# Earth Materials

## Second Edition

**Kevin Hefferan**
Marine Science, Safety and Environmental Protection Department
Massachusetts Maritime Academy

**John O'Brien**
Department of Earth and Environmental Sciences
New Jersey City University (Retired)

**WILEY** Blackwell

This second edition first published 2022
© 2022 John Wiley & Sons Ltd

*Edition History*
© 2010 by Kevin Hefferan and John O'Brien

All rights reserved. No part of this publication may be reproduced, stored in a retrieval system, or transmitted, in any form or by any means, electronic, mechanical, photocopying, recording or otherwise, except as permitted by law. Advice on how to obtain permission to reuse material from this title is available at http://www.wiley.com/go/permissions.

The right of Kevin Hefferan and John O'Brien to be identified as the authors of this work has been asserted in accordance with law.

*Registered Offices*
John Wiley & Sons, Inc., 111 River Street, Hoboken, NJ 07030, USA
John Wiley & Sons Ltd, The Atrium, Southern Gate, Chichester, West Sussex, PO19 8SQ, UK

*Editorial Office*
9600 Garsington Road, Oxford, OX4 2DQ, UK

For details of our global editorial offices, customer services, and more information about Wiley products visit us at www.wiley.com.

Wiley also publishes its books in a variety of electronic formats and by print-on-demand. Some content that appears in standard print versions of this book may not be available in other formats.

*Limit of Liability/Disclaimer of Warranty*
The contents of this work are intended to further general scientific research, understanding, and discussion only and are not intended and should not be relied upon as recommending or promoting scientific method, diagnosis, or treatment by physicians for any particular patient. In view of ongoing research, equipment modifications, changes in governmental regulations, and the constant flow of information relating to the use of medicines, equipment, and devices, the reader is urged to review and evaluate the information provided in the package insert or instructions for each medicine, equipment, or device for, among other things, any changes in the instructions or indication of usage and for added warnings and precautions. While the publisher and authors have used their best efforts in preparing this work, they make no representations or warranties with respect to the accuracy or completeness of the contents of this work and specifically disclaim all warranties, including without limitation any implied warranties of merchantability or fitness for a particular purpose. No warranty may be created or extended by sales representatives, written sales materials or promotional statements for this work. The fact that an organization, website, or product is referred to in this work as a citation and/or potential source of further information does not mean that the publisher and authors endorse the information or services the organization, website, or product may provide or recommendations it may make. This work is sold with the understanding that the publisher is not engaged in rendering professional services. The advice and strategies contained herein may not be suitable for your situation. You should consult with a specialist where appropriate. Further, readers should be aware that websites listed in this work may have changed or disappeared between when this work was written and when it is read. Neither the publisher nor authors shall be liable for any loss of profit or any other commercial damages, including but not limited to special, incidental, consequential, or other damages.

*Library of Congress Cataloging-in-Publication Data applied for*

Paperback ISBN: 9781119512172

Cover Design: Wiley
Cover Images: © Kevin Hefferan, Earth image © NASA

Set in 11/12pt Sabon by Straive, Pondicherry, India

Printed in Singapore
M090087_030222

# Contents

# Preface

Earth Materials encompass the minerals, rocks, soil, and water that constitute our planet and the physical, chemical, and biological processes that produce them. Since the expansion of computer technology in the last two decades of the twentieth century, many universities have compressed or eliminated individual course offerings such as mineralogy, optical mineralogy, igneous petrology, sedimentology, and metamorphic petrology and replaced them with Earth materials courses. Earth materials courses have become an essential curricular component in the fields of geology, geoscience, Earth science, engineering geology, environmental geology, and many related areas of study. This textbook is designed to address the needs of a one or two semester Earth materials course. It can also serve as a text for those individuals who want or need an expanded background in minerals, rocks, soils, and water resources.

The chapters featured in this textbook illuminate key topics in Earth materials including their:

1  Properties, origin, and classification
2  Associations and relationships in the context of Earth's major tectonic, petrologic, hydrologic, and biogeochemical systems
3  Uses as resources and their fundamental role in our lives and the global economy
4  Relation to natural and human induced hazards
5  Impact on health and the environment.

This textbook provides:

1  Comprehensive descriptive analysis of Earth materials
2  Color graphics and insightful texthe in a logical integrated format
3  Field examples and regional relationships with graphics that illustrate concepts discussed
4  Examples of how concepts discussed can be used to address real world issues
5  Contemporary references from current scientific journals related to developments in Earth materials research
6  Summative discussions of how Earth materials are interrelated with other science and non-science fields of study.

Chapter 1 contains an introduction to Earth materials and an overview that includes a discussion of Earth's interior and plate tectonics. This introductory chapter provides a global framework for following chapters. Chapters 2–6 constitute the minerals portion of the textbook. Chapter 2 addresses mineral chemistry, bonding mechanisms, and classification. Chapter 3 explores fundamentals of crystal chemistry, including ionic substitution, phase diagrams, and isotope geochemistry. Chapter 4 highlights the principles of crystallography including symmetry operations, crystal lattices, and crystal systems. Chapter 5 examines mineral formation, macroscopic mineral properties, and the major groups of rock-forming minerals. Chapter 6 focuses on the microscopic optical properties of minerals and petrographic microscopic techniques.

Chapters 7–10 encompass the igneous portion of the textbook. Chapter 7 discusses the composition, textures, and classification of igneous rocks. Chapter 8 describes the origin and evolution of magmas and plutonic structures. Chapter 9 vividly presents volcanic processes and landforms in all their wonder. Chapter 10 explores igneous rock associations and their relations to plate tectonics.

Chapters 11–15 focus on sedimentary rocks and processes. Chapter 11 addresses weathering, sediment production, and soils. Chapter 12 discusses the sedimentary cycle, the agents, and processes by which sediments are eroded, transported, and deposited and how they imprint sediments deposited in different environments. Chapters 13 and 14 examine the composition, textures, classification, and origin of detrital sedimentary rocks and biochemical sedimentary rocks that include carbonates, evaporites, siliceous rocks, iron formations, phosphates, and carbon-rich sedimentary materials that include coal, petroleum, and natural gas.

Chapters 15–18 address metamorphic rocks and processes. Chapter 15 introduces metamorphic agents, processes, protoliths, and types of metamorphism. Chapter 16 discusses metamorphic structures, stresses, and deformation processes. Chapter 17 investigates metamorphic rock textures and classification. Chapter 18 concentrates on metamorphic zones, facies, facies series, and metamorphic trajectories in relationship to plate tectonics.

Lastly, Chapter 19 explores ores minerals, industrial minerals, and gems as well as environmental and health issues related to Earth materials.

In addition to information presented in this textbook, additional resources, including detailed descriptions of major rock-forming minerals and keys for identifying minerals using macroscopic and/or optical methods are available on the website that supports this text at:

www.wiley.com/go/hefferan/earthmaterials.

Our overall goal is to produce an innovative, visually appealing, informative, and readable textbook that addresses the full spectrum of Earth materials. We present equal treatment to minerals as well as igneous, sedimentary, and metamorphic rocks and demonstrate their impact on our personal lives and the global environment on this planet. We hope you enjoy this text and use it to further your knowledge of Earth materials.

# Acknowledgments

The authors thank all those at John Wiley & Sons who worked with them on the second edition. Special thanks are due Rosie Hayden, Antony Samy, Anandan Bommen, Umar Saleem and Andrew Harrison. Their first edition greatly benefited from guidance provided by Ian Francis, Kelvin Matthews, Jane Andrews, Delia Sandford, Camille Poire, Catherine Flack and, once again, Rosie Hayden. The authors are grateful to Anita O'Brien who provided the index for both editions. We once again thank reviewers for the first edition who greatly improved the textbook, while in no way being responsible for its shortcomings. These individuals include Malcolm Hill, Stephen Nelson, Lucian Platt, Steve Dutch, Duncan Heron, Jeremy Inglis, Maria Luisa Crawford, Barbara Cooper, Alec Winters, David H. Eggler, Cin-Ty Lee, Samantha Kaplan, Penelope Morton and Ellen D'Andrea.

The authors truly appreciate many individuals and publishers who generously permitted reproduction of their figures and images from published work, educational websites and/or personal collections. The authors are particularly indebted to Doug Moore, Steve Dutch, Stephen Nelson, Kurt Hollacher, Gregory Finn, Patrice Rey, Neil Heywood, the U.S. Geological Survey, the Geological Survey of Canada and the Geological Society of America.

Kevin Hefferan would especially like to thank his wife Sherri and children Kaeli, Patrick, Sierra, Keegan and Quintin and parents Patrick and Catherine for their love, laughter and never wavering enthusiasm. Kevin is also grateful for the support of the Department of Marine Science, Safety and Environmental Protection at Massachusetts Maritime Academy of Buzzards Bay, MA. Kevin expresses his appreciation to all his collaborative colleagues through the years especially Jeff Karson, Tony Rathburn, William Hubbard, Heather Burton, Abderrahmane Soulaimani, Hassan Admou, Ali Saquaque, Nasrrrddine Youbi, Lucian Platt, Neil Heywood, Keith Rice, Doug Moore and Samantha Kaplan. Kevin was an undergraduate student of John O'Brien at New Jersey City University and greatly benefited from his wisdom as well as that of John Marchisin, Howard Parish and Barry Perlmutter.

John O'Brien would like to thank his wife Anita, his sons Tyler and Owen and granddaughter Scarlett for their love and support. John worked for 41 years in the Department of Geoscience-Geography (now Earth and Environmental Sciences) at New Jersey City State University. He is now retired and living in the San Francisco Bay area. Sabbaticals from the University in 2005 and again in 2011, gave John the time required to complete this project. John wishes to acknowledge his indebtedness to Ansel Gooding, Charles Martin, Wayne Martin, John Pope, Robert Webb, John Crowell, Don Runnels, Joe Clark and Barry Perlmutter for encouraging, supporting and challenging him at critical times in his career.

The authors believe that an understanding of Earth materials is more important than ever in this twenty-first century in order to appropriately utilize essential resources, for mitigating geohazards and dealing with environmental change. Universities and government agencies would be wise to revitalize the Earth science and geology curricula in the United States and other countries around the world. We must all renew efforts to use our resources wisely and to engage the green economy.

# About the Companion Website

This book is accompanied by a companion website.

www.wiley.com/go/hefferan/earthmaterials2

This website includes:

- Figures and Tables from the book
- Appendices and additional resources, including detailed descriptions of major rock-forming minerals and keys for identifying minerals using macroscopic and/or optical methods

# Chapter 1

# Earth materials and the geosphere

## 1.1   EARTH MATERIALS

This book concerns the nature, origin, evolution, and significance of Earth materials. Earth is composed of a variety of naturally occurring and synthetic materials whose composition can be expressed in many ways. These include their chemical, mineral, and rock composition. In simple terms, atoms combine to form minerals and minerals combine to form rocks. Discussion of the relationships between atoms, minerals, and rocks is fundamental to an understanding of Earth materials, their properties, and the processes that produce them.

## 1.2   MINERALS AND MINERALOIDS

The term **mineral** is used in a number of ways. For example, the chemical elements, such as calcium, iron, and potassium, listed on your breakfast cereal box, your bottle of vitamin supplements or your bag of fertilizer are called minerals. Coal, oil, and gas are referred to as mineral resources. All of these fall under a broad use of the term mineral. In a stricter sense used by many, but not all geologists, minerals are defined by the following properties:

1   Minerals are **solid,** so do not include liquids and gases. Minerals are solid because the atoms in them are held together in fixed positions by forces called chemical bonds (Chapter 2).
2   Minerals are **naturally occurring,** meaning that they occur naturally within the Earth. This definition excludes synthetic solids that are produced only by technologies in laboratories or factories. It does include solid Earth materials that are produced by both natural

*Earth Materials*, Second Edition. Kevin Hefferan and John O'Brien.
© 2022 John Wiley & Sons Ltd. Published 2022 by John Wiley & Sons Ltd.
Companion website: www.wiley.com/go/hefferan/earthmaterials2

and synthetic processes, such as natural and synthetic diamonds and the solid materials synthesized in high temperature and high pressure laboratory experiments that are thought to be analogous to real minerals that occur only in the deep interior of Earth.

3  Each mineral species has a **specific chemical composition** which may vary only within well-defined limits; that is to say that each mineral possesses a chemical composition that can be expressed by a chemical formula. An example is common table salt or halite which is composed of sodium and chlorine atoms in a 1 : 1 ratio (NaCl). Chemical compositions may vary within well-defined limits because minerals incorporate impurities, have atoms missing, or otherwise vary from their ideal compositions. In addition some types of atoms may substitute freely for one another when a mineral forms generating a well-defined range of chemical compositions. For example, magnesium (Mg) and iron (Fe) may substitute freely for one another in the mineral olivine whose composition is expressed as $(Mg,Fe)_2SiO_4$. The parentheses are used to indicate the variable amounts of Mg and Fe that may substitute for each other in olivine group minerals (Chapter 3).

4  Every mineral species **possesses a long-range, geometric arrangement of constituent atoms or ions.** This implies that the atoms in minerals are not randomly arranged. Instead minerals crystallize in geometric patterns so that the same pattern is repeated throughout the mineral. In this sense, minerals are like three-dimensional wall paper. A basic pattern of atoms, a motif, is repeated systematically to produce the entire geometric design. This long range pattern of atoms characteristic of each mineral species is called its **crystal structure**. All materials that possess geometric crystal structures are **crystalline materials**. They are minerals, in the narrow sense, if they are naturally occurring, inorganic solids with a well-defined chemical composition. Solid materials that lack a long-range crystal structure are **amorphous** materials, where amorphous means without form and without a long-range geometric order.

Many would add a fifth property that requires minerals to sometimes be **formed by inorganic processes.** It is certainly true that the vast majority of minerals conform to this property and that the vast majority of organically formed crystalline solids are not considered to be minerals. However, many solid Earth materials that form by both inorganic and organic processes are considered minerals, especially if they are important constituents of naturally formed rocks. For example, the mineral calcite is also precipitated as shell material by organisms such as clams, snails, and corals and is the major constituent of the rock limestone (Chapter 14).

Over 5500 minerals have been discovered to date (www.mindat.com) and each is distinguished by a unique combination of chemical composition and crystal structure. Strictly speaking, naturally occurring, solid materials that lack one of the properties described above are commonly referred to as **mineraloids**. Common examples include **amorphous** materials such as volcanic glass in which the atoms lack long-range order and amber or ivory which are formed only by organic processes.

### 1.2.1  Rocks

Earth is largely composed of various types of **rock**. A rock is an aggregate of mineral crystals and/or mineraloids. Scarce **monomineralic** rocks consist of multiple crystals of a single mineral. Examples include the sedimentary rock quartz sandstone which may consist entirely of quartz grains held together by quartz cement and the igneous rock dunite which can consist entirely of olivine crystals. The vast majority of rocks are **polymineralic;** they are composed of many types of mineral crystals. For example, granite commonly contains quartz, potassium feldspar, plagioclase, hornblende and/or biotite, and various other minerals in small amounts.

Mineral composition is one of the major defining characteristics of rocks. Rock textures and structures are also important defining characteristics. It is not surprising that the number of rock types is very large indeed, given the large number of different minerals that occur in nature, the different conditions under which they form, and the different proportions in which they can combine to form aggregates with various textures and structures. Helping students to understand the properties, classification, origin, and significance of minerals and rocks is the major emphasis of this text.

## 1.3   THE GEOSPHERE

Earth materials can occur anywhere on or within the **geosphere**, the portion the Earth from its surface to its center, whose radius is approximately 6370 km (Figure 1.1). In static standard models of the geosphere, Earth is depicted with a number of roughly concentric layers. Some of these layers are distinguished primarily on the basis of differences in composition and others by differences in their state or mechanical properties. These two characteristics by which the internal layers of Earth are distinguished are not totally independent, because differences in chemical, mineralogical and/or rock composition influence mechanical properties.

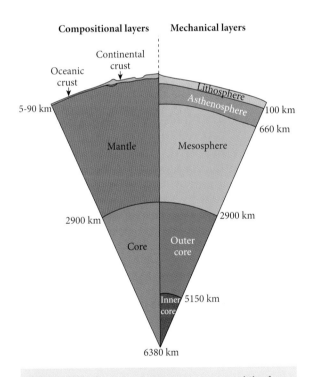

**Figure 1.1** Standard cross-section model of the geosphere. Major compositional layers are shown on the left: core (red shade), mantle (brown shade), and continental and oceanic crust (blue shades). Major mechanical layers are shown on the right: inner core and outer core (red shades), lithosphere (light brown), mesosphere (dark brown), and asthenosphere (burnt orange).

### 1.3.1   Compositional layers

The layers within Earth that are defined largely on the basis of chemical composition (Figure 1.1; left side) include the: (1) **crust**, which is subdivided into **continental** and **oceanic** crust, (2) **mantle**, and (3) **core**. Each of these layers has a distinctive combination of chemical, mineral, and rock compositions that distinguishes it from the others, as described in the next section. The thin crust typically ranges from 5 to 85 km thick and occupies <1% of Earth's volume. The much thicker mantle has an average radius of ~2885 km and occupies ~83% of Earth's volume. The core has a radius of ~3470 km and comprises ~16% of Earth's volume.

### 1.3.2   Mechanical layers

The layers within Earth defined principally on the basis of mechanical properties (Figure 1.1; right side) include: (1) a relatively strong **lithosphere** of variable thickness to an average depth of ~100 km that includes all of the crust and the upper part of the mantle, (2) a weaker **asthenosphere** that extends to depths between ~100 and 660 km and includes a **transition** *zone* from ~400 to 660 km, and (3) a **mesosphere** or **lower mantle** from ~660 to 2900 km. The underlying core is divided into a liquid **outer core** (~2900–5150 km) and a solid **inner core** below ~5150 km to the center of Earth. Each of these layers is distinguished from the layers above and below by its unique mechanical properties. The major features of each of these layers are summarized in the next section.

## 1.4   DETAILED MODEL OF THE GEOSPHERE

### 1.4.1   Earth's crust

The outermost layer of the geosphere, **Earth's crust**, is extremely thin; in some ways it is analogous to the very thin skin on an apple. The crust is separated from the underlying mantle by a boundary called the **Mohorovičić (Moho) discontinuity**. Two major types of crust occur.

*Oceanic crust*

**Oceanic crust** is composed largely of dark colored, basic (45–52% $SiO_2$) rocks

(Chapter 7) enriched in oxides of magnesium, iron, and calcium (MgO, FeO, and CaO) relative to average crust. The elevated iron (Fe) content is responsible for the both the dark color and elevated density of oceanic crust. Oceanic crust is thin; the depth to the Moho averages 5–7 km. Under some oceanic islands, its thickness reaches 18 km. The elevated density and small thickness of oceanic crust cause it to be less buoyant than continental crust, so that it occupies areas of lower elevation on Earth's surface. As a result, most oceanic crust of normal thickness is below sea level and covered by sea water to a depth of several thousand meters. Oceanic crust consists principally of basic igneous rocks such as basalt and gabbro composed largely of the minerals pyroxene and calcic plagioclase. These dark-colored, mafic igneous rocks comprise layers 2 and 3 of oceanic crust and are commonly topped with sediments that comprise layer 1 (Table 1.1). An idealized profile of typical ocean crust consists of these three main layers, each of which can be subdivided into sublayers which are briefly discussed later in this chapter.

Oceanic crust is young relative to the age of the Earth (~4.55 Ga = 4550 Ma). The oldest ocean crust in the major ocean basins, less than 190 million years old (190 Ma), occurs along the western and eastern borders of the Atlantic Ocean and in the Western Pacific Ocean. Recently, still older oceanic crust that may be 340 Ma has been discovered in the eastern Mediterranean Sea (Granot 2016). Still older oceanic crust has largely been destroyed by subduction, but fragments of such crust are preserved on land in the form of ophiolites. Ophiolites contain slices of ocean crust thrust onto continental margins and provide evidence for the existence of Precambrian oceanic crust. The age of the oldest true ophiolites of Precambrian age remains controversial (Chapter 18).

### Continental crust

**Continental crust** has a much more variable composition than oceanic crust. Continental crust can be generalized as "granitic" in composition, and is enriched in $K_2O$, $Na_2O$, and $SiO_2$ relative to average crust. Although igneous and metamorphic rocks of granitic composition are fairly common in the upper portion of continental crust, lower portions contain more rocks of intermediate dioritic and even basic gabbroic composition. Granites and related rocks tend to be light colored, lower density felsic rocks rich in quartz and potassium and sodium feldspars. Continental crust is generally much thicker than oceanic crust; depth to the Moho averages 30–40 km. Under areas of very high elevation, such as the Himalayas, its thickness approaches 85 km. The greater thickness and lower density of continental crust make it more buoyant than oceanic crust. As a result, the top of continental crust is generally located at higher elevations and the surfaces of continents with normal crustal thicknesses are above sea level. The distribution of Earth's land and sea is largely dictated by the distribution of continental and oceanic crust. Only the thinnest portions of continental crust, most frequently along thinned continental margins and in rifts, have surfaces below sea level.

**Table 1.1**  Characteristics of oceanic and continental crust: a comparison.

| Properties | Oceanic crust | Continental crust |
| --- | --- | --- |
| Composition | Dark colored, mafic rocks enriched in MgO, FeO, and CaO | Complex; many lighter colored felsic rocks |
|  |  | Enriched in $K_2O$, $Na_2O$, and $SiO_2$ |
|  | Averages ~50% SiO2 | Averages ~60% SiO2 |
| Density | Higher; less buoyant | Lower; more buoyant |
|  | Average 2.9–3.1 g/cm³ | Average 2.6–2.9 g/cm³ |
| Thickness | Thinner; average 5–7 km thickness | Thicker; average 30 km thickness |
|  | Up to 15 km under islands | Up to 80 km under mountains |
| Elevation | Low surface elevation; mostly submerged below sea level | Higher surface elevations; mostly emergent above sea level |
| Age | Up to 190 Ma for in-place crust | Up to more than 4000 Ma |
|  | ~3.5% of Earth history | 85–90% of Earth history |

Whereas modern oceans, with the exception of a small area in the Mediterranean Sea, are underlain by oceanic crust younger than 190 Ma, the oldest well-documented continental crust includes 4.03 Ga rocks from the Northwest Territories of Canada (Stern and Bleeker 1998). Approximately 4 Ga rocks also occur in Greenland and Australia. Greenstone belts (Chapter 18) may date back as far as 4.28 Ga (O'Neill et al. 2008) which suggests that continental crust began forming within 300 million years of Earth's birth. Individual detrital zircon grains, derived from the erosion of older continental crust, occur in metamorphosed sedimentary rocks in Australia. These zircons have been dated at 4.4 Ga (Wilde et al. 2001) an age recently confirmed by Valley et al. (2014). These data suggest that continental crust may have existed no more than 150 Ma after Earth formed. The great age of some continental crust results from its relative buoyancy. In contrast to ocean crust, continental crust is largely preserved as its density is generally too low for it to be subducted on as large a scale. Table 1.1 summarizes the major differences between oceanic and continental crust.

## 1.4.2  Earth's Mantle

The **mantle** is thick (~2900 km) relative to the radius of Earth (~6370 km) and constitutes ~83% of Earth's total volume. The mantle is distinguished from the crust by being very rich in MgO (30–40%) and, to a lesser extent, in FeO. It contains an average of approximately 40–45% $SiO_2$ which means it has an **ultrabasic composition** (Chapter 7). Some basic rocks such as eclogite occur in smaller proportions. In the upper mantle (depths to 400 km), the Mg-rich silicate minerals olivine and pyroxene dominate; spinel, plagioclase and garnet are locally common. These minerals combine to produce generally dark colored **ultramafic rocks** such as peridotite, the dominate group of rocks in the upper mantle. Under the higher pressure conditions deeper in the mantle similar chemical components combine to produce dense minerals with tightly packed crystal structures. These high-pressure minerals are produced by transformations that are largely indicated by changes high pressure in seismic wave velocity, which reveal that the mantle contains a number of sublayers (Figure 1.2) as discussed below.

### Upper Mantle and Transition Zone

The uppermost part of the mantle and the crust together constitute the relatively rigid **lithosphere** which is strong enough to rupture in response to Earth stresses. Because the lithosphere can rupture in response to stress, it is the site of most earthquakes and is broken into large fragments called plates, as discussed later in this chapter.

A discrete **low velocity zone (LVZ)** occurs within most areas of the upper mantle at depths of ~100–250 km below the surface. The top of low velocity zone marks the contact between the strong lithosphere and the underlying, weak asthenosphere (Figure 1.3). The **asthenosphere** is more plastic than the lithosphere and flows slowly, rather than rupturing, when subjected to stress. The anomalously low rigidity of the LVZ has been explained by small amounts of partial melting (Anderson et al. 1971). This is supported by laboratory studies that suggest peridotite should be very near its melting temperature at these depths due to the high temperature. This is especially likely if it contains small amounts of water or water-bearing minerals. Below the base of the low velocity zone (250–410 km), seismic wave velocities increase (Figure 1.2) indicating that the materials are more rigid solids. These materials are still part of the relatively weak asthenosphere which extends to the base of the transition zone at 660 km.

Seismic discontinuities marked by increases in seismic velocity occur within the upper mantle at depths of ~410 and ~660 km (Figure 1.2). This interval (~410–660 km) is called the **transition zone** between the upper and lower mantle. The sudden jumps in seismic velocity record sudden increases in rigidity and incompressibility. Laboratory studies suggest that the minerals in peridotite undergo transformations into new minerals at these depths. At approximately 410 km depth (pressures of ~14 GPa), olivine ($Mg_2SiO_4$) is transformed into more rigid, incompressible beta spinel (β-spinel), also known as wadleysite ($Mg_2SiO_4$). Within the transition zone, wadleysite is transformed into the higher pressure mineral ringwoodite ($Mg_2SiO_4$). At approximately 660 km depth (~24 GPa), ringwoodite and garnet are converted to very rigid, incompressible perovskite [(Mg,Fe,Al) $SiO_3$], also known as bridgmanite (Tschauner

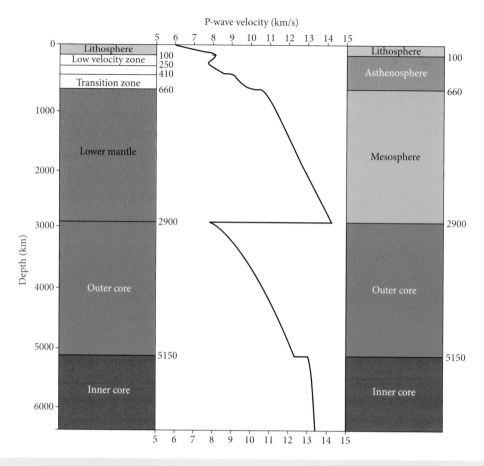

**Figure 1.2** Major layers and seismic (p-wave) velocity changes within Earth; showing details of upper mantle layers. Colors are as for Figure 1.1.

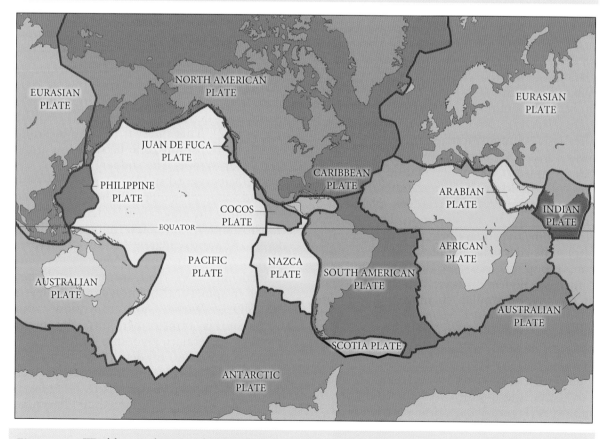

**Figure 1.3** World map showing the distribution of major plates separated by boundary segments that end in triple junctions. *Source*: From USGS.

et al. 2014) and oxide phases such as periclase (MgO). The mineral phase changes from olivine to wadleysite and from ringwoodite to perovskite are inferred to be largely responsible for the seismic wave velocity changes that occur at 410 and 660 km respectively (Ringwood 1975; Condie 1982; Anderson 1989). Inversions of pyroxene to garnet and garnet to minerals with ilmenite and perovskite structures may also be involved. The base of the transition zone at 660 km marks the base of the asthenosphere in contact with the underlying mesosphere or lower mantle (Figure 1.2).

*The lower mantle (mesosphere)*

The **lower mantle**, also called the **mesosphere**, extends from depths of 660 km to the core–mantle boundary at approximately 2900 km. Based on high pressure, high temperature laboratory studies, bridgmanite [(Mg,Fe,Al)$SiO_3$], ferropericlase [(Mg,Fe)O], magnesiowustite [(Mg,Fe)O], stishovite ($SiO_2$), and calcium-rich ferrite $(Ca,Na,Al)Fe_2O_4$ are thought to be the major minerals in the lower mantle. Our knowledge of the deep mantle continues to expand, largely based on high temperature, high pressure laboratory studies and on anomalous seismic signals deep within the Earth. A deeper layer has been proposed at about 1600 km depth where the rigidity of the mantle may increase considerably (Miyagi and Marquardt 2015). Anomalous seismic velocities are particularly common in a complex zone, of variable thickness, near the core–mantle boundary called the *D″* **layer**. The D″ discontinuity ranges from ~130 to 340 km above the core–mantle boundary. Williams and Garnero (1996) proposed an ultra-low velocity zone (ULVZ) in the lowermost mantle on seismic evidence. These sporadic ultra-low velocity zones may be related to the formation of deep mantle plumes within the lower mantle. Other areas near the core–mantle boundary are characterized by anomalously fast velocities. Hutko et al. (2006) detected subducted lithosphere which had sunk all the way to the D″ layer and may be responsible for the anomalously fast velocities. Deep subduction and deeply rooted mantle plumes support some type of whole mantle convection and may play a significant role in the evolution of a highly heterogeneous mantle, but these concepts are still controversial (Foulger et al. 2005).

### 1.4.3    Earth's core

Earth's **core** consists primarily of iron (~85%), with smaller, but significant amounts of nickel (~5%) and lighter elements (~8–10%) such as oxygen, sulfur and/or hydrogen. A dramatic decrease in P-wave velocity and the termination of S-wave propagation occurs at the 2900 km discontinuity which is **Gutenberg discontinuity** or **core–mantle boundary (CMB)**. Because S-waves are not transmitted by nonrigid substances such as fluids, the **outer core** is inferred to be a fluid. Geophysical studies suggest that the Earth's outer core is a highly compressed liquid with a density of ~10–12 g/cm³. Slowly circulating molten, iron-rich, very viscous liquids in the outer core are believed to be responsible for the production of most of Earth's magnetic field.

The outer/inner core boundary, the **Lehman discontinuity** at 5150 km, is marked by a rapid increase in P-wave velocity and the reemergence of low velocity S-waves. This suggests that the **inner core** is rigid. The inner core is solid and has a density of ~13 g/cm³. Density and magnetic studies suggest that the Earth's inner core also consists of largely of iron, with nickel and less oxygen, sulfur, and/or hydrogen than the outer core. Seismic studies have shown that the inner core is seismically anisotropic; that is seismic velocity in the inner core is faster in one direction than in others. This has been interpreted to result from the parallel alignment of iron-rich crystals or from a core consisting of a single crystal with a fast velocity direction. Recent discoveries suggest that the inner core is divided into two layers with the inner layer more rigid than the outer one and with a different orientation of its fast seismic wave direction (Ishii and Dziewonski 2002; Wang et al. 2015).

In this section, we have discussed the major layers of the geosphere, their composition, and their mechanical properties. This model of a layered geosphere provides us with a spatial context in which to visualize where the processes that generate earth materials occur. In the following sections we will examine the ways in which all parts of the geosphere interact to produce global tectonics. The ongoing story of global-scale tectonics is one of the most fascinating tales of scientific discovery in the last century and new discoveries continue to be made in this one.

## 1.5   GLOBAL TECTONICS

### 1.5.1   Introduction

Plate tectonic theory has profoundly changed the way geoscientists view Earth and provides an important theoretical and conceptual framework for understanding the origin and global distribution of igneous, sedimentary, and metamorphic rocks (Chapters 7–18). It also helps to explain the distribution of diverse phenomena that include faults, earthquakes, volcanoes, mountain belts, mineral deposits, and even the evolution of life and the evolving composition of the atmosphere.

The fundamental tenet of **plate tectonics** (Le Pichon 1968; Isacks et al. 1968) is that the lithosphere is broken along major fault systems into large, relatively rigid pieces called plates that move relative to one another. The existence of the strong, breakable lithosphere permits plates to form. Most plates are huge, with areas of $10^5$–$10^8$ km$^2$ and thicknesses that average ~$10^3$ km; some plates are smaller and microplates are smaller still. The fact that they overlie a weaker, slowly flowing asthenosphere permits them to move very slowly. Each plate is separated from adjacent plates by **plate boundary segments** that end in **triple junctions** (McKenzie and Morgan 1969) where three plates are in contact (Figure 1.3).

The relative movement of plates with respect to the boundary that separates them defines three major types of plate boundary segments (Figure 1.4) and two hybrids: (1) divergent plate (2) convergent, (3) transform, (4) divergent-transform hybrid, and (5) convergent-transform hybrid.

Each type of plate boundary produces a characteristic suite of features and Earth materials. This relationship between the kinds of Earth materials formed and the plate tectonic settings in which they are produced provides a major theme of the chapters that follow.

### 1.5.2   Divergent plate boundaries

**Divergent plate boundaries** occur where two plates are moving apart relative to their boundary (Figure 1.4a). Such areas are characterized by horizontal extension and vertical thinning of the lithosphere. Horizontal extension in continental lithosphere is marked by **continental rift systems** and in oceanic lithosphere by the **oceanic ridge system**.

*Continental rifts*

**Continental rifts** form where large-scale horizontal extension occurs in continental lithosphere (Figure 1.5). In such regions, the lithosphere is progressively stretched and thinned. A candy bar being very slowly stretched in two is a crude metaphor. This stretching occurs by brittle, normal, and detachment faulting near the cooler surface and by ductile flow at deeper, warmer levels.

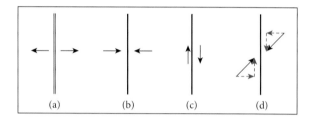

**Figure 1.4**   Principal types of plate boundaries: (a) divergent; (b) convergent; (c), transform; thick lines represent plate boundaries and black arrows indicate relative motion between plates; (d) hybrid convergent-transform boundary; red arrows show components of convergent and transform relative motion.

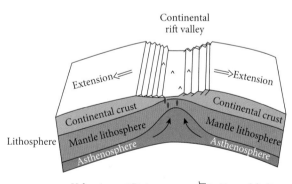

**Figure 1.5**   Major features of continental rifts include rift valleys, thinned continental crust (blue) and lithosphere (brown) with volcanic-magmatic activity from melts generated in rising asthenosphere (burnt orange).

Long-term extension is accompanied by uplift of the surface as the hot asthenosphere rises under the thinned lithosphere. Rocks near the surface of the lithosphere that rupture along normal and detachment faults produce **continental rift valleys**. The East African Rift, the Rio Grande Rift in the United States and the Dead Sea Rift in the Middle East are modern examples of continental rift valleys.

If horizontal extension and vertical thinning occur for a sufficient period of time, the continental lithosphere may be completely rifted into two separate continents. **Complete continental rifting** is the diachronous process by which supercontinents such as Pangea and Rodinia were broken into smaller continents such as those we see on Earth's surface at present. When this happens, a new and growing ocean basin begins to form between the two continental fragments by the process of **sea floor spreading** (Figure 1.6). The most recent example of this occurred when the Saudi Arabian Peninsula separated from Africa to produce the Red Sea basin some 5 Ma. Older examples include the separation of India from Africa to produce the northwest Indian Ocean basin (c. 115 Ma) and the separation of the Americas from Africa to produce the Atlantic Ocean basin (beginning c. 190 Ma). Once the continental lithosphere has rifted completely, the divergent plate boundary is no longer situated within continental lithosphere. Its position is instead marked by a portion of the oceanic ridge system where oceanic crust is produced and grows by sea floor spreading (Figure 1.6).

### Oceanic ridge system

The **oceanic ridge system (ridge)** is Earth's largest mountain range and covers roughly 20% of Earth's surface (Figure 1.7). The ridge is >65 000 km long, averages ~1500 km in width and rises to a crest with an average elevation of ~3 km above the surrounding sea floor. A moment's thought will show that the ridge system is only a broad swell on the ocean floor, whose slopes, on average, are very gentle. Since it rises only 3 km over a horizontal distance of 750 km, the average slope is 3 km/750 km which is about 0.004; the average slope is less than half a degree. We often exaggerate the vertical dimension on profiles

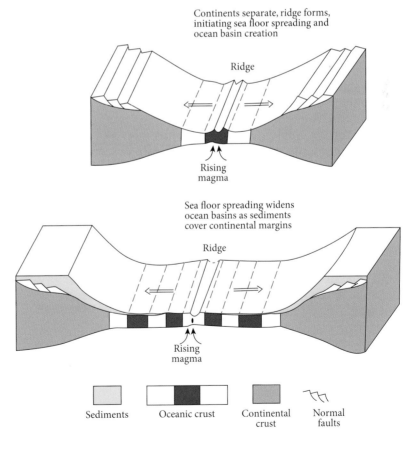

**Figure 1.6** Model showing the growth of ocean basins by sea floor spreading from the ridge system following the complete rifting of continental lithosphere along a divergent plate boundary. Separated continental crust (light blue, topped with yellow sediments) and recently formed oceanic crust (dark blue striped pattern) are produced.

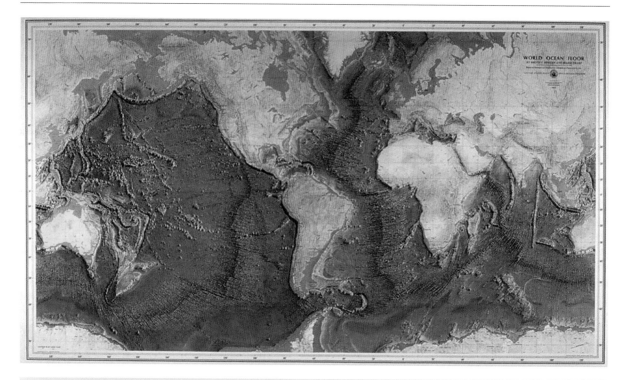

**Figure 1.7** Map of the ocean floor showing the distribution of the oceanic ridge system. *Source*: World Ocean Floor Manuscript Map; drawn by Berann, H.C., US Library of Congress, public domain after Heezan, Bruce C. and Tharpe, Marie.

and maps in order to make the subtle stand out. Still there are differences in relief along the ridge system. In general, warmer, faster spreading portions of the ridge such as the East Pacific Rise (~6–18 cm/yr) have gentler slopes than colder, slower spreading portions such as the Mid-Atlantic Ridge (~2–4 cm/yr). The central or axial portion of the ridge system is commonly marked by a **rift valley**, especially along slower spreading segments. This marks the position of a divergent plate boundary in oceanic lithosphere (Figure 1.7).

One of the most significant discoveries of the twentieth century (Dietz 1961; Hess 1962) was that oceanic crust forms along the axis of the ridge system, then spreads away from it in both directions, causing ocean basins to grow through time. The details of this process are illustrated by Figure 1.8. As the lithosphere is thinned, the asthenosphere rises toward the surface, generating basaltic-gabbroic melts. Melts that crystallize in magma bodies well below the surface form basic plutonic rocks such as gabbro that become layer 3 in oceanic crust. Melts intruded into near vertical fractures above the chamber form the

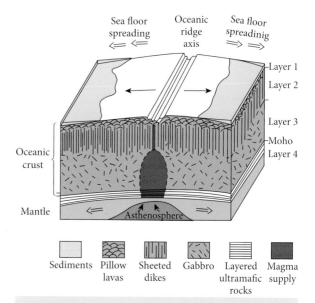

**Figure 1.8** The formation of oceanic crust along the ridge axis generates layer 2 pillow basalts and dikes and layer three gabbros of the oceanic crust (blue) and layer 4 mantle peridotites (gray). Sediment deposition atop these rocks produces layer 1 (yellow) of the crust. Sea floor spreading (black arrows) carries these laterally away from the ridge axis in both directions.

gabbroic-basaltic parallel "sheeted" dikes that become layer 2b. Lavas that flow onto the ocean floor commonly form basaltic pillow and sheet lavas that become layer 2a. The marine sediments of layer 1 are deposited atop the basalts as they spread away from the ridge axis. In this way layers 1, 2, and 3 of the oceanic crust are formed. The underlying mantle consists of ultramafic rocks (layer 4). Layered ultramafic rocks form by differentiation near the base of the basaltic-gabbroic magma bodies, whereas the remainder of layer 4 represents the unmelted, refractory residue that accumulates below the magma bodies.

Because the ridge axis marks a divergent plate boundary, the new sea floor on one side moves away from the ridge axis in one direction and the new sea floor on the other side moves in the opposite direction relative to the ridge axis. More melts rise from the asthenosphere and the process is repeated, sometimes over >100 Ma. In this way ocean basins grow by sea floor spreading as though new sea floor was being added to two slowly moving conveyor belts that carry older sea floor in opposite directions away from the ridge where it forms (Figure 1.8). Because most oceanic lithosphere is produced along divergent plate boundaries marked by the ridge system, these boundaries are also called **constructive** plate boundaries.

As sea floor spreads away from the ridge axis, the crust thickens from above by the accumulation of marine sediments and the lithosphere thickens from below by a process called **underplating** that occurs as the solid, unmelted portion of the asthenosphere spreads laterally and cools through a critical temperature below which it becomes strong enough to fracture. As the entire lithosphere cools, it contracts, becomes denser, and sinks, so that the floors of the ocean gradually deepen away from the thermally elevated ridge axis. As explained in the next section, if the density of oceanic lithosphere exceeds that of the underlying asthenosphere, subduction occurs.

The formation of oceanic lithosphere by sea floor spreading implies that the age of oceanic crust should increase systematically away from the ridge in opposite directions. Crust produced during a period of time characterized by normal magnetic polarity should split

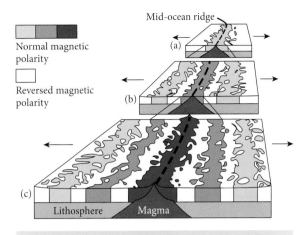

**Figure 1.9**    Model depicts the production of alternating normal (colored) and reversed (white) magnetic bands in oceanic crust by progressive sea floor spreading and alternating normal and reversed periods of geomagnetic polarity (a–c). The age of such bands should increase away from the ridge axis. *Source*: Courtesy of USGS.

in two and spread away from the ridge axis. New crust formed during the subsequent period of reversed magnetic polarity will form between the two areas of normally polarized crust and the reversely magnetized crust will also split in two. As indicated by Figure 1.9, repetition of this splitting process produces oceanic crust with bands (**linear magnetic anomalies**) of alternating normal and reversed magnetism whose age increases systematically away from the ridge, as initially explained by Vine and Matthews (1963).

Sea floor spreading was convincingly demonstrated in the middle to late 1960s by paleomagnetic studies and radiometric dating which showed that the age of ocean floors systematically increases in both directions away from the ridge axis, as predicted by sea floor spreading (Figure 1.10).

Hess (1962), and those who followed, realized that sea floor spreading causes the outer layer of Earth to grow substantially over time. If Earth's circumference is relatively constant and Earth's lithosphere is growing and being extended horizontally at divergent plate boundaries over long periods of time, then there must be places where it is undergoing long-term horizontal shortening of similar magnitude. As ocean lithosphere ages and

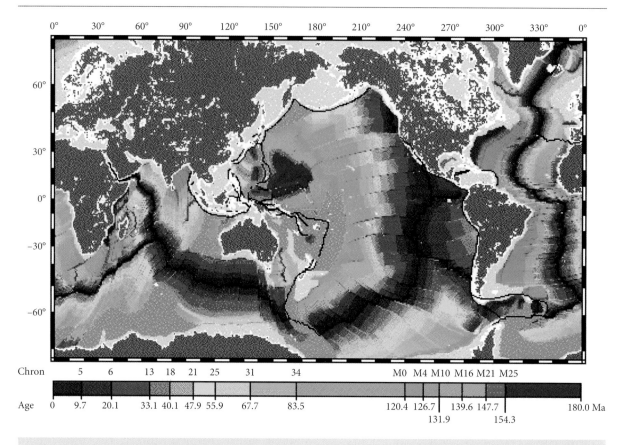

**Figure 1.10**   World map showing the age of oceanic crust; such maps confirmed the origin of oceanic crust by sea floor spreading. *Source*: From Lamont Doherty Earth Observatory.

continues to move away from oceanic spreading centers, it cools, subsides, and becomes denser over time. The increased density eventually causes the strong ocean lithosphere to become denser than the underlying, weak asthenosphere. As a result, a plate carrying old, cold, dense ocean lithosphere begins to sink downward into the asthenosphere under a more buoyant plate edge, creating a convergent plate boundary.

### 1.5.3   Convergent plate boundaries

**Convergent plate boundaries** occur where two plates are moving toward one another relative to their mutual boundary (Figure 1.11). The scale of such processes and the features they produce are truly awe inspiring.

*Subduction zones*

The process by which the leading edge of a denser lithospheric plate is forced downward

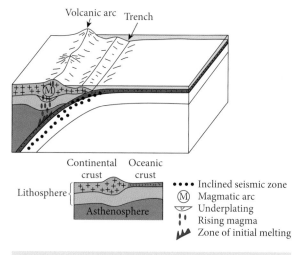

**Figure 1.11**   Convergent plate boundary, showing trench-arc system, inclined seismic zone and subduction of oceanic lithosphere.

into the underlying asthenosphere is called **subduction**. The downgoing plate is called the **subducted plate** or **downgoing slab**; the less dense plate is called the **overriding plate** or slab. The area where this process occurs is a **subduction zone**. The subducted plate, whose thickness averages 100 km, is generally composed of dense oceanic lithosphere. Subduction is the major process by which oceanic lithosphere is destroyed and recycled into the asthenosphere and deeper Earth at rates similar to its creation along the oceanic ridge system. For this reason, subduction zone plate boundaries are also called **destructive plate boundaries**.

The surface expressions of subduction zones are large **trench-arc systems** of the kind that encircle most of the shrinking Pacific Ocean (Isacks et al. 1968). **Trenches** are deep, elongate troughs in the ocean floors marked by water depths that can exceed 11 km. They are formed as the downgoing slab forces the overriding slab to bend downward forming a long trough along the boundary between them.

Because the asthenosphere is mostly solid, it resists the downward movement of the subducted plate to varying degrees. This produces stresses in the cool interior of the subducted lithosphere that generate earthquakes (Figure 1.11) along an **inclined seismic (Wadati-Benioff) zone** that marks the path of the subducted plate as it descends into the asthenosphere. The four largest magnitude earthquakes in the past 120 years occurred along inclined seismic zones beneath Chile (1909), Alaska (1964), Sumatra (2004), and Japan (2011). The latter two events produced the devastating 2004 Banda Aceh tsunami which killed some 300 000 people around the Indian Ocean region and the Fukushima tsunami which killed tens of thousands in eastern Japan and destroyed an atomic power plant.

What is the ultimate fate of subducted slabs? Earthquakes occur in subducted slabs to a depth of 660 km, so we know they reach the base of the asthenosphere transition zone. Earthquake records suggest that some slabs flatten out as they reach this boundary indicating that they may not penetrate into the lower mantle. Seismic tomography, which images three-dimensional variations in seismic wave velocity within the mantle, has shed some light on this question, while raising many others. A consensus has emerged (Grand 2002; Hutko et al. 2006) that some subducted slabs become dense enough to sink all the way to the core–mantle boundary where they contribute material to the D″ layer. These slab remnants may ultimately be involved in the formation of mantle plumes, as proposed by Jeanloz (1993).

Subduction zones produce a wide range of distinctive Earth materials. The increase in temperature and pressure within the subducted plate causes it to undergo significant metamorphism. The upper part of the subducted slab, in contact with the hot asthenosphere, releases volatile fluids as it undergoes metamorphism which lowers melting temperatures and triggers partial melting. A complex set of melts rise from this region to produce **volcanic-magmatic arcs**. These melts range in composition from basaltic–gabbroic through dioritic–andesitic and may differentiate or be contaminated to produce melts of granitic–rhyolitic composition. Granitic–rhyolitic melts may also be generated by partial melting of older continental crust heated by rising magma and volatiles. Melts that reach the surface produce **volcanic arcs** such as those that characterize the "ring of fire" of the Pacific Ocean basin. Mt. St. Helens in Washington, Mt. Pinatubo in the Philippines, Mt. Fuji in Japan and the many volcanoes in Central America and the Andes of South America are all examples of volcanic arc composite volcanoes that form over Pacific Ocean subduction zones. These volcanic arcs add to the volume of continental crust.

When magmas intrude the crust and solidify below the surface, they produce plutonic igneous rocks that add new continental crust to Earth. Most of the world's major **batholith belts** represent **plutonic magmatic arcs**, subsequently exposed by erosion of the overlying volcanic arc. In addition, many of Earth's most important ore deposits (Chapter 19) are produced in association with volcanic-magmatic arcs over subduction zones.

Many of the magmas generated over the subducted slab cool and crystallize at the base of the lithosphere, thickening it by underplating. Underplating and intrusion are two of the major sets of processes by which **new continental crust** is generated. Once produced, the low density of continental crust prevents most of it from being subducted. This helps to explain its preservation potential and the great age that continental crust can achieve (>4.0 Ga).

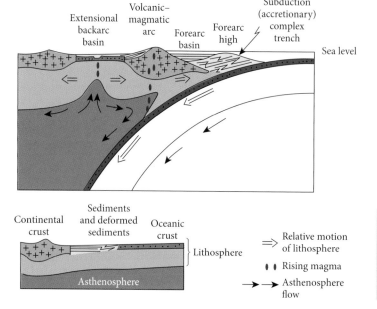

**Figure 1.12** Subduction zone depicting details of sediment distribution, sedimentary basins and volcanism in trench-arc system forearc and backarc regions.

Areas of significant relief, such as trench-arc systems are ideal sites for the production and accumulation of detrital (epiclastic) sedimentary rocks (Chapter 13). Huge volumes of detrital sedimentary rocks produced by the erosion of volcanic and magmatic arcs are deposited in forearc and backarc basins (Figure 1.12). They also occur with deformed abyssal sediments in the forearc subduction complex. As these sedimentary rocks are buried and deformed, they are commonly metamorphosed (Chapters 15 and 18).

*Continental collisions*

As ocean basins shrink by subduction, portions of the ridge system may be subducted. Once the ridge is subducted, growth of the ocean basin by sea-floor spreading ceases, the ocean basin continues to shrink, and the continents, microcontinents or arcs on either side are brought closer together as subduction proceeds. Eventually they converge to produce a continental collision.

When a **continental collision** (Dewey and Bird 1970) occurs, subduction eventually ceases. This occurs because most continental lithosphere is too buoyant to be subducted to great depths for prolonged periods of time. Small amounts may be subducted in this tectonic setting, likely because the density contrasts between the leading edges of the two plates are small. The continental lithosphere

involved in the collision may be part of a continent, a microcontinent or a volcanic-magmatic arc complex. Typically the collision occurs over millions to tens of millions of years as irregularly shaped ocean basins close at different times. As convergence continues, the margins of both continental plates are compressed and shortened horizontally and thickened vertically in a manner roughly analogous to what happens to two vehicles in a head-on collision. However, in the case of continents colliding at a convergent plate boundary, the convergence occurs over millions of years. The result is a severe horizontal shortening and vertical thickening which results in the progressive uplift of a mountain belt and/or extensive elevated plateau that mark the closing of an ancient ocean basin (Figure 1.13).

Long mountain belts formed along convergent plate boundaries are called **orogenic belts**. The increasing weight of the thickening orogenic belt causes the adjacent continental lithosphere to bend downward to produce **foreland basins** adjacent to the orogenic belt. Large amounts of detrital sediments derived from the erosion of the mountain belts are deposited in such basins. In addition, increasing temperatures and pressures within the thickening orogenic belt cause regional metamorphism of the rocks within it. If the temperatures become high enough, partial melting may occur to produce melts in the deepest

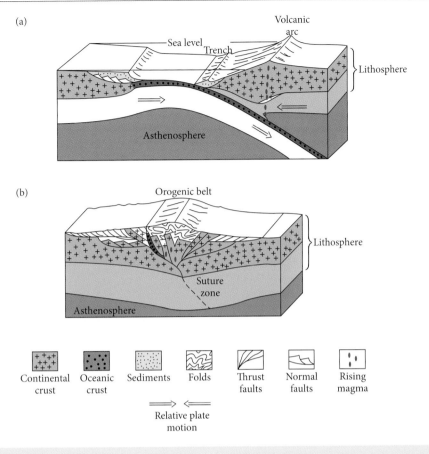

| | | | | | | |
|---|---|---|---|---|---|---|
| Continental crust | Oceanic crust | Sediments | Folds | Thrust faults | Normal faults | Rising magma |

Relative plate motion

**Figure 1.13**    (a) Ocean basin shrinks by subduction, as continents on two plates converge. (b) Continental collision produces a larger continent from two continents joined by a suture zone. Horizontal shortening and vertical thickening are accommodated by folds and thrust faults in the resulting orogenic belt.

parts of orogenic belts which rise to produce a variety of felsic igneous rocks.

A striking example of a modern orogenic belt is the Himalayan Mountain range formed by the collision of India with Eurasia over the past 40 Ma. The continued convergence of the Indian micro-continent with Asia has resulted in shortening and regional uplift of the Himalayan Mountain Belt along a series of major thrust faults and produced the Tibetan Plateau. Limestones near the summit of Mt. Everest, Earth's highest mountain, were formed on the floor of the Tethys Ocean that once separated India and Asia. They were then thrust to an elevation of nearly 9 km as that ocean was closed and the Himalayan Mountain Belt formed by continental collision. The collision has produced tectonic indentation of Asia, resulting in mountain ranges that wrap around India (Figure 1.14). The Ganges River in northern India flows approximately west-east in a trough that represents a modern foreland basin.

Continental collision inevitably produces a larger continent. It is now recognized that supercontinents such as Pangea and Rodinia were formed as the result of collisional tectonics. **Collisional tectonics** only requires converging plates whose leading edges are composed of lithosphere that is too buoyant to be easily subducted. In fact all the major continents display evidence of being composed of a collage of terranes that were accreted by collisional events at various times in their histories.

### 1.5.4    Transform plate boundaries

In order for plates to be able to move relative to one another, a third type of plate boundary is required. **Transform plate boundaries** are characterized by horizontal relative motion

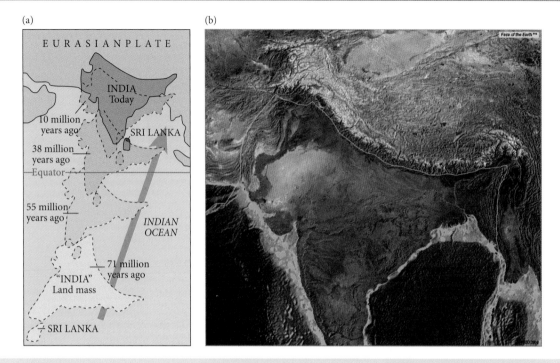

(a)                                    (b)

**Figure 1.14**   (a) Diagram depicting the convergence of India and Asia which closed the Tethys Sea. *Source*: Courtesy of NASA; (b) Satellite image of southern Asia showing indentation of Eurasia by India, the uplift of Himalayas and Tibetan Plateau and the mountains that "wrap around" India. *Source*: From UNAVCO.

along fault systems that is parallel to the plate boundary segment that separates two plates (Figure 1.4). Because the rocks on either side slide horizontally past each other, transform fault systems are a type of strike-slip fault system.

**Transform faults** were first envisioned by J.T. Wilson (1965) to explain the seismic activity along fracture zones in the ocean floor. **Fracture zones** are curvilinear zones of intensely faulted, fractured oceanic crust that are generally oriented nearly perpendicular to the ridge axis (Figure 1.15). Despite having been fractured by faulting along their entire length, earthquake activity is largely restricted to the **transform portion** of fracture zones that lies between offset ridge segments. Wilson (1965) reasoned that if sea floor was spreading away from two adjacent ridge segments in opposite directions, the portion of the fracture zone between the two ridge segments would be characterized by parallel relative motion in opposite directions. This would produce shear stresses that result in strike-slip faulting of the

lithosphere, frequent earthquakes and the development of a transform fault plate boundary. The exterior portion of fracture zones outside the ridge segments represents oceanic crust that was faulted and fractured when it was between ridge segments, then carried beyond the adjacent ridge segment by additional sea floor spreading. These exterior portions of fracture zones are appropriately called healed transforms or **transform scars**. They are no longer plate boundaries; instead, they are aseismic (lacking significant earthquakes) intraplate features because the seafloor on either side of them is spreading in the same direction (Figure 1.15). However, they do record the relative motion between two plates at the time they were actively forming.

**Continental transform plate boundaries** occur in continental lithosphere. The best known modern examples of continental transforms include the San Andreas Fault system in California (Figure 1.16), the Alpine Fault system in New Zealand, and the Anatolian Fault systems in Turkey and Iran. All of these are

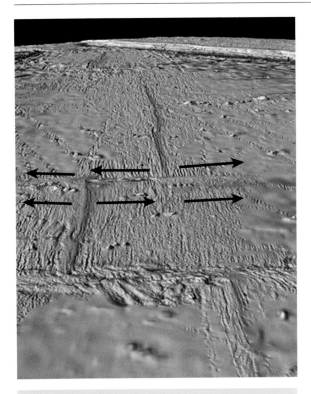

**Figure 1.15**   Transform faults offsetting ridge segments on the eastern Pacific Ocean floor off Central America. Arrows show directions of sea floor spreading away from the ridge. Oppositely directed arrows (black) indicate transform plate boundaries. Similarly directed arrows (red) indicate intraplate transform scars. *Source*: William Haxby, with the permission of Columbia University Earth Institute; Copyright Marine Geoscience Data System.

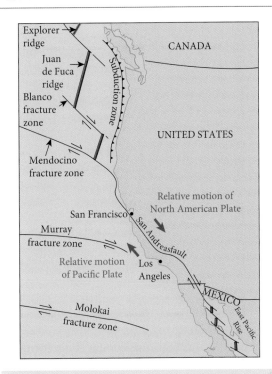

**Figure 1.16**   Fracture zones, transform faults and ridge segments in the eastern Pacific Ocean and western North America. The San Andreas Fault system is a continental transform fault plate boundary. *Source*: Courtesy of USGS.

characterized by active strike-slip fault systems of the type that characterize transform plate boundaries. In places where such faults bend or where their tips overlap, deep **pull-apart basins** may develop in which thick accumulations of sedimentary rocks accumulate rapidly.

Plates cannot simply diverge and converge; they must be able to slide past each other in opposite directions in order to move at all. Transform plate boundaries serve to accommodate this required sense of motion. Small amounts of igneous rocks form along transform plate boundaries, especially hybrid boundaries that have a component of divergence or convergence as well. They produce much smaller volumes of igneous and metamorphic rocks than are formed along

divergent and convergent plate boundaries. Because they neither create nor destroy large volumes of crust/lithosphere, these boundaries are sometimes referred to as **conservative plate boundaries**.

## 1.6   HOTSPOTS AND MANTLE CONVECTION

**Hotspots** (Wilson 1963) are long-lived areas in the mantle where anomalously large volumes of magma are generated. They occur beneath both oceanic lithosphere (e.g., Hawaii) and continental lithosphere (e.g., Yellowstone National Park, Wyoming) as well as along divergent plate boundaries (e.g., Iceland). Wilson pointed to **linear seamount chains** of volcanoes, such as the Hawaiian Islands (Figure 1.17), as surface expressions of hot spots, which he believed were fixed in one position for long periods. At any one

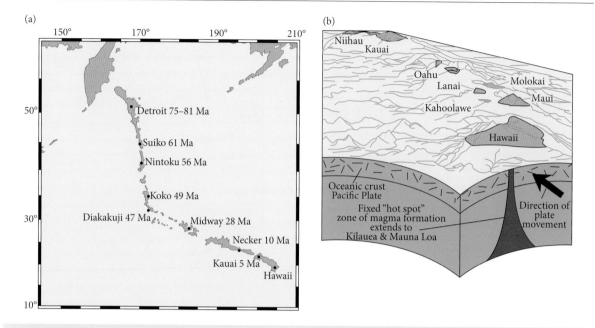

**Figure 1.17** (a) Linear seamount chain formed by plate movement over the Hawaiian hotspot and/or hot spot motion. *Source*: Tarduno et al. (2009). © The American Association for the Advancement of Science; (b) Mantle plume feeding surface volcanoes of Hawaiian Chain. *Source*: From USGS.

time, volcanism is largely restricted to that portion of the plate that lies above the hotspot. As the plate continues to move, older volcanoes are carried away from the hotspot and new volcanoes are formed above it. As a result, the age of these seamount chains increases systematically away from the hotspot in the direction of plate motion. For the Hawaiian chain, the data suggested a west-northwest direction of plate motion for the last 47 Ma. However, a change in orientation of the seamount chain to just a few degrees west of north for older volcanoes suggested that sea floor was spreading over the hotspot in a more northerly direction prior to 47 Ma. A similar trend of hotspot volcanism of increasing age over the past 15 Ma extends southwestward from the Yellowstone caldera. Some recent work suggests that the Hawaiian hotspot is not precisely fixed and some southward migration has been documented (Torsvik et al. 2017). Other work suggests that the amount of hotspot drift has been small (Wang et al. 2017). Stay tuned!

In the early 1970s, Morgan (1971) and others suggested that hotspots were the surface expression of fixed, long-lived mantle plumes. **Mantle plumes** were hypothesized to be columns of warm material which rose from near the core–mantle boundary. Some plumes appear to develop plume heads, as they spread outward near the base of the lithosphere (Griffiths and Campbell 1990). These evolving plume heads may be the cause of the apparent drift of hot spots, depending on how they spread out beneath the lithosphere. Later workers hypothesized that deep mantle plumes originate in the ultra-low velocity zone (LVZ) of the D″ layer at the base of the mantle and may represent the dregs of subducted slabs warmed sufficiently by contact with the outer core to become buoyant enough to rise. Huge **superplumes** (Larson 1991) were hypothesized to be significant players in extinction events, the initiation, and location (Arndt and Davaille 2013; Condie 2015) of continental rifting, and in the **supercontinent cycle** (Sheridan 1987) of rifting and collision that has caused supercontinents to form and rift apart numerous times during Earth's history. Eventually most intraplate volcanism and magmatism was linked to hotspots and mantle plumes.

The picture has become considerably muddled in the twenty-first century. Many Earth scientists have offered significant evidence that mantle plumes do not exist (Foulger et al. 2005). Others have suggested that mantle plumes exist, but are not fixed (Nataf 2000; Koppers et al. 2001; Tarduno et al. 2009). Still others (Nolet et al. 2006) suggest on the basis of fine-scale thermal tomography that some of these plumes originate near the core–mantle boundary, others at the base of the transition zone (660 km) and others at around 1400 km in the mesosphere. They suggest that the rise of some plumes from the deep mantle is interrupted by the 660 km discontinuity, whereas other plumes seem to cross this discontinuity. This is reminiscent of the behavior of subducted slabs, some of which spread out above the 660 km discontinuity, whereas others penetrate it and apparently sink to the core–mantle boundary. Recent advances in new imaging methods that use powerful supercomputers have suggested that plumes originating near the base of the mantle do exist beneath many hotspots (French and Romanowicz 2015; Nelson and Grand 2018; Sanni et al. 2019) including Yellowstone, Hawaii, and Iceland, even though they are not always vertical. Wang et al. (2017) demonstrated that most groups of hotspots migrate very slowly, if at all, over time. It is very likely that hot spots are generated by a variety of processes related to mantle convection patterns, but these are still not well understood. Deep Earth tomography will continue to be an exciting area of Earth research over the coming decade.

In this chapter, we have attempted to provide a spatial and tectonic context for the processes which form Earth materials. One part of this context involves the location of compositional and mechanical layers within the geosphere where Earth materials form. Ultimately, however, the geosphere cannot be viewed as a group of static layers. Plate tectonics implies significant horizontal and vertical movement of the lithosphere with compensating motion of the underlying asthenosphere and deeper mantle. Global tectonics suggests significant lateral heterogeneity within layers and significant vertical exchange of material between layers caused by processes such as convection, subduction and mantle plumes.

Helping students to understand how variations in composition, position within the geosphere and tectonic processes interact on many scales to generate distinctive Earth materials is the fundamental task of this book. We hope you will find what follows is both exciting and meaningful.

## CONTENT ASSESSMENT

1   What properties distinguish the following zones of Earth's interior? Elaborate.
    a.  continental crust, oceanic crust, and mantle
    b.  lithosphere, asthenosphere, and mesosphere
    c.  low velocity zone (LVZ), transition zone, and D″ layer
    d.  core, outer core, and inner core
2   Detail the processes by which oceanic crust is created and grows through time and contrast these with the processes by which it shrinks and is "destroyed."
3   Explain why the age of oceanic crust generally increases systematically away from the ridge system axis in both directions and the major reasons why there are so many local exceptions to this rule.
4   Describe the three major types of plate boundaries and the features that are associated with and produced by each.
5   Explain how transform faults between two ridge segments form and how, over time, they can generate long fracture zones in oceanic crust. In addition, contrast the earthquake activity on transforms with that on (the external portions of) fracture zones and explain the major reason for this contrast.
6   What is the major process are involved in "collisional tectonics"? Detail the features are produced by and that record such collisional events.

## REFERENCES

Anderson, D.L. (1989). *Theory of the Earth*. Oxford, UK: Blackwell Scientific Publications 366 pp.

Anderson, D.L., Sammis, C., and Jordan, T. (1971). Composition and evolution of the mantle and core. *Science 171*: 1103–1112.

Arndt, N. and Davaille, A. (2013). Episodic earth evolution. *Tectonophysics 609*: 661–674.

Condie, K.C. (1982). *Plate Tectonics and Crustal Evolution*. Oxford, UK: Pergamon Press 476 pp.

Condie, K.C. (2015). *Earth as an Evolving Planetary System*. Cambridge, MA: Academic Press 415 pp.

Dewey, J.F. and Bird, J.M. (1970). Mountain belts and new global tectonics. *Journal of Geophysical Research 75*: 2625–2647.

Dietz, R. (1961). Continental and ocean basin evolution by spreading of the sea floor. *Nature 190*: 854–857.

Foulger, G.R., Natland, J.H., Presnall, D.C., and Anderson, D.L. (eds.) (2005). *Plates, plumes, and paradigms*, vol. 388. Geological Society of America Special 861 pp.

French, S. and Romanowicz, B. (2015). Broad plumes rooted at the base of the Earth's mantle beneath major hot spots. *Nature 525*: 95–99.

Grand, S.P. (2002). Mantle shear-wave tomography and fate of subducted slabs. *Philosophical Transactions of the Royal Society of London: Mathematics, Physical and Engineering Sciences 360*: 2475–2491.

Granot, R. (2016). Palaeozoic oceanic crust preserved beneath the eastern Mediterranean. *Nature Geoscience 9*: 701–705.

Griffiths, R.W. and Campbell, I.H. (1990). Stirring and structure in mantle starting plumes. *Earth and Planetary Science Letters 99*: 66–78.

Hess, H.H. (1962). History of the ocean basins. In: *Petrologic Studies: A Volume in Honor of A. F. Buddington* (eds. A.D.J. Engel, H.L. James and B.F. Leonard), 599–620. Geological Society of America Publications.

Hutko, A.R., Lay, T., Garnero, D.J., and Revenaugh, J. (2006). Seismic detection of folded, subducted lithosphere at the core–mantle boundary. *Nature 441*: 333–336.

Isacks, B., Oliver, J., and Sykes, L.R. (1968). Seismology and the new global tectonics. *Journal of Geophysical Research 73*: 5855–5899.

Ishii, M. and Dziewonski, A.M. (2002). The innermost core of the earth: evidence for a change in anisotropic behavior at the radius of 300 km. *Proceedings of the Natironal Academy of Science 99*: 14026–14030.

Jeanloz, R. (1993). The mantle in sharper focus. *Nature 365*: 110–111.

Koppers, A.A.P., Morgan, J.P., Morgan, J.W., and Staudiget, H. (2001). Testing the fixed hotspot hypothesis using $^{40}Ar/^{39}Ar$ age progressions along seamount trails. *Earth and Planetary Science Letters 185*: 237–252.

Larson, R.V. (1991). Geological consequences of superplumes. *Geology 19*: 963–966.

Le Pichon, X. (1968). Sea-floor spreading and continental drift. *Journal of Geophysical Research 73*: 3661–3697.

McKenzie, D.P. and Morgan, W.J. (1969). Evolution of triple junctions. *Nature 224*: 125–133.

Miyagi, I. and Marquardt, H. (2015). Strength under pressure. *Nature Geoscience 8*: 255–256.

Morgan, W.J. (1971). Convection plumes in the lower mantle. *Nature 230*: 42–43.

Nataf, H.C. (2000). Seismic imaging of mantle plumes. *Annual Review of Earth and Planetary Science Letters 28*: 391–417.

Nelson, P.L. and Grand, S.P. (2018). Lower-mantle plume beneath the Yellowstone hot spot revealed by core waves. *Nature Geoscience 11*: 280–284.

Nolet, S.-I., Karato, S.I., and Montelli, R. (2006). Plume fluxes from seismic tomography. *Earth and Planetary Science Letters 248*: 685–699.

O'Neil, J., Carlson, R.W., Francis, D., and Stevenson, R.K. (2008). Neodynium-142 evidence for Hadean mafic crust. *Science 321*: 1828–1831.

Ringwood, A.E. (1975). *Composition and Petrology of the Earth's Mantle*. New York: McGraw-Hill 618 pp.

Sanni, T., Luttinen, A.V., Heinonen, J.S., and Jamai, D.L. (2019). Luehna picrites, Central Mozambique, messengers from a mantle plume source of Karoo continental flood basalts. *Lithos*: 346–347.

Sheridan, R. (1987). Pulsation tectonics as the control of long-term stratigraphic cycles. *Paleoceanography 2*: 97–118.

Stern, R.A. and Bleeker, W. (1998). Age of the world's oldest rocks refined using Canada's SHRIMP: the Acasta Gneiss Complex, Northwest Territories, Canada. *Geoscience Canada 25*: 27–31.

Tarduno, J., Bunge, H.-P., Sleep, N., and Hansen, U. (2009). The bent Hawaiian–Emperor hotspot track: inheriting the mantle wind. *Science 324*: 50–53.

Torsvik, T., Doubrovine, P.V., Steinberger, B. et al. (2017). Pacific plate motion change caused the Hawaiian-Emperor Bend. *Nature Communications 8* https://doi.org/10.1038/ncomms15660.

Tschauner, O., Ma, C., Beckett, J.R. et al. (2014). Discovery of bridgmanite, the most abundant mineral in Earth, in a shocked meteorite. *Science 346*: 1100–1102.

Valley, J.W., Cavosie, A.J., Ushikubo, T. et al. (2014). Hadean age for a post-magma-ocean zircon confirmed by atom probe tomography. *Nature Geoscience 7*: 219–223.

Vine, F.J. and Matthews, D.H. (1963). Magnetic anomalies over ocean ridges. *Nature 199*: 947–949.

Wang, T., Song, X., and Han, H.X. (2015). Equatorial anisotropy in the inner part of Earth's core from autocorrelation of earthquake coda. *Naure Geoscience 3*: 224–227.

Wang, C., Gordon, R.C., and Zhang, T. (2017). Bounds of geologically current rates of motion of groups of hot spots. *Geophysical Research Letters* 44: 6048–6056.

Wilde, S., Valley, J.W., Peck, W.H., and Graham, C.M. (2001). Evidence from detrital zircons for the existence of continental crust and ocean on the Earth 4.4 Ga ago. *Nature 409*: 175–178.

Williams, Q. and Garnero, E.J. (1996). Seismic evidence for partial melt at the base of Earth's mantle. *Science 273*: 1528–1530.

Wilson, J.T. (1963). Evidence from islands on the spreading of ocean floors. *Nature 197*: 536–538.

Wilson, J.T. (1965). A new class of faults and their bearing on continental drift. *Nature 207*: 343–347.

# Chapter 2

# Atoms, elements, bonds, and coordination polyhedra

If we zoom in on any portion of Earth, we will see that it is composed of progressively smaller entities. At very high magnification, we will be able to discern very small particles called atoms. Almost all Earth materials are composed of atoms which, in turn, strongly influence their material properties. Understanding the ways in which these basic chemical constituents combine to produce larger scale Earth materials is essential to understanding our planet.

In this chapter we will consider the fundamental chemical constituents that bond together to produce Earth materials such as minerals and rocks. We will discuss the nuclear and electron configurations of atoms and then discuss the role these play in determining both atomic and mineral properties and the conditions under which minerals form. This information will provide a basis for understanding how and why minerals, rocks and other Earth materials have the following characteristics:

1 They possess specific properties that characterize and distinguish them.
2 They provide benefits and hazards through their production, refinement and use.
3 They form in response to particular sets of environmental conditions and processes.
4 They record information about the conditions and processes that produce them.
5 They permit us to infer significant events in Earth's history.

## 2.1 ATOMS

Earth materials are composed of smaller entities. At very small scales, such as viewed through a tunneling electron microscope, we

*Earth Materials*, Second Edition. Kevin Hefferan and John O'Brien.
© 2022 John Wiley & Sons Ltd. Published 2022 by John Wiley & Sons Ltd.
Companion website: www.wiley.com/go/hefferan/earthmaterials2

are able to discern particles called **atoms** whose effective diameters are a few angstroms (1 Å = $10^{-10}$ m, about a million times smaller than the width of a human hair). These tiny atoms, in turn, consist of three main particles – electrons, protons, and neutrons – which were discovered between 1895 and 1902. The major properties of electrons, protons, and neutrons are summarized in Table 2.1.

**Protons (p⁺)** and **neutrons (n⁰)** each have a mass of ~1 atomic mass unit (amu) and are clustered together in a small, positively charged, central region of the atom called the **nucleus** (Figure 2.1). Protons possess a positive electric charge and neutrons are electrically neutral. The nucleus is surrounded by a vastly larger,

mostly "empty", region called the **electron cloud**. The electron cloud represents the area in which the **electrons (e⁻)** in the atom move about the nucleus in patterns called orbitals (Figure 2.1). Electrons have a negative electric charge and an almost negligible mass of 0.000 005 amu. Knowledge of these three fundamental particles in atoms is essential to understanding how minerals and other materials form, how they can be used as resources and how we can deal with their hazardous effects that can be seen occasionally.

### 2.1.1 The nucleus, atomic number, atomic mass number, and isotopes

The nucleus of atoms is composed of positively charged protons and uncharged neutrons bound together by a strong force. Ninety-two fundamentally different kinds of atoms called elements have been discovered in the natural world. More than 25 additional elements have been created synthetically in laboratory experiments during the past century. Each **element** is characterized by the number of protons in its nucleus. The number of protons in the nucleus is called the **atomic number (Z)**. The atomic number is typically represented by a subscript number to the lower left of the element symbol. The 92 naturally occurring elements range from hydrogen (Z = 1) through uranium (Z = 92). Hydrogen (₁H) is characterized by having one proton in its nucleus. Every atom of uranium (₉₂U) contains 92 protons in its nucleus. The atomic number is the unique property that distinguishes the atoms of each element from atoms of all other elements.

Every atom also possesses mass that largely results from the protons and neutrons in its nucleus. The mass of a particular atom is called its **atomic mass number**, and is expressed in atomic mass units (amu). As the mass of both protons and neutrons is ~1 amu, the atomic mass number is closely related to the total number of protons plus neutrons in its nucleus. The simple formula for atomic mass number is: atomic mass number = number of protons plus number of neutrons (#p⁺ + #n⁰). The atomic mass number is indicated by a superscript number to the upper left of the element symbol. For example, most oxygen atoms have eight

**Table 2.1** Major properties of electrons, protons, and neutrons.

| Particle type | Electric charge | Atomic mass (amu)ᵃ |
| --- | --- | --- |
| Proton (p⁺) | +1 | 1.00728 |
| Neutron (n⁰) | 0 | 1.00867 |
| Electron (e⁻) | −1 | 0.0000054 |

ᵃ amu = atomic mass unit = 1/12 mass of an average carbon atom.

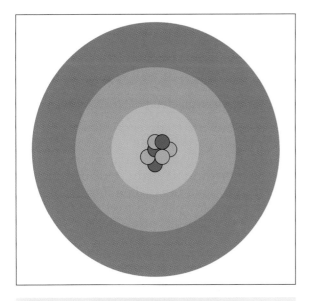

**Figure 2.1** Simplified model atom with nucleus that contains positively charged protons (dark blue) and electrically neutral neutrons (light blue) surrounded by an electron cloud (red shades) in which negatively charged electrons move in orbitals about the nucleus.

protons and eight neutrons so their atomic mass number is written $^{16}O$.

Although each element has a unique atomic number, many elements are characterized by atoms with different atomic mass numbers. Atoms of the same element that possess different atomic mass numbers are called **isotopes**. For example, three different isotopes of hydrogen exist (Figure 2.2a). All hydrogen ($_1H$) atoms have an atomic number of 1. The common form of hydrogen atom, sometimes called **protium**, has one proton and no neutrons in the nucleus; therefore protium has an atomic mass number of 1, symbolized as $^1H$. A less common form of hydrogen called **deuterium**, used in some nuclear reactors, has an atomic mass number of 2, symbolized by $^2H$. This implies that it contains one proton and one neutron in its nucleus ($1p^+ + 1n^0$). A rarer isotope of hydrogen called **tritium** has an atomic mass number of 3, symbolized by $^3H$. The nucleus of tritium has one proton and two neutrons. Similarly oxygen occurs in three different isotopes: $^{16}O$, $^{17}O$, and $^{18}O$. All oxygen atoms contain eight protons but neutron numbers vary between $^{16}O$, $^{17}O$, and $^{18}O$, which contain eight, nine, and ten neutrons, respectively (Figure 2.2a). The average atomic mass for each element is the weighted average for all the isotopes of that element. This helps

to explain why the listed atomic masses for each element do not always approximate the whole numbers produced when one adds the number of protons and neutrons in the nucleus of a particular isotope.

The general isotope symbol for the nucleus of an atom expresses its atomic number to the lower left of its symbol, the number of neutrons to the lower right and the atomic mass number (number of protons + number of neutrons) to the upper left. For example, the most common isotope of uranium has the symbolic nuclear configuration of 92 protons + 146 neutrons and an atomic mass number of 238:

$$^{238}_{92}U_{146}$$

**Stable isotopes** have stable nuclear configurations that tend to remain unchanged; they retain the same number of protons and neutrons over time. On the other hand, **radioactive isotopes** have unstable nuclear configurations (numbers of protons and neutrons) that spontaneously change over time via radioactive decay processes, until they achieve stable nuclear configurations and become stable isotopes of another element. Both types of isotopes are extremely useful in solving geological and environmental problems, as discussed in Chapter 3. Radioactive isotopes are used in many medical treatments, but also present serious environmental hazards (Chapter 19).

### 2.1.2 The electron cloud

Electrons are enigmatic entities, with properties of both particles and wave energy, that move very rapidly around the nucleus in ultimately unpredictable paths. Our depiction of the electron cloud is based on the probabilities of finding a particular electron at a particular place. The wave-like properties of electrons help to define the three-dimensional shapes of their probable locations, known as **orbitals**. The size and shape of the electron cloud defines the chemical behavior of atoms and ultimately the composition of all the Earth materials they combine to form. Simplified models of the electron cloud depict electrons distributed in spherical orbits around the nucleus (Figure 2.3); the reality is much more complex. Because the electron cloud largely determines the chemical behavior of atoms and how they combine to

(a)

Protium      Deuterium      Tritium
$^1H$          $^2H$         $^3H$

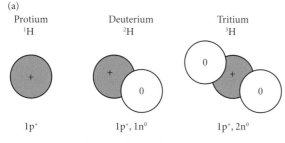

$1p^+$        $1p^+, 1n^0$      $1p^+, 2n^0$

Nuclei of the three hydrogen isotopes

(b)

$^{16}O$      $^{17}O$      $^{18}O$

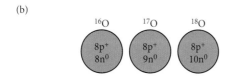

Nuclei of the three oxygen isotopes

**Figure 2.2** (a) Nuclear configurations of the three common isotopes of hydrogen. (b) Nuclear configurations of the three common isotopes of oxygen.

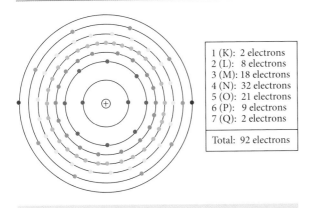

| 1 (K): | 2 electrons |
| 2 (L): | 8 electrons |
| 3 (M): | 18 electrons |
| 4 (N): | 32 electrons |
| 5 (O): | 21 electrons |
| 6 (P): | 9 electrons |
| 7 (Q): | 2 electrons |
| Total: | 92 electrons |

**Figure 2.3** Distribution of electrons in the principal quantum levels ("electron shells") of uranium: K-shell electrons violet; L-shell blue; M-shell bluish green; N-shell green; O-shell yellow; P-shell orange and Q-shell red.

produce Earth materials, it is essential to understand some fundamental concepts about it.

Every electron in an atom possesses a unique set of properties that distinguishes it from all the other electrons in that atom. An individual electron's identity is given by four properties that include its (1) principal quantum number, (2) azimuthal quantum number, (3) magnetic quantum number, and (4) spin number. Each electron in the electron cloud possesses a unique combination of the four quantum properties.

The **principal quantum number** (n) signifies the **principal quantum energy region**, sometimes called "**level**" or "**shell**" in which a particular electron occurs. It is related to its distance from the nucleus. Principle quantum regions are numbered in order of increasing electron energies **1, 2, 3, 4, 5, 6 or 7** or alternatively lettered **K, L, M, N, O, P or Q**. These are arranged from low principal quantum number for low energy regions closer to the nucleus to progressively higher quantum number for higher energy regions farther away from the nucleus.

Each principal quantum region or level contains electrons with one or more **azimuthal or angular momentum quantum numbers** which signify the directional quantum energy region, sometimes called "**subshell**", in which the electron occurs. This is related to the angular momentum of the electron and the shape of its orbital. Azimuthal quantum numbers or subshells are labeled **s, p, d, and f**. As discussed

below, the number of electrons in the s and p "subshells" of the atom's highest principal quantum level largely determines the chemical behavior of elements. All four azimuthal quantum numbers are important for understanding the layout of the periodic table of the elements which is discussed below. The other two quantum properties, the **magnetic quantum number** (m) and the **spin number** (±½) define the orientation of the quantum probability region in which the electron is located and its spin relative to a reference framework. Table 2.2 summarizes the quantum properties of the electrons that can exist in principle quantum regions or "shells" 1–7. To summarize, each electron in an atom possesses a unique set of the four principle quantum properties.

Atomic nuclei were created largely during the "big bang," by subsequent fusion reactions between protons and neutrons in the interior of stars, and in supernova. When elements are formed, electrons are added to the lowest available quantum level in numbers equal to the number of protons in the nucleus. Electrons are added to the atoms in a distinct

**Table 2.2** Quantum designations of electrons in the 92 naturally occurring elements. The numbers refer to the principal quantum region occupied by the electrons within the electron cloud; small case letters refer to the subshell occupied by the electrons.

| Principal quantum number | Subshell description | Number of electrons |
|---|---|---|
| 1 (K) | 1s | 2 |
| 2 (L) | 2s | 2 |
| | 2p | 6 |
| 3 (M) | 3s | 2 |
| | 3p | 6 |
| | 3d | 10 |
| 4 (N) | 4s | 2 |
| | 4p | 6 |
| | 4d | 10 |
| | 4f | 14 |
| 5 (O) | 5s | 2 |
| | 5p | 6 |
| | 5d | 10 |
| | 5f | 14 |
| 6 (P) | 6s | 2 |
| | 6p | 6 |
| 7 (Q) | 7s | 2 |
| | Total = 92 | |

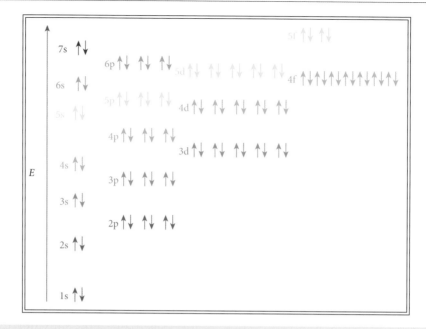

Figure 2.4 The quantum properties of electrons in the 92 naturally occurring elements, listed with increasing quantum energy (E) from bottom to top; K-shell electrons violet; L-shell blue; M-shell bluish green; N-shell green; O-shell yellow; P-shell orange and Q-shell red.

sequence, from lowest quantum level electrons to highest quantum level electrons. The relative quantum energy of each electron is shown in Figure 2.4.

The **diagonal rule** is a simple rule or memory device that predicts, with very few exceptions, the sequence in which electrons are added to the electron cloud. The order in which electrons are added to shells is depicted by a series of diagonal lines from 1s to 7p (Figure 2.5).

Table 2.3 shows the ground state electron configurations for the elements. One can write the electron configuration of any element in a sequence from lowest to highest energy electrons. For example, calcium (Z = 20) possesses the electron configuration $1s_2$, $2s_2$, $2p_6$, $3s_2$, $3p_6$, $4s_2$. Elements with principal quantum levels (shells) or azimuthal quantum s- and p-subshells that are completely filled (that is they contain the maximum number of electrons possible) possess very stable electron configurations. These elements include the noble gas elements such as helium (He), neon (Ne), argon (Ar), and krypton (Kr) which, because of their stable configurations, tend not to react with or bond to other elements. For elements other than helium, the highest quantum level stable

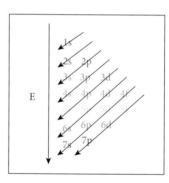

Figure 2.5 The diagonal rule for determining the sequence in which electrons are added to the electron cloud; K-shell electrons violet; L-shell blue; M-shell bluish green; N-shell green; O-shell yellow; P-shell orange and Q-shell red.

configuration is $s_2$, $p_6$, sometimes referred to as the "**stable octet.**"

## 2.2 THE PERIODIC TABLE

The naturally occurring and synthetic elements discovered to date display certain periodic traits; that is, several elements with different atomic numbers display similar chemical behavior. Tables that attempt to portray the periodic behavior of the elements are called **periodic tables**. It is now well known that the periodic behavior of the elements is related to their electron configurations. In most modern periodic tables (Table 2.3) the elements are arranged in **seven rows or periods** and **eighteen columns or groups**. Two sets of elements, the **lanthanides** and the **actinides**, which belong to the sixth and seventh rows, respectively, are listed separately at the bottom of such tables to allow all the elements to be shown conveniently on a printed page of standard dimensions. For a rather different approach to organizing the elements in a periodic table for Earth scientists, readers are referred to Railsback (2003).

## 2.2.1   Rows (periods) on the periodic table

On the left-hand side of the periodic table the row numbers 1–7 indicate the highest principle quantum level in which electrons occur in the elements in that row. Every element in a given horizontal row has its outermost electrons in the same principle energy region or level. Within each row, the number of electrons increases with the atomic number from left to right. The number of elements in each row varies, and reflects the sequence in which electrons are added to various quantum levels as the atoms are formed. For example, row 1 has only two elements because the first quantum level can contain only two 1s electrons. The two elements are hydrogen ($1s_1$) and helium ($1s_2$).

**Table 2.3**   Periodic table of the naturally-occurring elements displaying atomic symbols, atomic number (Z), average mass, ground state electron configuration, common valence states and electronegativity of each element.

Simplified periodic table of the elements showing symbols atomic numbers and mass numbers

| IA | IIA | III B | IVB | VB | VIB | VIIB | --------VIIIB--------- | | | IB | IIB | III A | IVA | VA | VIA | VIIA | VIIIA |
| 1 | 2 | 3 | 4 | 5 | 6 | 7 | 8 | 9 | 10 | 11 | 12 | 13 | 14 | 15 | 16 | 17 | 18 |
|---|---|---|---|---|---|---|---|---|---|---|---|---|---|---|---|---|---|
| 1<br>H<br>1.008 | | | | | | | | | | | | | | | | | 2<br>He<br>4.003 |
| 3<br>Li<br>6.941 | 4<br>Be<br>9.012 | | | | | | | | | | | 5<br>B<br>10.810 | 6<br>C<br>12.011 | 7<br>N<br>14.007 | 8<br>O<br>15.999 | 9<br>F<br>18.998 | 10<br>Ne<br>20.179 |
| 11<br>Na<br>22.990 | 12<br>Mg<br>24.305 | | | | | | | | | | | 13<br>Al<br>26.962 | 14<br>Si<br>28.086 | 15<br>P<br>30.974 | 16<br>S<br>32.060 | 17<br>Cl<br>35.453 | 18<br>Ar<br>39.948 |
| 19<br>K<br>39.098 | 20<br>Ca<br>40.080 | 21<br>Sc<br>44.956 | 22<br>Ti<br>47.900 | 23<br>V<br>50.942 | 24<br>Cr<br>51.996 | 25<br>Mn<br>54.938 | 26<br>Fe<br>55.847 | 27<br>Co<br>58.933 | 28<br>Ni<br>58.700 | 29<br>Cu<br>63.546 | 30<br>Zn<br>65.380 | 31<br>Ga<br>69.720 | 32<br>Ge<br>72.590 | 33<br>As<br>74.922 | 34<br>Se<br>78.960 | 35<br>Br<br>79.904 | 36<br>Kr<br>83.800 |
| 37<br>Rb<br>85.468 | 38<br>Sr<br>87.620 | 39<br>Y<br>88.906 | 40<br>Zr<br>91.220 | 41<br>Nb<br>92.906 | 42<br>Mo<br>95.940 | 43<br>Tc<br>(98) | 44<br>Ru<br>101.07 | 45<br>Rh<br>102.91 | 46<br>Pd<br>106.40 | 47<br>Ag<br>107.87 | 48<br>Cd<br>112.41 | 49<br>In<br>114.82 | 50<br>Sn<br>118.69 | 51<br>Sb<br>121.75 | 52<br>Te<br>127.60 | 53<br>I<br>126.90 | 54<br>Xe<br>131.30 |
| 55<br>Cs<br>132.91 | 56<br>Ba<br>137.33 | 57<br>La<br>138.91 | 72<br>Hf<br>178.48 | 73<br>Ta<br>180.95 | 74<br>W<br>183.85 | 75<br>Re<br>186.21 | 76<br>Os<br>190.20 | 77<br>Ir<br>192.22 | 78<br>Pt<br>195.09 | 79<br>Au<br>196.97 | 80<br>Hg<br>200.59 | 81<br>Tl<br>204.37 | 82<br>Pb<br>208.98 | 83<br>Bi<br>208.98 | 84<br>Po<br>(209) | 85<br>At<br>(210) | 86<br>Rn<br>(222) |
| 87<br>Fr<br>(223) | 88<br>Ra<br>226.03 | 89<br>Ac<br>227.03 | | | | | | | | | | | | | | | |

| Lanthanides | 58<br>Ce<br>140.12 | 59<br>Pr<br>140.91 | 60<br>Nd<br>144.24 | 61<br>Pm<br>(145) | 62<br>Sm<br>150.40 | 63<br>Eu<br>151.96 | 64<br>Gd<br>157.25 | 65<br>Tb<br>158.93 | 66<br>Dy<br>162.50 | 67<br>Ho<br>164.93 | 68<br>Er<br>167.29 | 69<br>Tm<br>168.94 | 70<br>Yb<br>173.04 | 71<br>Lu<br>174.97 |
|---|---|---|---|---|---|---|---|---|---|---|---|---|---|---|
| Actinides | 90<br>Th<br>232.04 | 91<br>Pa<br>231.04 | 92<br>U<br>238.03 | | | | | | | | | | | |

**Table 2.3**   Continued

Simplified periodic table of the elements showing symbols and electron configurations

| IA | IB | IIIB | IVB | VB | VIB | VIIB | --------XIIIB-------- | | | IB | IIB | IIIA | IVA | VA | VIA | VIIA | VIIIA |
|---|---|---|---|---|---|---|---|---|---|---|---|---|---|---|---|---|---|
| **1** | **2** | **3** | **4** | **5** | **6** | **7** | **8** | **9** | **10** | **11** | **12** | **13** | **14** | **15** | **16** | **17** | **18** |
| $1s_1$<br>**H** | | | | | | | | | | | | | | | | | $1s_2$<br>**He** |
| He +<br>**Li**<br>$2s_1$ | $2s_2$<br>**Be** | | | | | | | | | | | $2s_2 2p_1$<br>**B** | $2s_2 2p_2$<br>**C** | $2s_2 2p_3$<br>**N** | $2s_2 2p_4$<br>**O** | $2s_2 2p_5$<br>**F** | $2s_2 2p_6$<br>**Ne** |
| Ne +<br>**Na**<br>$3s_1$ | $3s_2$<br>**Mg** | | | | | | | | | | | $3s_2 3p_1$<br>**Al** | $3s_2 3p_2$<br>**Si** | $3s_2 3p_3$<br>**P** | $3s_2 2p_4$<br>**S** | $3s_1 3p_5$<br>**Cl** | $3s_2 3p_6$<br>**Ar** |
| Ar +<br>**K**<br>$4s_1$ | $4s_2$<br>**Ca** | $4s_2$<br>**Sc**<br>$3d_1$ | $4s_2$<br>**Ti**<br>$3d_2$ | $4s_2$<br>**V**<br>$3d_3$ | $4s_1$<br>**Cr**<br>$3d_5$ | $4s_2$<br>**Mn**<br>$3d_5$ | $4s_2$<br>**Fe**<br>$3d_6$ | $4s_2$<br>**Co**<br>$3d_7$ | $4s_2$<br>**Ni**<br>$3d_8$ | $4s_1$<br>**Cu**<br>$3d_{10}$ | $4s_2$<br>**Zn**<br>$3d_{10}$ | $4s_2 4p_1$<br>**Ga**<br>$3d_{10}$ | $4s_2 4p_2$<br>**Ge**<br>$3d_{10}$ | $4s_2 4p_3$<br>**As**<br>$3d_{10}$ | $4s_2 4p_4$<br>**Se**<br>$3d_{10}$ | $4s_2 4p_5$<br>**Br**<br>$3d_{10}$ | $4s_2 4p_6$<br>**Kr**<br>$3d_{10}$ |
| Kr +<br>**Rb**<br>$5s_1$ | $5s_2$<br>**Sr** | $5s_2$<br>**Y**<br>$4d_1$ | $5s_2$<br>**Zr**<br>$4d_2$ | $5s_1$<br>**Nb**<br>$4d_4$ | $5s_1$<br>**Mo**<br>$4d_5$ | $5s_2$<br>**Tc**<br>$4d_5$ | $5s_1$<br>**Ru**<br>$4d_7$ | $5s_1$<br>**Rh**<br>$4d_8$ | $5s_2$<br>**Pd**<br>$4d_8$ | $5s_1$<br>**Ag**<br>$4d_{10}$ | $5s_2$<br>**Cd**<br>$4d_{10}$ | $5s_2 5p_1$<br>**In**<br>$4d_{10}$ | $5s_2 5p_2$<br>**Sn**<br>$4d_{10}$ | $5s_2 5p_3$<br>**Sb**<br>$4d_{10}$ | $5s_2 5p_4$<br>**Te**<br>$4d_{10}$ | $5s_2 5p_5$<br>**I**<br>$4d_{10}$ | $5s_2 5p_6$<br>**Xe**<br>$4d_{10}$ |
| Xe +<br>**Cs**<br>$6s_1$ | $6s_2$<br>**Ba** | $6s_2$<br>**La**<br>$5d_1$ | $6s_2$<br>**Hf**<br>$4f_{14}5d_2$ | $6s_2$<br>**Ta**<br>$4f_{14}5d_3$ | $6s_2$<br>**W**<br>$4f_{14}5d_4$ | $6s_2$<br>**Re**<br>$4f_{14}5d_5$ | $6s_2$<br>**Os**<br>$4f_{14}5d_6$ | $6s_2$<br>**Ir**<br>$4f_{14}5d_7$ | $6s_1$<br>**Pt**<br>$4f_{14}5d_9$ | $6s_1$<br>**Au**<br>$4f_{14}5d_{10}$ | $6s_2$<br>**Hg**<br>$4f_{14}5d_{10}$ | $6s_2 6p_1$<br>**Tl**<br>$4f_{14}5d_{10}$ | $6s_2 6p_2$<br>**Pb**<br>$4f_{14}5d_{10}$ | $6s_2 6p_3$<br>**Bi**<br>$4f_{14}5d_{10}$ | $6s_2 6p_4$<br>**Po**<br>$4f_{14}5d_{10}$ | $6s_2 6p_5$<br>**At**<br>$4f_{14}5d_{10}$ | $6s_2 6p_6$<br>**Rn**<br>$4f_{14}5d_{10}$ |
| Rn +<br>**Fr**<br>$7s_1$ | $7s_2$<br>**Ra** | $7s_2$<br>**Ac**<br>$6d_1$ | | | | | | | | | | | | | | | |

| Lanthanides | $6s_2$<br>**Ce**<br>$5d_1 4f_1$ | $6s_2$<br>**Pr**<br>$5d_1 4f_2$ | $6s_2$<br>**Nd**<br>$5d_1 4f_3$ | $6s_2$<br>**Pm**<br>$5d_1 4f_4$ | $6s_2$<br>**Sm**<br>$5d_1 4f_5$ | $6s_2$<br>**Eu**<br>$5d_1 4f_6$ | $6s_2$<br>**Gd**<br>$5d_1 4f_7$ | $6s_2$<br>**Tb**<br>$5d_1 4f_8$ | $6s_2$<br>**Dy**<br>$5d_1 4f_9$ | $6s_2$<br>**Ho**<br>$5d_1 4f_{10}$ | $6s_2$<br>**Er**<br>$5d_1 4f_{11}$ | $6s_2$<br>**Tm**<br>$5d_1 4f_{12}$ | $6s_2$<br>**Yb**<br>$5d_1 4f_{13}$ | $6s_2$<br>**Lu**<br>$5d_1 4f_{14}$ |
|---|---|---|---|---|---|---|---|---|---|---|---|---|---|---|
| Actinides | $7s_2$<br>**Th**<br>$6d_1 5f_1$ | $7s_2$<br>**Pa**<br>$5d_1 4f_2$ | $7s_2$<br>**U**<br>$5d_1 4f_3$ | | | | | | | | | | | |

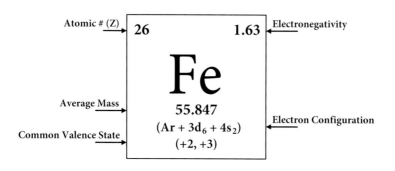

Atomic # (Z) → **26**   **1.63** ← Electronegativity

**Fe**

Average Mass → **55.847**

Common Valence State → **(Ar + 3d₆ + 4s₂)**   ← Electron Configuration

**(+2, +3)**

**Table 2.3**  Continued

Simplified periodic table of the elements electronegativities common and valence states

| IA | IIA | IIIB | IVB | VB | VIB | VIIB | --- | VIIIB | --- | IB | IIB | IIIA | IVA | VA | VIA | VIIA | VIIIA |
|---|---|---|---|---|---|---|---|---|---|---|---|---|---|---|---|---|---|
| 1 | 2 | 3 | 4 | 5 | 6 | 7 | 8 | 9 | 10 | 11 | 12 | 13 | 14 | 15 | 16 | 17 | 18 |
| 2.20 | | | | | | | | | | | | | | | | | ---- |
| **H** | | | | | | | | | | | | | | | | | **He** |
| ±1 | | | | | | | | | | | | | | | | | 0 |
| 0.98 | 1.57 | | | | | | | | | | | 2.04 | 2.55 | 3.04 | 3.44 | 3.95 | ---- |
| **Li** | **Be** | | | | | | | | | | | **B** | **C** | **N** | **O** | **F** | **Ne** |
| +1 | +2 | | | | | | | | | | | +3 | +4, 0 | many | -2 | -1 | 0 |
| 0.93 | 1.31 | | | | | | | | | | | 1.61 | 1.91 | 2.19 | 2.58 | 3.16 | ---- |
| **Na** | **Mg** | | | | | | | | | | | **Al** | **Si** | **P** | **S** | **Cl** | **Ar** |
| +1 | +2 | | | | | | | | | | | +3 | +4 | many | -2,+6 | -1 | 0 |
| 0.82 | 1.00 | 1.36 | 1.54 | 1.63 | 1.66 | 1.55 | 1.63 | 1.85 | 1.91 | 1.90 | 1.65 | 1.81 | 2.01 | 2.18 | 2.55 | 2.96 | ---- |
| **K** | **Ca** | **Sc** | **Ti** | **V** | **Cr** | **Mn** | **Fe** | **Co** | **Ni** | **Cu** | **Zn** | **Ga** | **Ge** | **As** | **Se** | **Br** | **Kr** |
| +1 | +2 | +3 | +4,+2 | many | +3,+6 | many | +2,+3 | +2,+3 | +2 | +1,+2 | +2 | +3 | +4 | many | -2,+6 | -1 | 0 |
| 0.82 | 0.95 | 1.22 | 1.33 | 1.60 | 2.16 | 1.90 | 2.20 | 2.28 | 2.20 | 1.93 | 1.69 | 1.78 | 1.96 | 2.05 | 2.10 | 2.66 | ---- |
| **Rb** | **Sr** | **Y** | **Zr** | **Nb** | **Mo** | **Tc** | **Ru** | **Rh** | **Pd** | **Ag** | **Cd** | **In** | **Sn** | **Sb** | **Te** | **I** | **Xe** |
| +1 | +2 | +3 | +4,+3 | many | many | many | many | many | +2,+4 | +1 | +2 | +3 | +4,+2 | +5,+3 | -2,+6 | -1 | 0 |
| 0.79 | 0.89 | 1.10 | 1.30 | 1.50 | 2.36 | 1.90 | 2.12 | 2.20 | 2.28 | 2.54 | 2.00 | 2.04 | 2.33 | 2.02 | 2.00 | 2.20 | ---- |
| **Cs** | **Ba** | **La** | **Hf** | **Ta** | **W** | **Re** | **Os** | **Ir** | **Pt** | **Au** | **Hg** | **Tl** | **Pb** | **Bi** | **Po** | **At** | **Rn** |
| +1 | +2 | +3 | +4 | +5 | many | many | many | many | +2,+4 | +1,+3 | +2,+1 | +3,+1 | +4.+2 | +5.+3 | +4,+2 | many | 0 |
| 0.70 | 0.87 | 1.10 | | | | | | | | | | | | | | | |
| **Fr** | **Ra** | **Ac** | | | | | | | | | | | | | | | |
| +1 | +2 | +3 | | | | | | | | | | | | | | | |

| | 1.12 | 1.13 | 1.14 | 1.13 | 1.17 | 1.20 | 1.20 | 1.20 | 1.22 | 1.23 | 1.24 | 1.25 | 1.10 | 1.27 |
|---|---|---|---|---|---|---|---|---|---|---|---|---|---|---|
| Lanthanides | **Ce** | **Pr** | **Nd** | **Pm** | **Sm** | **Eu** | **Gd** | **Tb** | **Dy** | **Ho** | **Er** | **Tm** | **Yb** | **Lu** |
| | +3,+4 | +3.+4 | +3 | +3 | +3,+2 | +3.+2 | +3 | +3.+4 | +3 | +3 | +3 | +3.+2 | +3,+2 | +3 |
| | 1.30 | 1.30 | 1.38 | | | | | | | | | | | |
| Actinides | **Th** | **Pa** | **U** | | | | | | | | | | | |
| | +4 | +5,+4 | many | | | | | | | | | | | |

Row 2 contains eight elements that reflect the progressive addition of 2s, then 2p electrons during the formation of lithium (helium + 2s$_1$) through neon (helium + 2s$_2$, 2p$_6$). Row 3 contains eight elements that reflect the filling of the 3s and 3p quantum regions respectively during the addition of electrons in sodium (neon + 3s$_1$) through argon (neon + 3s$_2$, 3p$_6$) as indicated in Table 2.3. Row 4 contains 18 elements which reflects the addition of 10 3d-subshell electrons after the 4s electrons and before the 4p elections. Row 5 contains 32 elements because of the addition of 14 3f electrons after the two 5s electrons prior to the addition 10 4d subshell electrons and the six 5p electrons (Figure 2.5). The process continues through rows 6 and 7 ending with uranium. The rows that contain up to 14 f-subshell electrons are too long to fit conveniently in a table, so the elements (lanthanides and actinides) to which f-subshell electrons have been added are shown separately at the bottom of the table. In summary, elements are grouped into rows on the periodic table according to the highest ground state principle quantum level (1–7) occupied by their electrons. Their position within each row depends on the distribution and numbers of electrons within the principle quantum levels.

### 2.2.2  Ionization

The periodic table not only organizes the elements into rows based on their electron properties, but also into vertical columns based upon their tendency to gain or lose electrons to become more stable (Table 2.3). Ideal atoms are electrically neutral because they contain the same numbers of positively charged protons and negatively charged electrons ($p^+ = e^-$). Many atoms are not electrically neutral; instead they are electrically charged particles called **ions**. The process by which they acquire their charge is called **ionization** (Box 2.1). In order for an ion to form, the number of positively charged protons and negatively charged electrons must become unequal. **Cations** are

## Box 2.1 Ionization energy

**Ionization energy (IE)** is the amount of energy required to remove an electron from its electron cloud. Ionization energies are periodic as illustrated for 20 elements in Table B2.1.

The first ionization energy is the amount of energy required to remove one electron from the electron cloud; the second ionization energy is the amount required to remove a second electron and so forth. Ionization energies are lowest for electrons that are weakly held by the nucleus and higher for electrons that are strongly held by the nucleus or are in stable configurations. Ionization energies decrease down the periodic table because the most weakly held outer electrons are shielded from the positively charged nucleus by a progressively larger number of intervening electrons. Elements with relatively low first ionization energies are called **electropositive elements** because they tend to lose one or more electrons and become positively charged cations. Most elements with high first ionization energies are **electronegative elements** because they tend to add electrons to their electron clouds and become negatively charged anions. Since opposite charges tend to attract, you can imagine the potential such ions have for combining to produce other Earth materials. The arrangement of elements into vertical columns or groups within the periodic table helps us to comprehend the tendency of specific atoms to lose, gain or share electrons. For example, on the periodic table (see Table 2.3), column 2 (IIA) elements commonly exist as divalent (+2) cations because the first and second ionization energies are fairly similar and much lower than the third and higher ionization energies. This permits two electrons to be removed fairly easily from the electron cloud, but makes the removal of additional electrons much more difficult. Column 13 (IIIA) elements commonly exist as trivalent (+3) cations (see Table 2.3). These elements have somewhat similar first, second and third ionization energies, which are much smaller than the fourth and higher ionization energies. The transfer of electrons is fundamentally important in the understanding of chemical bonds and the development of mineral crystals.

**Table B2.1** Ionization energies for hydrogen through calcium (units in kJ/mole).

| Element | Ionization energy | | | | | | | |
|---|---|---|---|---|---|---|---|---|
| | First | Second | Third | Fourth | Fifth | Sixth | Seventh | Eighth |
| H | 1312 | | | | | | | |
| He | 2372 | 5250 | | | | | | |
| Li | 520 | 7297 | 11810 | | | | | |
| Be | 899 | 1757 | 14845 | 21000 | | | | |
| B | 800 | 2426 | 3659 | 25020 | 32820 | | | |
| C | 1086 | 2352 | 4619 | 6221 | 37820 | 47260 | | |
| N | 1402 | 2855 | 4576 | 7473 | 9452 | 53250 | 64340 | |
| O | 1314 | 3388 | 5296 | 7467 | 10987 | 13320 | 71320 | 84070 |
| F | 1680 | 3375 | 6045 | 8408 | 11020 | 15150 | 17860 | 91010 |
| Ne | 2080 | 3963 | 6130 | 9361 | 12180 | 15240 | — | — |
| Na | 496 | 4563 | 6913 | 9541 | 13353 | 16610 | 20114 | 26660 |
| Mg | 737 | 1451 | 7733 | 10540 | 13630 | 17995 | 21703 | 25662 |
| Al | 578 | 1817 | 2745 | 11575 | 14830 | 18376 | 23292 | — |
| Si | 787 | 1577 | 3231 | 4356 | 16091 | 19784 | 23783 | — |
| P | 1012 | 1903 | 2912 | 4956 | 6273 | 22233 | 25397 | — |
| S | 1000 | 2251 | 3361 | 4564 | 7012 | 8495 | 27105 | — |
| Cl | 1251 | 2297 | 3822 | 5160 | 6540 | 7458 | 11020 | — |
| Ar | 1520 | 2665 | 3931 | 5570 | 7238 | 8781 | 11995 | — |
| K | 418 | 3052 | 4220 | 5877 | 7975 | 9590 | 11343 | 14944 |
| Ca | 590 | 1145 | 4912 | 6491 | 8153 | 10496 | 12270 | 14206 |

positively charged ions because they have more positively charged protons than negatively charged electrons ($p^+ > e^-$). Their charge is equal to the number of excess protons ($p^+ - e^-$). Cations form when electrons are lost from the electron cloud. Ions that have more negatively charged electrons than positively charged protons, such as the ion chlorine ($Cl^-$), will have a negative charge and are called **anions**. The charge of an anion is equal to the number of excess electrons ($e^- - p^+$). Anions form when electrons are added to the electron cloud during ionization.

### 2.2.3  Ionization behavior of columns (groups) on the periodic table

The elements in every column or group on the periodic table (Table 2.3) share a similarity in their electron configuration that distinguishes them from elements in every other column. This shared property causes the elements in each group to behave in a similar manner during chemical reactions. As will be seen later in this chapter and throughout the book, knowledge of these patterns is fundamental to understanding and interpreting the formation and behavior of minerals, rocks, and other Earth materials. The tendency of atoms to form cations or anions is indicated by the location of elements in columns of the periodic table.

**Metallic elements** have relatively low first ionization energies (<900 kJ/mol) and tend to give up one or more weakly held electrons rather easily. Column 1 and 2 (group IA and group IIA) elements, the **alkali metals** and **alkali earths**, respectively, tend to display the most metallic behaviors. Most of the elements in columns 3–12 (groups IIIB through IIB), called the transition metals, also display metallic tendencies.

**Nonmetallic elements** have high first ionization energies (>900 kJ/mol) and tend not to release their tightly bound electrons. With the exception of the stable, nonreactive, very nonmetallic Noble elements in column 18 (group VIIIA), nonmetallic elements tend to be electronegative and possess high electron affinities. Column 16–17 (group VIA and group VIIA) elements, with their especially high electron affinities, display a strong tendency to capture additional electrons to fill their highest principle quantum levels. They provide the best examples of highly electronegative, nonmetallic elements. A brief summary of the characteristics of the columns and elemental groups is presented below and in Table 2.4.

**Table 2.4**  Common ionization states for common elements in columns on the periodic table.

| Column (group) | Ionic charge | Description | Examples |
|---|---|---|---|
| 1 (IA) | +1 | Monovalent cations due to low first ionization energy | $Li^{+1}$, $Na^{+1}$, $K^{+1}$, $Rb^{+1}$, $Cs^{+1}$ |
| 2 (IIA) | +2 | Lose two electrons due to low first and second ionization energy | $Be^{+2}$, $Mg^{+2}$, $Ca^{+2}$, $Sr^{+2}$, $Ba^{+2}$ |
| 3–12 (IIIB–IIB) | +1 to +7 | Transition elements; lose variable numbers of electrons depending upon environment | $Cu^{+1}$, $Fe^{+2}$, $Fe^{+3}$, $Cr^{+2}$, $Cr^{+6}$, $W^{+6}$, $Mn^{+2}$, $Mn^{+4}$, $Mn^{+7}$ |
| 13 (IIIA) | +3 | Lose three electrons due to low first through third ionization energy | $B^{+3}$, $Al^{+3}$, $Ga^{+3}$ |
| 14 (IVA) | +4 | Lose four electrons due to low first through fourth ionization energy; may lose a smaller number of electrons | $C^{+4}$, $Si^{+4}$, $Ti^{+4}$, $Zr^{+4}$, $Pb^{+2}$, $Sn^{+2}$ |
| 15 (VA) | +5 to −3 | Lose up to five electrons or capture three electrons to achieve stability | $N^{+5}$, $N^{-3}$, $P^{+5}$, $As^{+3}$, $Sb^{+3}$, $Bi^{+4}$ |
| 16 (VIA) | −2 | Generally gain two electrons to achieve stability; gain six electrons in some environments | $O^{-2}$, $S^{-2}$, $S^{+6}$, $Se^{-2}$, |
| 17 (VIIA) | −1 | Gain one electron to achieve stable configuration | $Cl^{-1}$, $F^{-1}$, $Br^{-1}$, $I^{-1}$ |
| 18 (VIIIA) | 0 | Stable electron configuration; neither gain nor lose electrons | He, Ne, Ar, Kr |

- **Column 1 (IA)** metals are the only elements with a single s-electron in their highest quantum levels. Elements in column 1 (IA) achieve the stable configuration of the next lowest quantum level when they lose their single s-electron from the highest principal quantum level. For example, if sodium (Na) with the electron configuration $(1s_2, 2s_2, 2p_6, 3s_1)$ loses its single 3s electron ($Na^{+1}$), its electron configuration becomes that of the stable noble element neon $(1s_2, 2s_2, 2p_6)$ with the "stable octet" in the highest principle quantum level. Similar behavior is true for $Li^{+1}$, $K^{+1}$, and the other univalent (+1) cations in column 1 (IA).
- **Column 2 (IIA)** metals are the only elements with only two electrons in their highest quantum levels in their electrically neutral states. Column 2 (IIA) elements achieve stability by the removal of two s-electrons from the outer electron shell to become divalent (+2) cations.
- **Columns 3–12 (IIIB through IIB)** transition elements are situated in the middle of the periodic table. Column 3 (IIIB) elements tend to occur as trivalent (+3) cations by giving up three of their electrons $(s_2, d_1)$ to achieve a stable electron configuration. The other groups of transition elements, from column 4 (IVB) through column 12 (IIB) are cations that occur in a variety of ionization states. Depending on the chemical reaction in which they are involved, these elements can give up as few as one s-electron as in $Cu^{+1}$, $Ag^{+1}$, and $Au^{+1}$ or as many as six or seven electrons, two s-electrons, and four or five d-electrons as in $Cr^{+6}$, $W^{+6}$, and $Mn^{+7}$. An excellent example of the variable ionization of a transition metal is iron (Fe). In environments where oxygen is relatively scarce, iron commonly gives up two electrons to become $Fe^{+2}$ or ferrous iron. In other environments, especially those where oxygen is abundant, iron gives up three electrons to become smaller $Fe^{+3}$ or ferric iron.
- **Column 13 (IIIA)** elements such as $Al^{+3}$ commonly exist as trivalent (+3) cations by losing three electrons $(s_2, p_1)$.
- **Column 14 (IVA)** elements such as $Si^{+4}$ commonly exist as tetravalent (+4) cations by losing four electrons. The behavior of the heavier elements in this group is some-

what more variable than in those groups discussed previously. It depends on the chemical reaction in which the elements are involved. Tin (Sn) and lead (Pb) behave in a similar manner to silicon and germanium in some chemical reactions, but in other reactions they only lose the two s-electrons in the highest principal quantum level to become divalent cations.
- **Column 15 (VA)** elements commonly have a wide range of ionization states from tetravalent (+5) cations through trivalent (−3) anions. These elements are not particularly electropositive, nor are they especially electronegative. Their behavior depends on the other elements in the chemical reaction in which they are involved. For example, in some chemical reactions, with electropositive elements, nitrogen attracts three additional electrons to become the trivalent anion $N^{-3}$. In other chemical reactions, with electronegative elements, nitrogen releases as many as five electrons in the second principal quantum level to become the pentavalent cation $N^{+5}$. In still other situations, nitrogen gives up or attracts smaller numbers of electrons to form a cation or anion of smaller charge. All the other elements in group VA exhibit analogous situational ionization behaviors. Phosphorous, arsenic, antimony and bismuth all have ionic states that range from +5 to −3.
- **Column 16 (VIA)** nonmetallic elements commonly exist as divalent (−2) anions. These elements attract two additional electrons into their highest principal quantum levels to achieve a stable electron configuration. For example, oxygen adds two electrons to become the divalent anion $O^{-2}$. With the exception of oxygen, however, the column 16 elements display other ionization states as well, especially when they react chemically with oxygen, as will be discussed later in this chapter. Sulfur and the other VIA elements are also quite electronegative, with strong electron affinities, so that they tend to attract two electrons to achieve a stable configuration and become divalent anions such as $S^{-2}$. However, in the presence of highly electronegative oxygen these elements may lose electrons and become cations such as $S^{+6}$.

- **Column 17 (VIIA)** nonmetallic elements such as Cl⁻ and F⁻¹ commonly exist as monovalent (−1) anions. Because electrons are very difficult to remove from their electron clouds, these elements tend to attract one additional electron into their highest principal quantum level to achieve a stable electron configuration.
- **Column 18 (VIIIA)** noble gas elements such as He, Ar, and Ne contain complete outer electron shells ($s_2$, $p_6$) and do not commonly combine with other elements to form minerals. Instead, they tend to exist as monatomic (composed of single atoms) gases.

The periodic table is a highly visual and logical way in which to illustrate patterns in the electron configurations of the elements. Elements are grouped in rows or classes according to the highest principal quantum level in which electrons occur in the ground state. Elements are grouped into columns or groups based on similarities in the electron configurations in the higher principal quantum levels; those that are farthest from the nucleus and involved in most chemical reactions. More thorough explanations of the periodic table and the properties of elements are available in many standard texts in chemistry and physics.

From the discussion above, it should be clear that during the chemical reactions that produce Earth materials, elements display behaviors that are related to their electron configurations. Group 18 (VIIIA) elements in the far right column of the periodic table have stable electron configurations and tend to exist as uncharged atoms. Metallic elements toward the left side of the periodic table are strongly electropositive and tend to give up one or more electrons to become positively charged particles called cations. Nonmetallic elements toward the right side of the periodic table, especially in groups 16 (VIA) and 17 (VIIA), are strongly electronegative and tend to attract electrons to become negatively charged particles called anions. Elements toward the middle of the periodic table are somewhat electropositive and tend to lose various numbers of electrons to become cations with various amounts of positive charge. These tendencies are summarized in Table 2.4.

### 2.2.4  Atomic and ionic radii

**Atomic radii** are defined as half the distance between the nuclei of identical, bonded neighboring atoms. Because the electrons in higher quantum levels are farther from the nucleus, the effective radius of electrically neutral atoms generally increases from the top to bottom (row 1–row 7) in the periodic table (see Table 2.3). However, atomic radii generally decrease within rows from left to right (Figure 2.6). This occurs because the addition

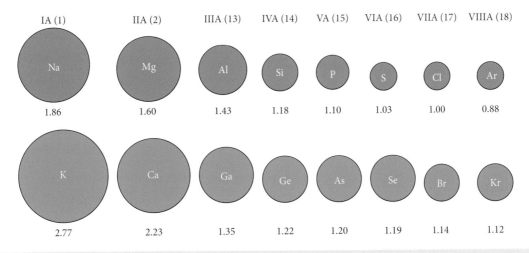

**Figure 2.6** Trends in variation of atomic radii (in angstroms; 1Å = 10⁻¹⁰ m) with their position on the periodic table, illustrated by rows 3 and 4. With few exceptions, radii tend to decrease from left to right and from bottom to top.

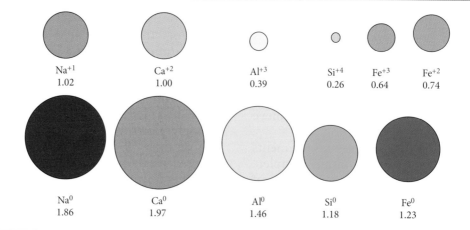

| Na$^{+1}$ | Ca$^{+2}$ | Al$^{+3}$ | Si$^{+4}$ | Fe$^{+3}$ | Fe$^{+2}$ |
| 1.02 | 1.00 | 0.39 | 0.26 | 0.64 | 0.74 |

| Na$^0$ | Ca$^0$ | Al$^0$ | Si$^0$ | Fe$^0$ |
| 1.86 | 1.97 | 1.46 | 1.18 | 1.23 |

**Figure 2.7**  Radii (in angstroms) of some common cations in relationship to the atomic radius of the neutral atoms.

of electrons to a given quantum level does not significantly increase atomic radius, while the increase in the number of positively charged protons in the nucleus causes the electron cloud to contract as electrons are pulled closer to the nucleus. Atoms with large atomic numbers and large electron clouds include cesium (Cs), rubidium (Rb), potassium (K), barium (Ba), and uranium (U). Atoms with small atomic numbers and small electron clouds include hydrogen (H), beryllium (Be), and carbon (C).

Electrons in the outer, higher energy electron levels are least tightly bound to the positively charged nucleus. This weak attraction results because these electrons are farthest from the nucleus and because they are shielded from the nucleus by the intervening electrons that occupy lower quantum level positions closer to the nucleus. These outer electrons or **valence electrons** are the electrons that are involved in a wide variety of chemical reactions, including those that produce minerals, rocks, and a wide variety of synthetic materials. The loss or gain of these valence electrons produces anions and cations, respectively.

When atoms are ionized by the loss or gain of electrons, their **ionic radii**, invariably change. This results from the electrical forces that act between the positively charged protons in the nucleus and the negatively charged electrons in the electron clouds. The ionic radii of cations tend to be smaller than the atomic radii of the same element (Figure 2.7). As electrons are lost from the electron cloud during cation formation, the positively charged protons in the nucleus

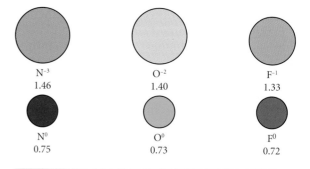

| N$^{-3}$ | O$^{-2}$ | F$^{-1}$ |
| 1.46 | 1.40 | 1.33 |

| N$^0$ | O$^0$ | F$^0$ |
| 0.75 | 0.73 | 0.72 |

**Figure 2.8**  Radii (in angstroms) of some common anions in relationship to the atomic radius of the neutral atoms.

tend to exert a greater force on each of the remaining electrons. This force draws electrons closer to the nucleus, reducing the effective radius of the electron cloud as it contracts. The larger the charge on the cation, the more its radius is reduced by the excess positive charge in the nucleus. This is well illustrated by the radii of the common cations of iron (Figure 2.8). Ferric iron (Fe$^{+3}$) has a smaller radius (0.64 Å) than does ferrous iron (Fe$^{+2}$ = 0.74 Å). Both iron cations possess much smaller radii than neutral iron (Fe$^0$ = 1.23 Å) in which there is no excess positive charge in the nucleus.

The ionic radii of anions are significantly larger than the atomic radii of the same neutral (uncharged) element (Figure 2.8). When electrons are added to the electron cloud during anion formation, the positively charged protons in the nucleus exert a smaller force on

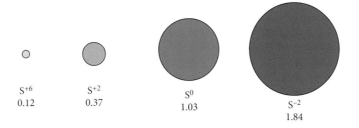

S⁺⁶
0.12

S⁺²
0.37

S⁰
1.03

S⁻²
1.84

**Figure 2.9**  Radii (in angstrom units) of some common anions and cations of sulfur in relationship to the neutral atom radius.

each of the electrons. This allows the electrons to move farther away from the nucleus, which causes the electron cloud to expand, increasing the effective radius of the anion. The larger the charge on the anion, the more its effective radius is increased.

The expansion of anions and the contraction of cations are well illustrated by the common ions of sulfur (Figure 2.9). The divalent sulfur ($S^{-2}$) anion possesses a relatively large average radius of $1.84\,\text{Å}$. In this case, the two electrons gained during the formation of a divalent sulfur anion produce a large deficit between positive charges in the nucleus and negative charges in the electron cloud. This leads to a significant increase in the effective ionic radius compared to that of electrically neutral sulfur ($S^0 = 1.03\,\text{Å}$). Neutral sulfur in turn is much larger than the divalent sulfur cation ($S^{+2} = 0.37\,\text{Å}$) and the very small, highly charged hexavalent sulfur cation ($S^{+6} = 0.12\,\text{Å}$). Keep in mind that the effective radius of a particular anion does vary somewhat. As we will see in the following sections, it depends on the environment in which bonding occurs, the number of nearest neighbors and the type of bond that forms.

## 2.3  CHEMICAL BONDS

### 2.3.1  The basics

Atoms in minerals, rocks, and other Earth materials are held together by forces or mechanisms called **chemical bonds**. The nature of these bonds strongly influences the properties and behavior of these materials. The nature of the bonds is, in turn, strongly influenced by the electron configuration of the elements that combine to produce the mineral, rock or other material.

Five principle bond types and many hybrids occur in minerals. The three most common bond types are (1) ionic, (2) covalent, and (3) metallic. They can be modeled based on the behavior of valence electrons in the outer quantum levels of atoms. During bonding, valence electrons display varying tendencies to change position based on their periodic properties. In discussing chemical bonds, it is useful to divide elements into those that are metallic and those that are nonmetallic.

**Ionic bonds** involve the linking together of metallic and nonmetallic elements, **covalent bonds** involve the linking of two nonmetallic elements, and **metallic bonds** involve the linking of two metallic elements. Hybrids between these bond types are common. Minerals with such hybrid or transitional bonds commonly possess combinations of features characteristic of each bond type. Other bond types include van der Waals and hydrogen bonds. Chemical bonding is a very complicated process; the models used below are simplifications designed to make this complex process easier to understand to a reasonable degree.

Another useful concept for understanding chemical bonds, developed originally by Linus Pauling (1929), is the concept of electronegativity (see Table 2.3). **Electronegativity (En)** is an empirical measure that expresses the tendency of an element to attract electrons when atoms bond. Highly electronegative elements ($En > 3.0$) have a strong tendency to attract electrons and become anions during bonding. Many column 16 (group VIA) and column 17 (group VIIA) elements are highly electronegative, requiring capture of two or one electrons, respectively, to achieve a stable electron configuration. Elements with low electronegativity ($En < 1.5$) are electropositive, metallic elements with a tendency to give up electrons to more electronegative elements during bonding to become positively charged cations. Highly electropositive elements include column 1 (group IA) and column 2 (group IIA) elements that tend to release one or two electrons, respectively, to achieve a stable electron configuration. Electronegativity is a very helpful concept in discussions of how atoms bond to produce larger molecules and minerals.

### 2.3.2 Ionic (electrostatic) bonds

When very metallic atoms bond with very nonmetallic atoms, an **ionic bond,** also called an **electrostatic bond,** is formed. Because the very metallic atoms (e.g., columns 1 and 2) are electropositive, they have a strong tendency to give up one or more electrons to achieve a stable configuration in their highest principal quantum level. In doing so, they become positively charged cations, whose charge is equal to the number of electrons each has lost. At the same time, very nonmetallic atoms (columns 16 and 17) are electronegative and have a strong tendency to gain one or more electrons in order to achieve a stable configuration in their highest principal quantum level. In doing so, they become

negatively charged anions, with a charge equal to the number of electrons each has gained. When very metallic and very nonmetallic atoms bond, the metallic atoms give up or donate their valence electrons to the nonmetallic atoms that capture them. It is like a tug-of-war in which the electronegative side always wins the battle for electrons. In the electron exchange process, the atoms of both elements develop stable noble element electron configurations while becoming ions of opposite charge. Because particles of opposite charge attract, the cations and anions are held together by the electrostatic attraction between them that results from their opposite charges. Larger clusters of ions form as additional ions exchange electrons and are bonded and crystals begin to grow.

The most frequently cited example of ionic bonding is the bonding between sodium ($Na^{+1}$) and chloride ($Cl^{-1}$) ions in the mineral halite (NaCl) (Figure 2.10). As a column 1 (group IA) element, sodium is very metallic and electropositive, with a rather low electronegativity (0.93). Sodium has a strong tendency to give up one electron to achieve a stable electron configuration. On the other hand, chlorine, as a column 17 (group VIIA)

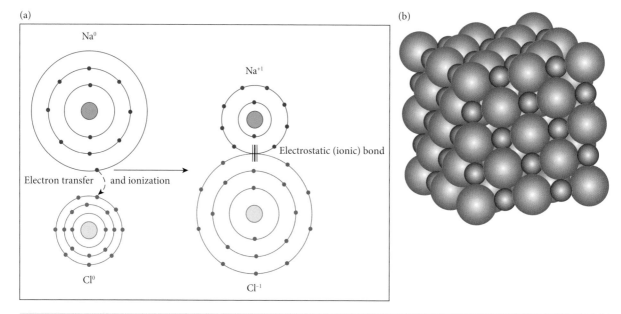

**Figure 2.10** (a) Ionic bonding develops between highly electronegative anions and highly electropositive cations. When neutral sodium ($Na^0$) atoms (red) release an electron to become cations ($Na^{+1}$) their ionic radius decreases. When neutral chlorine ($Cl^0$) atoms (blue) capture an electron to become anions ($Cl^{-1}$) their ionic radius increases. (b) Ions of opposite charge attract to form crystals such as sodium chloride (NaCl).

element, is very nonmetallic, has a strong affinity for electrons and has a high electronegativity (3.5). It has a strong tendency to gain one electron to achieve a stable electron configuration. When sodium and chlorine atoms bond, the sodium atoms release one electron to become smaller sodium cations ($Na^{+1}$) with the "stable octet" electron configuration (neon), while the chlorine atoms capture one electron to become larger chloride anions ($Cl^{-1}$) with "stable octet" electron configurations. As we shall see, the "exchange" is incomplete. The two atoms are then joined together by the electrostatic attraction between particles of opposite charge to form the compound NaCl. In macroscopic mineral specimens of halite, many millions of sodium and chloride ions are bonded together, each by the electrostatic or ionic bond described above. Note that the numbers of chloride anions and sodium cations must be equal if the electric charges are to be balanced so that the mineral is electrically neutral. Other group IA (1) and group VIIA (17) elements bond ionically to produce minerals such as sylvite (KCl).

Ionic bonds also form when group IIA and group VIA elements combine. In the mineral periclase (MgO), magnesium ($Mg^{+2}$) and oxygen ($O^{-2}$) ions are bonded together to form MgO. In this case, electropositive, metallic magnesium atoms from group IIA tend to donate two valence electrons to become stable, smaller divalent magnesium cations ($Mg^{+2}$) while highly electronegative, nonmetallic oxygen atoms from group VIA capture two valence electrons to become stable, larger divalent oxygen anions ($O^{-2}$). The two oppositely charged ions are then held together by virtue of their opposite charges by an electrostatic or ionic bond. Once again, the number of magnesium cations ($Mg^{+2}$) and oxygen anions ($O^{-2}$) in periclase (MgO) must be the same if electrical neutrality is to be conserved. A slightly more complicated example of ionic bonding involves the formation of the mineral fluorite ($CaF_2$). In this case, electropositive, metallic calcium atoms from class IIA release two electrons to become stable divalent cations ($Ca^{+2}$). At the same time, two nonmetallic, strongly electronegative fluorine atoms from class VIIA each accept one of these electrons to become stable univalent anions ($F^{-1}$). Pairs of $F^{-1}$ anions bond to each $Ca^{+2}$ cation to form ionic bonds in electrically neutral fluorite ($CaF_2$).

In ideal ionic bonds, ions can be modeled as spheres of specific ionic radii in contact with one another (Figure 2.10), as though they were ping-pong balls or marbles in contact with each other. This approximates real situations because the attractive force between ions of opposite charge (Coulomb attraction) and the repulsive force (Born repulsion) between the negatively charged electron clouds are balanced when the two ions approximate spheres in contact (Figure 2.11). If they were moved farther apart, the electrostatic attraction between ions of opposite charge would move them closer together. If they were moved closer together, the repulsive forces between the negatively charged electron clouds would move them farther apart. It is when they behave as approximate spheres in contact that these attractive and repulsive forces are balanced.

Bonding mechanisms play an essential role in contributing to material properties. Crystals with ionic bonds are generally characterized by the following:

1  Variable hardness that increases with increasing electrostatic bonding forces
2  Brittle at room temperatures.
3  Quite soluble in polar substances (such as water).
4  Intermediate melting temperatures.
5  Absorb relatively small amounts of light, producing translucent to transparent minerals with light colors and vitreous to subvitreous luster in macroscopic crystals.

### 2.3.3  Covalent (electron-sharing) bonds

When nonmetallic atoms bond with other nonmetallic atoms they tend to form **covalent bonds**, also called **electron-sharing bonds**. Because the elements involved are highly electronegative they each tend to attract electrons; neither gives them up easily. This is a little bit like a tug-of-war in which neither side can be moved so neither side ends up with sole possession of the electrons needed to achieve stable electron configurations. In simple models of covalent bonding, the atoms involved share valence (thus covalent) electrons (Figure 2.12). By sharing electrons, each atom gains the electrons necessary to achieve a more stable electron configuration in its highest principal quantum level.

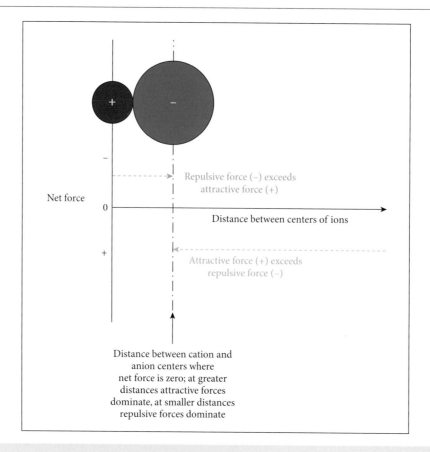

**Figure 2.11**  Relationship between attractive and repulsive forces between ions produces a minimum net force when the near spherical surfaces of the ions are in contact.

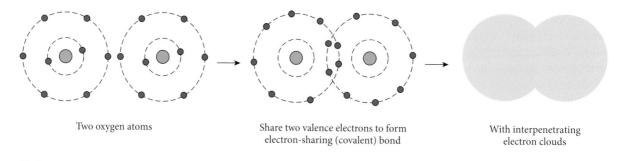

**Figure 2.12**  Covalent bonding in oxygen ($O_2$) by the sharing of two electrons from each atom.

A relevant nonmineral example is oxygen gas ($O_2$) molecules. Each oxygen atom from column 16 (group VIA) requires two electrons to achieve a stable electron configuration in its highest principal quantum level. Since both oxygen atoms have an equally large electron affinity and electronegativity, they tend to share two electrons in order to achieve the stable electron configuration. This sharing is modeled as interpenetration or overlapping of the two electron clouds (Figure 2.12). Interpenetration of electron clouds due to the sharing of valence electrons forms a strong covalent or electron-sharing bond. Because the bonds are localized in the region where the electrons are "shared" each atom has a larger probability of electrons in the area of the bond than it does elsewhere in its electron cloud. This causes each atom to become electrically polarized with a more negative charge in the vicinity of the bond and a less negative charge away from the bond. Polarization of atoms during covalent bonding is accentuated when covalent bonds form between different atoms with different electronegativities. This causes the electrons to be more tightly held by the more electronegative atom which in turn distorts the shape of the atoms so that they cannot be as effectively modeled as spheres in contact.

Other diatomic gases with covalent bonding mechanisms similar to oxygen include the column 17 (group VIIA) gases chlorine ($Cl_2$), fluorine ($F_2$), and iodine ($I_2$) in which single electrons are shared between the two atoms to achieve a stable electron configuration. Another gas that possesses covalent bonds is nitrogen ($N_2$) from column 15 (group V) where three electrons from each atom are shared to achieve a stable electron configuration. Nitrogen is the most abundant gas (>79% of the total) in Earth's lower atmosphere. In part because the two atoms in nitrogen and oxygen gas are held together by strong electron-sharing bonds that yield stable electron configurations, these two molecules are the most abundant constituents of Earth's lower atmosphere.

The best known mineral with covalent bonding is diamond, which is composed of carbon (C). Because carbon is a column 14 atom, it must either lose four electrons or gain four electrons to achieve a stable electron configuration. In diamond, each carbon atom in the structure is bonded to four nearest neighbor carbon atoms that share with it one of their electrons (Figure 2.13). In this way, each carbon atom attracts four additional electrons, one from each of its neighbors, to achieve the stable noble electron configuration. The long-range crystal structure of diamond is a pattern of carbon atoms in which every carbon atom is covalently bonded to four other carbon atoms.

Covalently bonded minerals are generally characterized by the following:

1   Hard and brittle at room temperature.
2   Insoluble in polar substances such as water.
3   Crystallize from melts.
4   Moderate to high melting temperatures.
5   Absorb very little light, producing transparent to translucent minerals with light colors and vitreous to sub-vitreous lusters in macroscopic crystals.

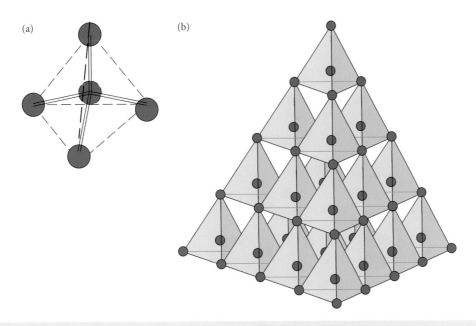

(a)            (b)

**Figure 2.13**   (a) Covalent bonding (double lines) in a carbon tetrahedron with the central carbon atom bonded to four carbon atoms that occupy the corners of a tetrahedron (dashed lines). (b) A larger scale diamond structure with multiple carbon tetrahedra. *Source*: Courtesy of Steve Dutch.

### 2.3.4 Metallic bonds

When metallic atoms bond with other metallic atoms, a **metallic bond** is formed. Because very metallic atoms have low first ionization energies, are highly electropositive and possess low electronegativities they do not tend to hold their valence electrons strongly. In such situations, each atom releases valence electrons to achieve a stable electron configuration. The positions of the valence electrons fluctuate or migrate between atoms. Metallic bonding is difficult to model, but is usually portrayed as positively charged partial atoms (nuclei plus the strongly held inner electrons) in a matrix or "gas" of "delocalized" valence electrons that are only temporarily associated with individual atoms (Figure 2.14). The weak attractive forces between positive partial atoms and valence electrons bond the atoms together. Unlike the strong electron-sharing bonds of covalently bonded substances, or the frequently strong electrostatic bonds of ionically bonded substances, metallic bonds are rather weak, less permanent and easily broken and reformed. Because the valence electrons are not strongly held by any of the partial atoms, they are easily moved in response to stress or in response to an electric field or thermal gradient.

Excellent examples of metallic bonding exist in the native metals such as native gold (Au), native silver (Ag), and native copper (Cu). Such materials are excellent conductors of electricity and heat. When materials with metallic bonds are subjected to an electric potential or field, delocalized electrons flow toward the positive anode, which creates and maintains a strong electric current. Similarly, when a thermal gradient exists, thermal vibrations are transferred by delocalized electrons, making such materials excellent heat conductors. When metals are stressed, the weakly held electrons tend to flow, which helps to explain the ductile behavior that characterizes native copper, silver, gold, and other metallically bonded substances.

Minerals containing metallic bonds are generally characterized by the following features:

1 Fairly soft to moderately hard minerals.
2 Deform plastically; malleable and ductile.
3 Excellent electrical and thermal conductors.
4 Frequently high specific gravity.
5 Excellent absorbers and reflectors of light; so are commonly opaque with a metallic luster in macroscopic crystals.

### 2.3.5 Transitional (hybrid) bonds

**Transitional** or **hybrid bonds** display combinations of ionic, covalent and/or metallic bond behavior. Some transitional bonds can be modeled as ionic–covalent transitional,

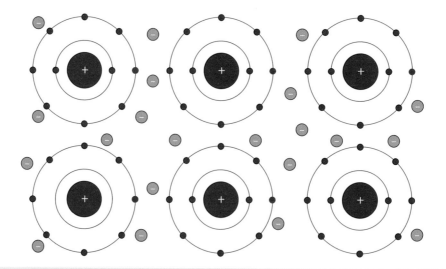

**Figure 2.14** A model of metallic bonds with delocalized electrons (dark red) surrounding positive charge centers that consist of tightly held lower energy electrons (light red dots) surrounding individual nuclei (blue).

others as ionic–metallic or covalent–metallic transitional. A detailed discussion of all the possibilities is beyond the scope of this book, but because most bonds in Earth materials are transitional, it is a subject worthy of mention. The following discussion also serves to illustrate once again the enigmatic behavior of which electrons are capable.

Earlier in this chapter, we defined electronegativity in relation to the periodic table. Linus Pauling developed the concept of electronegativity (En) to help model transitional ionic–covalent bonds. In models of such bonds, electrons are partially transferred from the more metallic, more electropositive element to the less metallic, more electronegative element to produce a degree of ionization and electrostatic attraction typical of ionic bonding. At the same time, the electrons are partially shared between the two elements to produce a degree of electron sharing associated with covalent bonding. Such bonds are best modeled as hybrids or transitions between ionic and covalent bonds. Materials that possess such bonds commonly display properties that are transitional between those of ionically bonded substances and those of covalently bonded substances. Using the **electronegativity difference** – the difference between the electronegativities of the two elements sharing the bond – Pauling was able to predict the percentages of covalent and ionic bonding, that is, the percentages of electron sharing and electron transfer that characterize ionic–covalent transitional bonds. Figure 2.15 illustrates the relationship between electronegativity difference and the percentages of ionic and covalent bond character that typify the transitional ionic–covalent bonds.

Where electronegativity differences in transitional ionic–covalent bonds are smaller than 1.68, the bonds are primarily electron-sharing covalent bonds. Where electronegativity differences are larger than 1.68, the bonds are primarily electron-transfer ionic bonds. Calculations of electronegativity and bond type lead to some interesting conclusions. For example, when an oxygen atom with En = 3.44 bonds with another oxygen atom with En = 3.44 to form $O_2$, the electronegativity difference (3.44 – 3.44 = 0.0) is zero and the resulting bond is 100% covalent. The valence electrons are completely shared by the two oxygen atoms. This will be the case

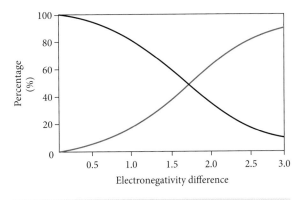

**Figure 2.15**   Graph showing the electronegativity difference and bond type in covalent–ionic bonds. Percent covalent bonding is indicated by the black line and percent ionic bonding by the blue line.

whenever two highly electronegative, nonmetallic atoms of the same element bond together. On the other hand, when highly electronegative, nonmetallic atoms bond with strongly electropositive, metallic elements to form ionically bonded substances, the bond is never purely ionic. There is always at least a small degree of electron sharing and covalent bonding. For example, when sodium (Na) with En = 0.93 bonds with chlorine (Cl) with En = 3.6 to form sodium chloride (NaCl), the electronegativity difference (3.6 – 0.93 = 2.67) is 2.67 and the bond is only 83% ionic and 17% covalent. Although the valence electrons are largely transferred from sodium to chloride and the bond is primarily electrostatic (ionic), a degree of electron sharing (covalent bonding) exists. Even in this paradigm of ionic bonding, electron transfer is incomplete and a degree of electron sharing occurs. The bonding between silicon (Si) and oxygen (O), so important in silicate minerals, is very close to the perfect hybrid since the electronegativity difference is 3.44 – 1.90 = 1.54 and the bond is 45% ionic and 55% covalent.

This simple picture of transitional ionic–covalent bonding does not hold in bonds that involve transition metals. For example, the mineral galena (PbS) has properties that suggest its bonding is transitional between metallic and ionic. In this case some electrons are partially transferred from lead (Pb) to sulfur (S) in the manner characteristic of ionically bonded substances, but some electrons are

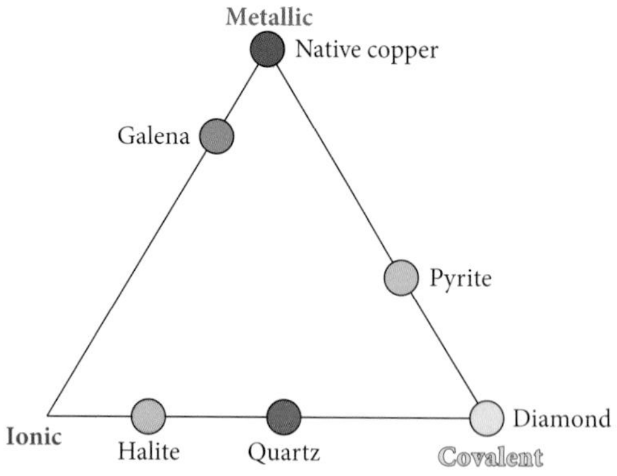

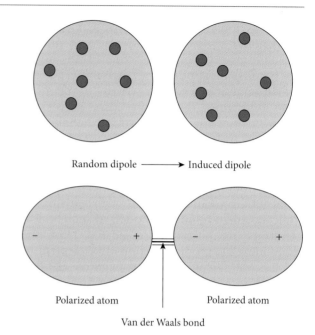

**Figure 2.16**   Triangular diagram representing the bond types of some common minerals.

weakly held in the manner characteristic of metallic bonds. As a result, galena displays both ionic properties (brittle and somewhat soluble) and metallic properties (soft, opaque and a metallic luster). Figure 2.16 utilizes a triangle, with pure covalent, ionic and metallic bonds at the apices, to depict the pure and transitional bonding characteristic of selected minerals, including those discussed above.

### 2.3.6   Van der Waals and hydrogen bonds

Because the distributions of electrons in the electron cloud are probabilistic and constantly changing, they may be, at any moment, asymmetrically distributed within the electron cloud. This asymmetry gives rise to weak electric dipoles on the surface of the electron cloud; areas of excess negative charge concentration where the electrons are located and areas of negative charge deficit (momentary positive charge) where they are absent. Areas of momentary positive charge on one atom attract electrons in an adjacent atom, thus inducing a dipole in that atom. The areas of excess negative charge on one atom are attracted to the areas of positive charge on an adjacent atom to form a very weak bond that holds the atoms together (Figure 2.17). Bonds that result from weak electric dipole forces that are caused by the asymmetrical distribution of electrons in the electron cloud are called **van der Waals bonds**. The presence of very weak van der Waals bonds helps to explain why minerals such as graphite and talc are extremely soft and have a "greasy" feel (Chapter 5).

**Figure 2.17**   Van der Waals bonding occurs when one atom becomes dipolar as the result of the random concentration of electrons in one region of an atom. The positively charged region of the atom attracts electrons in an adjacent atom causing it to become dipolar. Oppositely charged portions of adjacent dipolar atoms are attracted creating a weak van der Waals bond. Larger structures result from multiple bonds.

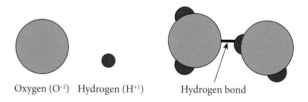

Oxygen ($O^{-2}$)   Hydrogen ($H^{+1}$)         Hydrogen bond

**Figure 2.18**   Diagram showing two water molecules joined by a hydrogen bond that links the hydrogen in one molecule to the oxygen in the other molecule.

**Hydrogen bonds** exist between electropositive hydrogen and electronegative ions such as oxygen in molecules such as water or hydroxyl ions. Because of the profound importance of water ($H_2O$) and hydroxyl ion ($OH^{-1}$), in both organic and inorganic compounds, this type of bond has been given its own separate designation (Figure 2.18). Hydrogen bonds are relatively weak bonds that occur in hydrated (water-bearing) or hydroxide (hydroxyl-bearing) minerals.

Atoms are held together by a variety of chemical bonds. The type of bond that forms depends largely on the electron configurations of the combining elements, as expressed by their electronegativities, although environmental factors also play a role. Each bond type imparts certain sets of properties to Earth materials that contain those bonds. In the following section we will discuss factors that determine the three-dimensional properties of the molecular units that result from such bonding. In Chapter 4 we will elaborate on the long-range crystalline structures that form when these molecular units combine to produce crystals. Remember: it all starts with atoms, their electron properties and the way they bond together to produce crystals.

## 2.4 PAULING'S RULES AND COORDINATION POLYHEDRA

### 2.4.1 Pauling's rules and radius ratios

Linus Pauling (1929) established five rules, now called **Pauling's rules**, which describe cation–anion relationships in ionically bonded substances and are paraphrased below:

- **Rule 1:** A polyhedron of anions is formed about each cation, with the distance between a cation and an anion determined by the sum of their radii (radius sum). The number of coordinated anions in the polyhedron is determined by the cation : anion radius ratio.
- **Rule 2:** An ionic structure is stable when the sum of the strengths of all the bonds that join the cation to the anions in the polyhedron equals (balances) the charge on the cations and on the anions. This rule is called the **electrostatic valency rule**.
- **Rule 3:** The sharing of edges and particularly faces by adjacent anion polyhedral elements decreases the stability of an ionic structure. Similar charges tend to repel. If they share components, adjacent polyhedra tend to share corners, rather than edges, to maximize cation spacing.
- **Rule 4:** Cations with high valence charges and small coordination number tend not to share polyhedral elements. Their large positive charges tend to repel.

- **Rule 5:** The number of different cations and anions in a crystal structure tends to be small. This is called the **rule of parsimony**.

Pauling's rules provide important tools for understanding crystal structures. Especially important is the rule concerning radius ratio and coordination polyhedra. Coordination polyhedra provide a powerful means for visualizing crystal structures and their relationship to crystal chemistry. In fact, they provide a fundamental link between the two. When atoms and ions combine to form crystals, they bond together into geometric patterns in which each atom or ion is bonded to a number of nearest neighbors. The number of nearest neighbor ions or atoms is called the **coordination number (CN)**. Clusters of atoms or ions bonded to other coordinating atoms produce **coordination polyhedron** structures. Polyhedrons include triangles, cubes, octahedra, tetrahedra, and other geometric forms.

When ions of opposite charge combine to form minerals, each cation attracts as many nearest neighbor anions as can fit around it as approximate "spheres in contact". In this way, the basic units of crystal structure are formed which grow into crystals as multiples of such units are added to the existing structure. One can visualize crystal structures in terms of different coordinating cations and coordinated anions that together define a simple three-dimensional polyhedron structure. As detailed in Chapters 3 and 4, complex polyhedral structures develop by linking of multiple coordination polyhedra.

The number of nearest neighbor anions that can be coordinated with a single cation "as spheres in contact" depends on the **radius ratio ($RR = Rc/Ra$)** which is the radius of the smaller cation ($Rc$) divided by the radius of the larger anion ($Ra$). For very small, highly charged cations coordinated with much larger, highly charged anions, the radius ratio ($RR$) and the coordination number ($CN$) are small. This is analogous to fitting basketballs as spheres in contact around a small marble. Only two basketballs can fit as spheres in contact with the marble. For cations of smaller charge coordinated with anions of smaller charge, the coordination number is larger. This is analogous to fitting golf balls around a larger marble. One can fit a larger number of

**Table 2.5** Relationship between radius ratio, coordination number, and coordination polyhedra.

| Radius ratio (Rc/Ra) | Coordination number | Coordination type | Coordination polyhedron |
|---|---|---|---|
| <0.155 | 2 | Linear | Line |
| 0.155–0.225 | 3 | Triangular | Triangle |
| 0.225–0.414 | 4 | Tetrahedral | Tetrahedron |
| 0.414–0.732 | 6 | Octahedral | Octahedron |
| 0.732–1.00 | 8 | Cubic | Cube |
| >1.00 | 12 | Cubic or hexagonal closest packed | Cubeoctahedron complex |

golf balls around a large marble as spheres in contact because the radius ratio is larger.

The general relationship between radius ratio, coordination number and the type of coordination polyhedron that results is summarized in Table 2.5. For radius ratios less than 0.155, the coordination number is 2 and the "polyhedron" is a line. The appearance of these coordination polyhedra is summarized in Figure 2.19.

When predicting coordination number using radius ratios, several caveats must be kept in mind.

1 The ionic radius and coordination number are not independent. As illustrated by Table 2.6, effective ionic radius increases as coordination number increases.
2 Since bonds are never truly ionic, models based on spheres in contact are only approximations. As bonds become more covalent and more highly polarized, radius ratios become increasing less effective in predicting coordination numbers.
3 Radius ratios do not successfully predict coordination numbers for metallically bonded substances.

The great value of the concept of coordination polyhedra is that it yields insights into the fundamental patterns in which atoms bond during the formation of crystalline materials. These patterns most commonly involve three-fold (triangular), fourfold (tetrahedral), six-fold (octahedral), eightfold (cubic) and, to a lesser extent, 12-fold coordination polyhedra or small variations of such basic patterns. Other coordination numbers and polyhedron types exist, but are rare in inorganic Earth materials.

Another advantage of using spherical ions to model coordination polyhedra is that it allows one to calculate the size or volume of the resulting polyhedron. In a coordination polyhedron of anions, the cation–anion distance is determined by the **radius sum ($R_s$)**. The radius sum is simply the sum of the radii of the two ions (Rc + Ra); that is, the distance between their respective centers. Once this is known, the size of any polyhedron can be calculated using the principles of geometry. Such calculations are beyond the scope of this book but are discussed in Wenk and Bulakh (2016) and Klein and Dutrow (2007).

### 2.4.2 Electrostatic valency

An important concept related to the formation of coordination polyhedra is **electrostatic valency (EV)**. In a stable coordination structure, the total strength of all the bonds that reach a cation from all neighboring anions is equal to the charge on the cation. This is another way of saying that the positive charge on the cation is neutralized by the electrostatic component of the bonds between it and its nearest neighbor anions. Similarly, every anion in the structure is surrounded by some number of nearest neighbor cations to which it is bonded, and the negative charges on each anion are neutralized by the electrostatic component of the bonds between it and its nearest neighbor cations. For a cation of charge Z bonded to a number of nearest neighbor anions (CN), the electrostatic valency of each bond is given by the charge of the cation divided by the number of nearest neighbors to which it is coordinated:

$$EV = Z/CN$$

For example, in the case of the silica tetrahedron $(SiO_4)^{-4}$ each $Si^{+4}$ cation is coordinated with four $O^{-2}$ anions (Figure 2.20). The electrostatic valency of each bond is given by

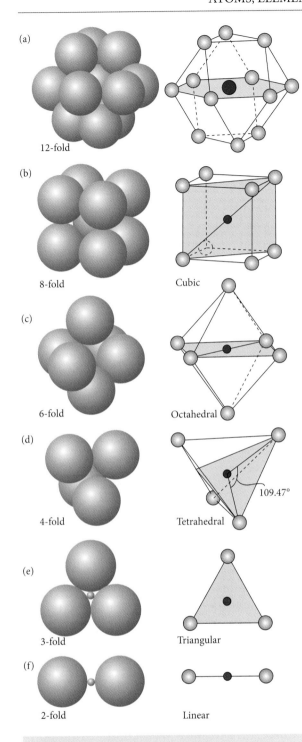

**Figure 2.19** Common coordination polyhedra: (a) cubic closest packing, (b) cubic, (c) octahedral, (d) tetrahedral, (e) triangular, (f) linear. *Source*: Wenk and Bulakh (2004). © Cambridge University Press.

**Table 2.6** Variations in ionic radius (in angstroms) with coordination number (CN) for some common cations.

| Ion | CN = 4 | CN = 6 | CN = 8 |
|-----|--------|--------|--------|
| $Na^{+1}$ | 0.99 | 1.02 | 1.18 |
| $K^{+1}$ | | 1.38 | 1.51 |
| $Rb^{+1}$ | | 1.52 | 1.61 |
| $Cs^{+1}$ | | 1.67 | 1.74 |
| $Mg^{+2}$ | 0.57 | 0.72 | |
| $Al^{+3}$ | 0.39 | 0.48 | |
| $Si^{+4}$ | 0.26 | 0.40 | |
| $P^{+5}$ | 0.17 | 0.38 | |
| $S^{+6}$ | 0.12 | 0.29 | |

EV = Z/CN = +4/4 = +1. What this means is that each bond between the coordinating silicon ion ($Si^{+4}$) and the coordinated oxygen ions ($O^{-2}$) balances a charge of +1. Another way to look at this is to say that each bond involves an electrostatic attraction between ions of opposite charge of one charge unit. Since there are four Si–O bonds, each balancing a charge of +1, the +4 charge on the silicon ion is fully neutralized by the four nearest neighbor anions to which it is bonded. However, although the +4 charge on the coordinating silicon ion is fully satisfied, the –2 charge on each of the coordinated ions is not. Since each has a –2 charge, a single bond involving an electrostatic attraction of one charge unit neutralizes only half their charge. They must attract and bond to one or more additional cations, with an additional total electrostatic valency of one, in order to have their charges effectively neutralized. So it is that during mineral growth, cations attract anions and anions attract additional cations of the appropriate charge and radius which in turn attract additional anions of the appropriate charge and radius as the mineral grows. In this manner minerals retain their essential geometric patterns and their ions are neutralized as the mineral grows. In the following section we will introduce the major mineral groups and see how their crystal chemistry forms the basis of the mineral classification.

## 2.5 THE CHEMICAL CLASSIFICATION OF MINERALS

The formation and growth of most minerals can be modeled by the attractive forces

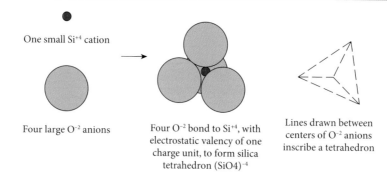

One small Si⁺⁴ cation

Four large O⁻² anions

Four O⁻² bond to Si⁺⁴, with
electrostatic valency of one
charge unit, to form silica
tetrahedron (SiO4)⁻⁴

Lines drawn between
centers of O⁻² anions
inscribe a tetrahedron

**Figure 2.20**   A silica tetrahedron is formed when four oxygen ions ($O^{-2}$) bond to one silicon ion ($Si^{+4}$) in the form of a tetrahedron. The electrostatic valency of each silicon–oxygen bond in the silica tetrahedron is one charge unit, which fully neutralizes the charge on the central silicon ion (four = four), while leaving the charge on the oxygen ions only partially neutralized (one is one-half of two).

**Table 2.7**   Mineral classification based on the major anion groups.

| Mineral group | Major anion groups | Mineral group | Major anion groups |
|---|---|---|---|
| Native elements | None | Nitrates | $(NO_3)^{-1}$ |
| Halides | $F^{-1}$, $Cl^{-1}$, $Br^{-1}$ | Borates | $(BO_3)^{-3}$ and $(BO_4)^{-5}$ |
| Sulfides | $S^{-2}$, $S^{-4}$ | Sulfates | $(SO_4)^{-2}$ |
| Arsenides | $As^{-2}$, $As^{-3}$ | Phosphates | $(PO_4)^{-3}$ |
| Sulfarsendies | $As^{-2 \text{ or }} As^{-3}$ and $S^{-2 \text{ or }} S^{-4}$ | Chromates | $(CrO4)^{-5}$ |
| Selenides | $Se^{-2}$ | Arsenates | $(AsO_4)^{-3}$ |
| Tellurides | $Te^{-2}$ | Vanadates | $(VO_4)^{-3}$ |
| Oxides | $O^{-2}$ | Molybdates | $(MO_4)^{-2}$ |
| Hydroxides | $(OH)^{-1}$ | Tungstates | $(WO_4)^{-2}$ |
| Carbonates | $(CO_3)^{-2}$ | Silicates | $(SiO_4)^{-4}$ |

between cations and anions, the formation of coordination polyhedra with unsatisfied negative charges and the attraction of additional ions to build additional coordination polyhedra *ad infinitum*, until the conditions for growth cease to exist. It is useful to visualize minerals in terms of major anions and anion groups and/or radicals bonded to various cations that effectively neutralize their charge during the formation and growth of minerals. One common way to group or classify minerals is to do so in terms of the major anion group in the mineral structure. Those that contain $(SiO_4)^{-4}$ silica tetrahedra, discussed in the previous section, are silicate minerals, by far the most common minerals in Earth's crust and upper mantle. Those that do not contain silica tetrahedra are nonsilicate minerals and are further subdivided on the basis of their major anions. Table 2.7 summarizes the common mineral groups according to this

classification system. These groups are discussed in more detail in Chapter 5.

Oxygen (O) and silicon (Si) are the two most abundant elements in Earth's continental crust, oceanic crust and mantle. Under the relatively low pressure conditions that exist in the crust and the upper mantle, the most abundant rock-forming minerals (Chapter 5) are silicate minerals. **Silicate minerals**, characterized by the presence of silicon and oxygen that have bonded together to form silica tetrahedra, are utilized here to show how coordination polyhedra are linked to produce larger structures with the potential for the long-range order characteristic of all minerals.

### 2.5.1   The basics: silica tetrahedral linkage

Silica tetrahedra are composed of a single, small, tetravalent silicon ion ($Si^{+4}$) in fourfold, tetrahedral coordination with four larger, divalent oxygen ions $(O^{-2})$. These silica tetrahedra may be thought of as the basic building blocks, the LEGO®, of silicate minerals. Because the electrostatic valency of each of the four Si–O bonds in the tetrahedron is one (EV = 1), the +4 charge of the silicon ion is effectively neutralized. However, the −2 charges on the oxygen $(O^{-2})$ ions are not neutralized. Each oxygen ion possesses an unsatisfied charge of −1 which it can only neutralize by bonding with one or more additional cations in the mineral structure. Essentially, as a crystal forms, oxygen anions can bond to another silicon $(Si^{+4})$ ion to form a second bond with an electrostatic valency of 1 or it can bond to some other combination of cations (e.g., $Al^{+3}$, $Mg^{+2}$, $Fe^{+2}$, $Ca^{+2}$, $K^{+1}$, $Na^{+1}$) with a total electrostatic valency of 1.

Many factors influence the type of silica tetrahedral structure that develops when silicate minerals form; the most important is the relative availability of silicon and other cations in the environment in which the mineral crystallizes. Environments with abundant silicon (and therefore silica tetrahedra) tend to favor the linkage of silica tetrahedra through shared oxygen ions. Environments depleted in silicon tend to favor the linkage of the oxygen ions in silica tetrahedra to cations other than silicon. In such situations, silica tetrahedra tend to link to coordination polyhedral elements other than silica tetrahedra.

If none of the oxygen ions in a silica tetrahedron bond to other silicon ions in adjacent tetrahedra, the silica tetrahedron will occur as an isolated tetrahedral unit in the mineral structure. If all the oxygen ions in a silica tetrahedron bond to other silicon ions of adjacent tetrahedra, the silica tetrahedra form a three-dimensional framework structure. If some of the oxygen ions in the silica tetrahedra are bonded to silicon ions in adjacent tetrahedra and others are bonded to other cations in adjacent coordination polyhedra, a structure that is intermediate between totally isolated silica tetrahedra and three-dimensional frameworks of silica tetrahedra will develop.

Six major silicate groups (Figure 2.21) are distinguished based upon the linkage patterns of silica tetrahedra. These are: (1) nesosilicates, (2) sorosilicates, (3) cyclosilicates, (4) inosilicates, (5) phyllosilicates, and (6) tectosilicates. **Nesosilicates** ("island" silicates) are characterized by isolated silica tetrahedra that are not linked to other silica tetrahedra through shared oxygen ions. **Sorosilicates** ("bow-tie" silicates) contain pairs of silica tetrahedra linked through shared oxygen ions. In **cyclosilicates** ("ring" silicates), each silica tetrahedron is linked to two other tetrahedra through shared oxygen ions into ring-shaped structural units. In **single-chain inosilicates,** each silica tetrahedron is linked through shared oxygen anions to two other silica tetrahedra in the form of a long, one-dimensional chain-like structure. When two chains are linked through shared oxygen anions a **double-chain inosilicate** structure is formed. When chains are "infinitely" linked to one another through shared oxygen anions, a two-dimensional sheet of linked silica tetrahedra is formed which is the basic structural unit of **phyllosilicates** ("sheet" silicates). Finally, when silica tetrahedra are linked to adjacent silica tetrahedra by sharing all four oxygen anions, a three-dimensional framework of linked silica tetrahedra results, which is the basic structure of **tectosilicates** ("framework" silicates).

Because the silicate groups constitute the most significant rock-forming minerals in Earth's crust and upper mantle they are discussed more fully in Chapter 5. In Chapter 3, we will further investigate significant aspects of mineral chemistry, including substitution solid solution and the uses of isotopes and phase stability diagrams in understanding Earth materials.

## CONTENT ASSESSMENT

1 Explain the difference between:
 (a) An *isotope* and an *ion*.
 (b) A *cation* and an *anion* and the reason the latter two types of ions exist.
 (c) A *stable isotope* and an *unstable isotope* and how the latter evolve through time.

2 Detail the major factors that determine the effective radii of different atoms and ions.

3 What is the *diagonal rule* and how does it help to predict the electron configuration of most major elements whose atomic number is known? Use the diagonal rule

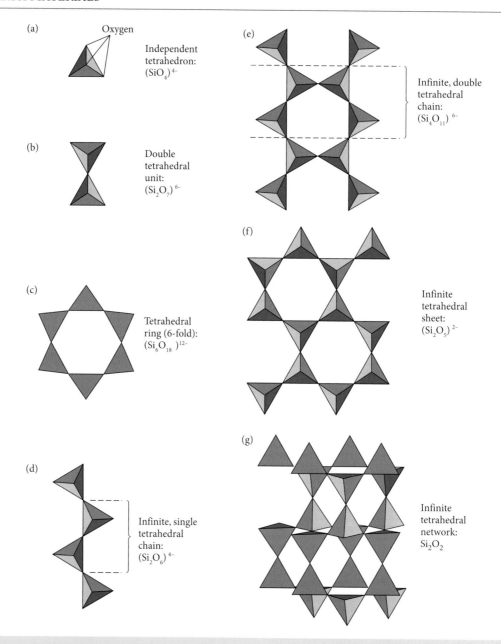

**Figure 2.21** Major silicate structures: (a) nesosilicate, (b) sorosilicate, (c) cyclosilicate, (d) single-chain inosilicate, (e) double chain inosilicate, (f) phyllosilicate, (g) tectosilicate. *Source*: Wenk and Bulakh (2004). © Cambridge University Press.

to write the "ground state" electron configurations for the following elements.

(a) helium (H)
(b) carbon (C)
(c) oxygen (O)
(d) aluminum (Al)
(e) argon (Ar)
(f) iron (Fe).

4  Referring to the periodic table of the elements (Table 2.3), which of the reactive (non-Noble) elements is the most metallic (electropositive) and which of these is the most nonmetallic (electronegative). Use their electron configurations and position on the periodic table to explain why.

5  Explain the differences between ionic, covalent, metallic, and transitional

(hybrid) bonds and the properties that commonly characterize materials that possess that type of bond.

6   Using the periodic table of the elements (Table 2.3), calculate the electronegativity difference and then predict the bond type and properties for each of the following minerals:

   a.   Native platinum (Pt)
   b.   Periclase (MgO)
   c.   Sylvite (KCl)
   d.   Pyrite (FeS2)
   e.   Sphalerite (ZnS).

7   Explain the concepts of *radius ratio* and *electrostatic valency* and use them to explain why silica tetrahedra tend to link by sharing oxygen ions. Then describe and explain the basic differences between nesosilicate, sorosilicate, cyclosilicate, nesosilicate, phyllosilicate and tectosilicate structures.

## REFERENCES

Klein, C. and Dutrow, B. (2007). *Manual of Mineral Science (Manual of Mineralogy)*, 23e. New York: Wiley 704 pp.

Pauling, L. (1929). The principles determining the structure of complex ionic structures. *Journal American Chemical Society 51*: 1010–1026.

Railsback, L.B. (2003). An earth scientist's periodic table of the elements and their ions. *Geology 31*: 737–740.

Wenk, H.R. and Bulakh, A. (2004). *Minerals: Their Constitution and Origin*, 3e. Cambridge, UK: Cambridge University Press 646 pp.

Wenk, H.R. and Bulakh, A. (2016). *Minerals: Their Constitution and Origin*, 3e. Cambridge, UK: Cambridge University Press 621 pp.

# Chapter 3

# Atomic substitution, phase diagrams, and isotopes

## 3.1   ATOMIC (IONIC) SUBSTITUTION

Minerals are composed of atoms or ions that occupy structural sites in a crystal structure (Chapter 2). Different ions can occupy the same structural site if (1) they have similar size, (2) have similar charge, and (3) are available in the environment in which the mineral is forming. This process of one ion replacing another ion is called **ionic substitution**. In mineral formulas, ions that commonly substitute for one another are generally placed within a single set of parentheses. In the olivine group, iron and magnesium can freely substitute for one another in the sixfold, octahedral site. As a result, the formula for olivine is commonly written as $(Mg,Fe)_2SiO_4$.

Substitution is favored for ions of similar ionic radius. In general, cation substitution at surface temperatures and pressures is limited when the larger cation radii exceed the smaller by 10–15% and becomes negligible for differences greater than 30%. Such ions are "too big" or "too small" to easily substitute for one another (Figure 3.1a), while ions of similar size are "just right." Substitution of ions of significantly different radii distorts coordination polyhedra and decreases the stability of crystals. However, at higher temperatures, where the crystal structure is expanded, ions with larger differences in radius may more easily substitute for one another.

Substitution is favored for ions of similar charge. Where substitutions occur in only one coordination site, substitution is largely limited to ions with the same charge (Figure 3.1b). This enables the mineral to remain electrically neutral, which increases its stability. However, where substitution can occur in multiple coordination sites, ions of different charge may

*Earth Materials*, Second Edition. Kevin Hefferan and John O'Brien.
© 2022 John Wiley & Sons Ltd. Published 2022 by John Wiley & Sons Ltd.
Companion website: www.wiley.com/go/hefferan/earthmaterials2

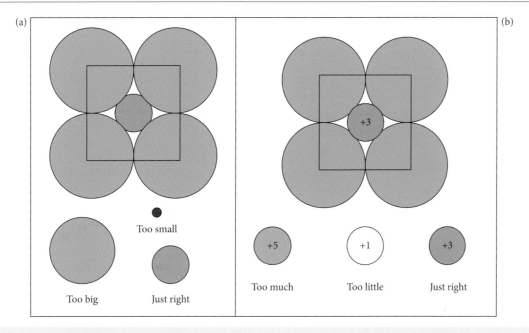

**Figure 3.1**   Criteria for substitution are (a) similar size, (b) similar charge and (not shown) availability.

substitute for one another in one site so long as this charge difference is balanced by a second substitution of ions of different charge in a second coordination site.

Substitution is favored for ions that are widely available in the environment in which the mineral is growing (Figure 3.1). As minerals grow, coordination sites will preferentially select ions with the appropriate radii and charge that are available in the vicinity of the growing crystal. The ions that occupy a coordination site in a mineral provide vital clues to the chemical composition of the system and environmental conditions under which crystallization occurred.

### 3.1.1   Simple ionic substitution

**Simple, complete substitution** exists when two or more ions of similar radii and the same charge may substitute for one another in a coordination site in any proportions. In such cases, it is convenient to define **end members** or **components** that have only one type of ion in the structural site in question. The olivine group illustrates complete substitution. In the olivine group, $(Mg,Fe)_2SiO_4$, $Mg^{+2}$ (radius = 0.66Å) and $Fe^{+2}$ (radius = 0.74Å) can substitute for one another in the octahedral site in any proportion. The two end members are the pure

magnesium silicate component called **forsterite** $[(Mg)_2SiO_4]$ and the pure iron silicate component called **fayalite** $[(Fe)_2SiO_4]$. Since these two end members can substitute for one another in any proportion in olivine, a **complete solid solution series** exists between them. As a result, the composition of any olivine can be expressed in terms of the proportions of forsterite (Fo) and/or fayalite (Fa). Simple two-component, complete solid solution series are easily represented by a number line called a **tie line** between the two end members (Figure 3.2).

Compositions of any olivine can be represented in a number of different ways. For example, pure magnesium olivine can be represented by (1) a formula $(Mg_2SiO_4)$, (2) a name (forsterite), (3) its position on the tie line (far right) or (4) the proportion of either end member ($Fo_{100}$ or $Fa_0$). Similarly, pure iron olivine can be represented by a formula $(Fe_2SiO_4)$, a name (fayalite), its position on the tie line (far left) or the proportion of either end member ($Fo_0$ or $Fa_{100}$). Any composition in the olivine complete solid solution series can be similarly represented. For example, the composition of an olivine with equal amounts of the two end member components can be represented by the formula $[(Mg_{0.5}, Fe_{0.5})_2SiO_4]$, its position on the tie line (halfway between the ends), or the proportions of either end member

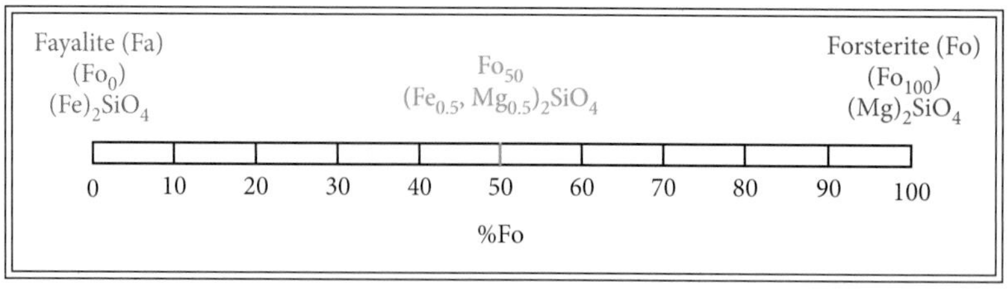

Fayalite (Fa)
(Fo$_0$)
(Fe)$_2$SiO$_4$

Fo$_{50}$
(Fe$_{0.5}$, Mg$_{0.5}$)$_2$SiO$_4$

Forsterite (Fo)
(Fo$_{100}$)
(Mg)$_2$SiO$_4$

| 0 | 10 | 20 | 30 | 40 | 50 | 60 | 70 | 80 | 90 | 100 |

%Fo

**Figure 3.2**   Olivine complete substitution solid solution series.

(Fo$_{50}$ or Fa$_{50}$). Typically the forsterite component is used (e.g., Fo$_{50}$) and the fayalite (Fa) component (100 − Fo) is implied.

In cases where three ions substitute freely for one another in the same coordination site, it is convenient to define three end member components. Each of these end member components contains only one of the three ions in the structural site in which substitution occurs. For example, ferrous iron (Fe$^{+2}$), magnesium (Mg$^{+2}$), and manganese (Mn$^{+2}$) can all substitute for one another in any proportions in the cation site of rhombohedral carbonates. The general formula for such carbonate minerals can be written as (Fe,Mg,Mn)CO$_3$. The three end member components are the "pure" minerals siderite (FeCO$_3$), magnesite (MgCO$_3$), and rhodochrosite (MnCO$_3$). On a three-component diagram, the three pure end member components are plotted at the three apices of a triangle (Figure 3.3).

Points on the apices of the triangle represent "pure" carbonate minerals with only one end member component. Percentages of any component decrease systematically from 100% at the apex toward the opposite side of the triangle where its percentage is zero. Each side of the triangle is a tie line connecting two end members. Points on the sides represent carbonate solid solutions between two end member components. Point A on Figure 3.3 lies on the side opposite the magnesite apex and so contains no magnesium. Because it lies half-way between rhodochrosite and siderite, its composition may be written as Rc$_{50}$Sd$_{50}$ or as (Mn$_{0.5}$, Fe$_{0.5}$)CO$_3$. Any point that lies within the triangle represents a solid solution that contains all three end member components. The precise composition of any three-component solid solution can be determined by the distance from the point to the three apices of the triangle. Point B in Figure 3.3 lies closest

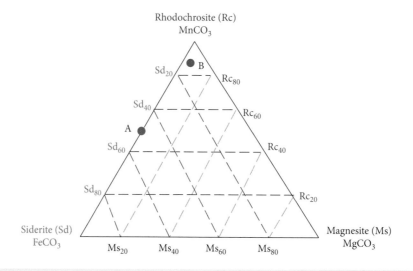

**Figure 3.3**   Compositions of carbonate minerals expressed in terms of the proportions of iron, magnesium, and manganese; that is of the three components: siderite (Sd), magnesite (Ms), and rhodochrosite (Rc) plotted on a ternary diagram.

to the rhodochrosite apex and farthest from the magnesite apex and so clearly contains more Mn than Fe and more Fe than Mg. Its precise composition can be expressed as $Sd_{10}Ms_2Rc_{88}$ or as $(Fe_{0.10},Mg_{0.02},Mn_{0.88})CO_3$. Many other examples of three-component systems with complete solid solution exist; all may be represented in a similar fashion by their position on a triangular diagram.

### 3.1.2 Coupled (paired) ionic substitution

**Coupled (paired) ionic substitution** involves the simultaneous substitution of ions of different charges in two different structural sites in a way that preserves the electrical neutrality of the crystal lattice (Figure 3.4). The substitution of ions of different charge in one structural site changes the electric charge of the crystal lattice; this requires a second set of substitutions of ions in a second structural site to balance that change in charge. Many examples of coupled ionic substitution exist; none are more important than those that occur in the plagioclase feldspars, the most abundant mineral group in Earth's crust.

In the plagioclase feldspars, similar size ions of sodium ($Na^{+1}$) and calcium ($Ca^{+2}$) can substitute for one another in any proportion in the large cation coordination site. However, when calcium ($Ca^{+2}$) substitutes for sodium ($Na^{+1}$), the positive charge of the crystal lattice is increased, and when the reverse occurs, the positive charge of the lattice is decreased. These changes in charge are balanced by a second set of substitutions. This second set of substitutions occurs in the small tetrahedral cation coordination site where aluminum ($Al^{+3}$) and silicon ($Si^{+4}$) substitute for one another. When a sodium ($Na^{+1}$) ion is added to the large cation coordination site, a silicon ($Si^{+4}$) ion is added to the small cation structural site. The two sites together contain a total charge of +5 that is balanced by the anions in the plagioclase structure. When a calcium ($Ca^{+2}$) is added to the large cation site,

an aluminum ($Al^{+3}$) is added to the small cation site. Once again, the two sites together contain a total charge of +5 which is balanced by the anions in the plagioclase structure. The two substitutions are paired. If a sodium ($Na^{+1}$) ion replaces a calcium ($Ca^{+2}$) ion in the first coordination site, a silicon ($Si^{+4}$) ion must simultaneously replace an aluminum ($Al^{+3}$) ion in the second structural site for the two sites to total +5, so that the electrical neutrality of the crystal lattice is maintained. Ideally all substitutions are paired and any change in the proportion of sodium to calcium (Na/Ca) in the large ion site is balanced by a similar change in the proportion of silicon to aluminum (Si/Al) in the small ion site. As a result, the general composition of plagioclase can be represented by the formula $(Na,Ca)(Si,Al)AlSi_2O_8$ to emphasize the nature of coupled ionic substitutions.

The plagioclase group can be represented as a two-component system with coupled ionic substitution (Figure 3.4). The two end members are the "pure" sodium plagioclase called **albite (Ab)**, whose formula can be written as $(Na)(Si)AlSi_2O_8$ (or $NaAlSi_3O_8$), and the "pure" calcium plagioclase called **anorthite (An)**, whose formula can be written as $(Ca)(Al)AlSi_2O_8$ (or $CaAl_2Si_2O_8$). Since a complete solid solution series exists between these two end members, any plagioclase composition can also be represented by its position on a tie line between the end member components or by the proportions of albite (Ab) and/or anorthite (An). In Figure 3.4, the composition of "pure" sodium plagioclase can be represented by (1) its position on the left end of the tie line, (2) $Ab_{100}$, (3) $An_0$, or (4) the formula $[(Na_{1.0},Ca_{0.0})(Si_{1.0},Al_{0.0})AlSi_2O_8] = NaAlSi_3O_8$. Similarly, the composition of "pure" calcium plagioclase can be represented by (1) its position on the right end of the tie line, (2) $Ab_0$, (3) $An_{100}$, or (4) the formula $[(Na_{0.0},Ca_{1.0})(Si_{0.0},Al_{1.0})AlSi_2O_8] = CaAl_2Si_2O_8$. Plagioclase compositions are generally expressed in terms of anorthite proportions, with the implication

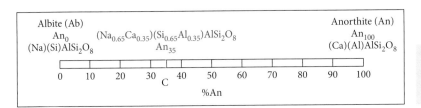

**Figure 3.4** Coupled ionic substitution in the plagioclase solid solution series.

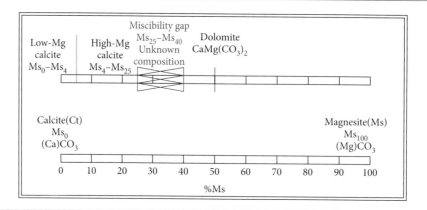

**Figure 3.5** Limited substitution and miscibility gap in calcium–magnesium carbonates with compositional ranges of low-Mg calcite, high-Mg calcite, and dolomite.

that the proportion of albite is $(100 - An_x)$. An intermediate plagioclase solid solution such as the composition marked C on the tie line in Figure 3.4 has a composition that is represented by its position on the tie line, which can be represented as $An_{35}$ $(=Ab_{65})$. $An_{35}$ can also be expressed as $(Na_{0.65}, Ca_{0.35})(Si_{0.65}, Al_{0.35})$ $AlSi_2O_8$. This indicates that 65% of the large cation site is occupied by sodium $(Na^{+1})$ ions and 35% by calcium $(Ca^{+2})$ ions with a coupled substitution of 65% silicon $(Si^{+4})$ and 35% aluminum $(Al^{+3})$ existing in the small cation site.

### 3.1.3 Limited ionic substitution

As noted previously, substitution is limited by significant differences in the ionic radii or charge of substituting ions. Ions of substantially different size limit the amount of substitution so that only a **limited solid solution** can exist between end member components. This situation can be illustrated in the rhombohedral carbonates by the limited solid solution series that exists between calcite $(CaCO_3)$ and magnesite $(MgCO_3)$. Once again, the potential solid solution series can be represented as a line between the two end members, and the composition of any calcium–magnesium-bearing, rhombohedral carbonate may be represented by a formula, by its position on the tie line or by the proportion of an end member component (calcite = Ct or magnesite = Ms). However, because calcium cations $(Ca^{+2})$ are more than 30% larger than magnesium $(Mg^{+2})$ cations, the substitution between the two end members is limited. Because the amount of substitution

is limited, many potential compositions do not exist in nature. Such gaps in a solid solution series are called **miscibility gaps** by analogy with immiscible liquids that do not mix in certain proportions. In this series, a miscibility gap exists between approximately $Ms_{25} = Ct_{75}$ and $Ms_{40} = Ct_{60}$ (Figure 3.5). To the left of this miscibility gap, a partial solid solution series exists between $Ms_0 = (Ca_{1.0}, Mg_{0.0})CO_3$ and $Ms_{25} = (Ca_{0.75}, Mg_{0.25})CO_3$. Many organisms secrete shells in this compositional range (Chapter 14). Within this range, we can define **low magnesium calcite** and **high magnesium calcite** in terms of their proportions of calcite (Ct) and magnesite (Ms) end members. Low magnesium calcites generally contain less than 4% magnesium $(Mg^{+2})$ substituting for calcium $(Ca^{+2})$ in this structural site and so have compositions in the range $Ct_{96-100} = Ms_{0-4}$ (Figure 3.5). High magnesium calcites have more than 4% magnesium substituting for calcium and therefore have compositions in the range $Ct_{75-96} = Ms_{4-25}$. Some workers further subdivide these compositions into medium magnesium and high magnesium calcite with a boundary at 10% magnesium (James and Jones 2016). Compositions from $Ms_{40-55} = Ct_{45-60}$ actually have a different structure – that of the double carbonate mineral **dolomite** whose average composition is $CaMg(CO_3)_2$. Many other examples exist of limited substitution series with miscibility gaps. The importance of mineral compositional variations that result from variations in substitution can be more fully understood in the context of phase stability diagrams, as discussed in the following section.

## 3.2 PHASE STABILITY (EQUILIBRIUM) DIAGRAMS

The behavior of materials in Earth systems can be modelled using thermodynamic calculations and/or empirical results from laboratory investigations. The results of such calculations and/or investigations are commonly summarized on phase stability diagrams. A **phase** is a mechanically separable part of the system. **Phase stability (equilibrium) diagrams** display the stability fields (conditions under which a phase is stable) for various phases in a system of specified composition. These fields are separated by **phase stability boundary lines** that represent the conditions under which phase changes from one phase to another occur. Phase stability diagrams related to igneous systems and processes summarize relationships between liquids (melts) and solids (crystals) in a system. Such diagrams usually have temperature increasing upward on the vertical axis and composition shown on the horizontal axis. At high temperatures the system is completely melted. The stability field for 100% liquid is separated from the remainder of the phase diagram by a phase boundary line called the **liquidus** that represents the temperature above which the system exists as 100% melt and below which it contains some crystals. The low temperature stability field for 100% solid crystals is separated from higher temperature conditions by a phase boundary line called the **solidus**. At intermediate temperatures between the solidus and liquidus, the system consists of two types of stable phases in equilibrium, both liquid and solid crystals. Phase equilibrium diagrams, based on both theoretical and laboratory analyses, exist for a variety of multicomponent systems. A one-component and five representative two-component systems related to the discussion of igneous rocks and processes (Chapters 7–10) are discussed below. Metamorphic phase diagrams are discussed in Chapter 18. For discussions of systems that are beyond the scope of this text, including three- and four-component systems, the reader is referred to mineralogy books by Wenk and Bulakh (2016), Nesse (2016), Dyer et al. (2008) and Klein and Dutrow (2007) and to petrology books by Frost and Frost (2019), Philpotts and Ague (2009), Winter (2009) and Best (2000). Some of the more important terms you will encounter in this discussion are defined in Table 3.1.

**Table 3.1** A list of some common terms used in discussing phase diagrams.

| Terms | Definitions |
| --- | --- |
| Liquidus | Phase boundary (line) that separates the all-liquid (melt) stability field from stability fields that contain at least some solids (crystals) |
| Solidus | Phase boundary (line) that separates the all-solid (crystal) stability field from stability fields that contain at least some liquid (melt) |
| Eutectic | Condition under which liquid (melt) is in equilibrium with two different solids |
| Peritectic | Condition under which a reaction occurs between a pre-existing solid phase and a liquid (melt) to produce a new solid phase |
| Phase | A mechanically separable part of the system; may be a liquid, gas or solid with a discrete set of mechanical properties and composition |
| Invariant melting | Occurs when melts of the same composition are produced by melting rocks of different initial composition |
| Incongruent melting | Occurs when a solid mineral phase melts to produce a melt and a different mineral with a different composition from the initial mineral |
| Discontinuous reaction | Mineral crystals and melt react to produce a completely different mineral; negligible solid solution exists between the minerals |
| Continuous reaction | Mineral crystals and melt react to continuously and incrementally change the composition of both; requires a mineral solid solution series |
| Solvus | Phase boundary (line) that separates conditions in which complete solid solution occurs within a mineral series from conditions under which solid solution is limited |

### 3.2.1 The phase rule

The **phase rule** (Gibbs 1928) governs the number of phases that can coexist in equilibrium in any system and can be written as:

$$P = C + 2 - F$$

where

**P** represents the number of phases present in a system. **Phases** are mechanically separable varieties of matter that can be distinguished from other varieties based on their composition, structure and/or state. Phases in igneous systems include minerals of various compositions, and crystal structures, amorphous solids (glass) and fluids such as liquids or gases. All phases are composed of one or more of the components used to define the composition of the system.

**C** designates the minimum number of chemical **components** required to define the phases in the system. These chemical components are usually expressed as proportions of oxides. The most common chemical components in igneous reactions include $SiO_2$, $Al_2O_3$, FeO, $Fe_2O_3$, MgO, CaO, $Na_2O$, $K_2O$, $H_2O$, and $CO_2$. All phases in the system can be made by combining components in various proportions.

**F** refers to the number of **degrees of freedom** or variance. Variance means the number of independent factors that can vary, such as temperature, pressure, and the composition of each phase, without changing the phases that are in equilibrium with one another. We will use the first phase diagram in the next section to show how the phase rule can be applied to understanding phase diagrams. A discussion of the phase rule and of phase diagrams related to metamorphic processes is presented in Chapter 18.

### 3.2.2   One component phase diagram: silica polymorphs

Pure silica ($SiO_2$) occurs as a number of different mineral phases, each characterized by a different crystal structure. These silica minerals include low quartz (alpha quartz), high quartz (beta quartz), tridymite (alpha and beta), cristobalite (alpha and beta), coesite, and stishovite. Each polymorph of silica is stable under a different set or range of temperature and pressure conditions. A phase stability diagram (Figure 3.6), where pressure increases upward and temperature increases to the right, shows the stability fields for the silica minerals. The stability fields represent the temperature and pressure conditions under

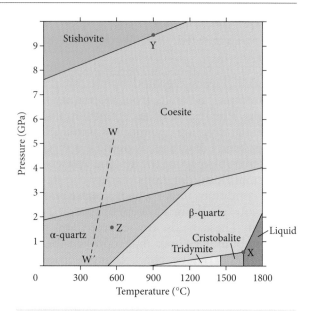

**Figure 3.6**   Phase diagram for silica depicting the temperature–pressure stability fields for the major polymorphs and the liquid phase. *Source*: Adapted from Wenk and Bulakh (2004). © John Wiley & Sons.

which each mineral phase is stable. Each stability field is bounded by phase boundaries, lines that define the limits of the stability field as well as the conditions under which phases in adjoining fields can coexist in equilibrium. Where three phase boundaries intersect, a unique set of conditions is defined under which three stable phases can coexist simultaneously.

The positions of the stability fields show that **stishovite** and **coesite** are high pressure varieties (polymorphs) of silica and that **tridymite** and **cristobalite** are high temperature/low pressure minerals. In this diagram both high temperature/low pressure polymorphs are the beta varieties. The high pressure polymorphs, coesite and stishovite, occur in association with meteorite impact and thermonuclear bomb sites, and stishovite is very likely a constituent of the deep mantle. The diagram shows that quartz is the stable polymorph of silica over a broad range of temperature–pressure conditions common in Earth's crust. This wide stability range and an abundance of silicon and oxygen help to explain why quartz is such an abundant rock-forming mineral in the igneous, sedimentary and metamorphic rocks of Earth's crust.

Figure 3.6 also shows that alpha quartz (low quartz) is generally more stable at lower temperatures than beta quartz (high quartz). Lastly, the diagram shows a phase stability boundary (liquidus or melting curve) on the far right that separates the lower temperature/pressure conditions under which silica is solid from the higher temperature/pressure conditions under which it is a liquid. The phase rule permits a deeper understanding of the relationships portrayed in the diagram. Places where three phase boundaries intersect represent unique temperature and pressure conditions where three stable mineral phases can coexist. For example, at point X at ~1650°C and ~0.04 GPa high quartz, cristobalite and liquid coexist because the high quartz/liquid, high quartz/cristobalite, and cristobalite/liquid phase boundaries intersect. Because there are three phases and one component, the phase rule ($P = C + 2 - F$) yields $3 = 1 + 2 - F$, so that F must be 0. This simply means that the temperature and pressure cannot be varied if three phases are to coexist. There are no degrees of freedom. Figure 3.6 shows additional triple points where three phases, all of them solid, coexist under a unique set of temperature and pressure conditions. If either temperature or pressure is varied, the system moves to a place on the diagram where one or more phases are no longer stable. At triple points, there are no degrees of freedom; the system is **invariant**.

In other situations depicted in Figure 3.6, two phases coexist under the conditions marked by phase stability boundary lines rather than points. As a result, the phase rule ($P = C + 2 - F$) yields $2 = 3 - F$, so that F must be 1. For example, under the conditions at point Y (900°C, 9.2 GPa), both coesite and stishovite can coexist. If the temperature increases the pressure must also increase, and vice versa, in order for the system to remain on the phase stability boundary line where these two phases coexist. There is only one independent variable or 1 degree of freedom. The temperature and pressure cannot be changed independently. In a similar vein, two phases, one solid and one liquid, can coexist anywhere on the melting curve that separates the liquid and a single solid stability fields.

However, for any point within a phase stability field (e.g., point Z) only one phase is stable (e.g., low quartz). The phase rule ($P = C + 2 - F$) yields $1 = 1 + 2 - F$, so that F must be 2. This means that the temperature and the pressure can change independently without changing the phase composition of the system. For point Z, the temperature and pressure can increase or decrease in many different ways without changing the phase that is stable, as long as they remain within the stability field. There are two independent variables and 2 degrees of freedom. All points to the right of the melting curve in the liquid field represent the stability conditions for a single phase, liquid silica.

One can also use this diagram to understand the sequence of mineral transformations that might occur as Earth materials rich in silica experience different environmental conditions. From a liquid silica system cooling at a pressure of 0.3 GPa cristobalite will begin to crystallize at ~1650°. As the system continues to cool, it will reach the cristobalite/tridymite phase boundary (~1460°C), where cristobalite will be transformed into tridymite. Ideally, the system will continue to cool until it reaches the tridymite/high quartz phase boundary. Here it will be transformed into high quartz, then cool through the high quartz field until it reaches the low quartz/high quartz phase boundary, where it will be converted to low quartz and continue to cool. Two phases will coexist only at phase boundaries during phase transformations that take finite amounts of time to complete (Chapter 4).

Similarly, a system undergoing decompression and cooling as it slowly rises toward the surface might follow line W–W′ on the phase diagram. It will start as coesite and be converted into alpha quartz (low quartz) as it crosses the phase boundary that separates them. Note that low quartz is the common form of quartz in low temperature, low pressure Earth materials.

### 3.2.3 Two component phase diagram: plagioclase

Figure 3.7 is a phase stability diagram for plagioclase, the most abundant mineral group in Earth's crust. One critical line on the phase stability diagram is the high temperature, convex upward **liquidus line**, above which is the all liquid (melt) stability field that comprises

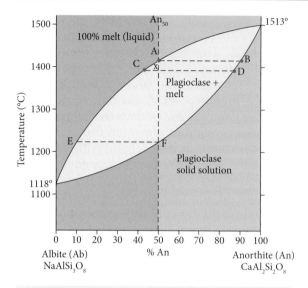

**Figure 3.7** Plagioclase phase stability diagram at atmospheric pressure, with a complete solid solution between the two end member minerals albite (Ab) and anorthite (An).

the conditions under which the system is 100% liquid (melt). A second critical line is the lower temperature, convex downward **solidus line**, below which is the all solid stability field that comprises the conditions under which the system is 100% solid (plagioclase crystals). A third stability field occurs between the liquidus and the solidus. This is the melt plus solid field where conditions permit both crystals and liquid to coexist simultaneously.

To examine the information that can be garnered from the plagioclase phase stability diagram, let us examine the behavior of a system, with equal amounts of the two end member components albite and anorthite. whose composition can be expressed as $An_{50}$ (Figure 3.7). On the phase diagram, the system is located on the vertical $An_{50}$ composition line. This line is above the liquidus (100% liquid) at high temperatures, between the liquidus and solidus (liquid + solid) at intermediate temperatures and below the solidus (100% solid) at low temperatures. If this system is heated sufficiently, it will be well above the liquidus temperature for $An_{50}$ and will be 100% melt, much like an ideal magma. Now let us begin to cool the $An_{50}$ system until it reaches the liquidus temperature (1420 °C) at point A. Once the system moves incrementally below A, it

moves into the melt plus solid field. This means that crystallization of the melt begins at point A. To determine the composition of the first crystals, a horizontal line (A–B), called a tie line, is constructed between the liquidus and the solidus. The tie line represents the composition of the two phases (liquid and solid solution) in equilibrium with each other at that temperature. The intersection of the tie line with the liquidus (point A) represents the composition of the liquid (~$An_{50}$), because the melt has just begun to crystallize. The tie line intersection with the solidus (point B) represents the composition of the first solid solution mineral (~$An_{90}$) to crystallize from the melt.

As the system continues to cool, the composition of the melt continues to change incrementally down the liquidus line (e.g., to point C) while the composition of the crystalline solid solution simultaneously changes composition as it moves incrementally down the solidus line (e.g., to point D). This process continues as liquid compositions evolve down the liquidus and solid compositions evolve down the solidus until the latter reaches the vertical system composition line where it intersects the solidus at point F. Any further cooling brings the system into the 100% solid field. The tie line E–F at this temperature indicates that the last drops of liquid in the system have the composition ~$An_{10}$, whereas the final solid crystals will be the same as the system composition ($\rightarrow An_{50}$).

Clearly the percentage of crystals must increase (from 0 to 100) and the percentage of melt must decrease (from 100 to 0) as cooling proceeds. Meanwhile the composition of the melt and the crystals continuously changes down the liquidus and solidus lines, respectively. How does this happen? As the system cools through the melt plus solid field, two phenomena occur simultaneously. First, ideally the melt and the existing crystals continuously react with one another so that crystal compositions are progressively converted into more albite-rich crystals (lower An) stable at progressively lower temperatures. Second, newly formed (lower An) crystals of the stable composition form and earlier formed crystals continue to grow as they react with the melt, so that the percentage of crystalline material increases progressively at the expense of melt. Crystal compositions evolve down the solidus line toward more albite-rich compositions (decreasing An) as temperature decreases. Liquid compositions evolve down the liquidus, also toward more albite-rich composition (decreasing An), as temperature decreases because the additional crystals that separate from the melt are always enriched in anorthite relative to the instantaneous melt composition.

The precise proportion of melt and solid at any temperature can be determined by the lever rule. The **lever rule** states that the proportion of the tie line on the solidus side of the system composition represents the proportion of liquid in the system, whereas the proportion of the tie line on the liquidus side of the system composition represents the proportion of crystals in the system. In Figure 3.7, the proportion of tie line A–B on the solidus side of the system composition line is ~100% and the proportion on the liquidus side of the system composition line is ~0%. This makes sense because crystallization has just begun. So tie line A–B indicates that, just as crystallization begins, ~0% solids of composition $An_{90}$ coexist with ~100% liquid of composition $An_{50}$. As the system cools (1) the percentage of crystals increases at the expense of the melt; (2) crystal composition evolves down the solidus; and (3) liquid composition evolves down the liquidus during continuous melt–crystal reaction and additional crystallization.

We can check this by drawing tie lines between the liquidus and the solidus for any temperature in which melt coexists with solids. Tie line C–D provides an example. In horizontal (An) units, this tie line is ~45 units long ($An_{86} - An_{41} = 45$). The proportion of the tie line on the liquidus side of the system composition (x) that represents the percentage of crystals is 20% (9/45). The proportion of the tie line on the solidus side (y) that represents the percentage of liquid is 80% (36/45). The system is 20% crystals of composition $An_{86}$ and 80% liquid of composition $An_{41}$. As the system cooled from temperature A–B to temperature C–D, existing crystals reacted continuously with the melt and new crystals continued to separate from the melt. Therefore, the percentage of crystals progressively increased as crystal composition evolved incrementally down the solidus line and melt composition evolved incrementally down the liquidus line. When the system has cooled to the solidus temperature (1225 °C), the

proportion of the tie line (E–F) on the liquidus side approaches 100% indicating that the system is approaching 100% solid and the proportion on the solidus side approaches 0%, implying that the last drop of liquid of composition $An_{10}$ is reacting with the remaining solids to convert them into $An_{50}$. We can use the albite–anorthite phase diagram to trace the progressive crystallization of any composition in this system. The lever rule can be used for compositions and temperatures other than those specifically discussed in this example.

The crystallization behavior of plagioclase in which An-rich varieties crystallize at high temperatures and react continuously with the remaining melt to form progressively lower temperature Ab-rich varieties forms the basis for understanding the meaning of the continuous reaction series of Bowen's reaction series, as discussed in Chapter 8. Phase stability diagrams summarize what happens when equilibrium conditions are obtained. In the real world, disequilibrium conditions are common so that incomplete reactions between crystals and magmas occur. These are discussed in the section of Chapter 8 that deals with fractional crystallization.

In addition, phase diagrams permit the melting behavior of minerals to be examined by raising the temperature from below the solidus. Let us do this with the same system we examined earlier ($An_{50}$). As the system is heated to the solidus temperature (1225 °C), it will begin to melt. The lever rule (line E–F) indicates that the first melts ($An_{10}$) will be highly enriched in the albite component. As the temperature increases, the percentage of melt increases and the percentage of remaining crystals decreases as the melt and crystals undergo the continuous reactions characteristic of systems with complete solid solution. The melt continues to be relatively enriched in the albite (lower temperature) component, but progressively less so, as its composition evolves incrementally up the liquidus. Simultaneously, the remaining solids become progressively enriched in the anorthite (higher temperature) component as the composition of the solids evolves up the solidus. The lever rule allows us to check this at 1400 °C where tie line C–D provides an example. The proportion of the tie line on the liquidus side of the system composition that represents the percentage of crystals is 20% (9/45), whereas the proportion of the tie line on the solidus side that represents the percentage of liquid is 80% (36/45). The system is 20% crystals of composition $An_{86}$ and 80% liquid of composition $An_{41}$. Complete equilibrium melting of the system occurs at 1420°C (point A), where the last crystals of $An_{90}$ melt to produce 100% liquid with the composition of the original system ($An_{50}$).

Why are phase diagrams important in understanding igneous processes? Several important concepts concerning melting in igneous systems are illustrated in the plagioclase phase diagram.

1  *All partial melts are enriched in low temperature components,* in this case albite, relative to the composition of the original rock.
2  The smaller the amount of partial melting that occurs in a system, the more enriched are the melts in low temperature constituents such as albite.
3  Progressively larger percentages of partial melting progressively dilute the proportion of low temperature constituents.
4  If melts separate from the remaining solids, the solids are enriched in high temperature, refractory constituents.

During crystallization, the liquidus indicates the temperature at which a system of a given composition (An content) begins to crystallize; and the stable composition of any liquid in contact with crystals in the melt plus solid field. During crystallization, the solidus represents the stable composition of any solid crystals that are in contact with liquid in the melt plus solid field as crystallization continues and the temperature of final crystallization for a system of given composition.

It might be useful to briefly note that olivine group minerals exhibit behavior that is similar to that of plagioclase in that there is complete substitution solid solution between the two end-members, high-temperature forsterite ($Mg_2SiO_4$) and fayalite ($Fe_2SiO_4$). In this case only one substitution, $Mg^{+2}$ for $Fe^{+2}$ and vice versa, occurs (Chapter 2). Olivine exhibits continuous chemical reactions between solids and melts, similar to those discussed above with plagioclase group minerals. During cooling below the liquidus, crystals are enriched in high temperature, Mg-rich forsterite, relative to system composition, and liquids are progressively enriched in low temperature, Fe-rich fayalite. Eventually, the melt has completely crystallized and the system crosses the solidus. Similarly, with increasing temperature, as the system crosses the solidus, early melts are enriched in low temperature, Fe-rich fayalite and residual solids are progressively enriched in high temperature, Mg-rich forsterite. More detailed descriptions of this system are available in the references cited above.

Phase stability diagrams deliver quantitative information regarding the behavior of melts and crystals during both melting and crystallization. This provides simple models for understanding such significant processes as anatexis (partial melting) and fractional crystallization, which strongly influence magma composition and the composition of igneous rocks. All these topics are explored in the context of igneous rock composition, magma generation, and magma evolution in Chapters 7 and 8. Phase stability diagrams are also important in understanding the conditions that produce sedimentary minerals and rocks (Chapters 11–14) and the reactions that generate metamorphic minerals and rocks (Chapters 15–18). Let us now consider two-component systems with distinctly different end members, between which no solid solution exists, using the diopside–anorthite binary phase diagram.

### 3.2.4  Two component phase diagram: diopside–anorthite

Figure 3.8 illustrates a simple type of two-component or binary phase stability diagram in which the two end members possess entirely different mineral structures so that there is no solid solution between them. The two components are the calcic plagioclase **anorthite** ($CaAl_2Si_2O_8$), a tectosilicate mineral, and the calcium-magnesium clinopyroxene **diopside** ($CaMgSi_2O_6$), a single-chain inosilicate mineral. The right margin of the diagram represents 100% anorthite component and the left margin represents 100% diopside component. Compositions in the system are expressed as weight % anorthite component; the weight % diopside component is 100% minus the weight % anorthite component. Temperature

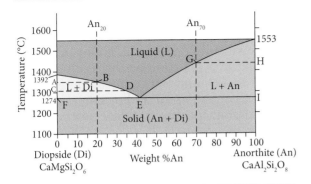

**Figure 3.8** Diopside–anorthite phase diagram at atmospheric pressure.

(°C) increases upward on the vertical axis. Because anorthite-rich plagioclases and diopside-bearing clinopyroxenes are the major minerals in mafic/basic igneous rocks, this phase diagram yields insights into their formation.

The diopside–anorthite phase stability diagram illustrates the temperature–composition conditions under which systems composed of various proportions of diopside and anorthite end member components exist as 100% melt, as melt plus solid crystals and as 100% solid crystals. At high temperatures all compositions of the system are completely melted. The stability field for 100% liquid (red) is separated from the remainder of the phase diagram by the *liquidus*. The liquidus temperature increases in both directions away from a minimum value for $An_{42}$ ($Di_{58}$), showing that either a higher anorthite (An) or a higher diopside (Di) content requires higher temperatures to maintain 100% melt. The phase diagram also shows that at low temperatures the system is completely crystallized. The stability field for 100% solid (blue) is separated from the remainder of the phase diagram by the *solidus*. For compositions of $An_{100}$ ($Di_0$) and $Di_{100}$ ($An_0$), which behave as one-component systems, the solidus temperature is the same as the liquidus temperature so that the solidus and liquidus intersect at 1553 and 1392 °C, respectively. For all intermediate two-component compositions, the solidus temperature is a constant 1274 °C.

The liquidus and solidus lines define a third type of stability field that is bounded by the two lines. This stability field represents the temperature–composition conditions under which both melt and crystals coexist; a liquid

of some composition coexists with a solid of either pure anorthite or pure diopside. Two melt plus solid fields are defined: (1) a melt plus diopside field for compositions of <42% anorthite by weight (yellow), and (2) a melt plus anorthite field (green) for compositions of >42% anorthite by weight. The liquidus and the solidus intersect where these two fields meet at a temperature of 1274 °C and a composition of 42% anorthite by weight ($An_{42}$). This point defines a temperature trough in the liquidus where it intersects the solidus and is called a **eutectic point** (E in Figure 3.8). Let us use a couple of examples, one representative of compositions of <42% anorthite by weight and the other of compositions of >42% anorthite by weight, to illustrate how this system works.

To investigate crystallization behavior, we'll start with a system rich in diopside component with a composition of $An_{20}$ ($Di_{80}$). We will start at a temperature above the liquidus temperature for this composition (Figure 3.8). As the system cools it will eventually intersect the liquidus at a temperature of ~1350 °C for this system composition (point B in Figure 3.8). To determine the composition of the first crystals, a horizontal tie line (A–B) may be drawn between the liquidus and the solidus. The intersection of the tie line with the liquidus (point B) represents the composition of the liquid (~$An_{20}$) because the melt has just begun to crystallize and its intersection with the solidus (point A) indicates the composition of the first crystals (diopside). As the system continues to cool, diopside crystals continue to form and grow. This increases the percentage of solid diopside crystals in the system while incrementally decreasing its proportion while increasing the proportion of anorthite in the remaining melt as the percentage of melt decreases. As the system continues to cool to 1315 °C (tie line C–D), the composition of the melt continues to change incrementally down the liquidus line (to point D) while the composition of the crystalline solid remains pure diopside (point C). As cooling continues, liquid compositions evolve down the liquidus and solid compositions evolve down the solidus until the vertical system composition line intersects the solidus at point E, after which any further cooling brings the system into the 100% solid (diopside plus anorthite) field. As the system approaches 1274 °C (tie line E–F), it contains a large proportion of diopside crystals and a smaller proportion of melt with the composition $\sim An_{42}$. When the system reaches the eutectic point at 1274 °C, where the liquidus and solidus intersect, the remaining melt crystallizes completely by isothermal, eutectic crystallization of diopside and anorthite until all the melt has been crystallized. Cooling of the system below 1274 °C causes it to enter the all-solid diopside plus anorthite field.

The percentage of crystals must increase (from 0 to 100) and the percentage of melt must decrease (from 100 to 0) as cooling proceeds. During this process, the composition of the melt continuously changes down the liquidus and the solids are crystallized in the sequence all diopside prior to the eutectic and diopside plus anorthite at the eutectic. Can we quantify these processes? In Figure 3.8, the proportion of tie line A–B on the solidus side of the system composition line is $\sim 100\%$ and the proportion on the liquidus side of the system composition line is $\sim 0\%$. This makes sense because crystallization has just begun. So tie line A–B indicates that $\sim 0\%$ solid diopside coexists with $\sim 100\%$ melt of composition $An_{20}$, at the moment crystallization begins. As the system cools, the percentage of crystals should increase at the expense of the melt as liquid composition evolves down the liquidus, with increasing An content caused by the continuous crystallization of diopside crystals. We can check this by drawing tie lines between the liquidus and the solidus for any temperature in which melt coexists with solids. Tie line C–D provides an example. In horizontal (An) units, this tie line is $\sim 35$ units long ($An_{35} - An_0 = 35$). The proportion of the tie line on the liquidus side of the system composition that represents the percentage of crystals is $\sim 43\%$ (15/35), whereas the proportion of the tie line on the solidus side is $\sim 57\%$ (20/35). The system is 43% diopside crystals ($An_0$) and 57% liquid of composition $An_{35}$. As the system cools from temperature A–B to temperature C–D, existing diopside crystals grow and new crystals continue to separate from the melt so that the percentage of crystals progressively increases. During this time, melt composition evolves incrementally down the liquidus line toward more anorthite-rich compositions. When the system approaches the eutectic temperature, the tie line (E–F) is $\sim 42$ An units long. The

proportion of the tie line (E–F) on the liquidus side approaches 52% (22/42), indicating that the system contains 52% diopside crystals, and the proportion on the solidus side is $\sim 48\%$ (20/42) liquid of composition $An_{42}$. At the eutectic temperature, diopside and anorthite simultaneously crystallize isothermally until the remaining melt is depleted. The proportion of crystals that form during eutectic crystallization of the remaining melt (48% of the system) is given by the lever rule as 42% (42/100) anorthite crystals and 58% (58/100) diopside crystals. The composition of the final rock is given by the proportions of the tie line between the solid diopside and solid anorthite that lie to the right and left of the system composition line. For this system, with a composition of $An_{20}$, the lever rule yields a final rock composition of 20% anorthite and 80% diopside.

The specific example related above is representative of the behavior of all compositions in this system between $An_0$ and $An_{42}$. When the system cools to the liquidus, diopside begins to crystallize, and as the system continues to cool, diopside continues to crystallize and grow. This causes the composition of the increasingly An-rich remaining melt to evolve down the liquidus toward the eutectic. Separation of crystals from the melt causes melt composition to change. When the system reaches the eutectic composition, isothermal crystallization of diopside and anorthite occurs simultaneously until no melt remains.

For compositions between $An_{42}$ and $An_{100}$ (e.g., $An_{70}$), the system diopside–anorthite behaves differently. For these compositions, when the system cools to intersect the liquidus, the first crystals formed are anorthite crystals (tie line G–H). Continued separation of anorthite crystals from the cooling magma causes the melt to be depleted in anorthite component (and enriched in diopside component) so that the melt composition evolves down the liquidus line to the left. Tie lines can be drawn and the lever rule can be used for any temperature in the anorthite plus liquid field. When the system cools to approach the eutectic temperature (tie line E–I), it contains a proportion of anorthite crystals in equilibrium with a liquid of composition $\sim An_{42}$. At the eutectic temperature, both diopside and anorthite simultaneously crystallize isothermally in eutectic proportions (58% diopside, 42% anorthite) until no melt remains.

Several important concepts emerge from studies of the equilibrium crystallization of two-component eutectic systems such as diopside–anorthite:

1  Which minerals crystallize first from magma depends on the specifics of melt composition
2  Separation of crystals from the melt generally causes melt composition to change
3  Multiple minerals can crystallize simultaneously from a magma.

This means that no standard reaction series, such as Bowen's reaction series (Chapter 8), can be applicable to all magma compositions because the sequence in which minerals crystallize or whether they crystallize at all is strongly dependent on magma composition, as well as on other variables. It also means that the separation of crystals from liquid during magma crystallization generally causes magma compositions to change or evolve through time. These topics are discussed in more detail in Chapter 8, which deals with the origin, crystallization and evolution of magmas.

Phase diagrams can also provide simple models for rock melting and magma generation. To do this, we choose a composition to investigate starting at subsolidus temperatures low enough to ensure that the system is 100% solid, and then gradually raise the temperature until the system reaches the solidus line where partial melting begins. As temperature continues to rise, we can trace the changes in the composition and proportions of melts and solids, using the lever rule, until the system composition reaches the liquidus, which implies that it is 100% liquid. Let us examine such melting behavior, using the two compositions previously used in the discussion of crystallization. A solid system of composition 20% anorthite ($An_{20}$) and 80% diopside ($Di_{80}$) will remain 100% solid until it has been heated to a temperature of 1274 °C where it intersects the solidus. Further increase in temperature causes the system to enter the melt plus diopside field as indicated by tie line E–F. The composition of the initial melt is given by the intersection of the tie line with the liquidus (point E), so that first melts have the eutectic composition ($An_{42}$), and

the composition of the remaining, unmelted solids is indicated by the intersection of the tie line with the solidus (point F = $An_0$ = $Di_{100}$). As the system is heated incrementally above the eutectic, the tie line (E–F) is 42 An units long and the proportion of the tie line on the liquidus side is ~52% (22/42) indicating that the system contains 52% diopside crystals, and the proportion on the solidus side is ~48% (20/42), indicating that all the anorthite and some of the diopside have melted at the eutectic to produce a liquid of composition $An_{42}$. At the eutectic temperature, both diopside and anorthite simultaneously melt isothermally until the remaining anorthite is completely melted. The proportion of crystals that melt during eutectic melting (48% of the system) is given by the lever rule and is 42% anorthite crystals and 58% diopside crystals as reflected in the melt composition. Further increases in temperature cause more diopside to melt. This increases the amount of melt and changes the melt composition toward less An-rich compositions as melt composition. As temperature continues to increase, melt composition evolves up the liquidus toward progressively diopside-enriched, anorthite-depleted compositions. When the temperature approaches the liquidus temperature for the bulk composition ($An_{20}$) of the system, the lever line (A–B) clearly indicates that the system consists of nearly 100% melt ($An_{20}$) and nearly 0% diopside ($An_0$) as the last diopside is incorporated into the melt. For the composition $An_{70}$, the initial also have the eutectic composition ($An_{42}$).

Several important concepts emerge from an examination of melting behavior in two-component systems such as diopside–anorthite:

1  The composition of first melts in such systems is the same – is invariant – for a wide range of system compositions
2  Melt compositions depend on the proportion of melting so that increasing degrees of partial melting cause liquid compositions to change
3  Changes in liquid composition depend on the composition of the crystals being incorporated into the melt.

**Invariant melting** helps to explain why some magma compositions (e.g., basaltic magmas) are more common than others, because some magma compositions can be generated by partial melting of a wide variety of available source rock compositions. The dependence of melt composition on the degree of partial melting suggests that it might be an important influence on ultimate melt composition. The ways in which magma composition depends on the incorporation of constituents from crystals in contact with the melt is also discussed in Chapter 8 in conjunction with a discussion of magma origin and evolution.

### 3.2.5 Two-component phase diagram: albite–orthoclase

Mineral compositions may offer vital clues to the conditions under which they were produced. This is well illustrated by the temperature-dependent substitution of potassium ($K^{+1}$) and sodium ($Na^{+1}$) in the alkali feldspars $(Na,K)AlSi_3O_8$, as illustrated by the **albite–orthoclase phase diagram** (Figure 3.9). Because albite-rich plagioclases and potassium feldspars are abundant in felsic/acidic igneous rocks, a significant component of continental crust, this diagram is useful in understanding their formation.

At high temperatures (>~620 °C at 1 atm pressures) a complete substitution solid solution series exists between the two end members. These are the potassium feldspar **orthoclase** ($KAlSi_3O_8$) and the sodium plagioclase feldspar, **albite** ($NaAlSi_3O_8$). Feldspar

crystals that form at high temperatures can have any proportions of orthoclase (Or) or albite (Ab) end member. Actual proportions depend largely on the composition of the system; that is, the availability of potassium and sodium ions. Because a complete solid solution exists between the two end members, crystallization and melting in this system share many similarities with the albite–anorthite system (see Figure 3.7) discussed earlier. For systems with <40% Or, initial crystals are rich in the albite plagioclase component. As plagioclase crystals continue to separate on cooling, they react continuously with the melt so that crystal composition changes down the solidus as the remaining liquid changes composition down the liquidus, both toward increasing Or content until no melt remains.

The result is a solid composed of sodic plagioclase, with a potassic orthoclase component in solid solution. For systems with >40% Or, initial crystals are relatively enriched in the potassium feldspar (orthoclase) component. As such crystals continue to separate on cooling, they react continuously with the melt so that crystal composition changes down the solidus as the remaining liquid changes composition down the liquidus, both toward decreasing Or content until all the melt is used up. The result is a rock composed of a feldspar solid solution. For systems with >40% Or, these crystals may be thought of as potassic orthoclase crystals with an albite component in solid solution. All solid solutions between the two end members are stable at high temperatures (and low pressures) after they begin to cool below the solidus temperature.

However, at lower temperatures (<~620 °C), the solid solution between orthoclase and albite becomes limited and a miscibility gap exists in which the solid solution between the two end members is unstable. The lower the temperature, the more limited the solid solution and the larger the miscibility gap becomes. As high temperature potassium-sodium feldspar solid solutions cool, they eventually reach the **solvus** temperature (Figure 3.9), a phase stability boundary that separates the conditions under which a complete solid solution is stable from conditions under which solid solutions are unstable. The solvus temperature is generally highest for compositions with large amounts of both end members. Below the solvus temperature, the original complete

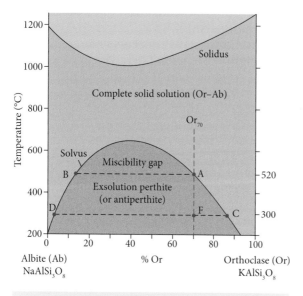

**Figure 3.9** Albite–orthoclase phase diagram at atmospheric pressure.

solid solution becomes unstable and begins to unmix or exsolve into an intergrowth of two distinct feldspars, one enriched in Ab component, the other in Or component.

Let us examine a potassium-rich feldspar (line $Or_{70}$ in Figure 3.9) that is a complete solid solution of composition $Or_{70}$ ($Ab_{30}$) as it cools below the solidus temperature. As this feldspar cools it eventually intersects the convex-up **solvus** curve at point A, at a temperature of 520°C, below which the solid solution becomes unstable. At temperatures below the solvus, the original solid solution unmixes or **exsolves** into two stable, but distinctly different, feldspars whose compositions lie on the solvus line that borders the miscibility gap. In this case, plagioclase of composition $Ab_{70}$ ($Or_{30}$) begins to unmix (exsolve) from the potassic feldspar as the solid solution becomes limited and a miscibility gap is created. Because the solid solution becomes increasingly limited and the miscibility gap widens as the temperature decreases, more plagioclase exsolves from the potassic feldspar and becomes increasingly sodic (Ab rich) as the crystals cool. As a result, the composition of the exsolved plagioclase evolves down the solvus to the left toward increasing Ab enrichment. Because albite component is exsolving from the potassic feldspar, the latter's orthoclase content progressively increases as its composition evolves down the solvus to the right. The lever rule can be used to trace the proportions and the composition of the exsolved plagioclase and the potassic feldspar at any temperature. Tie line C–D (~85 Or units long) between the two feldspar compositions on the solvus can be used for this purpose. On cooling to 300°C, the potassic feldspar component (point C) is ~$Or_{88}$ and the plagioclase component (point D) is ~$Or_3$. The percentage of exsolved Ab-rich plagioclase, given by line segment C–F, is ~21% (18/85), and the percentage of potash feldspar, given by line segment D–F, is ~79% (67/85). Progressive unmixing (exsolution) produces one feldspar increasingly enriched in potassium (Or) and another feldspar increasingly enriched in sodium (Ab). For initially potassium-rich feldspar solid solutions, the result is a specimen of potassium-rich feldspar that contains sodium-rich feldspar blebs, stringers or patches. A potassium feldspar crystal that contains sodium feldspar blebs, stringers or patches produced by the **exsolution** of two distinct feldspars is called **perthite**. Look closely at most potash feldspar crystals (e.g., orthoclase, microcline or sanidine) and you will see the generally less transparent blebs and stringers of plagioclase produced by exsolution. For initially albite-rich compositions (e.g., $Ab_{80}$), the result of exsolution can be plagioclase crystals that contain blebs, patches and/or stringers of exsolved orthoclase in albite and are called **antiperthite**. Antiperthite is less common than perthite because calcium-rich plagioclase does not form a solid solution series with orthoclase or any other potassic feldspar.

### 3.2.6   Two component phase diagram: nepheline–silica

The **nepheline–silica phase diagram** (Figure 3.10) illustrates a type of two-component system in which there is an **intermediate**

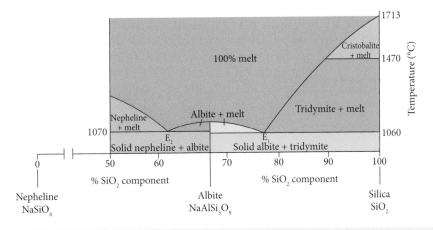

**Figure 3.10** Phase diagram for the system nepheline–silica with the intermediate compound albite, at atmospheric pressure.

**compound** whose composition can be produced by combining the compositions of the two end member components. In this case **silica (SiO2)** and **nepheline (NaAlSiO4)** are the two end member components. The intermediate compound formed by combining one molecular unit of nepheline and two of silica [NaAlSiO$_4$ + 2(SiO$_2$)] is the plagioclase mineral **albite (NaAlSi3O8)**. No solid solution exists between nepheline, albite, and silica minerals. Compositions are expressed on the horizontal axis in terms of molecular percent silica (SiO$_2$) component, so that the percentage of nepheline component is %Ne = 100% − %SiO$_2$ component. The composition of the intermediate compound albite is two-thirds SiO$_2$ component. Temperature increases on the vertical axis; pressure is 1 atm. The polymorphs of silica (see Figure 3.6) that crystallize in this system are the high-temperature, low-pressure minerals cristobalite and tridymite, but the more common polymorph of silica at slightly higher pressures is quartz. All these silica minerals have the same chemical composition, but different crystal structures.

Several important concepts are illustrated and reinforced by the nepheline–silica phase diagram (Figure 3.10). The most important is the notion of **silica saturation**, which is fundamental to igneous rock classification (Chapter 7). When there is sufficient silica component (more than two-thirds) so that each molecular unit of nepheline component can be converted into albite by adding a molecular unit of silica with an additional silica component remaining, the system is said to be **oversaturated with respect to silica**. Evidence for silica oversaturation is the presence of a silica mineral, such as tridymite or quartz formed from the excess silica, along with plagioclase feldspar in the final rock. When there is insufficient silica component (less than two-thirds) to convert each molecular unit of nepheline into albite by adding a molecular unit of silica, the system is said to be **undersaturated with respect to silica**. Evidence for silica undersaturation is the presence of a low silica feldspathoid mineral such as nepheline in the final rock. Only when the silica component is exactly two-thirds is there precisely the amount of silica component required to convert each molecular unit of nepheline into albite. Such systems are said to be exactly **saturated with respect to silica**. Evidence for exact silica saturation is the presence of feldspar and the absence of both silica and feldspathoids from the final rock, which in the system nepheline–silica consists of 100% albite. The International Union of Geological Sciences (IUGS) classification of igneous rocks (Chapter 7) is largely based on the concept of silica saturation; rocks in the upper triangle contain quartz and feldspar, whereas those in the lower triangle contain feldspathoids and feldspar. Rocks that lie on the line or join between the two triangles contain feldspar but neither quartz nor feldspathoids and are ideally saturated with respect to silica.

The nepheline–silica phase diagram shows some similarity to the diopside–anorthite phase diagram. The most significant difference is the presence of two eutectic points where troughs in the liquidus intersect the solidus. For many purposes, this diagram may be interpreted as two side-by-side eutectic diagrams: one diagram for undersaturated compositions (less than two-thirds silica component), with a eutectic point at 1070°C and ~62% silica component, and a second diagram for oversaturated compositions (over two-thirds silica component), with a eutectic point at 1060°C and ~77% silica component. A brief discussion of the crystallization and melting behaviors for these two compositional ranges follows.

Let us first examine the behavior of silica oversaturated systems with between 78 and 100% silica component. If a cooling melt with silica content between 89 and 100% intersects the liquidus, the first crystals to separate are composed of the high temperature silica polymorph called cristobalite. With continued cooling (Figure 3.10), more cristobalite separates from the melt, causing its composition to evolve down the liquidus toward lower percentages of silica as the proportion of melt decreases. As the system reaches a temperature of 1470°C, cristobalite becomes unstable and inverts isothermally to the stable, low temperature polymorph of silica called tridymite. This ideal inversion temperature is shown by the phase boundary between cristobalite and tridymite in the silica plus melt field. With continued cooling below 1470°C, more tridymite separates from the melt, and melt compositions continue to evolve (with progressively lower silica concentrations) down the liquidus toward the eutectic (E$_1$) at 1060°C. Upon reaching the eutectic, both albite and tridymite crystallize simultaneously until the melt is used up and the system enters the solid albite plus tridymite field. For compositions between 78 and 89% silica component, the behavior is similar except that the first crystals to form are tridymite. Final rock compositions can be calculated using the lever rule.

For compositions between ~67% and 78% $SiO_2$ (Figure 3.10), albite crystallizes when the system cools to the liquidus temperature. Continued separation of albite on cooling causes the liquid composition to move down the liquidus toward increasing silica content. As the system cools to the eutectic temperature of 1060 °C, albite and tridymite crystallize simultaneously and isothermally until the melt is used up. The final rock contains both albite and a silica mineral in proportions that can be determined by the lever rule.

Let us now examine the behavior of so-called silica undersaturated systems with between 0 and 67% silica component. For compositions between 62 and 67% silica, cooling of the system to the liquidus temperature causes albite crystals to separate from the melt (Figure 3.10). Continued cooling below the liquidus temperature causes further separation of albite from the melt which causes melt compositions to change down the liquidus to the left toward decreasing silica content. As the eutectic temperature ($E_2$) is reached at 1070 °C, both albite and nepheline crystallize isothermally until the melt is used up. The final rock contains percentages of both albite and nepheline that can be determined by the lever rule. Lastly, for those compositions with 50–62% silica component addressed in the diagram (additional complexities, not shown, exist for systems with lower amounts of silica component), the first crystals to separate are nepheline crystals. Continued separation of silica-poor nepheline causes melt compositions to change down the liquidus toward the eutectic at 1070 °C. At the eutectic, both albite and nepheline crystallize isothermally until the melt is used up. Once again the final rock is composed of albite and nepheline, and their percentages can be calculated using the lever rule.

### 3.2.7 Two component phase diagram: forsterite–silica

Another set of mineral relationships is well illustrated by the two-component system **forsterite–silica** (Figure 3.11). Forsterite ($Mg_2SiO_4$)

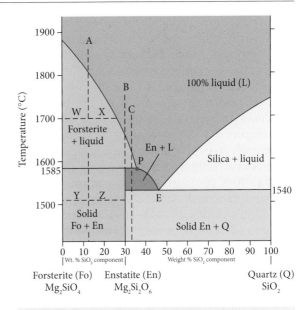

**Figure 3.11** Phase diagram for the system forsterite–silica with the intermediate compound enstatite, at atmospheric pressure.

is the magnesium end member of the olivine solid solution series, and the silica mineral is commonly quartz ($SiO_2$). As in the nepheline–silica system discussed above, this system contains an **intermediate compound**, in this case the orthopyroxene mineral **enstatite** ($MgSi_2O_6$), that can be thought of as being composed of one molecular unit of each of the two end member components ($MgSiO_4 + SiO_2 = MgSi_2O_6$). The horizontal axis in this phase diagram is weight % silica end member component (weight % $SiO_2$), rather than the molecular proportions used in the nepheline–silica diagram. Temperature increases on the vertical axis. Six phase stability fields are defined: (1) 100% melt, (2) melt + quartz, (3) melt + enstatite, (4) melt + forsterite, (5) forsterite + enstatite, and (6) enstatite + quartz. No solid solution exists between the three minerals in this system (forsterite, enstatite, and quartz). Instead, a **discontinuous reaction** occurs between forsterite and enstatite in which early formed minerals react with the melt to produce new minerals at a specific temperature. These reactions occur when the system reaches point P on the liquidus line, the **peritectic point** at 1585 °C and 35% silica component. There is also a *eutectic point* (E) located in the trough in the liquidus where it intersects the solidus at 1540 °C and 46% silica. Let us examine four selected compositions in this system during crystallization, each of which demonstrates different behaviors and/or results. For compositions of >46% *silica* component by weight, the system behaves as a simple eutectic system. As melts cool to the liquidus, silica (quartz) begins to separate from the melt and continues to separate as the system cools further (Figure 3.11). This causes the composition of the melt to evolve down the liquidus toward lower silica contents. Upon reaching the eutectic at 1540 °C, both quartz and enstatite crystallize simultaneously until the melt is used up. For compositions of ~35–46% silica that are richer in silica component than the peritectic (P) composition, the system also behaves as a simple eutectic. The only change is that enstatite crystallizes first, causing the liquid to evolve down the solidus toward increasing silica content until it reaches the eutectic. There enstatite and quartz crystallize simultaneously until the system is 100% crystalline. The lever rule can be used to determine phase percentages and compositions for any composition in which two phases coexist.

Systems of between 0 and 35% silica component behave somewhat differently because they pass through the *peritectic point* where reactions occur between forsterite, enstatite, and melt. During cooling and crystallization, three fundamentally different situations can be recognized. For example, with a composition of *12% silica* component (dashed line A, Figure 3.11), the system cools to the liquidus at 1810 °C, where forsterite begins to separate. Continued cooling causes additional forsterite to separate from the melt, which causes the melt composition to evolve down the liquidus toward the peritectic. At 1700 °C, the system consists of 50% forsterite crystals (line segment x) and 50% melt (line segment w) with a composition of ~24% silica component as inferred from the lever rule. Further cooling and separation of forsterite crystals cause the melt composition to approach the peritectic point at 1585 °C, where the lever rule shows that the system contains ~66% forsterite and ~34% melt of the peritectic composition 35% silica component. Below this temperature the system enters the 100% solid forsterite plus enstatite field with ~60% forsterite olivine line, as shown by tie-line segment z) and ~40% enstatite, as shown by tie-line segment y). So what happens when the melt reaches the peritectic? The percentage of

solids increases as the melt is used up, and the percentage of solid forsterite decreases while the percentage of solid enstatite increases dramatically. The percentage of forsterite decreases because some of the forsterite reacts with some of the remaining melt to produce enstatite. Simultaneously the percentage of enstatite increases dramatically because as some olivine is converted to enstatite, new enstatite crystallizes simultaneously from the remaining melt until it is used up. More generally, for all compositions of <30% silica component (enstatite composition), the equilibrium behavior is (1) forsterite crystallization as the melt cools below the liquidus; (2) increasing proportions of olivine and decreasing proportions of melt as the melt cools; (3) evolution of the remaining silica-enriched melt down the liquidus toward the peritectic; and (4) isothermal conversion of some forsterite to enstatite by discontinuous reaction with the remaining melt at the peritectic accompanied by additional isothermal crystallization of enstatite until the melt is used up. Some forsterite always remains because there is insufficient silica component to convert all of it into the intermediate compound enstatite. This indicates that the system was *undersaturated* with respect to silica. The peritectic reaction that converts the olivine mineral forsterite to the pyroxene mineral enstatite, as note previously, is called a **discontinuous reaction**. This phase diagram provides an excellent example of how early formed crystals can react with remaining melt to produce an entirely different mineral. These reactions are characteristic of the minerals in the discontinuous reaction series of Bowen's reaction series (Chapter 8).

Systems with exactly 30% silica component have the composition of the intermediate compound enstatite. They are precisely saturated with respect to silica. Line B (Figure 3.11) indicates the cooling behavior of such a system. Initial crystallization at the liquidus produces forsterite crystals. Continued cooling, accompanied by addition, separation, and growth of forsterite crystals, causes the remaining melt to evolve down the liquidus toward the peritectic with forsterite and melt percentages and melt compositions given by the lever rule. When this system reaches the peritectic, it contains 14% forsterite crystals and 86% melt with 35% silica component. Discontinuous reaction between the remaining melt and the forsterite crystals converts all the forsterite crystals to enstatite, while simultaneous crystallization of enstatite causes the melt to be used up. The final rock is 100% enstatite and is neither undersaturated (lacks forsterite) nor oversaturated (lacks quartz), but is **saturated with respect to silica.**

Systems with silica component contents of 30–35% silica are *oversaturated* with respect to silica and thus exhibit another set of behaviors. Line C (33% silica component) is representative of these behaviors (Figure 3.11). The system cools to the liquidus at 1650 °C where forsterite crystals begin to separate. Continued separation of forsterite causes melt composition to evolve down the liquidus toward the peritectic. As the melt composition reaches the peritectic, the lever rule shows that the system contains ~6% (2/35) forsterite crystals and ~94% (33/35) melt with ~35% silica component. At the peritectic, all the forsterite is converted to enstatite with the remaining melt and additional enstatite crystallizes, but additional melt remains. The lever rule shows that as the system leaves the peritectic and enters the enstatite plus liquid stability field it contains ~40% (2/5) enstatite and ~60% (3/5) melt. Further cooling leads to additional crystallization of enstatite (30% silica component), which causes the remaining melt to evolve down the liquidus toward the eutectic (at 46% silica component). As the system reaches the eutectic, it contains ~81% enstatite (13/16) and ~19% (3/16) melt with 46% silica component. At the eutectic, enstatite and quartz crystallize simultaneously until the melt is used up. The final rock contains 96% (67/70) enstatite and 4% (3/70) quartz and records a melt that was *oversaturated* with respect to silica.

Compositions in this system between 30 and 35% silica component clearly show that minerals, such as forsterite, that are undersaturated with respect to silica can crystallize from magmas that are oversaturated with respect to silica. If equilibrium conditions between these crystals and the melt are maintained, they will eventually be converted to the intermediate compound and therefore will not be preserved. However, if disequilibrium conditions exist, of the kinds that commonly occur during fractional crystallization, early formed crystals may well be preserved in the final rock. In addition, because silica in the remaining melt was not used to convert forsterite to enstatite, the melt will be more enriched in silica than would otherwise be the case. As discussed in Chapter 8, such concepts are very important in understanding the evolution of magma composition.

The system forsterite–silica (Figure 3.11) clearly illustrates the concept of silica saturation. Compositions of >30% silica ($SiO_2$) end member component by weight are *oversaturated* with respect to silica, so that there is sufficient silica to convert all the forsterite into enstatite, with additional silica remaining. Equilibrium crystallization in such silica-rich systems produces the intermediate compound enstatite with excess silica to form quartz. As discussed in connection with the nepheline–silica diagram (see Figure 3.10), quartz forms by equilibrium crystallization of melts that are oversaturated with respect to silica. On the other hand, compositions of <30% silica component by weight are undersaturated with respect to silica, so that there is insufficient silica to convert all the forsterite into enstatite. Equilibrium crystallization in such silica-poor systems produces forsterite plus as much of the intermediate compound enstatite as can be formed at the peritectic. Forsterite forms by equilibrium crystallization of melts that are undersaturated with respect to silica. As detailed in Chapter 7, both forsterite-rich olivine and feldspathoids suggest crystallization from systems undersaturated with respect to silica. Systems with exactly 30% silica component are exactly saturated with respect to silica because they possess precisely the amount of silica component required to convert forsterite into enstatite without excess silica remaining.

One can also investigate melting behaviors in this system. For compositions of >35% silica, the system behaves as a simple eutectic, producing first melts with an invariant composition of 46% silica. Melts possess this composition until either quartz (for systems 35–46% silica component) or enstatite (for systems >46% silica component) is completely melted. Subsequent melting of the remaining mineral causes the liquid to change composition up the liquidus. These behaviors once again demonstrate the ways in which *melt compositions depend both on the percentage of partial melting and on the composition of the original rock.* For compositions of <35% silica, melting involves peritectic reactions. In these systems, whenever the system reaches the peritectic, some or all of the enstatite remaining in the solid fraction melts to produce both forsterite and melt at the peritectic (35% silica component). This behavior is essentially the reverse of what happens when systems cool through the peritectic, and melt plus olivine yields enstatite. Such behavior, in which the melting of one crystalline material produces both a new crystalline material and a melt of different composition, is called **incongruent melting**. It also illustrates how silica oversaturated melts might be obtained from the partial melting of silica undersaturated, forsterite-rich rocks such as ultramafic peridotites in the mantle.

## 3.3   ISOTOPES

This section provides a brief introduction to the uses of some radioactive isotopes and stable isotopes important in the understanding of Earth materials and processes. Isotope studies provide powerful insights concerning the age, behavior and history of Earth materials. In geology, a thorough understanding of both stable and radioactive isotopes is essential for determining the ages and origin of minerals and rocks. Isotope ratios, determined by mass spectroscopy, are also instrumental in understanding a variety of other phenomena discussed in this book, including the determination of:

1   Source rocks from which magmas are derived.
2   Origin of water on Earth's surface.
3   Timing of mountain building events involving igneous intrusions and metamorphism.
4   Timing of unroofing of such rocks and the dispersal of their erosional products by sedimentary agents.
5   Source rocks for petroleum and natural gas.
6   Changes in ocean water temperatures, biological productivity and circulation.
7   History of ice age glacial expansions and contractions.
8   Climate change.

### 3.3.1   Stable isotopes

**Stable isotopes** contain nuclei that do not tend to change spontaneously. Instead, their nuclear configurations (number of protons and neutrons) remain constant over time. Many elements occur in the form of multiple stable isotopes with different atomic mass numbers. In many cases, these isotopes, because of their different mass, exhibit subtly different behaviors in Earth environments. These differences in behavior are recorded as differences in the ratios between isotopes that can be used to infer the conditions under which the isotopes were selectively incorporated into Earth materials. We will use oxygen and carbon isotopes to illustrate the uses of stable isotope ratios to increase our understanding of Earth materials and processes. Other stable isotopes that are commonly utilized in such studies include those of sulfur, nitrogen, and helium (Chapter 13, Box 13.2).

*Oxygen isotopes*

Three isotopes of oxygen occur in Earth materials (Chapter 2): oxygen-18 ($^{18}O$), oxygen-17 ($^{17}O$), and oxygen-16 ($^{16}O$). Each oxygen isotope contains eight protons in its nucleus; the remaining mass results from the number of neutrons (10, 9, or 8 respectively) in the nucleus. $^{16}O$ constitutes >99.7% of the oxygen on Earth, $^{18}O$ constitutes ~0.2%, and $^{17}O$ is relatively rare. The ratio $^{18}O/^{16}O$ is widely used to infer important information concerning Earth history.

During evaporation, water with lighter $^{16}O$ is preferentially evaporated relative to water with heavier $^{18}O$. During the evaporation of ocean water, water vapor in the atmosphere is enriched in $^{16}O$ relative to $^{18}O$ (lower $^{18}O/^{16}O$) while the remaining ocean water is preferentially enriched in $^{18}O$ relative to $^{16}O$ (higher $^{18}O/^{16}O$). Initially (Epstein and Mayeda 1953), these ratios were related to temperature because evaporation rates are proportional to temperature. It was proposed that higher $^{18}O/^{16}O$ ratios in ocean water record higher temperatures, which cause increased evaporation and preferential removal of lighter $^{16}O$. It was quickly understood that organisms using oxygen to make calcium carbonate ($CaCO_3$) shells could preserve this information as carbonate sediments accumulated on the sea floor over time. Such sediments would have the potential to record changes in water temperature over time; especially when the changes are large and the signal is clear (see Box 3.1).

However, it was soon realized that small, short-term temperature signals could be largely obliterated by a second set of processes. These involve changes in global ice volumes associated with the expansion and contraction of continental glaciers, e.g., during ice ages. Glaciers expand when more snow accumulates each year than is ablated (Chapter 12). This produces a net growth in glacial ice volume. Because atmospheric water vapor largely originates by evaporation, the snow (eventually converted to ice) is enriched in $^{16}O$ and has a low $^{18}O/^{16}O$ ratio. As glaciers expand, they store huge volumes of water with low $^{18}O/^{16}O$ ratios, causing the $^{18}O/^{16}O$ ratio in ocean water to progressively increase. As a result, periods of maximum glacial ice volume correlate with global periods of maximum $^{18}O/^{16}O$ in marine sediments. Prior to the use of oxygen isotopes, the record of Pleistocene glaciation was known largely from glacial till deposits on the continent, and only four periods of major Pleistocene glacial expansion had been established. Subsequently, the use of oxygen isotope records from marine

## Box 3.1   The Paleocene–Eocene thermal maximum

In the mid-nineteenth century, scientists recognized a rapid change in mammalian fossils that occurred early in the Tertiary era. The earliest Tertiary epoch, named the Paleocene (early life), was dominated by archaic groups of mammals that had mostly been present during the preceding Mesozoic Era. The succeeding period, marked by the emergence and rapid radiation of modern mammalian groups, was called the Eocene (dawn of life). The age of the Paleocene–Eocene boundary is currently judged to be 55.8 Ma. Later workers noted that the boundary between the two epochs was also marked by the widespread extinction of major marine groups, most prominently deep-sea benthic foraminifera (Pinkster 2002; Ivany et al. 2018). The cause of these sudden biotic changes initially remained unknown. Oxygen and carbon isotope studies have given us some answers.

Kennett and Stott (1991) reported a rapid rise in $\delta^{18}O$ at the end of the Paleocene, which they interpreted as resulting from a rapid rise in temperature, since they believed that no prominent ice sheets existed at this time. Subsequent work (e.g., Zachos et al. 1993; Rohl et al. 2000; Gehler et al. 2016; Ivany et.al. 2018) has confirmed that temperatures rose ~6–8 °C at high latitudes and ~3–5 °C at low latitudes over a time interval not longer than 10 000 years. Rapid global warming, in this case the Paleocene–Eocene thermal maximum (PETM), has apparently occurred in the past, with significant implications for life on Earth. Researchers have also shown that the higher temperatures *lasted for* approximately 100 Ka (Pinkster 2002; Ivany et al. 2018). How long will the current period of global warming last?

What caused the rapid global warming? Researchers studying carbon isotopes have shown that the sudden increase in temperature implied by rising $\delta^{18}O$ values was accompanied by sudden decreases in $\delta^{13}C$. Several hypotheses have been suggested for this, most of which involve warming and the release of large quantities of $^{12}C$ from organic carbon reservoirs. Two rapid spikes in negative $\delta^{13}C$, each occurring over time periods of less than 1000 years, suggest that some releases were extremely rapid. The currently favored hypothesis involves the melting of frozen clathrates in buried ocean floor sediments. Clathrates consist of frozen water in which methane, methanol, and other organic carbon molecules are trapped. The hypothesis is that small amounts of warming caused clathrates to melt, releasing large volumes of methane to the atmosphere in sudden bursts. This would account for the sudden negative $\delta^{13}C$ spikes. Because methane ($CH_4$) is a very effective greenhouse gas (10–20 times more effective than $CO_2$), this theory also accounts for the sudden warming of Earth's surface and the extinction and mammalian radiation events that mark the Paleocene–Eocene boundary. Of course scientists wonder if the current episode of global warming that already involves an order of magnitude larger release of $CO_2$ (Ivany et al. 2018) might be accelerated by triggering a sudden release of clathrates, and how long the effects of such releases might linger.

sediments and ice ($H_2O$) cores in Greenland and Antarctica has established a detailed record that involves dozens of glacial ice volume expansions and contractions during the Pliocene and Pleistocene.

$^{18}O/^{16}O$ ratios are generally expressed with respect to a standard in terms of $\delta^{18}O$. One standard is the $^{18}O/^{16}O$ ratio in a belemnite from the Cretaceous Pee Dee Formation of South Carolina, called PDB. $\delta^{18}O$ is usually expressed in parts per thousand (mils) and calculated from:

$$\delta^{18}O = \frac{\left[ ^{18}O/^{16}O_{SAMPLE} - {^{18}O/^{16}O_{PDP}} \right]}{^{18}O/^{16}O_{PDB}} \times 1000$$

Because the Cretaceous was an unusually warm period in Earth's history, with high evaporation rates, PDB has an unusually high $^{18}O/^{16}O$ ratio. As a result, most Pliocene–Pleistocene samples have a negative $\delta^{18}O$, with small negative numbers recording maximum glacial ice volumes and larger negative numbers recording minimum glacial ice volumes. Because different organisms selectively fractionate $^{18}O$ and $^{16}O$, a range of organisms must be analyzed and the results averaged when determining global changes in $^{18}O/^{16}O$. Nothing is ever as easy as it first seems.

It should be noted that many $\delta^{18}O$ analyses have used a different standard. This standard is the average $^{18}O/^{16}$ ratio in ocean water known as standard mean ocean water (SMOW). Of course, that has been complicated by global warming, that generally increases evaporation rates, changing the ratios in natural waters.

Because the original SMOW and PDB standards have been used up in comparative analyses, yet another standard, Vienna standard mean ocean water (VSMOW), is also used. This name is misleading as the Vienna standard is actually a pure water sample with no dissolved solids. There is currently much discussion concerning the notion of which standards are most appropriate and how $\delta^{18}O$ and other isotope values should be reported.

Oxygen isotope analyses, as well as carbon isotopes discussed below, contribute essential data regarding Earth's paleoclimate as well as dynamic climate changes currently affecting Earth.

### Carbon isotopes

Three isotopes of carbon occur naturally in Earth materials: carbon-12 ($^{12}C$), carbon-13 ($^{13}C$) and the radioactive carbon-14 ($^{14}C$), used in radiocarbon dating. Each carbon isotope contains six protons in its nucleus; the remaining mass results from the number of neutrons (six, seven or eight) in the nucleus. $^{12}C$ constitutes >98.9% of the stable carbon on Earth, and $^{13}C$ constitutes most of the other 1.1%.

When organisms synthesize organic molecules, they selectively utilize $^{12}C$ in preference to $^{13}C$ so that organic molecules have lower than average $^{13}C/^{12}C$ ratios. Enrichment of the organic material in $^{12}C$ causes the $^{13}C/^{12}C$ in the water column to increase. Ordinarily, there is a rough balance between the selective removal of $^{12}C$ from water during organic synthesis and its release back to the water column by bacterial decomposition, respiration, and other processes. Mixing processes produce a relatively constant $^{13}C/^{12}C$ ratio in the water column. However, during periods of stagnant circulation in the oceans or other water bodies, disoxic–anoxic conditions develop in the lower part of the water column and/or in bottom sediments. These conditions inhibit bacterial decomposition (Chapters 11 and 14) and lead to the accumulation of $^{12}C$-rich organic sediments. These sediments have unusually low $^{13}C/^{12}C$ ratios. As they accumulate, the remaining water column, depleted in $^{12}C$, develops a higher ratio of $^{13}C/^{12}C$. However, any process, such as the return of vigorous circulation and oxidizing conditions, that rapidly release the $^{12}C$-rich carbon from organic sediments, is associated with a rapid decrease in $^{13}C/^{12}C$. By carefully plotting changes in $^{13}C/^{12}C$ ratios over time, paleooceanographers have been able to document both local and global changes in oceanic circulation. In addition, because different organisms selectively incorporate different ratios of $^{12}C$ to $^{13}C$, the evolution of new groups of organisms and/or the extinction of old groups of organisms can sometimes be tracked by rapid changes in the $^{13}C/^{12}C$ ratios of carbonate shells in marine sediments or organic materials in terrestrial soils.

$^{13}C/^{12}C$ ratios are generally expressed with respect to a standard in terms of $\delta^{13}C$. The standard once again is the $^{13}C/^{12}C$ ratio of the

Pee Dee Belemnite, or PDB. $\delta^{13}C$ is usually expressed in parts per thousand (mils) and calculated from:

$$\delta^{13}C = \frac{\left[ {}^{13}C/{}^{12}C_{SAMPLE} - {}^{13}C/{}^{12}C_{PDP} \right]}{{}^{13}C/{}^{12}C_{PDB}} \times 1000$$

Box 3.1 illustrates an excellent example of how oxygen and carbon isotope ratios can be used to document Earth history, in this case a period of sudden global warming that occurred 55 million years ago.

### 3.3.2   Radioactive isotopes

**Radioactive isotopes** possess unstable nuclei whose nuclear configurations tend to be spontaneously transformed by **radioactive decay**. Radioactive decay occurs when the nucleus of an unstable **parent isotope** is transformed into that of a **daughter isotope**. Daughter isotopes have different atomic numbers and/or different atomic mass numbers from their parent isotopes. Three major radioactive decay processes (Figure 3.12) have been recognized: alpha ($\alpha$) decay, beta ($\beta$) decay, and electron capture.

**Alpha decay** involves the ejection of an alpha ($\alpha$) particle plus gamma ($\gamma$) rays and heat from the nucleus. An **alpha particle** consists of two protons and two neutrons, which is the composition of a helium ($^4He$) nucleus. The ejection of an alpha particle from the nucleus of a radioactive element reduces the atomic number of the element by two ($2p^+$) while reducing its atomic mass number by four ($2p^+ + 2n^0$). The spontaneous decay of uranium-238 ($^{238}U$) into thorium-234 ($^{234}Th$) is but one of many examples of alpha decay.

**Beta decay** involves the ejection of a beta ($\beta$) particle plus heat from the nucleus. A **beta particle** is a high-speed electron ($e^-$). The ejection of a beta particle from the nucleus of a radioactive element converts a neutron into a proton ($n^0 - e^- = p^+$) increasing the atomic number by one while leaving the atomic mass number essentially unchanged. The spontaneous decay of radioactive rubidium-87 (Z = 37) into stable strontium-87 (Z = 38) is one of many examples of beta decay.

**Electron capture** involves the addition of a high-speed electron to the nucleus with the release of heat in the form of gamma rays. It

can be visualized as the reverse of beta decay. The addition of an electron to the nucleus converts one of the protons into a neutron ($p^+ + e^- = n^0$). Electron capture decreases the atomic number by one while leaving the atomic mass number unchanged. The decay of radioactive potassium-40 (Z = 19) into stable argon-40 (Z = 18) is a useful example of electron capture. It occurs at a known rate, which allows the age of many potassium-bearing minerals and rocks to be determined. Only about 9% of radioactive potassium decays into argon-40; the remainder decays into calcium-40 ($^{40}Ca$) by beta emission.

The heat released by radioactive decay is called **radiogenic heat**. Radiogenic heat is a major source of the heat generated within Earth. It is an important driver of global tectonics and many of the rock-producing processes discussed in this text, including magma generation and metamorphism.

The time required for one half of the radioactive isotope to be converted into a new isotope is called its **half-life** and may range from seconds to billions of years. Radioactive decay processes continue until a stable nuclear configuration is achieved and a stable isotope is formed. The radioactive decay of a parent isotope into a stable daughter isotope may involve a sequence of decay events. Table 3.2 illustrates the 14-step process required to convert radioactive uranium-238 ($^{238}U$) into the stable daughter isotope lead-206 ($^{206}Pb$). All of the intervening isotopes are unstable, so that the radioactive decay process continues. The first step, the conversion of $^{238}U$ into thorium-234 by alpha decay, is slow, with a half-life of 4.47 billion years (4.47 Ga). Because many of the remaining steps are relatively rapid, the half-life of the full decay, sequence is just over 4.47 Ga. As the rate at which $^{238}U$ atoms are ultimately converted into $^{206}Pb$ atoms is known, the ratio $^{238}U/^{206}Pb$ can be used to determine the crystallization ages for minerals, especially for those formed early in Earth's history, as explained below.

Those isotopes that decay rapidly, beginning with protactinium-234 and radon-222, produce large amounts of decay products in short amounts of time. Radioactive decay products, especially alpha particles, can produce notable damage to crystal structures and significant tissue damage in human populations (Box 3.2). Radioactive isotopes also have significant applications in medicine, especially in cancer treatments. Radioactive decay provides a significant energy source through nuclear fission in reactors. Radioactive materials remain important global energy resources, even though the radioactive isotopes in spent fuel present long-term hazards, expecially with respect to their disposal. As noted above, radioactive decay is also the primary heat engine within Earth and

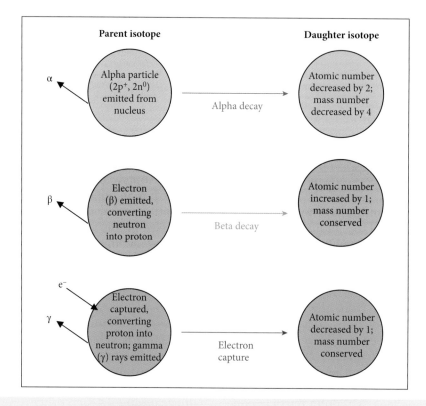

**Figure 3.12**  Three types of radioactive decay: alpha decay, beta decay, and electron capture (gamma decay) and the changes in nuclear configuration that occur as the parent isotope decays into a daughter isotope.

**Table 3.2**  The 14-step radioactive decay sequence that occurs in the conversion of the radioactive isotope $^{238}$U into the stable isotope $^{206}$Pb.

| Parent isotope | Daughter isotope | Decay process | Half-life |
|---|---|---|---|
| **Uranium-238** | Thorium-234 | Alpha | $4.5 \times 10^9$ years |
| Thorium-234 | Protactinium-234 | Beta | 24.5 days |
| Protactinium-234 | Uranium-234 | Beta | 1.1 minutes |
| Uranium-234 | Thorium-230 | Alpha | $2.3 \times 10^5$ years |
| Thorium-230 | Radium-226 | Alpha | $8.3 \times 10^4$ years |
| Radium-226 | Radon-222 | Alpha | $1.6 \times 10^3$ years |
| Radon-222 | Polonium-218 | Alpha | 3.8 days |
| Polonium-218 | Lead-214 | Alpha | 3.1 minutes |
| Lead-214 | Bismuth-214 | Beta | 26.8 minutes |
| Bismuth-214 | Polonium-214 | Beta | 19.7 minutes |
| Polonium-214 | Lead-210 | Alpha | $1.5 \times 10^{-4}$ seconds |
| Lead-210 | Bismuth-210 | Beta | 22.0 years |
| Bismuth-210 | Polonium-210 | Beta | 5.0 days |
| Polonium-210 | **Lead-206** | Alpha | 140 days |

## Box 3.2    Radon and lung cancer

Inhalation of radon gas is the second largest cause of lung cancer worldwide, second only to cigarette smoking. In the 1960s, underground uranium miners began to show unusually high incidences of lung cancer. The cause was shown to be related to the duration of the miner's exposure to radioactive materials. To cause lung cancer, the radioactive material must enter the lungs as a gas. It then causes progressive damage to the bronchial epithelium or lining of the lungs. What is the gas and how does it originate? Table 3.2 shows the many radioactive isotopes that are produced by the decay of the common isotope of uranium ($^{238}U$). Uranium miners would be exposed to all of these, but which one would they inhale into their lungs? Because radon possesses a stable electron configuration, it tends not to combine with other elements. Like most noble elements, under normal near surface conditions, it tends to exist as separate atoms in the form of a gas. In the confined space of poorly ventilated underground mines, radioactive decay in the uranium series produces sufficient concentrations of radon to significantly increase the incidence of lung cancer. The other property that makes radon-222 so dangerous is its short half-life (3.825 days). Within days, most of the radon inhaled by miners decays into polonium-218 with the emission of alpha particles ($^4$He nuclei). Subsequently, most of the radioactive $^{218}Po$ decays within hours into lead-210 with the release of more alpha particles. Lung damage leading to lung cancer largely results from continued rapid release of alpha particles over long periods of exposure. Scientific studies on radon exposure have been complicated by the fact that many miners were also smokers. It turns out that smoking and radon exposure act synergistically to multiply the risk of developing lung cancer.

Is the general public at risk of radon exposure? Uranium is ubiquitous in the rocks of Earth's crust, and so therefore is radon production. Potassium feldspar-bearing rocks such as granites and gneisses, black shales, and phosphates contain higher uranium concentrations (>100 ppm) than average crustal rocks (<5 ppm). They therefore pose a greater threat. Radon gas occurs in air spaces and is quite soluble in water; think of the dissolved oxygen that aqueous organisms use to respire or the carbon dioxide dissolved in carbonate beverages. Groundwater circulating through uranium-rich rocks can dissolve substantial amounts of radon gas and concentrate radium, another carcinogenic isotope. Ordinarily, this is not a problem. The gas rises and is released to the atmosphere, where it is dispersed and diluted to very low levels. But if radon gas is released into a confined space such as a home, especially one that is well insulated and not well ventilated, radon gas concentrations can reach hazardous levels. Most radon gas enters the home through cracks in the walls and foundations, either as gas or in water from which the gas is released. Most of the remainder is released when water from radon-contaminated wells is used, again releasing radon into the home atmosphere. The problem is especially bad in winter and spring months when homes are heated, basements flooded, and ventilation poor. As warm air in the home rises, air is drawn from the soil into the home, increasing radon concentrations. The insulation that increases heat efficiency also increases radon concentrations. What can be done to reduce the risk? Making sure that basements and foundation walls are well sealed and improving ventilation can reduce radon concentrations to acceptable levels, even in homes built on soils with high concentrations of uranium. Radon test kits can be purchased from hardware stores. If indoor radon levels exceed 4 pCi/l, remediation is recommended by the installation of indoor air pumps and ventilation pipes to remove gases from beneath basement floors. Radon remediation typically costs $1500 and is highly recommended as a health measure.

is partly responsible for driving plate tectonics and core–mantle convection. Without radioactive heat, Earth would be a very different kind of home.

In the following sections we have chosen a few examples, among the many that exist, to illustrate the importance of radioactive isotopes and decay series in the study of Earth materials.

*Age determinations using radioactive decay series*

Table 3.3 lists several radioactive parent to stable daughter transformations that can be used to determine the formation ages of Earth materials. All these are based on the principle that after the radioactive isotope is incorporated into Earth materials, the ratio of

**Table 3.3**   Systematics of radioactive isotopes important in age determinations in Earth materials.

| Decay series | Decay process | Decay constant ($\lambda$) | Half-life | Applicable dating range |
|---|---|---|---|---|
| $^{14}C \rightarrow {}^{14}N$ | Beta decay | $1.29 \times 10^{-4}$/year | 5.37 Ka | <60 Ka |
| $^{40}K \rightarrow {}^{40}Ar$ | Electron capture | $4.69 \times 10^{-10}$/year | 1.25 Ga | 25 Ka to >4.5 Ga |
| $^{87}Rb \rightarrow {}^{87}Sr$ | Beta decay | $1.42 \times 10^{-11}$/year | 48.8 Ga | 10 Ma to >4.5 Ga |
| $^{147}Sm \rightarrow {}^{143}Nd$ | Alpha decay | $6.54 \times 10^{-12}$/year | 106 Ga | 200 Ma to >4.15 Ga |
| $^{232}Th \rightarrow {}^{208}Pb$ | Beta and alpha decays | $4.95 \times 10^{-11}$/year | 14.0 Ga | 10 Ma to >4.5 Ga |
| $^{235}U \rightarrow {}^{207}Pb$ | Beta and alpha decays | $9.85 \times 10^{-10}$/year | 704 Ma | 10 Ma to >4.5 Ga |
| $^{238}U \rightarrow {}^{206}Pb$ | Beta and alpha decays | $1.55 \times 10^{-10}$/year | 4.47 Ga | 10 Ma to >4.5 Ga |

radioactive parent isotopes to stable daughter isotopes decreases through time by radioactive decay. The rate at which such parent: daughter ratios decrease depends on *the rate of decay*, which is given by the **decay constant ($\lambda$)**, the proportion of the remaining radioactive atoms that will decay per unit of time. One useful formula that governs decay series states that the number of radioactive atoms remaining at any given time (N) is equal to the number of radioactive atoms originally present in the sample ($N_0$) multiplied by a negative exponential factor ($e^{-\lambda t}$) that increases with the rate of decay ($\lambda$) and the time since the sample formed (t), that is, its age. These relationships are given by:

$$N = N_0 e^{-\lambda t}$$

It should be clear from the formula that when t = 0, N = $N_0$, and that N becomes smaller through time as a function of the rate of decay given by the decay constant; rapidly for a large decay constant, more slowly for a small one. Figure 3.13 illustrates a typical decay curve, showing how the abundance of the radioactive parent isotope decreases exponentially over time while the abundance of the daughter isotope increases in a reciprocal manner. The two curves cross where the number of radioactive parent and stable daughter atoms is equal. The time required for this to occur is called the half-life of the decay series and is the time required for one half of the radioactive isotopes to decay into stable daughter isotopes.

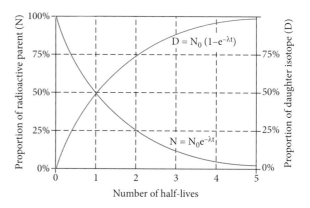

**Figure 3.13**   Progressive change in the proportions of radioactive parent (N) and daughter (D) isotopes over time, in terms of number of half-lives.

More generally, the age of any sample may be calculated from the following equation:

$$t = (1/\lambda)\ln(d/p + 1)$$

where t = age, $\lambda$ = the decay constant, d = number of stable daughter atoms and p = number of radioactive parent atoms. Where stable daughter atoms were present in the original sample, a correction must be made to account for them, as explained below.

Many radioactive to stable isotope decay series, each with a unique decay constant, can be used, often together, to determine robust formation ages for Earth materials, especially for older materials. Table 3.3 summarizes some common examples. Three of these are discussed in more detail in the sections that follow.

### Uranium–lead systematics

Uranium (U) occurs in two radioactive isotopes, both of which decay in steps to different stable isotopes of lead (Pb). The closely related actinide element thorium (Th) decays in a similar fashion to yet another stable isotope of lead. The essential information on the radioactive and stable isotopes involved for these three decay series ($^{238}U \rightarrow \, ^{206}Pb$, $^{235}U \rightarrow \, ^{207}Pb$, and $^{232}Th \rightarrow \, ^{208}Pb$) is summarized in Table 3.3. This includes the types of decay processes, their half-lives, and the useful age range for dating by each of these methods.

Of the three decay series, the most commonly used is $^{238}U \rightarrow \, ^{206}Pb$. This is because $^{238}U$ is much more abundant than the other two radioactive isotopes and thus it and its daughter product are easier to measure accurately. Typically, heavy minerals, such as zircon, sphene, and/or monazite, are used for analyses because they contain substantial actinides, are relatively easy to separate and resist chemical alteration.

A problem arises because some $^{206}Pb$ occurs naturally in Earth materials, without radioactive decay, so that measured $^{238}U/^{206}Pb$ ratios include both daughter and nondaughter $^{206}Pb$. This must be corrected if an accurate age is to be determined. The correction can be accomplished because yet another lead isotope, $^{204}Pb$, is not produced by radioactive decay. The ratio of $^{204}Pb/^{206}Pb$ in meteorites, formed at the time Earth formed, is known and is taken to be the original ratio in the rock whose age is being determined. The correction is then accomplished by measuring the amount of $^{204}Pb$ in the sample and subtracting the meteoritic proportion of $^{206}Pb$ from total $^{206}Pb$ to arrive at the amount of $^{206}Pb$ that is the daughter of $^{238}U$.

In most analyses, the ratio of $^{235}U/^{207}Pb$ is also determined. The $^{235}U/^{207}Pb$ decay series has a much shorter half-life than the $^{238}U/^{206}Pb$ series, so that the U/Pb ratios are much smaller for a given sample age. Once the ratios of both parent uranium isotopes to the corrected daughter isotopes ($^{238}U/^{206}Pb$, $^{235}U/^{207}Pb$) have been determined, sample ages are generally

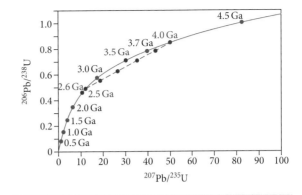

**Figure 3.14** Uranium–lead concordia plot (blue) showing sample ages as a function $^{206}Pb/^{238}U$ and $^{207}Pb/^{235}U$ ratios and samples with lower $^{206}Pb/^{238}U$ and $^{207}Pb/^{235}U$ ratios (red), reset during a 2.5 Ga metamorphic event.

determined on a **concordia plot** (Figure 3.14) that displays the expected relationship between the reciprocal ratios $^{206}Pb/^{238}U$ and $^{207}Pb/^{235}U$ as a function of sample age.

Either decay series is capable of producing a sample age on its own. For example, a $^{206}Pb/^{238}U$ ratio of 0.5 indicates an age of approximately 2.6 Ga (Figure 3.14), while a $^{207}Pb/^{235}U$ ratio of 40 indicates an age of approximately 3.7 Ga. One advantage of these decay series is that the measurement of multiple, closely related isotopes permits ages to be checked against each other and provides robust sample ages when both methods yield a closely similar age. For example, if a sample possesses a $^{206}Pb/^{238}U$ ratio of 0.7 and a $^{207}Pb/^{235}U$ ratio of 32, a sample age near 3.5 Ga (Figure 3.14) is highly likely. When two age determinations are in agreement, their ages are said to be **concordant**. Another way to state this is that samples that plot on the concordia line when their $^{206}Pb/^{238}U$ versus $^{207}Pb/^{235}U$ ratios are plotted yield concordant ages. Such dating techniques are especially powerful in determining robust crystallization ages of igneous rocks that crystallized from magmas and lavas (Chapters 8 and 9).

A significant problem arises from the fact that subsequent events can alter the geochemistry of rocks, so that age determinations are no longer concordant. For example, when rocks undergo metamorphism, lead is commonly mobilized and lost from the minerals or rocks in question. This results in lower than expected amounts of daughter lead isotopes, lower Pb/U ratios and points that fall below the concordant age line on $^{206}$Pb/$^{238}$U versus $^{207}$Pb/$^{235}$U diagrams. When samples of similar real ages, affected differently during a single metamorphic event, are plotted on a concordia diagram, they fall along a straight line below the concordant age line. One interpretation of such data is that the right end of the dashed straight line intersects the concordant age line at the original age of the sample (4.0 Ga in Figure 3.14) and that the left end intersects the concordant age line at the time of metamorphism (2.5 Ga in Figure 3.14). Wherever possible, such interpretations should be tested against other data so that a robust conclusion can be drawn. One such independent radiometric decay series, with widespread application, is described in the section that follows.

*Rubidium–strontium systematics*

Several isotopes of the relatively rare element rubidium (Rb) exist; some 27% of these are radioactive $^{87}$Rb. Most Rb$^{+1}$ is concentrated in continental crust, especially in substitution for potassium ion (K$^{+1}$) in potassium-bearing minerals such as potassium feldspar, muscovite, biotite, sodic plagioclase, and amphibole. Radioactive $^{87}$Rb is slowly (half-life = 48.8 Ga) transformed by beta decay into strontium-87 ($^{87}$Sr). Unfortunately, the much smaller strontium ion (Sr$^{+2}$) does not easily substitute for potassium ion (K$^{+1}$) and therefore tends to migrate into minerals in the rock that contain calcium ion (Ca$^{+2}$) for which strontium easily substitutes. This makes using $^{87}$Rb/$^{86}$Sr for age dating a much less accurate method than the U/Pb methods explained previously.

One basic concept behind rubidium–strontium dating is that the original rock has some initial amount of $^{87}$Rb, some initial ratio of $^{87}$Sr/$^{86}$Sr and some initial ratio of $^{87}$Rb/$^{86}$Sr. These ratios evolve through time in a predictable manner. Over time, the amount of $^{87}$Rb decreases and the amount of $^{87}$Sr increases by radioactive decay so that the $^{87}$Sr/$^{86}$Sr ratio increases. At the same time, the $^{87}$Rb/$^{86}$Sr ratio decreases by an amount proportional to sample age. A second basic concept is that the initial amount of $^{87}$Rb varies from mineral to mineral, being highest in potassium-rich minerals. As a result, the rate at which the $^{87}$Sr/$^{86}$Sr ratio increases depends on the individual

mineral. For example, in a potassium-rich (rubidium-rich) mineral, the $^{87}$Sr/$^{86}$Sr ratio will increase rapidly, whereas for a potassium-poor mineral it will increase slowly. For a mineral with no $^{87}$Rb substituting for K, the $^{87}$Sr/$^{86}$Sr ratio will not change; it will remain the initial $^{87}$Sr/$^{86}$Sr ratio. The $^{87}$Rb/$^{86}$Sr ratio of the whole rock however decreases at a constant rate that depends on the decay constant.

In a typical analysis, the amounts of $^{87}$Rb, $^{87}$Sr, and $^{86}$Sr in the whole rock and in individual minerals are determined by mass spectrometry, and the $^{87}$Rb/$^{86}$Sr and $^{87}$Sr/$^{86}$Sr ratios are calculated for each. These are plotted on an $^{87}$Sr/$^{86}$Sr versus $^{87}$Rb/$^{86}$Sr diagram (Figure 3.15). At the time of formation, assuming no fractionation of strontium isotopes, the $^{87}$Sr/$^{86}$Sr in each mineral and in the whole rock was a constant initial value, while the $^{87}$Rb/$^{86}$Sr values varied from relatively high for rubidium-rich minerals such as biotite and potassium feldspar to zero for minerals with no rubidium. These initial $^{87}$Sr/$^{86}$Sr and $^{87}$Rb/$^{86}$Sr values are shown by the horizontal line in Figure 3.15. As the rock ages, $^{87}$Rb progressively decays to $^{87}$Sr, which causes the $^{87}$Sr/$^{86}$Sr ratio to increase at rates proportional to the initial amount of $^{87}$Rb, while the $^{87}$Rb/$^{86}$Sr ratio decreases at a constant rate. Over time, the $^{87}$Sr/$^{86}$Sr ratios and $^{87}$Rb/$^{86}$Sr ratios for each mineral and the whole rock evolve along paths shown by the arrowed lines in Figure 3.15. If each mineral acts as a closed system, points representing the current $^{87}$Sr/$^{86}$Sr versus $^{87}$Rb/$^{86}$Sr ratios will fall on a straight line whose slope increases through time (Figure 3.15). The slope of the best-fit line, called an **isochron** (line of constant age), yields the age of the sample. The y-intercept of any isochron yields the initial $^{87}$Sr/$^{86}$Sr ratio, which is unchanging for a theoretical sample that contains no $^{87}$Rb. The **initial 87Sr/86Sr** is especially important in identifying the source regions from which magmas are derived in the formation of igneous rocks (Chapter 8).

*Potassium–argon systematics*

Potassium-40 ($^{40}$K) is a relatively rare radioactive isotope of potassium. Approximately 9.1% of $^{40}$K atoms decay to the daughter isotope Argon-40 ($^{40}$Ar), almost entirely by electron capture that converts a proton into a

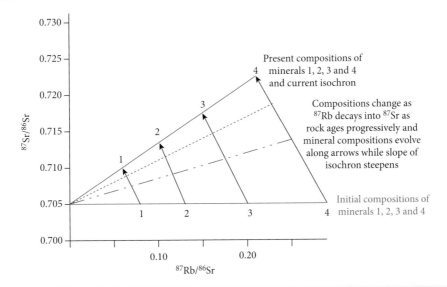

**Figure 3.15** Rubidium–strontium systematics, showing evolution in the composition of four representative minerals (1–4) from initial composition (blue line) to current composition (red line) as $^{87}Rb$ decays into $^{87}Sr$ over time. Whole rock compositions would lie somewhere between minerals 1 and 4 depending on the specifics of mineral composition and their proportions in the rock. Slope of red line yields provisional age of rock.

neutron. This lowers the atomic number by one without changing the atomic mass number. Argon-40 is produced only from the decay of potassium-40, with a half-life of 1.248 Ga. The other 88.9% of the radioactive $^{40}K$ atoms decay into the daughter isotope Calcium-40 ($^{40}Ca$) by β particle emission which converts a neutron into a proton. This increases the atomic number by one without changing the atomic mass number. Because calcium-40 from other sources is abundant in variable amounts in most rocks its ratio with $^{40}K$ cannot be used to obtain rock ages.

However, because argon-40 is a Noble element it generally occurs as a gas and therefore rarely occurs in minerals at the time they form. Therefore, argon-40 that exists in minerals is likely the product of the radioactive decay of the potassium-40. Assuming that there is no loss of this argon-40 from the mineral and no addition from other sources, the ratio of $^{40}K/^{40}Ar$ in the mineral should increase over time and yield reliable ages for the minerals and/or rocks in which it occurs. This is especially true for volcanic rocks because, at high temperatures, argon is a gas that escapes easily from the lava into the atmosphere (where it is the third most abundant gas, after nitrogen and oxygen). On the other hand,

when the lava crystallizes to form potassium-bearing minerals, argon-40 produced by the decay of potassium-40 tends to be trapped in the crystal lattice because its radius is larger than the spacing between atoms. Ideally, this sets the stage for using $^{40}K/^{40}Ar$ to date such rocks, but as we shall see, many challenges remain. This ratio is most useful for dating samples that formed more than 100 Ka in which enough time has elapsed for accurately measurable argon-40 to accumulate, although some dates as young as 25 Ka have been reported.

Three isotopes of potassium exist and tend to occur in a known fixed ratio in mineral-forming environments. The stable isotopes potassium-39 and potassium-41 constitute 93.25815 and 6.73025% of all potassium atoms. Radioactive potassium-40 contributes only 0.0117% of all potassium atoms. The rarity of potassium-40 means that its initial abundance in minerals or rocks must generally calculated from its known ratio to the other two isotopes that are much easier to measure accurately.

The abundance of argon-40 in the sample is determined by mass spectrometry. The amount of potassium-40 remaining in the sample is calculated by subtracting the amount of argon-40 that has accumulated and the corresponding amount of calcium-40 that would have formed in the double decay of potassium-40. On the assumption that no argon-40 existed in the sample at the time it was formed and that the mineral has behaved as a closed system with no gain or loss of argon-40, the age of the sample can be calculated. Of course this assumption is not always valid. Two processes that produce excess argon-40 and therefore anomalously old ages are (1) the incorporation of mantle xenoliths and xenocrysts that contain ungassed argon-40 and (2) younger lavas that contain argon-40 bearing gas-bubble/vesicles. Processes that cause argon-40 to leak from rocks and minerals, which produces anomalously young ages, include subsequent heating and alteration. For these reasons, care is taken to select samples that are unaltered and unfractured. This often requires extensive sample searching and preparation. Potassium feldspar (sanidine, orthoclase, and microcline) is the most commonly used mineral group, but biotite, muscovite, and illite clays can also be dated.

In recent decades, many techniques have been developed for the purpose of producing more accurate and refined age dates. Of these, the most significant has been the evolution of $^{40}Ar/^{39}Ar$ dating methods which compare the ratios of these two argon isotopes from a small portion of a sample to avoid the inaccuracies inherent in inhomogeneous samples such as whole rocks, parts of which may not be representative. In this method, the sample and a standard of known age containing potassium-40 are bombarded with neutrons in a nuclear reactor to produce argon-39 which does not exist naturally. The amount of argon-39 produced under a standard set of conditions is a proxy for the amount of potassium-40 in the sample. From this information, the $^{40}K/^{40}Ar$ ratio can be calculated and the age of the sample determined. On the whole, $^{40}Ar/^{39}Ar$ dating methods appear to be more accurate than conventional $^{40}K/^{40}Ar^{40}$ methods and can date samples as young as 25 Ka, but they do require access to a nuclear reactor.

There are many other isotope series utilized in determining rock ages, the history of magmatic source rocks, the conditions of sedimentation and/or the age of metamorphic events. Some of these are discussed in the chapters that follow in contexts where they are especially important.

## CONTENT ASSESSMENT

1  What three conditions favor the *complete substitution* of different atoms/ions for one another in a particular coordination site as a mineral grows? What three conditions limit such substitutions? Why are some substitutions *simple* and other substitutions *paired*?

2  For an initial solid plagioclase of composition $An_{60}$, use Figure 3.4 to:
   (a) Describe the proportions of $Na^+$ and $Ca^{+2}$ in the large cation substitution site.
   (b) Describe the proportions of $Si^{+4}$ and $Al^{+3}$ in the small cation substitution site.
   (c) Write a formula for this plagioclase in the form $(Na_xCa_x)$ $(Si_xAl_x)AlSi_2O_8$, similar to the formula for $An_{35}$ shown in the figure.

3  In addition to the β-quartz—cristobalite—liquid triple point show by the letter X in Figure 3.6, identify the *two other triple points* shown in the figure in terms of the *three phases that exist in equilibrium* and the unique *temperature* and *pressure* at which the three phases are equilibrium. Which of these sets of coexisting minerals might be expected to exist deep below the surface and which might be more likely to occur much nearer the surface? How might you recognize a rock that formed under *triple point* conditions?

4  Starting with a melt of composition $An_{30}$, trace the crystallization history as the melt cools, crystallizes and solidifies by answering the following questions.
   (a) At what temperature does the melt begin to crystallize?
   (b) What is the composition of the first crystals?
   (c) In general, what happens as the system continues to cool?
   (d) At a temperature of 1250°, what is the composition and percentage of crystals?
   (e) What is the composition and percentage of the remaining melt?
   (f) How has the composition and percentage of each changed as the system cooled from the temperature of initial crystallization?

(g) Explain in terms of continuous reactions how this happened?

(h) At what temperature does the system become solid rock and what is its final composition?

5 Starting with a solid of composition $An_{70}$ at 1200°, trace the melting history as the system is heated progressively.

(a) At what temperature does it begin to melt?

(b) What is the composition of the first melt?

(c) In general, what happens to the system as the temperature continues to increase.

(d) At a temperature of 1400°, what is the composition and percentage of the remaining melt?

(e) What is the composition of the remaining crystals?

(f) How has the percentage of each changed and how can this be explained in terms of continuous reactions?

(g) At what temperature would the melting become complete and what would be its composition?

6 For feldspar of composition $Or_{55}$, please answer the following questions using Figure 3.4.

(a) At what temperature did this mineral crystallize?

(b) As it cools, at what temperature would it reach the *solvus* and begin to undergo exsolution (unmixing)?

(c) What is the composition of the first exsolved component?

(d) At 400°C, what is the composition and percentage of the exsolved component? What is the composition and percentage of the host phase?

(e) What is the best name for the texture of this feldspar after exsolution has occurred?

7 Explain what is meant by *silica undersaturation*, *silica oversaturation* and *silica saturation*. What modal mineral group is indicative of silica oversaturation? What two modal mineral groups are generally suggestive of silica undersaturation? Using the phase diagram for nepheline-silica (Figure 3.10), describe in terms of $\%SiO_2$ component the compositions that are undersaturated, oversaturated and precisely saturated with respect to $SiO_2$ and the minerals that would form from

each during equilibrium crystallization of a melt. Do the same for the system forsterite-silica in terms of weight $\%SiO_2$, noting the mineral that forms from a system that is precisely saturated.

8 Summarize the differences between $^{16}O$ and $^{18}O$? Then explain the effects of temperature and global ice volumes on the ratio of these two isotopes ($^{18}O/^{16}O$) in ocean water and how they can be used to estimate fluctuations in temperatures and glacial ice volumes through time.

9 What is a radioactive element? What are the three major types of radioactive decay and how does each type account for the specific changes that occur when a parent isotope is converted into a daughter isotope?

10 Explain the meaning of the term *half-life* of a radioactive decay process. What percentage of parent isotopes has been converted into the daughter isotope after one half-life? After 2 half-lives? After 3, 4, 5 and 6 half-lives?

## REFERENCES

Best, M.G. (2000). *Igneous Petrology*. New York: Wiley-Blackwell 455 pp.

Dyer, M.D., Gunter, M.E., and Tasa, D. (2008). *Mineralogy and Optical Mineralogy*. Chantilly, VA: Mineralogical Society of America 708 pp.

Epstein, S. and Mayeda, T. (1953). *Variation of O18 content of waters from natural sources. Geochimica et Cosmochimica Acta* 4: 213–224.

Frost, B.R. and Frost, C.D. (2019). *Essentials of Igneous and Metamorphic Petrology*, 2e. Cambridge University Press 321 pp.

Gehler, A., Gingerich, P.D., and Pack, A. (2016). Temperature and atmospheric $CO_2$ concentration estimates through PETM using triple oxygen isotope analysis of mammalian bioapatite. *Proceedings of the National Academy of Science*. 113 (28): 7739–7744.

Gibbs, J.W. (1928). *The Collected Works of J. Willard Gibbs, Vol. I. Thermodynamics*. New Haven, CT: Yale University Press 438 pp.

Ivany, L., Pietsch, C., Handley, J.C. et al. (2018). Little lasting impact of the Paleocene–Eocene thermal maximum on shallow water molluscan faunas. *Science Advances* 4 (9): 1–9.

James, N.P. and Jones, B. (2016). *Origin of Carbonate Sedimentary Rocks*. Chichester, UK: Wiley 446 pp.

Kennet, J.P. and Stott, L.D. (1991). Abrupt deep-sea warming, paleooceanographic changes and benthic extinctions at the end of the Palaeocene. *Nature 353*: 225–229.

Klein, C. and Dutrow, B. (2007). *Manual of Mineral Science (Manual of Mineralogy)*, 23e. New York: Wiley 704 pp.

Nesse, W.D. (2016). *Introduction to Mineralogy*, 3e. New York: Oxford University Press 512 pp.

Philpotts, A. and Ague, J. (2009). *Principles of Igneous and Metamorphic Petrology*. Cambridge University Press 684 pp.

Pinkster, L.M. (2002). Sudden warming in the past. *Geotimes 30*: 77.

Rohl, V., Bralower, T.J., and Norris, R.D. (2000). New chronology for the late Paleocene thermal maximum and its environmental implications. *Geology 28*: 927–930.

Wenk, H.-R. and Bulakh, A. (2004). *Minerals: Their Constitution and Origin*. Cambridge, UK: Cambridge University Press 646 pp.

Wenk, H.-R. and Bulakh, A. (2016). *Minerals: Their Constitution and Origin*, 3e. Cambridge, UK: University Press 621 pp.

Winter, J.D. (2009). *Principles of Igneous and Metamorphic Petrology*, 2e. New York: Pearson 720 pp.

Zachos, J.C., Lohmann, K.C., Walker, J.C.G., and Wise, S.W. (1993). Abrupt climate change and transient climates during the Paleogene – a marine perspective. *Journal of Geology 101*: 191–213.

# Chapter 4

# Crystallography

## 4.1 CRYSTALLINE SUBSTANCES

Crystallography emphasizes the long-range order or crystal structure of crystalline substances. It focuses on the symmetry of crystalline materials and on the ways in which their long-range order is related to the three-dimensional repetition of fundamental units of pattern during crystal growth. In minerals, the fundamental units of pattern are molecular clusters of coordination polyhedra or stacking sequences (Chapter 2). The ways in which these basic units can be repeated to produce crystal structures with long-range order are called symmetry operations. Crystallography is also focused on the description and significance of planar features in crystals including planes of atoms, cleavage planes, crystal faces, and the forms of crystals. In addition, it is concerned with crystal defects which are local imperfections in the long-range order of crystals.

### 4.1.1 Crystals and crystal faces

Mineral crystals are one of nature's most beautiful creations. Many crystals are enclosed by flat surfaces called crystal faces. **Crystal faces** are formed when mineral crystals grow, and enclose crystalline solids when they stop growing. Perfectly formed crystals are notable for their remarkable symmetry (Figure 4.1). The external symmetry expressed by crystal faces allowed us to infer the geometric patterns of atoms in mineral crystal structures.

*Earth Materials*, Second Edition. Kevin Hefferan and John O'Brien.
© 2022 John Wiley & Sons Ltd. Published 2022 by John Wiley & Sons Ltd.
Companion website: www.wiley.com/go/hefferan/earthmaterials2

(a)

(b)

**Figure 4.1**   Representative mineral crystals: representative mineral crystals: (a) pyrite. *Source*: Courtesy of Doug Moore. (b) quartz. *Source*: Robert Lavinsky. iRocks.com—CCBYSA 3.0, https:// commons.wikimedia.org/wiki/File:Quartz-153455.jpg; last accessed 09/24/2020.

These patterns inferred from external symmetry have been confirmed by advanced analytical techniques such as X-ray diffraction (XRD) and atomic force microscopy (AFM).

Mineralogists have developed language to describe the symmetry of crystals and the crystal faces that enclose them. Familiarizing students with the concepts and terminology of crystal symmetry and crystal faces is one of the primary goals of this chapter. A second goal of this chapter is to build connections between crystal chemistry (Chapters 2 and 3) and crystallography. This involves an explanation of the relationships between chemical composition, coordination polyhedra and the crystal structures, crystal faces, and crystal forms that develop as crystals grow.

*Motifs and nodes*

When minerals begin to form, atoms or ions bond together, so that partial or complete coordination polyhedra develop (Chapter 2). Because the ions on the corners and edges of coordination polyhedra have unsatisfied electrostatic charges, they tend to bond to additional ions available in the environment as the mineral grows. Eventually, a small cluster of coordination polyhedra is formed that contains all the coordination polyhedra characteristic of the mineral and its chemical

composition. In any mineral, we can recognize a small cluster of coordination polyhedra that contains the mineral's fundamental composition and unit of pattern or motif. As the mineral continues to grow, additional clusters of the same pattern of coordination polyhedra are added to form a mineral crystal with a three-dimensional geometric pattern – a long-range, three-dimensional crystal structure. Clusters of coordination polyhedra are added, often one atom or ion at a time, as (1) the crystal nucleates, (2) it becomes a microscopic crystal, and, if growth continues, (3) it becomes a macroscopic crystal. Growth continues in this manner until the environmental conditions that promote growth change and growth ceases.

Long-range, geometric arrangements of atoms and/or ions in crystals are produced when a fundamental array of atoms, a unit of pattern or motif, is repeated in three dimensions to produce the crystal structure. A **motif** is the smallest unit of pattern that, when repeated by a set of symmetry operations, will generate the long-range pattern characteristic of the crystal. In minerals, the motif is composed of one or more coordination polyhedra. In wallpaper, it is a basic set of design elements that are repeated to produce a two-dimensional pattern, whereas in a brick wall the fundamental motif is that of a single brick that is repeated in space to form the three-dimensional structure. The repetition of these fundamental units of pattern by a set of rules called **symmetry operations** can produce a two- or three-dimensional pattern with long-range order. When several different motifs could be repeated by a similar set of symmetry operations, we may wish to emphasize the general rules by which different motifs may be repeated to produce a particular type of long-range order. In such cases, it is useful to represent motifs by using a point. A point used to represent any motif is called a **node**. The pattern or array of atoms about every node must be the same throughout the pattern the nodes represent.

## 4.2 SYMMETRY OPERATIONS

### 4.2.1 Simple symmetry operations

Symmetry operations may be simple or compound. **Simple symmetry operations** produce repetition of a unit of pattern or motif using a single type of operation. **Compound symmetry operations** involve motif repetition using a combination of two symmetry operations. Simple symmetry operations, discussed in this section, include (1) translation, by specific distances in specified directions, (2) rotation, about a specified set of axes, (3) reflection, across a mirror plane, and (4) inversion, through a point called a center. These operations are discussed below and provide useful insights into the geometry of crystal structures and the three-dimensional properties of such crystals.

*Translation*

The symmetry operation called **translation** involves the periodic repetition of nodes or motifs by systematic linear translation. Two-dimensional translation of basic design elements generates a row of similar elements (Figure 4.2a). The translation is defined by the **unit translation vector (t)**, a specific length, and direction of systematic displacement by which the pattern is repeated. Motifs other than triangles, any motif, could be translated by the same unit translation vector to produce a one-dimensional pattern. In minerals, the motifs are clusters of atoms or coordination polyhedra that are repeated by translation.

**Two-dimensional translations** are defined by **two unit translation vectors** ($t_a$ and $t_b$ or $t_1$ and $t_2$, respectively). The translation in one direction is represented by the length and direction of $t_a$ or $t_1$. Translation in the second direction is represented by the length and direction of $t_b$ or $t_2$. The pattern generated depends on the length of the two unit translation vectors and the angles between their directions. The result of any two-dimensional translation is a **plane lattice** or **plane mesh**. A plane lattice is a two-dimensional array of motifs or nodes in which every node has an environment similar to every other node in the array (Figure 4.2a, b).

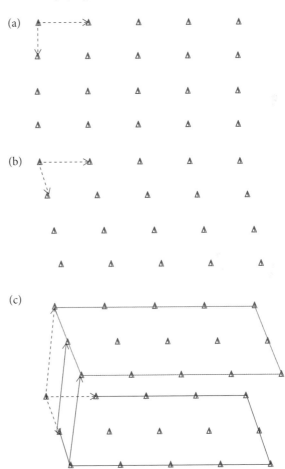

**Figure 4.2** (a) Two-dimensional translation at right angles ($t_1$ and $t_2$) to generate a two-dimensional mesh of motifs or nodes. (b) Two-dimensional translation ($t_1$ and $t_2$) not at right angles to generate a two-dimensional mesh or lattice. (c) Three-dimensional translation ($t_1$, $t_2$, and $t_3$) to generate a three-dimensional space lattice.

**Three-dimensional translations** are defined by **three unit translation vectors** *(t_a, t_b, and t_c* or *t_1, t_2, and t_3*, respectively). The translation in one direction is represented by the length and direction of $t_a$ or $t_1$, the translation in the second direction is represented by $t_b$ or $t_2$ and the translation in the third direction is represented by $t_c$ or $t_3$. The result of any three-dimensional translation is a **space lattice**. A space lattice is a three-dimensional array of motifs or nodes in which every node has an environment similar to every other node in the array. Since crystalline substances such as minerals have long-range, three-dimensional order and since they may be thought of as motifs repeated in three dimensions, the resulting array of motifs is a **crystal lattice**. Figure 4.2c illustrates a space lattice produced by a three-dimensional translation of nodes or motifs.

*Rotation*

Motifs can also be repeated by non-translational symmetry operations. Many patterns can be repeated by rotation (n). **Rotation (n)** is a symmetry operation that involves the rotation of a pattern about an imaginary line or axis, called an **axis of rotation (A)**, in such a way that every component of the pattern is perfectly repeated one or more times during a complete 360° rotation. The symbol "n" denotes the number of repetitions that occur during a complete rotation. Figure 4.3 uses triangle motifs to depict the major types of rotational symmetry (n) that occur in minerals and other crystals. The axis of rotation for each motif is perpendicular to the page. Table 4.1 summarizes the major types of rotational symmetry.

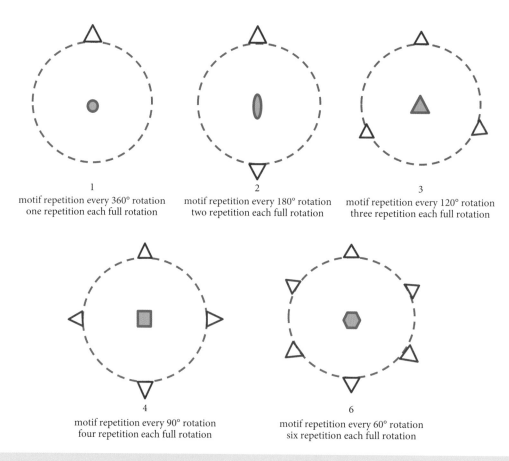

| | | |
|---|---|---|
| **1** | **2** | **3** |
| motif repetition every 360° rotation<br>one repetition each full rotation | motif repetition every 180° rotation<br>two repetition each full rotation | motif repetition every 120° rotation<br>three repetition each full rotation |

| | |
|---|---|
| **4** | **6** |
| motif repetition every 90° rotation<br>four repetition each full rotation | motif repetition every 60° rotation<br>six repetition each full rotation |

**Figure 4.3** Five major types of rotational symmetry operations, viewed looking down rotational axes marked by blue symbols in the center of each circle: dot marks axis of onefold rotation (1), oval marks axis of twofold rotation (2), triangle marks axis of threefold rotation (3), square marks axis of fourfold rotation (4), and hexagon marks axis of sixfold rotation (6).

**Table 4.1** Five common axes of rotational symmetry in minerals.

| Type | Symbolic notation | Description |
|------|-------------------|-------------|
| Onefold axis of rotation | (1 or $A_1$) | Any axis of rotation about which the motif is repeated only once during a 360° rotation (Figure 4.3 (1)) |
| Twofold axis of rotation | (2 or $A_2$) | Motifs repeated every 180° or twice during a 360° rotation (Figure 4.3 (2)) |
| Threefold axis of rotation | (3 or $A_3$) | Motifs repeated every 120° or three times during a complete rotation (Figure 4.3 (3)) |
| Fourfold axis of rotation | (4 or $A_4$) | Motifs repeated every 90° or four times during a complete rotation (Figure 4.3 (4)) |
| Sixfold axis of rotation | (6 or $A_6$) | Motifs repeated every 60° or six times during a complete rotation (Figure 4.3 (5)) |

*Reflection*

Reflection is as familiar to us as our own reflections in a mirror or that of a tree in a still body of water. It is also the basis for the concept of bilateral symmetry that characterizes many organisms (Figure 4.4). Yet it is a symmetry operation that is somewhat more difficult for most people to visualize than rotation. **Reflection** is a symmetry operation in which every component of a pattern is repeated by reflection across a plane called a **mirror plane** (m). Reflection occurs when each component is repeated by equidistant projection perpendicular to the mirror plane. Reflection retains all the components of the original motif but changes its "handedness"; the new motifs produced by reflection across a mirror plane are mirror images of each other (Figure 4.4). Symmetry operations that change the handedness of motifs are called **enantiomorphic operations**.

One test for the existence of a mirror plane of symmetry is that all components of the motifs on one side of the plane are repeated at equal distances on the other side of the plane along projection lines perpendicular to the plane. If this is not true, the plane is not a plane of mirror symmetry.

*Inversion*

Inversion is perhaps the most difficult of the simple symmetry operations to visualize. **Inversion** involves the repetition of motifs by inversion through a point called a **center of inversion** (i). Inversion occurs when every component of a pattern is repeated by equidistant projection through a common point or center of inversion. The two "letters" in

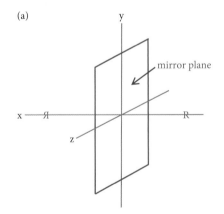

**Figure 4.4** Two- and three-dimensional motifs that illustrate the concept of reflection across a plane of mirror symmetry (m). (a) Mirror image of the letter "R". (b) Bilateral symmetry of a butterfly; the two halves are nearly, but not quite, perfect mirror images of each other. *Source*: Image from butterfly-website.com. © Mikula Web Solutions.

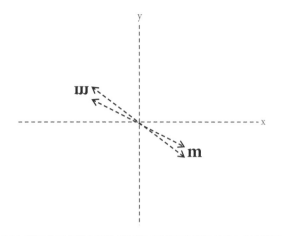

**Figure 4.5** Inversion through a center of symmetry (i) illustrated by the letter "m" repeated by inversion through a center (inversion point).

Figure 4.5 illustrates this enantiomorphic symmetry operation and shows the center through which inversion occurs. In some symbolic notations centers of inversion are symbolized by (c) rather than (i).

One test for the existence of a center of symmetry is that all the components of a pattern are repeated along straight lines that pass through a common center and are repeated at equal distances from that center. If this is not the case, the pattern does not possess a center of symmetry.

### 4.2.2   Compound symmetry operations

Three other symmetry operations exist. However, unlike those discussed so far, they are *compound symmetry operations* that combine two simple symmetry operations. **Glide reflection** (g) is a symmetry operation that combines translation (t) parallel to a mirror plane (m) with reflection across the mirror plane to produce a glide plane (Figure 4.6).

**Rotoinversion** (n⁻) is an operation that combines rotation about an axis with inversion through a center to produce an axis of rotoinversion (Figure 4.7). Figure 4.7b illustrates an axis of fourfold rotoinversion ($\bar{4}$) in which the motif is repeated after 90° rotation followed by inversion through a center so that it is repeated four times by rotoinversion during a 360° rotation. Axes of twofold rotoinversion ($\bar{2}$) are unique symmetry operations, whereas axes of threefold rotoinversion ($\bar{3}$) are equivalent to a threefold axis of rotation and a center of inversion (3i), and axes of sixfold rotoinversion ($\bar{6}$) are equivalent to a threefold axis of rotation perpendicular to a mirror plane (3/m).

**Screw rotation** ($n_a$) is a symmetry operation that combines translation parallel to an axis with rotation about the axis. This is similar to what occurs when a screw is inserted into a wall (Figure 4.7c). Much more detailed treatments of the various types of compound symmetry operations can be found in Klein and Dutrow (2007), Wenk and Bulakh (2016) or Nesse (2016).

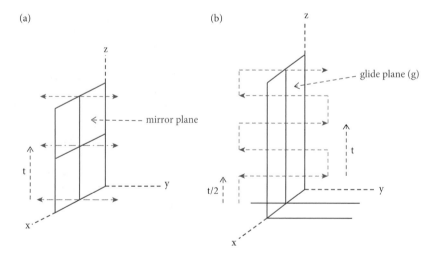

**Figure 4.6**   (a) Mirror plane (m) with the translation vector (t), contrasted with (b) a glide plane (g) with the translation vector (t/2) combined with mirror reflection.

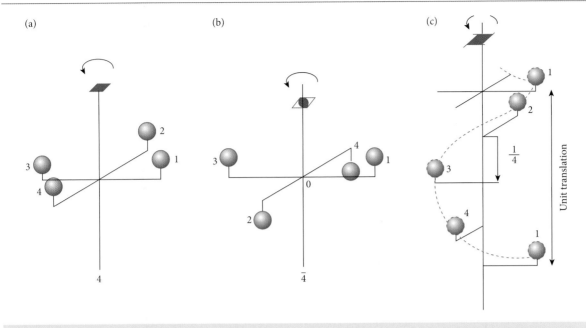

**Figure 4.7** (a) An axis of fourfold rotation (4). This contrasts with (b) an axis of fourfold rotoinversion ($\bar{4}$) that combines rotation with inversion every 90°, and (c) a fourfold screw axis ($4_1$) that combines translation with 90° rotations every one-fourth translation. *Source*: Wenk and Bulakh (2004). © Cambridge University Press.

## 4.3 TWO-DIMENSIONAL MOTIFS AND LATTICES (MESHES)

The symmetry of three-dimensional crystals can be quite complex. Understanding symmetry in two dimensions provides an excellent basis for understanding the increased complexity that characterizes three-dimensional symmetries. It also provides a basis for learning to visualize planes of constituents within three-dimensional crystals. Being able to visualize and reference lattice planes is of the utmost importance in describing cleavage and crystal faces and in the identification of minerals by X-ray diffraction methods.

### 4.3.1 Plane point groups

As discussed earlier, any fundamental unit of two-dimensional pattern, or motif, can be repeated by various symmetry operations to produce a larger two-dimensional pattern. All two-dimensional motifs that are consistent with the generation of long-range two-dimensional arrays can be assigned to one of **ten plane point groups** based on their unique plane point group symmetry (Figure 4.8). Using the symbolic language discussed in the

previous section on symmetry, the ten plane point groups are **1, 2, 3, 4, 6, m, 2mm, 3m, 4mm, and 6mm.** The numbers refer to axes of rotation that are perpendicular to the plane (or page); the m refers to mirror planes perpendicular to the page. The first m refers to a set of mirror planes that is repeated by the rotational symmetry and the second m to a set of mirror planes that bisects the first set. Note that the total number of mirror planes is the same as the number associated with its rotational axis (e.g., 3m has three mirror planes and 6mm has six mirror planes).

### 4.3.2 Plane lattices and unit meshes

As discussed previously, any motif can be represented by a point called a node. Points or nodes can be translated some distance in one direction by a unit translation vector $t_a$ or $t_1$ to produce a line of nodes or motifs. Nodes can also be translated some distances in two directions $t_a$ and $t_b$ or $t_1$ and $t_2$ to produce a two-dimensional array of points called a plane mesh or plane net. Simple translation of nodes in two directions produces five basic types of two-dimensional patterns (Figure 4.9). The smallest units of such meshes, which contain

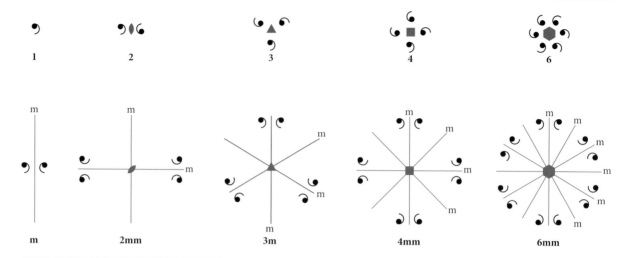

**Figure 4.8** The 10 plane point groups defined by rotational and reflection symmetry. *Source*: Klein and Hurlbut (1985). © John Wiley & Sons.

the unit translation vectors and at least one node, are called **unit meshes** (**unit nets**). Unit meshes contain all the information necessary to produce the larger two-dimensional pattern. They contain only translation symmetry information. The five basic types of unit mesh are classified on the basis of (1) the unit translation vector lengths (equal or unequal), the angles between them (90°, 60°, and 120° or none of these) and (2) whether they have nodes only at the corners (primitive = p) or have an additional node in the center (c) of the mesh.

**Square unit meshes** (Figure 4.9a) are primitive and have equal unit translation vectors at 90° angles to each other (p, $t_a = t_b$, $\gamma = 90°$). **Primitive rectangular unit meshes** (Figure 4.9b) differ in that, although the unit translation vectors intersect at right angles, they are of unequal lengths (p, $t_a \neq t_b$, $\gamma = 90°$). **Diamond unit meshes** have equal unit translation vectors that intersect at angles other than 60°, 90° or 120°. Diamond lattices can be produced and represented by **primitive diamond unit meshes** (p, $t_a = t_b$, $\gamma \neq 60°$, 90° or 120°). They can also be produced by the translation of **centered rectangular unit meshes** (Figure 4.9c) in which the two unit mesh sides are unequal, the angle between them is 90°, and there is a second node in the center of the mesh (c, $t_a \neq t_b$, $\gamma = 90°$). In a centered rectangular mesh there is a total content of two nodes = two motifs. If one looks closely, one

may see evidence for glide reflection in the centered rectangular mesh and/or the larger diamond lattice. The **hexagonal unit mesh** (Figure 4.9d) is a special form of the primitive diamond mesh because, although the unit translation vectors are equal, the angles between them are 60° and 120° (p, $t_a = t_b$, $\gamma = 120°$). Rotation through 120° generates three such unit meshes which combine to produce a larger pattern with hexagonal symmetry. **Oblique unit meshes** (Figure 4.9e) are primitive and are characterized by unequal unit translation vectors that intersect at angles that are not 90°, 60° or 120° (p, $t_a \neq t_b$, $\gamma \neq 90°$, 60° or 120°) and produce the least regular, least symmetrical two-dimensional lattices. The arrays of nodes on planes within minerals always correspond to one of these basic patterns.

### 4.3.3 Plane lattice groups

When the ten plane point groups are combined with the five unit meshes in all ways that are compatible, a total of **17 plane lattice groups** are recognized on the basis of the total symmetry of their plane lattices. Note that these symmetries involve translation-free symmetry operations that include rotation and reflection, translation and compound symmetry operations such as glide reflection. Table 4.2 summarizes the 17 plane lattice groups and their symmetries.

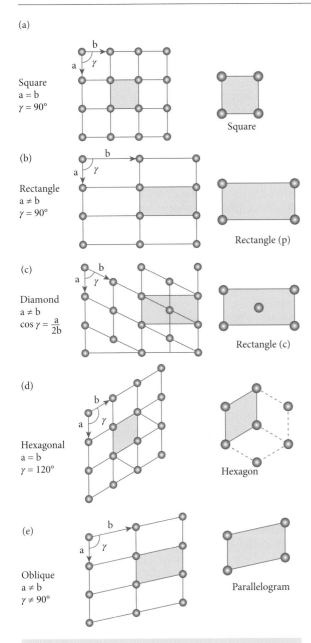

**(a)**

Square
$a = b$
$\gamma = 90°$

Square

**(b)**

Rectangle
$a \neq b$
$\gamma = 90°$

Rectangle (p)

**(c)**

Diamond
$a \neq b$
$\cos \gamma = \dfrac{a}{2b}$

Rectangle (c)

**(d)**

Hexagonal
$a = b$
$\gamma = 120°$

Hexagon

**(e)**

Oblique
$a \neq b$
$\gamma \neq 90°$

Parallelogram

**Figure 4.9** The five principal types of meshes or nets and their unit meshes (shaded gray): (a) square, (b) primitive rectangle, (c) diamond or centered rectangle, (d) hexagonal, (e) oblique. *Source*: Nesse (2016). © Oxford University Press.

Primitive lattices are denoted by "P" and centered lattices by "C." Axes of rotation for the entire pattern perpendicular to the plane are noted by 1, 2, 3, 4, and 6. Mirror planes perpendicular to the plane are denoted by "m"; glide planes perpendicular to the plane are denoted by "g."

**Table 4.2** The 17 plane lattice groups and the unique combination of point group and unit mesh that characterizes each.

| Lattice | Point group | Plane group |
|---|---|---|
| Oblique (P) | 1 | P1 |
| | 2 | P2 |
| Rectangular | m | Pm |
| (P and C) | | Pg |
| | | Cm |
| | 2mm | P2mm |
| | | P2mg |
| | | P2gg |
| | | C2mm |
| Square (P) | 4 | P4 |
| | 4mm | P4mm |
| | | P4gm |
| Hexagonal (P) | 3 | P3 |
| (rhombohedral) | 3m | P3m1 |
| | | P3lm |
| Hexagonal (P) | 6 | P6 |
| (hexagonal) | 6mm | P6mm |

The details of plane lattice groups are well documented (see for example, Klein and Dutrow 2007), but are beyond the introductory material in this text.

## 4.4  THREE-DIMENSIONAL MOTIFS AND LATTICES

Minerals are three-dimensional Earth materials with three-dimensional crystal lattices. The fundamental units of pattern in any three-dimensional lattice are **three-dimensional motifs** that can be classified according to their translation-free symmetries. These three-dimensional equivalents of the two-dimensional plane point groups are called **space point groups**.

Space point groups can be represented by nodes. These nodes can be translated to produce three-dimensional patterns of points called **space lattices**. Space lattices are the three-dimensional equivalents of plane nets or meshes. By analogy with unit meshes or nets, we can recognize the smallest three-dimensional units, called unit cells, which contain all the information necessary to produce the three-dimensional space lattices. In this section, we will briefly describe the space point groups, after which we will introduce Bravais lattices, unit cells, and their relationship to the six (or seven) major crystal systems to which minerals belong.

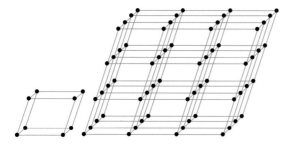

**Figure 4.10**  A primitive unit cell and a long-range space point lattice that results from its repetition by symmetry operations in three dimensions.

### 4.4.1  Space point groups

In minerals, the fundamental motifs are parts of clusters of three-dimensional coordination polyhedra sufficient to establish the composition of the mineral. When these are repeated in three dimensions during mineral growth, they produce the long-range order characteristic of crystalline substances (Figure 4.10). Like all fundamental units of pattern, these three-dimensional motifs can be classified on the basis of their translation-free symmetries.

Only 32 different three-dimensional motif symmetries exist. These define **32 space point groups**, each with unique space point group symmetry. In minerals, the **32 crystal classes** – to one of which all minerals belong – correspond to the 32 space point group symmetries of the mineral's three-dimensional motif. That the crystal classes were originally defined on the basis of the external symmetry of mineral crystals is another example of the fact that the external symmetry of minerals reflects the internal symmetry of their constituents. The 32 crystal classes belong to 6 (or 7) crystal systems, each with its own characteristic symmetry. Table 4.3 summarizes the crystal systems, the symmetries of the 32 space point groups or crystal classes and their names, which are based on general crystal forms. It is important to remember that a crystal cannot possess more symmetry than that of the motifs of which it is composed. However, it can possess less, depending on how the motifs are arranged and how the crystal developed during growth.

### 4.4.2  Bravais lattices, unit cells, and crystal systems

As noted earlier, any motif can be represented by a point called a node. Nodes, and the motifs they represent, can also be translated in three directions ($t_a$, $t_b$, and $t_c$) to produce three-dimensional space point lattices and unit cells (Figure 4.10).

Unit cells are the three-dimensional analogs of unit meshes. A **unit cell** is a parallelepiped whose edge lengths and volume are defined by the three unit translation vectors ($t_a$, $t_b$, and $t_c$). The unit cell is the smallest unit that contains all the information necessary to reproduce the mineral by three-dimensional symmetry operations. Unit cells may be **primitive (P)**, in which case they have nodes only at their corners and a total content of one node (=one motif). **Non-primitive cells** are multiple because they contain extra nodes in one or more faces (A, B, C or F) or in their centers (I) and possess a total unit cell content of more than one node or motif.

Unit cells bear a systematic relationship to the coordination polyhedra and packing of atoms that characterize mineral structures, as illustrated by Figure 4.11.

In 1850, Bravais recognized that only 14 basic types of three-dimensional translational point lattices exist; these are known as the **14 Bravais space point lattices** (Klein and Hurlbut 1985) and define 14 basic types of unit cells. The 14 Bravais lattices are distinguished on the basis of (1) the magnitudes of the three unit translation vectors $t_a$, $t_b$, and $t_c$ or more simply a, b, and c, (2) the angles (alpha, beta, and gamma) between them, where ($\alpha = b \wedge c$; $\beta = c \wedge a$; $\gamma = a \wedge b$), and (3) whether they are primitive lattices or some type of multiple lattice. Figure 4.12 illustrates the 14 Bravais space point lattices.

The translational symmetry of every mineral can be represented by one of the 14 basic types of unit cells. Each unit cell contains one or more nodes that represent motifs and contains all the information necessary to characterize chemical composition. Each unit cell also contains the rules according to which motifs are repeated by translation; the repeat distances, given by $t_a$ = a, $t_b$ = b, $t_c$ = c, and directions, given by angles $\alpha$, $\beta$, and $\gamma$. The 14 Bravais lattices can be grouped into **crystal systems** on the basis of the relative dimensions

**Table 4.3** The six crystal systems and 32 crystal classes, with their characteristic symmetry and crystal forms.

| System | Crystal class | Class symmetry | Total symmetry |
|---|---|---|---|
| Isometric | Hexoctahedral | $4/m\overline{3}2/m$ | $3A_4, 4\overline{A}_3, 6A_2, 9m$ |
| | Hextetrahedral | $\overline{4}3m$ | $3A_4, 4A_3, 6m$ |
| | Gyroidal | $432$ | $3A_4, 4A_3, 6A_2$ |
| | Diploidal | $2/m\overline{3}$ | $3A_2, 3m, 4A_3$ |
| | Tetaroidal | $23$ | $3A_2, 4A_3$ |
| Tetragonal | Ditetragonal–dipyramidal | $4/m2/m2/m$ | $i, 1A_4, 4A_2, 5m$ |
| | Tetragonal–scalenohedral | $\overline{4}2m$ | $1A_4, 2A_2, 2m$ |
| | Ditetragonal–pyramidal | $4mm$ | $1A_4, 4m$ |
| | Tetragonal–trapezohedral | $422$ | $1A_4, 4A_2$ |
| | Tetragonal–dipyramidal | $4/m$ | $i, 1A_4, 1m$ |
| | Tetragonal–disphenoidal | $\overline{4}$ | $\overline{A}_4$ |
| | Tetragonal–pyramidal | $4$ | $1A_4$ |
| Hexagonal (hexagonal) | Dihexagonal–dipyramidal | $6/m2/m2/m$ | $i, 1A_6, 6A_2, 7m$ |
| | Ditrigonal–dipyramidal | $6m2$ | $1A_6, 3A_2, 3m$ |
| | Dihexagonal–pyramidal | $6mm$ | $1A_6, 6m$ |
| | Hexagonal–trapezohedral | $622$ | $1A_6, 6A_2$ |
| | Hexagonal–dipyramidal | $6/m$ | $i, 1A_6, 1m$ |
| | Trigonal–dipyramidal | $\overline{6}$ | $1A_6$ |
| | Hexagonal–pyramidal | $6$ | $1A_6$ |
| Hexagonal (rhombohedral or trigonal) | Hexagonal–scalenohedral | $\overline{3}2/m$ | $1A_3, 3A_2, 3m$ |
| | Ditrigonal–pyramidal | $3m$ | $1A_3, 3m$ |
| | Trigonal–trapezohedral | $32$ | $1A_3, 3A_2$ |
| | Rhombohedral | $\overline{3}$ | $1A_3$ |
| | Trigonal–pyramidal | $3$ | $1A_3$ |
| Orthorhombic | Rhombic–dipyramidal | $2/m2/m2/m$ | $i, 3A_2, 3m$ |
| | Rhombic–pyramidal | $mm2$ | $1A_2, 2m$ |
| | Rhombic–disphenoidal | $222$ | $3A_2$ |
| Monoclinic | Prismatic | $2/m$ | $i, 1A_2, 1m$ |
| | Sphenoidal | $2$ | $1A_2$ |
| | Domatic | $m$ | $1m$ |
| Triclinic | Pinacoidal | $\overline{1}$ | $i$ |
| | Pedial | $1$ | None |

(a)   (b)   (c)

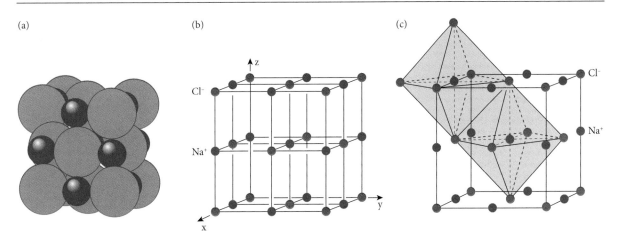

**Figure 4.11** Relationship between (a) atomic packing, (b) a unit cell, and (c) octahedral coordination polyhedra in halite (NaCl). *Source*: Wenk and Bulakh (2016). © Cambridge University Press.

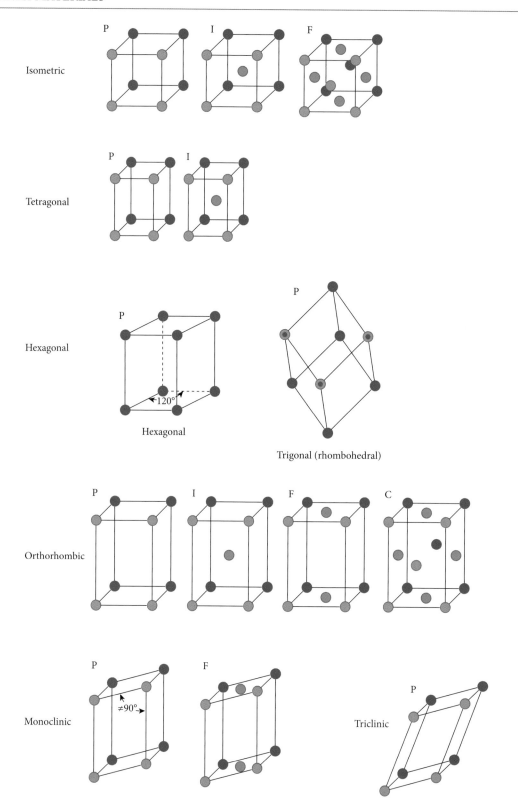

**Figure 4.12** The 14 Bravais lattices and the six (or seven) crystal systems they represent. *Source*: Courtesy of Steve Dutch.

**Table 4.4** Major characteristics of Bravais lattice cells in the major crystal systems.

| Crystal system | Unit cell edge lengths | Unit cell edge intersection angles | Bravais lattice types |
|---|---|---|---|
| Isometric (cubic) | (a = b = c) Preferred format for edges of equal length is ($a_1 = a_2 = a_3$) | $\alpha = \beta = \gamma = 90°$ | Primitive (P) Body centered (I) Face centered (F) |
| Tetragonal | ($a_1 = a_2 \neq a_3$) or a = b ≠ c | $\alpha = \beta = \gamma = 90°$ | Primitive (P) Body centered (I) |
| Hexagonal (hexagonal) | ($a_1 = a_{2} \neq c$) (a = b ≠ c) | ($\alpha = \beta = 90° \neq \gamma = 120°$) | Primitive (P) |
| Hexagonal (trigonal or rhombohedral) | ($a_1 = a_2 = a_3$) | $\alpha = \beta = \gamma \neq 90°$ | Primitive (P) |
| Orthorhombic | a ≠ b ≠ c | ($\alpha = \beta = \gamma = 90°$) | Primitive (P) Body centered (I) End centered (A, B, C) Face centered (F) |
| Monoclinic | a ≠ b ≠ c | ($\alpha = \gamma = 90° \neq \beta$) | Primitive (P) End centered (C) |
| Triclinic | a ≠ b ≠ c | ($\alpha, \beta,$ and $\gamma \neq 90°$) | Primitive (P) |

of the unit cell edges (a, b, and c) and the angles between them ($\alpha$, $\beta$, and $\gamma$). These six (or seven if the hexagonal system is divided into trigonal and hexagonal) systems in which all minerals crystallize. These include the **isometric (cubic), tetragonal, orthorhombic, monoclinic, triclinic,** and **hexagonal systems.** The latter is subdivided into the **hexagonal division or system and the trigonal (rhombohedral) division or system.** Table 4.4 summarizes the characteristics of the Bravais lattices in the major crystal systems.

## 4.5 CRYSTAL SYSTEMS

Imagine yourself, if you can, the crystals on a web site, in a mineral shop or a museum. Such crystals are partially or completely bounded by planar crystal faces that are produced when minerals grow. Many other mineral specimens are partially or completely bounded by flat, planar cleavage faces produced when minerals break along planes of relatively low total bond strength. The shapes of the crystals, the number and orientation of the crystal faces, and the nature of the cleavage depend on the crystal structure of the mineral. That is, they depend on the basic motif and the symmetry operations that produce the three-dimensional crystal lattice. The nature of the crystal forms and cleavage surfaces depends on the crystal system and crystal class in which the mineral crystallized.

### 4.5.1 Crystallographic axes

To identify, describe and distinguish between different planes in minerals, including cleavage planes, crystal faces, and X-ray diffraction planes, a comprehensive terminology has been developed that relates each set of planes to the three **crystallographic axes** (Figure 4.13). For all but the rhombohedral (trigonal) division or system, the three crystallographic axes, designated a, b, and c, are chosen to correspond to the three unit cell translation vectors ($t_a$, $t_b$, and $t_c$). With the exception noted, the three crystallographic axes have lengths and angular relationships that correspond to those of the three sets of unit cell edges (Table 4.4). The rules for the three crystallographic axes are specific to each system; some systems have multiple sets of rules labeling. For details, see Klein and Hurlbut (1985).

When referencing crystallographic planes to the crystallographic axes, a **standard set of orientation rules** is used (Table 4.5). To indicate their similarity, crystallographic axes with the same length are labeled $a_1$, $a_2$ and/or $a_3$ instead of a, b and/or c. In the isometric, tetragonal, and orthorhombic systems (Figure 4.14), the b-axis (or $a_2$-axis) is oriented from left to right, with the right end designated as the positive end of the axis (b or $a_2$) and the left end designated as the negative end of the axis ($\bar{b}$ or $\bar{a}_2$). The a-axis (or $a_1$-axis) is oriented so that it trends from back to front toward the observer. The end of the a-axis toward the observer is designated as the positive end of the axis (a) and the end of the a-axis away from the observer is designated as the negative end ($\bar{a}$). The c-axis is oriented vertically with the top end designated as the positive end of the axis (c) and the bottom end designated as the negative end ($\bar{c}$). There are small exceptions to these rules in the hexagonal, monoclinic, and triclinic systems that result from the fact that unit cell edges are not perpendicular to one another so that not all crystallographic axes intersect at right angles. Even in the hexagonal and monoclinic systems, the b-axis is oriented horizontally with the positive end to the right, and in all systems the c-axis is vertical with the positive end toward the top and the negative end toward the bottom. The orientations, lengths, and intersection angles between crystallographic axes in each of the major crystal systems are illustrated in Figure 4.14. The characteristics of the crystallographic axes in each system and the standard rules for orienting them are summarized in Table 4.5.

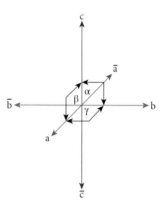

**Figure 4.13** Conventional labeling of crystallographic axes, illustrating the positive and negative ends of the three crystallographic axes and the angles between the axes for crystals in the orthorhombic system.

**Table 4.5** The relationships of crystallographic axes and the rules for orienting crystals in each of the crystal systems. Note that the trigonal system (division) is listed independently of the hexagonal system (division) in this table.

| Crystal system | Verbal description | Symbolic description |
|---|---|---|
| Isometric (cubic) | Three mutually perpendicular axes ($a_1$, $a_2$, $a_3$) of equal length that intersect at right angles | ($a_1 = a_2 = a_3$) ($\alpha = \beta = \gamma = 90°$) |
| Tetragonal | Three mutually perpendicular axes; axes ($a_1$, $a_2$) are of equal length; the c axis may be longer or shorter | ($a_1 = a_2 \neq c$) ($\alpha = \beta = \gamma = 90°$) |
| Orthorhombic | Three mutually perpendicular axes of different lengths (a, b, c); two axial length ratios have been used to identify the axes: c > b > a (older) or b > a > c (newer) | ($a \neq b \neq c$) ($\alpha = \beta = \gamma = 90°$) |
| Monoclinic | Three unequal axes lengths (a, b, c) only two of which are perpendicular. The angle ($\beta$) between a and c is not 90°. The a-axis is inclined toward the observer. The b-axis is horizontal and the c-axis is vertical | ($a \neq b \neq c$) ($\alpha = \gamma = 90°$; $\beta \neq 90°$) |
| Triclinic | Three unequal axes, none of which are generally perpendicular. The c axis is vertical and parallel to the prominent zone of crystal faces | ($a \neq b \neq c$) ($\alpha \neq \beta \neq \gamma \neq 90°$) |
| Hexagonal | Four crystallographic axes; three equal horizontal axes ($a_1$, $a_2$, $a_3$) intersecting at 120°. One longer or shorter axis (c) perpendicular to the other three. $a_1$ oriented to front left of observer; $a_2$ to right; $a_3$ to back left; c vertical. Sixfold axis of rotation or rotoinversion | ($a_1 = a_2 = a_3 \neq c$) ($\alpha = \beta = 90°$; $\gamma = 120°$) |
| Trigonal (or rhombohedral) | Axes and angles are similar to the hexagonal system; crystal symmetry is different with the c-axis a threefold axis of rotation or rotoinversion | ($a_1 = a_2 = a_3 \neq c$) ($\alpha = 120°$; $\beta = \gamma = 90°$) |

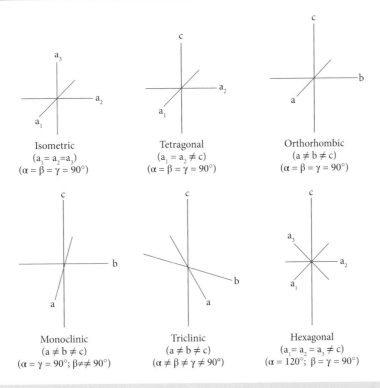

**Figure 4.14** Crystallographic axes (positive ends labeled) and intersection angles for the major crystal systems: isometric, tetragonal, orthorhombic, monoclinic, triclinic, and hexagonal systems.

## 4.5.2 Crystal forms

Each of the crystal systems has an associated set of common (and rarer) crystal forms. **Crystal forms** consist of a three-dimensional set of one or more crystal faces that possess similar spatial relationships to the crystallographic axes. Some natural crystals possess only one crystal form; others possess multiple or combined crystal forms. Crystal forms can be subdivided into two major groups: closed forms and open forms.

**Closed crystal forms** can completely enclose a mineral specimen and therefore can exist alone in perfectly formed (euhedral) crystals which are enclosed completely by the crystal form. Common closed forms include all the forms in the isometric system and many forms in the tetragonal, hexagonal, trigonal, and orthorhombic systems. The pyritohedron (Figure 4.15) is a typical closed form, common in the mineral pyrite. Each closed form possesses a different shape that is related to the number and shape of faces in the form and to their angular relationships to the crystallographic axes. Figure 4.16 illustrates common dipyramid (double pyramid) closed forms in the trigonal, tetragonal, and hexagonal systems.

**Figure 4.15** A pyritohedron, a closed form in which all faces have the same general relationship to the crystallographic axes.

**Open crystal forms** (Figure 4.17) cannot completely enclose a mineral specimen and so must occur in combination with other open or closed crystal forms in order to completely enclose a crystal. Common open forms include: (1) **pedions**, which consist of a single face, (2) **pinacoids**, a pair of parallel faces, (3) **prisms**, three or more faces parallel to an axis, (4) **pyramids**, three or more faces that intersect an axis, (5) **domes**, a pair of faces symmetrical about a mirror plane, and (6)

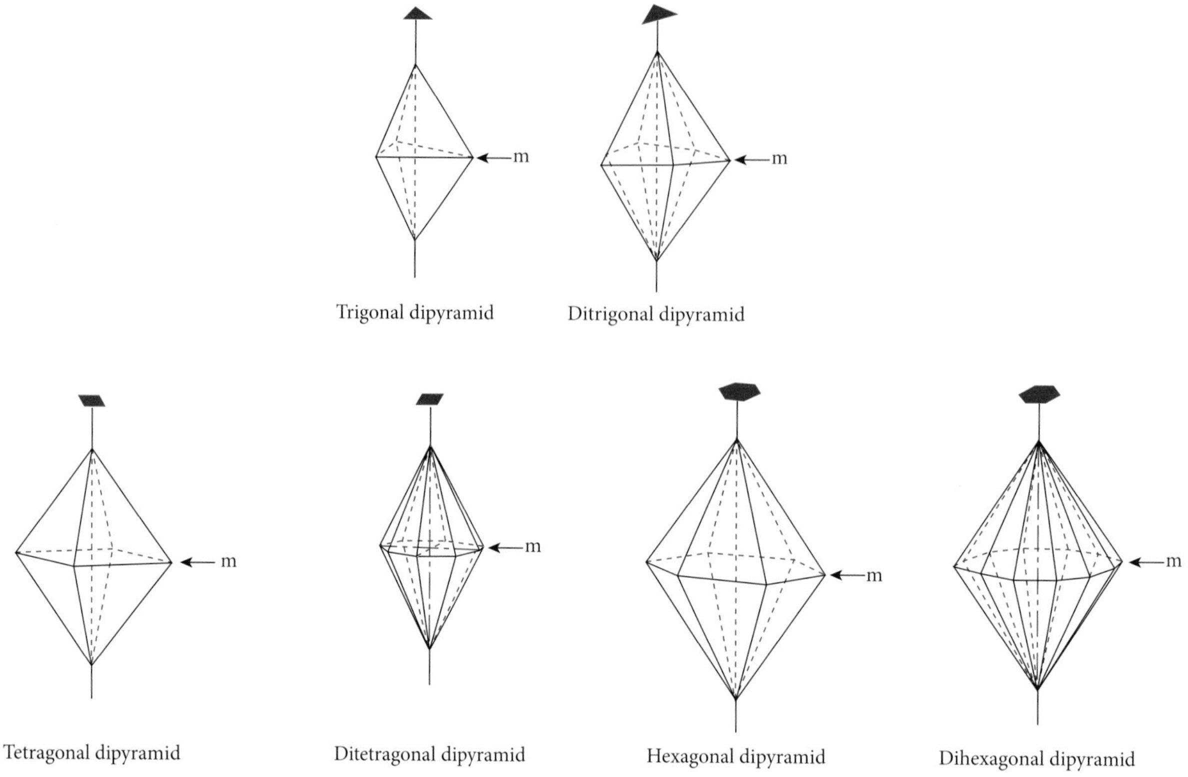

Trigonal dipyramid          Ditrigonal dipyramid

Tetragonal dipyramid          Ditetragonal dipyramid          Hexagonal dipyramid          Dihexagonal dipyramid

**Figure 4.16**   Different types of dipyramid forms in the trigonal, tetragonal, and hexagonal systems. *Source*: Klein and Hurlbut (1985). © John Wiley & Sons.

**sphenoids**, a pair of faces symmetrical about an axis of rotation. Figure 4.17b illustrates the kinds of prisms that occur in the trigonal, tetragonal, and hexagonal crystal systems.

The most common crystal forms in each system are discussed later in this chapter, after we have presented the language used to describe them. More detailed discussions are available in Nesse (2016) and Klein and Dutrow (2007).

### 4.6   INDEXING PLANES IN CRYSTALS

#### 4.6.1   Axial ratios

Whatever their respective lengths, the proportional lengths or **axial ratios** of the three crystallographic axes (a : b : c) can be calculated. The standard method for expressing axial ratios is to express their lengths relative to the length of the b-axis (or $a_2$-axis) which is taken to be unity so that the ratio is expressed as a : b : c; b = 1. This is accomplished by dividing the lengths of all three axes by the length of the b-axis (a/b: b/b: c/b). An example from

the monoclinic system, the pseudo-orthorhombic mineral staurolite, will illustrate how axial ratios are calculated. In staurolite, the unit cell edges have average dimensions expressed in angstrom (Å) units of: a = 7.87 Å, b = 16.58 Å, and c = 5.64 Å. The axial ratios are calculated from a/b : b/b : c/b = 7.87/16.58 Å : 16.58/16.58 Å : 5.64/16.58 Å. The average axial ratios of staurolite are 0.47 : 1.00 : 0.34.

Axial ratios are essential to understanding how crystallographic planes and crystal forms are described or indexed by reference to the crystallographic axes as discussed in the section that follows.

#### 4.6.2   Crystal planes and crystallographic axes

Crystalline substances such as minerals have characteristic planar features that include: (1) crystal faces that develop during growth, (2) cleavage surfaces that develop during breakage, and (3) crystal lattice planes that reflect X-rays and other types of electromagnetic radiation. All these types of planes possess a number of shared properties.

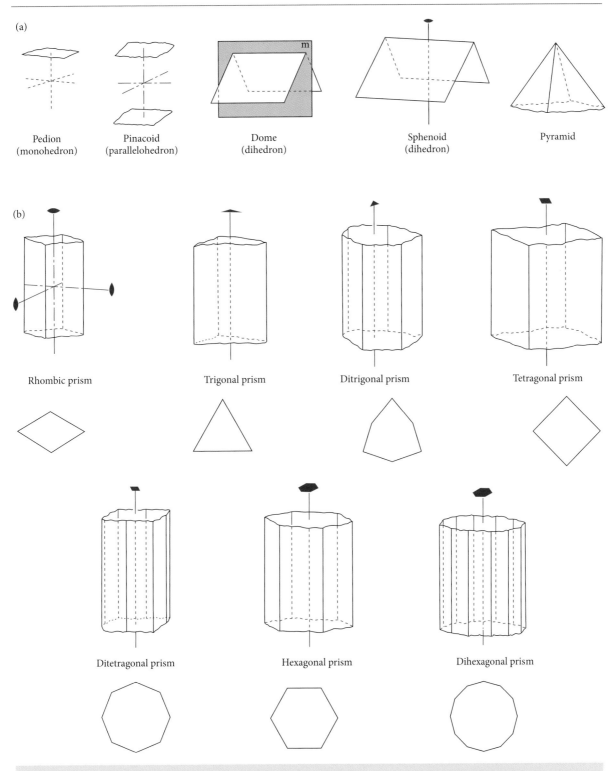

**Figure 4.17** (a) Common open forms: pedions, pinacoids, domes, sphenoids, and pyramids. (b) Different types of prisms that characterize the orthorhombic, trigonal, tetragonal, and hexagonal systems. The illustrated prisms are bounded by pinacoids at the top and bottom. *Source*: Klein and Hurlbut (1985). © John Wiley & Sons.

Each type of plane is part of a large set of parallel lattice planes of which it is representative. As a mineral with a particular crystal form grows freely it may be bounded by a sequence of planar faces. When it stops growing, it is bounded by crystal faces that are parallel to many other lattice planes that bounded the mineral as it grew over time. When a mineral cleaves, it breaks along a specific set of parallel planes of relative weakness, but these cleavage planes are parallel to large numbers of planes of weakness or potential cleavage surfaces in the mineral structure along which the mineral did not happen to rupture. When X-rays are reflected from a reflecting plane, they are reflected simultaneously from all the planes in the crystal that are parallel to one another to produce a "reflection peak" that is characteristic of the mineral and can be used to identify it.

In addition, any set of parallel planes in a crystal is ideally characterized by a particular molecular content; all the parallel planes in the set possess closely similar molecular units, spacing, and arrangement. A molecular image of one of these planes is sufficient to depict the general molecular content of all the planes that are parallel to it. All planes in a set of parallel planes have the same general spatial relationship to the three crystallographic axes. This means that they can be collectively identified in terms of their spatial relationship to the three crystallographic axes. This is true for crystal faces, for cleavage surfaces, for X-ray reflecting planes or for any set of parallel crystallographic planes that we wish to identify. A universally utilized language has evolved that uses the relationship between the planar features in minerals and the crystallographic axes to identify different sets of planes. A discussion of this language and its use follows.

Figure 4.18 depicts several representative crystal planes with different relationships to the three crystallographic axes. Some crystal planes, or sets of parallel planes, intersect one crystallographic axis and are parallel to the other two (Figure 4.18a, b). Alternatively, a set of crystal planes may intersect two crystallographic axes and be parallel to the third (Figure 4.18c, d). Still other sets of planes intersect all three crystallographic axes (Figure 4.18e, f). No other possibilities exist in Euclidean space; sets of planes in crystalline substances must intersect one, two, or three axes and be parallel to those they do not intersect.

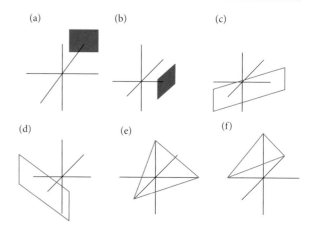

**Figure 4.18** Representative crystal faces that cut one, two or three crystallographic axes. See text for further discussion of parts (a–f).

Of course, some sets of planes or their projections intersect the positive ends of crystallographic axes (Figure 4.18b, c, and e). Others, with different spatial orientations with respect to the axes, intersect the negative ends of crystallographic axes (Figure 4.18a). Still others, with yet different orientations, intersect the positive ends of one or more axes and the negative ends of other crystallographic axes (Figure 4.18d, f). Given the myriad possibilities, a simple language is needed that allows one to visualize and communicate to others the relationship and orientation of any set of crystal planes to the crystallographic axes. The language for identifying and describing crystallographic planes involves the use of symbols called Miller indices, which has been employed since the 1830s and is explained in the following sections.

### 4.6.3 Unit faces or planes

In any crystal, the three crystallographic axes have a characteristic axial ratio, typically grounded in the cell edge lengths of the unit cell. No matter how large the mineral becomes during growth, even if it experiences preferred growth in a particular direction or inhibited growth in another, the axial ratio remains constant and corresponds to the axial ratio implied by the properties of the unit cell.

In the growth of any mineral one can imagine the development of a crystal face (or any plane parallel to it) that intersects the positive ends of all three axes at lengths that correspond

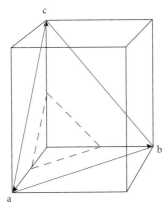

**Figure 4.19** Unit face (outlined in solid blue) in an orthorhombic crystal with three unequal unit cell edges and crystallographic axes that intersect at right angles. All parallel faces (e.g., outlined in dotted red) will have the same general relationship to the crystallographic axes and the same atomic content and properties.

to the axial ratio of the mineral (Figure 4.19). For crystals with a center (i), such faces would intersect each axis at a distance from the center of the crystal that corresponds to the axial ratio. For the monoclinic (pseudo-orthorhombic) mineral staurolite discussed in the previous section, such a face could cut the a-axis at 0.47 mm, the b-axis at 1.00 mm, and the c-axis at 0.34 mm from the center, or, for a larger crystal, it could cut the three axes at 0.47, 1.00, and 0.34 cm (or 0.235, 0.50, and 0.17 cm) from the center. All these faces would be parallel to one another. Any face or plane that intersects all three axes at distances from the center that correspond to the axial ratio of the mineral is a **unit face** or **unit plane**. It is part of a set of parallel planes all of which are unit planes because they intersect the three crystallographic axes at lengths that correspond to the axial ratios.

### 4.6.4 Weiss parameters

**Weiss parameters** provide a method for describing the orientation of sets of crystal faces or planes in relationship to the crystallographic axes. They are always expressed in the sequence a : b : c, where a represents the relationship of the planes to the a-axis (or $a_1$-axis), b represents the relationship between the planes and the b-axis (or

$a_2$-axis) and c depicts the relationship between the planes and the c-axis (or $a_3$-axis). A unit face or plane that cuts all three crystallographic axes at ratios that correspond to their axial ratios has the Weiss parameters (1 : 1 : 1). Mathematically, if we divide the actual lengths at which the face or plane intercepts the three axes by the corresponding axial lengths, the three intercepts have the resulting ratio 1 : 1 : 1, or unity, which is why such planes are called unit planes. Again, using the pseudo-orthorhombic mineral staurolite as an example, if we divide the actual intercept distances by the axial lengths, the resulting ratios are 0.47/0.47 : 1.00/1.00 : 0.34/0.34 = 1 : 1 : 1. Even if we utilize the magnitudes of the dimensions, the three resulting numbers have the same dimensional magnitude, and so their ratio reduces to 1 : 1 : 1.

As discussed in the section on unit faces and planes, many planes intersect all three crystallographic axes. Unit faces or planes (1 : 1 : 1), such as those in Figure 4.19, intersect all three axes at lengths that correspond to their axial ratios. Other sets of planes, however, intersect one or more axes at lengths that do not correspond to their axial ratios. A face or plane that intersects all three crystallographic axes at different lengths relative to their axial ratios is called a **general face**. The Weiss parameters of such a face or plane will be three rational numbers that describe the fact that each axis is intersected at a different proportion of its axial ratio. There are many sets of general planes. For example, a general plane with the Weiss parameters (1 : 1/2 : 1/3), shown in Figure 4.20, would be a plane that intersects the c-axis at one-third the corresponding length of the c-axis and the b-axis at one-half the corresponding length of the b-axis. Since many general faces are possible, for example (1 : 1/2 : 1/3) or (1/2 : 1 : 1/4), it is possible to write a general notation for all the faces that intersect the three crystallographic axes as (h : k : l) where h, k, and l express intercepts with a, b, and c, respectively.

Many crystal planes intersect the negative ends of one or more crystallographic axes. The location and/or orientation of these planes are not the same as those of planes that intersect the positive ends of the same axes. Planes that intersect the negative ends of one or more crystallographic axes are indicated by placing a bar over their Weiss parameters. For example, the dashed plane in Figure 4.20 has the Weiss parameters $(1 : \frac{1}{4} : \frac{1}{2})$.

Mineral planes may be parallel to one or two crystallographic axes. How do we determine the Weiss parameters for such faces and planes? The Weiss parameters of any face or plane that is parallel to a crystallographic axis are infinity ($\infty$) because the plane never intersects the axis in question. If a set of planes is parallel to two crystallographic axes and intersects the third, it is assumed to intersect that axis at unity. Planes that are parallel to the a- and b-axes and intersect the c-axis have the Weiss parameters ($\infty : \infty : 1$). Planes parallel to the b- and c-axes that intersect the a-axis have the Weiss parameters ($1 : \infty : \infty$). Planes that

cut the b-axis and are parallel to the a- and c-axes (Figure 4.21a) have the Weiss parameters ($\infty : 1 : \infty$). Each set of planes, with its unique relationship to the crystallographic axes possesses its own unique Weiss parameters. If a set of planes intersects two axes and is parallel to the third, only one of the Weiss parameters will be infinity. The other two will be one if, and only if, the two axes are intersected at lengths corresponding to their axial ratios. Therefore, the Weiss parameters ($1 : \infty : 1$) represent planes that parallel the b-axis and intersect the a-and c-axes at unit lengths. Similarly the Weiss parameters ($1 : 1 : \infty$) are those of planes that cut the a- and b-axes at unit lengths and are parallel to the c-axis (Figure 4.21b).

Having begun to master the concepts of how Weiss parameters can be used to represent different sets of crystal planes with different sets of relationships to the crystallographic axes, students are generally thrilled to find that crystal planes are commonly referenced, not by Weiss parameters, but instead by Miller indices.

### 4.6.5 Miller indices

The **Miller indices** of any face or set of planes are the *reciprocals of its Weiss parameters*. They are calculated by inverting the Weiss parameters and multiplying by the lowest common denominator. Because of this reciprocal relationship, large Weiss parameters become small Miller indices. For planes parallel to a crystallographic axis, the Miller index is zero. This is because when the large Weiss parameter infinity ($\infty$) is inverted it becomes the Miller index $1/\infty \rightarrow 0$.

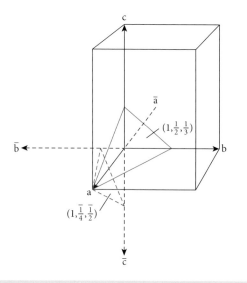

**Figure 4.20** Faces with different Weiss parameters on an orthorhombic crystal.

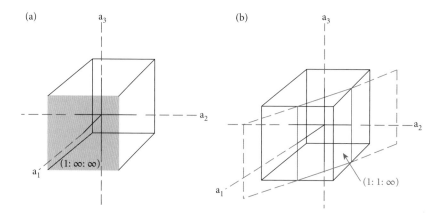

**Figure 4.21** (a) The darkened front crystal face possesses the Weiss parameters: ($1 : \infty : \infty$). (b) The face outlined in green possesses the Weiss parameters ($1 : 1 : \infty$).

The Miller index of any face or set of planes is, with a few rather esoteric exceptions, expressed as three integers hkl in a set of parentheses (**hkl**) that represent the reciprocal intercepts of the face or planes with the three crystallographic axes (a, b, and c) respectively.

We can use the example of the general face cited in the previous section (Figure 4.20), where a set of parallel planes intercepts the a-axis at unity, the b-axis at half unity and the c-axis at one-third unity. The Weiss parameters of such a set of parallel planes are 1 : 1/2 : 1/3. If we invert these parameters they become 1/1, 2/l, and 3/1. The lowest common denominator is one and multiplying by the lowest common denominator yields 1, 2, and 3. The Miller indices of such a face are (123). These reciprocal indices should be read as representing all planes that intersect the a-axis at unity (1) and the b-axis at one-half unity (reciprocal is 2), and then intercept the c-axis at one-third unity (reciprocal is 3) relative to their respective axial ratios. Every parallel plane in this set of planes has the same Miller indices.

As is the case with Weiss parameters, the Miller indices of planes that intersect the negative ends of one or more crystallographic axes are denoted by the use of a bar placed over the indices in question. We can use the example from the previous section in which a set of planes intersect the positive end of the a-axis at unity, the negative end of the b-axis at twice unity and the negative end of the c-axis at three times unity. If the Weiss parameters of each plane in the set are 1, $\bar{2}$ / 3, and $\bar{1}$ / 2, inversion yields 1/1, $\bar{3}$ / 2, and $\bar{2}$ / 1. Multiplication by two, the lowest common denominator, yields 2/1, $\bar{6}$ / 2, and $\bar{4}$ / 1 so that the Miller indices are ($2\bar{3}\bar{4}$). These indices can be read as indicating that the planes intersect the positive end of the a-axis and the negative ends of the b- and c-crystallographic axes with the a-intercept at unity and the b-intercept at two-thirds unity and the c-intercept at one-half unity relative to their respective axial ratios.

A simpler example is the cubic crystal shown in Figure 4.22. Each face of the cube intersects one crystallographic axis and is parallel to the other two. The axis intersected is indicated by the Miller index "1" and the axes to which it is parallel are indicated by the Miller index "0". Therefore the six faces of the cube have the Miller indices (100), ($\bar{1}$00), (010), (0$\bar{1}$0), (001), and (00$\bar{1}$) for the front, back, right, left, top, and bottom faces respectively.

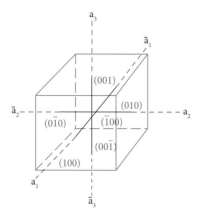

**Figure 4.22**   Miller indices of various crystal faces on a cube depend on their relationship to the crystallographic axes.

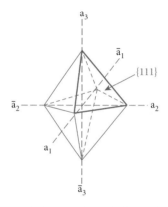

**Figure 4.23**   Isometric octahedron outlined in blue possesses eight faces; the form face {111} is outlined in bold blue.

Miller indices are a symbolic language that allows us to represent the spatial relationship of any crystal face, cleavage face or crystallographic plane with respect to the crystallographic axes.

### 4.6.6   Form indices

Every face in a form has the same general relationship to the crystallographic axes and therefore the same general Miller index, yet every face in a form has a different specific relationship to the crystallographic axes and therefore has a different Miller index. These statements can be clarified by using an example. Figure 4.23 shows the common eight-faced isometric form called the octahedron.

Each face in the octahedron has the same general relationship to the three crystallographic axes in that each intersects the three crystallographic axes at unity. The Miller indices of each face are some form of (111). However, only the top, right front face intersects the positive ends of all three axes. The bottom, left back face intersects the negative ends of all three axes, and the other six faces intersect some combination of positive and negative ends of the three crystallographic axes. None of the faces are parallel to one another; each belongs to a different set of parallel planes within the crystal. The Miller indices of these eight faces and the set of planes to which each belongs are (111), ($\bar{1}\bar{1}1$), ($1\bar{1}1$), ($11\bar{1}$), ($\bar{1}11$), ($\bar{1}1\bar{1}$), ($1\bar{1}\bar{1}$), and ($\bar{1}\bar{1}\bar{1}$). Their unique Miller indices allow us to distinguish each of the eight faces and the sets of planes to which they belong. However, they are all parts of the same form because they all have the same general relationship to the crystallographic axes. It is cumbersome and often unnecessary to have to recite the indices of every face within a form. To represent the general relationship of the form to the crystallographic axes, the indices of a single face, called the form face, are chosen and placed in brackets to indicate that they refer to the form indices. The rule for choosing the **form face** is generally to select the top face if there is one, or the top right face if there is one, or the top right front face if there is one. In the case of the octahedron, the top right front face is the face that intersects the positive ends of the $a_1$-axis (front), the $a_2$-axis (right), and the $a_3$-axis (top) and has the Miller indices (111). The **form indices** for all octahedral crystals are the Miller indices of the form face placed between curly brackets {111}. Similarly the form indices for the cube (see Figure 4.22), in which the

faces intersect one axis and are parallel to the other two are {001}, the Miller indices of the top face, whereas the form indices for the dodecahedron, in which each face intersects two axes at unity and is parallel to the third is {011}, the indices for the top, right face.

Many other forms exist. Every crystal form has a form index, which is the Miller index of the form face placed in brackets. Each form consists of one or (generally) more faces and each face possesses a Miller index different from that of every other face in the form. Every crystal system has a characteristic suite of forms that reflect the unique characteristics of the crystal lattice of the system, especially the relative lengths of the three crystallographic axes that directly or indirectly reflect the lengths of the unit cell edges. The forms that are characteristic of each class (space point group) in each crystal system are beyond the scope of this text (see Klein and Hurlbut 1985). However, a brief review of some common forms in each crystal system is appropriate.

### 4.6.7 Common crystal forms in each system

*Isometric (cubic) system forms*

All forms in the isometric system are closed forms. Common crystal forms in the isometric system include the cube, octahedron, dodecahedron, tetrahedron, and pyritohedron (Figure 4.24). These forms may occur alone or in combination with each other. Common isometric minerals, their crystal forms, and form indices are summarized in Table 4.6.

These form indices are also used to describe cleavage in isometric minerals. These minerals include halite and galena, which possess cubic cleavage {001} with three orientations of cleavage at right angles; fluorite, which

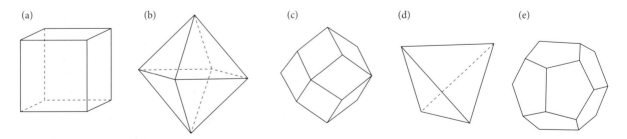

**Figure 4.24** Five common forms in the isometric system: (a) cube, (b) octahedron, (c) dodecahedron, (d) tetrahedron, (e) pyritohedron.

**Table 4.6**   Common isometric crystal forms, form indices, form descriptions, and minerals.

| Crystal form | Form indices | Form description | Minerals that commonly exhibit crystal form |
|---|---|---|---|
| Cube | {001} | Six square faces | Halite, galena, pyrite, fluorite, cuprite, perovskite, analcite |
| Octahedron | {111} | Eight triangular faces | Spinel, magnetite, chromite, cuprite, galena, diamond, gold, perovskite |
| Dodecahedron | {011} | 12 diamond-shaped faces | Garnet, sphalerite, sodalite, cuprite |
| Tetrahedron | {1$\bar{1}$1} | Four triangular faces | Tetrahedrite, sphalerite |
| Pyritohedron | {h0l} | 12 pentagonal faces | Pyrite |

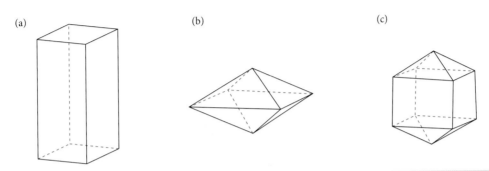

**Figure 4.25**   Common crystal forms in the tetragonal crystal system: (a) tetragonal prism in combination with a pinacoid, (b) tetragonal dipyramid, (c) tetragonal dipyramid in combination with a tetragonal prism.

**Table 4.7**   Common tetragonal crystal forms, form indices, form descriptions, and minerals.

| Crystal forms | Form indices | Form description | Minerals that commonly exhibit crystal form |
|---|---|---|---|
| Tetragonal dipyramid | {111} {hh1} {011} {0kl} and variations | Eight triangular faces; top four separated from bottom four by mirror plane | Zircon, rutile, cassiterite, scheelite, wulfenite, vesuvianite, scapolite |
| Tetragonal prism | {010} {110} and variations | Four rectangular faces parallel to c-axis | Zircon, scheelite, vesuvianite, rutile, malachite, azurite, cassiterite, scapolite |
| Tetragonal disphenoid | {0kl} | Four triangular faces; alternating pairs symmetrical about c-axis | Chalcopyrite |
| Basal pinacoid | {001} | Pair of faces perpendicular to c-axis | Vesuvianite, wulfenite |

possesses octahedral cleavage {111} with four cleavage orientations; and sphalerite, which possesses dodecahedral cleavage {011} with six orientations of cleavage.

*Tetragonal system forms*

Tetragonal crystals can possess closed and/or open forms. Common closed crystal forms in the tetragonal crystal system include different eight-sided dipyramids. Common open forms include four-sided and eight-sided prisms and pyramids, as well as pinacoids and pedions.

Typical crystal forms and associated minerals in the tetragonal crystal system are shown in Figure 4.25 and Table 4.7.

*Hexagonal system (hexagonal division) forms*

Common crystal forms in the hexagonal system include 6–12-sided prisms, dipyramids, and pyramids. Pinacoids and pedions are also common. Some selected examples of common crystal forms and minerals in the hexagonal system are illustrated in Figure 4.26 and Table 4.8.

*Trigonal system (hexagonal system, trigonal division) forms*

Common closed crystal forms in the trigonal system include the six-sided rhombohedron, the 12-sided scalenohedron, six-sided hexagonal dipyramids and three-sided trigonal pyramids. Three- and six-sided prisms, and pinacoids and pedions are also common. Many forms common in the hexagonal division also occur in trigonal crystals, but not vice versa. Common crystal forms and representative minerals in the trigonal crystal system are summarized in Figure 4.27 and Table 4.9.

*Orthorhombic crystal system*

Common crystal forms in the orthorhombic system include four-sided rhombic prisms and pyramids, and eight-sided dipyramids. Pinacoids are the dominant form and pedions are also common. Common crystal forms and associated minerals in the orthorhombic crystal system are indicated in Figure 4.28 and Table 4.10.

*Monoclinic crystal system*

Because of its lower symmetry, the only crystal forms in the monoclinic crystal system are four-sided prisms, two-sided domes, sphenoids and pinacoids, and pedions. More complex

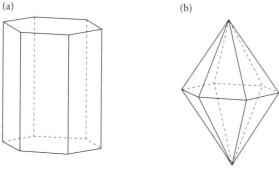

**Figure 4.26** Common crystal forms in the hexagonal crystal system (hexagonal division): (a) hexagonal prism closed by a basal pinacoid, (b) hexagonal dipyramid.

**Table 4.8** Common hexagonal crystal forms and associated hexagonal and trigonal minerals.

| Crystal forms | Common form indices | Form description | Minerals that commonly exhibit crystal form |
|---|---|---|---|
| Hexagonal dipyramid | $\{11\bar{2}1\}$ and variations | 12 triangular faces inclined to c-axis; top six separated from bottom six by mirror plane | Apatite, zincite |
| Hexagonal prism | $\{11\bar{2}0\}$ and variations | Six rectangular faces parallel to c-axis | Apatite, beryl, quartz, nepheline, corundum, tourmaline |
| Basal pinacoid | $\{0001\}$ | Pair of faces perpendicular to c-axis | Apatite, beryl, corundum |

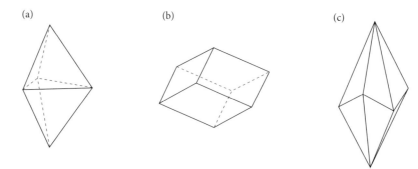

**Figure 4.27** Common crystal forms in the trigonal system: (a) trigonal dipyramid, (b) rhombohedron, (c) scalenohedron.

**Table 4.9** Common trigonal crystal forms and associated minerals.

| Crystal forms | Form indices | Form description | Minerals that commonly exhibit crystal form |
|---|---|---|---|
| Rhombohedron | $\{10\bar{1}1\}$ $\{h0\bar{h}l\}$ | Six parallelogram faces inclined to c-axis | Dolomite, calcite, siderite, rhodochrosite, quartz, tourmaline; chabazite |
| Trigonal Scalenohedron | $\{hkil\}$ and variations | 12 scalene triangle faces inclined to c-axis | Calcite |
| Trigonal prism | $\{hki0\}$ and variations | Three rectangular faces parallel to c-axis | Tourmaline, calcite, quartz |
| Trigonal dipyramid | $\{hkil\}$ and variations | Six triangular faces; top three separated from bottom three by a mirror plane | Tourmaline |

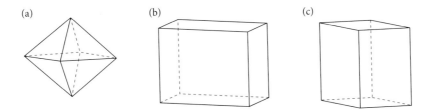

**Figure 4.28** Common crystal forms in the orthorhombic crystal system: (a) rhombic dipyramid, (b) front, side, and basal pinacoids, (c) rhombic prism with a pinacoid.

**Table 4.10** Common crystal forms, form indices, and minerals in the orthorhombic system.

| Crystal forms | Form indices | Form description | Minerals that commonly exhibit crystal form |
|---|---|---|---|
| Rhombic dipyramids | $\{111\}$ $\{hkl\}$ and variations | Eight triangular faces; top four separated from bottom four by a mirror plane | Topaz, aragonite, witherite, olivine |
| Rhombic prisms; first, second, and third orders | $\{011\}$ $\{0kl\}$ $\{101\}$ $\{h0l\}$ $\{011\}$ $\{0kl\}$ | Four rectangular faces parallel to a single crystallographic axis | Stibnite, aragonite, barite, celestite, topaz, enstatite, andalusite, cordierite, epidote, olivine |
| Pinacoids; front, side and basal | $\{001\}$ $\{010\}$ $\{001\}$ | Two parallel faces perpendicular to a-, b- or c-axis | Barite, celestite, olivine, andalusite, topaz, hemimorphite |

forms cannot exist in systems with low symmetry in which some crystallographic axes do not all intersect at right angles. Common crystal forms and associated minerals in the monoclinic crystal system are depicted in Figure 4.29 and Table 4.11.

*Triclinic crystal system*

The only crystal forms in the triclinic system, with its extremely low symmetry, are pinacoids and pedions. Common forms and minerals in the triclinic system are illustrated in Figure 4.30 and listed in Table 4.12.

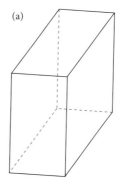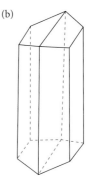

**Figure 4.29** Monoclinic crystal forms: (a) front, side, and basal pinacoids, (b) two monoclinic prisms and a side pinacoid.

**Table 4.11**    Common crystal forms, form indices, and minerals in the monoclinic system.

| Crystal form | Form indices | Form description | Minerals that commonly exhibit crystal form |
|---|---|---|---|
| Monoclinic prisms; first, third, and fourth orders | {011} {0kl} {110} {hk0} {hkl} | Four rectangular faces | Gypsum, staurolite, clinopyroxenes, amphiboles, orthoclase, sanidine, sphene (titanite), borax |
| Pinacoids; front, side, and basal | {001} {010} {001} | Pair of rectangular faces perpendicular to a-, b-, or c-axis | Gypsum, staurolite, sphene (titanite), epidote, micas, clinopyroxenes, amphiboles |

**Table 4.12**    Common crystal forms, form indices, and minerals in the triclinic system.

| Crystal forms | Form indices | Form description | Minerals that commonly exhibit crystal form |
|---|---|---|---|
| Pinacoids | {001}{010}{001} {0k1} {hk1} and variations | Two parallel faces | Kyanite, plagioclase, microcline, amblygonite, rhodonite, wollastonite |
| Pedions | {hk1} | Single face | Similar |

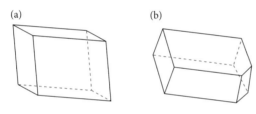

(a)                    (b)

**Figure 4.30**    Triclinic crystal forms: (a) front, side, and basal pinacoids, (b) various pinacoids and a pedion to the lower right.

## 4.7    TWINNED CRYSTALS

Many mineral specimens display **twinned crystals** that contain two or more related parts called twins. **Twins** have the following characteristics: (1) they possess different crystallographic orientations, (2) they share a common surface or plane, and (3) they are related by a symmetry operation such as reflection, rotation or inversion (Figure 4.31). Because twins are related by a symmetry operation, twinned crystals are not random intergrowths.

A **twin law** describes the symmetry operation that produces the twins and the plane (hkl) or axis involved in the operation. For example, swallowtail twins in gypsum (Figure 4.31a) are related by reflection across a plane (001), which is not a mirror plane in single gypsum crystals. Carlsbad twins in

potassium feldspar (Figure 4.31f) are related by a twofold axis of rotation that is parallel to the c-axis (001), which is not a rotational axis in single potassium feldspar crystals.

The surfaces along which twins are joined are called **composition surfaces**. If the surfaces are planar, they are called **composition planes**, which may or may not be equivalent to twin planes. Other composition surfaces are less regular. Twins joined along composition planes are called **contact twins** and do not appear to penetrate one another. Good examples of contact twins are shown in Figure 4.31a and b. Twins joined along less regular composition surfaces are usually related by rotation and are called **penetration twins** because they appear to penetrate one another. Good examples of penetration twins are shown in Figure 4.31c–f.

Twinned crystals that contain only two twins are called **simple twins**, whereas **multiple twins** are twinned crystals that contain more than two twins. If multiple twins are repeated across multiple parallel composition planes, the twins are called **polysynthetic twins**. Polysynthetic albite twins (Figure 4.31b) are repeated by the albite twin law, reflection across (010), and are very common in plagioclase. They produce small ridges and troughs on the cleavage surfaces of plagioclase, which the eye detects as linear striations – a key to hand-specimen identification of plagioclase. They also produce the

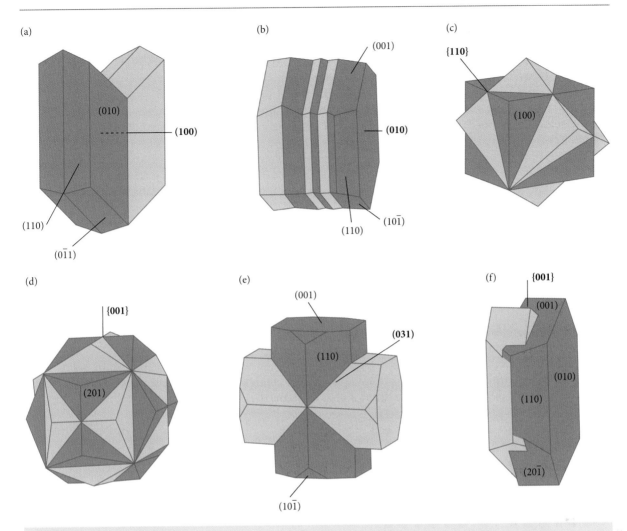

(a)

(010)

(100)

(110)

(0$\bar{1}$1)

(b)

(001)

(010)

(10$\bar{1}$)

(110)

(c)

{110}

(100)

(d)

{001}

(201)

(e)

(001)

(031)

(110)

(10$\bar{1}$)

(f)

{001}

(001)

(010)

(110)

(20$\bar{1}$)

**Figure 4.31** Examples of twinned crystals: (a) swallowtail twins in gypsum; (b) polysynthetic albite twins in plagioclase; (c) penetration twins in galena; (d) penetration twins in pyrite; (e) penetration twins in staurolite; (f) Carlsbad twins in potassium feldspar. *Source*: Wenk and Bulakh (2004). © Cambridge University Press.

characteristic "zebra stipe" pattern observed in many plagioclases when viewed under crossed polars with a petrographic microscope (Chapter 6).

Most twins are **growth twins** that form during mineral crystallization. Less commonly, twins result from displacive mineral transformations or from deformation. Deformation twins are called **mechanical twins**. The common mineral calcite typically develops mechanical twins (102) during deformation, and their development can play a significant role in the deformation of calcitic marbles and other metamorphic rocks, especially at low temperatures. This helps to explain why marbles deform more plastically than many other

rocks during lower grades of metamorphism (Chapter 16).

## 4.8 CRYSTAL DEFECTS

Ideally, crystals are perfectly formed with no defects in their lattice structures (Figure 4.32a). However, nearly all crystals contain small-scale impurities or imperfections that cause mineral composition and/or structure to vary from the ideal. These local-scale inhomogeneities are called **crystal defects**. Crystal defects have some profound effects on the properties of crystalline material that belie their small scale (Box 4.1). A convenient way to classify crystal defects is in terms of their dimensions.

## Box 4.1    Frenkel and Schottky defects

**Frenkel defects** (Figure B4.1a) are formed when the ions in question move to an interstitial site, leaving unoccupied structural sites or holes behind. Frenkel defects combine omission and interstitial defects. Because the ion has simply moved to another location, the overall charge balance of the crystal is maintained, but local lattice distortions occur in the vicinities of both the holes and the extra ions. **Schottky defects** (Figure B4.1b) are omissions formed when the ions migrate out of the crystal structure or were never there. Schottky defects create a charge imbalance in the crystal lattice. Such charge imbalances may be balanced by the creation of a second hole in the crystal structure; for example, an anion omission may be created to balance a cation omission. They may also be balanced by the substitution of ions of appropriate charge difference elsewhere in the structure. The highly magnetic mineral **pyrrhotite** ($Fe_{1-x}S$) provides a good example. When a ferrous iron ($Fe^{+2}$) ion is omitted from a cation site in the crystal structure, leaving a charge deficit of 2, two ferric iron ($Fe^{+3}$) ions can substitute for ferrous iron ($Fe^{+2}$) ions to increase the charge by 2 and produce an electrically neutral lattice (Figure B4.1b). The formula for pyrrhotite ($Fe_{1-x}S$) reflects the fact that there are fewer iron (total $Fe^{+2}$ and $Fe^{+3}$ ions than sulfur ($S^{-2}$) ions in the crystal structure due to the existence of a substantial number of such Schottky omission defects.

Point defects can occur on still smaller scales. In some cases electrons are missing from a quantum level, which produces an electron hole in the crystal structure. In others, an electron substitutes for an anion in the crystal structure. As with other point defects, the existence of electron holes plays an important role in the properties of the crystalline materials in which they occur. In most minerals, as temperature increases, the number of omission defects tends to increase. This allows minerals to deform more readily in a plastic manner at higher temperatures.

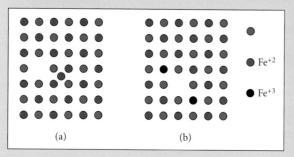

Fe$^{+2}$

Fe$^{+3}$

(a)                    (b)

**Figure B4.1**    (a) Frenkel defect, with a vacancy due to an ion displaced to the interstitial site. (b) Shottky defect in pyrrhotite ($Fe_{1-x}S$) where a vacancy (absent $Fe^{+2}$) is balanced by the substitution of $Fe^{+3}$ for $Fe^{+2}$ in two lattice sites.

### 4.8.1    Point defects

**Point defects** involve individual atoms and therefore do not have longer range extent; they are considered to be **zero-dimensional defects**. Many types of point defects exist, and they are important in explaining the properties of minerals as well as other materials that include steel, cement, glass products, semiconductors, and superconductors. These include the following:

1   **Substitution defects** (Figure 4.32b) form when anomalous ions of inappropriate size and/or charge substitute for ions of appropriate size and/or charge in a structural site. These anomalous ions tend to distort the crystal lattice locally and to be somewhat randomly distributed within the crystal lattice.

2   **Interstitial defects** (Figure 4.32c) occur when anomalous ions occupy the spaces between structural sites. Such "extra" ions are trapped in the interstices between the "normal" locations of ions in the crystal lattice.

3   **Omission defects** (Figure 4.32d) form when structural sites that should contain ions are unoccupied. In such cases, ions that should occur within the ideal crystal structure are omitted from the crystal lattice leaving a "hole" in the ideal crystal structure.

### 4.8.2 Line defects

Line defects are called **dislocations**. As lines, they possess extent in one direction and are therefore **one-dimensional defects.**

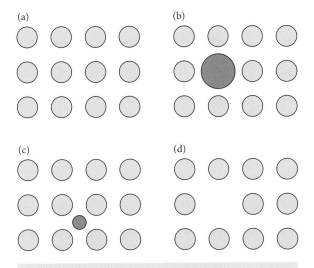

**Figure 4.32** (a) Perfect crystal lattice; (b) substitution defect; (c) interstitial defect; (d) omission defect.

Dislocations commonly result from shearing stresses produced in crystals during deformation. These stresses cause atomic planes to shift position, producing distortions in the crystal lattice that can be represented by a line called a **dislocation line**. Two major types of dislocations are recognized: **edge dislocations** (Figure 4.33a) and **screw dislocations** (Figure 4.33b).

Dislocations are extremely important in the plastic deformation of crystalline materials that leads to changes in rock shape and volume without macroscopic fracturing. Dislocations permit rocks to flow plastically at very slow rates over long periods of time. Detailed discussions are available in many books on mineralogy (e.g., Wenk and Bulakh 2016) and structural geology (e.g., van der Pluijm and Marshak 2004; Davis et al. 2011). Figure 4.34 shows how an edge dislocation can migrate through a crystal by breaking a single bond at a time. The result of dislocation migration is a crystal shape change accomplished without rupture. Dislocations are important plastic deformation processes allowing rocks to change shape without visible rupture (Chapter 16).

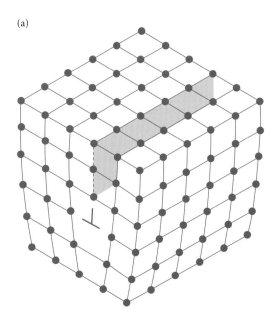

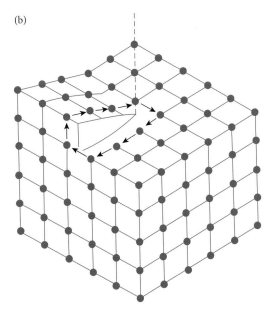

**Figure 4.33** (a) Edge dislocation with an extra half plane of atoms; this is a line defect because the base of the half plane can be represented by a dislocation line (⊥). (b) Screw dislocation, where a plane of atoms has been rotated relative to the adjacent plane. *Source*: Klein and Hurlbut (1985). © John Wiley & Sons.

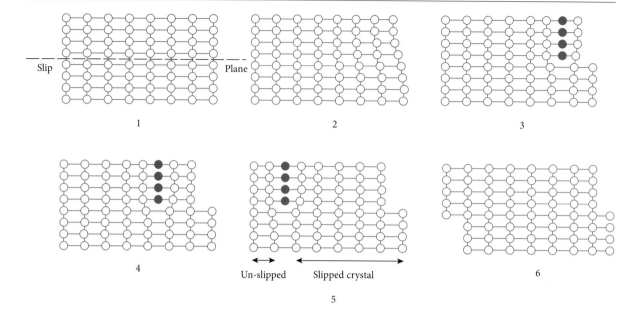

**Figure 4.34** Two-dimensional depiction of how an edge dislocation created by slip due to shear can migrate through a crystal by breaking one bond at a time, so that no fractures develop as the crystal changes shape during deformation (steps 1–6). *Source*: Adapted from Hobbs et al. (1976). © John Wiley & Sons.

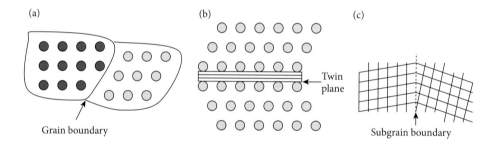

**Figure 4.35** Three types of planar defect (shown in two dimensions): (a) intergranular grain boundary between two different crystals; (b) intragranular mechanical twin boundary resulting from mechanical slip; (c) intragranular subgrain boundary within a crystal, separated by a wall of dislocations. Imagine each extending in a second dimension perpendicular to the page and note how (b) and (c) accommodate changes in crystal shape.

### 4.8.3   Planar defects

**Planar defects** (Figure 4.35) extend in two dimensions and are therefore **two-dimensional defects**. They are places within a crystal where the crystal structure changes across a distinct planar boundary. Examples include: (1) the boundaries between exsolution lamellae, for example between albite and potassium feldspar in perthites; (2) the subgrain boundaries between twins in twinned crystals; (3) the subgrain boundaries within crystals between out-of-phase crystal structures generated during ordering transformations; (4) grain boundaries between different crystals; and (5) extra atomic planes or missing planes called stacking faults.

Point defects, line defects, and planar defects are critically important in the study of deformed rocks, particularly in the elastic and plastic deformation processes discussed in Chapter 16 (Box 4.2).

## 4.9 POLYMORPHS AND PSUEDOMORPHS

### 4.9.1 Polymorphs

As noted earlier (Chapters 1 and 3), different minerals can have the same chemical

---

## Box 4.2  Defects and plastic deformation in crystals

You may recall from earlier courses in which folds, faults, and metamorphic foliations were discussed that when stresses are applied to rocks, they experience changes in shape and/or volume. These changes in shape and/or volume that occur in response to stress are called strains. They are analogous to the strains that occur in bones and muscles when they change shape in response to stress. Non-elastic strains are subdivided into those in which rocks break along fractures such as faults or joints and those in which shape changes are accomplished without macroscopic fracturing. Irreversible strains that involve visible fracturing are called rupture; those that do not are called plastic strains and accommodate plastic deformation. Rupture is favored by rapid strain rates (think how fast the bone changes shape as it fractures), low confining pressures, and low temperatures. On the other hand, plastic strain is favored by very low strain rates, high confining pressures, and high temperatures (Figure B4.2a). Under such conditions, deep below the surface, rocks respond very slowly to stress in a manner more like Playdough® or modeling clay than like the rigid rocks we see at Earth's surface. How can rocks undergo significant strain without rupturing? A major key lies in the large number of defects that the minerals in rocks contain.

Plastic deformation at high temperatures and low strain rates largely results from two significant types of diffusion creep (Figure B4.2a) that are dependent on the existence of omission defects in minerals: (1) Coble (grain boundary diffusion) creep, and (2) Herring–Nabarro (volume diffusion) creep. Elevated temperatures are associated with elevated molecular vibration in an expanded crystal lattice. Such vibrations lower bond strength and increase the number of omission defects (also called holes or vacancies) in the crystal structure. As holes are created, adjacent atoms can migrate into the vacancy by breaking only one weak bond a time. The movement of the ions in one direction causes the holes or vacancies to migrate in the opposite direction (Figure B4.2b).

Under conditions of differential stress, ions tend to be forced toward the direction of least compressive stress $(\sigma_3)$, which tends to lengthen the crystal in that direction. Simultaneously, holes tend to migrate toward the direction of maximum compressive stress $(\sigma_1)$ until they reach the surface of the crystal where they disappear, causing the crystal to shorten in this direction (Figure B4.2b). In Coble creep, the vacancies and ions migrate near grain boundaries to achieve the strain, whereas in Herring–Nabarro creep, the vacancies and ions migrate through the interior of the crystals. Since thousands of omission defects are created over long periods of time, even in small crystals, the long-term summative effects of plastic strain as each crystal changes shape by diffusion creep can be very large indeed.

At higher strain rates related to higher differential stresses, dislocation creep processes become dominant (Figure B4.1). In these environments edge dislocations and screw dislocations migrate through the crystal structure, once again breaking only one bond at a time, while producing plastic changes in shape. Because such dislocations result from strain, large numbers are produced in response to stress, and their migration accommodates large amounts of plastic strain. Imagining the summative plastic changes in shape that can be accomplished by the migration of thousands of diffusing vacancies and/or migrating dislocations in a small crystal or $10^{20}$ dislocations migrating through the many crystals in a large mass of rock offers insight into the power of crystal defects to accommodate plastic deformation on scales that range from microcrystals to regionally metamorphosed mountain ranges.

*Continued*

**Box 4.2** *Continued*

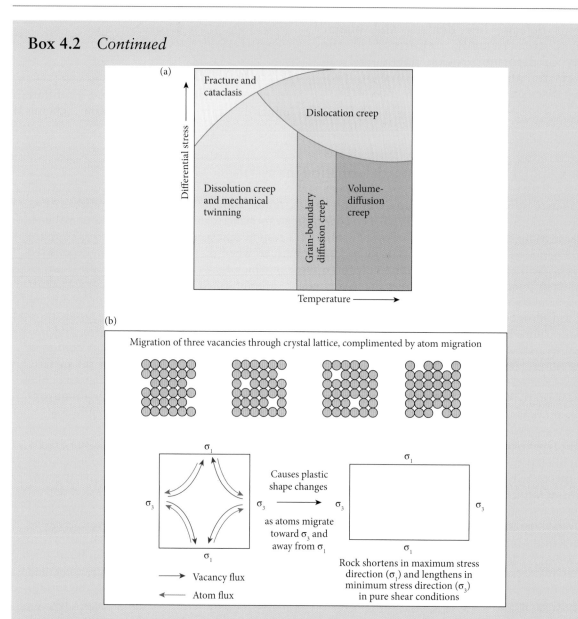

**Figure B4.2** (a) Deformation map showing the significant role of omission defects and dislocations in the high temperature plastic deformation of crystals. *Source*: Modified from Davis et al. (1996). © John Wiley & Sons. (b) Diagrams showing how the existence of omission defects permits adjacent ions to move into their former locations, effectively causing the omission or hole to migrate in one direction as the ions migrate in the other. The flux of atoms (blue arrows) toward regions of least compressive stress ($\sigma_3$) and of vacancies (black arrows) toward areas of maximum compressive stress ($\sigma_1$) cause crystals to change shape.

composition, but different crystal structures. This ability for a specific chemical composition to occur in multiple minerals, each with a different crystal structure is called **polymorphism**. The resulting minerals are called **polymorphs**. In most cases, the crystal structure or form taken by the mineral is strongly influenced by the environment in which it forms. Polymorphs therefore record important information concerning the environments that produced them. Many polymorphs belong to very common and/or economically significant mineral groups, such as the examples summarized in Table 4.13.

The polymorphs of carbon can be used to illustrate how environmental conditions during growth determine which crystal structure a chemical compound possesses. Figure 4.36 is a phase stability diagram for systems composed of pure carbon. This phase stability diagram clearly indicates that diamond is the high pressure polymorph of carbon, whereas graphite is the low pressure polymorph. If we add **geotherms**, lines showing the average temperature of Earth at any depth, to this diagram, we can infer that diamonds are the stable polymorph of carbon at pressures of more than 3.5 GPa, corresponding to depths of more than 100 km below the surface whereas graphite is the stable polymorph of carbon at all shallower depths. Inferences must be tempered by the fact that Earth's interior is not pure carbon and temperature distributions with depth are not constant. Nonetheless, it is widely believed that most natural diamonds originate at high pressures far below the surface of old continental shields in which they most commonly occur. If graphite is the stable polymorph of carbon at low pressures, why do diamonds occur in deposits at Earth's surface where pressures are low? Obviously, as diamonds rise toward Earth's surface into regions of substantially lower pressure, something keeps the carbon atoms from rearranging into the graphite structure. What keeps the transformation from unstable diamond to stable graphite from occurring?

*Reconstructive transformations*

**Reconstructive transformations** involve the conversion of one polymorph (or mineral) into another by processes that require bond breakage so that a significant change in structure occurs. Such transformations require large amounts of energy, and this requirement tends to slow or inhibit their occurrence. In the transformation of diamond to graphite, a

**Table 4.13**  Important rock-forming mineral polymorphs.

| Chemical composition | Common polymorphs |
| --- | --- |
| Calcium carbonate ($CaCO_3$) | Calcite and aragonite |
| Carbon (C) | Diamond and graphite |
| Silica ($SiO_2$) | α-quartz, β-quartz, tridymite, cristobalite, coesite, stishovite |
| Aluminum silicate ($AlAlOSiO_4$) | Andalusite, kyanite, sillimanite |
| Potassium aluminum silicate ($KAlSi_3O_8$) | Orthoclase, microcline, sanidine |
| Iron sulfide ($FeS_2$) | Pyrite, marcasite |

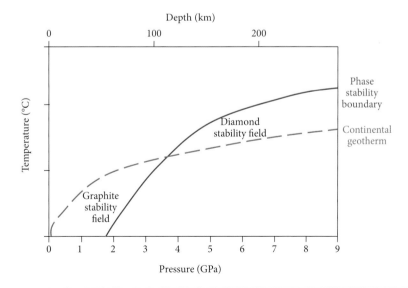

**Figure 4.36**  Phase stability diagram showing the conditions under which graphite, the low pressure polymorph of carbon, and diamond, the high pressure polymorph of carbon, are stable beneath continental lithosphere.

large amount of energy is required to break the strong bonds that hold carbon atoms together in the isometric diamond structure, so that they can rearrange into the more open, hexagonal structure of graphite. This inhibits the transformation of diamonds into graphite as diamonds find themselves in lower pressure and lower temperature environments near Earth's surface. Minerals such as diamond that exist under conditions where they are not stable are said to be **metastable**. All polymorphs that require reconstructive transformations have the potential to exist outside their normal stability ranges as metastable minerals. This allows them to preserve important information about the conditions under which they, and the rocks in which they occur, were formed and, in the case of diamonds, to grace the necks and fingers of people all over the world.

*Displacive transformations*

Some polymorphs are characterized by structures that, while different, are similar enough that the conversion of one into the other requires only a rotation of the constituent atoms into slightly different arrangements without breaking any bonds. Transformations between polymorphs that do not require bonds to be broken and involve only small rotations of atoms into the new structural arrangement are called **displacive transformations** and tend to occur very rapidly under the conditions predicted by laboratory experiments and thermodynamic calculations. Polymorphs involved in displacive transformations rarely occur as metastable minerals far outside their normal stability ranges and so may preserve less information about the conditions under which they and the rocks in which they occur originally formed.

Alpha quartz (low quartz) is generally stable at lower temperatures than beta quartz (high quartz). Although α- and β-quartz have different structures, the structures are so similar (Figure 4.37) that the conversion of one to the other is a displacive transformation. It is not at all unusual, especially in volcanic rocks, to see quartz crystals with the external crystal form of β-quartz but the internal structure of α-quartz. These quartz crystals are interpreted to have crystallized at the elevated temperatures at which β-quartz is stable and to have

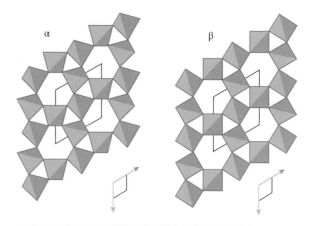

**Figure 4.37**  The closely similar structures of α- and β-quartz. *Source*: Courtesy of Bill Hames.

been diplacively transformed into the α-quartz structure as they cooled, while retaining their original external crystal forms.

Other transformations between silica polymorphs are reconstructive. For example, the transformations between the high-pressure minerals stishovite and coesite and between coesite and quartz are reconstructive. Therefore, both stishovite and coesite can be expected to exist as metastable phases at much lower pressures than those under which they are formed. Their preservation in rocks at low pressures allows them to be used to infer high pressure conditions, such as meteorite impacts, long after such conditions have ceased to exist.

*Order–disorder transformations*

Many polymorphs differ from one another only in terms of the degree of regularity in the distribution of certain ions within their respective crystal structures. Their structures can range from perfectly ordered to a random distribution of ions within structural sites (Figure 4.38). The potassium feldspar minerals ($KAlSi_3O_8$) provide many examples of such variation in regularity or order in the distribution of aluminum ions within the structure. In the feldspar structure, one in every four tetrahedral sites is occupied by aluminum ($Al^{+3}$), whereas the other three are occupied by silicon ($Si^{+4}$). In the potassium feldspar **high sanidine**, the distribution of aluminum cations is completely random (high disorder); the

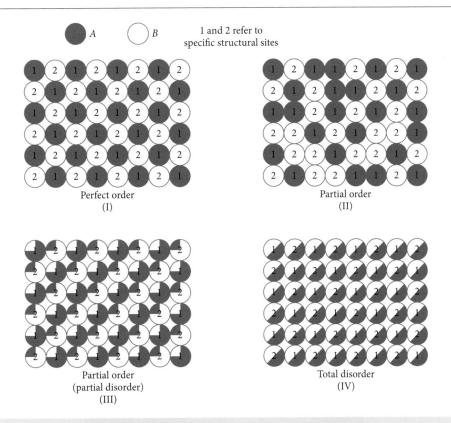

**Figure 4.38**    Variations in the order of minerals. *Source*: Klein and Hurlbut (1985). © John Wiley & Sons.

probability of finding an aluminum cation in any one of the four sites is equal. Crystal structures with such random distributions of cations are highly disordered and are favored by high temperatures and low pressures of formation. On the other hand, in the potassium feldspar **low microcline**, the distribution of aluminum cations is highly ordered (low disorder), with the aluminum distributed regularly in every fourth tetrahedral site. The probability of finding an aluminum cation in these sites approaches 100%, and the probability of finding one in the other three sites approaches zero. Crystal structures with such regular distributions of cations possess very low disorder, and their formation is favored by low temperatures and/or high pressures of formation. Intermediate degrees of order exist within the potassium feldspar group. High sanidine, with its high degree of disorder, crystallizes in the monoclinic system, and is common in volcanic rocks formed at high temperatures and low near surface pressures, whereas low microcline, with its low degree of disorder, crystallizes in the triclinic system,

and is common in rocks formed at higher pressures, and in some cases lower temperatures, below the surface.

### 4.9.2    Pseudomorphs

Minerals that take the crystal form of another, pre-existing mineral are called **pseudomorphs** and are said to be pseudomorphic after the earlier mineral (Figure 4.39). Pseudomorphs can be produced in many ways. All require that the original crystal possessed a significant number of crystal faces (was euhedral or subhedral) at the time it formed. Some pseudomorphs are produced by **replacement** in which the atoms in a pre-existing mineral are replaced by the atoms of a new mineral that retains the external crystal form of the original crystal. A common example is the replacement of pyrite ($FeS_2$) crystals by goethite ($FeOOH$) to produce goethite pseudomorphs after pyrite. Another common example is quartz ($SiO_2$) pseudomorphs after fluorite ($CaF_2$). Some pseudomorphs are **casts** produced by dissolution of the old mineral

**Figure 4.39** (a) Hematite replacing pyrite; (b) chalcedony encrusting aragonite; (c) quartz cast filling an aragonite solution cavity. *Source*: Photo courtesy of Stan Celestian, Maricopa Community College.

followed by precipitation of the pseudomorph to fill the cavity left behind. Other pseudomorphs are produced by the **loss of a constituent** from the original crystals. For example, the dissolution of carbonate ion from crystals of the copper carbonate mineral azurite $[Cu_3(CO_3)_2(OH)_2]$ can produce native copper (Cu) pseudomorphs after azurite. Still other pseudomorphs are produced when the new mineral forms a thin layer or crust over the original crystal. The **encrustation** of the original mineral by the new mineral allows the new mineral to mimic the crystal form of the original mineral. Still other pseudomorphs form by **inversion** as when β quartz crystals are transformed into α quartz, as described in the preceding section.

The properties of minerals and other crystalline materials are strongly influenced by their crystal structures and chemical compositions. These properties and the minerals that possess them are the subjects of Chapter 5.

## CONTENT ASSESSMENT

1 Discuss the difference between a *motif* and a *node* and explain why both are important in illustrating and understanding how long-range patterns of atoms/ions can be generated by symmetry operations.

2 What are *symmetry operations* and how are they related to crystal structures? How do *simple* and *compound* symmetry operations differ?

3 Carefully define the following symmetry operations and the symbols used to denote each one.
   (a) translation and unit translation vector
   (b) rotation and axis of rotation
   (c) reflection and mirror plane
   (d) inversion and center
   (e) glide reflection
   (f) rotoinversion
   (g) screw rotation

4 Discuss the properties and differences between the following types of *plane meshes or nets*.
   (a) square
   (b) rectangle
   (c) diamond
   (d) hexagonal (rhombic)
   (e) oblique

5 Define the essential properties of each of the *seven* (if hexagonal and trigonal are separated) *crystal systems* in terms of the properties of their *Bravais lattices* and their *crystallographic axes* (lengths and intersection angles).

6 In *Weiss parameters* and *Miller indices*, what do the three sequential parameters or indices represent? Convert the following Weiss parameters into the equivalent Miller indices.
   (a) $2, \infty, 1$   (b) $\frac{1}{2}, 1, 2$   (c) $2, \infty, \bar{2}/3$
   (d) $2, 2, 1$   (e) $1/3, \bar{1}/2, 1$

7 Explain the meaning of the following Miller indices in terms of crystallographic axial intercepts with respect to axial ratios
   (a) $(111)$   (b) $\{11\underline{\bar{1}}\}$   (c) $(11\bar{2})$
   (d) $\{hkl\}$   (e) $(hh\bar{1})$

8 What are *twinned crystals* and what are their chief defining characteristics? What is the essential difference between *contact* and *penetration* twins? Give a couple of good examples of each.

9 What are *polymorphs*? What is the difference between *reconstructive* and *displacive* polymorphs? Give an example or two of each. Which is most likely to produce long term metastable polymorphs?

## REFERENCES

Davis, G.H., Reynolds, S.J., and Kluth, C.F. (2011). *Structural Geology of Rocks and Regions*, 3e. New York: Wiley 864 pp.

Hobbs, B.E., Means, W.D., and Williams, P.F. (1976). *An outline of structural geology*. New York: Wiley 571 pp.

Klein, C. and Dutrow, B. (2007). *Manual of Mineral Science (Manual of Mineralogy)*, 23e. New York: Wiley 704 pp.

Klein, C. and Hurlbut, C.S. Jr. (1985). *Manual of Mineralogy*, 20e. New York: Wiley 644 pp.

Nesse, W.D. (2016). *Introduction to Mineralogy*, 3e. New York: Oxford University Press 512 pp.

van der Pluijm, B.A. and Marshak, S. (2004). *Earth Structure: An Introduction to Structural Geology*, 2e. New York: W.W. Norton 672 pp.

Wenk, H.-R. and Bulakh, A. (2016). *Minerals: Their Constitution and Origin*, 3e. Cambridge, UK: University Press 621 pp.

# Chapter 5

# Mineral properties and rock-forming minerals

## 5.1 MINERAL FORMATION

Minerals are ephemeral in the sense that they have finite, if often very long, life spans. They represent atoms that have bonded together to form crystalline solids whenever and wherever environmental conditions permit. Ice is a familiar example. It forms whenever temperature and pressure conditions permit hydrogen and oxygen atoms to bond together to form crystals with a hexagonal structure. When temperatures increase or pressures decrease sufficiently, ice ceases to exist because the atoms separate into the partially bonded arrays that characterize liquid water. Ice, like all minerals, is ephemeral; more ephemeral under Earth's surface conditions than are most minerals. Another good example of the impermanence of minerals is the transition between the polymorphs of carbon. At relatively low pressures, carbon atoms combine to form graphite. If the pressure increases sufficiently, the carbon atoms are rearranged so that graphite is transformed into diamond. Likewise, diamond can be transformed back into graphite if pressures decrease. The reaction kinetics of such transformations can be very slow which is why diamonds can exist as a metastable mineral under low pressure conditions at Earth's surface.

Minerals form via natural environmental processes that cause atoms to bond together to form solids. These include

1 **Precipitation from solution.** Solutions from which minerals precipitate include
   a **Surface water** in springs, rivers, lakes, and oceans.
   b **Groundwater** in soils and underground aquifers.

*Earth Materials*, Second Edition. Kevin Hefferan and John O'Brien.
© 2022 John Wiley & Sons Ltd. Published 2022 by John Wiley & Sons Ltd.
Companion website: www.wiley.com/go/hefferan/earthmaterials2

  c  **Hydrothermal solutions**, which are warm, aqueous solutions that have been heated at depth and/or by proximity to a body of magma.

2  **Sublimation from a gas.** Sublimation, mineral precipitation from a gas, occurs where volcanic gases are vented at Earth's surface or where gas phases separate from solution in the subsurface.

3  **Crystallization from a melt or other liquid:**
  a  **Lava flows** at the surface that form volcanic minerals and rocks.
  b  **Magma** bodies in the subsurface that produce plutonic minerals and rocks.

4  **Solid state growth.** In solid state growth, new mineral crystals grow from the constituents of preexisting minerals. This is especially common during the formation of metamorphic minerals and rocks.

5  **Solid–liquid or solid–gas reactions.** In such reactions, atoms are exchanged between solid minerals and the liquid or gas phase with which they are in contact, producing a new mineral. These solid–liquid or solid–gas reactions are common in mineral-forming processes that include weathering, hydrothermal alteration, vein formation, and metamorphism.

This list illustrates the most important processes by which minerals form. These processes continually alter and destroy preexisting arrays of atoms to form new minerals in ways that depend on the conditions and processes in the environment in which they form. It should be noted that the environmental conditions that produce minerals have evolved in complex ways over the 4.55 Ga of Earth's history. As a result, the minerals that have been produced by these processes have evolved with them. Hazen et al. (2008) introduced the concept of mineral evolution which traces how Earth's mineralogy has evolved in response to the evolution of Earth's mineral forming environments. They argued that the number of minerals generated in Earth environments evolved from about 250 species at the time it formed to roughly 5400 at the present time. Major episodes of new mineral formation occurred during major changes in Earth environments related to the (1) Hadean–Archaean initiation of partial melting that generated oceanic crust and then continental crust (~3.5–4.4 Ga), (2) initiation of plate tectonics during the late Archaean (>3.0 Ga),

(3) early Proterozoic (2.5–1.9 Ga) Great Oxygenation Event (GOE) that radically changed the composition of Earth's atmosphere, and (4) Phanerozoic (<542 Ma) explosion of biomineralization by "modern" organisms that further changed oceanic and atmospheric chemistry. To this, many would add the large number of crystalline solids produced by humans during the "Anthropocene," though many do not consider these minerals, as discussed earlier. Since 2008, new minerals have been discovered, leading to the revised number of roughly 5500, some of which are no longer produced, because the environmental conditions that generated them no longer exist.

## 5.2   CRYSTAL HABITS

### 5.2.1   Habits of individual crystals

Minerals start small. Each mineral crystal begins with the bonding of a few atoms into a three-dimensional geometric pattern. Initial growth leads to the formation of small "seed crystals" called **nuclei**. If the appropriate atoms are available and the environmental conditions are suitable for growth, the nuclei will continue to attract appropriate atoms or ions and grow into larger mineral crystals. When it stops growing, the mineral can be bounded by crystal faces that reflect its internal crystal structure. Since minerals frequently occur as well-formed crystals and since crystal habits reflect the crystal structure of the mineral in question, being able to recognize crystal habits is very useful in mineral identification.

Single crystals can be described using a variety of terminology. The simplest terminology is based on the relative proportions of the crystals in three mutually perpendicular directions (a, b, and c) where a $\geq$ b $\geq$ c. Table 5.1 and Figure 5.1 summarize the terminology used for individual crystal habits and illustrate examples of each.

### 5.2.2   Habits of mineral crystal aggregates

Where environmental conditions are appropriate for the nucleation and growth of a single mineral crystal, they tend to be suitable for the formation of multiple crystals of the same mineral. The production and growth of multiple crystals located adjacent to one another

**Table 5.1** Crystal Habits of Individual Crystals: Images Source: Equant, filiform, prismatic, bladed; courtesy of Kurt Hollacher, Union College; tabular (NMNH, Smithsonian Institute open source file), platy (Aram Dulyan, Wikimedia commons, public domain).

| Crystal habit | Colloquial description | Crystal dimensions | Image |
|---|---|---|---|
| Equant | Equal dimensions; shape may approach that of cube or sphere | $a = b = c$ | |
| Tabular | Tablet or diskette-like | $a = b > c$; c thin | |
| Platy | Thin, sheet-like, a thinner version of tabular | $a \approx b >> c$; c very thin | |
| Prismatic or columnar | Pillar-like or column-like; slender to stubby | $a > b = c$; a long | |
| Bladed | Blade-like or knife-like | $a > b > c$; a long, c thin | |

| Acicular | Needle-like; somewhat thicker than filiform, thinner than prismatic | a >>> b = c; b and c very thin | |
| Capillary or filiform | Hair-like | a >>> b = c; b and c extremely thin | |

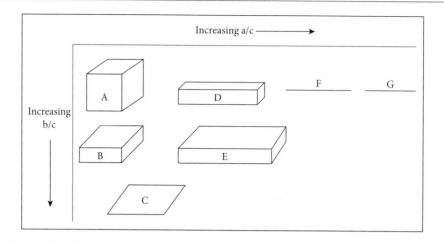

**Figure 5.1** Individual crystal habits: A. equant, B. tabular, C. platy, D. prismatic or columnar, E. bladed, F. acicular, G. capillary or filiform.

produces an assemblage of similar crystals called a **crystal aggregate**. In samples where crystal aggregates occur, at least two sets of crystal habits exist: one set describes the habit of individual crystals (Table 5.1) and the second set describes the habit of the aggregate. Examples of crystal aggregate habits are illustrated in Figures 5.2 and 5.3 and summarized in Table 5.2.

**Geodes** are crystal aggregates produced by the partial or complete filling of a subspherical cavity as crystallization proceeds from the walls of the cavity inward. Precipitation of outer layers of microscopic crystals produces most of the bands. Precipitation of larger crystals produces the druzy, divergent, or reticulated crystals that line the centers of many geodes (Figures 5.3 and 5.4). If the geode material is more resistant to weathering than the host rock, the geodes weather out as subspherical crystal aggregates whose concentric layering is revealed when they are broken or sawed in two. Another intriguing type of concentrically layered crystal aggregate is called a **concretion**. Concretions grow outward from a central nucleus within a rock body and incorporate preexisting mineral material as they do so. Thus, concretionary aggregates consist of subspherical bodies that include both newly precipitated crystals and preexisting material. Concretions are discussed in more detail and illustrated in Chapter 13.

Minerals with individual and aggregate crystal habits also possess a variety of other macroscopic and microscopic properties. In the following section, we will consider the macroscopic physical properties used in the identification of minerals. The microscopic properties used to identify minerals are discussed in Chapter 6.

## 5.3   MACROSCOPIC MINERAL PROPERTIES

Every mineral species possesses a unique set of properties that results from its unique combination of chemical composition, chemical bonding mechanisms, and crystal structure. Minerals are identified on the basis of their distinctive properties. In the field, correct mineral identification generally depends on the accurate identification of macroscopic properties that distinguish each mineral group and species. This often begins with such obvious features as color, luster, and the presence or absence of flat surfaces. In the following section, we will begin by considering static and mechanical properties followed by optical and electromagnetic properties.

### 5.3.1   Static and mechanical properties

Several common mineral properties depend on the fundamental static properties of minerals and/or how minerals respond mechanically to stress.

*Hardness*

**Hardness (H)** is the resistance of a smooth mineral surface to scratching or abrasion by a

**Figure 5.2**  Crystal aggregate habits, clockwise from upper left: top row: fibrous serpentine and radiating wavellite. Second row: divergent aragonite and reticulate barite. Third row: roseiform barite and drusy quartz. Bottom row: dendritic pyrolusite and foliated serpentine. *Source*: Courtesy of Doug Moore.

sharp tool. Several scales are used to express the hardness of minerals. One of the oldest was developed by Frederic Mohs in 1824 (Klein and Hurlbut 1985). Mohs recognized that some minerals could scratch other minerals. This indicated that some minerals were harder than others. Mohs did not quantify these relationships but created a relative

**Figure 5.3** Crystal aggregate habits (left to right); top row: massive kaolinite clay and granular green olivine; second row: concentrically banded agate and colloform (globular) prehnite; third row: stalactitic to colloform malachite and amydaloidal zeolites in vesicular basalt; bottom row: pisolitic bauxite and oolitic calcite. *Sources*: First (anonymous, CC, https://www.wpclipart.com/rocks_minerals/K/ Kaolin__clay_mineral.png.html), second through fifth courtesy of Doug Moore; sixth by John O'Brien, seventh (courtesy of Scott Horvath, USGS public domain) and eighth (Marek Novotnak, CCBYSA https://commons.wikimedia.org/wiki/File:Basalt_-_amygdaloidal_structure.jpg, last visited 9/13/2020).

**Table 5.2**    Summary of the habits of crystal aggregates.

| Aggregate habit | Colloquial description |
| --- | --- |
| Fibrous | Parallel arrangement of acicular or filiform crystals |
| Radiating | Acicular–filiform crystals radiating outward from a central point |
| Divergent | Prismatic crystals diverging from a common area |
| Reticulated | Lattice work of tabular to bladed crystals |
| Roseiform | Petal-like arrangement of tabular or bladed crystals |
| Drusy | Surface lined with very small (<2–3 mm), goosebump-like crystals |
| Dendritic, arborescent | Tree-like, branching network of crystals |
| Lamellar/foliated | Subparallel layers of minerals |
| Massive | Aggregate of very small crystals with a fine-grained appearance |
| Granular | Subequant, macroscopic crystal aggregate with a granular appearance |
| Banded | Parallel layers of same mineral with different color; as in agate |
| Concentric | Spherical to subspherical layers about a common center |
| Botryoidal/colloform | Rounded, mound-like aggregates; kidney-like |
| Oolitic | Spherical, concentrically layered, sand-sized (<2 mm) grain aggregates |
| Pisolitic | Spherical, concentrically layered, gravel-sized aggregates |
| Amygdaloidal | Spherical to ellipsoidal gas vesicles infilled with secondary minerals |

**Figure 5.4**    Crystals with various percentages of bounding crystal faces. From left: euhedral twinned orthoclase crystals, completely enclosed by crystal faces; subhedral aragonite crystal, partially bounded by crystal faces; anhedral rose quartz, lacking crystal faces. *Source*: Courtesy of Doug Moore.

hardness scale that bears his name. Using 10 readily available minerals, he assigned each an integral value from 1 for the softest mineral, talc, to 10 for the hardest mineral, diamond (Table 5.3). Minerals with higher Mohs hardness numbers (H) can scratch minerals with lower Mohs hardness numbers. This is a relative hardness scale rather than a linear scale. Diamond (H = 10) is not twice as hard as apatite (H = 5) nor 10 times harder than talc (H = 1).

Simple scratch tests can be used to determine the relative Mohs hardness of any other mineral. The flat surface of one mineral is scratched with the sharp point of the other mineral while applying firm pressure. With a set of indenting tools or points composed of

**Table 5.3**    Comparison between the Mohs and Knoop hardness scales; the former is a relative hardness scale, whereas the latter is an absolute hardness scale.

| Mineral | Mohs hardness | Knoop hardness |
| --- | --- | --- |
| Talc | 1 | 1 |
| Gypsum | 2 | 32 |
| Calcite | 3 | 135 |
| Fluorite | 4 | 163 |
| Apatite | 5 | 430 |
| Orthoclase | 6 | 560 |
| Quartz | 7 | 820 |
| Topaz | 8 | 1340 |
| Corundum | 9 | 1800 |
| Diamond | 10 | 7000 |

the minerals on the **Mohs hardness scale,** the relative Mohs hardness of any mineral can be quickly ascertained. For example, a mineral such as hornblende that can scratch apatite but is scratched by orthoclase possesses a Mohs hardness between 5 and 6. Implements commonly used for approximate hardness determinations include a fingernail ($H \cong 2.5$), a penny ($H = 3+$), a steel nail or knife blade ($H = 5–5.5$), and a glass plate ($H = 5.5$). Using these simple tools, minerals with hardness $\leq 3$ are considered to be **soft minerals.** Minerals with a hardness of 3–5.5 possess **intermediate hardness.** **Hard minerals** possess hardness greater than 5.5. In many situations, knowledge of the approximate hardness is sufficient to allow a mineral to be identified. For example, fluorite is commonly purple, translucent and vitreous, as is the variety of quartz called amethyst. But quartz is a hard mineral ($H = 7$) whereas fluorite is an intermediate mineral ($H = 4$), so a scratch test using a nail, knife blade, or glass plate quickly distinguishes the two minerals.

More quantitative measures of the resistance of a mineral surface to abrasion exist. The **Knoop hardness scale** is an example of indentation hardness and uses an absolute hardness scale related to the stress required to indent a polished surface. Knoop hardness is calculated from data collected using a diamond tool to indent a polished mineral surface. The tool is held against the specimen under a controlled force for a controlled period of time and the size of the indentation is then measured microscopically. The Knoop hardness is then calculated from the ratio between the force and the time over which it was applied to the size of the indentation; the larger the indentation, the softer the mineral. For example, the indentation is 32 times smaller for gypsum than it is for talc, so gypsum is 32 times more resistant to indentation than is talc. Table 5.3 compares the relative hardness of the minerals on the Mohs hardness scale with the absolute hardness on the Knoop hardness scale. The indentation in talc produced by the Knoop hardness test is 7000 times larger than that for diamond, so on this scale diamond is 7000 times harder (more resistant to indentation) than talc.

What accounts for the variation in hardness between minerals? When a mineral is scratched or abraded, a fine powder is produced. This implies that many of the bonds that held the atoms together in the mineral have been broken. Hardness depends first and foremost on bond strength and second on how densely concentrated the bonds are in the crystal structure. The stronger and more numerous the bonds are, the harder the mineral. Diamond is characterized by closely spaced, very strong bonds that result in the hardest known mineral. Talc, on the other hand, possesses some areas with a small number of very weak bonds, which helps to explain why it is "soft enough for a baby's skin" and used in cosmetics.

### Density, specific gravity, and weight

**Density ($\rho$)** is the mass per unit volume of a material. It is expressed in units of mass divided by units of volume. These units are expressed in kilograms per cubic meter ($kg/m^3$) or, more commonly in the case of minerals, grams per cubic centimeter ($g/cm^3$). The density of a mineral is proportional to the number of atoms per unit volume, which is called the **packing index,** and to their **atomic mass number.** Minerals with very high density, such as native gold (Au), tend to have crystal structures with very closely spaced atoms (many atoms per volume) that have very high atomic mass numbers ($Au = 197$). Minerals with very low density, such as ice ($H_2O$), tend to have crystal structures with widely spaced atoms (few atoms per volume) and low atomic mass numbers ($H = 1$, $O = 16$).

**Specific gravity (SG)** is the ratio between the density of a material and the density of pure water at standard temperature and pressure (temperature = $3.9\,°C$ and pressure = 1 atmosphere). It is a dimensionless quantity. Since the density of pure water at standard temperature and pressure (STP) is $1.0\,g/cm^3$, the specific gravity of a mineral will have the same numerical value as its density but will be expressed without dimensions. An example will illustrate this point. The lead sulfide (PbS) mineral galena is quite dense because it contains very massive lead atoms; it has an average density of $7.5\,g/cm^3$. It has a specific gravity of 7.5. Mathematically, the number is obtained from the ratio between the density of the mineral ($7.5\,g/cm^3$) and the density of water at STP ($1.0\,g/cm^3$):

$$SG = \frac{\text{density of mineral}}{\text{density of water}} = \frac{7.5 \text{ g/cm}^3}{1.0 \text{ g/cm}^3} = 7.5$$

The **weight** of a mineral or of any other material is simply its total mass accelerated by gravity. For example, when an object is placed on a scale, its mass is accelerated downward by gravity to produce a downward force on the scale, which determines the object's weight. The total mass of the object, in grams (g) or kilograms (kg), is the total mass of all of the atoms it contains. Total mass therefore is proportional to the density of the material multiplied by its volume:

$$
\begin{aligned}
\text{Total mass}(m) \;&= \text{density} \times \text{volume} \\
&= \text{g/cm}^3 \times \text{cm}^3 \\
&= \text{grams (or kilograms)}
\end{aligned}
$$

The downward acceleration that creates weight is the acceleration due to gravity (g) which varies from place to place but averages about $9.8 \text{ m/s}^2$ on Earth's surface. Weight (w) then is simply the total mass (m) of an object multiplied by the acceleration due to gravity (g). Mathematically, w = mg. Weight is a downward force and is expressed in force units called *Newtons* (N), which have the dimensions of $\text{kg m/s}^2$, since

$$w = mg = \text{kg} \cdot \text{m/s}^2 = \text{Newtons (N)}$$

The concept of weight and its relationship to density, specific gravity, and volume can be illustrated using native gold as an example. Native gold, because of its closely packed, massive atoms has a density of $20 \text{ g/cm}^3$. Its specific gravity is 20 because the ratio of its density to the density of pure water at STP is

$$
\begin{aligned}
\text{Specific gravity (SG)} \;&= \frac{\text{density of sample}}{\text{density of water}} \\
&= \left(20 \text{ g/cm}^3\right) / \left(1.0 \text{ g/cm}^3\right) \\
&= 20
\end{aligned}
$$

A cubic centimeter ($\text{cm}^3$) of gold has a small total mass of $20 \text{ g}$, since

$$
\begin{aligned}
\text{Mass}(m) \;&= \text{density} \times \text{volume} \\
&= 20 \text{ g/cm}^3 \times 1 \text{cm}^3 \\
&= 20 \text{ g}
\end{aligned}
$$

How much will this amount of gold weigh at Earth's surface? Since weight (w) = mg,

$$
\begin{aligned}
\text{Weight}(w) \;&= mg = 20 \text{ g} \times 9.8 \text{ m/s}^2 \\
&= 196 \text{ g} \cdot \text{m/s}^2 = 0.196 \text{ kg} \cdot \text{m/s}^2 \\
&= 0.196 \text{ N}
\end{aligned}
$$

A larger volume of gold, though its density and specific gravity are constant, will weigh more because it contains more atoms and possesses more mass.

*Tenacity*

**Tenacity** is defined as the manner in which minerals respond to short-term stresses. Mineral tenacity depends on environmental factors such as temperature and pressure, but in the context of mineral identification is typically described for behavior at normal surface temperatures and pressures. Thus, it is a subset of **rheology**, the way materials generally respond to stress, which is discussed in detail in Chapter 16.

Many specimens of minerals in the mica group can be bent when stressed but return to their original shape when the stress is released. Such minerals are said to be **elastic**. If stressed sufficiently, thin sheets of some minerals demonstrate flexibility. Such specimens can be bent without breaking when stressed but, if stressed enough, do not return to their original shape. These minerals are said to be **flexible**.

Some minerals respond to stress by deforming elastically by a small percentage and then undergo significant plastic strain in which they change shape and/or volume without visibly fracturing. Minerals such as native metals are so plastic at surface temperatures and pressures that they can be hammered into thin sheets. They are said to be **malleable**. Minerals that can be drawn into a thin wire are **ductile**. This allows minerals such as native gold, silver, and copper to be formed into jewelery by cold working.

**Brittle** materials respond to stress by deforming elastically by a small percentage before fracturing after little or no plastic strain. Most minerals at normal surface conditions fracture when stressed; their tenacity is therefore brittle. A few minerals, with a closely spaced set of cleavage planes, can be cut into thin shavings, and these otherwise brittle minerals have a tenacity called **sectile**. Since most minerals are brittle, it is the unusual tenacities such as elastic, flexible, and malleable which are especially useful in mineral identification.

*Growth surfaces and breakage surfaces*

All mineral specimens possess external surfaces that enclose the mineral specimen and separate it from its surroundings. **Growth surfaces** are the surfaces that enclosed the mineral when ceased to grow. **Breakage surfaces** result when the specimen is broken, either from its host rock or at a subsequent time. Let us first consider growth surfaces.

Crystal faces
As discussed in Chapter 4, **crystal faces** are relatively flat, geometric surfaces generated by mineral growth. As crystals grow, atoms are added in the geometric patterns that characterize the mineral in question. When the mineral stops growing, it is bounded by growth surfaces. If flat crystal faces are present, they represent an external expression of the mineral's internal crystal structure. The beautiful crystals housed in many museums are testimony to the ability of mineral growth surfaces to express their geometric internal crystal structures (Figure 5.4). Mineral crystals completely enclosed by crystal faces are said to be **euhedral**. Euhedral crystals develop under conditions that allow them to be completely enveloped by relatively flat, geometric crystal faces. Mineral crystals that are only partially enclosed by crystal faces are said to be **subhedral**. Subhedral crystals form under conditions that allow them to be only partially bounded by relatively flat crystal. Mineral crystals that possess no crystal faces are said to be **anhedral** and record growth conditions that did not permit the development of crystal faces. What factors determine whether a mineral will be bounded by crystal faces? Many crystals nucleate (begin to grow) on a preexisting surface. The shape of that side of the mineral will reflect the shape of the preexisting surface, rather than the crystal structure of the mineral. When mineral growth fills a restricted space, for example, one enclosed by other minerals, the mineral surfaces will commonly reflect the shape of the space, rather than the internal geometry of the mineral's crystal structure. Only when a mineral nucleates and is free to grow in every direction, may it be completely enveloped by crystal faces. This happens most commonly when minerals nucleate in a fluid but also occurs when minerals grow around and/or incorporate preexisting minerals during growth.

Most mineral specimens are broken from the rock in which they occurred. Breaking the mineral produces breakage surfaces, which include cleavage, fracture, and parting surfaces, as discussed below.

Cleavage surfaces
Some minerals invariably break along flat, planar, light-reflecting surfaces. These flat breakage surfaces are called **cleavage surfaces**. They are related to the presence of sets of parallel planes of relatively weak bonding in the crystal structure, along which the mineral cleaves preferentially. Fresh cleavage surfaces can be recognized by a combination of features that may include (1) flat, planar surfaces, (2) strong reflection of light from these surfaces, and (3) the repetition of flat surfaces as sets of parallel surfaces on the mineral. These flat, parallel, light-reflecting cleavage surfaces may occur on opposite sides of the mineral specimen or they may occur as smaller parallel surfaces on one side of the specimen analogous to parallel steps on a stairway. In the latter instance, the parallelism of the step-like surfaces can be demonstrated by rotating the specimen in the light and noting that all the steps reflect light to the eye at the same moment in a manner analogous to the flash of light from the flat surface of a mirror. Since all the parallel cleavage surfaces on one or both sides of a mineral crystal have the same orientation, they together constitute one set, orientation, or direction of cleavage. In minerals that possess more than one set or orientation of cleavage (because they possess more than one set of relatively weakly bonded planes), the cleavage surfaces intersect each other at specific angles. When describing cleavage, one should carefully note the number of sets or directions of cleavage and the intersection angles between them. It is often sufficient simply to note whether or not the cleavage sets intersect at right angles.

What causes minerals to possess cleavage? Since cleavage results from the existence of

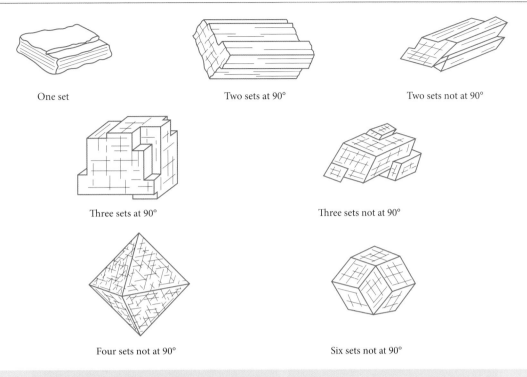

One set

Two sets at 90°

Two sets not at 90°

Three sets at 90°

Three sets not at 90°

Four sets not at 90°

Six sets not at 90°

**Figure 5.5**  Types of cleavage exhibited by minerals.

repeated sets of relatively weakly bonded planes in the crystal structure, a set of cleavages reflects the existence of a set or direction of such weak planes in the crystal structure. The number of sets or directions of cleavage equals the number of sets or directions of relatively weakly bonded planes in the mineral's structure. The angles of intersection between sets of cleavage correspond to the angles between these weakly bonded planes in the crystal structure. What produces planes of relative weakness in minerals? These planes possess some combination of weaker bonds or fewer bonds, that is to say they possess lower **total bond strength**, than do other planes within the mineral. It should be noted that even very hard minerals, such as diamond, can possess cleavage if there are directions in which there are fewer bonds, so that total bond strengths are weaker relative to those in other directions. Minerals that do not possess parallel planes of relative weakness, that is lower total bond strength, in their crystal structure do not possess cleavage.

Cleavage produces smaller mineral fragments of a predictable and repeatable shape that reflect the orientations of planes of relative weakness in the crystal structure. Minerals can have one, two, three, four, or six sets of cleavage (Figures 5.5 and 5.6). Like all crystallographic planes, cleavages can be symbolically represented by their Miller indices, as discussed in Chapter 4.

Minerals, such as those in the mica group, that possess only one set of planes of low total bond strength in their crystal structure tend to break repeatedly along those planes and so possess one direction of cleavage. Feldspar minerals such as orthoclase possess two sets of cleavage that intersect at nearly right (90°) angles, whereas amphiboles such as hornblende possess two sets of cleavage that intersect at angles of 57° and 123°. Still other minerals possess three sets or orientations of cleavage. In halite and galena, the three sets of relatively weak planes intersect at 90° angles, producing cubic cleavage, whereas in calcite and dolomite they do not intersect at right angles and instead produce rhombic cleavage. Fluorite possesses four sets of cleavage (octahedral cleavage). The largest number of cleavage plane sets is six (dodecahedral cleavage), as exemplified by the mineral sphalerite (see Figure 5.5).

It is important to remember that mineral cleavage is a property of single crystals. In crystal aggregates, the characteristic cleavage of the mineral can often be obscured. The

**Figure 5.6**   Examples of mineral cleavage (left to right). Top row: one set of cleavage, biotite and two sets of cleavage not at right angles, hornblende; row 2: two sets of cleavage at right angles, orthoclase and three sets of cleavage not at right angles, calcite; bottom row: three sets of cleavage at right angles (cubic), halite and four sets of cleavage (octahedral), fluorite. *Source*: Courtesy of Doug Moore.

three varieties of the hydrated calcium sulfate mineral gypsum provide an instructive example. Gypsum possesses one set of especially weak planes in its internal crystal structure that are weaker than all others and therefore possesses one excellent direction of cleavage (it does cleave less perfectly in two other directions). The variety of gypsum called selenite consists of relatively large single crystals. A single crystal of pure selenite possesses one excellent direction of cleavage that is generally easy to discern. However, when selenite occurs in aggregates of multiple crystals with different orientations, the aggregate may appear to have more than one set of cleavage orientations, unless one is careful to observe only the cleavage of each single crystal. A second variety of gypsum, called satinspar, is composed of a fibrous aggregate with numerous very thin, needle-like crystals that grow parallel to one another during its formation. Because satinspar is a fibrous aggregate of very thin, needle-like crystals, the cleavage cannot be discerned macroscopically. A third common variety of gypsum is alabaster. Alabaster is a massive form of gypsum that is an aggregate of numerous very small, randomly oriented gypsum crystals that have been compacted together. Because the crystals are so small and their cleavage planes are randomly oriented, alabaster does not display the perfect cleavage observed in larger single crystals like selenite.

Fracture surfaces

When broken, many mineral specimens display non-flat, commonly less reflective, non-parallel surfaces called **fracture surfaces.** Four types of fracture are described below: conchoidal, irregular, splintery, and hackly (Figure 5.7). Fracture surfaces may completely enclose a mineral specimen, or they may occur in combination with cleavage surfaces or crystal faces. Note that two sets of flat, planar surfaces cannot enclose a volume; therefore, broken mineral specimens with fewer than three sets or orientations of cleavage surfaces will possess fracture surfaces (and/or crystal faces) as well. For example, orthoclase has two sets of cleavage planes at right angles, and the remaining breakage surfaces exhibit irregular fracture (Figure 5.6).

**Conchoidal fracture** is a distinctive type of fracture marked by smooth, curved surfaces and occurs in both minerals and rocks. Common minerals such as quartz, chalcedony, opal, and garnet are often recognized because of their lack of cleavage and the presence of well-defined conchoidal fracture. Conchoidal fracture develops in materials that have similar total bond strengths in all directions and is characteristic of glass, such as the volcanic rock obsidian, and microcrystalline rocks, such as chert and flint. Such materials lack the long-range order necessary for planes of weakness and cleavage. When smooth, curved conchoidal fracture surfaces intersect, they have the potential to produce very sharp edges or points. Hard Earth materials with conchoidal fracture, such as obsidian, chert, and flint, were prized for their use as stone scrapers, blades, and arrowheads by "stone age" societies.

The most common type of fracture observed in mineral specimens is **uneven or irregular fracture.** As the names suggest, these fracture surfaces have a rather nondescript appearance. Some irregular fractures are produced when single mineral crystals break in directions that do not possess preferred planes of weakness in the crystal structure. In many

**Figure 5.7** Fracture types, clockwise from upper left: conchoidal fracture in chalcedony; splintery fracture in serpentine asbestos; uneven or irregular fracture in glauconite; hackly fracture in native copper. *Source*: Courtesy of Doug Moore; courtesy of Kurt Hollander.

instances, however, uneven or irregular fracture is produced in fine-grained, randomly oriented mineral aggregates such as alabaster gypsum or finely granular olivine, hematite, and limonite specimens.

A third type of fracture is characteristic of fibrous aggregates. Many minerals grow to produce aggregates of numerous, thin, parallel, hair- or needle-like crystals. When these fibrous aggregates break, they tend to separate between the fibers to produce **splintery fracture**. The fibrous variety of gypsum called satinspar is an excellent example of a mineral that possesses splintery fracture, as are the fibrous serpentine and amphibole minerals known as asbestos.

Some minerals, especially the malleable native metals, break in such a way as to produce ragged, sharp edges. These breakage surfaces are characteristic of **hackly fracture**.

Parting surfaces
Some minerals break along flat surfaces that are related to previous stress or twinning. Surfaces produced when minerals break along planes of weakness produced by twinning and/or by earlier stresses are called **parting surfaces**. These are not cleavage surfaces because they are not controlled by planes of weakness inherent in the mineral's essential crystal structure. Instead, the planes of weakness are produced when the mineral's structure is changed by twinning (Chapter 4) or in response to stress after its initial formation. For example, garnet does not possess cleavage, but stressed garnets commonly break along flat surfaces. Stressed corundum crystals also display well-defined parting surfaces not related to cleavage.

*Other static or mechanical properties*

Certain minerals display additional static or mechanical properties that enable them to be identified.

Striations
Some specimens of minerals, such as plagioclase, pyrite, and calcite, exhibit parallel sets of linear features called **striations** that appear as thin ridges and/or grooves on mineral surfaces. In the case of pyrite, the striations are due to the intersection of pyritohedral crystal faces on the face of cubic crystals (Hurlbut and Sharp, 1998). In the minerals calcite and plagioclase, striations develop due to

crystallographic twinning, as described in Chapter 4. The presence of striations is the primary means by which plagioclase feldspar is distinguished from potassium feldspar. Because feldspars are the most abundant mineral group in Earth's crust and their proportions are central to the classification of igneous rocks (Chapter 7), this distinction is of real significance.

Taste
Certain minerals possess a characteristic **taste**. The salty taste characteristic of halite, the bitter taste of sylvite, and the sweet, alkaline taste characteristic of borax are good examples. However, because many minerals are harmful if ingested, it is not a good idea to taste unknown samples in the field or the laboratory unless you are certain that it is a non-toxic mineral.

Feel
Very soft minerals, such as talc, graphite, and molybdenite, possess a characteristic "greasy" or "soapy" **feel**. The greasy feel results from planes of very weak van der Waals bonds that allow the minerals to be broken into soft, dust-like particles that "lubricate" the surface when the specimen is rubbed with the fingers.

Smell
Native sulfur (S) and many sulfide minerals such as marcasite ($FeS_2$) and sphalerite (ZnS) produce a sulfur **smell** when crushed or powdered. Arsenic-bearing minerals such as arsenopyrite (FeAsS) and realgar (AsS) possess a garlicky smell.

*Additional mineral properties*

Some carbonate minerals, such as calcite, aragonite, witherite, and rhodochrosite, possess the property of **effervescence**. When a drop of dilute hydrochloric acid ($HCl_{(a)}$) touches the mineral's surface, a chemical reaction releases carbon dioxide to produce a bubbling reaction. Other carbonate minerals, such as dolomite, effervesce only when the specimen is powdered or the hydrochloric acid is heated to promote the chemical reaction.

Other static mineral properties, such as melting and crystallization temperature, solubility, and thermal and electrical conductivity, are discussed in relationship to chemical bonding (Chapter 2) and in the chapters on magma generation and igneous rocks (Chapters 7 and 8), weathering and soils (Chapter 11), and ore deposits (Chapter 19).

### 5.3.2   Optical and electromagnetic properties

Many mineral properties result from interactions with light. Light is that part of the electromagnetic spectrum that can be sensed by the normal human eye. It is a very small portion of the total electromagnetic spectrum (Chapter 6), restricted to electromagnetic radiation with wavelengths between 700 nm (violet) and 300 nm (red). Light that strikes the surface of a mineral may be

1   Reflected by the mineral, either from its surface or internally.
2   Transmitted through the mineral, in whole or in part.
3   Absorbed by the mineral, either completely or selectively.
4   Dispersed, scattered, or reradiated by the mineral.

A fuller analysis of the potentially complex ways in which minerals interact with light and with other wavelengths of the electromagnetic spectrum is given in Chapter 6, which deals with optical microscopy.

*Diaphaneity (opacity)*

**Diaphaneity** or opacity depends on the amount of light transmitted by a mineral specimen (Figure 5.8). Mineral specimens that do not transmit any light are **opaque**. When light strikes these minerals, no light is transmitted through it. Instead the light is reflected, absorbed or both, even by a thin specimen of the mineral. Opaque minerals generally possess a dark gray to black streak. Even a finely powdered sample of the mineral does not transmit light. Mineral specimens that transmit a significant portion of the incident light are **transparent**. Such specimens transmit enough light that an image can be transmitted through them. In this way, they are analogous to a glass window that is transparent enough to clearly transmit an image. Whether or not a mineral specimen is transparent may depend on its thickness; thin specimens tend to be more transparent than thick ones because their smaller number of atoms scatter and absorb less light. Minerals that are macroscopically transparent often possess a white streak. Mineral specimens that transmit some light, but not enough to transmit an image, are **translucent**. Translucent minerals range from barely translucent to quite translucent or almost transparent. Diaphaneity is closely related to the property of luster, as explained below.

**Figure 5.8**   Examples of mineral diaphaneity, top-down. Left column: opaque minerals; galena and pyrite; center-left column: somewhat translucent minerals: perthite and milky quartz; center-right column: quite translucent minerals: tourmaline (rubellite) and trona; right column: transparent minerals: calcite and halite. *Sources*: Courtesy of Doug Moore; Andrew Silver, Brigham Young University mineral collection, public domain. © John Wiley & Sons.

Most mineral specimens are translucent. Translucent minerals commonly possess either white streaks or colored streaks that are neither gray nor black. One way to ascertain the diaphaneity or opacity of a specimen is to hold the specimen up to a light source and peer through the thinnest edge you can find. In this manner, you can determine whether no light, some light, or enough light to transmit an image is passing through the mineral. But another way is to use the streak and luster of the mineral to test your hypothesis about diaphaneity. This will be clearer when you have finished reading the sections that follow.

*Luster*

**Luster** is the appearance of a mineral surface in reflected light. Because it involves appearance, it is a rather subjective property. The luster we perceive when looking at a macroscopic specimen is principally the result of the amount of light reflected from its surface, the scattering of light from the surface and internally, and the amount of light absorbed by the mineral. Figure 5.9 relates the different lusters possessed by minerals to these three major variables. Luster is broadly subdivided into metallic and nonmetallic lusters (Figure 5.9).

Metallic and related lusters
Many minerals have a luster that is similar to the bright luster of metals, such as freshly polished silver, brass, or chrome. We normally see shiny, opaque materials as having the luster associated with metals, thus the term metallic luster (Figure 5.10). This **metallic luster** is characteristic of minerals that reflect and/or reradiate large amounts of minimally scattered, coherent light and absorb the rest. They are therefore shiny but opaque (or nearly so). As a result, they tend to have a gray to black streak. Galena, pyrite, and magnetite are excellent examples of minerals with metallic luster that are also opaque and possess a gray or black streak.

Some metallic minerals, such as chromite, magnetite, and bornite, may possess a **submetallic luster**. This usually results from a larger amount of scattering of the light or from the mineral being not quite opaque. Some opaque mineral specimens with a gray to black streak reflect little or no coherent light and therefore appear to have a **dull (earthy) luster**. Such specimens are often finely granular aggregates, so that they scatter light and do not reflect coherent light in a particular direction. Other nearly opaque minerals, such as rutile, reflect extremely bright coherent light, which gives them an unusually resplendent quality called an **adamantine luster**.

Nonmetallic lusters
Most minerals transmit some light and therefore possess **nonmetallic lusters** (Figure 5.11). These minerals commonly possess white streaks or colored streaks other than gray to black. Many of these minerals reflect light from a transparent or translucent background in a manner analogous to a mirror or a pane of glass. These minerals have the bright luster of shiny glass, which is formally called **vitreous** from the Latin word for glass. It is

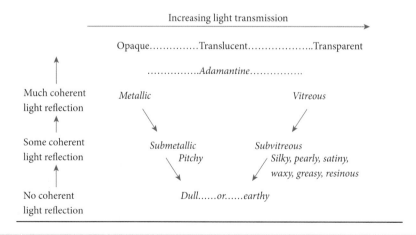

**Figure 5.9**   General relationships between light transmission (diaphaneity), light reflection/reradiation and luster. Light transmission increases from left to right and reflectivity increases from bottom to top.

**Figure 5.10**  Metallic and related lusters (opaque and nearly opaque minerals); left to right: metallic luster, pyrite (*Source*: Courtesy of Doug Moore. © John Wiley & Sons); submetallic luster, sphalerite (*Source*: Photo by Andreas Fruh, Wikimedia commons CCBYSA, https://commons.wikimedia.org/wiki/File:Sphalerite4.jpg, last accessed 9/13/2020) and dull (earthy) luster, limonite *Source*: Anonymous, USGS, public domain.

important to note that both light-colored (e.g. quartz, calcite, fluorite, and feldspars) and dark-colored minerals (amphiboles and many pyroxenes) possess vitreous lusters. A few transparent to translucent minerals with cleavage, such as sphalerite, reflect very bright light that produces the resplendent surfaces that characterize adamantine luster. Many gem mineral specimens, such as diamond, emerald, ruby, and sapphire, possess an adamantine luster which helps to give them their visual appeal. Many specimens of minerals that transmit light do not reflect coherent light and therefore have a **dull (earthy) luster**. As with other dull or earthy specimens, these specimens are commonly fine-grained aggregates of multiple crystals that randomly scatter the available reflected light.

Several different types of **subvitreous luster** exist. These are generally produced by partial scattering of light waves by the mineral's surfaces or by internal scattering of light. All of these scattering effects produce lusters that are somewhat subdued when compared to those of vitreous and adamantine specimens. **Silky luster** is characteristic of minerals with a fibrous habit that consist of parallel fibers with capillary or acicular habits. The parallel fibers reflect light in a manner reminiscent of silk threads. **Waxy** and **resinous lusters** are quite similar, having the subdued luster typical of floor wax and of tree resin or amber, respectively. **Greasy, pearly**, and **satiny** are also subdued subvitreous lusters that are similar to those of grease, pearls, and satin, respectively. In practice, it is not generally necessary to distinguish between the various

subvitreous lusters. In most mineral identification, it is only necessary to recognize that the specimen has a somewhat subdued, but not dull, luster and that it is translucent.

*Streak*

**Streak** is the color of the mineral powder and is typically obtained by scratching a mineral specimen on an unglazed porcelain plate called a streak plate. Since most streak plates have a hardness of about 6.5, only minerals which are somewhat softer than 6 will leave a powder when scratched on the porcelain plate. Harder minerals will not leave a powder on such plates but can be powdered using a harder material for the streak plate or grinding the material using a mortar and pestle or a grinding device. Because a mineral's powder generally has a fairly consistent color, a streak plate is very useful for identifying minerals, especially those that produce a non-white colored streak.

The color of the streak is largely the product of which wavelengths of light are transmitted by the mineral powder. Minerals that transmit a mixture of all wavelengths of light have a white streak. Such minerals are very common, so their streak is not a particular useful for identification. Opaque minerals absorb all wavelengths of light, producing a gray to black streak. However, many minerals selectively absorb certain wavelengths of light while transmitting others. These minerals possess a characteristic colored streak that is neither white nor gray to black. For example, hematite possesses a diagnostic dark brick-red streak.

**Figure 5.11**   Nonmetallic lusters (nonopaque minerals) left to right; top row: vitreous amethyst and subvitreous spodumene: second row: resinous amber and waxy chalcedony; third row: silky satinspar gypsum and pearly talc; last row: earthy limonite. *Sources*: Courtesy of Doug Moore. © John Wiley & Sons, except resinous amber (*Source*: Anonymous, freeimage.com) and pearly talc. *Source*: Michael C. Rygel, CCBYSA. https://commons.wikimedia.org/wiki/File:Schist_detail.jpg, last accessed September 14, 2020.

This is because hematite absorbs nearly all wavelengths of light except for a narrow band in the red part of the spectrum. It is nearly opaque, with its characteristic brick-red streak sometimes accompanied by a dark gray streak as well.

Similarly, azurite possesses a distinctive blue streak because it transmits blue wavelengths while absorbing the red and yellow parts of the spectrum. The closely related mineral malachite possesses a green streak because it transmits blue and yellow (blue + yellow = green) wavelengths while absorbing the red end of the visible spectrum. Sphalerite has a yellow streak because it transmits yellow wavelengths while absorbing the red and blue ends of the spectrum. The determination of a characteristically colored streak is frequently the test required to clinch the macroscopic identification of mineral specimens.

*Color*

The **color** of a macroscopic mineral specimen is the result of a complex interplay among reflection, absorption, transmission, refraction, scattering, and dispersion of light as it interacts with the mineral's chemical and structural components. This interplay is strongly influenced by chemical impurities and structural irregularities called defects (Chapter 4) that are common in all naturally occurring solids. Minerals characterized by a relatively constant shade of color are said to be **idiochromatic**, which means they are "self-colored." Idiochromatic minerals possess essentially the same color, independent of any impurities and/or defects they contain. Color is a diagnostic property of idiochromatic minerals and can be used as a criterion for their identification. Excellent examples of idiochromatic minerals include azurite, which is always blue, sulfur, which is always yellow, and galena, which is always gray.

Other minerals, however, are characterized by colors that vary from one specimen to another or even within the same specimen. These minerals are said to be **allochromatic**, which means "foreign colored." Their color is strongly influenced by impurities and/or defects so that different specimens possess different colors that depend on the types of impurities, most commonly metallic elements, and/or defects they possess. A good example of an allochromatic mineral is quartz which occurs in several different varieties that include colorless rock crystal, white milky quartz, pink rose quartz, honey-brown to dark brown smoky quartz, yellow citrine, blue to green aventurine, and purple amethyst that are explained later in this chapter (see Table 5.8). Color is not a diagnostic property of quartz or of any other allochromatic mineral. Many other minerals are allochromatic. For example, calcite can be colorless, white, gray, blue, green, yellow, pink, or red. In the case of calcite, however, the different colors are not generally given varietal names.

If all naturally occurring solids have chemical impurities and/or defects in their crystal structures, why are some mineral species idiochromatic whereas others are allochromatic? Generally, if a mineral possesses a strong color in its pure state, it will be idiochromatic. The small amounts of chemical impurities and/or crystal defects permissible in the definition of a mineral are insufficient to affect its color significantly. On the other hand, if a mineral is colorless in its pure state, small amounts of impurities or defects will cause the selective absorption of light and the mineral will transmit or reflect selected wavelengths that give it color. Such minerals have a tendency to be allochromatic.

Craftspeople have long used the knowledge of impurities and color to dope glass with the appropriate impurities that impart color to this otherwise colorless material. Stained glass, with its gorgeous array of colors, often in artistic designs, bears witness to the concept of how impurities in colorless materials produce allochromatic effects. For a more detailed analysis of color, using the principles of crystal field theory, the reader is referred to Wenk and Bulakh (2016) and Zoltai and Stoudt (1984).

*Play of colors*

Many minerals exhibit colors that change as the angle at which the light strikes them changes, as when the mineral is rotated in incident light. Such changes in color as the angle of incident light changes are collectively referred to as a **play of colors**. Several specific types have been recognized. **Asterism** (Figure 5.12a) is caused by inclusions oriented according to the host mineral's hexagonal crystal structure, which produces a six-sided, star-like pattern as light is scattered by the inclusions at specific incident angles. A closely related phenomenon is **chatoyancy** (Figure 5.12b). Chatoyancy is characteristic of certain fibrous minerals, that when rotated produce a band of light that moves perpendicular to the fibers, especially if the fibers are curved. Many minerals display a play of colors called **iridescence** in which the scattering of light from zones of contrasting composition or structure within the mineral produces changes in color as the mineral is

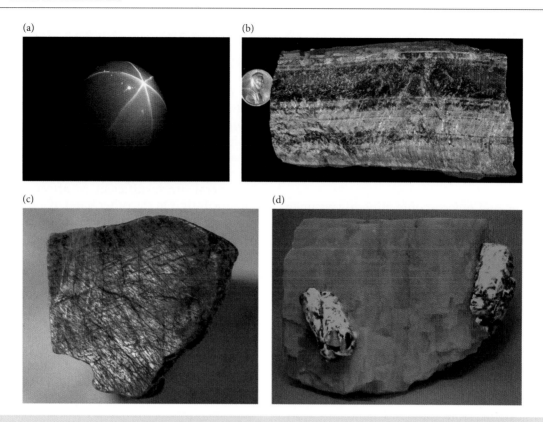

**Figure 5.12** Clockwise from upper left: asterism in the "Star of India" sapphire, Sri Lanka, Smithsonian Museum (*Source*: Daniel Torres, Jr., CCBY, https://commons.wikimedia.org/wiki/File:Star_of_India_Gem. JPG, last accessed September 10, 2020; Smithsonian Institution, with permission); chatoyancy in tiger eye; fluorescence in calcite (red) and willemite (green), Franklin, NJ (*Source*: Rob Lavinsky, Irocks.com. CC-BY-SA, wikimedia.org/wiki/File:Willemite-Calcite-225254.jpg, last accessed January 17, 2021); labradorescence in calcic plagioclase (*Source*: Courtesy of Doug Moore).

rotated. An excellent example of iridescence is the **labradorescence** (Figure 5.12c) that occurs in some specimens of plagioclase. The iridescence exhibited by fire opals is referred to as **opalescence**.

*Luminescence*

Many minerals emit light or luminesce when subjected to an external source of energy. **Luminescence** is best observed in darkness, where such minerals appear to glow in the dark. Many minerals luminesce when subjected to some type of short wavelength radiation, such as gamma rays, X-rays, or ultraviolet waves. Luminescence is produced when electrons are temporarily energized by an energy source. In their excited state, they release heat and visible light as they return to their original energy state or ground state. Specific types

of luminescence are summarized in Table 5.4. An example of mineral fluorescence caused by ultraviolet radiation is shown in Figure 5.12d.

*Magnetism*

Many minerals display some degree, often negligible, of **magnetism**, that is, a response to an external magnetic field. The three major types of magnetism displayed by minerals are indicated in Table 5.5.

**Ferromagnetic** and **ferrimagnetic** minerals have the ability to become magnetized parallel to an external magnetic field and are responsible for rock magnetization. The most magnetic minerals are ferromagnetic and include magnetite ($FeFe_2O_4$) and pyrrhotite ($Fe_{1-x}S$). These two minerals are strongly attracted by small magnets. Magnetite that has been strongly magnetized by an external

**Table 5.4**   Four major types of luminescence exhibited by minerals.

| Luminescence | Description |
| --- | --- |
| Fluorescence | Fluorescence occurs when materials are subjected to short wavelength radiation such as gamma rays, X-rays, or ultraviolet waves. The color produced during fluorescence depends on the wavelength of visible light emitted |
| Phosphorescence | Material emits visible light after it is no longer subjected to the incident radiation. This is used in glow-in-the-dark toys, which continue to emit light after removal of the light source |
| Thermoluminescence | Materials emit visible light when heated to 50–475 °C |
| Triboluminescence | Materials emit visible light in response to stress induced by rubbing or crushing the specimen |

**Table 5.5**   Different varieties of magnetism exhibited by minerals.

| Type of magnetism | Description |
| --- | --- |
| Dimagnetic minerals | Not attracted to even very powerful magnets; in fact they are slightly repelled by them |
| Paramagnetic minerals | Weakly attracted to strong magnets, become magnetized in an external magnetic field, but lose their magnetization when the external field is removed |
| Ferromagnetic–ferrimagnetic minerals | Strongly attracted to even weak magnets and can retain magnetization for long periods of time |

field becomes a magnet in its own right, as exemplified by the strongly magnetic variety of magnetite called **lodestone**. Prisms of lodestone cut parallel to its magnetization provided the first compass needles used by mariners more than a thousand years ago.

*Electrical properties*

Minerals exhibit a variety of electrical properties. **Pyroelectricity** is a phenomenon in which an increase in temperature induces an electric current that flows from one end of the crystal to the other. **Piezoelectricity** is similar but is produced by a pressure or stress applied to one end of the mineral. Both of these properties are characteristic of anisotropic minerals such as quartz and tourmaline that lack a center of symmetry. The fact that one end of the crystal is different from the other allows an electric potential to be created across the crystal. These electrical properties are widely utilized in the expanding semiconductor, heat sensor, power generation, and nuclear fusion industries, often using synthetically produced "designer" crystalline solids.

Using a combination of macroscopic properties, most minerals can be tentatively identified with reasonable accuracy. Tables summarizing the macroscopic properties used to identify significant rock-forming and/or economic minerals are available as downloadable files (appendices) from the website that supports this text. Hypothetical identifications can be tested using a variety of microscopic and chemical/structural analytical techniques (Chapter 6). Detailed descriptions of the minerals briefly described in the following sections are available in the appendices that accompany this book. These include their macroscopic properties, their optical properties, and their uses. They are accompanied by tables that permit rather straightforward methods for identifying each mineral.

## 5.4   SILICATE MINERALS

The widespread occurrence of silicate minerals makes them by far the most important group of rock-forming minerals; one with which every Earth materials scientist needs to be familiar. As discussed in Chapter 2, the unifying characteristic of silicate minerals is the presence of silica tetrahedra $(SiO^4)^{-4}$ composed of silicon $(Si^{+4})$ in tetrahedral coordination with oxygen $(O^{-2})$. Silicate minerals are subdivided according to the degree to which silica tetrahedra are linked and the pattern in which they are linked through shared oxygen ions in the crystal structure. Because oxygen and silicon are the two most abundant elements in Earth's crust and mantle, silicate minerals are abundant and widespread, comprising more than 90% by volume of the minerals in Earth's crust and upper mantle.

## 5.4.1 Nesosilicates (orthosilicates)

**Nesosilicates**, also known as **orthosilicates**, are silicate minerals characterized by isolated silica tetrahedra that are not linked through shared oxygen ions to other silica tetrahedra in the structure (Figure 5.13). The ratio of silicon ($Si^{+4}$) ions to oxygen ($O^{-2}$) ions in the tetrahedral sites of such minerals is 1 : 4. This ratio is reflected in the formulas of nesosilicate minerals, which always contain a $(SiO_4)^{-4}$ structural component that implies the existence of isolated tetrahedra. For example the *olivine group*, the most abundant mineral group in the upper mantle, has the formula $(Mg,Fe)_2SiO_4$. The formula indicates that it is a nesosilicate $(SiO_4)$ in which the silica tetrahedra are isolated and linked through oxygen to different polyhedral elements. In this case, these are sixfold, octahedral elements that contain magnesium ($Mg^{+2}$) and/or iron ($Fe^{+2}$) cations, which electrically neutralize the $-4$ silica tetrahedral component. As discussed in Chapter 3, olivine consists of two end member minerals in a complete solid solution series in which iron and magnesium substitute for one another. Forsterite is the magnesium-rich end member, while fayalite is the iron-rich end member.

A second important tectosilicate mineral group, the **garnet group**, is widespread and abundant in metamorphic rocks. It also forms in igneous rocks and is an important constituent of the mantle. Garnet group minerals have variable compositions due to multiple substitutions and have the general formula $A_3B_2(SiO_4)_3$. The A in the formula signifies cations in a structural site in which oxygen ions are in coordination with a variety of $+2$ cations that may include $Fe^{+2}$, $Ca^{+2}$, $Mg^{+2}$, and $Mn^{+2}$, among others. The B in the formula signifies a structural site for cations, an octahedral site in which oxygen atoms are in sixfold coordination with small $+3$ cations that may include $Al^{+3}$, $Fe^{+3}$, or $Cr^{+3}$, among others. The general formula for garnet could be written as $(Fe,Ca,Mg,Mn)_3(Al,Fe,Cr)_2(SiO_4)_3$. However, $A_3B_2(SiO_4)_3$, also written $X_3Y_2(SiO_4)_3$, is preferred, not only for its simplicity but because real garnets have specific compositions that depend on which cations were available in the environment in which they formed. Both ways of writing the garnet formula, however, clearly show that garnets are nesosilicates with three principal polyhedral or structural sites into which cations of the appropriate radius ratio will fit. Names given to the principal end member varieties of garnet and the occurrences of garnets dominated by specific end member components are summarized in Table 5.6.

A third important nesosilicate group is the **aluminum silicate group**. This group comprises three polymorphs that are common in metamorphic rocks, especially in pelitic assemblages produced by the metamorphism of shales and mudrocks. The three polymorphs of aluminum silicate ($AlAlOSiO_4$) are the low pressure

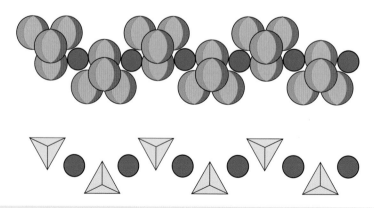

**Figure 5.13** Basic nesosilicate (orthosilicate) structure with isolated tetrahedra linked to other polyhedral elements in the structure. *Source*: Courtesy of Steve Dutch.

**Table 5.6**   Composition and occurrence of the major varieties of garnet.

| Variety | Chemical formula | Common occurrence |
|---|---|---|
| Almandine | $Fe_3Al_2(SiO_4)_3$ | Abundant in pelitic metamorphic rocks, including schists, gneisses, and granulites and occurs in some aluminum-rich pegmatites |
| Andradite | $Ca_3Fe_2(SiO_4)_3$ | In regionally metamorphosed carbonate rocks and skarns |
| Grossularite | $Ca_3Al_2(SiO_4)_3$ | In regionally metamorphosed carbonate rocks and skarns |
| Pyrope | $Mg_3Al_2(SiO_4)_3$ | In ultrabasic rocks, including mantle peridotites and kimberlites |
| Spessartine | $Mn_3Al_2(SiO_4)_3$ | Scarcer mineral, in skarns |
| Uvarovite | $Ca_3Cr_2(SiO_4)_3$ | Scarce mineral, in chromium-enriched ultrabasic rocks |

polymorph **andalusite**, the high-pressure polymorph **kyanite**, and the high-temperature polymorph **sillimanite**. The metamorphic conditions that produce each of these important minerals are detailed in Chapter 18.

Several other nesosilicates, including chloritoid, staurolite, topaz, titanite (sphene), and zircon, are significant rock-forming or economically important minerals.

### 5.4.2   Sorosilicates (disilicates)

**Sorosilicate**, also called **disilicate**, minerals possess the minimal amount of linkage possible between linked silica tetrahedra. In sorosilicates, pairs of silica tetrahedra are linked together through a single shared oxygen ($O^{-2}$) ion to form a basic unit that resembles a "bow tie" (Figure 5.14). The ratio of silicon ($Si^{+4}$) ions to oxygen ($O^{-2}$) ions in the linked pair of silica tetrahedra is 2 : 7. This is reflected in the formulas of sorosilicate minerals that always contain an $(Si_2O_7)^{-6}$ element, representing the linked pairing of tetrahedra. For example, the mineral **hemimorphite** has the formula $Zn_4(Si_2O_7)(OH)_2 \bullet H_2O$. The formula indicates that hemimorphite is a sorosilicate ($Si_2O_7$) with paired silica tetrahedra. The Zn indicates that zinc occupies a second coordination site. The (OH) indicates the presence of a hydroxyl ($OH^{-1}$) anion in addition to oxygen ($O^{-2}$) anions and the $H_2O$ signifies the presence of neutral, but dipolar, water molecules within the structure. The formula is written in a way that provides information on the chemical composition, the number of structural sites, and the common elements that occupy each structural site in the mineral.

The most important group of sorosilicate minerals is the **epidote group**. The epidote group consists of five species. Several these can display extensive substitution solid solution. They include the monoclinic minerals epidote, clinozoisite, and allanite. The rarer manganese-rich monoclinic mineral piemontite and the orthorhombic mineral zoisite are the other members of the group. Epidote is an especially important metamorphic mineral in the greenschist and epidote–amphibolite facies, as discussed in Chapter 18. Other common and/or economically important sorosilicate minerals include lawsonite and vesuvianite (idocrase), which also occur largely in metamorphic rocks.

### 5.4.3   Cyclosilicates

When multiple silica tetrahedra link together through shared oxygen ($O^{-2}$) ions, they may form rings or chains of linked silica tetrahedra (Figure 5.15). In both cases, each silica tetrahedron is linked to two additional silica tetrahedra through the sharing of two of its oxygen ions. The ratio of silicon ($Si^{+4}$) ions to oxygen ($O^{-2}$) ions is 1 : 3 in both cases. In **cyclosilicates**, each silica tetrahedra is linked to two adjacent silica tetrahedra through shared oxygen ($O^{-2}$) ions to form a ring-shaped structural element (Figure 5.15).

Three basic **ring structures** exist. In a few minerals, three silica tetrahedra link together through shared oxygen ions to form a ring element in the shape of a triangle. This ring

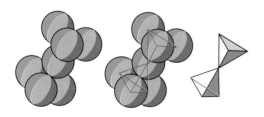

**Figure 5.14**   Basic unit of sorosilicate structure; pairs of silica tetrahedra are linked through a shared oxygen ion. *Source*: Courtesy of Steve Dutch.

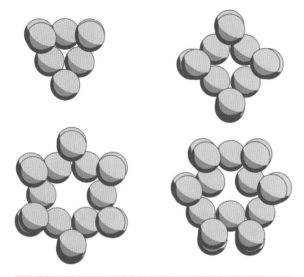

Figure 5.15 Triangular, square, and hexagonal ring structures in cyclosilicates and the hexagonal net characteristic of beryl. *Source*: Courtesy of Steve Dutch.

element has the formula ($Si_3O_9$), as in the rare cyclosilicate mineral **benitoite**, whose formula is $BaTiSi_3O_9$. This formula shows that, in addition to the triangular rings ($Si_3O_9$), benitoite has two other polyhedral elements: a large cation structural site that is occupied mostly by barium ($Ba^{+2}$) cations and a small cation structural site that is occupied mostly by titanium ($Ti^{+4}$) cations.

In other cyclosilicates, four silica tetrahedra join through shared oxygen ions to form a square-shaped ring element. This ring element has the formula $Si_4O_{12}$, as in the scarce cyclosilicate mineral **axinite**. Its chemical formula can be written as $(Ca,Fe,Mn)_3Al_2(BO_3)(Si_4O_{12})$ (OH). The formula shows that axinite possesses a square ring ($Si_4O_{12}$) structure, a structural site for divalent cations of moderate size ($Ca^{+2},Fe^{+2},Mn^{+2}$), and a third structural site for smaller cations that is commonly occupied by aluminum ($Al^{+3}$) cation. The borate $(BO_3)^{-1}$ and hydroxyl $(OH)^{-1}$ anions occupy spaces within the open rings of the structure.

In the most common group of cyclosilicates, six silica tetrahedra link through shared oxygen ions to form six-sided, hexagonal ring elements. These ring components have the formula $Si_6O_{18}$, as in beryl, whose chemical formula can be written as $Be_3Al_2Si_6O_{18}$. As in the previous examples of cyclosilicates, beryl has three principal polyhedral structural sites: (1) a tetrahedral site occupied by $Si^{+4}$ linked in the form of hexagonal rings ($Si_6O_{18}$), (2) a second tetrahedral site occupied by small beryllium ($Be^{+2}$) cations, and (3) an octahedral site occupied by slightly larger aluminum ($Al^{+3}$) cations. Tourmaline and cordierite are other common cyclosilicates with hexagonal ring structures.

### 5.4.4 Inosilicates

**Inosilicate** is the formal term for silicates in which silica tetrahedra are linked together through shared oxygen ions into one-dimensional chains of long-range extent. Because these chain structures extend from one side of the mineral crystal to the other, chain silicate structures are classified as **one-dimensional structures**. Many different types of inosilicate or chain structures exist, the most common of which are single-chain and double-chain inosilicate structures.

*Single-chain inosilicates: pyroxenes and pyroxenoids*

In **single-chain inosilicates**, each silica tetrahedron is linked to two adjacent silica tetrahedra through shared oxygen ions so that, as in the cyclosilicates, the Si : O ratio is 1 : 3. There are many different ways in which tetrahedra can be linked into single chains. In the **pyroxene group**, the most abundant group of single-chain inosilicate minerals, the tetrahedra alternate on opposite sides of the axis of the chain (Figure 5.16). One can visualize this structure as consisting of a basic unit of two doubly linked silica tetrahedra ($Si_2O_6$), on opposite sides of the axis, being repeated infinitely along the axis of the chain. This is reflected in the

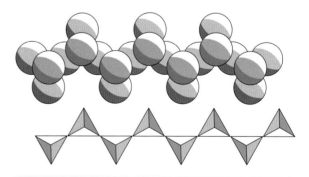

Figure 5.16 Single-chain inosilicate structure in pyroxene. *Source*: Courtesy of Steve Dutch.

general formula for the minerals of the pyroxene group which is $XY(Si_2O_6)$. The $(Si_2O_6)$ infers a single-chain inosilicate and the X and Y represent two other coordination sites. In real pyroxenes, the X represents a transitional octahedral–cubic structural site for cations. When it contains larger cations such as $Ca^{+2}$ and $Na^{+1}$, as in the clinopyroxenes augite and diopside, it is somewhat distorted with many of the properties of an eightfold site. When it contains only smaller cations, such as $Fe^{+2}$, $Mg^{+2}$, and $Mn^{+2}$, as in the orthopyroxenes hypersthene and enstatite, it resembles a normal sixfold site. The Y represents a normal octahedral structural site for smaller cations, such as $Fe^{+3}$, $Al^{+3}$, and $Ti^{+4}$, as well as $Fe^{+2}$, $Mg^{+2}$, and $Mn^{+2}$. The general formula for pyroxene group minerals can be written as $(Ca^{+2},Na^{+1},Fe^{+2},Mg^{+2},Mn^{+2})(Fe^{+3},Al^{+3},Ti^{+4},Fe^{+2},Mg^{+2},Mn^{+2})(Si_2O_6)$, but there are advantages in the simpler form $XY(Si_2O_6)$. As indicated by the assortment of elements in the X and Y coordination sites, widespread substitution occurs among elements of similar charge and atomic radii. Which elements are actually added to the site also depends on what is available in the environment in which the mineral crystallizes and also the temperature at which the mineral forms.

In **pyroxenoids**, the tetrahedra are distributed about the chain axis in a different fashion, so that the repeat distance between tetrahedra occupying similar positions is larger. For example, in wollastonite ($Ca_3Si_3O_9$) the repeat distance is every third tetrahedron, whereas in rhodonite ($Mn_5Si_5O_{15}$) it is every fifth tetrahedron, as indicated by the manner in which their formulas are written. The ratio of Si : O remains at 1 : 3 that is characteristic of single-chain inosilicates.

*Double-chain inosilicates: amphibole group*

In the **double-chain inosilicates**, two single chains are linked together through additional shared oxygen ions to form a double chain, commonly with a Si/O ratio of 4 : 11 (1 : 2.75). In the abundant **amphibole group**, the basic structural unit consists of eight linked silica tetrahedra, four on each side of the axis of a double chain. These are repeated along the chain axis to form a long-range double chain (Figure 5.17). The basic unit has the formula $Si_8O_{22}$. Many of the oxygen atoms in the double chain are not linked to other silica tetrahedra and thus have unsatisfied charges (–1) that must be satisfied by bonding to other cations in other coordination polyhedra. This is reflected in the simplified general formula for minerals in the amphibole group, which is $X_2Y_5(Si_8O_{22})(OH)_2$. Once again, the $(Si_8O_{22})$ signifies a double-chain inosilicate from the amphibole group, and the X and the Y denote the occurrence of two additional types of structural sites or coordination polyhedra. The $(OH)_2$ signifies the presence of hydroxyl ($OH^{-1}$) anion in addition to oxygen ($O^{-2}$) anions in the structure. It conveys the important information that amphibole group minerals are hydrous silicates. The X structural site commonly contains larger cations, such as $Fe^{+3}$, $Ca^{+2}$, and $Na^{+1}$ in addition to $Fe^{+2}$, $Mg^{+2}$, and $Mn^{+2}$. The Y structural sites are octahedral, sixfold sites that contain slightly smaller cations, such as $Fe^{+2}$, $Mg^{+2}$, $Mn^{+2}$, and $Al^{+3}$. As with the pyroxenes and many other silicates, formulas may be written in different ways to emphasize the structure of the mineral, its chemistry or both. Hornblende is the most commonly encountered amphibole mineral, but with so many possible substitutions it is not surprising that there are a plethora of others.

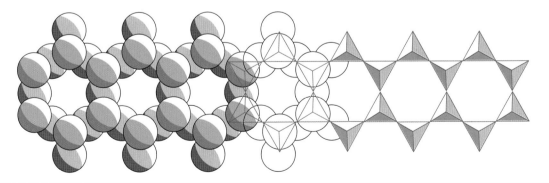

**Figure 5.17**  Double-chain silicate structure in amphiboles. *Source*: Courtesy of Steve Dutch.

Minerals with triple- and quadruple-chain elements exist but are not common. Like other inosilicates, these chain silicates link silica tetrahedra into one-dimensional structures that possess long-range extent only in the direction parallel to the chain axis (Figure 5.17).

### 5.4.5   Phyllosilicates

When multiple chains of silica tetrahedra are linked through shared oxygen ions in a direction at a large angle to the chain axis, the chains combine to form a sheet of linked silica tetrahedra with long-range extent in two directions. Such two-dimensional sheet structures are characteristic of **phyllosilicate** minerals (Figure 5.18). Minerals in this group, often called sheet silicates, commonly have a Si/O ratio of 2 : 5 or 4 : 10. Their chemical formulas are often complicated by the fact that aluminum ($Al^{+3}$) cations substitute in limited, and often variable, amounts for silicon ($Si^{+4}$) cation in the "silica" tetrahedra to form aluminum tetrahedra.

Because one out of every four oxygen ions is not linked to another silica tetrahedron, one out of every four oxygen ions must bond to other cations in order to be electrically neutralized. Typically these cations occur in octahedral sites that alternate in some way with the tetrahedral layers to which they are linked. The two most common octahedral sites are those that contain magnesium ($Mg^{+2}$) bonded to oxygen ($O^{-2}$) and hydroxyl ($OH^{-1}$) ions and others that contain aluminum ($Al^{+3}$) bonded to oxygen and hydroxyl ions. The magnesium octahedral sites are called **brucite sites** (**b**) after the mineral brucite [$Mg(OH)_2$], and the aluminum sites are called **gibbsite sites** (**g**) after the mineral gibbsite [$Al(OH)_3$]. Iron frequently substitutes partially for the magnesium in brucite octahedral sites.

Common rock-forming phyllosilicate mineral groups include (1) serpentine, (2) talc, (3) chlorite, (4) mica, and (5) clay. Other moderately common phyllosilicate minerals include apophyllite, prehnite, and stilpnomelane.

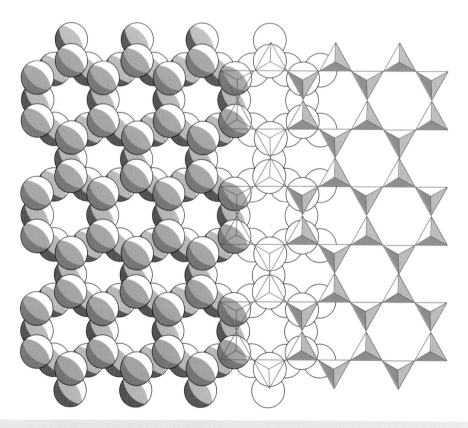

**Figure 5.18**   Two-dimensional sheet structure typical of phyllosilicate minerals. *Source*: Courtesy of Steve Dutch.

## Serpentine, talc, and chlorite group minerals

Serpentine, talc, and chlorite provide an excellent basis for discussing the major phyllosilicate structures. **Serpentine group** minerals (Figure 5.19a) are composed of alternating tetrahedral (t) layers $(Si_2O_5)^{-2}$ and octahedral (o) layers that give rise to a basic two-layer (t-o) structure. In serpentines, the octahedral layers are typically brucite layers (b), and the chemical formula $(Mg,Fe)_3Si_2O_5(OH)_4$ reflects the two-layer (t-o or t-b) structure typical of serpentine minerals. On the other hand, **talc** (Figure 5.19b) is composed of one octahedral (o) or brucite layer (b) sandwiched between two tetrahedral (t) layers, giving rise to a basic three-layer (t-o-t or t-b-t) structure that is reflected in its mineral formula $Mg_3(Si_4O_{10})(OH)_2$. **Pyrophyllite** $[(Al_2(Si_4O_{10})(OH)_2]$ possesses a similar three-layer structure, but with a gibbsite layer (g), rather than a brucite layer, sandwiched between the two tetrahedral layers (t-g-t). **Chlorite group** minerals possess an extra brucite layer so that the basic structure is four layers (t-b-t-b), as reflected in its formula $(Mg,Fe)_3(OH)_6 \bullet (Mg,Fe,Al)_3(Si,Al)_4O_{10}(OH)_2$, where the portion before the stop represents the extra brucite layer. A more detailed discussion of phyllosilicate structures, especially clay minerals, occurs in Chapter 11, where they are discussed and illustrated in conjunction with weathering and soils. Serpentine-, chlorite-, and talc-group minerals, common in metamorphic rocks, are discussed further in Chapters 15–18. These minerals are all relatively soft and possess one set of excellent cleavage because the fundamental, sheet-like structural units are held together by planes of weak bonds.

## Mica group minerals

**Mica group** minerals are widespread and nearly ubiquitous in their occurrence. These minerals are significant rock-forming minerals in both igneous and metamorphic rocks and are not uncommon in sedimentary rocks. Micas are three-layer (t-o-t) phyllosilicates with two tetrahedral layers in which one out of every four silica tetrahedra has an aluminum ($Al^{+3}$) cation substituting for the silicon ($Si^{+4}$) cation as reflected in the ($AlSi_3O_{10}$) component of their formulas. **Biotite** $[K(Mg,Fe)_3(AlSi_3O_{10})(OH)_2]$ and phlogopite $[K(Mg)_3(AlSi_3O_{10})(OH)_2]$ possess brucite octahedra sandwiched between the two tetrahedral layers (t-b-t). The potassium ions are weakly held in interlayer spaces between the basic structural units. **Muscovite** $[(KAl_3AlSi_3O_{10}(OH)_2)]$ and **lepidolite** $[(K(Li,Al)_3AlSi_3O_{10}(OH,F)_2)]$ possess gibbsite-related octahedra sandwiched between two tetrahedral layers (t-g-t). The potassium ions occur in interlayer spaces between the basic structural units (Chapter 11). The potassium ($K^{+1}$) ions balance the negative charge that results from the substitution of aluminum ($Al^{+3}$) ions for silicon ($Si^{+4}$) ions in the tetrahedral sites. The micas are characterized by the elasticity of their sheets, one set of perfect cleavage parallel to the sheet structure, and relatively low hardness.

## Clay group minerals

**Clay group** minerals commonly occur as microscopic crystals (<4 μm) in soils, sedimentary rocks, and low-temperature hydrothermal and metamorphic rocks, where they form by the low-temperature alteration of

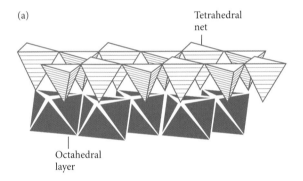

(a)

Tetrahedral net

Octahedral layer

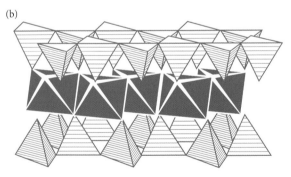

(b)

**Figure 5.19** (a) Two-layer (t-o or t-b) structure of serpentine. (b) Three-layer (t-o-t or t-b-t) structure of talc. Both are viewed perpendicular to the sheets, with repeated stacking of the basic structural units to produce a three-dimensional crystal lattice. *Source*: Wenk and Bulakh (2004). © 2004 Cambridge University Press.

aluminum silicate minerals, such as feldspars, micas, and amphiboles. The clay minerals can be subdivided into two structural groups. The two-layer (t-o or t-g) **kandite group** *clays* include minerals, such as **kaolinite** $[(Al_2Si_2O_5)]$, that possess serpentine-type structures. The three-layer (t-o-t) clays possess talc-type structures and belong to the **illite group** (t-g-t), with the approximate formula $KAl_3AlSi_3O_{10}(OH)_2$, and the **smectite group** (t-b-t), which includes minerals such as **montmorillonite** with the formula $\sim$(Ca,Na) $(Mg,Fe,Al)_3AlSi_3O_{10}(OH)_2 \bullet nH_2O$. Smectites are called expansive clays due to their ability to absorb and release large amounts of water held in interlayer sites and thus to change volume significantly. As discussed in Chapter 11, these expansive clays are hazardous for use in construction foundation due to their shrink and swell properties. Clay minerals are widely used in industry as absorbents and fillers (Chapter 19). Their structures, compositions, and properties are discussed in detail and illustrated in Chapter 11.

### 5.4.6 Tectosilicates

**Tectosilicate** minerals are composed principally of silica tetrahedra linked through all their oxygen anions to adjacent silica tetrahedra to form three-dimensional framework structures (Figure 5.20). These silicates, often called **framework silicates**, have a Si/O ratio of 1 : 2, unless aluminum substitutes for some of the silicon ions in the tetrahedral sites, in which case the (Si±Al)/O ratio is 1 : 2. This is because all four oxygen $(O^{-2})$ anions are shared with adjoining Si±Al tetrahedra. This implies that each silica (or aluminum) tetrahedron utilizes one-half of each of the charge of four oxygen anions, making the Si±Al/O ratio $1 : 4 \times 0.5 = 1 : 2$. Tectosilicates constitute an extremely important group of rock-forming minerals, comprising nearly 75% of the minerals in Earth's crust, and occur widely in igneous, metamorphic, and sedimentary rocks.

The two most abundant groups of tectosilicate minerals are the pure $SiO_2$ **silica group** and the aluminum silicate **feldspar group**; the latter is the most abundant group of minerals in Earth's crust. Other important tectosilicate mineral groups include (1) the silica-poor, aluminum-rich **feldspathoid group**, (2) the aluminum-rich, hydrated **zeolite group**, and (3) the **scapolite group**. All but the scapolite group are discussed in the sections that follow. Details concerning the macroscopic and optical properties of these minerals and the criteria for their recognition are available in the online appendices that support this book.

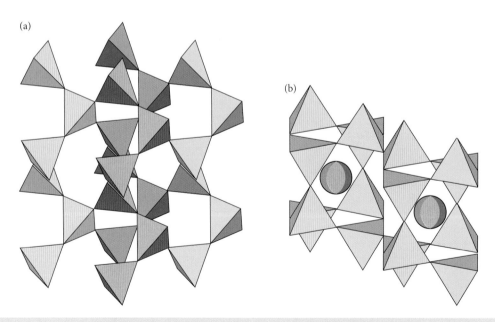

(a)

(b)

**Figure 5.20** Three-dimensional framework structures typical of tectosilicates: (a) quartz and (b) potassium feldspar, with the blue-gray atoms representing potassium. *Source*: Courtesy of Steve Dutch.

## Silica group

A classic and abundant example of tectosilicates is the **polymorphs of silica**, all of which have the chemical formula $SiO_2$. This implies that in their pure states they consist solely of silica tetrahedra linked together through shared oxygen anions to form a three-dimensional framework structure. The silica polymorphs (Chapter 3) include the high-pressure polymorphs **coesite** and **stishovite**, the high-temperature/low pressure polymorphs **tridymite** and **cristobalite**, and the quartz polymorphs **alpha quartz** ($\alpha$-quartz) and **beta quartz** ($\beta$-**quartz**).

The phase stability diagram for the silica group is discussed in Chapter 3. Coesite is the stable polymorph of silica at pressures of more than 20 kbar (2 GPa) and less than 75 kbar (7.5 GPa) which corresponds to burial depths of ~60–250 km. Stishovite is a very high-pressure polymorph of silica, with silicon and oxygen in sixfold coordination, which is stable only at pressures >75 kbar (7.5 GPa) or at depths that exceed ~250 km. These high-pressure polymorphs, coesite and stishovite, also occur in association with meteorite impact and thermonuclear bomb sites, and stishovite is likely a significant constituent of the deep mantle. Tridymite and cristobalite are the stable polymorphs of silica under high-temperature/low pressure conditions. Tridymite and cristobalite are relatively common in silica-rich volcanic rocks produced under high-temperature/low pressure conditions. The phase stability diagram (see Figure 3.6) illustrates that quartz is the stable polymorph of silica over a broad range of temperature–pressure conditions common in Earth's crust. This wide stability range and the abundance of silicon and oxygen help to explain why quartz is such an abundant rock-forming constituent of common igneous, sedimentary, and metamorphic rocks. The diagram also illustrates that $\alpha$-quartz (low quartz) is generally the stable form of quartz at normal near-surface temperatures and pressures. Quartz is an important economic mineral, widely used in the manufacture of glass and optical fibers and as a source of silicon for the manufacture of microprocessors.

**Opal** ($SiO_2 \bullet nH_2O$) is an amorphous, hydrated form of silica with a hardness of 5.5–6, conchoidal fracture, and a waxy luster.

Electron microscopy has revealed that it is composed of minute (1500–8000 Å) spheres with fairly regular packing (Wenk and Bulach 2016). This arrangement of spheres acts as a diffraction grating for light, which produces the opalescence characteristic of gem opals such as fire opals. Opal is also an important constituent of microscopic shells of organisms such as diatoms and radiolaria that accumulate to form siliceous sediments on ocean floors, as discussed in Chapter 14, in the section on siliceous sedimentary rocks. The major properties and occurrences of the silica polymorphs and opal are summarized in Table 5.7.

Macroscopic quartz (Figure 5.20a) is easily recognized by its combination of hardness (H = 7), vitreous–greasy luster, lack of cleavage, conchoidal fracture, and/or hexagonal prismatic crystals. Quartz is a classic example of an allochromatic mineral. Because quartz is colorless and transparent in its pure state, small amounts of impurities or defects may cause its color to change significantly. Because macrocrystalline quartz is so abundant, each of the major color varieties has its own name. The major varieties of macrocrystalline quartz, the cause of their colors and their common occurrences, are summarized in Table 5.8.

Several varieties of **microcrystalline to cryptocrystalline quartz** are quite common, especially as sedimentary rocks. Members of the **chert group** are composed of an aggregate of microcrystalline quartz crystals that are fairly equant, if not all the same size. As an aggregate of mineral crystals, chert qualifies as a rock, and the term is frequently used in that way. Members of the chert group include chert (white–medium gray), flint (dark gray–black), jasper (red–yellow), and prase (green). Chert group members are hard (H = 7), lack visible crystals, and are smooth, with dull lusters and excellent conchoidal fracture. Members of the **chalcedony group** are characterized by aggregates or sheaves of radiating microscopic silica crystals that are often water bearing. The properties of chalcedony are similar to those of chert, but chalcedony is often slightly more transparent than chert and possesses a somewhat waxy or resinous luster, rather than a dull one. Major varieties of chalcedony include chalcedony (gray), carnelian (red), sard (yellow–brown), and chrysoprase (green). Banded varieties are called agate

**Table 5.7** Colors and common occurrences of the major varieties of macroscopic quartz.

| Variety | Colors | Cause of color | Common occurrences | Photo |
|---|---|---|---|---|
| Amethyst | Purple–violet | $Fe^{+3}$ impurities | In open fractures and cavitites | |
| Aventurine | Green | Chrome mica inclusions | | |
| Citrine | Yellow–yellow–brown | $Fe^{+3}$ impurities | In open fractures and cavitites | |
| Milky quartz | White | Water bubble inclusions | Widespread in veins | |
| Rock crystal | Clear, colorless | "Pure" quartz | In open cavities and veins | |
| Rose quartz | Pink | Inclusions of dumortierite nannocrystals | In pegmatites and veins | |
| Smoky quartz | Brown–black | $Al^{+3}$ substitution for $Si^{+4}$ | Widespread in silicic igneous rocks and pegmatites | |

**Table 5.8**   Colors and common occurrences of the major varieties of macroscopic quartz.

| Variety | Colors | Cause of color | Common occurrences |
|---|---|---|---|
| Amethyst | Purple–violet | $Fe^{+3}$ impurities | In open fractures and cavities |
| Aventurine | Green | Chrome mica inclusions | |
| Citrine | Yellow to yellow–brown | $Fe^{+3}$ impurities | In open fractures and cavities |
| Milky quartz | White | Water bubble inclusions | Widespread in veins |
| Rock crystal | Clear, colorless | | In open cavities and veins |
| Rose quartz | Pink | Inclusions of dumortierite nannocrystals | In pegmatites and veins |
| Smoky quartz | Brown–black | $Al^{+3}$ substitution for $Si^{+4}$ | Widespread in silicic igneous rocks and pegmatites |

(concentric bands) and onyx (nonconcentric bands). Chert and chalcedony can be difficult to distinguish, not the least because they are often intergrown in a single rock, yielding specimens with intermediate characteristics. Opal, the amorphous form of silica, is often grouped with chert and chalcedony because of its similar properties, including smooth surface texture and conchoidal fracture. With its higher water content and unique internal structure, opal tends to be a little softer than chert and chalcedony and to have a decidedly waxy luster. Like chert and chalcedony, it also occurs in a wide variety of colors.

*Feldspar group*

Because aluminum ($Al^{+3}$) cations can substitute to some degree for silicon ($Si^{+4}$) cations in the tetrahedral site, many tectosilicate minerals have somewhat more complex formulas than the members of the silica group. This is largely because for each aluminum ($Al^{+3}$) cation that replaces a silicon ($Si^{+4}$) cation in the tetrahedral

site, a charge deficiency of –1 is created in the structure. This charge deficiency must be balanced by the addition of sufficient cations to create a neutrally charged mineral. This is well illustrated in the most abundant group of minerals in Earth's crust, the **feldspar group**.

The **potassium feldspar subgroup** includes several polymorphs including **orthoclase**, **microcline**, and **sanidine**, whose simplified formula can be written as $KAlSi_3O_8$. The formula conveys the information that potassium feldspars are tectosilicates in which one out of every four silica tetrahedra contains an aluminum ($Al^{+3}$) cation instead of a silicon ($Si^{+4}$) cation, so that the $(Al+Si)/O$ ratio is $1 : 2$ or more precisely $4 : 8$. The substitution of the $Al^{+3}$ for the $Si^{+4}$ creates a charge deficiency of –1 which is balanced by the incorporation of potassium ($K^{+1}$) cation (blue and gray atoms in Figure 5.20b) into the crystal structure to produce an electrically neutral mineral.

The polymorphs of potassium feldspar share several macroscopic properties (Table 5.9). They tend to be hard ($H = 6.0–6.5$), have

**Table 5.9**   Macroscopic properties and occurrences of the major potassium feldspars.

| Mineral | Crystallography | Other distinguishing properties[a] | Common occurrences |
|---|---|---|---|
| Microcline | Triclinic; stubby crystals of low symmetry | The variety amazonite is bright green; but also white–gray and pink–red | Felsic plutonic igneous rocks and pegmatites; metamorphic schists and gneisses; sedimentary arkoses |
| Orthoclase | Monoclinic; stubby prismatic with evident symmetry | White–gray–pink–red–green, but rarely bright green | Felsic plutonic igneous rocks; metamorphic schists and gneisses; sedimentary arkoses |
| Sanidine | Monoclinic; tabular with evident symmetry | More transparent than others; typically colorless to light gray | High temperature potassium feldspar, usually in felsic volcanic rocks |

[a] Feldspars are best distinguished using the analytical techniques discussed in Chapter 6.

similar specific gravity (2.5–2.6), possess colors that range from white to gray to green to pink to red, and two sets of excellent cleavage at right angles. Potash feldspars are commonly perthitic as the result of exsolved albite (Chapter 3). Unfortunately, potash feldspars possess overlapping macroscopic properties and are easier to distinguish in thin section using optical techniques (Chapter 6) than in hand specimens.

The other group of feldspar minerals is the **plagioclase group**, which has a general formula that can be written as $(Ca,Na)(AlSi)AlSi_2O_8$. As discussed in Chapter 3, this formula reflects the complete coupled ionic substitution series between the plagioclase end members **albite** $(NaAlSi_3O_8)$ and **anorthite** $(CaAl_2Si_2O_8)$ that characterizes the plagioclase group. Recall also that the plagioclase phase diagram is discussed in Chapter 3 as an example of a two-component system with complete solid solution between end member components. Plagioclase compositions are typically expressed in terms of the proportion of the anorthite end member (%An). The albite (%Ab) component is simply (100% – %An). Traditionally, the plagioclase solid solution series is divided into six compositional ranges that are given specific names (Table 5.10), but it is more precise to simply express the plagioclase composition in terms of percent anorthite end member. The anorthite content (Table 5.10) and varieties of plagioclase are difficult to distinguish macroscopically but can be distinguished using analytical techniques that include optical mineralogy, X-ray crystallography, and advanced chemical analysis. Calcic

plagioclases have more than 50% anorthite (An) and include labradorite (50–70% An), bytownite (70–90% An), and anorthite (>90% An). Sodic plagioclases contain less than 50% anorthite and include andesine (30–50% An), oligoclase (10–30% An) and albite (<10% An). All plagioclase varieties possess similar hardness (H = 6), specific gravity (2.6–2.8, increasing with %An), cleavage (two sets, one perfect, one good, near 90°), color (white–green–gray), and luster (vitreous–pearly). All crystallize in the triclinic system. The tendency of plagioclase to display parallel striations that result from twinning on one set of cleavage surfaces is particularly useful in identifying plagioclase and distinguishing it from potash feldspars.

*Feldspathoid group*

Like feldspars, **feldspathoids** are aluminum-bearing tectosilicates. However, the feldspathoids possess lower silica contents, higher aluminum contents, and higher contents of the alkali cations such as potassium, sodium, and calcium, required to make them electrically neutral. Feldspathoids are relatively scarce minerals that occur primarily in silica-poor (silica-undersaturated) and alkali-rich (peralkaline) igneous rocks. They are excellent indicator minerals for such silica-undersaturated rocks, which occupy fully half of the International Union of Geological Sciences (IUGS) standard classification chart for igneous rocks. The classification, occurrence, and origin of alkaline igneous rocks are discussed in Chapters 7–10. The macroscopic properties and occurrences of the major feldspathoid minerals are summarized in Table 5.11.

**Table 5.10**  Major varieties of plagioclase by anorthite (An) and albite (Ab) content.

| Plagioclase variety | Chemical formula | An content | Ab content |
|---|---|---|---|
| Albite | $(Ca_{0-0.1}Na_{0.9-1.0})(Al_{0-0.1}Si_{0.9-1.0})AlSi_2O_8$ | $An_{0-10}$ | $Ab_{90-100}$ |
| Oligoclase | $(Ca_{0.1-0.3}Na_{0.7-0.9})(Al_{0.1-0.3}Si_{0.7-0.9})AlSi_2O_8$ | $An_{10-30}$ | $Ab_{70-90}$ |
| Andesine | $(Ca_{0.3-0.5}Na_{0.5-0.7})(Al_{0.3-0.5}Si_{0.5-0.7})AlSi_2O_8$ | $An_{30-50}$ | $Ab_{50-70}$ |
| Labradorite | $(Ca_{0.5-0.7}Na_{0.3-0.5})(Al_{0.5-0.7}Si_{0.3-0.5})AlSi_2O_8$ | $An_{50-70}$ | $Ab_{30-50}$ |
| Bytownite | $(Ca_{0.7-0.9}Na_{0.1-0.3})(Al_{0.7-0.9}Si_{0.1-0.3})AlSi_2O_8$ | $An_{70-90}$ | $Ab_{10-30}$ |
| Anorthite | $(Ca_{0.9-1.0}Na_{0.0-0.1})(Al_{0.9-1.0}Si_{0-0.1})AlSi_2O_8$ | $An_{90-100}$ | $Ab_{0-10}$ |

*Zeolite group*

**Zeolite group** minerals (Table 5.12) are hydrous silicates that form as secondary minerals at temperatures of 100–250 °C as discussed in Chapter 18. Zeolites commonly occur as amygdules or cavity fillings in altered basalts and related rocks or as veins and alteration products in volcanic pyroclastic and glassy rocks. Upon heating, zeolites expel their water while their crystal structures remain intact. This allows such dehydrated zeolites to act as "sponges" and/or "molecular sieves" that can selectively absorb dissolved constituents such as hydrocarbons, heavy metals, or other contaminants from water. Zeolite minerals and their synthetic counterparts are extensively utilized for the purposes of sewage treatment, water softening (by removal of $Ca^{+2}$), and water purification. They are also used in the catalysis of high octane lead-free gasoline, the removal of radioactive

**Table 5.11** Macroscopic properties and common occurrences of significant feldspathoid minerals.

| Mineral/formula | Crystallography | Other distinguishing properties | Common occurrences |
|---|---|---|---|
| Cancrinite $Na_6Ca_2(AlSiO_4)_6(CO_3)_2 \bullet nH_2O$ | Hexagonal; prismatic; rare | Yellow–rose–blue; greasy–waxy luster; granular–massive | By alteration of nepheline in silica-undersaturated, feldpathoidal igneous rocks |
| Lazurite $Na_3Ca(AlSiO_4)_3SO_4S_2$ | Isometric; equant; rare | Deep azure blue–greenish blue; massive–granular | Scarce mineral in metamorphosed limestones/skarns |
| Leucite $KAlSi_2O_6$ | Hexagonal; stubby prismatic; pseudo-isometric | White–gray; vitreous to dull luster | In silica-undersaturated, potassium-rich volcanic rocks |
| Nepheline $Na_3K(AlSiO_4)_4$ | Hexagonal; prismatic; rare | White–gray; greasy luster; massive–granular habit | In silica-undersaturated, feldpathoidal igneous rocks |
| Scapolite* $(Na,Ca)_4(Al_{1-2}Si_{2-3}O_8)_3(CO_3,SO_4,Cl)$ | Tetragonal; prismatic; square sections | Prismatic; square sections | In medium–high-grade metamorphic carbonates/skarns and pelitic schists, gneisses |
| Sodalite $Na_4(AlSiO_4)_3Cl$ | Isometric; equant; rare | Blue; also gray–green; massive–granular | In silica-undersaturated, feldpathoidal igneous rocks |

* Scapolite is not a feldspathoid but is a closely related mineral.

**Table 5.12** Macroscopic properties of common zeolite minerals.

| Mineral/formula | Crystallography | Other distinguishing properties |
|---|---|---|
| Analcime $NaAlSi_2O_6 \bullet 6H_2O$ | Isometric; equant; trapezohedra | Colorless to white; no visible cleavage |
| Chabazite $(Ca,Na,K)_4Al_4Si_8O_{24} \bullet 12H_2O$ | Trigonal; equant; pseudocubic rhombohedra | White–yellow; sometimes pink; poor cleavage |
| Clinoptilolite $(Ca,Na,K)_6Al_6Si_{30}O_{72} \bullet 20H_2O$ | Monoclinic; platy–capillary; scaly–fibrous | White; one cleavage, usually not visible |
| Heulandite $(Ca,Na,K)_9Al_9Si_{27}O_{72} \bullet 24H_2O$ | Monoclinic; prismatic–platy | Colorless–white–pale yellow; one perfect cleavage |
| Laumontite $Ca_4Al_8Si_{16}O_{48} \bullet 18H_2O$ | Monoclinic; prismatic | White; two cleavages not at 90° |
| Natrolite $Na_2Al_2Si_3O_{10} \bullet 2H_2O$ | Orthorhombic; prismatic–acicular; radiated–fibrous | Colorless–white–pale yellow; two perfect cleavages at 90° |
| Stilbite $(Ca,Na,K)_9Al_9Si_{27}O_{72} \bullet 28H_2O$ | Monoclinic; tabular–platy; close radiated, sheaf-like groups | White–yellow–brown; one perfect cleavage set |

isotopes from nuclear waste water, and the scrubbing of pollutants from industrial stacks. Zeolite group minerals share several physical properties, including diaphaneity (translucent–transparent), luster (vitreous–pearly, silky in fibrous varieties), low specific gravity (2.1–2.3), and streak (white). Their hardness is intermediate, ranging between 3.5 and 5.5.

To summarize, silicate minerals, the predominant minerals in Earth's crust and upper mantle (depth < 400 km), consist of fundamental units called silica tetrahedra linked to various degrees to produce the six major groups of silicate structures whose properties are summarized in Figure 2.21. Because of the abundance of oxygen and silicon in Earth's crust and mantle, these minerals are the major group of rock-forming minerals with which Earth scientists must deal. Many are of economic importance as well. Knowledge of silicate minerals is essential to working in most areas of the Earth sciences. More detailed descriptions and information about the occurrences, macroscopic and microscopic properties, and uses of the silicate minerals introduced in this section are available in the online appendices to this book.

## 5.5   NONSILICATE MINERALS

With the exception of native elements, nonsilicate minerals are classified by their major anion or anionic group. Although nonsilicate minerals are far less abundant in Earth's crust and mantle than are silicate minerals, many are of great economic and social value. This makes knowledge of nonsilicate minerals extremely important to practitioners of Earth science. Details concerning the macroscopic and optical properties of these minerals and the criteria for their recognition may be found in the online appendices that support this text.

### 5.5.1   Native elements

Minerals composed of a single essential element are called **native elements**. This group also includes several minerals that are composed of two or more closely related elements that possess very similar chemical characteristics. Fewer than 20 elements occur in their native state. Most native element minerals are quite rare; many have only been discovered with the advent of sophisticated instrumentation for examining Earth materials. Together they comprise less than 0.00002% of Earth's crust by weight (Wenk and Bulach 2016). Native elements are divided into three subgroups based on the chemical behavior of the elements: (1) metals, (2) semi-metals, and (3) nonmetals.

### Native metals

The **native metals** are composed of metallic elements. All native metals crystallize in the isometric system, and almost all are characterized by cubic closest packing with face-centered cubic crystal structures (Chapter 4). As a result, substitution solid solution of metals of similar radii (e.g. gold and silver, iron and nickel) is common. Native metals are subdivided into three groups: the gold group, the platinum group, and the iron group. These groups are distinguished on the basis of the chemical nature of the metallic elements and the properties they confer on their members.

The **gold group metals** include gold (Au), silver (Ag), and copper (Cu). Excellent metallic bonding (Chapter 2) causes these minerals to be soft, malleable, and ductile and to be excellent thermal and electrical conductors. The metallic bonding of gold group metals also causes them to be opaque and to possess metallic luster, hackly fracture, and low melting points. High atomic mass numbers and cubic closest packing produce minerals with high specific gravity. The members of the gold group, together with such elements as mercury, lead, and palladium, are so similar in crystallographic and chemical properties that they readily substitute for one another.

The **platinum group metals (PGM)** include several rare, but valuable, isostructural minerals that contain platinum (Pt), palladium (Pd), iridium (Ir), and osmium (Os). Solid solutions of these metals are common. Platinum group metals possess bonds that are less metallic than those of the gold group and so are generally harder and have higher melting points than gold group metals. Platinum group metals are excellent thermal and electrical conductors, are opaque, and possess metallic luster and high specific gravity. They are important as catalysts for many chemical processes.

The **iron group metals** include minerals composed of iron and/or nickel. Native iron is rare, but two iron–nickel minerals, iron-rich kamacite and nickel-rich taenite, are very important constituents of iron-rich meteorites.

*Semi-metals*

The **native semi-metals** are composed of semi-metallic elements such as arsenic (As), antimony (Sb), and bismuth (Bi). Because they possess metallic–covalent transitional bonds, they are brittle and much poorer thermal and electrical conductors than the native metals. This more directional bond type also results in lower symmetry, so that most semi-metals crystallize in the hexagonal system.

*Nonmetals*

The **native nonmetals** are composed of non-metallic elements, chiefly sulfur and carbon. Native sulfur forms by sublimation from gases at volcanic vents (Chapter 9) and by bacterial reduction of sulfate minerals occurring for example in the caprock of salt domes (Chapter 14). Some atoms in sulfur are bound together by covalent bonds, while others are bound by van der Waals forces. This accounts for sulfur's translucency, brittle nature, and low hardness.

In contrast, one of the two polymorphs of native carbon is diamond which is the hardest mineral on Earth due to short, strong covalent bonds that bind atoms tightly in its cubic closest packing structure. On the other hand, the other common polymorph of carbon, graphite, is very soft because some of its atoms are bound together by van der Waals forces. This property allows graphite to be used for soft pencils and as an important lubricant. The loosely held electrons involved in these bonds also interact with light, which accounts for graphite's opacity and submetallic luster. As discussed in Chapter 3, diamond is a high-pressure polymorph formed at depth within the mantle, whereas graphite is a low pressure polymorph formed under near-surface conditions. The formation and occurrence of diamonds in kimberlite pipes are discussed in Chapter 10.

Many native elements are of significant economic value. They are the principle sources for such elements as gold, platinum, iridium, and osmium and for the carbon minerals graphite and diamond.

### 5.5.2 Halides

**Halide** minerals are characterized by large, highly electronegative, monovalent anions such as fluorine ($F^{-1}$), chlorine ($Cl^{-1}$), bromine ($Br^{-1}$), and iodine ($I^{-1}$). These elements, called the halogens, occur in row 17 (class VIIA) of the periodic table (Chapter 2). In halide minerals, they are ionically bonded with electropositive, metallic cations such as sodium ($Na^{+1}$), potassium ($K^{+1}$), and calcium ($Ca^{+2}$). The halides ionic bonding causes them to be brittle, translucent, and quite soluble. They tend to possess low to moderate hardness and moderate to high melting temperatures, and to be poor conductors of heat and electricity. More than 80 halide minerals exist. However, only three halides: (1) halite (NaCl), (2) fluorite ($CaF_2$) and (3) sylvite (KCl) are common. A fourth halide mineral, cryolite ($Na_3AlF_6$), was formerly essential to the refining of aluminum ores such as bauxite. Rarer halides typical of evaporite deposits are discussed further in Chapter 14, in the section on biochemical sedimentary rocks.

### 5.5.3 Sulfides

**Sulfide** minerals are composed of metallic and semi-metallic elements bonded with sulfide ($S^{-2}$) anions. A common example is pyrite ($FeS_2$). Closely related are the arsenides ($As^{-2}$), selenides ($Se^{-2}$), and tellurides ($Te^{-2}$), in which another element takes the place of sulfur. Nearly, 500 sulfide and related minerals have been reported (Wenk and Bulach 2016); only the common and/or economically important examples are discussed here. These minerals exhibit a variety of crystal structures that depend on ionic/atomic radii and on bond types which range from ionic–metallic to covalent–metallic. Most sulfides crystallize in the cubic, tetragonal, or hexagonal systems, reflecting the high degree of symmetry of their crystal lattices.

The variety of bonding mechanisms in sulfides and related minerals leads to a great variety of characteristics, making generalizations difficult. Many sulfides with a significant component of metallic bonding are opaque with metallic luster, distinctive colors, and characteristic streaks. Nonopaque examples tend to possess extremely high refractive indices and transmit light only on thin edges. Most sulfides are relatively soft and are good electrical conductors, which reflects the metallic component of their bonds.

The sulfides and related minerals are economically significant as the major source of many metallic elements important to civilization, including copper (chalcopyrite, $CuFeS_2$; bornite, $CuFe_5S_4$; chalcocite, $Cu_2S$), lead (galena, $PbS$), zinc (sphalerite, $ZnS$), silver (argentite, $AgS$), nickel (nickeline, $NiAsS$), cobalt (cobaltite, $CoAsS$), molybdenum (molybdenite, $MoS_2$), and mercury (cinnabar, $HgS$). The description and origins of many of these economically important mineral deposits are detailed in Chapter 19.

### 5.5.4 Oxides

**Oxide** minerals contain metals or semi-metals ionically bonded with oxygen anions ($O^{-2}$) in a diverse range of fairly symmetrical, closely packed structures. As a result, most oxide minerals are relatively hard and dense, possess relatively high melting temperatures, and crystallize in the isometric, tetragonal, or hexagonal systems. Oxide minerals are subdivided into **simple oxides** that contain a single metal (X) bonded with oxygen and **complex oxides** in which more than one metal (X, Y) is bonded with oxygen. Simple oxides can be further subdivided into: (1) the **cuprite group** ($X_2O$), (2) the **periclase group** (XO), (3) the **rutile group** ($XO_2$), and (4) the **corundum group** ($X_2O_3$), where X represents the single metal ion in the mineral. Complex oxides include the **ilmenite group** ($XYO_2$), **perovskite group** ($XYO_3$), and **spinel group** ($XY_2O_4$).

Oxides are widespread and abundant and are often concentrated in economically valuable ore deposits. Notable examples include deposits of the iron oxide minerals hematite ($Fe_2O_3$) and magnetite ($FeFe_2O_4$), which are mined from iron-rich sedimentary and metasedimentary deposits, detailed in Chapter 14, and from magmatic and hydrothermal deposits, discussed in Chapter 19. Other oxide minerals yield economic deposits of manganese (pyrolusite, $MnO_2$), copper (cuprite, $Cu_2O$), titanium (ilmenite, $FeTiO_2$), tin (cassiterite, $SnO_2$), chromium (chromite, $FeCr_2O_4$), and

uranium (uraninite, $UO_2$). Spinels are a common component of the upper mantle (Chapter 1).

### 5.5.5 Hydroxides and oxyhydroxides

The defining components of **hydroxide** minerals are metallic elements in combination with hydroxyl ($OH^{-1}$) ions. Oxygen is an additional essential component in a subgroup of hydroxide minerals called **oxyhydroxides**. Hydroxide minerals tend to be softer than oxides and to possess a somewhat lower specific gravity. Hydroxide minerals are commonly formed by the weathering and alteration of other minerals under near-surface conditions.

The most economically important hydroxide/oxyhydroxide minerals are those that belong to the **bauxite group**, which is the major source of aluminum ore. They include boehmite (AlOOH), diaspore (AlOOH), and gibbsite [$Al(OH)_3$]. Bauxite group minerals form during intense weathering of aluminum-bearing rocks in tropical environments (Chapter 11). Another significant oxyhydroxide mineral is the iron-bearing goethite [FeO(OH)], which is an important iron ore mineral in lateritic soils (Chapter 11) and in bog iron deposits (Chapter 14). **Limonite** is a term used for finely granular, sometimes amorphous, mixtures of iron oxyhydroxide and hydrated iron oxyhydroxide minerals with close affinities to goethite. Such granular masses are typically soft, with a characteristic rusty yellow to yellow–brown color. Other important hydroxide/oxyhydroxide minerals include the manganese minerals brucite [$Mn(OH)_2$], manganite [$MnMnO_2(OH)_2$], and romanechite [$BaMnMn_9O_{20}\bullet 3H_2O$] and the mixed, often amorphous, manganese equivalent of limonite that is called **wad** (Chapter 14).

### 5.5.6 Carbonates

All **carbonate** minerals contain carbonate [$(CO_3)^{-2}$] anions bonded with metallic or semi-metallic cations. Carbonate minerals are characterized by their relative softness and varying degrees of solubility in dilute HCl (hydrochloric acid). HCl breaks bonds in the carbonate mineral and releases carbon dioxide gas ($CO_2$), causing the mineral to effervescence. Carbonates are significant rock-forming minerals. They occur widely in biochemical sedimentary rocks, where they are the major constituents of limestones and dolostones (Chapter 14). They also are major constituents of metamorphic rocks such as marble and skarns (Chapter 15). Carbonatites are rare igneous rocks containing carbonate minerals crystallized from carbonate magmas and lavas (Chapter 10).

Approximately 70 carbonate minerals have been reported. The most common carbonate minerals belong to one of three groups. **Calcite group** carbonates constitute an isostructural group composed of small cations bonded to carbonate ions in trigonal (rhombohedral) crystal structures that display rhombohedral cleavage. Calcite group minerals include calcite ($CaCO_3$), magnesite ($MgCO_3$), siderite ($FeCO_3$), rhodochrosite ($MnCO_3$), and smithsonite ($ZnCO_3$). **Aragonite group** carbonates constitute a second group composed of larger cations bonded to carbonate ions in orthorhombic structures. Aragonite group minerals include aragonite ($CaCO_3$), cerrusite ($PbCO_3$), strontianite ($SrCO_3$), and witherite ($BaCO_3$). The important rock-forming mineral dolomite [$CaMg(CO_3)_2$] bears some resemblance to the minerals in the calcite group. **Hydroxycarbonates**, such as the minor copper ore minerals azurite [$Cu_3(CO_3)_2(OH)_2$] and malachite [$Cu_2CO_3(OH)_2$], contain hydroxyl ion and/or water and commonly exhibit monoclinic structures. Carbonate minerals such as calcite and aragonite are economically important as the major raw materials in cement products. Both limestone, and to a lesser extent, dolostone are widely used as building and dimension stone.

### 5.5.7 Borates

The fundamental building blocks of **borate** minerals are $(BO_3)^{-3}$ triangles and, less commonly, $(BO_4)^{-5}$ tetrahedra. These are bonded with metal cations, most commonly sodium and/or calcium. The triangular and tetrahedral coordination polyhedra commonly link through shared oxygen ions into larger structural units such as rings and chains, in a manner similar to the linkage of silica tetrahedra in silicate minerals. Most borate minerals also contain appreciable hydroxyl ($OH^{-1}$) ion and/or water ($H_2O$) in their structures. Borates are particularly abundant in enclosed lake basins of the Mojave Desert region in California, where their

formation is attributed to evaporitic concentration of boron derived from volcanic hot springs and fumarole activity. More than 100 borate minerals have been described. Four commercially important examples are borax [$Na_2B_4O_5(OH)_4 \bullet 8H_2O$], colemanite [$CaB_3O_4(OH)_3 \bullet H_2O$], kernite [$Na_2B_4O_6(OH)_2 \bullet 3H_2O$], and ulexite [$NaCaB_5O_6(OH)_6 \bullet 5H_2O$]. Once widely used in soaps, detergents, and water softeners, boron is currently used primarily as an additive in the production of fiberglass for insulation and to strengthen glass and ceramics products. It is also used to ignite rocket fuels and in the production and the ignition of fireworks.

### 5.5.8 Sulfates

**Sulfates** are minerals that contain sulfate anion $(SO_4)^{-2}$ combined with one or more metals or semi-metals. One major group of sulfate minerals is the **hydrated sulfates**, of which the most common example is gypsum ($CaSO_4 \bullet 2H_2O$). A second important group is the **anhydrous sulfates**, of which the most common example is anhydrite ($CaSO_4$). As discussed in section "Growth Surfaces and Breakage Surfaces," gypsum varieties include **selenite**, which is composed of large macroscopic crystals that display good cleavage; **satinspar**, which consists of fibrous aggregates of parallel acicular–capillary crystals; and **alabaster**, which consists of masses of randomly oriented, microscopic crystals. Gypsum, anhydrite, and several other sulfate minerals are significant components of sedimentary evaporite deposits (Chapter 14). Gypsum is an essential ingredient in plaster and sheetrock that is widely used in the construction industry.

Other sulfate minerals include alunite [$KAl_3(SO_4)_3(OH)_6$], anglesite ($PbSO_4$), barite ($BaSO_4$), celestite ($SrSO_4$), and several evaporite minerals that include epsomite ($MgSO_4 \bullet 7H_2O$) and polyhalite [$K_2Ca_2Mg(SO_4)_4 \bullet 2H_2O$].

### 5.5.9 Phosphates

**Phosphate** minerals contain $(PO_4)^{-3}$ anions bonded with metal cations. Although the phosphate group is large, only two minerals, apatite and monazite, are relatively common. **Apatite** [$Ca_5(PO4)_3(Cl,F,OH)$] forms in crystalline igneous rocks and in a marine sedimentary rock called phosphorite (Chapter 14). In addition to its use as fertilizer, apatite is the main component of human teeth. **Monazite** [(Ce,Y,La,Th)PO$_4$], the other important phosphate mineral, is a principal source of rare Earth minerals (REE) such as thorium and numerous lanthanides (Chapter 2). The value of rare Earth has increased dramatically because they are essential constituents in the manufacture of cell phones, computers, catalytic converters, rechargeable batteries, and DVDs. Economic quantities of monazite are recovered from beach deposits along with other heavy, resistant minerals such as zircon, magnetite, rutile, ilmenite, and garnet. Amblygonite (LiAlFPO$_4$) occurs in some pegmatites and is mined as a source of lithium used in refractory glass and metallic alloys. Turquoise [CuAl$_6$(PO$_4$)$_4$(OH)$_8$•4H$_2$O] is a significant gemstone, widely used in jewelery.

### 5.5.10 Tungstates and molybdates

**Tungstate group** minerals are characterized by the tungstate [(WO$_4$)$^{-2}$] anion group bonded to metallic cations. Some 20 tungstate minerals have been identified. All are relatively rare, but **scheelite** (CaWO$_4$) and **wolframite** [(Fe,Mn)WO$_4$] are economically important sources of tungsten. Tungsten's high melting point and steel hardening properties make it an important element in the manufacture of hard, refractory steel alloys for use in high-speed cutting tools, incandescent lamps, electrical contacts, crucibles, and spark plugs.

**Molybdates** are characterized by the molybdate [(MoO$_4$)$^{-2}$] anion group bonded to metallic cations. The only significant molybdate mineral is **wulfenite** (PbMoO$_4$), which is a source of molybdenum used with iron in the manufacture of high-quality steel alloys.

### 5.5.11 Other nonsilicate minerals

Other mineral groups, based on their principal anion group, are represented by relatively rare minerals and are not detailed in this text. These include **nitrates** [(NO$_3$)$^{-1}$] such as nitratite (NaNO$_3$) and niter (KNO$_3$) used in fertilizers, **chromates** [(CrO$_4$)$^{-2}$] such as crocoite

(PbCrO$_4$), **vanadates** [(VO$_4$)$^{-3}$] such as vanadinite [Pb$_5$(VO$_4$)$_3$Cl] and carnotite [K$_2$(UO$_2$)$_2$(VO$_4$)$_2$•3H$_2$O] which is a significant uranium ore, and **arsenates** [(AsO$^4$)$^{-3}$] such as erythrite [Co$_2$(AsO$_4$)$_2$•8H$_2$O].

In this chapter, we have focused on the macroscopic properties of minerals. Chapter 6 details the microscopic techniques used to identify minerals and summarizes a few of the many advanced analytical methods used by scientists who seek a deeper understanding of Earth materials. Because of space constraints, the macroscopic and microscopic properties of significant minerals used to identify them and their uses are detailed in the online appendices that support this text.

## CONTENT ASSESSMENT

1  What is the meaning of the term *hardness*? What two principal factors determine the hardness of minerals and how does each influence it?

2  Define the term *density*. What two factors determine the density of minerals? Explain the role played by each.

3  Distinguish between *density, specific gravity, and weight*.

4  Define the term *cleavage* in minerals. What three criteria best permit cleavage surfaces to be recognized? What determines the number of sets or orientations of cleavage and the intersection angles between them in a given mineral?

5  What are the differences between *idiochromatic* and *allochromatic* minerals? Explain why some minerals are allochromatic and others are idiochromatic and give three good examples of each.

6  Distinguish between the silica minerals: (a) *macroscopic quartz*, (b) the *chert* group, and (c) the *chalcedony* group and *opal*.

7  Distinguish between the following types of *phyllosilicate* minerals: (a) two-layer (t-o), three-layer (t-o-t), four-layer (t-o-t-o). Then discuss the major minerals represented in each group (serpentine, talc,

chlorite, biotite, phlogopite, muscovite, kandites, illites, and smectites, using the terminology that distinguishes tetrahedral (t), gibbsite (g) and brucite (b) layers.

8   What is the chief compositional characteristic of each of the following *nonsilicate* mineral groups?

  a   halides
  b   sulfides
  c   oxides
  d   hydroxides
  e   oxyhydroxides
  f   carbonates
  g   sulfates
  h   phosphates and nitrates
  i   tungstates and molybdates

## REFERENCES

Hazen, R.M., Papineau, D., Bleeker, W. et al. (2008). Mineral evolution. *American Mineralogist.* 93: 1693–1720.

Hurlbut, C.S.J. and Sharp, W.E. (1998). *Dana's Minerals and How to Study Them*, 4e. New York: Wiley 328pp.

Klein, C. and Hurlbut, C.S. Jr. (1985). *Manual of Mineralogy*, 20e. New York: Wiley 644pp.

Wenk, H.R. and Bulakh, A. (2004). *Minerals: Their Consititution and Origin.* Cambridge, UK: Cambridge University Press 646pp.

Wenk, H.R. and Bulakh, A. (2016). *Minerals: Their Constituion and Origin* (2nd edn). Cambridge, UK: Cambridge University Press 640pp.

Zoltai, T. and Stoudt, J.H. (1984). *Mineralogy: Concepts and Principles.* Minneapolis: Burgess Publishing 506pp.

# Chapter 6

# Optical identification of minerals

In our efforts to analyze Earth materials from as many different perspectives as possible, we have developed a variety of microscopic, imaging, and analytical techniques that permit more precise and accurate characterization of such materials. Due to space constraints, only the optical methods are discussed here. A brief discussion of other imaging and analytical techniques is available in an appendix that accompanies this text. Many of these techniques involve insights adapted from physics and chemistry to investigate the ways in which various forms of electromagnetic radiation interact with materials. These interactions reveal the detailed structure and chemical compositions of both natural and synthetic materials. This, in turn, has stimulated the increasingly sophisticated production of synthetic materials which possess the desired structural and chemical properties for specific applications that has marked the post-industrial revolution in materials science. Because many of these techniques involve the interaction of Earth materials with electromagnetic radiation, a brief review of the major concepts of electromagnetic radiation is presented here.

## 6.1 ELECTROMAGNETIC RADIATION AND THE ELECTROMAGNETIC SPECTRUM

### 6.1.1 Electromagnetic radiation

**Electromagnetic radiation**, most familiar to us in the form of visible light energy, includes a much broader spectrum of energies that are discussed below. Electromagnetic energy is generated by (1) the acceleration of electrically charged particles, e.g., electrons, protons

*Earth Materials*, Second Edition. Kevin Hefferan and John O'Brien.
© 2022 John Wiley & Sons Ltd. Published 2022 by John Wiley & Sons Ltd.
Companion website: www.wiley.com/go/hefferan/earthmaterials2

or ions, or (2) changing magnetic fields. Once generated, electromagnetic radiation moves away from the place where it was produced carrying both energy and momentum. In a vacuum, such energy moves (propagates) at the speed of light ($c = 3.0 \times 10^8$ m/s); in matter it always propagates at less than the speed of light, but still "lightening fast."

Electromagnetic radiation is characterized by both an **electrical component** and a **magnetic component**. The two components are mutually self-generating; the magnetic component can be generated by the electrical component and *vice versa*. Changes in one component are accompanied by changes in the other.

Electromagnetic radiation possesses the properties of both particles and waves. In **particle models**, electromagnetic radiation is interpreted in terms of **photons** which are discrete packets of energy that interact with atoms at the subatomic level. Photons have zero mass and carry an amount of energy characteristic of the type of electromagnetic radiation in question. Particle models work best to explain very small-scale (nanometer-scale) phenomena of short (nanosecond) duration. Scientists are attempting to find models that link these behaviors to those of larger-scale phenomena.

For most larger-scale phenomena, a wave model generally provides the most appropriate visualization. In **wave models**, electromagnetic radiation is portrayed as a series of sinusoidal, transverse waves with features characteristic of the type electromagnetic radiation in question. The **electrical component (E)** of the radiation vibrates in a plane (Y or E) perpendicular to the direction of propagation (X) while the **magnetic component (B)** of the radiation vibrates in a plane (Z or B) perpendicular to both (Y and X). These relationships are shown in Figure 6.1. For reasons of efficiency, in many of the discussions later in this chapter only the electrical vibration direction is considered.

Each wave series is characterized by **crests** and **troughs** that represent the maximum up and down or side to side, that is positive and negative, displacements from the propagation direction as the wave energy oscillates transverse to it. This maximum displacement from the propagation direction is called the **wave**

**amplitude (A)**. The **wavelength ($\lambda$)** of a single wave is the distance between successive wave crests or any other corresponding parts of successive wave forms. **Wave frequency (f)** is defined by the number of waves in the propagating wave series that pass a point in one second. The unit of wave frequency is the number per second (#/s) or cycles per second and is called a **Herz (Hz)**. For example, an electromagnetic wave with a frequency of 20 000 Hz (20 kHz) would be a wave in which 20 000 crests passed a point during a single second of wave propagation. Wavelength ($\lambda$) and frequency (f) are related to velocity ($\upsilon$) by the following equation

$$\upsilon = \lambda f \qquad \text{(equation 6.1)}$$

For a wave propagating at a constant velocity, for example the speed of light in a vacuum (c), increasing wavelength is associated with decreasing frequency and *vice versa*.

### 6.1.2 The electromagnetic spectrum

The electromagnetic spectrum constitutes the full range of electromagnetic energy (Figure 6.2). Every part or **band** of the full spectrum is characterized by three properties: (1) **energy (E)** expressed in **electron volts (Ev)**, (2) frequency (f) expressed in cycles per second or Herz (Hz), and (3) wavelength ($\lambda$). These three properties are closely related. Energy values are proportional to frequency

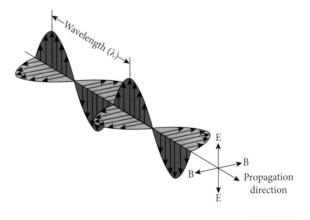

**Figure 6.1** The electrical (E, in red) and magnetic (B, in yellow) components of electromagnetic waves vibrate perpendicular to each other and to the propagation direction (X).

and frequency (equation 6.1) is inversely proportional to wavelength. Figure 6.2 illustrates the major, sometimes overlapping, parts of the electromagnetic spectrum and their characteristic energies, wavelengths, and frequencies. Radio and TV waves have the longest wavelengths, smallest frequencies, and least energy, whereas gamma rays and x-rays have the shortest wavelengths, highest frequencies, and most energy. This brief introduction to the electromagnetic spectrum is necessary to understand optical crystallography and many other advanced analytical techniques.

## 6.2 ESSENTIALS OF OPTICAL CRYSTALLOGRAPHY

Optical crystallography is the study of crystals based on how they interact with visible light incident upon their surfaces. It cannot be practiced well, unless some basic concepts about light-crystal interactions are clearly understood.

### 6.2.1 Light and crystals

Whenever light strikes (is incident to) the surface of a solid substance, a number of phenomena may occur. Light may be reflected, refracted, dispersed, transmitted, absorbed, scattered and/or reradiated by the substance. Such phenomena depend primarily on details of the substance's chemical composition and crystal structure. Because each mineral possesses a unique combination of chemical composition and crystal structure (including defects), the manner in which each mineral interacts with light is unique. Once these interactions are known, they can be used to identify minerals or any other crystalline substance. **Ordinary**

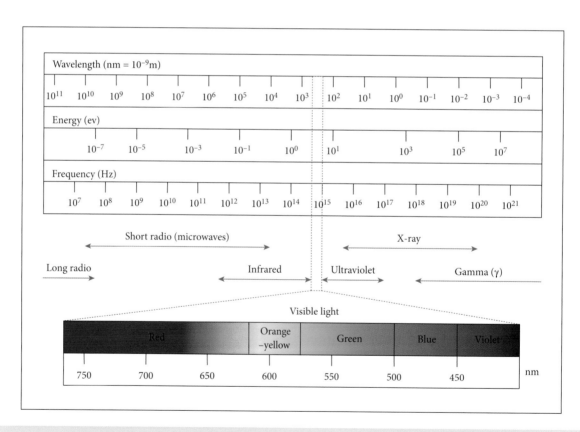

**Figure 6.2** The continuous electromagnetic spectrum showing the wavelengths, frequencies and energies associated with each part or band of the spectrum. The narrow part of the spectrum visible to humans is shown along with the approximate wavelengths characteristic of each major color of the visible spectrum.

light (Figure 6.3) vibrates in all directions perpendicular to the direction of propagation (ray path). **Plane polarized light** (Figure 6.3) is constrained by a polarizing lens to vibrate in a single plane perpendicular to the direction of propagation.

*Reflection, refractive index, refraction, and dispersion*

When light strikes (is incident upon) the surface of a solid, part of the light is reflected from the surface, much like light from a mirror.

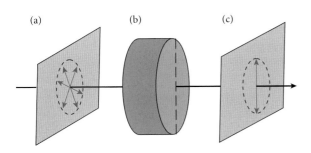

(a)      (b)      (c)

**Figure 6.3**   Ordinary light vibrates in all directions perpendicular to the direction of propagation. When ordinary light (a) is passed through a polarizing filter (b), it emerges as plane polarized light (c), constrained to vibrate in a single plane perpendicular to the propagation direction.

The **angle of reflection** is the complement of the **angle of incidence** *(i)*. Only incident light that is perpendicular to the boundary between two media is reflected back in the direction from which it came. All other light is reflected at the complementary angle (Figure 6.4). This is why light flashes to your eyes from a mirror or the cleavage surfaces of a mineral at a particular angle as the objects are rotated.

Light that is not reflected from the surface continues into the solid as **refracted light**. Both the velocity and the propagation direction of the refracted light change by an amount that depends on the (1) wavelength of visible light in question, (2) the angle of incidence, and (3) the properties of the solid through which it is being propagated (Figure 6.4). The change in velocity is expressed by the **refractive index (n or RI)** which is the ratio between the velocity of light in a vacuum ($V_v$) and the velocity of light in the material ($V_m$) and is given by:

$$RI = V_v / V_m \qquad \text{(equation 6.2)}$$

Refractive index is inversely proportional to the velocity of light in the material. It increases as the velocity of light in the material decreases. Because light slows down as it interacts with the electrical fields in atoms, the velocity of light in substances is always less than the velocity of light in a vacuum, and the refractive index of a substance is therefore always more

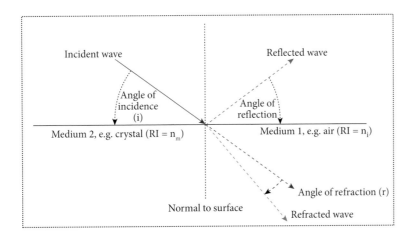

**Figure 6.4**   Incident light is reflected (blue arrow) from a surface between two media, for instance air and a crystalline substance. It is also refracted (blue arrow) as it moves into the crystalline substance with a new velocity, given by the refractive index (n), and in a new direction, given by the angle of refraction (r), as related by Snell's Law.

than 1.0. For example, the refractive index of air, with its low proportion of atoms per volume, at standard temperature and pressure is 1.0008. Most minerals, with their much higher concentration of atoms per volume to slow the passage of light, possess refractive indices between 1.4 and 2.0 meaning that light slows down from 20 to 50% as in passes from air into most crystals. Refractive index is critical in mineral identification.

The change in direction across the boundary between two media is called the **angle of refraction (r)**. This angle is related to the angle of incidence (i) and the refractive indices of the two materials ($n_i$ and $n_m$), as given by Snell's Law:

$$n_i \sin(i) = n_m \sin(r) \quad \text{(equation 6.3)}$$

Snell's Law predicts that incident light waves propagating in a direction not perpendicular to the surface will always be refracted or bent toward the medium with the lowest velocity and highest refractive index. In Figure 6.4, the light is bent toward the crystalline material as it enters from the air, because the crystal has a higher refractive index and a lower velocity. The amount of refraction increases with decreasing angle of incidence and with increases in refractive index (or velocity) across the boundary.

In a vacuum, all wavelengths of the magnetic spectrum are propagated at the velocity of light (c). In Earth materials, the velocity varies with wavelength. For example higher energy, shorter wavelengths toward the blue-violet end of the visible spectrum slow less than lower energy, longer wavelengths nearer the red end. As a result, violet-blue wavelengths (~400 μm) are refracted less than red wavelengths (~700 μm). As white light, a mix of all wavelengths in the visible spectrum, is refracted through a crystal, the light is dispersed into separate wavelengths that travel different paths due to different amounts of refraction. This phenomenon in which light is separated into its component wavelengths is called **dispersion**. Good examples are the dispersion of light by raindrops to produce a rainbow or the dispersion of light by a prism (Figure 6.5).

*Diaphaneity and color*

Some materials, including many native metals, metallic sulfides, and metallic oxides, are opaque. **Opaque** substances do not transmit light, even through very thin samples. Instead, light is absorbed as photons interact with electrons. When the energy of a photon is the energy required to excite an electron and cause it to "jump" a higher quantum level (energy shell), the photon is absorbed as its energy is transferred to the electron. If all wavelengths (photons) of light are absorbed, no light is transmitted and the substance is opaque.

**Translucent** and **transparent** substances transmit some light and many of them transmit a distinct **color** of light. These colors result from the selective absorption and transmission of specific wavelengths (or particles) of light. For example, if the red wavelengths (red photons) are absorbed, and the blue and yellow wavelengths are transmitted, the mineral will appear green, as in mineral malachite. Many minerals possess the property of **pleochroism**, in which color varies with crystallographic orientation. Pleochroism occurs because these minerals selectively absorb and transmit light differently in different crystallographic directions in which the light passes through different arrays of atoms.

*Isotropic and anisotropic substances*

**Isotropic** substances transmit light at the same velocity in every mutually perpendicular direction; they possess a single refractive index which is the same in every direction. Minerals that crystallize in the isometric system (where $a_1 = a_2 = a_3$) and amorphous substances such as fluids and glass are isotropic substances. When a ray of plane polarized light enters an **isotropic medium**, in which the distribution of atoms is statistically the same in every mutually perpendicular direction, it changes velocity and passes through the medium as a single light ray that vibrates in the same direction. Certain wavelengths may be selectively absorbed, but the light that is transmitted leaves the mineral as a single light ray vibrating in the same direction as when it passed through the polarizer.

All other crystals are **anisotropic**; they transmit light at different velocities in different directions. They possess a direction of maximum velocity, a direction of minimum velocity, and many directions with intermediate velocities. As a result, they possess a minimum refractive

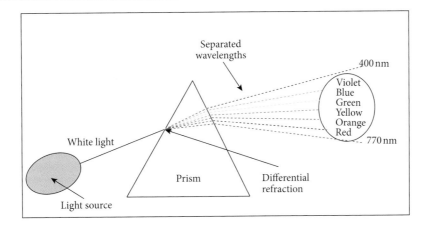

**Figure 6.5** The dispersion of white light into different parts of the visible spectrum as it passes through a crystalline prism. Note that the amount of refraction increases as wavelength decreases, causing the separation of wavelengths into separate colors.

index in one direction, a maximum refractive index in a second direction, and intermediate refractive indices in between. Minerals that crystallize in the tetragonal, hexagonal, orthorhombic, monoclinic, and triclinic systems (where $a \neq c$) are all anisotropic crystals.

**Birefringence (B or δ)** is the difference between the maximum refractive index (in the slowest direction) and the minimum refractive index (in the fastest direction) in a crystal. It is given by

$$B = RI_{max} - RI_{min} \quad \text{(equation 6.4)}$$

Birefringence is zero in isotropic materials and has a finite value in all anisotropic crystals. This value increases with increases in the difference between the maximum and minimum refractive indices. It can be best observed and measured directly in crystals oriented so that both the minimum and maximum refractive index directions are visible and measurable.

When plane polarized light enters an anisotropic mineral, the light generally changes significantly. In most orientations, **two light rays** are generated that vibrate perpendicular to each other, but that generally travel in different directions with different velocities. It is these two different light rays that are responsible for the **double refraction** seen in minerals such as calcite where refracted images are transmitted separately along two different ray paths. As the two rays pass through the mineral, the **slow ray** (traveling in the high refractive index direction) lags behind the **fast ray** (traveling in the low

refractive index direction). The amount (measured in number of wave lengths) by which the slow ray lags behind the fast ray is called the **retardation (Δ)**. The retardation is proportional to the difference between the two refractive indices, i.e. the birefringence, and to the distance (d) traveled through the specimen, roughly its thickness. Retardation (Δ) is given by:

$$\Delta = d\left(RI_s - RI_f\right) = d\left(B\right) \quad \text{(equation 6.5)}$$

Where d is the distance traveled or "thickness," $RI_s$ is the refractive index of the slow ray, $RI_f$ is the refractive index of the fast ray and $(RI_s - RI_f)$ is the birefringence. For thin-sections where the thickness is relatively constant (30 μm), the retardation is largely a function of birefringence.

### 6.2.2 The petrographic microscope

The light-transmitting microscope used in most optical investigations of rocks, minerals, and other Earth materials is called a **petrographic microscope**. It is an essential tool in **petrography** which involves the careful description of rock compositions, textures, and small-scale structures. Figure 6.6 shows a typical, if somewhat older, model of a petrographic microscope with its major components labeled. Newer models often combine some of the features described and are less useful for the fuller description that follows.

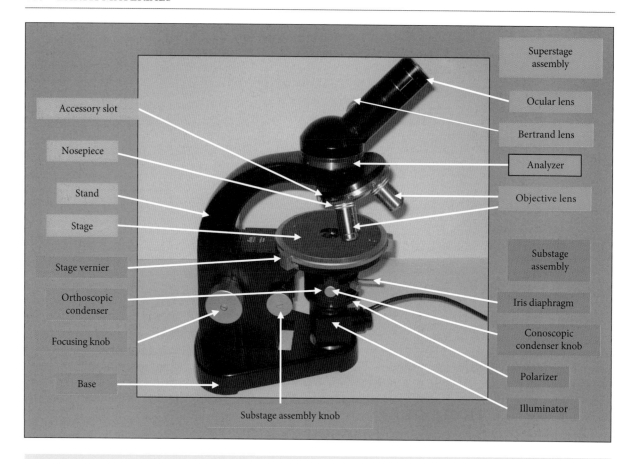

Accessory slot

Nosepiece

Stand

Stage

Stage vernier

Orthoscopic
condenser

Focusing knob

Base

Substage assembly knob

Superstage
assembly

Ocular lens

Bertrand lens

Analyzer

Objective lens

Substage
assembly

Iris diaphragm

Conoscopic
condenser knob

Polarizer

Illuminator

**Figure 6.6** A standard petrographic microscope with the major components identified.

### The illuminator (light source)

Our discussion of the components of petro-graphic microscopes proceeds from the bottom up. Such microscopes rest on a flat **base** that contains a **light source** or **illuminator**. The illuminator consists of a power source, an incandescent bulb, and a blue filter. Light from the bulb is directed upward by lenses or mirrors and the blue filter passes light that approximates sunlight. Many models have a **rheostat** that permits the intensity of the light source to be adjusted.

### The substage assembly

Between the illuminator and the stage, where the sample is placed, is the **substage assembly.** This assembly is designed to modify the light from the illuminator that illuminates the sample from below.

The **polarizer** (also called the **lower polar** or **lower Nicol**) is a lens that causes random light from the illuminator to be polarized so that it is passed upward as light vibrating in a single plane. In most microscopes, the polarizer contains a polarizing film that only passes light vibrating in an east–west direction. All other components of light from the illuminator are extinguished. Once light leaves the polarizer on its way to the overlying stage, it is **plane-polarized light**, constrained to vibrate in a single plane.

Above the polarizer are two condensing lenses. The first is a fixed condensing lens, the **orthoscopic condenser**, which focuses the light on the area of the sample to be viewed. This lens provides what is called **orthoscopic illumination** and is the illumination under which most samples are initially viewed.

Above the orthoscopic condenser is an auxiliary condensing lens, called the **conoscopic lens** or **condenser**. It is auxiliary in the sense that it can be swung into the optical path of orthoscopic light when needed by means of an attached lever. The conoscopic lens produces a strongly convergent cone of light, **cono-scopic illumination**, which is focused on only a tiny area of the sample on the stage. It is used for viewing interference figures in **cono-scopic mode**, as discussed later in the chapter.

The final component of the substage assembly, which is placed between or below the two condensing lenses, is an **iris diaphragm**. The diaphragm can be opened or closed by means of an attached lever to vary the amount of light that reaches the sample. For most work, the diaphragm is kept in the full open position to maximize the amount of available light.

*The microscope stage*

The freely rotating, **circular stage** possesses a circular hole in the middle through which light from the substage assembly passes. Glass slides with samples are placed over the circular hole. The periphery of the stage is a **stage goniometer** subdivided into 360° for measuring rotation angles of the sample. Some models also possess a **stage vernier** for more precise angular measurements. Although the stage is ordinarily free to rotate, its position can be fixed by tightening a **thumb screw** on the stage margin. Additional screw holes on the stage permit instruments to be mounted on the microscope stage. These range from **stage clips** for holding sample slides in place to **mechanical stages** for counting the proportions of components as they are randomly identified.

*The superstage assembly*

Between the stage and the eyepiece at the top of the microscope is the **superstage assembly**. At the bottom of this assembly are three (or more) **objective lenses**. These lenses provide initial magnification of the sample image. One of these should always be in the optical path of the microscope. The **low power objective** typically provides 2.5–4× magnification; the **medium power objective** typically provides 8–10× magnification; the **high power objective** provides ~40× magnification. The objectives are mounted to a circular **nosepiece** that can be rotated to switch objective lenses. *Always change objectives by holding onto and rotating the nosepiece. Never change objectives by holding on to the objectives.* The latter will inevitably cause the objectives to become uncentered, so that samples will not

remain in the field of view as the stage is rotated. This will make proper identification impossible. Objective lenses may be recentered using two centering screws, but the process is time-consuming for all but the very experienced.

Above the nosepiece on most microscopes is the **accessory slot**, used for the insertion of **accessory plates**, also called **compensator plates**. These plates change the quality of light from the objective lenses in ways that permit sophisticated mineral identification. A very important component of the superstage assembly is the **analyzer** (also called the **upper polar** or **upper Nicol**). The analyzer is an auxiliary lens that can be swung into position by a lever. The analyzer lens is similar to the polarizer or lower polar except that its polarization direction is oriented perpendicular to that of the polarizer. Whereas the polarizer (lower polar) typically passes plane polarized light vibrating in an east–west plane, the analyzer (upper polar) only passes light vibrating in a north–south plane. That the upper polar can pass any light at all is testimony to the ability of anisotropic samples to change the vibration direction of the plane polarized light that enters them from the lower polar. When the analyzer is not inserted, the sample is said to be viewed in plane-polarized light or **plane light mode**. When the analyzer is inserted, the sample is said to be viewed under **crossed-polars mode** (crossed-Nicols mode) because the vibration directions of the two polarizers are at right angles. Plane light mode and crossed-polars mode are two of the major ways of viewing samples using petrographic microscopes. One can quickly switch back and forth between them by swinging the analyzer in and out of position.

Above the analyzer, commonly in the microscope tube, is the **Bertrand lens**. The Bertrand lens is yet another auxiliary lens that can be swung into the optic path by an attached lever. It is used only for **conoscopic mode** viewing using the high power objective with both the conoscopic condenser and the analyzer inserted into the optic path. It is in this mode that interference figures, magnified and focused on the eyepiece by the Bertrand lens, may be examined and interpreted by those doing advanced optical work. For most preliminary work, in plane light and under crossed-polars, the Bertrand lens is not used.

*The eyepiece or ocular head*

In all major modes, the sample is directly viewed through the **eyepiece** or **ocular lens** that fits into the top of the microscope tube. The ocular lens provides additional image magnification that ranges from 5×–12× (most commonly 8×–10×). The **total magnification** of the sample is given by the product of the objective magnification and the eyepiece magnification. A sample viewed using a 10× objective and a 10× eyepiece is magnified to 100× its true size. All eyepieces contain **reticule markings** that include **cross-hairs**. These should ordinarily be oriented north–south and east–west, that is parallel to the vibration direction of the analyzer (upper polar) and of the polarizer (lower polar) respectively. The cross-hairs are used to center a feature of interest in the field of view. They also divide the field of view into quadrants for referencing the location of features relative to the center of the field of view. Some microscopes also contain **eyepiece micrometers**. These are scales that are used to measure sizes of and distances between objects in the sample being viewed. They must be calibrated with a stage micrometer for each objective and the correction factors recorded, after which the eyepiece micrometer can be used to measure distances and sizes accurately. An adjustable **focusing ring** on the circumference of the ocular can be used to focus the reticule markings. This should be done before starting work. **Stereographic microscopes** have two eyepieces, one of which has an adjustable focusing ring. **Trinocular microscopes** possess a third tube to which a camera or computer monitor may be attached and through which photomicrographs (microscopic images) and video images may be captured.

*The focusing knobs and free working distance*

The distance between the objective and the sample on the stage is called the **free working distance (FWD)** and can be changed by moving the stage up or down using knurled **focusing knobs** on the microscope arm. The larger knob is for coarse focusing; the smaller knob for fine focusing. Moving the stage up, decreases the free working distance; moving it down increases the free working distance. The specimen will be in focus at a particular free working distance that depends on the objective used and the eyesight of the investigator. Note that the three objective lenses are of different lengths. The shorter, lower power lenses are typically the first lenses used. When using the high-power lens, it is necessary to lower the stage assembly before using the focusing to decrease the free working distance so as to avoid damaging the glass slide (whose cover slip should always be on top).

### 6.2.3   Modes of optical investigation

Petrographic microscopes permit the investigation of rocks and minerals using a variety of techniques. These techniques are described below.

*Grain mounts*

One method for mineral identification involves preparing one or more **grain mounts**. This involves grinding the sample, for example with a mortar and pestle or a crusher, into small grains which are passed through sieves to produce a sample with grains between 0.075 and 0.105 mm (Nesse 2012). The finer material can be saved and used for other types of analyses such as x-ray diffraction. Several dozen grains are randomly scattered onto a glass slide to achieve a variety of orientations, covered with a thin cover slip, and then examined under a petrographic microscope. The grains may be mounted to the slide with a mounting medium or cement of known refractive index or may be immersed in an oil of known refractive index. In the **oil immersion method**, a sequence of samples is immersed in oils of different refractive indices. The refractive indices of the mineral are compared to the known refractive indices of the oils until they have been matched. Identification is then made on the basis of the mineral's refractive indices and other properties. This method is generally time-consuming, but offers a very useful way to learn the theory and practice of optical mineralogy in an integrated manner. Further information on these methods can be found in Nesse (2012).

*Thin-sections*

**Thin-sections** are thin slices of solid materials, typically rocks, mounted to a glass slide. They are the "bread-and-butter" of optical investigations in petrography. Rock thin-sections permit multiple minerals to be magnified and identified and their textural relationships to be revealed. Most "ground-up" work with rocks involves examining thin-sections at some stage in the research process. Thin-sections are prepared by first cutting a small piece or "chip" from a larger specimen using a rock saw. One flat area of the chip must be smaller than the glass slide on which it is to be mounted. This flat side of the chip is smoothed and polished, typically on a grinding wheel, and that flat side of the chip is carefully mounted to a glass slide (Figure 6.7a). The traditional mounting medium was Canada balsam with a refractive index of 1.537; this expensive medium has largely been replaced by epoxies with similar indices of refraction. The mounted chip is then cut on a special rock saw to produce a thin slice of even thickness (Figure 6.7b). The inverted sample is then ground to a thickness of approximately 50 μm (0.05 mm) prior to a final polishing that produces a second smooth surface on a thin-section of rock whose thickness should be 30 μm

(0.03 mm). Any dies that stain minerals for the purposes of easier identification are applied after this stage. A thin cover slip is mounted atop the thin-section using epoxy (Figure 6.7c). The thin-section is now ready for analysis. Preparing a good thin-section with a constant thickness of 30 μm is a skill not mastered by everyone. Many commercial enterprises are available to prepare thin-sections to order at a price.

Both grain mounts and thin-sections can be examined in three principle modes using the petrographic microscope. These are, in order of increasing complexity: (1) plane polarized light (plane light) mode, (2) crossed-polars (crossed-Nicols) mode, and (3) conoscopic mode. Each mode reveals information about how the minerals in the sample transmit light and permit progressively more accurate mineral identification. The first two are especially useful modes for investigating rock textures and microstructures. *When examining thin-sections, make sure that the thin specimen and its cover slip are on top, so that the thin specimen will come into focus, as the free working distance is decreased, before lens reaches it. If the slide is upside down, with the cover glass on the bottom, the slide can easily be broken by the lens as the working distance is decreased, before the thin specimen comes into focus, especially when using the high power lens.*

*Plane polarized light mode*

The initial investigation of most thin-sections involves a combination of plane polarized light (plane light) mode and crossed-polars (crossed-Nicols) mode using the low and medium power objectives. Plane light investigations reveal how the sample transmits plane polarized light from the polarizer (low polar) when none of the auxiliary lenses, including the analyzer (upper polar), conoscopic lens, and Bertrand lens, have been inserted into the optical path. What is observed through the eyepiece is plane polarized light that may have changed as it passed through the sample. Plane light mode is the preferred mode for observing (1) diaphaneity, (2) color, (3) pleochroism, (4) cleavage and fracture, (5) relief and Becke lines (relative refractive indices), and (6) some inclusions and alteration products. The combination of plane light properties is a starting point for distinguishing

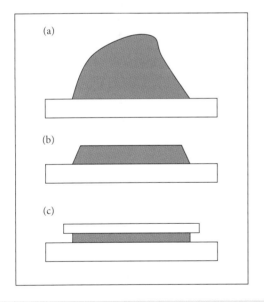

**Figure 6.7** The essential steps in thin-section preparation. (a) A chip with a flat surface is cut from the specimen, ground, polished and mounted to a slide. (b) The chip is cut and ground to a constant thickness. (c) The chip is polished to a thickness of 30 μm and covered with a thin cover slip.

between different minerals. It allows one to develop hypotheses that can be tested by further observations. Plane light properties of some common rock-forming minerals are summarized in Table 6.1.

Because they do not transmit light, **opaque minerals** can be recognized in plane light mode by the fact that they appear black in all orientations. Because they reflect light differently, opaque minerals can be distinguished from one another using a **reflecting light microscope** in which a polished specimen is illuminated from above by a strong light source. **Nonopaque minerals** transmit some light, so appear white or colored in plane light. The **transmitted light color** of minerals is best observed under plane light and depends on the wavelengths of visible light that are transmitted through the specimen. Many minerals, including common nonferromagnesian rock-forming silicates, including quartz, feldspars, and muscovite, are colorless in plane light. This is because all wavelengths are transmitted by the mineral. Many other minerals, including most ferromagnesian rock-forming silicates, are colored to some degree

**Table 6.1** Plane light and crossed-polars properties of some common rock-forming minerals.

| | Mineral plane light properties | | | Crossed-polars properties | | |
|---|---|---|---|---|---|---|
| | Color and pleochoism | Relief | Cleavage | Birefringence | Extinction | Twinning and others |
| Quartz | Colorless; clear | Low | None | Low | Often undulatory | Clear; little alteration |
| Orthoclase | Colorless; "dusty" | Low | Two near 90° | Low | | Carlsbad twins, if any |
| Microcline | Colorless; "dusty" | Low | Two near 90° | Low | | Gridiron or scotch-plaid twins |
| Plagioclase | Colorless; "dusty" | Low | Two near 90° | Low | Albite twins alternate | Albite, Carlsbad, or pericline twins |
| Muscovite | Colorless | Moderate | One | Moderate to high | Almost parallel to cleavage | Birdseye extinction |
| Biotite | (Red) brown to yellow; some green | Moderate | One | Moderate to high | Almost parallel to cleavage | Birdseye extinction |
| Chlorite | Colorless to light green | Moderate | One | Low | Almost parallel to cleavage | Anomalous blue-violet interference colors |
| Hornblende | Brown to green | Moderate | Two not 90° | Low to moderate | Large angles | Some simple paired twins |
| Augite | Pale green | Moderate | Two near 90° | Moderate | Large angles | Some simple paired twins |
| Enstatite/ hypersthene | Pale pink or pale yellow | Moderate | Two near 90° | Low to barely moderate | Parallel or symmetrical in some orientations | |
| Olivine | Colorless | Moderate to high | None | Moderate to high | | Curved fractures; alteration |
| Calcite/ dolomite | Colorless | Low (−) to moderate (+) | Three not 90° | Very high | Symmetrical to cleavage | Mottled extinction |
| Gypsum | Colorless | Low | One perfect | Low | | |
| Garnet | Pale pink to pale red | High | None | None; isotropic | In all positions | Equant dodecaheral crystals |
| Staurolite | Colorless to yellow | High | Poor | Low to barely moderate | Parallel to long axis | |

because of the tendency of iron and other metallic elements to absorb and transmit light selectively. Isotropic minerals transmit and absorb light equally in all directions and therefore have constant colors. However, anisotropic minerals tend to absorb light differently in different crystallographic directions and often display **pleochroism**, colors that change as the stage is rotated and the plane light passes through different atomic arrays. For example, biotite shows pleochroism in shades of brown, yellow, red-brown and/or green. Hornblende is characterized by green to brown pleochroism in most orientations.

Cleavage and fracture are observed best in plane light mode (Figure 6.8). They are most clearly seen when the Iris diaphragm is partially closed to reduce light and increase contrast. **Cleavage** appears as sets of parallel fractures in the mineral. Where the cleavage is perfect, the parallel fractures are easy to recognize, as are their intersection angles in some orientations. Where the cleavage is less perfect, the fractures may only extend over small

**Figure 6.8** Plane polarized light photomicrograph of plutonic igneous rock (granitoid). Gray-green crystals with relatively high relief and two cleavages not at right angles are hornblende. Slightly dusty, colorless crystals with two cleavages at right angles and low relief are plagioclase. The light to darker brown crystals, with moderate relief and one cleavage are biotite. *Source*: Courtesy of Kurt Hollacher, Union College.

parts of the mineral and careful observation under medium to high power is required to recognize its existence. Noncleavage fractures will appear irregular or curved. It should be noted that a cleavage will not be visible when it is parallel to the stage.

Another property that is clearly visible in plane light is relief. **Relief** results from the difference between the refractive index (R.I.) of a substance and the refractive index of the mounting medium or adjacent grains. It is seen under plane light as the sharpness or boldness of the grain boundary against the adjacent medium (Figure 6.8). The higher the relief is, the bolder the grain outline that results due to the larger the difference in refractive index. When the R.I. difference between the mineral and its surroundings approaches zero, the relief becomes so low that the mineral boundaries are practically invisible. Among common rock forming minerals, quartz, and feldspars generally possess low relief because their refractive indices approach that of common mounting media (~1.537). On the other hand, minerals such as olivine, garnet, and zircon possess much higher relief because their refractive indices are much higher than those of the mounting medium. Some minerals (e.g., calcite and dolomite) display relief that varies from high to low as the stage is rotated, producing a "twinkling" effect because in some orientations the refractive index is much higher than the mounting medium, whereas in others it approaches that of the mounting medium. Table 6.1 describes the relief of some common rock-forming minerals.

Crystals possess **positive relief** when they possess a higher refractive index than the mounting medium. They are said to possess **negative relief** when their refractive index is lower than that of the mounting medium. The sign of the relief can be determined by observing the behavior of **Becke Lines**, named after Friedrich Becke who discovered this phenomenon. Becke lines are bright lines produced by the refraction and dispersion of light as it crosses the boundary between the crystal and its surroundings. In such cases, light is refracted toward the slower medium with the higher refractive index which generates a bright line near the boundary. Becke lines are best observed by focusing on the grain boundary under medium to high power with the Iris diaphragm partially closed down. As the working distance is very slowly increased (stage lowered), the Becke line will appear to move toward the medium with the highest refractive index. If the Becke line moves toward the mineral, the relief is positive. If it moves toward the adjacent medium, the relief is negative. When the refractive index of a mineral closely matches that of the mounting or immersion medium, its relief is extremely low and two Becke lines appear due to dispersion; a blue line which moves toward the medium of slightly higher refractive index and a yellow-orange line that moves away from it. This is very useful when determining a mineral's refractive indices by matching the refractive index of grain mounts to that of the mounting medium or oil in which it is immersed.

Inclusions and alteration products are best observed using a combination of plane light and crossed-polars modes. Because they affect the appearance of minerals in plane light, they are introduced here. **Fluid inclusions** are incorporated into many crystals as they grow, especially in crystals formed in the presence of meteoric, hydrothermal, magmatic or metasomatic fluids (Chapters 11 and 19). These inclusions tend to scatter light and make the mineral less transparent than it would otherwise be. **Solid inclusions** represent smaller mineral grains incorporated into larger crystals as they grew. **Alteration products** involve the partial or complete alteration of a previously formed crystal into new materials. Generalizations can be dangerous, but in many cases alteration causes altered minerals tend to be less transparent in plane light than are unaltered ones. For example, most quartz appears quite clear (hydrothermal vein quartz with many fluid inclusions is an exception) under plane light because quartz is generally quite resistant to alteration. On the other hand, feldspars, especially plagioclases, are more susceptible to alteration which makes them appear "dusty" or "dirty" under plane light. When scanning colorless, low relief minerals in plane light, the combination of cleavage and alteration can allow one to distinguish feldspars from relatively unaltered

quartz with its lack of cleavage. These preliminary ideas can be tested using crossed-polars techniques.

To summarize, plane light investigations allow students of Earth materials to make preliminary distinctions between minerals based on combinations of diaphaneity, color, pleochroism, cleavage, fracture, relief, inclusions, and alteration products (Table 6.1). Many of these preliminary identifications can be further refined and often confirmed in crossed-polars mode.

*Crossed-polars mode*

The transition from plane light to **crossed-polars mode** is accomplished by rotating the upper polar or analyzer into the optic path by means of a lever. Once this is done the light that reaches the eyepiece must pass through both the polarizer (lower polar), which polarizes light from the illuminator in an east–west direction, and the analyzer which resolves only components of light from the specimen that vibrate in a north–south direction. The two polarizers are oriented perpendicular to each other and the sample is being viewed under crossed-polars. It is routine for an investigator to switch back and forth between plane light and crossed-polars modes by alternately retracting and inserting the analyzer lens. This allows one to rapidly catalog both the plane light and crossed-polars properties of the crystals being examined. Low and medium power objectives are suitable for most of this work. As Table 6.1 makes clear, this approach is often sufficient to tentatively distinguish between all the major minerals in a rock, as well as to characterize its textures and microstructures.

When plane polarized light enters an **isotropic medium** it slows down, but passes through the medium as a single light ray that vibrates in the direction of polarization. Certain wavelengths may be selectively absorbed, but the light that is transmitted leaves the mineral as a single light ray vibrating in the same east–west direction as when it passed through the polarizer. Since there is only one polarized ray and no birefringence, no retardation occurs. Such isotropic minerals do not transmit any of the polarized light that vibrates at right angles to the vibration direction of the analyzer (upper Nicol) and thus appear dark under crossed-polars (when the analyzer is inserted) and remain so during a complete 360° stage rotation.

However, when plane polarized light enters an **anisotropic medium** in which the crystal structure varies with direction it is split into **two light rays** that vibrate perpendicular to each other, but that generally travel in different directions with different velocities. As the two rays pass through the mineral, the **slow ray**, vibrating in the high refractive index direction lags behind the **fast ray**, vibrating in the low refractive index direction. This causes retardation to occur between the two waves. The amount of retardation is proportional to the difference between the two refractive indices, the birefringence, and the distance traveled through the specimen, roughly the thickness. For thin-sections where the thickness is relatively constant, the retardation is largely a function of birefringence.

When the fast ray and later the slow ray leave the mineral, they head toward the analyzer with whatever retardation they had when they left the crystal (no further retardation occurs in air, an isotropic medium). When they enter the analyzer, of necessity the two vibration directions are resolved so that only their north–south components are transmitted through the polarizer. Figure 6.9 summarizes this process.

For some wavelengths and retardations of visible light, the fast ray and the slow ray are canceled destructively by the analyzer and are not transmitted to the eyepiece. For other wavelengths and retardations of visible light, the two rays are added constructively, so that light passes through the analyzer and upward toward the eyepiece. These are seen as colors under crossed-polars that, because they result from the interference of the two waves, are called **interference colors**. Several orders of interference colors can be produced, the number increasing with retardation, essentially with distance (specimen thickness) and birefringence. For a more detailed treatment of the theory behind these phenomena, the reader is referred to books by Nesse (2011, 2012). The relationship between interference colors, thickness, and birefringence can be summarized on a color chart such as that produced by Michel–Levy (Figure 6.10).

The orders of interference colors increase with retardation. If the thickness of the grain is closely known, as it is for well-prepared thin-sections, the interference colors can be predicted from the birefringence or the birefringence can be estimated from the interference colors. Let us use this chart by reading across the horizontal line that represents a thin-section thickness of 30 μm (0.03 mm). Lines that represent birefringence values

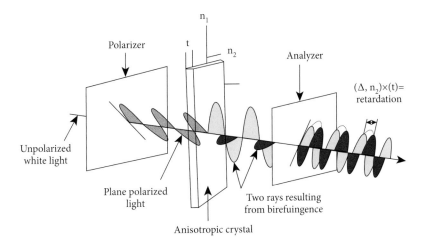

**Figure 6.9** Plane polarized light from the polarizer is split into a fast ray and a slow ray by anisotropic (birefringent) crystals which are then resolved by the analyzer to produce a new polarized ray whose properties depend on the retardation of the two rays.

radiate from the origin and birefringence values are labeled at the end of each line on the margins of the diagram. Isotropic materials, such as isometric minerals and glass, that have zero birefringence produce no retardation and will appear dark under crossed-polars. Minerals with **low birefringence** values ($B = n_f - n_s < 0.10$) will show first-order colors that range from grays and whites through pale yellow. Minerals with **moderate birefringence** ($B = 0.10–0.30$) will display second-order colors, including brighter blues, yellows, and purples. In practice, one must examine several grains to determine birefringence, because the maximum birefringence and interference colors for such minerals will only be observed when both the fast and slow ray vibration directions are parallel to the stage. Minerals with **higher birefringence** ($B > 0.30$) may display third- and fourth-order colors. These can be recognized by their paler appearance and less intense hues. Among common rock forming minerals, garnet is isotropic, while quartz, feldspars, and chlorite possess low birefringence. Orthopyroxenes, hornblende, and most clinopyroxenes have moderate birefringence and biotite, muscovite, olivine have relatively high birefringence. Figure 6.11 depicts several minerals under cross-polars.

There are specific directions in anisotropic minerals where only one ray travels through the mineral. These directions are called the **optic axes** of the mineral. For light waves propagating in the direction of the optic axes (and vibrating perpendicular to them), the refractive index is constant. Minerals in the tetragonal and hexagonal systems are **uniaxial minerals**. They have a single optic axis parallel to the crystallographic c-axis. A ray traveling along the c-axis vibrates perpendicular to it and therefore encounters the same crystal structure ($a_1 = a_2$) with the same refractive index. Minerals in the orthorhombic, monoclinic, and triclinic systems are **biaxial minerals**. They have two optic axes; directions in which the ray encounters the same crystal structure with the same refractive indices. This topic is developed more fully in the section that follows.

Another important property observed under crossed-polars is **extinction**. This occurs whenever a nonopaque mineral appears completely dark under crossed-polars because no light is being transmitted to the eyepiece. In this case, all light from the illuminator has been extinguished, and the mineral is said to be at extinction. Under crossed-polars, **isotropic substances** remain in the extinction position through an entire 360° rotation of the stage. This occurs because east–west vibrating plane polarized light produced by the polarizer (lower polar) that passes through the mineral with the same vibration direction is completely extinguished by the analyzer (upper polar) because it only passes light that

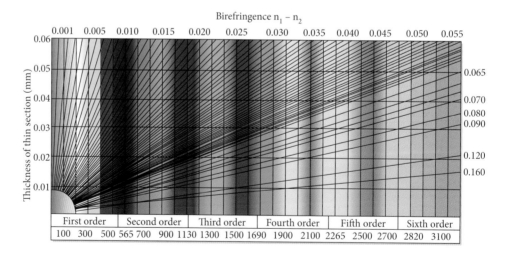

**Figure 6.10** A modified version of the Michel–Levy Color Chart for interference colors viewed under crossed-polars. Orders of color increase to right as hue intensities decrease. *Source*: Courtesy of Olympus Microscopy.

**Figure 6.11**  Photomicrograph of a thin-section of gabbro viewed in cross-polars mode. Plagioclase is striped and has low birefringence (low first-order white and gray interference colors). Brightly-colored augite has moderate-high birefringence (well-developed second-order interference colors in shades of blue, purple, red, and orange). *Sources*: Siim Sepp, CCBYSA; https://commons.wikimedia.org/wiki/File:Gabbro_pmg_ss_2006.jpg; last accessed 9/27/2020; Lapidary Lab, School of Oceanography and Earth Sciences and Technology, University of Hawaii.

has some north–south component of vibration. Since no light passes the analyzer, the mineral remains in extinction in all orientations. The quickest way to recognize a nonopaque, isotropic mineral or glass is when all crystals remain in extinction as the stage is rotated 360° under crossed-polars.

There is one situation in which **anisotropic crystals** mimic the behavior of isotropic materials. When anisotropic crystals are viewed looking down an optic axis (with the optic axis perpendicular to the stage) only one ray is viewed moving up the optic axis, vibrating east–west, and the mineral remains at extinction through a complete 360° rotation because the analyzer cannot pass east–west vibrating light. On the other hand, in all other orientations, such anisotropic minerals do not stay in the extinction position. Instead they go to extinction every 90° or four times during a 360° rotation of the stage. The extinction positions correspond to those crystal orientations where the vibration directions of the fast ray and the slow ray correspond to the orientation of the analyzer and the polarizer (upper and low polars) so that no light is transmitted to the eyepiece (Figure 6.12).

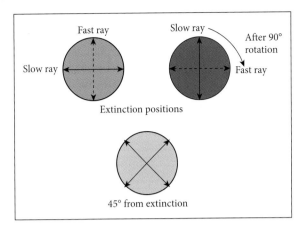

**Figure 6.12**  Diagram shows the correspondence between extinction positions and the vibration directions of the upper and lower polars (analyzer and polarizer lenses). These are parallel to the cross-hairs when they are oriented N–S and E–W. At 45° from the extinction position, the vibration directions of the two rays are oriented NE–SW and NW–SE respectively.

When anisotropic minerals are rotated to 45° from extinction, they display maximum brightness because they are in a position that maximizes the amount of light that passes through the analyzer after the two rays recombine.

**Extinction angles** (Table 6.1) are angles between extinction positions and mineral properties such as a cleavage, crystallographic axis, long axis or twin plane. To measure an extinction angle, the specimen is rotated until the property in question, for example a prominent cleavage, is parallel to the N–S cross-hair. Its position, read from the stage goniometer, is recorded. The specimen is then slowly rotated clockwise until it goes to maximum extinction. The new position is read from the stage goniometer and the difference between the two angles recorded is the extinction angle.

**Parallel extinction** occurs when the extinction angle is zero; that is when the mineral is at extinction when a cleavage, long dimension or other feature is parallel to the cross-hairs. Micas, such as biotite and muscovite have one prominent cleavage with very small extinction angles (<5°); their extinction is commonly described as near-parallel to cleavage. Quartz crystals are commonly elongated parallel to the c-axis (the vibration direction of the slow ray); they show extinction parallel to the long dimension of such crystals. Micas such as muscovite and biotite and carbonates such as calcite and dolomite display an unusual, characteristic, extinction phenomenon, variously referred to as **mottled or birdseye extinction.** Birdseye extinction occurs when, in the maximum extinction position, many small points of light can be seen. **Symmetrical extinction** characterizes minerals that go to extinction when a feature such as a pair of cleavages or crystal faces is bisected into symmetrical halves by the cross-hair. Triclinic minerals with their low symmetry, never display symmetrical extinction. **Uniform extinction** occurs when the entire crystal goes to extinction at one time. **Nonuniform extinction** occurs when different parts of a crystal go to extinction at different times. The major causes of nonuniform extinction are (1) twinning of crystals (2) chemical zoning in crystals, and (3) strain distortion of a crystal's structure.

**Twinned crystals** are easily recognized under crossed-polars. Because each twin component of the larger crystal possesses a different crystallographic orientation, each goes to extinction at a different stage of rotation; each has its own extinction angle. Simple **composition twins** are common in potassium feldspars such as orthoclase and sanidine, and in ferromagnesian minerals such as hornblende and augite. The twinning can be observed only in thin-sections cut across the twin plane and many crystals of these minerals are not twinned. In those that are, as the stage is rotated, one twin goes to extinction while the other does not; as the stage is rotated further their appearances are reversed as the second twin goes to extinction (Figure 6.13). These extinction angles between composition planes and extinction positions can be measured using the stage goniometer.

**Microcline** can generally be distinguished from other potassium feldspars such sanidine and orthoclase by its complex twinning. The twins appear as spindle-shaped crystals that are intergrown at a high angle giving rise to what is variously called **gridiron or scotch-plaid twinning** (Figure 6.14)

**Plagioclase** displays a variety of twin types. The most prominent in many cases is polysynthetic **albite twinning**. Albite twins consist of crystals that possess two sets of orientations that alternate with one another (Figure 6.15). As the stage is rotated, one set of alternating crystals goes to extinction at one time; the second set goes to extinction at another time which produces alternating dark and light "stripes," a bit like zebra stripes. The extinction angles at which the two extinctions occur can be used, along with other characteristics,

**Figure 6.13** Sanidine crystal (clear, low birefringence) showing simple composition (Carlsbad) twin crystal with the upper twin at extinction and the lower twin not at extinction. *Source*: Courtesy of Kurt Hollacher, Union College.

**Figure 6.14** Crossed-polars photomicrograph of microcline crystal that displays gridiron twinning and low birefringence (first-order gray and white colors) that make this mineral easy to recognize under crossed-polars. *Source*: Photo by permission of Kurt Hollacher, Union College.

**Figure 6.15** Crossed-polars photomicrograph of diorite (see Figure 6.8 for plane light view). Plagioclase crystal in the center (low birefringence, "stripes" due alternate extinction of albite twins with pericline twins at nearly right angles), hornblende (higher relief, moderate birefringence at lower left and right corners, two sets of cleavage not at right angles in left-hand crystal, and biotite (thin greenish crystal with moderately-high birefringence and one cleavage orientation. *Source*: Photo by permission of Kurt Hollacher, Union College.

**Figure 6.16** Crossed-polars photomicrograph of a volcanic rock. Large plagioclase phenocryst with low birefringence, simple Carlsbad twins, narrower albite twins, and chemical zoning shown by the shadows which indicate that the darker parts are near extinction and the lighter part farther from the extinction position. The phenocryst is set in a fine groundmass of microscopic crystals and some cryptocrystalline or glassy material. *Source*: Photo by permission of Kurt Hollacher, Union College.

to closely estimate the anorthite (An) content of the plagioclase. Albite twins sometimes occur with **Carlsbad twinning** which consists of simple paired composition twins like those in sanidine or orthoclase. Spindle-shaped **pericline twins** also occur, usually at an angle to the other twin types (Figure 6.15). These twin combinations can be used to closely approximate plagioclase compositions, but other methods now yield more accurate results.

Minerals that are part of a solid solution series may change composition as they grow. In such cases the crystals may display **chemical zoning** in which their composition changes from the early-formed inside of the crystal toward the later-formed margins of the crystal. Plagioclases and clinopyroxenes, especially in volcanic and shallow plutonic environments, are commonly zoned (Figure 6.16) as is garnet in many metamorphic rocks.

Nonuniform extinction can result from the deformation of crystals associated with strain. This distorts the crystal structure in a manner that causes different parts of the crystal to go to extinction at different times as the stage is rotated. Strained quartz (Figure 6.17) and micas crystal show **undulatory extinction** in which the extinction shadow sweeps across the crystal as the stage is rotated through several degrees.

**Intergrowths** between minerals can be observed most readily under crossed-polars. **Perthitic intergrowths,** caused by the exsolution of plagioclase from potash-rich feldspars as they cool below the solvus temperature (Chapter 3), occur as blebs, patches, and stringers of sodic plagioclase in a potassium feldspar host. The perthitic intergrowths of plagioclase go to extinction at different times than the host potassium feldspar (Figure 6.18). Similar exsolution intergrowths between orthopyroxene and clinopyroxene are common and produce parallel intergrowths referred to as **Schiller structure**.

Table 6.1 summarizes the optical characteristics of some common rock forming minerals in plane light and crossed-polar modes.

The more advanced conoscopic mode techniques discussed in the following section are not

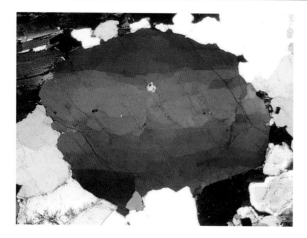

**Figure 6.17** Crossed-polars photomicrograph of a quartz crystal near the extinction position. Strain has produced nonuniform, undulatory extinction, so that the lower and upper parts of the crystal are at extinction while the central parts are not. The extinction shadow would sweep across the crystal toward the center as the stage is rotated. *Source*: Photo by permission of Kurt Hollacher, Union College.

**Figure 6.18** Crossed-polars photomicrograph of perthite. Exsolved plagioclase blebs and stringers trend upper right to lower left across potassium feldspar crystal. Note the low birefringence and first-order colors characteristic of both feldspars. *Source*: Photo by permission of Kurt Hollacher, Union College.

required for preliminary work with thin-sections including initial rock identification and interpretation. What they do provide are multiple methods for testing one's hypothetical identifications and for making more specific and accurate mineral identifications.

### 6.2.4 Conoscopic mode

A third important mode of observation available to investigators of Earth materials using petrographic microscopes is **conoscopic mode**. In this mode the sample is viewed (1) using the high power objective, (2) with the auxiliary condensing lens (conoscopic lens) in the substage assembly and (3) the Bertrand lens in the superstage assembly rotated into the optical path. The conoscopic lens receives plane polarized light that has passed though the fixed condensing lens and focuses it as a cone of light (thus conoscopic lens) onto a very small portion of the sample. The light that passes through the sample is no longer orthoscopic light, but conoscopic light which behaves in a rather different manner. One significant difference is that the all the light waves but one in a cone of light strike the sample at an angle less than 90°. Several new features are observed as conoscopic light interacts with the sample, the most significant of which are shadow-like figures called **interference figures**. Differential retardation of cones of light also produce rainbow-like colored lines, often of multiple orders, called **isochromatic lines or isochromes**. For thin-sections, the maximum number of orders of isochromatic lines is directly proportional to birefringence. Interpretation of interference figures and isochromes permits hypothetical identifications based on plane light and crossed-polars characteristics to be tested and confirmed, modified or rejected. These interpretations commonly involve the use of accessory plates that are placed into the accessory slot just above the objective nosepiece while the specimen is being observed in conoscopic mode.

Using conoscopic mode techniques, investigators can quickly determine whether an anisotropic material is uniaxial (tetragonal or hexagonal systems) or biaxial (orthorhombic, monoclinic or triclinic systems). With a bit more time and effort, one can determine which one of the two rays is the faster ray and which one is the slower ray. This allows one to determine whether the mineral is optically positive or optically negative. For biaxial minerals, the acute angle between the two optic axes (2V) can be closely estimated. The primary goals of conoscopic mode investigations are to improve the precision and accuracy of mineral identifications by determining (1) whether a crystal is uniaxial or biaxial, (2) whether it is optically positive or optically negative and, (3) for biaxial minerals the angle 2V between the two optic axes. Armed with knowledge of (1) the plane light characteristics, (2) the crossed-polars characteristics, (3) the conoscopic mode characteristics, and (4) a good reference book (Philpotts 1989; Nesse 2012; Barker 2014; Mackenzie et al. 2017) most minerals can be accurately identified by an experienced investigator. Determinative tables for the identification of rock-forming minerals using optical techniques are available in the appendices that support this text. MacKenzie et al. (2017) and Barker (2014) provide, in addition, comprehensive color photomicrographs.

## 6.3    THE OPTICAL INDICATRIX, INTERFERENCE FIGURES AND OPTIC SIGN DETERMINATIONS

In this section, the theory and methods used to determine the optic sign of uniaxial and biaxial minerals are discussed. The intention is that the discussion will permit students to make such determinations using a standard petrographic microscope. This will take practice.

### 6.3.1    The optical indicatrix

An **optical indicatrix** is a three-dimensional geometric figure used to represent the refractive indices in a crystal in all directions. The insights gained from understanding an optical indicatrix are essential for a clear understanding of optical microscopy. *Radial distances from the center of the indicatrix to its margins are proportional to the refractive index of light that vibrates in that direction.* The optical indicatrix is a sphere for isotropic materials in which the speed of light and the refractive index is the same in every direction. It is an ellipsoid for anisotropic materials which have a fast ray and slow ray directions perpendicular to each other and intermediate velocities and refractive indices in all other directions. Since ray paths are perpendicular

to the direction of vibration, the ray paths associated with a particular refractive index are tangent to the surface of the indicatrix at the point where the radial distance representing the refractive index intersects it. A two-dimensional representation of these relationships is shown below in a central section through a typical indicatrix (Figure 6.19). Note that the semi-major axis of the ellipse corresponds to the higher refractive index and therefore to the vibration direction of the slow ray. In addition, note that the semi-minor axis of the ellipse corresponds to the lower refractive index and therefore to the vibration direction of the fast ray. Remember: Light waves vibrate perpendicular to their propagation vector and its direction.

### 6.3.2    The isotropic indicatrix

The **isotropic indicatrix** is a sphere; an indicatrix of constant radius. This is because isotropic media have the same refractive index in every direction; because the velocity of light moving through an isotropic medium is the same in every direction. Every central section through the isotropic indicatrix is a circular section (Figure 6.20). No matter what the

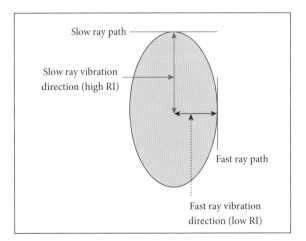

**Figure 6.19**   A cross-section through a typical ellipsoidal indicatrix showing the relationship between the long axis (with higher RI) of the elliptical cross-section and the associated slow ray vibration direction (red) and the short axis (with lower RI) of the elliptical cross-section and the associated fast ray vibration direction (blue).

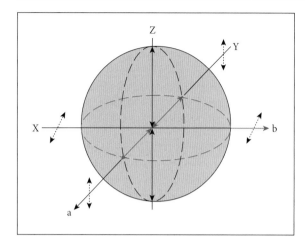

**Figure 6.20** Isotropic indicatrix shows sample ray paths "a" (parallel to Y-axis) and "b" (parallel to X-axis) with vibration directions perpendicular to ray paths. Two sample circular sections contain the potential vibration directions for ordinary light traveling along "a" (YZ section) and "b" (XY section) are shown as are potential vibration directions (double arrows) for plane polarized light.

direction of the ray path, its velocity and inversely proportional refractive index are invariant. There is only one refractive index, called **omega** ($n_\omega$) for any isotropic material.

When light enters an isotropic medium, no double refraction occurs; the light does not split into two rays. This explains why: (1) as plane polarized light enters an isotropic medium, it continues to vibrate in the same plane as it passes; (2) all light that leaves the isotropic medium is extinguished by the analyzer so that isotropic materials remain at extinction under crossed-polars in all positions of the stage.

### 6.3.3 The uniaxial indicatrix

A **uniaxial indicatrix** is an ellipsoid of revolution about the crystal's c-axis that represents the refractive index for waves vibrating in any direction within the crystal. Radial distances from the center of the ellipsoid to its surface are proportional to the refractive index for waves vibrating in that direction and being propagated at right angles to it. The semi-axis parallel to the c-axis has a length proportional to the refractive index (n) parallel to c and is called

**epsilon** ($n_\varepsilon$). The second semi-axis, perpendicular to the c-axis (and thus in the $a_1a_2$ plane) has a length proportional to the refractive index (n) perpendicular to c and is called **omega** ($n_\omega$). The **maximum birefringence (B)** is given by the absolute value of the difference between the two refractive indices ($B = |n_\varepsilon - n_\omega|$).

A uniaxial indicatrix (ellipsoid of revolution) has a unique **circular section** that is perpendicular to the c-axis and therefore perpendicular to the epsilon refractive index ($n_\varepsilon$). In standard representations of the uniaxial indicatrix, the unique circular section is horizontal and has a radius equal to the omega refractive index ($n_\omega$).

Elliptical **principal sections** through the center of the indicatrix that contain the c-axis are perpendicular to the unique circular section. In standard representations, principle sections have one semi-axis parallel to the c-axis with a length proportional to the epsilon refractive index ($n_\varepsilon$), and a second semi-axis perpendicular to the c-axis with a length proportional to the omega refractive index ($n_\omega$). If $n_\varepsilon > n_\omega$ the principal sections are prolate (extended) ellipses and the indicatrix is a prolate ellipsoid with one long axis parallel to the c-axis and two equal short axes in the circular section (Figure 6.21a). If $n_\varepsilon < n_\omega$ the principal sections are oblate (flattened) ellipses and the indicatrix is an oblate ellipsoid with one short axis parallel to the c-axis and two equal long axes in the circular section (Figure 6.21b).

All other sections through the uniaxial indicatrix that are neither parallel nor perpendicular to the c-axis and the epsilon refractive index are called **random sections**. Random sections have one semi-axis equal to the omega refractive index ($n_\omega$) and a second semi-axis equal to a refractive index called **epsilon prime** ($\varepsilon'$) that possesses a refractive index between that of omega and epsilon (Figure 6.22)

*The ordinary and extraordinary rays*

A uniaxial indicatrix, with one optic axis, is appropriate for crystals in the tetragonal and hexagonal systems. When light enters such crystals where $a_1 = a_2 \neq c$, in any direction not parallel to the optic axis (c-axis), it doubly refracts into two rays that move in different directions. Each wave vibrates in a plane perpendicular to its ray path and the two waves vibrate in mutually perpendicular planes.

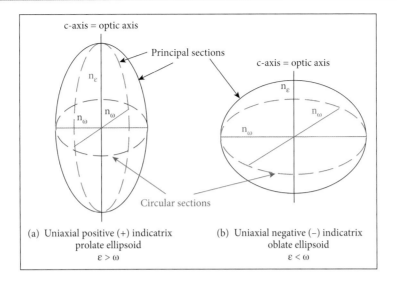

**Figure 6.21** (a) Prolate uniaxial indicatrix with a vertical long axis (blue) parallel to the c-axis, the optic axis and the epsilon (n$_\varepsilon$) refractive index. These are perpendicular to the two horizontal short axes (red) of the ellipsoid which are parallel to the omega (n$_\omega$) refractive index. (b) Oblate uniaxial ellipsoid with a vertical short axis (red) parallel to the c-axis, the optic axis and the epsilon (n$_\varepsilon$) refractive index. These are perpendicular to the two equal long axes (blue) of the ellipsoid which are parallel to the omega (n$_\omega$) refractive index.

One ray, called the **ordinary ray (o-ray = omega ray)** always moves parallel to the c-axis, which is parallel to the crystal's optic axis. The ordinary wave vibrates in directions perpendicular to the c-axis, which is in the a$_1$a$_2$ plane in which atomic configurations are uniform in every opposing direction. As a result, the ordinary wave has a constant velocity and is associated with a constant refractive index (n) called **omega (ω or n$_\omega$)**. Omega is proportional to the radius of the **circular section** perpendicular to the c-axis and to the optic axis in the uniaxial indicatrix.

The second wave, called **the extraordinary ray (e-ray = epsilon)**, moves in another direction that may be anywhere from perpendicular to nearly parallel to the ordinary wave, depending on its angle of refraction. Its velocity varies with direction because atomic configurations vary with e-ray direction. The velocity of the unique extraordinary wave that propagates perpendicular to the ordinary wave is associated with a refractive index called epsilon (ε or n$_\varepsilon$) which is perpendicular to omega (n$_\omega$). The epsilon vibration direction is parallel to the c-axis, and is contained in all principle sections through the indicatrix. Intermediate velocities of epsilon that are associated with

other e-ray directions possess intermediate refractive indices referred to as **epsilon prime (ε′ or n$_\varepsilon$′)** and are associated with random sections through the indicatrix.

Figure 6.22 depicts the three major types of central sections through a uniaxial indicatrix. The *unique circular section* (Figure 6.22a) is perpendicular to the optic axis, the c-crystallographic axis and ordinary ray path (labeled "b"). When light enters a uniaxial crystal traveling parallel to the optic axis, it does not split into two rays. Instead, it travels through the crystal as an ordinary wave with a constant refractive index equal to the radius of the circular section for light vibrating perpendicular to the optic axis. Crystals viewed looking down the optic axis (c-axis) with the circular section parallel to the stage appear isotropic because only the ordinary wave (omega = ω) vibrates in this plane and birefringence in this unique section is zero (Figure 6.22b). Such crystals remain at extinction under crossed-polars as the stage is rotated.

On the other hand, *principal sections* (Figure 6.22c) are viewed when the optic axis, parallel to the extraordinary wave vibration direction, is parallel to the stage. In such orientations, light passing through the crystal refracts into two rays. The ordinary ray moves

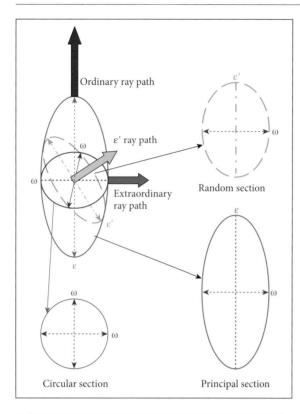

**Figure 6.22** Uniaxial positive indicatrix and three major types of section. (a). The unique circular section (red) is perpendicular to the optic axis and ordinary (fast) ray path (red arrow) and parallel to the ordinary ray vibration direction; no birefringence is displayed when such sections are parallel to the stage; (b) When random sections (green) are parallel to the stage, minerals display an intermediate apparent birefringence $|\varepsilon' - \omega|$. (c) principal sections (blue) are viewed when the extraordinary ray path (red arrow) is parallel to the stage as is its vibration direction the maximum birefringence $|\varepsilon - \omega|$ is observed in this orientation only.

parallel to the optic axis and vibrates perpendicular to it in the principle section with the refractive index omega ($\omega$). At the same time, the extraordinary wave moves perpendicular to the optic axis and vibrates parallel to it within the principle section with the refractive index epsilon ($\varepsilon$). Since both the minimum and maximum refractive indexes lie in the plane of principal sections parallel to the stage, the maximum birefringence between the two rays $|\varepsilon - \omega|$ is observed in this orien-

tation only. It is in these sections only that maximum birefringence can be determined since both maximum and minimum refractive indices can be measured.

When any other *random sections* are parallel to the stage, two rays pass through the crystal. The ordinary ray moves along the optic axis and vibrates in the circular section with the refractive index omega ($\omega$) However, the extraordinary ray moves perpendicular to the random section. It vibrates at right angles to the ordinary ray in the random section and has some intermediate refractive index epsilon prime ($\varepsilon'$). These sections display an intermediate apparent birefringence $|\varepsilon' - \omega|$.

### The optic sign

The **optic sign** of a uniaxial mineral is determined by the relative velocities and associated refractive indices of the ordinary ray and the extraordinary ray. A crystal is **uniaxial positive (+)** if the refractive index of epsilon is more than omega ($\varepsilon > \omega$); that is, if epsilon is the slow ray associated with the larger refractive index and omega is the fast ray associated with the smaller refractive index. In such cases, the uniaxial indicatrix is a prolate ellipsoid (Figure 6.23). The long axis is labeled Z and is the optic axis parallel to the c-axis. The two short axes are radii of the circular section, so are equal in length and are labeled X and Y.

A crystal is **uniaxial negative (−)** if epsilon is less than omega ($\varepsilon < \omega$); that is, if epsilon is the fast ray associated with the lower refractive index and omega is the slow ray associated with the higher refractive index. In such cases the uniaxial indicatrix is an oblate ellipsoid (Figure 6.24). The short axis is labeled Z and is the optic axis parallel to the c-axis. The two long axes are radii of the circular section, so are equal in length and are labeled X and Y.

### Uniaxial interference figures

Interference figures result from complex interference phenomena that occur between the fast and slow rays as they pass through the sample in conoscopic light and are processed by the analyzer. Detailed treatments of their origin are available in advanced books on optical mineralogy (Kerr 1977; Nesse 2004, 2012). *To obtain an interference figure*, it is necessary to follow several steps very carefully:

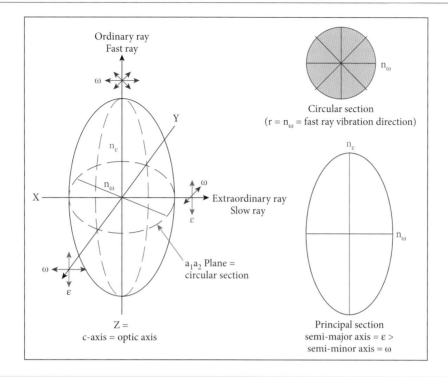

**Figure 6.23** Uniaxial positive indicatrix. The crystal c-axis is vertical and parallel to the epsilon vibration direction. The $a_1$ and $a_2$ crystallographic axes are horizontal and lie in the unique circular section parallel to the omega vibration direction. The ordinary ray, traveling parallel to the c-axis and vibrating perpendicular to it, is the fast ray (blue) with the lower refractive index ($n_\omega$). The extraordinary ray, traveling perpendicular to the c-axis (e.g., along X or Y) and vibrating parallel to it, is the slow ray (red) with the higher refractive index ($n_\varepsilon$). One of an infinite number of possible prolate elliptical principal sections that contains the vibration directions for both the fast ray ($\omega$) and the slow ray ($\varepsilon$) is shown. Random sections (with $\varepsilon'$ and $\omega$) are not shown.

(1) the crystal must be carefully focused under the high power objective, being sure not to touch the thin-section (cover glass on top!) with the microscope lens, (2) the analyzer must be inserted so that the crystal is being viewed under crossed-polars and (3) both the conoscopic lens and the Bertrand lens must be swung into the optical path and focused. The type of interference figure observed will depend on the orientation of the crystal and its indicatrix with respect to the rotating stage.

Uniaxial optic axis figure and optic sign determination
**The centered uniaxial optic axis (O.A.) figure** is observed when the optic axis is perpendicular to the stage so that the circular section ($n_\omega$) is parallel to the stage. To obtain a centered optic axis figure, one should select from a mineral population a crystal that shows minimum birefringence; one that appears to be nearly isotropic so that it remains dark under crossed-polars as the stage is rotated. A centered optic axis figure (Figure 6.25a) appears as a cross that consists of two **isogyres**, shadow-like features that are tapered toward the center of the field of view. The center of the area where the two isogyres cross is called the **melatope** and represents the position of the vertical optic axis or the c-axis along which the ordinary ray is traveling. The isogyres are produced by the extinction of conoscopic light as it passes through the analyzer. One isogyre represents light vibrating parallel to the polarizer which does not pass through the analyzer; the other represents light vibrating in the preferred direction of the analyzer that does not pass through the polarizer. These extinctions produce the **"uniaxial cross"** as the centered optic axis figure is sometimes known (Figure 6.25a). Any crystal that displays a uniaxial cross in conoscopic

mode is a uniaxial mineral that crystallized in either the tetragonal or the hexagonal system.

In minerals with reasonably high birefringence ($n_s - n_f > 0.10$), colored lines will appear in concentric circles around the melatope. The order of these lines increases with distance from the melatope and their hue intensities decrease in a manner consistent with the Michel–Levy color chart (Figure 6.10). These lines of constant color are called **isochromatic lines or isochromes** (Figure 6.25b). They represent cones of equal retardation produced when conoscopic light passes through the sample. When conoscopic light passes through the sample, light in the center of the cone travels the shortest distance and experiences minimum retardation whereas light near the margins of the cone travel the longest distance and experience maximum retardations. Retardation increases outward from the optic axis and is observed as isochromes of increasing order outward from the melatope.

Because conoscopic light strikes the sample at an angle that decreases with distance from the center of the cone, not all the light travels up the optic axis, even when the latter is perpendicular to the stage. Therefore, both an ordinary ray and an extraordinary wave are generated by refraction. To determine the optic sign of a uniaxial mineral, one must determine the relative velocities (inverse to refractive indices) of the ordinary and extraordinary rays. If the extraordinary ray is the slow ray so that the epsilon refractive index is more than omega ($\varepsilon > \omega$), the mineral is positive. If the extraordinary ray is the fast ray because the epsilon refractive index is less than omega ($\varepsilon < \omega$), the mineral is negative. A uniaxial centered optic axis figure (uniaxial cross) is ideal for this determination.

The vibration directions of the ordinary and extraordinary rays are always perpendicular to each other. However, their propagation directions change with orientation. In (+) minerals, the ordinary wave always vibrates tangent to the isochromes, whereas the extraordinary ray always vibrates perpendicular to the isochromes. At 45° from the isogyres, the extraordinary ray always bisects the quadrants so that it is oriented from NE–SW in the NE–SW quadrants and from NW–SE in

the NW–SE quadrants. In (–) minerals, these relationships are reversed (Figure 6.24).

For minerals of low birefringence, a **gypsum accessory plate** (or 1λ plate) is generally used to make sign determinations. The gypsum plate has a uniform retardation of 550 µm; that is, one order of interference on the Michel–Levy color chart. The gypsum plate is cut parallel to a principal section. Because gypsum is anisotropic, it has a slow ray direction that is marked on the plate and is usually perpendicular to the plate long axis (Figure 6.26) and is oriented from NE–SW when the plate is inserted into the optical path. The vibration direction of the fast ray is perpendicular to that of the slow ray and is generally parallel to the gypsum plate long axis which is oriented NW–SE. To use the gypsum plate, insert it into the accessory slot. When the slow ray direction of the gypsum plate is parallel to the slow ray direction of the crystal all colors will increase by one full order, the retardation of the gypsum plate. First-order grays and whites plus the retardation of the gypsum plate will add to produce shades of second-order blues (Figure 6.26a). When the slow ray direction of the gypsum plate is parallel to the fast ray direction of the crystal the colors will subtract (partially cancel). First-order grays and whites will increase by less than the retardation of the gypsum plate to produce first-order yellows (Figure 6.26b).

In **uniaxial positive minerals**, the extraordinary ray is the slow ray, its vibration direction epsilon is associated with the highest refractive index and $\varepsilon > \omega$. If the extraordinary ray associated with the epsilon vibration direction is the slow ray, when the gypsum plate is inserted, **addition** will occur in the **NE–SW quadrants**. In this case, the slow ray vibration direction of the mineral is parallel to the slow ray vibration direction of the gypsum plate, addition will occur, and first-order grays and white will turn **blue** when the plate is inserted. **Subtraction** will occur in the **NW–SE quadrants** where the slow ray vibration direction of the mineral is perpendicular to the slow ray vibration direction of the gypsum plate and first-order grays and whites will change to **yellows**. The dark gray-black isogyres will change to a reddish hue (Figure 6.26a).

In **uniaxial negative minerals**, the extraordinary ray is the fast ray, its vibration direction epsilon is associated with the lowest refractive

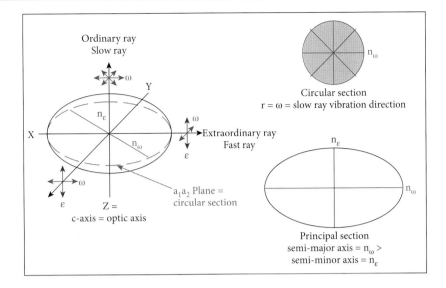

**Figure 6.24**  Uniaxial negative indicatrix. The crystal c-axis is vertical and parallel to the epsilon ($\varepsilon$) vibration direction. The $a_1$ and $a_2$ crystallographic axes are horizontal and lie in the unique circular section parallel to the omega ($\omega$) vibration direction. Note that the ordinary ray, traveling parallel to the c-axis and vibrating perpendicular to it, is the slow ray (red) with the higher refractive index ($n_\omega$). The extraordinary ray, traveling perpendicular to the c-axis (e.g., along X or Y) and vibrating parallel to it, is the fast ray (blue) with the lower refractive index ($n_\varepsilon$). One of many elliptical principle sections that contain the vibration directions for both the fast ray ($\omega$) and the slow ray ($\varepsilon$) is shown. Random sections (with $\varepsilon'$ and $\omega$) are not shown.

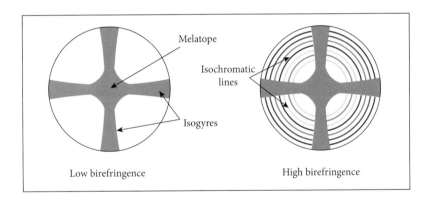

**Figure 6.25**  Centered optic axis figures for uniaxial minerals display two isogyres in the form of a cross, the center of which is the trace of the optic axis or melatope. (a) Minerals with low birefringence show first-order grays and whites in the quadrants between the isogyres; (b) Minerals with higher birefringence display isochromatic lines concentric about the melatope, whose numbers increase with retardation; that is with birefringence for a sample of constant thickness.

index, and $\varepsilon < \omega$. If the extraordinary ray associated with the epsilon vibration direction is the fast ray, when the gypsum plate is inserted, **subtraction** will occur in the **NE–SW quadrants**. In this case, the fast ray vibration direction of the mineral is parallel to the slow ray vibration direction of the gypsum plate

and first-order grays and white will turn **yellow** when the plate is inserted. *Addition* will occur in the **NW–SE quadrants** where the slow ray vibration direction of the mineral is perpendicular to the slow ray vibration direction of the gypsum plate and first-order grays and whites will change to **blue**. The dark

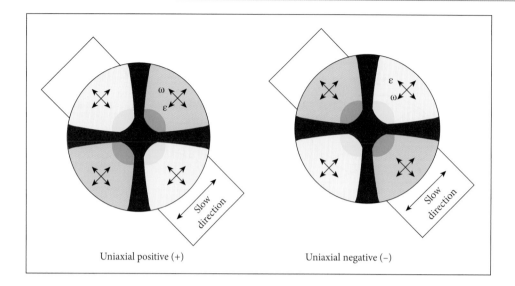

Uniaxial positive (+)                    Uniaxial negative (−)

**Figure 6.26** Use of the gypsum plate for optic sign determinations of uniaxial minerals. When the gypsum plate is inserted, addition (blues) occurs in the quadrants where the slow vibration direction is parallel to that of the gypsum plate and subtraction (yellows) occurs where the slow vibration direction is perpendicular to that of the gypsum plate. Blue in the NE–SW quadrants means that the extraordinary ray is the slow ray associated with a higher refractive index so that $\varepsilon > \omega$ and the mineral is uniaxial positive. Yellow in the NE–SW quadrants indicates that the extraordinary ray is the fast ray associated with a lower refractive index (epsilon) so that $\varepsilon < \omega$ and the mineral is uniaxial negative. *Source*: Courtesy of Gregory Finn.

gray-black isogyres will change to a reddish hue (Figure 6.26b).

If the gypsum plate is used to determine the sign of higher birefringent minerals, care must be taken to examine the color changes in those portions of each quadrant immediately adjacent to the melatope where the colors were first-order grays or whites prior to the insertion of the gypsum plate. Then the rules about blue from addition and yellow from subtraction hold true. Blue in the NE–SW quadrants indicates a uniaxial positive mineral, and yellow in the NE–SW quadrants signifies a uniaxial negative mineral.

The behavior of isochromatic lines associated with an optic axis figure can be also used to determine optic sign. In this case, a **quartz wedge compensation plate** is inserted into the accessory slot and the movement of the isogyres is observed as the wedge is slowly inserted. As with the gypsum plate, the slow direction of the quartz wedge is NE–SW and the fast direction is NW–SE. Because the wedge is tapered, inserting it progressively increases the retardation and therefore the amount of addition or subtraction that occurs

in each quadrant. If addition occurs, the isogyres appear to move inward toward the melatope. This occurs because first-order isochromes become higher-order isochromes during insertion so that higher-order isochromes move closer to the melatope. If subtraction occurs, the isogyres appear to move outward away from the melatope.

To summarize, addition in the NE–SW quadrants (isochromes move in) and subtraction in the NW–SE (isochromes move out) indicates that the mineral is uniaxial positive (Figure 6.27a). Subtraction in the NE–SW quadrants (isochromes move out) and addition in the NW–SE quadrants (isochromes move in) indicates that the mineral is uniaxial negative (Figure 6.27b).

When the optic axis is inclined more steeply than 60°–70° with respect to the stage, so that a random section is parallel to the stage, an **off-centered optic axis figure** may be observed. As the stage is rotated the trace of the optic axis, the melatope, will move in the field of view. As long as the melatope remains in the field of view throughout rotation, quadrants can be recognized. In such cases, the quadrants

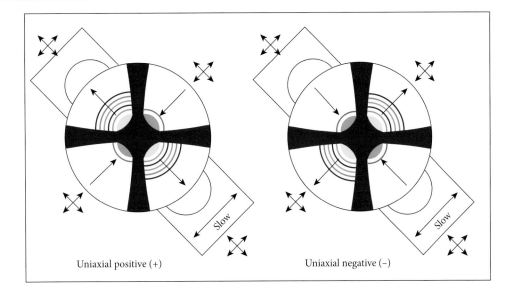

Uniaxial positive (+)          Uniaxial negative (−)

**Figure 6.27**  Use of the quartz wedge for optic sign determinations in minerals with high birefringence and well-developed isochromatic lines. If isochromes move toward the melatope in the NE–SW quadrants and away from the melatope in the NW–SE quadrants, the mineral is positive. If isochromes move away from the melatope in the NE–SW quadrants and toward the melatope in the NW–SE quadrants, the mineral is negative. *Source*: Courtesy Gregory Finn.

of addition and subtraction can be determined using an accessory plate and the optic sign can be determined. When the angle of inclination of the optic axis with respect to the stage is less than 60°, the optic axis figure is so off-centered that the melatope leaves the field of view as isogyres sweep through it every 90°. In such cases, it is best to look for another crystal of the same mineral, with lower birefringence, to find a more centered optic axis figure in order to make an accurate determination of optic sign.

Optic normal or flash figure
When the optic axis lies within the plane of the stage so that a principle section is being viewed, both the omega and epsilon vibration directions are being observed (Figure 6.22). In this orientation the crystal will show maximum birefringence under crossed-polars. In conoscopic mode, **a uniaxial flash figure** or **optic normal figure** (Figure 6.28) will be observed each time the optic axis is parallel to one of the two polarizing lenses. As the stage is rotated, an interference figure appears every 90°, in the form of a broad cross that nearly fills the field of view. With a few degrees of additional rotation, it breaks up into two curved isogyres that

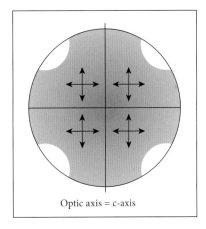

Optic axis = c-axis

**Figure 6.28**  The orientation of the uniaxial indicatrix (epsilon and omega vibration directions parallel to the stage) that produces uniaxial flash figures and the resulting flash figure observed every 90° during rotation when the optic axis is parallel to one of the polarizing lenses. *Source*: Courtesy of Gregory Finn.

leave the field of view in the quadrants that contain the optic axes. It appears, and then disappears in a "flash" – now you see it, now you don't – as the stage is rotated through a few degrees. The flash figure appears when both the extraordinary ray vibration direction (parallel to the optic axis) and the ordinary ray vibration direction (perpendicular to the optic axis) are parallel to the cross-hairs. The flash figure is not easy to use to determine optic sign, but it does alert the viewer to the fact that a principle section is essentially parallel to the stage. In such orientations, both the omega and epsilon vibration directions lie in the plane of the stage. In grain mount investigations with immersion oils, these orientations permit omega and epsilon to be measured and birefringence ($|\varepsilon-\omega|$) to be accurately determined.

### 6.3.4   The biaxial indicatrix

As their name implies, **biaxial crystals** possess two optic axes. You may recall that all minerals that crystallize in the orthorhombic, monoclinic, and triclinic systems where $a \neq b \neq c$ (except incidentally) are biaxial. The optical features of biaxial minerals can be illustrated using a biaxial indicatrix.

A **biaxial indicatrix** (Figure 6.29) is a **triaxial ellipsoid** that represents the refractive indices of a biaxial crystal in every direction. It possesses three mutually perpendicular axes that can be labeled X, Y, and Z. **X is the short axis** of the ellipsoid which is parallel to the vibration direction of the fast ray and is associated with the **minimum refractive index called alpha** ($n_\alpha$ or $\alpha$). The **long axis is the Z axis** which is parallel to the vibration direction of the slow ray and is associated with the **maximum refractive index called gamma** ($n_\gamma$ or $\gamma$). The **intermediate axis is called Y** and is associated with the vibration direction of an intermediate ray that possesses an **intermediate refractive index called beta** ($n_\beta$ or $\beta$). For every biaxial mineral, alpha is less than beta which is less than gamma ($\alpha < \beta < \gamma$). The elliptical XY plane is called the **optic plane** (Figure 6.29) because it contains the two optic axes that characterize biaxial minerals. It also contains the vibration directions of the fast ray and slow ray and therefore the maximum (gamma) and minimum (alpha) refractive indices. As a

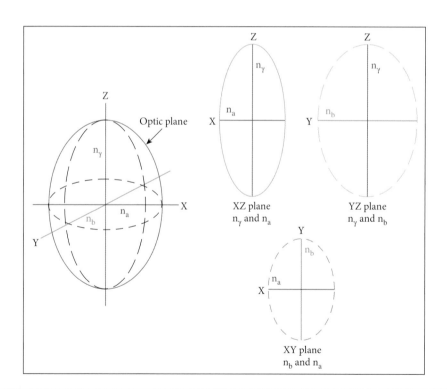

**Figure 6.29**   The general relationships between the three axes (X, Y, and Z) of the biaxial indicatrix and refractive indices ($\alpha$, $\beta$, and $\gamma$) associated with each axis. The XZ plane is called the optic plane because it contains the two optic axes. The three principle planes, each of which contains two semi-axes of the triaxial ellipsoid and two refractive indices are also shown. *Source*: Courtesy Gregory Finn.

result, biaxial minerals exhibit maximum birefringence ($|\alpha - \gamma|$) when the optic plane is parallel to the stage.

All triaxial ellipsoids such as the biaxial indicatrix contain two **circular sections** (Figure 6.30). Each of these circular sections has a radius equal to the beta ($\beta$) refractive index. Light rays moving perpendicular to the circular sections have vibration directions that lie in the circular sections that possess the same refractive index in every direction. The directions perpendicular to the two circular sections are the **two optic axes**. The angle between the two optic axes, measured in the optic plane, is called **optic angle (2V)**. The angle bisected by the X axis is called **2Vx** and the angle bisected by the Z-axis is called **2Vz**. If just one angle 2V is stated, it is the acute angle between the two optic axes (2Vz in Figure 6.30, but 2Vx in other situations). As with light moving parallel to the optic axis and perpendicular to the circular section in uniaxial minerals, light moving along either of the two optic axes in biaxial minerals does not doubly refract. The apparent birefringence in these two directions is zero. The circular

sections are also called **optic normal sections** because each is perpendicular to one of the two optic axes. Note that intersection of the two circular sections is a line parallel to the Y-axis of the indicatrix. This line is an **optic normal line** and is parallel to the intermediate ray vibration direction with a refractive index of Beta ($\beta$). Biaxial minerals exhibit minimum (zero) birefringence when an optic normal (circular) section is parallel to the stage and an optic axis is perpendicular to it.

A line in the optic plane that bisects the acute angle between the optic axes (optic angle or 2V) is called an **acute bisectrix (Bxa)**. Depending on mineral optics, the acute bisectrix may bisect 2Vx or 2Vz. The complimentary line that bisects the obtuse angle between the optic axes is the **obtuse bisectrix (Bxo)**. In Figure 6.30, the acute bisectrix (Bxa) bisects 2Vz and the obtuse bisectrix (Bxo) bisects 2Vx.

Light that enters biaxial minerals in any direction not parallel to one of the optic axes refracts into **two extraordinary rays.** Unless the rays are vibrating parallel to one of the axes of the indicatrix, their refractive indices

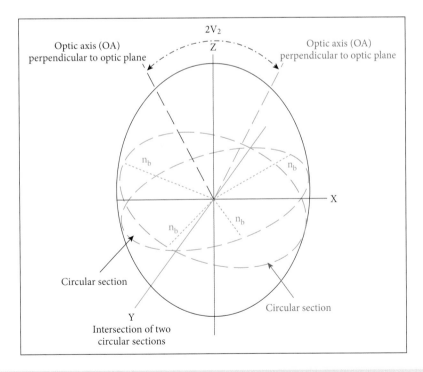

**Figure 6.30** Diagram that depicts two circular sections (with refractive index $\beta$) perpendicular to which are two optic axes separated by an acute angle (2V) that is contained in the XZ (optic) plane. *Source*: Courtesy of Gregory Flinn.

will be given by alpha prime (nα′ or α′) and gamma prime (nγ′ or γ′). Birefringence of such grains will vary from near minimum when the optic axis is almost perpendicular to the stage, to near maximum when the circular sections are nearly perpendicular to the stage and the optic section and the two optic axes are nearly parallel to it.

The **optic sign** of biaxial minerals may be either positive or negative (or rarely neutral). Minerals are **biaxial positive [biaxial (+)]** when (1) Z is the acute bisectrix, so that (2) gamma (γ) is the acute bisectrix, and (3) the slow ray vibration direction (perpendicular to the slow ray path) is the acute bisectrix. The acute bisectrix is 2Vz and the obtuse bisectrix is 2Vx (Figure 6.31a). Minerals are **biaxial negative [biaxial (−)]** when these relationships are reversed (Figure 6.31b). In this case (1) X is the acute bisectrix associated with (2) the alpha (α) refractive index and (3) the fast ray vibration direction (perpendicular to the fast ray path) is the acute bisectrix. The acute bisectrix is 2Vx and the obtuse bisectrix is 2Vz. Biaxial minerals are **optically neutral** when their 2V equals 90° so that the acute bisectrix

and the obtuse bisectrix are undefined. Note that if 2V were to become zero, the mineral would possess only one optic axis and be uniaxial.

The **optic orientation** is defined by the relationship of the optical indicatrix to the crystallographic axes (Nesse 2012). For **orthorhombic minerals**, the three mutually perpendicular crystallographic axes (a, b, and c) correspond in some way to the axes of the indicatrix X, Y, and Z (α, β, and γ refractive indices). Which crystallographic axis corresponds to which axis of the indicatrix depends on how the crystallographic axes were originally defined. In the **monoclinic system**, the b crystallographic axis (which is perpendicular to the ac plane) corresponds to one of the axes of the indicatrix (X, Y or Z). However, because the crystallographic axis and the c crystallographic axis are inclined with respect to each other, they do not correspond to axes of the indicatrix, except by chance. In the **triclinic system**, where the crystallographic axes do not intersect at right angles, there is no direct correspondence between the crystallographic axes and the axes of the indicatrix, except by chance.

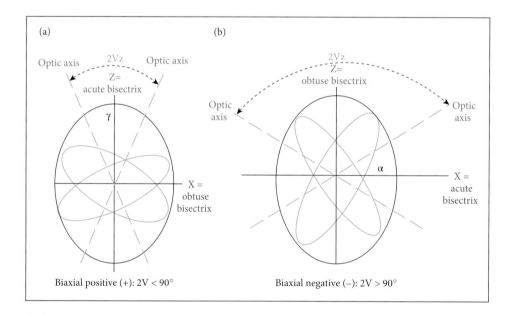

**Figure 6.31** Depiction of simplified versions of the positive and negative biaxial indicatrix. (a) A positive indicatrix in which the acute bisectrix (Bxa) is Z and the slow ray vibration direction gamma (γ). (b) Depicts a negative indicatrix in which the acute bisectrix (Bxa) is X and the fast ray vibration direction alpha (α).

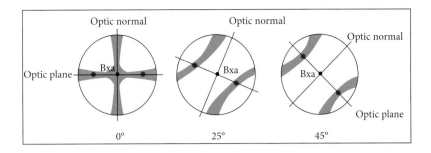

**Figure 6.32** An acute bisectrix figure for a biaxial mineral with a moderate 2V. When the optic normal, parallel to thick isogyre, and the optic plane, parallel to thin isogyre that contains the two melatopes (red dots) are parallel to the cross hairs, a crude cross is formed. As the stage is rotated, the optic normal and optic plane rotate and the cross splits into to two curved isogyres with the melatopes located at the points of maximum curvature. At 45° of rotation, the two isogyres reach maximum separation. When the Bxa figure is centered, the Bxa (black dot) remains in the center of the field of view during rotation. *Source*: Modified after Gregory Finn, with permission.

### 6.3.5 Biaxial interference figures

As with uniaxial interference figures, **biaxial interference figures** are the products of complex extinction of light by the analyzer (upper polarizing lens) when crystals are viewed in conoscopic mode. To obtain any interference figure, it is necessary to follow several steps very carefully: (1) the crystal must be carefully focused under the high power objective being sure the lens does not touch the sample, (2) the analyzer must be inserted so that the crystal is being viewed under crossed-polars, and (3) both the conoscopic lens and the Bertrand lens must be swung into the optical path. The interference figure will change as the stage and the orientations of the fast, intermediate and slow rays are rotated. The type of interference figure observed will depend on the orientation of the crystal and its indicatrix with respect to the stage.

*Acute bisectrix figures*

An **acute bisectrix figure (Bxa figure)** is observed when the acute bisectrix is perpendicular to the stage. If the acute bisectrix is perpendicular to the stage, the optic normal parallel to the Y-axis of the bisectrix and the Bxo within the optic plane both lie in the plane of the stage at right angles to one another. As the stage is rotated the trace of the optic normal (with refractive index β) remains perpendicular to the trace of the optic plane. The acute bisectrix figure consists of two

isogyres that change as the stage is rotated (Figure 6.32). When the optic plane is parallel to the E–W cross-hair and the optic normal is parallel to the N–S cross-hair or *vice versa*, a crude cross centered on the acute bisectrix is observed (Figure 6.32a). The **thicker isogyre** of the cross is **parallel to the optic normal** and the **thinner isogyre** is **parallel to the trace of the optic plane.** The two melatopes which represent the traces of the two optic axes occur in the optic plane on either side of the Bxa where the thinner isogyre is the thinnest (Figure 6.32a). During every 90° rotation a cross will form; four times during a complete rotation. The orientations of the thick and thin isogyres will alternate between N–S and E–W every 90° as the orientations of the optic normal and the optic plane rotate through 90°. As the stage is rotated from these positions, the cross begins to separate into two curved isogyres (Figure 6.32b) that move relatively slowly away from each other through 45° of rotation, and then begin to converge again to reform the cross after 90°. For a centered Bxa figure in minerals with an optic angle (2V) less than 55–60° (<55°), both isogyres remain in field of view throughout rotation. The maximum separation of the melatopes increases from zero for a 2V of zero (think uniaxial cross where no melatope separation occurs because there is only one optic axis) to progressively larger separations in the 45° position. In the 45° position (Figure 6.32c), (1) the two melatopes are at the points of maximum curvature on the two isogyres, (2) the trace of the optic plane is a line through the two melatopes, (3) the Bxa lies on this line, half way between the two melatopes, (4) the Bxo lies along this line, outside the field of view, on the concave side of the isogyres (other side of the melatopes), and (5) the optic normal is along a line perpendicular to the trace of the optic plane half way between the two isogyres. For optic angles larger than 60°, the two melatopes leave the field of view before the 45° position is reached.

Recall that crystals are **biaxial positive** when (1) Z is the acute bisectrix, so that (2) gamma ($\gamma$) is the acute bisectrix, so that (3) the slow ray vibration direction is the acute bisectrix. In such cases the acute bisectrix is 2Vz and the XY plane is parallel to the stage. Between the melatopes, the optic plane trace (parallel to X vibration direction) is the fast ray vibration direction associated with the minimum refractive index alpha ($\alpha$), and the optic normal trace parallel to Y is the fast ray vibration direction associated with the intermediate refractive index beta ($\beta$). On the concave side of the melatopes, the optic plane trace is the slow ray vibration direction associated with a larger refractive index gamma prime ($\gamma'$) and the optic normal trace parallel to Y is the slow ray vibration direction associated with the intermediate refractive index beta ($\beta$). Crystals are biaxial negative when (1) X is the acute bisectrix associated with (2) the alpha ($\alpha$) refractive index, so that (3) the fast ray vibration direction is the acute bisectrix. In such cases the acute bisectrix is 2Vx and the YZ plane is parallel to the stage. Between the melatopes, the optic plane trace (parallel to X vibration direction) is the slow ray vibration direction associated with the maximum refractive index gamma ($\gamma$), and the optic normal trace is the fast ray vibration direction associated with the refractive index beta ($\beta$). On the concave side of the melatopes, the optic plane trace is the fast ray vibration direction associated with a smaller refractive index alpha prime ($\alpha'$), and the optic normal trace parallel to Y is the slow ray vibration direction associated with the intermediate refractive index beta ($\beta$).

If a gypsum plate is inserted into the accessory slot, the quadrants of addition and subtraction will allow the optic sign of the mineral to be determined. As shown in Figure 6.32

this is most easily accomplished in the 45° (or 135°) position. The slow ray of the gypsum plate is oriented NE–SW and its fast ray is oriented NW–SE. If the slow ray vibration direction of the mineral is oriented parallel to that of the gypsum plate, addition will occur, and first-order white-gray colors will be increased to second-order blue colors by the 550 µm retardation of the gypsum plate. If the slow ray vibration direction of the mineral is oriented perpendicular to that of the gypsum plate, subtraction will occur and first-order grays-whites will become shades of yellow.

For **biaxial positive minerals**, in the 135° position, **first-order white-grays will be converted to blues in the NE–SW quadrants** by addition. This occurs on the concave side of the isogyres in the 135° position because the optic plane that contains the slow ray vibration direction ($\gamma'$) is parallel to the slow ray direction of the gypsum plate in the 135° position. In the 135° position, **first-order white-grays will be converted to yellows in the NW–SE quadrants** by subtraction. Subtraction occurs between the isogyres because the optic normal contains the fast ray vibration direction ($\beta$) which is perpendicular to the fast ray of the gypsum plate. Figure 6.33a summarizes these relationships for several positions of the stage during rotation where 180° simply returns the situation to that seen in the 0° position. In the initial position, the optic plane parallel to X ($\alpha$) is E–W and the optic normal is N–S parallel to Y ($\beta$) is N–S. They have been rotated clockwise in increments of 45° between images.

For **biaxial negative minerals**, the same types of arguments can be made to demonstrate that **first-order white-grays will be converted to yellows in the NE–SW quadrants** because the slow ray vibration direction is perpendicular to the slow ray vibration direction of the gypsum plate so that subtraction occurs and **first-order white-grays will be converted to blues in the NW–SE quadrants** where the slow ray vibration direction is parallel to the slow ray vibration direction of the gypsum plate so that subtraction occurs These relationships are summarized in Figure 6.33b. Their similarity to those that relate quadrants and colors for uniaxial minerals is worth remembering.

Along the acute bisectrix, the two rays are vibrating perpendicular to each other. At 45°

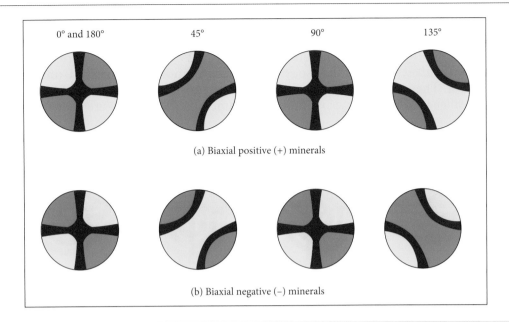

| 0° and 180° | 45° | 90° | 135° |

(a) Biaxial positive (+) minerals

(b) Biaxial negative (–) minerals

**Figure 6.33** (a) The appearance of a Bxa interference figure for a biaxial (+) mineral of low-moderate birefringence viewed in various orientations with the gypsum plate inserted. When the gypsum plate is inserted, positive minerals show white-gray changing to blue in the NE–SW quadrants and to yellow in the NW–SE quadrants. (b) A similar sequence that contrasts the appearance of a biaxial (–) mineral of similar birefringence where changes are to yellow in the NE–SW quadrants and to blue in the NW–SE quadrants.

addition occurs between isogyres because slow parallels slow and fast parallels fast. At 135°, subtraction occurs between isogyres because slow parallels fast and fast parallels slow! This is the key.

As is the case with uniaxial minerals, biaxial minerals with moderate to high birefringence display one or more orders of isochromatic lines. Because biaxial minerals have two optic axes and two melatopes, the pattern of isochromes is somewhat different. As with uniaxial minerals, isochromatic lines are concentrically arranged about the melatope, but because there are two melatopes, the pattern roughly resembles a figure eight (Figure 6.34).

For high birefringent minerals such as the one illustrated by Figure 6.34, optic sign determinations are most easily obtained by inserting the *quartz wedge* when the mineral is in the 45° or 135° position with the optic normal NW–SE or NE–SW. As with uniaxial minerals, optically **positive minerals** are characterized by addition in the NE–SW quadrants that causes isochromes to move inward toward the melatopes and by subtraction in the NW–SE quadrants that causes isochromes to move outward away from the melatopes. For optically **negative minerals**, the movements are reversed; isochromes move outward in the NE–SW quadrants and inward in the NW–SE quadrants.

*Optic axis figures*

A centered optic axis figure is an interference figure viewed when one of the optic axes is perpendicular to the stage. In such orientations, a circular section is parallel to the stage. In this orientation, only one refractive index ($\beta$) exists, so the mineral displays no birefringence. To obtain a centered optic axis figure, one should look for a biaxial mineral crystal that displays minimal birefringence, one that remains dark as the stage is rotated, and then examine it under conoscopic mode.

The **centered biaxial optic axis figure** consists of a single isogyre whose curvature is inversely proportional to the crystal's 2V (Figure 6.35). The **melatope** (trace of an optic axis) is located on the thinnest part of the isogyre at its point of maximum curvature. On rotation of the stage, the melatope will remain centered in the field of view as the isogyre rotates around it. The isogyre will be straight when it is parallel to one of the cross-hairs, and becomes concave to the NE (shown in Figure 6.34), SE, SW, and NW in the 45°, 135°, 225°, and 315° positions during a complete rotation. The curvature of the isogyres in these positions is inversely proportional to mineral's 2V. An isogyre that remains nearly

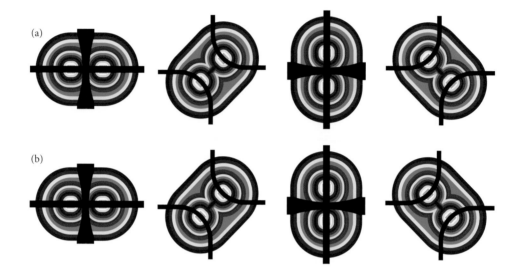

**Figure 6.34** (a) The appearance of a Bxa interference figure for a biaxial (+) mineral of moderate-high birefringence viewed in various orientations. When a quartz wedge is inserted into the accessory slot, positive minerals display isochromes that move in toward the melatopes in the NE–SW quadrants and outward in the NW–SE quadrants. (b) A similar sequence that contrasts the appearance of a biaxial (–) mineral of similar birefringence where changes are outward in the NE–SW quadrants and inward the NW–SE quadrants. *Source*: After Steven Dutch, with permission.

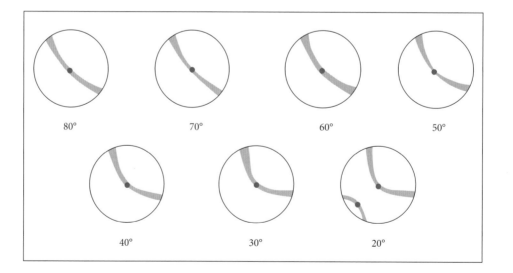

**Figure 6.35** Depicts centered optic axis figures in a position of maximum curvature, with the melatope (red dot). The amount of curvature in such positions is inversely proportional to the crystal's optic angle (2V), with no curvature corresponding to a 2V of 90°.

straight indicates a large 2V, while one that is strongly curved indicates a small 2V (Figure 6.35). Because 2V is the optic angle between the two optic axes, when 2V is less than 25°, a second isogyre may appear in the field of view (Figure 6.35). But only one of these will be a centered optic axis figure. One can imagine that if the 2V became zero, there would be only one optic axis and the two isogyres would "merge" into a uniaxial cross.

In addition to being useful for estimating the 2V of a crystal, biaxial optic axis figures can be used to determine its optic sign. The stage is rotated until the isogyre is in a position of maximum curvature, in which orientation it will appear to largely "enclose" a quadrant. The opposite quadrant will not be visible, so that only three quadrants are defined. The procedure is similar to that used for interpreting Bxa figures as is the explanation. Figure 6.36 illustrates both. For a *biaxial positive mineral*, the optic normal (β) is parallel to the tangent to the isogyre through the melatope. The acute bisectrix is Z and is located along the optic plane which passes through the melatope on the convex side of the isogyre. The obtuse bisectrix lies on the optic plane on the concave side of the isogyre.

On the convex side, the fast ray (α') is perpendicular to the optic normal (β) slow ray. On the concave side of the isogyre, the slow ray (γ') is perpendicular to the optic normal (β) fast ray. When the gypsum plate is inserted, its slow ray (NE–SW) is parallel to the crystal slow ray (γ') on the *concave side* so that addition occurs and *white-grays change to blues* (Figure 6.36a). On *the convex side* (toward the Bxa), the slow ray of the gypsum plate (NE–SW) is perpendicular to the crystal slow ray (β) so that *subtraction occurs and white-grays change to yellows*. Arguments similar to these can be made for all other orientations in which addition and subtraction occurs. The end result is that biaxial positive minerals show addition (white-gray changes to blue) in the NE–SW quadrants and subtraction (white-gray changes to yellows) in the NW–SE quadrants for all maximum curvature orientations of the centered optic axis figure. For biaxial negative minerals, the quadrants of addition and subtraction and therefore of color changes are reversed. Biaxial negative minerals show subtraction (white-gray changes to yellows) in the NE or SW quadrants and addition (white-gray changes to blues) in the NW–SE quadrants for all orientations of the centered optic

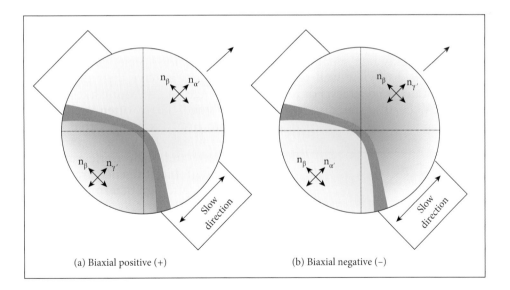

(a) Biaxial positive (+)          (b) Biaxial negative (−)

**Figure 6.36** Gypsum plate determination of optic sign, using a biaxial centered optic axis figure in a position of maximum curvature of the isogyre. If concave side of isogyre in SW (illustrated) or NE quadrant turns blue, and the convex side yellow, the mineral is biaxial positive (+). If concave side of isogyre in SW (illustrated) or NE quadrant turns yellow, and the convex side blue, the mineral is biaxial negative (−). *Source*: Courtesy of Gregory Finn. (a) Biaxial positive (+). (b) Biaxial negative (−).

axis figures in maximum curvature orientations (Figure 6.36b).

For minerals with high birefringence and many isochromatic lines around the melatope, centered optic axis figures can be analyzed using a quartz wedge to determine optic sign. The arguments are similar to those used previously. When the isogyre is in a position of maximum curvature (so that it "encloses" a quadrant), the quartz wedge is slowly inserted. If the mineral possesses a positive optic sign, isochromes will move in toward the melatope in the NE–SW quadrants and outward in the NW–SE quadrants. If the mineral has a negative optic sign, the movement of isochromes will be reversed to become outward in the NE–SW quadrants and inward in the NW–SE quadrants.

### Obtuse bisectrix figure

An **obtuse bisectrix figure (Bxo figure)** is observed when the Bxo is essentially perpendicular to the stage (Figure 6.37). In this orientation, the Bxa and optic normal lie in the plane of the stage. As the stage is rotated, a cross may form that superficially resembles that of an acute bisectrix figure. However, there are two important differences. First the isogyre that is parallel to the optic plane is thicker because it is farther away from the melatopes or optic axes in this orientation, so that the cross fills a larger proportion of the field of view. Secondly, as the stage is rotated, the cross breaks up quickly into separate isogyres that leave the field of view well before the stage has been rotated 45 degrees. For minerals with large 2V (>70°), the distinction

between an acute bisectrix figure and an obtuse bisectrix figure is difficult to make. Although the obtuse bisectrix figure can be used to determine sign, many workers prefer to find another crystal with an orientation appropriate for viewing an acute bisectrix or optic axis figure. If one does use an obtuse bisectrix figure to determine optic sign, the general rules for Bxa and optic axis figures are reversed. For positive minerals, subtraction (yellow, isochromes out) occurs in the NE–SE quadrants and addition (blue, isochromes in) occurs in the NW–SE quadrants. For negative minerals, addition (blue, isochromes in) occurs in the NE–SE quadrants and subtraction (yellow, isochromes out) occurs in the NW–SE quadrants. It is for this reason and the difficulty in distinguishing between Bxo and Bxa figures that many workers choose to look for clear examples of Bxa and/or optic axis figures when making sign determinations.

Biaxial minerals that appear isotropic and display no birefringence ($|\beta - \beta = 0|$) are minerals where one of the optic axes is perpendicular to the stage with one circular section parallel to the stage so that only one refractive index (beta) is viewed. These crystals should yield approximately centered optic axis figures, with four well-defined quadrants. This permits optic sign determination and estimates of the 2V. In addition, crystals within a population of the same mineral that exhibit slightly higher, but still low birefringence for the species are good candidates for crystals that have their Bxa perpendicular to the stage and the optic axes at a high angle to the stage. These often yield nearly centered acute bisectrix figures that are valuable for determining both the biaxial nature of the crystals and their optic sign. Crystals with very high birefringence are generally less useful for optic sign determinations.

### Biaxial flash (optic normal) figure

A **biaxial flash figure** is observed when the y-axis (optic normal) is perpendicular to the stage, so that the two circular sections are also perpendicular to the stage. As with the uniaxial flash figure, as the stage is rotated an interference figure appears every 90°, in the form of a broad cross that nearly fills the field of view. With a few degrees of additional rotation, it leaves the field of view. It appears, and

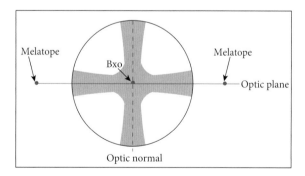

**Figure 6.37** Bxo figure and the orientation of the optic normal and optic plane when it is observed. *Source*: Courtesy of Gregory Finn.

then disappears in a "flash," as the stage is rotated through a few degrees. The flash figure is not easy to use to for optic sign determination, but it does alert the viewer to the fact that the optic plane is essentially parallel to the stage. In such orientations, Z ($\gamma$) and X ($\alpha$) and the two optic axes are parallel to the stage. In grain mount investigations with immersion oils, these orientations permit gamma and alpha to be measured and birefringence ($|\gamma - \alpha|$) to be determined.

Those who study Earth materials have at their disposal a "Pandora's Box" of ingenious instruments and methods that permit accurate and precise inferences concerning the previously hidden nature of Earth materials. For geoscientists, optical crystallography methods are supplemented, and often superseded, by a variety of chemical analytical techniques such as x-ray diffraction, mass spectrometry, cathodoluminescence, electron microprobe analysis, x-ray fluorescence, optical emission spectroscopy, optical absorption spectroscopy, scanning electron microscopy, atomic force microscopy and a variety of other specialized methods that together reveal the chemical composition and structure of Earth materials.

## CONTENT ASSESSMENT

1   Compare and contrast:
   (a) Wavelength and frequency
   (b) Ordinary and plane polarized light
   (c) Reflection and refraction
   (d) Isotropic and anisotropic substances
   (e) Refractive index (RI) and birefringence (B)
2   Explain the function(s) of the following elements of a petrographic microscope.
   (a) Lower polar (Nicol)
   (b) Orthoscopic condenser
   (c) Conoscopic lens (condenser)
   (d) Iris diaphragm
   (e) Upper polar (Nicol)
   (f) Objective lens
   (g) Ocular lens (eyepiece)
   (h) Eyepiece micrometer
3   In plane light mode, explain how the following distinguished from one another.
   (a) Opaque and nonopaque minerals
   (b) Color and pleochroism
   (c) Cleavage and fracture
   (d) High relief and low relief
   (e) Positive relief and negative relief
4   In crossed-polars mode, explain how the following are distinguished from one another.
   (a) Isotropic and anisotropic minerals
   (b) Uniform and nonuniform extinction
   (c) Symmetrical and parallel extinction
   (d) Quartz and feldspars
   (e) Plagioclase, orthoclase-sanadine, and microcline
   (f) Biotite, muscovite, and chlorite
   (g) Augite, enstatite, and hornblende
5   Discuss the major criteria by which isotropic, uniaxial, and biaxial minerals are recognized.
6   Discuss:
   (a) How to choose an appropriate crystal (grain) when searching for a centered uniaxial optic axis interference figure.
   (b) The best ways to distinguish between uniaxial positive (+) and uniaxial negative (–) minerals using the accessory plate and the quartz wedge compensation plate.
7   Discuss:
   (a) How to choose an appropriate crystal (grain) when searching for a centered biaxial optic axis interference figure.
   (b) The best ways to distinguish between uniaxial positive (+) and uniaxial negative (–) minerals using the gypsum accessory plate and the quartz wedge compensation plate with such a figure.
   (c) How to choose an appropriate crystal (grain) when searching for an acute bisectrix (Bxa) interference figure.
   (d) The best ways to distinguish between uniaxial positive (+) and uniaxial negative (–) minerals using the gypsum accessory plate and the quartz wedge compensation plate with such a figure.

## REFERENCES

Barker, A.J. (2014). *A Key for Identification of Rock-Forming Minerals in Thin Section.* London: CRC Press 182 pp.

Kerr, P.F. (1977). *Optical Mineralogy*, vol. 4. New York: McGraw-Hill 452 pp.

MacKenzie, W.S., Adams, A.E., and Brodie, K.H. (2017). *Rocks and Minerals in Thin-section*, 2e. London: CRC Press 242 pp.

Nesse, W.D. (2004). *An Introduction to Optical Mineralogy*, vol. 3. New York: Oxford University Press 370 pp.

Nesse, W. (2011). *Introduction to Mineralogy*, 2e. Oxford University Press 496 pp.

Nesse, W. (2012). *Introduction to Optical Mineralogy*, 4e. Oxford University Press 384 pp.

Philpotts, A.R. (1989). *Petrography of Igneous and Metamorphic Rocks.* New York: Prentice-Hall 178 pp.

# Chapter 7

# Igneous rock texture, composition, and classification

## 7.1 MAGMA AND IGNEOUS ROCKS

Igneous rocks derived from magma constitute the bulk of Earth's crust. **Magma** is molten material generated by partial melting of Earth's mantle and crust. Magma contains liquids, crystals, gases, and rock fragments in varying proportions that depend upon temperature, pressure, and chemical conditions. At temperatures over 1200 °C, above the crystallization temperature of most minerals, magma contains mostly liquids and dissolved gases. With cooling, magma becomes progressively more enriched in solid crystalline minerals at the expense of liquid melt.

Gas solubility in liquids is related to pressure. At high pressures in the lower crust and mantle, gases are dissolved in liquid magma. As magma rises toward Earth's surface, decompression causes gases to segregate from the melt as a separate phase. Opening a carbonated beverage bottle produces a similar reaction. When the bottle is capped, carbon dioxide gases are dissolved in the liquid under pressure. The removal of the bottle cap induces a pressure reduction such that gases segregate (**exsolve**) from the solution. This **exsolution** process creates gas bubbles that expand and rise toward the bottle's top as a separate "foam" phase, analogous to a volcanic eruption. Magma under high pressure conditions has a high volatile (dissolved gas) saturation point whereas magma under low pressure has a low volatile saturation point. We will consider these aspects later in this chapter when we consider noncrystalline rock textures.

Igneous rock texture relates to cooling history. Magma that solidifies within Earth produces **intrusive** or **plutonic** rocks with coarse-grained,

*Earth Materials*, Second Edition. Kevin Hefferan and John O'Brien.
© 2022 John Wiley & Sons Ltd. Published 2022 by John Wiley & Sons Ltd.
Companion website: www.wiley.com/go/hefferan/earthmaterials2

crystalline textures. Intrusive rocks develop from magma that cools slowly within Earth's interior to produce large, visible crystals in plutons. **Plutons** are magma bodies of various sizes, shapes, and depths that store magma within Earth. Recent research suggests that magma is largely stored in "mush reservoirs" where magma resides in the pore spaces between solid crystals and chemically reacts with the surrounding crystalline host rock material (Jackson et al. 2018).

As most magma is probably stored in rock fractures and pore spaces, plutons may contain relatively small amounts of melt at any given time. When rock fractures develop, magma migrates from rock pores into the open fissures and rises toward Earth's surface. Magma's buoyancy is due to its lower density relative to the surrounding rock. By further lowering magma density through expansion, gases provide buoyant forces propelling magma upward and producing violent, explosive volcanic eruptions. In a sense, dissolved gases within magma are analogous to fuel blasting a rocket from its launch pad. Magma that rises and erupts onto Earth's surface is called **lava**. Lava reaches the surface via vents generating lava flows and associated volcanic debris. **Volcanic** or **extrusive** igneous rocks form by rapid solidification of lava and volcanic debris on Earth's surface. As a result of rapid cooling, volcanic rocks are dominated by small crystals or noncrystalline, glassy materials of various sizes.

### 7.1.1 How do we classify igneous rocks?

Geologists classify igneous rocks based on composition and texture. Composition is determined by magma chemistry. Texture refers to the size, shape, arrangement, and degree of crystallinity of a rock's constituents. Together, composition and texture provide the means to classify rocks and to determine environmental conditions of rock formation.

Whenever possible, we classify rocks based upon identifiable minerals, if present. Plutonic rocks – such as gabbro, diorite, granodiorite, and granite – tend to have large crystals so that we can easily identify minerals simply by looking at the rock with our eyes or with the aid of a small magnifier or hand lens. Thus we can identify most plutonic rocks based on the relative proportion of minerals. Volcanic

rocks – such as basalt, andesite, dacite, and rhyolite – tend to have fine grained crystals that may be too small to be identified with the eye, even with a hand lens. With volcanic rocks, geologists commonly rely upon the petrographic microscope to identify very fine grained minerals under magnification. If the volcanic rock constituents cannot be determined even with microscopic techniques, laboratory geochemical analyses may be conducted to determine the rock's chemical composition focusing upon percent $SiO_2$ and major elements. Whereas crystalline igneous rocks are defined largely based on composition, noncrystalline igneous rocks are defined based solely on texture and color as crystalline minerals tend not to occur. A more detailed igneous classification system based upon texture and composition will be presented later in this chapter.

### 7.1.2 Percent silica

Nearly all magmas are **silicate magmas**, enriched in silicon and oxygen which bond together to form silica tetrahedra. Rare examples of nonsilicate magmas do exist; for example, carbonatite magmas, as their name suggests, are rich in carbonate minerals such as calcite. Silicate magmas contain anywhere from ~40% to over 75% silica ($SiO_2$). As silica is the dominant chemical component, magma and igneous rocks are classified as ultrabasic, basic, intermediate, and acidic based upon percent $SiO_2$ (Table 7.1). Acidic rocks are also referred to as **silicic**, based on their high $SiO_2$ content. Note that the terms acidic and basic are misleading in the sense that they have no reference to pH, acids or bases. However, the classification terminology continues to be widely used.

Magma chemistry determines the percentage of dark-colored and light-colored minerals.

**Table 7.1** Magma or igneous rock classification based on weight percent silica.

| Igneous classification | Weight percent silica ($SiO_2$) |
| --- | --- |
| Ultrabasic | <45% |
| Basic | 45–52% |
| Intermediate | 52–66% |
| Acidic (silicic) | >66% |

**Mafic** or **ultramafic** crystalline rocks are commonly enriched in dark-colored iron and magnesium (ferromagnesian) minerals that commonly impart black, gray, brown, or greenish hues. Light-colored **felsic** rocks, depleted in ferromagnesian minerals, are generally enriched in silicon, oxygen, potassium, and sodium and commonly appear white, light gray, pink or red. **Intermediate** rocks contain approximately equal amounts of light- and dark-colored minerals. Figure 7.1 illustrates some of the more common ferromagnesian and nonferromagnesian minerals.

### 7.1.3 Color index

Crystalline igneous rocks are broadly classified based on a **color index** (Table 7.2 and Figure 7.2). The **color index (CI)**, also called the mafic index, is the proportion of mafic minerals in the total population of felsic and mafic minerals given as:

$$CI = (\% \text{ mafic minerals} / (\% \text{ felsic minerals} + \% \text{ mafic minerals})) \times 100$$

The color index classification is based on the proportion of light-colored and dark-colored crystalline minerals. In field identification, extremely fine grain size may necessitate the use of the color index to tentatively classify volcanic rocks. The color index does not apply to noncrystalline solids such as glass. In all cases, color should be used with caution as rock color can vary due to weathering, glass content, and other factors. For example, silica rich rhyolites that are highly weathered or have high glass contents can be quite dark. While color index is used in the field, petrographic microscope, and geochemical analyses

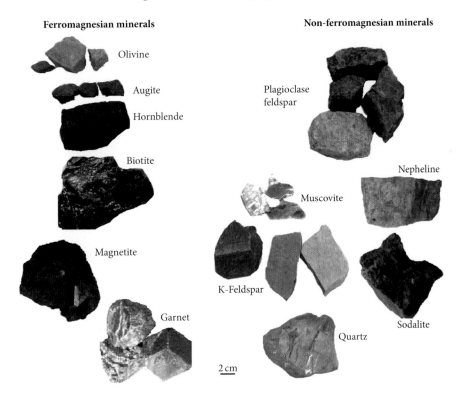

**Ferromagnesian minerals**

Olivine

Augite

Hornblende

Biotite

Magnetite

Garnet

**Non-ferromagnesian minerals**

Plagioclase feldspar

Nepheline

Muscovite

K-Feldspar

Sodalite

Quartz

2 cm

**Figure 7.1** Common igneous minerals. Ferromagnesian minerals include olivine, augite pyroxene, hornblende amphibole, biotite as well as minor minerals magnetite and garnet. Nonferromagnesian minerals include plagioclase, muscovite, K-feldspar (orthoclase, microcline, and sanidine), quartz, as well as minor minerals sodalite, nepheline and also garnet. The garnet group contains six mineral species, some of which tend to be darker due to iron (almandine) or magnesian (pyrope) and others that are lighter (grossular) because they lack ferromagnesian components. *Source*: Kevin Hefferan. © John Wiley & Sons.

**Table 7.2** Generalized description of rock color based on mineral composition.

| Color index | % Dark colored | Common minerals and rock color |
|---|---|---|
| Ultramafic | >90% dark-colored or greenish minerals | Olivine, pyroxene (augite) and amphibole (hornblende) minerals; Ultramafic rocks are dark gray, black or dark green. |
| Mafic | 70–90% dark-colored minerals | Pyroxene, plagioclase, amphibole, olivine, magnetite, garnet minerals; mafic rocks are commonly gray to black. |
| Intermediate | 30–70% dark-colored minerals | Plagioclase, amphibole, potassium feldspar, quartz, biotite minerals; Intermediate rocks appear as gray to salt and pepper-colored rocks. |
| Felsic | <30% dark-colored minerals | Feldspar, quartz, feldspathoid, muscovite minerals; Felsic rocks are commonly white, gray, pink or light red. |

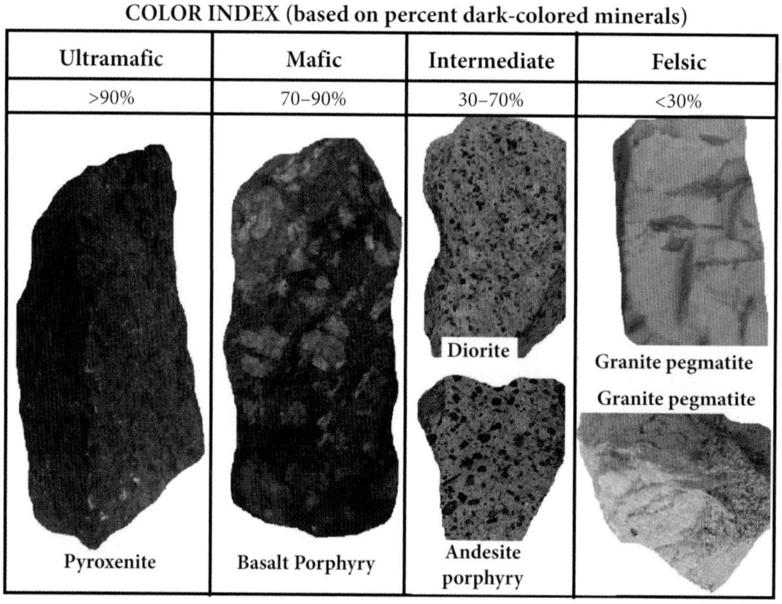

**COLOR INDEX (based on percent dark-colored minerals)**

| Ultramafic | Mafic | Intermediate | Felsic |
|---|---|---|---|
| >90% | 70–90% | 30–70% | <30% |
| Pyroxenite | Basalt Porphyry | Diorite / Andesite porphyry | Granite pegmatite / Granite pegmatite |

**Figure 7.2** Rock classification based upon percentage of dark-colored minerals (DCM) whereby ultramafic contains >90%, mafic 70–90%, intermediate 30–70% and felsic <30% DCM. *Source*: Kevin Hefferan. © John Wiley & Sons.

in the laboratory provide greater certainty as to appropriate rock name determination.

### 7.1.4 Texture

Texture provides the primary means to classify crystalline and noncrystalline igneous rocks. Figure 7.3 illustrates a simplified means by which some common crystalline (Figure 7.3a) and noncrystalline (Figure 7.3b) igneous rocks are identified based on texture, major minerals, percent $SiO_2$ and color:

1 Pyroxenite and peridotite are very dark-colored (ultramafic) rocks, depleted in $SiO_2$ (<45% $SiO_2$ = ultrabasic) and commonly enriched in coarse grained pyroxene, olivine, amphibole, and plagioclase minerals. Ultramafic rocks occur in Earth's mantle. Ultramafic volcanic rocks, such as olivine rich komatiites, are very rare.

2 Basalt and gabbro are dark-colored (mafic) rocks rich in plagioclase, pyroxene, and olivine. Although classified as $SiO_2$ poor (basic), basalt, and gabbro contain

(a)

**Phaneritic texture**

Peridotite  Gabbro  Diorite  Granodiorite  Granite

2 cm

Basalt  Andesite  Dacite  Rhyolite

**Aphanitic texture**

(b)

Ash fall tuff

2 cm

Ash flow tuff

2 cm

2 cm

Glassy obsidian

Volcanic breccia

2 cm

Vesicular scoria

2 cm

Vesicular pumice

2 cm

Plutonic breccia

2 cm

**Figure 7.3** (a) Common phaneritic (coarse crystals) and aphanitic (fine crystals) igneous rocks. (b) Noncrystalline igneous rock textures focus on the presence of vesicles, color, glass content, and pyroclast size. *Source*: Kevin Hefferan. © John Wiley & Sons.

~50% $SiO_2$. Fine grained basalt is a very common volcanic rock that forms at ocean spreading ridges rift valleys and hotspots. Coarse grained gabbro crystallizes slowly at depth. Together basalt and gabbro compose Earth's ocean basin crust that encompasses 70% of our planet.

3 Andesite and diorite are gray-colored (intermediate) to salt and pepper-colored rocks rich in hornblende, pyroxene, and plagioclase. Andesite and diorite contain ~50–63% $SiO_2$. Andesite is a common, fine grained volcanic rock above convergent plate boundaries such as around the Pacific Ring of Fire. Andesite volcanoes derive magma from dikes originating from underlying plutons containing coarse grained diorite.

4 Dacite and granodiorite are light-colored (felsic) rocks, containing ~63–67% $SiO_2$. Dacite and granodiorite are rich in plagioclase, alkali feldspar and quartz and may also contain hornblende and biotite. Dacite is a fine grained, volcanic rock that commonly occurs with andesite above convergent plate boundaries. Granodiorite is a coarse grained plutonic rock that underlies and feeds magma to surficial andesite–dacite volcanoes.

5 Rhyolite and granite are light-colored (felsic) rocks containing more than 67% $SiO_2$ (silicic or acidic) and rich in quartz, alkali feldspar with smaller percentages of plagioclase and biotite. Rhyolite is a fine grained volcanic rock that usually erupts over thick, continental crust. Granite plutons contain coarse grained crystals and commonly occurs in continental crust.

Noncrystalline rocks (Figure 7.3b), those characterized by the absence of crystals, include frothy, vesicular rocks such as pumice (light-colored) and some scoria (dark-colored). Other noncrystalline rocks include those with glassy textures such as obsidian or those enriched in glassy rock fragments. Fragmental, also known as pyroclastic, volcanic rocks include tuff (volcanic ash to gravel size) and breccia (gravel size and larger).

Igneous rock are broadly categorized based on whether they exhibit crystalline or noncrystalline textures. Igneous textures are further differentiated on the degree of crystallinity, completeness of crystal faces, grain size, shape, and variability.

## Degree of crystallinity

Igneous textures can be categorized on the **degree of crystallinity**. The degree of crystallinity varies between **holocrystalline** (100% crystals), **hypocrystalline** (mixture of crystals and glass) and **holohyaline** (100% noncrystalline glass). Stable conditions in which temperature, pressure, and gas content do not vary significantly over time allow silica tetrahedra to link together to form crystals. Holocrystalline textures predominate in plutonic rocks that cool slowly, producing stable crystalline mineral assemblages. Less stable conditions tend to produce textures that are not 100% crystalline. Hypocrystalline textures may develop when a partially crystallized magma experiences rapid upwelling as in a volcanic eruption so that crystalline minerals are embedded in a glassy matrix. In some cases, extremely rapid cooling or the sudden loss of gas results in instantaneous rock solidification without any crystals, producing a totally glassy holohyaline texture that characterizes some explosive volcanic eruptions.

## Crystal faces

Crystalline textures are based on completeness of crystal faces as well as crystal size. **Euhedral** minerals are enclosed by complete crystal faces (Figure 7.4). Euhedral igneous crystals typically develop as early mineral phases in the crystallization of magma. Under such conditions, the crystals have abundant free space for growth, enhancing the likelihood that perfectly formed faces develop. Later in the magma crystallization sequence, subhedral or anhedral crystals develop in the remaining void spaces between earlier formed crystals.

**Subhedral** minerals contain partially complete crystal forms in which at least one of the crystal faces is not fully developed (Figure 7.4). In subhedral textures, crystal face development may be inhibited by: (1) contact against previously formed minerals; (2) nucleation on preexisting surfaces such as early formed crystals or the margins of the magma chamber; (3) resorption in which preexisting euhedral crystals are partially remelted; (4) other secondary alteration processes that destroy preexisting mineral faces.

**Anhedral** minerals lack observable crystal faces. As crystallization progresses in magma, the space available for the development of euhedral and subhedral crystals diminishes.

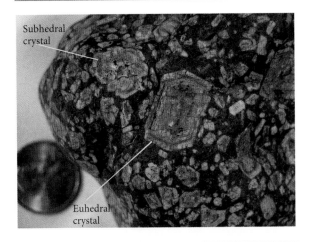

**Figure 7.4** Basalt porphyry containing euhedral and subhedral plagioclase crystals. Some euhedral crystals display zoning suggesting that crystals continued to react with the melt during crystallization. *Source*: Kevin Hefferan. © John Wiley & Sons.

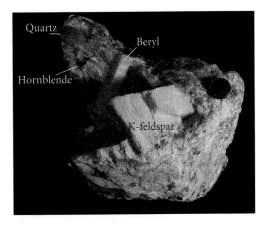

**Figure 7.5** Granite pegmatite with K-feldspar, beryl, quartz, and hornblende. Note 2 cm penny for scale. *Source*: Kevin Hefferan. © John Wiley & Sons.

As a result, anhedral crystal forms are commonly determined by the shape of the existing interstitial void space between early formed crystal grains. Anhedral textures may also develop by secondary alteration processes that obliterate preexisting mineral faces.

## 7.2 CRYSTALLINE IGNEOUS TEXTURES

Crystalline textures provide critical information as to whether the rock solidified in a plutonic or volcanic setting. Magma that cools deep within Earth produces plutonic rocks with crystals easily visible to the eye. Such plutons retain heat and allow for slow cooling which enhances crystal growth. Igneous textures typical of plutonic rocks include coarse grained pegmatitic, phaneritic, and phaneritic–porphyritic textures. Conversely, magma that cools in less stable conditions, such as very near Earth's surface (subvolcanic) or on Earth's surface lose heat quickly. As a result, small crystals develop producing fine grained aphanitic or aphanitic–porphyritic crystalline textures. These textures will be described below.

### 7.2.1 Pegmatitic textures

Pegmatitic texture is characterized by large crystals averaging more than 30 mm (3 cm) in diameter. Rocks with such textures are called pegmatites, and display large, early formed euhedral crystals surrounded by smaller euhedral to subhedral crystals. In naming rocks with pegmatitic texture, the textural term (e.g., pegmatite) is included in the rock name. Therefore, a rock with a pegmatitic texture and the composition of granite or granodiorite is a granite pegmatite or granodiorite pegmatite. Pegmatitic textures develop most commonly in granitic plutons (Figure 7.5) with high volatile contents. Gabbroic plutons less commonly display pegmatitic textures, partly due to the lower volatile gas content. Because of the large, well developed crystal forms, pegmatites are the source of many gemstones such as aquamarine and tourmaline. High volatile content also produces valuable ore deposits of metals such as gold, silver, and tin in pegmatite deposits.

### 7.2.2 Phaneritic textures

Phaneritic texture implies crystal diameters ranging from 1 to 30 mm (0.1–3 cm). Rocks with a phaneritic texture contain crystals visible to the eye (Figure 7.6). Early formed crystals tend to be euhedral, while later formed crystals are subhedral to anhedral. Phaneritic textures may be subdivided into fine (1–3 mm in diameter), medium (3–10 mm in diameter) or coarse grained (10–30 mm in diameter). Fine grained phaneritic textures commonly develop in shallow plutonic structures such as dikes and sills. Coarse grained textures are associated with larger or deeper intrusions. Rock names such as granite, diorite and gabbro imply a phaneritic texture so that we do not refer to "phaneritic granite" but simply "granite."

**Figure 7.6**   Coarse-grained phaneritic granite with early formed, euhedral to subhedral pink to white K-feldspar. Subhedral to anhedral gray quartz and black hornblende represent later, void filling minerals. *Source*: Kevin Hefferan. © John Wiley & Sons.

### 7.2.3   Aphanitic textures

Aphanitic textures contain small crystals less than 1 mm (<0.1 cm) in diameter that are not generally discernible to the unaided eye. With the use of a microscope or other analytical equipment, geologists can determine the composition and relative sequence of crystallization in the same manner as with phaneritic textures. Aphanitic textures are associated with volcanic and subvolcanic rocks that cool quickly on or near Earth's surface. Aphanitic textures may be subdivided into microcrystalline and cryptocrystalline varieties. **Microcrystalline** textures contain **microlite** crystals large enough to be identified with a petrographic microscope (Figure 7.7). Aphanitic textures in which the crystal size is too fine to be recognized even with a petrographic microscope are termed **cryptocrystalline** (MacKenzie et al. 1984). Rock names such as rhyolite, andesite, and basalt imply an aphanitic texture so that we do not refer to "aphanitic basalt" but simply basalt.

### 7.2.4   Porphyritic textures

Rocks with porphyritic textures consist of two distinctly different size crystals. Large crystals are referred to as **phenocrysts**; finer grained

**Figure 7.7**   Aphanitic dacite from Mt. St. Helens 2004 dome eruption. Dacite contains plagioclase, K-feldspar, and hornblende microlite crystals less than 0.3 mm in diameter. *Source*: Photo courtesy of the United States Geological Survey.

material constitutes the **groundmass**. In **porphyritic–phaneritic** textures, all crystals are visible to the eye (>1 mm), but the phenocrysts are distinctly larger than the groundmass crystals (Figure 7.8a). In rocks with **porphyritic–aphanitic**

(a)

(b)

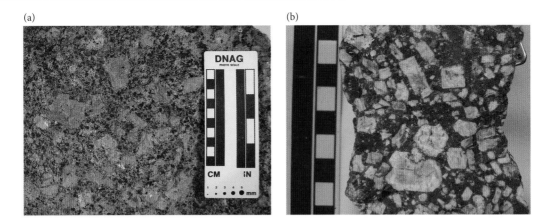

**Figure 7.8** (a) Porphyritic phaneritic texture with large pink K-feldspar phenocrysts and visible quartz and hornblende crystals in the groundmass. (b) Andesite porphyry displaying aphanitic porphyritic texture. Note the white, subhedral plagioclase phenocrysts and the gray, fine grained aphanitic groundmass. *Source*: Kevin Hefferan. © John Wiley & Sons.

textures, the larger, visible phenocrysts are embedded in an aphanitic groundmass composed largely of microcrystalline, cryptocrystalline or glassy material (Figure 7.8b).

Rocks with two distinctly different size crystals (porphyritic) are commonly explained by a two-stage cooling process. In two stage cooling processes, the larger phenocrysts form slowly at depth, while the finer grained groundmass crystals cool rapidly as magma approaches Earth's surface. These straightforward explanations for the development of crystalline textures provide simple models that do not always hold true. In addition to cooling rate and depth, other important factors influencing igneous textures include crystal nucleation rates, crystal growth rates, rate of magma undercooling, ion availability, chemical diffusion rates, viscosity, chemical composition, and volatile content of the magma. These factors will be discussed in the following section.

### 7.2.5   The origin of crystalline textures

In assessing the origin of crystalline textures, let us first consider magma crystallization using a simplified model that focuses on the solid and liquid states, ignoring the vapor state. At sufficiently high temperatures, an ideal magma exists in the liquid state. As temperatures decrease, the magma begins to crystallize so that liquid and crystals coexist. At sufficiently low temperatures, all the liquid has crystallized and the

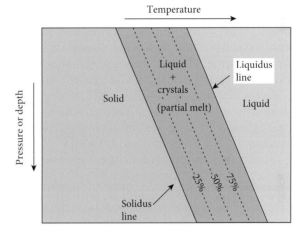

**Figure 7.9** Temperature–pressure relations depicting solid, solid plus liquid and liquid fields separated by a solidus line and a liquidus line. Dashed lines indicate % melt. *Source*: Courtesy of Stephen Nelson.

system becomes completely solid. On a phase-stability diagram, three fields are recognized, separated by lines that represent the initial crystallization temperature and pressure and the final crystallization temperature and pressure (Figure 7.9). The high temperature "all liquid" field is separated from the lower temperature "liquid+crystal" field by a **liquidus line**. The liquidus line represents the initial crystallization temperature/pressure below which crystals and liquids coexist. Below the liquidus conditions, crystals nucleate and continue to grow at **subliquidus** temperatures and pressures. Eventually, when the magma has cooled sufficiently, all remaining liquid crystallizes and the system crosses the solidus line at which point the magma has completely solidified. The **solidus line** separates a higher temperature field containing liquids and solid crystals from a lower temperature field in which only solid crystalline material exists. Temperatures below the solidus line correspond to **subsolidus** conditions.

The exact temperatures at which the ideal system changes from all melt to melt plus crystals (liquidus line), and from melt plus crystals to all solid (solidus line), depends on a variety of factors that include pressure and chemical composition. This means that for a given magma, we cannot define a single temperature at which melting or solidification occurs. Instead, these changes occur along a temperature/pressure range as represented by the inclined liquidus and solidus lines in Figure 7.9. Real magmatic systems are even more complicated, but this conceptual model approximates their behavior closely enough to be of value. We will now use this model to explain crystal nucleation and growth rates.

Below the liquidus line, crystal size depends primarily on the interplay between the crystal nucleation rate and the crystal growth rate (Swanson 1977; Špillar and Doleiš 2013). Crystal nucleation involves the formation of new crystals, called nuclei or "seed" crystals, large enough to persist and grow into larger crystals. The **crystal nucleation rate**, the number of new seed crystals that develop per volume per time, is commonly expressed as nuclei per cubic centimeter per second (nuclei/cm³/s). Crystal nuclei grow at rates that depend on the: (1) rate of undercooling; (2) availability of the necessary ions; and (3) ease with which these ions migrate to the crystal growth site.

In real magmas, the rate at which crystal nuclei form varies for each mineral. One might expect nucleation rates to peak at the liquidus temperature because the liquidus temperature represents conditions in which the maximum amount of liquid magma is available to crystallize. In reality, nucleation rates peak at temperatures well below the liquidus temperature, for reasons discussed below.

Subliquidus crystal nucleation represents undercooling of the magma. **Undercooling** occurs when liquids are cooled to temperatures below the liquidus line. One reason why nucleation requires undercooling is that crystal formation requires the development of bonds between ions, which produce a **heat of formation**. This heat increases the local temperature and remelts seed crystals. Only when the magma has been significantly undercooled below the liquidus temperature can incipient crystals nucleate, persist and grow (Brandeis et al. 1984). Therefore, nucleation rates are low near the liquidus temperature and increase at temperatures below the liquidus (undercooling). Nucleation rates decrease to approach zero at very large degrees of undercooling because increased magma viscosity impedes the assembly of nuclei. Very large degrees of undercooling suggest that the magma is approaching the solidus line where all liquid has solidified.

**Crystal growth rate**, expressed in millimeters per second (mm/s), is a measure of the increase in crystal radius over time. Špillar and Doleiš (2013) suggest that the largest growth rates occur at the beginning and end of crystallization. They further suggest that crystal growth rates are inversely proportional to the size of the pluton and that growth rates increase from the pluton interior toward the pluton walls, likely due to the loss of heat along the margins of the plutons. Crystal growth rates are also determined by the rate of undercooling as well as the availability of elements and magma viscosity. Magmas that experience small undercooling stay at temperatures just below the liquidus temperature for a long time. At such temperatures, the nucleation rate is low and the crystal growth rate is high producing a small number of large crystals (Swanson 1977). Prolonged undercooling conditions generate phaneritic, porphyritic or pegmatitic textures with euhedral to subhedral crystals (Figure 7.10).

Magmas that experience large undercooling at temperatures well below the liquidus temperatures experience higher nucleation rates and lower growth rates (Swanson 1977). As a result, a large number of small seed crystals produce aphanitic textures, some of which may display skeletal or dendritic textures. Extremely rapid temperature decreases prevent the growth of seed crystals. Failure to nucleate seed crystals produces noncrystalline glass such as obsidian that may contain small spherulites (Figure 7.10). **Spherulites** are

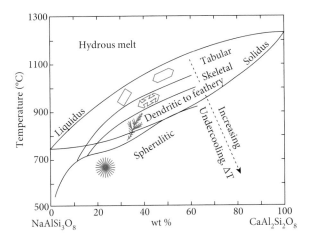

**Figure 7.10** Plagioclase phase diagram illustrating crystalline textures that can form in response to rates of undercooling. Laboratory data (Lofgren 1980, 1983) suggest that tabular, euhedral crystals form with cooling of a few °C per hour. Temperature changes of hundreds of °C per hour produce dendritic to spherulitic textures in lab experiments. *Source*: After Best (2003); with permission of Wiley-Blackwell Publishers.

small, spherical, radiating incipient crystals that were not able to fully develop. In summary, nucleation rates and growth rates are two independent processes which together determine the number and size of crystals.

**Ion availability** refers to the availability of ions that can fill specific ionic sites in a crystal lattice structure. If appropriate ions are readily available and can migrate to the lattice site, crystal growth is enhanced. The ion availability factor helps explain why minerals that require relatively rare trace elements tend to be small. For example, coarse grained, crystalline igneous rocks with large feldspar minerals may contain smaller apatite and zircon crystals. Apatite is composed of minor elements such as phosphorous. Zircon consists of the trace element zirconium. Apatite and zircon crystals tend to be small, even within most granite pegmatites, as their mineral growth rates are impeded by the limited availability of the minor and trace elements phosphorous and zirconium.

**Diffusion**, the rate at which elements migrate through magma, depends primarily on the viscosity of the melt. **Viscosity ($\eta$)**, defined as the resistance of a fluid to shear stress, is measured by the ratio of shear stress ($\tau$) to shear strain rate ($\dot{\varepsilon}$) given by $\tau = \eta \dot{\varepsilon}$. Shear stress is a force applied parallel to the surface of an object. Shear strain rate is a time dependent change in shape of an object in response to shear stress. We can rewrite the above formula as: $\eta = \tau/\dot{\varepsilon}$. In this modified formula, viscosity is proportional to shear stress and inversely proportional to shear strain rate. Low viscosity materials, such as water, flow readily. Higher viscosity materials, such as honey, flow slowly. For a given shear stress, the higher the viscosity, the lower the strain rate.

How does viscosity affect ion diffusion and crystal growth rates? Viscosity affects the ability of ions to diffuse through the magma to crystal growth sites. Low magma viscosity increases the rate of diffusion, allowing ions to migrate to molecular sites and increase crystal growth rate. Conversely, high magma viscosity reduces the diffusion rate, impedes ion migration and decreases crystal growth rate. Viscosity is largely determined by the $SiO_2$ content, temperature and the dissolved gases in the magma. As most magmas contain half to over two-thirds $SiO_2$, magma viscosity is strongly influenced by $SiO_2$ content which determines the degree of molecular linkage in magma.

Viscosity is strongly related to the degree to which molecules are bonded or linked together in the fluid: an increase in molecular linkages results in higher magma viscosity.

Molecular linkage is determined by the relative abundance of network formers and network modifiers. **Network formers** are elements that tend to increase molecular linkage, thereby increasing viscosity. Network formers in silicate melts include silicon and aluminum. Molecular linkage of the elements silicon and oxygen, creating the **silica tetrahedron** ($SiO_4$) structure, exerts primary control on magma viscosity. Silica tetrahedra behave as network formers that bond together through shared oxygen atoms to create polymerized chains of linked silica tetrahedra. The creation of these silica tetrahedron networks results in higher viscosity magmas. Under most conditions, magmas rich in $SiO_2$ have significantly higher viscosities than $SiO_2$ poor magmas.

Elements such as magnesium and iron are network modifiers. **Network modifiers** are elements that decrease molecular linkage, thereby reducing viscosity. In silicate magmas, network modifiers inhibit the linkage of silica tetrahedron polymerization by filling sites otherwise occupied by silicon and oxygen. Therefore, magmas enriched in iron and magnesium and depleted in silicon and oxygen tend to have lower viscosities. Decreased linkage of silica tetrahedra results in the lower viscosity (Figure 7.11a) that characterizes basic (45–52% $SiO_2$) and ultrabasic (<45% $SiO_2$) melts (Spera 2000; Lesher and Spera 2015).

In addition to composition, the degree of molecular bonding is also strongly influenced by temperature. Temperature is inversely proportional to viscosity and the degree of molecular bonding. As the temperature increases, molecular vibrations increase, bonds break and the amount of molecular linkage decreases thereby reducing viscosity. For example, honey is a high viscosity fluid at room temperature due to a relatively large number of organic molecular bonds. If honey is stored at cool temperatures, it tends to crystallize, developing organic molecular bonds resulting in greater viscosity. However, if you heat the crystallized honey, molecular bonds vibrate and break, drastically reducing viscosity. This example illustrates the inverse relationship of temperature and viscosity: as temperature increases, magma viscosity decreases.

Low magma viscosity also promotes high crystal growth rates. Crystal nucleation is an exothermic process that releases energy in the form of heat. Low viscosity magma aids crystal growth by increasing the rate at which the heat of formation can be diffused away from the surface of the growing crystal. This allows the crystal to be cool enough for additional growth to occur. It also increases the diffusion rates of necessary constituents to the surfaces of growing crystals.

Magma contains up to 7 wt. % volatile gases, largely consisting of $H_2O$, $CO_2$ and $SO_2$. Minor gases include N, H, S, F, Ar, CO, and Cl. Dissolved gases play a key role in magma viscosity by acting as network modifiers, reducing molecular bonding. Dissolved water vapor is a particularly potent network modifier (Figure 7.11b): an increase in water vapor pressure ($P_{H_2O}$) decreases magma viscosity, particularly for silicic magmas (Wallace and

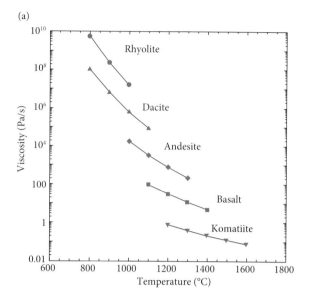

(a)

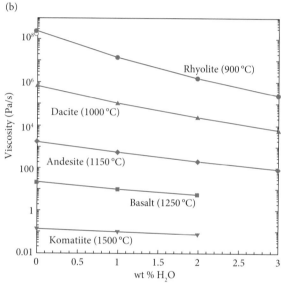

(b)

**Figure 7.11** (a) Graph depicts viscosity versus temperature for five different lavas as determined through experimentation. Note that low $SiO_2$, ultrabasic lavas (e.g. komatiite) have the lowest viscosity and as $SiO_2$ content increases, viscosity increases. The graph also indicates that viscosity is inversely proportional to temperature as discussed in the text. (b) The graph of viscosity versus weight percent dissolved water illustrates that silicic (dacite and rhyolite) magmas experience the greatest decrease in viscosity with an increase in water content. *Source*: Courtesy of Frank J. Spera.

Anderson 2000). High dissolved water vapor content is an effective network modifier because it partially bonds with the corners of silica tetrahedra: the higher the magma $SiO_2$ content, the greater the effect. As water vapor occupies tetrahedra bonding sites, the water vapor effectively inhibits adjacent silica compounds from bonding together. This reduces the molecular linkage in the magma and reduces viscosity. A reduction in magma viscosity increases diffusion rates and crystal growth rates. The result is that high $P_{H2O}$ produces a low nucleation rate and a high growth rate generating small numbers of large crystals.

The role of $P_{H2O}$ helps to explain an important observation about the relationship between texture and composition. The higher the $SiO_2$ content of the magma and the higher the confining pressure, the more water vapor can be dissolved. Silica poor (<52% $SiO_2$), basic magmas generally contain <1% dissolved water vapor. $P_{H2O}$ is only a little higher in basic magma under pressure at depth than in the shallow subsurface. As a result, $P_{H2O}$ has a very limited effect on crystal size in basic plutonic rocks such as gabbro. Phaneritic textures in gabbro develop by slow undercooling at depth while aphanitic textures in basalt develop by rapid undercooling on the surface. As a result, pegmatitic crystals are not common in basic and ultrabasic rocks (Wallace and Anderson 2000).

On the other hand, silicic magmas (>66% $SiO_2$) possess a large capacity to dissolve water vapor and other volatiles. The unusually high (3–7%) $P_{H2O}$ in silicic magmas enhances both diffusion rates and crystal growth rates, producing relatively few crystals which grow to a very large size (Wallace and Anderson 2000). As a result, granite pegmatites and granodiorite pegmatites are relatively common. Variations in $P_{H2O}$ also help explain why earlier formed granitic veins display pegmatitic textures while later formed, cross-cutting veins commonly display fine grained (aplitic) textures. As silicic magma rises toward the surface, the confining pressure in the magma decreases enhancing exsolution of gases from the magma. Exsolution allows gases such as water vapor to bubble out from the magma, drastically lowering magma $P_{H2O}$. $P_{H2O}$ reduction causes silica tetrahedra to link rapidly which accelerates magma viscosity. High viscosity inhibits crystal growth and helps to explain the very small crystal size typical of fine-grained granitic dikes (aplites) or aphanitic rocks such as rhyolites. As will be discussed in the next section, this also explains the abundance of noncrystalline, glassy textures in silicic volcanic rocks as opposed to basic volcanic rocks.

### 7.2.6  Textural equilibration

Crystal growth can be influenced by textural equilibration processes. Igneous minerals form through crystal nucleation and experience kinetic growth forces generating cumulate textures. **Cumulate** textures develop as crystals accumulate through floating or sinking in a pluton due to density differences with the remaining melt. Early formed cumulate crystals with a high surface area may contain excess surface energy, particularly if grains develop elongate shapes with high surface area to volume ratios. The excess surface energy can catalyze chemical reactions with surrounding crystals and interstitial fluids that mechanically modify initial cumulate textures. Chemical and mechanical reactions that result in igneous grain boundary modifications after initial crystal formation are referred to as **textural equilibration** processes. As magma cools, continued crystal growth and accumulation can produce grain compaction, deformation (dislocation creep), solution and reprecipitation of grains (diffusion creep) as well as grain recrystallization (Hunter 1996). Textural equilibration processes commonly result in crystal coarsening where smaller crystals are consumed by larger growing crystals, obliterating previously existing boundaries between grains (Figure 7.12). Various names have been applied to these textural equilibration processes such as Ostwald ripening, textural maturation, annealing, grain boundary migration, and textural coarsening (McBirney and Nicolas 1997; Higgins 2011). If magma crystallizes a single mineral at a temperature close to the liquidus of that mineral, then both kinetically controlled and mechanically modified textures will create equilibrium textures. If the magma contains several minerals, then their textural equilibration will be determined by the degree of undercooling for each mineral (Higgins 2011). Textural equilibration, which

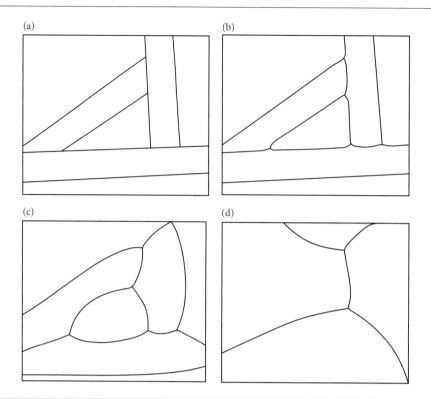

**Figure 7.12**   Grain boundary alteration and crystal coarsening is produced by textural equilibration. (a) Initial crystal growth boundaries generate primary interlocking crystal framework characterized by crystals with large surface areas relative to their volume. (b) Dihedral angles at corners of crystals are modified and curved. (c) Grain edge curvature produces migration of grain boundaries. (d) Continued textural equilibration results in smaller grains being consumed by larger grains. After equilibration, fewer but larger grains are characterized by rounded surfaces and more equant crystal forms. *Source*: Courtesy of Michael D. Higgins.

occurs in both magmatic and metamorphic processes, may be responsible for the growth of unusually large K-feldspar phenocrysts in granite porphyry and granite pegmatite rocks (Glazner and Johnson 2013).

In addition to the crystalline textural concepts described above, Table 7.3 lists other crystalline igneous textural terms that may be encountered in the literature. Together with rock composition, rock texture provides a keystone component in the classification of igneous rocks. In the following section, we will further investigate element and mineral abundances in igneous rocks. The final section of this chapter presents more detailed classification schemes based on textural and composition factors.

## 7.3   NONCRYSTALLINE TEXTURES

Many volcanic rocks contain noncrystalline (holohyaline) or a mixture of crystalline and noncrystalline (hypocrystalline) textures. The general name for noncrystalline rock material is glass. Common noncrystalline textures (Figure 7.3b) are discussed below.

### 7.3.1   Glassy textures

A **glass**, such as the volcanic rock obsidian, is an amorphous solid. Amorphous solids possess a disordered atomic array, lacking an ordered crystalline structure. Many glasses contain small amounts of very small microlites and/or cryptocrystalline material as well as glass. Glassy textures develop in lava that solidifies without experiencing significant crystallization. Glasses are essentially supercooled liquids that form by the near instantaneous solidification of melts, preserving their disordered structure. Instantaneous solidification results from two major mechanisms: quenching and rapid gas loss.

**Table 7.3**   Summary of igneous textural names and brief descriptions.

| Textural name | Textural description |
| --- | --- |
| Allotromorphic-granular | Rocks dominated by anhedral crystals |
| Antiperthitic | Stringers of K-feldspar in a sodic-plagioclase host rock |
| Aphyric | Volcanic rocks without phenocrysts |
| Chadocryst | Smaller randomly oriented crystals enclosed by oikocrysts |
| Equigranular | Rock characterized by all crystals approximately the same size |
| Hypidiomorphic-granular | Rocks dominated by subhedral crystals |
| Idiomorphic-granular | Rocks dominated by euhedral crystals |
| Inequigranular | Rock characterized by unequal crystal sizes |
| Intergranular | Wedge shaped spaces between a meshwork of lath shaped grains are filled with granular crystals. Common in basalt with lath-shaped plagioclase. |
| Intersertal | Wedge shaped spaces between a meshwork of lath shaped grains are filled with glassy or hypocrystalline material |
| Glomeroporphyritic | Porphyritic texture with phenocrysts clustered in aggregates |
| Granophyric | Microscopic scale intergrowth of alkali feldspar and quartz |
| Miarolitic | Irregularly shaped cavities infilled with euhedral crystals |
| Myrmekite | Plagioclase crystals intergrown with rod-like or vermicular (worm-like) inclusions of quartz |
| Oikocryst | Large enclosing crystal in poikilitic texture |
| Ophitic | Poikilitic texture in which large crystals of augite partially or entirely enclose laths of plagioclase |
| Orbicular | Concentric rings of rhythmically alternating mineral composition |
| Perthitic | Stringers of sodic plagioclase in a K-feldspar host rock |
| Phyric | Volcanic rocks with phenocrysts |
| Poikilitic | Large crystals of one mineral enclose numerous smaller crystals of one or more randomly oriented minerals. |
| Rapikivi | Visible, coarse-grained crystals of rounded, pink K-feldspar crystals rimmed by white sodic plagioclase in granites |
| Seriate | Primary mineral crystals display a range of sizes |
| Spherulitic | Spheroidal to elliptical aggregates of radiating, fibrous crystals |
| Sub-ophitic | Poikilitic texture with some plagioclase laths enclosed by augite; other plagioclase laths penetrate augite crystals |
| Symplectite | Intimate intergrowth of two minerals, one of which has a vermicular (worm-like) appearance |
| Trachytic | Sub-parallel orientation of microcrystalline lath-shaped feldspars in the groundmass of a holocrystalline or hypocrystalline rock |
| Variolitic | Radiating, fan like arrangement of acicular, hair-like crystals |

**Quenching** occurs when melts of any composition supercool when they come in contact with liquid water or air. Water rapidly absorbs heat from the melt, causing it to solidify before crystals have time to nucleate and grow. Most basic (low $SiO_2$) glasses quench when volcanoes erupt on the ocean floor or in lakes. Thin glassy zones also occur on lava flow tops that have been quenched by contact with the atmosphere. **Rapid loss of dissolved gas** from silicic melts can also generate glass. The rapid loss of dissolved water vapor causes melt viscosity to increase so rapidly that crystal nucleation and crystal growth are severely inhibited. This second model explains why glassy rocks, such as obsidian, are far more common in silicic rocks than in basic rocks. Unlike silicic magmas, basic magmas contain neither enough dissolved water nor sufficient silica tetrahedra to solidify rapidly due to loss of dissolved gases.

As noted earlier, many glasses do contain small microlites and cryptocrystalline minerals. Microlites and cryptocrystalline grains represent incipient crystal nucleation in a nearly solid magma of extremely high viscosity. Larger crystals may form by partial crystallization of the magma prior to its rapid solidification by quenching or gas loss. This can produce porphyritic glassy rocks called vitrophyres. **Vitrophyres** contain recognizable phenocrysts in a glassy groundmass and are said to have a **vitrophyric texture**. Over time, glasses may crystallize in the solid state by growing on preexisting microlitic or cryptocrystalline nuclei in a process called **devitrification**. Growth commonly occurs outward from existing crystal nuclei to produce rounded masses of radiating spherulite seed crystals. **Snowflake obsidian** is an excellent example in which cristobalite seed crystals grow into white snowflake forms within black glassy obsidian (Figure 7.13). **Perlites**, glassy $SiO_2$ rich volcanic rocks with higher water contents than obsidian, display a **perlitic texture** characterized by a cloudy appearance and curved or sub-spherical cooling cracks called perlitic cracks. Perlite is widely used as an aerating medium in potting soils.

### 7.3.2 Vesicular textures

Rocks with **vesicular** textures contain spherical to ellipsoidal void spaces called **vesicles**, which are analogous to holes in a household sponge (Figure 7.14). Vesicular textures develop due to exsolution and entrapment of gas bubbles in lava as it cools and solidifies. A model for the development of vesicular textures is illustrated in Figure 7.15. Plutons contain magmas at relatively high confining pressures such that gases are dissolved, and the magma is undersaturated in volatile content. As volatile gases are low density and buoyant, they ascend within the pluton and can saturate magma in the upper part of the pluton. In

conditions where magma rises toward the surface, the confining pressure decreases, and the ability of the magma to retain dissolved gases decreases. As a result, magma becomes supersaturated with volatiles so that it can no longer hold all the gas in solution. At a depth referred to as the **level of exsolution**, volatiles exsolve from the liquid as a separate phase. Above the level of exsolution, volatiles nucleate as small bubbles in a process called **vesiculation**. As the magma continues to rise, gas solubility continues to decrease and the number and size of the gas bubbles increase. Gas bubble expansion decreases magma density, resulting in gas driven, explosive volcanic eruptions. Since exsolution is time dependent, as in bubbles released from champagne, gases often continue to exsolve even after lavas are extruded. It is the entrapment of such gas bubbles in shallow plutons and erupted lava that produces vesicular textures (Sparks 1978; Wilson 1980; Cashman et al. 2000).

Vesicular rocks, defined as containing >30% vesicles by volume, include pumice and scoria. White to gray-colored pumice solidifies as a frothy glass from silicic lava. Pumice is widely used as an abrasive soap. Scoria, an iron rich brownish red- or black-colored vesicular rock, is used as a decorative landscape stone. Scoria is derived from basic to intermediate lava and may be partially crystalline (hypocrystalline). Highly vesicular rocks are characterized by very low specific gravity and may contain such a large volume of vesicles that they float in

**Figure 7.13** Snowflake obsidian displaying cristobalite seed crystals as well as conchoidal fracture. *Source*: Kevin Hefferan. © John Wiley & Sons.

**Figure 7.14** Cross-bedded vesicular pumice bomb ejected by the 1980 eruption of Mt. St. Helens and deposited in a "pumice plain." *Source*: Kevin Hefferan. © John Wiley & Sons.

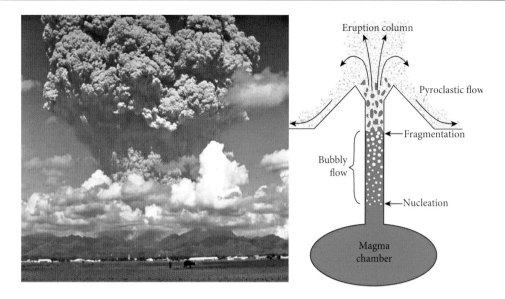

**Figure 7.15**   At depth within the magma pluton, gases are dissolved within the pressurized magma. As the magma rises toward the top of the pluton, pressure decreases and gases exsolve or separate from the magma. The separation of gases results in the generation of gas bubble growth, which produces a vesicular texture. *Source*: Courtesy of Ed Llewellin; US Geological Survey.

water. Rocks that contain smaller amounts (5–30%) of vesicles are named using a modifier such as vesicular basalt or vesicular andesite while those rocks with just a few vesicles (<5%) are given names such as vesicle-bearing basalt and andesite.

Hot fluids that flow through vesicular rocks may later precipitate secondary minerals in the void spaces of vesicles, producing **amygdules**. Common secondary minerals that infill preexisting vesicles include quartz, calcite, epidote, zeolites, and metals. Figure 7.16 depicts an amygdaloidal basalt in which quartz and epidote have precipitated in vesicles. Secondary fluid flow through rock can produce significant ore deposits of copper and other metals that precipitate in the void spaces. Amygdaloidal ore deposits are particularly important in rift basins such as the Keweenaw copper belt of North America.

### 7.3.3   Pyroclastic textures

Volcanic eruptions eject broken rock particles of varying sizes, known as **pyroclasts** (which means fiery fragment). Pyroclasts may be ejected into the atmosphere as airborne **tephra** or transported along Earth's surface as pyroclastic flows. Following deposition, these particles are compacted, cemented or welded together to produce volcanic rocks with pyro-

**Figure 7.16**   Amygdaloidal basalt in which vesicles have been infilled with quartz and epidote. This basalt has been altered by hydrothermal solutions which have changed the existing chemistry of the basalt and precipitated new (secondary) minerals in the vesicles. *Source*: Kevin Hefferan. © John Wiley & Sons.

clastic textures. Pyroclasts are classified (Figure 7.17) according to their composition, size, and shape (Table 7.4). Pyroclasts consist of several different types of materials. **Lithic** pyroclasts are rock fragments such as basalt, andesite or other rocks. **Vitric** pyroclasts are composed of glassy fragments called **shards**,

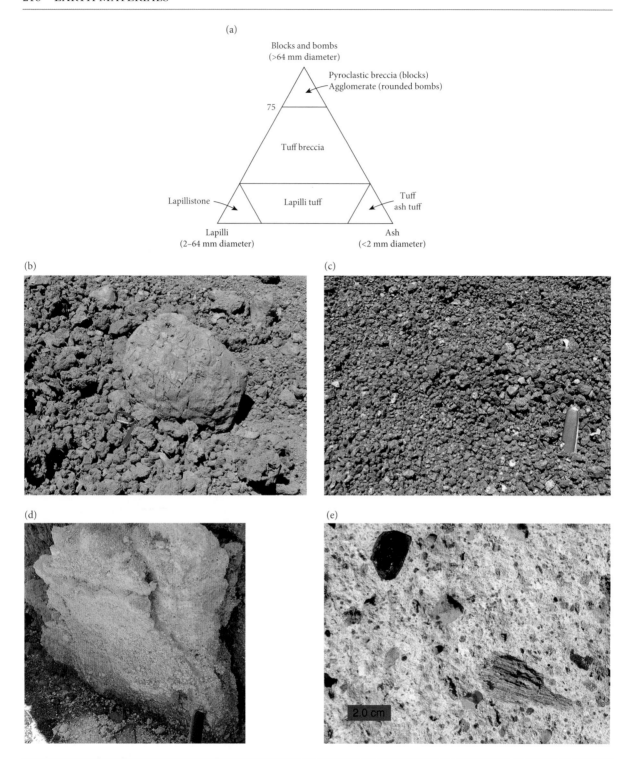

**Figure 7.17** (a) IUGS classification for pyroclastic rocks based on diameter clast size and relative abundance. *Source*: Based on Fisher (1961, 1966). © John Wiley & Sons. Inset Images: (b) Rounded bomb and angular blocks from Hawaii. *Source*: Kevin Hefferan. © John Wiley & Sons. (c) Lapilli from Hawaii. (d) Hawaiian ash tuff. *Source*: Kevin Hefferan. © John Wiley & Sons. (e) Tuff Breccia from hole in the wall, Mojave National Preserve, CA. *Source*: Photograph courtesy of Mark A. Wilson, Department of Geology, The College of Wooster.

**Table 7.4** Common fragmental (pyroclastic) textures classified by size and degree of roundness (Schmid 1981; Fisher and Schmincke 1984).

| Clast diameter size (mm) | Pyroclast name | Rock name |
| --- | --- | --- |
| >64 | Angular block | Breccia or tuff breccia |
| | Rounded bomb | Agglomerate |
| 2–64 | Lapilli | Lapillistone or lapilli tuff |
| <2 mm | Ash | Tuff or ash tuff |

*Source*: Based on Schmid (1981) and Fisher and Schmincke (1984). © John Wiley & Sons.

most commonly composed of pumice or scoria. **Crystal** pyroclasts contain mineral fragments.

Pyroclasts are further divided by average grain size diameters (Table 7.4 and Figure 7.17).

- Pyroclasts, greater than 64 mm in diameter, are called **blocks** if angular (Figure 7.17) and **bombs** if rounded. Angular blocks lithify as **breccias** and rounded blocks form **agglomerates.**
- Gravel-sized pyroclasts (2–64 mm diameter) are called **lapilli**. Rocks that consist largely of lapilli are called **lapillistones.**
- **Ash** consists of sand-sized and finer-sized pyroclasts (<2 mm diameter) which can be subdivided into coarse ash (1/16–2 mm) and fine ash (<1/16 mm) or **dust**. A rock composed of solidified volcanic ash is called **tuff**. Tuffs that contain significant

amounts of gravel-size lapilli are called **lapilli tuffs.**

Large clasts provide information about distance from source vents and degree of transport. **Breccias** are deposited proximal to the volcano vent and subjected to minimal transport such that the angular block edges are not abraded (Figure 7.18). **Agglomerates** are composed of volcanic bombs or fragments abraded and rounded during transport by volcanic or subsequent sedimentary processes.

**Tuffs** and **lapilli tuffs** are deposited by airfall and pyroclastic flows. Explosive mushroom shaped ash clouds descend to produce unwelded ashfall and lapilli tuff deposits. These **unwelded tuffs** display random shard orientations and spherical to ellipsoidal vesicles in pumice fragments (Figure 7.19). Due to the relative absence of heating and plastic deformation during compaction, unwelded tuffs tend to be soft, low density rocks with light colors and ashy lusters.

Pyroclastic flows consist primarily of hot gases, crystals, pumice shards, and rock fragments of variable sizes. Sufficient heat and gas buoyancy can allow pyroclastic flows to travel upslope as well as downslope. Pyroclastic flows may incorporate volcanic sediments and water during transport to produce a viscous cement like slurry. Relatively small or single event eruptions produce unwelded tuffs, similar to those generated by air fall. Increasing volumes of hot pyroclastic

**Figure 7.18** Angular volcanic blocks from explosive eruptions at Kilauea, Hawaii. *Source:* Kevin Hefferan. © John Wiley & Sons.

**Figure 7.19** Unwelded ashfall deposits from explosive volcanic eruptions. *Source:* Kevin Hefferan. © John Wiley & Sons.

**Figure 7.20** Partially welded ash flow tuff deposit. *Source*: Kevin Hefferan. © John Wiley & Sons.

**Figure 7.21** Densely welded tuff deposit. *Source*: Kevin Hefferan. © John Wiley & Sons.

debris produce more compaction and the generation of **partially welded tuffs**. **Welding** results as warm fragments become progressively fused together as porosity decreases during compaction. Compaction induces plastic flow deformation in which hot shards are flattened and rotated from a random to a more parallel orientation. As a result, partially welded tuffs display parallelism among shards, elongated and partially flattened pumice vesicles and draping of glassy pyroclasts around rigid fragments. Compared to unwelded tuffs, partially welded tuffs are harder, darker, and higher density rocks.

Multiple hot pyroclastic flows emplaced over short periods of time may form a single unit in which the deposits cool and compact together. While the exteriors of such units cool rapidly, the interior portions can remain hot allowing for crystal development. The variable cooling rate allows the shards and pumice fragments in the interior portion to remain plastic as the unit undergoes compaction. Plastic fragments tend to bend around or drape over rigid crystal and lithic fragments. The flattening of pumice fragments produces elongated vesicles during rotation and closure producing a foliated (layered) texture.

Large volcanic eruptions marked by the deposition of numerous, thick pyroclastic layers results in intense welding. Intense welding produces hard, **densely welded tuffs** with dark colors and glassy lusters that may resemble obsidian. In densely welded tuffs, shards show marked parallelism and flattening (Figure 7.21).

Pumice fragments are extremely elongated, with completely closed vesicles and pronounced draping of vitric pyroclasts. They often mimic obsidian or vitrophyres in their appearance (Figure 7.20).

As discussed above, rock texture is a keystone component in the classification of igneous rocks. In the following section, we will discuss chemical composition in igneous rocks. The final section of this chapter presents more detailed crystalline igneous rock classification based on textural and composition factors.

## 7.4 CHEMICAL COMPOSITION OF IGNEOUS ROCKS

In Section 7.1 we described how silica ($SiO_2$) content is fundamentally important in magma classification. In Section 7.2, we explained how $SiO_2$ has a profound effect on igneous textures. In this section, we will describe how magma and rock chemistry are described in terms of major, minor, and trace elements.

### 7.4.1 Major elements

**Major elements** have concentrations greater than 1 wt. % in Earth's crust. Note in Figure 7.22a that silicon and oxygen represent the bulk of Earth's crust by weight (74.3%) and volume (94.7%). The remaining 90 naturally occurring elements constitute only 25.7% by weight and 5.3% by volume of Earth's crust. In addition to silicon and oxygen, the other six major elements in Earth's

(a)

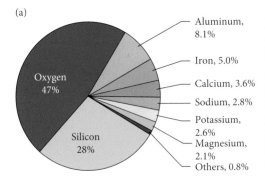

| | Ionic radius | | |
|---|---|---|---|
| Element | Volume % | (Å) | Ionic charge |
| Oxygen | 93.8 | 1.4 | −2 |
| Silicon | 0.9 | 0.4 | +4 |
| Aluminum | 0.5 | 0.5 | +3 |
| Iron | 0.4 | 0.7 | +2, +3 |
| Calcium | 1.0 | 1.0 | +2 |
| Sodium | 1.3 | 1.0 | +1 |
| Potassium | 1.8 | 1.4 | +1 |
| Magnesium | 0.3 | 0.7 | +2 |
| Total | 100.0 | | |

(b)

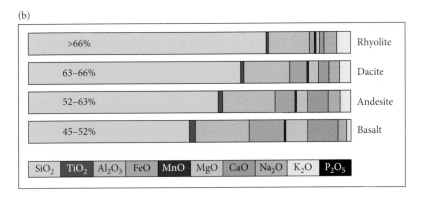

**Figure 7.22**    (a) The eight major elements in Earth's crust by weight percent, volume percent, radius, and common valence state (ionic charge). (b) Oxide compounds of major and minor elements that are common in volcanic rocks. *Source*: U.S. Geological Survey, United States Department of the Interior.

crust are aluminum, iron, calcium, sodium, potassium, and magnesium.

The eight major elements constitute the bulk of Earth's crust, encompassing 98.6% by weight and nearly 100% by volume. As silicon and oxygen bond together to form the compound $SiO_2$, it should be no surprise that most crustal minerals (~92%) are silicate minerals. $SiO_2$ concentration is the most important chemical factor in igneous rock classification.

Note that oxygen is the only anion (negatively charged ion) listed among the eight most common crustal elements. Therefore, the other major cation (positively charged ion) elements tend to bond with the anion oxygen. The eight major crustal elements – seven cations and one anion – form seven major oxide compounds in igneous rocks: $SiO_2$, $Al_2O_3$, $FeO$, $MgO$, $CaO$, $Na_2O$, and $K_2O$. In chemical analyses, geologists commonly express chemical components of igneous rocks in terms of oxide compounds. For example, the chemical components of K-feldspar ($KAlSi_3O_8$) can be expressed in terms of three different oxide compounds: $1/2(K_2O + Al_2O_3 + 3SiO_2)$.

The second most abundant oxide compound after $SiO_2$ is aluminum oxide. Aluminum oxide occurs in the feldspar mineral group, along with calcium, sodium, and potassium oxides. Ferromagnesian minerals such as the pyroxene, amphibole, olivine minerals, and biotite account for most of the iron oxide and magnesium oxide concentrations. Figure 7.22b illustrates how the relative abundance of major oxide compounds serves to distinguish four common volcanic rocks. As might be expected, the $SiO_2$ abundance is the primary determining factor in classifying these rocks. Note that rhyolite contains >66% $SiO_2$, dacite 63–66% $SiO_2$, andesite 52–63% $SiO_2$ and basalt 45–52% $SiO_2$. Also note the chemical trends depicted: as $SiO_2$ concentration increases, $FeO$, $MgO$, $CaO$, and $TiO_2$ decrease while $Na_2O$ and $K_2O$ increase. Figure 7.22 also displays minor oxide compounds such as titanium oxide, manganese oxide, and phosphate. These major and minor element variations are critically important in the determination of magma and mineral chemistry as well as in rock nomenclature.

## 7.4.2 Minor and trace elements

In addition to major elements, rocks also contain minor elements and trace elements. **Minor elements** consist of those elements that commonly occur in concentrations of 1.0–0.1% by weight. Chromium, manganese, phosphorous, hydrogen, and titanium are among the most common minor elements in igneous rocks; in some rocks, these minor elements occur in concentrations normally associated with major elements (>1 wt. %). Element concentrations vary somewhat among rock types. Minor element concentrations are related in part to $SiO_2$ concentrations. Basic igneous rocks with $SiO_2$ concentrations between 45 and 52% are relatively enriched in minor elements such as Cr, Ni, and Cu. Highly acidic igneous rocks with $SiO_2$ concentrations in excess of 66% tend to be enriched in minor elements such as Li, Be, and Ba.

Trace elements (Box 7.1) occur in crustal rock concentrations less than 0.1% by weight and are typically measured in parts per million (<1000 ppm). Many minor and trace elements are economically important, in part due to their scarcity in rocks. Although their crustal concentrations are infinitesimally small, trace elements can provide information on the genesis and

---

## Box 7.1   Trace elements

Trace elements include rare Earth elements, high field strength elements, and large ion lithophile elements. **Rare Earth elements (REE)**, also called the **lanthanides,** have atomic numbers ranging from 57 to 71. In Earth's crust, REE with odd atomic numbers are more abundant than REE with even atomic numbers. This odd–even imbalance makes graphing REE abundances difficult. To correct this imbalance, geologists calculate REE of Earth's crustal rocks and divide them by REE abundances in chronditic meteorite samples. Why use chrondritic meteorites as a comparison rock? Chrondritic meteorites formed at the same time as Earth (4.6 billion years ago) and contain the same relative proportions of elemental abundances. However, unlike meteorites, Earth's crustal rocks have experienced partial melting. The partial melting has altered the REE abundances from their original (primordial) concentration, due to the mobility of incompatible elements. Thus, dividing the REE concentration of an Earth crustal rock sample by the REE concentration in a chondritic meteorite is thought to be akin to comparing a sample rock to Earth's original REE concentration. This process standardizes the differences between even and odd atomic numbered REE concentrations and makes data presentation graphs easier to plot and comprehend. The net result typically produces a relatively flat REE concentration pattern. This process is referred to as a chondrite-normalized pattern (Hess 1989). REE are divided into light and heavy rare Earth elements. **Light rare Earth elements (LREE)** such as lanthanum (La), cerium (Ce), praseodymium (Pr), neodymium (Nd), and samarium (Sm) are situated on the left end of the periodic table. **Heavy rare Earth elements (HREE),** situated at the right end of the periodic table, include europium (Eu), gadolinium (Gd), terbium (Tb), dysprosium (Dy), holmium (Ho), erbium (Er), thulium (Tm), ytterbium (Yb), and lutetium (Lu). Although all REE are incompatible, LREE elements are more incompatible than HREE and for this reason we distinguish between the two groups of REE.

High field strength elements (HFS) are characterized as having a relatively high ionic charge (+3 or +4) for a given radius (Figure B7.1). These elements are considered "immobile", in that they tend not to be incorporated into the melt under limited partial melting. HFS elements, having a ionic charge/ ionic radius ratio greater than 2, include Th, Ti, V, Zr, Hf, Ce, Nb, Ta, and U. As immobile elements, HFS and HREE tend to remain with the parent or restite rock. As these elements are retained with the original parent source material, these elements have "long memories" and are useful in tracing mantle-related processes.

Trace elements, containing an ionic charge/ionic radius ratio less than 2, are referred to as **Large ion lithophile (LIL)** elements (Figure B7.1). These elements include Cs, Ba, Rb, Sr, Pb, K, and Eu. LIL elements tend to be mobile, incorporated into partial melts and depleted in the restite. LIL elements are useful in determining the role of hydrous fluid interaction and the parental source of the partial melt.

*Continued*

## Box 7.1 *Continued*

For example, a lithospheric magma source is indicated if an igneous rock is enriched in Sr and Nd (LREE); an asthenospheric magma source is suggested if Sr and Nd are depleted in an igneous rock.

Not all incompatible elements are incompatible to the same degree. Rb, Ba, and Cs are among the most incompatible elements. Sm and Hf are moderately incompatible. Sr is more compatible than Rb; Sm is more compatible than Nd and Lu is more compatible than Hf. Furthermore, an element's relative compatibility is dependent upon the mineral chemistry in the source rocks and within the magma. Eu is incompatible, except in the presence of plagioclase. In an oxidizing environment more of the Eu will be in the $Eu^{+3}$ state (vs. $Eu^{+2}$) and the radius of $Eu^{+3}$ closely matches the radius of Ca ions in feldspar. As a result, $Eu^{+3}$ is compatible in Ca-plagioclase. Likewise, chromium is compatible in the presence of pyroxene minerals, garnet or chromite. Thus, the relative abundance of the REE is related to the mineral chemistry of the magma. What is the significance of trace elements? As will be discussed in

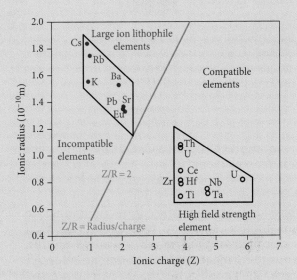

**Figure B7.1**  Large ion lithophile elements contain an ionic charge/ionic radius ratio less than 2 while high field strength elements contain an ionic charge/ionic radius ratio greater than 2.

Chapter 10, trace elements may be used to determine the magmatic source, tectonic environment, and the age of igneous rocks. $Sr^{87}/Sr^{86}$ ratios are used to determine the source region for magmas. Discrimination diagrams are particularly useful for ancient orogenic belts where tectonic environments are uncertain. For example, Faure (1977) calculated the mean $Sr^{87}/Sr^{86}$ ratios of rocks derived from different magmatic environments such as ocean floor (<0.7028), ocean islands (0.7039) and island arcs (0.7044). Cann (1970, 1971) and Pearce and Cann (1973) pioneered tectonic discrimination diagrams that display trace element and minor element concentrations. Using minor and trace element concentrations of Rb, Sr, Y, and Nb, Cann (1970) distinguished between mid ocean ridge basalt and ocean island basalt. Pearce and Cann (1973) discriminated mid-ocean ridge basalt, ocean island basalt, volcanic arc basalt and continental basalt by plotting concentrations of Ti, Zr, Y, and Sr. Discrimination diagrams have also been developed based on variations in minor and major element concentrations. Mullen (1983) used $MnO_2$, $TiO_2$ and $P_2O_5$ to identify five basalt environments. Pearce et al. (1975, 1977) distinguished between oceanic and continental basalts through variations in major element concentrations. Saccani (2015) proposed tectonic discrimination of ten different types of ophiolitic basalts based on high field strength elements Nb, Th, Ce, Dy, and Yb. Geochemical discrimination is particularly important for complexly faulted regions where the rocks have been displaced from their original site of development. In highly deformed tectonic zones, such as convergent and divergent margin settings, the original geologic setting and rock relationships have long since been disrupted by faulting and other deformational processes. Geochemical analysis can be very useful in revealing conditions for the rock origins. However, one must also be cognizant of the fact that these rocks have not existed in a closed system. Hydrothermal alteration and metamorphism can profoundly alter the original rock chemistry so that only the least altered samples containing immobile elements should be considered for analysis. The critical point of this discussion is that trace elements, which represent an infinitesimally small component of magma, play a profound role in allowing us to determine the genesis, history, and age of igneous rocks.

history of igneous rocks. Trace elements may be regarded as the DNA of magma. Trace elements are used as tectonic environmental indicators, essentially providing information on the origin and evolution of magma. How is this accomplished? Trace elements bear chemical characteristics that are said to be "compatible" versus "incompatible" or "mobile" versus "immobile."

### 7.4.3 Element compatibility

**Compatibility** is a measure of the ease with which an element fits into a crystal structure as a mineral grows, and is analogous to the description of a compatible or incompatible person. A compatible person might be described as someone who gets along well with other people, fits into social structures, tends to form long lasting bonds, and doesn't leave when conditions change. An incompatible person is someone who does not get along with other people, tends not to fit into social structures, moves around and quickly departs from relationships when things get hot.

The same descriptions aptly describe compatible and incompatible elements. **Compatible elements** tend to form long-lasting bonds and enter into crystal structures. Compatible elements are **immobile** in that they do not readily migrate from the crystal structure. **Incompatible elements** do not fit easily into crystal structures and their bonds are easily broken. Incompatible elements are **mobile** in that they tend to migrate from crystal structures into the melt when the rock is subjected to partial melting. Under conditions that lead to partial melting of a parent rock, compatible elements tend to remain within the solid crystalline structure and do not migrate with melt. The solid **residual rock**, or **restite** remaining after melt removal, has a different chemical composition than the original parent rock. The restite is enriched in compatible elements and depleted in incompatible elements. Conversely, the partial melts are enriched in incompatible elements and depleted in compatible elements.

Thus, the relative abundance of compatible and incompatible trace elements provides important information related to the melting history of igneous rocks. With small degrees of partial melting, incompatible elements such as K, Rb, Sr, and Ba are depleted in the restite and enriched in magmas that migrate upward toward Earth's surface. Rocks derived from partial melts are enriched in incompatible elements and depleted in compatible elements. The restite rock, from which the partial melts have been removed, tends to be enriched in compatible elements such as Fe, Mn, Zn, Ti, V, Cr, Co, Ni, Cu, and depleted in incompatible elements. For example, igneous rocks generated by partial melts, such as ocean ridge basalts, are incompatible element enriched compared to the restite mantle. Detailed studies of trace element concentrations are used by petrologists to infer melting histories and to indicate tectonic environments for the genesis of rocks at plate boundaries and intraplate settings. We will return to this discussion in Chapter 10 when we address igneous rock associations.

## 7.5  MINERAL COMPOSITION OF IGNEOUS ROCKS

Minerals may be regarded as primary or secondary, depending upon the time of their development. **Primary minerals** are those minerals that crystallize directly from magma over a range of relatively high temperatures. **Secondary minerals** form later in response to chemical processes that affect a preexisting rock. Secondary minerals replace primary minerals or fill voids (as amygdules) through alterations by hot solutions or other secondary alteration processes. For the purpose of this chapter we will discuss only primary minerals, which are subdivided into major minerals and accessory minerals.

### 7.5.1  Major minerals

The seven oxide compounds combine to form eight common **major or essential mineral** groups. Major mineral groups are those constituents that commonly occur in abundances greater than 5%. These include the quartz, K-feldspar, plagioclase feldspar, feldspathoid, mica, amphibole, pyroxene, and olivine mineral groups (Table 7.5). Common feldspar minerals include the plagioclase minerals (anorthite to albite) and K-feldspar minerals (microcline, orthoclase, and sanidine).

**Table 7.5** Common oxide compounds and major minerals in igneous rocks.

| Oxide compound | Major igneous minerals or mineral groups |
| --- | --- |
| MgO | Olivine, pyroxene, amphibole, biotite, phlogopite |
| FeO | Olivine, pyroxene, amphibole, biotite |
| $Al_2O_3$ | Plagioclase feldspar, potassium feldspar, and micas |
| CaO | Plagioclase feldspar |
| $Na_2O$ | Plagioclase feldspar, feldspathoids |
| $K_2O$ | Potassium feldspar, micas, feldspathoids |
| $SiO_2$ | Quartz group minerals and all the others |

Feldspathoid minerals are common in alkali rich, $SiO_2$ poor rocks and include leucite, nepheline, cancrinite, sodalite, and kalsilite. Mica minerals include muscovite, biotite, phlogopite, and lepidolite. Amphibole group minerals include hornblende, riebeckite, and richterite. Pyroxene group minerals include augite, diopside, pigeonite, aegerine, hypersthene, enstatite, and bronzite. Olivine minerals range from forsterite to fayalite. The major minerals are not of great economic value; their importance rests in the fact that they are the most common minerals in igneous rocks.

### 7.5.2 Accessory minerals

Less common minerals are considered accessory minerals, which typically occur in concentrations less than ~5%. Accessory minerals consist largely of oxide, sulfide and silicate minerals. Accessory oxide minerals include magnetite, hematite, ilmenite, spinel, sphene, rutile, chromite, corundum, uraninite, columbite, and cassiterite. Sulfides include pyrite, chalcopyrite, molybdenite, pentlandite, and pyrrhotite. Silicate minerals include zircon, garnet, melilite, monazite, epidote, allanite, tourmaline, and topaz. The fluoride mineral fluorite is also common. While accessory minerals are minor constituents in most igneous rocks, some of these are critical minerals of strategic importance in our society. Accessory minerals are valued as metallic ore deposits, gems, abrasives and for many other useful applications.

## 7.6 MINERAL COMPOSITION AND TERMINOLOGY

Whenever possible, Earth scientists name and describe rocks based upon their major mineral content. Mineral content yields important clues regarding rock chemistry, particularly with respect to major oxide compounds such as silica and aluminum oxide. Mineral composition and rock texture also help us understand the conditions in which the rock formed. Different sets of terminology have been developed to describe mineral components in igneous rocks.

### 7.6.1 Determination of modal and normative mineral composition in igneous rocks

For rocks in which minerals are visible to either the unaided eye or with a petrographic microscope, one can determine the mineral percentages using modal composition. For very fine grained or glassy rocks in which minerals are either too small to identify or not present, Earth scientists use a system referred to as normative classification. Both techniques will be discussed below. Normative classifications can be used for any igneous rock, whereas modal classifications are used only for rocks in which the minerals can be identified.

*Modal composition*

The most straightforward approach to determining rock mineralogy involves visually identifying the minerals and determining their percentages by volume. The composition of a rock determined by actual mineral identification is referred to as its **modal composition** or **mode**. In coarse grained phaneritic and pegmatitic rocks, a modal composition can be estimated by visual inspection of a rock, perhaps with the aid of a hand lens. For finer grained crystalline rocks, modal composition is determined using a petrographic microscope.

A much more accurate mode can be calculated for any coarse grained rock by doing a **point count analysis** on a thin-section. Accurate point count analysis requires use of a petrographic microscope, whereby a thin section is moved incrementally on a grid system such that at least 400 mineral points are tabulated. For each individual point, the scientist identifies the mineral and keeps a running total of all the major mineral points using a counter. A counter is similar to the device used by a baseball umpire to count balls, strikes, and outs. However, in this case, the "umpire" is counting quartz, plagioclase, and other modal minerals. The mineralogic point count results are then tabulated and expressed in volume percent. Modal analysis is a direct measuring technique in which constituent mineral grains are identified, and the volume percentages determined. The modal composition approach works well for crystalline rocks in which minerals are readily identifiable but fails for very fine grained or glassy igneous rocks in which individual minerals cannot be identified or do not exist.

*Normative composition*

Normative mineralogy (Cross et al. 1902; Kelsey 1965) is an indirect technique using data derived from the chemical analysis of a rock sample. The first norm classification was devised by C̲ross, I̲ddings, P̲irsson, and W̲ashington (Cross et al. 1902), and is referred to as the C.I.P.W. norm classification in their honor. Normative classification systems are commonly used for aphanitic or glassy volcanic rocks, in which a rock's modal mineral composition cannot be accurately determined.

In normative calculations, specific rules are used to convert rock bulk chemical composition, expressed as oxide %, into a hypothetical suite of **normative** minerals. Normative minerals represent those minerals geochemists believe might be observed if the rock texture was coarse grained. A **norm** calculation is a process that measures a rock bulk chemical composition and creates a hypothetical set of representative minerals based on that rock's chemical dataset. This method involves assumptions. First, one hopes that the hypothetical set of minerals chosen closely resembles the actual minerals in the rock or potential minerals that would have formed. Second, normative mineralogy is also calculated on the assumption that the magma crystallized near Earth's surface at low pressure, anhydrous conditions that approximate those under which many volcanic rocks form. Thus, standard CIPW normative calculations do not consider the role of high pressure or hydrous

mineral phases such as micas and amphiboles. Given that hydrous igneous minerals include the mica and amphibole mineral groups, this is a major omission. To address this shortcoming, special non-CIPW normative calculations have been developed to calculate normative minerals for rocks thought to be enriched in hydrous or other high pressure minerals.

Normative values are calculated using a multistep procedure utilizing whole rock chemical analysis of a rock sample. The chemical composition data are expressed in oxide weight percents (for example, $FeO$, $Al_2O_3$, etc.). These oxide weight percents are determined by powdering a rock and deriving geochemical results from an analytical spectrometer. Standard CIPW normative minerals include common anhydrous igneous minerals such as: quartz, orthoclase, albite, anorthite, nepheline, magnetite, ilmenite, apatite, corundum, diopside, enstatite, hypersthene, and olivine.

### 7.6.2 Descriptive terminology based on chemical composition

Numerous approaches have been developed to subdivide igneous rocks into distinctive groups based on major, minor, and trace element abundances. Remember that the major elements of greatest abundance in Earth's crust include oxygen, silicon and aluminum which bond together to form silica and aluminum oxide. The most useful terminology for the chemical composition of igneous rocks naturally involves these three elements.

*Abundance of silica ($SiO_2$)*

As silicon and oxygen are the primary chemical constituents in magma, the percentage of $SiO_2$ in rocks is an important means by which we classify magma and igneous rocks.

**$SiO_2$ saturation** classification is a concept that grew out of CIPW normative calculations. In such calculations, all oxides reported in the chemical analysis are used to make minerals according to a set of more than twenty steps that usually involves combining other oxides (e.g. $Al_2O_3$) with $SiO_2$. Three normative minerals provide useful examples of the concept of $SiO_2$ saturation (Table 7.6):

1   Normative orthoclase ($KAlSi_3O_8$) forms by combining $\frac{1}{2} K_2O + \frac{1}{2} Al_2O_3 + 3SiO_2$;

**2** Normative albite ($NaAlSi_3O_8$) crystallizes by combining $\frac{1}{2} Na_2O + \frac{1}{2} Al_2O_3 + 3SiO_2$;

**3** Normative enstatite ($Mg_2Si_2O_6$) is produced by combining $2MgO + 2SiO_2$.

The concept of $SiO_2$ saturation (Table 7.6) is based on the presence or absence of three mineral groups: quartz, feldspars, and feldspathoids. **$SiO_2$ oversaturation** implies that when all available cation oxides have been used to make normative minerals, additional $SiO_2$ remains available to generate normative quartz. The excess $SiO_2$ is indicated by the presence of "free quartz." All quartz normative rocks are therefore oversaturated with $SiO_2$. Most rocks with significant modal quartz are likely to also be quartz normative and therefore oversaturated with $SiO_2$. Granitic rocks are almost always oversaturated with $SiO_2$.

**$SiO_2$ saturation** exists in a CIPW normative calculation when exactly enough $SiO_2$ exists to consume all the other oxides and no excess $SiO_2$ remains. $SiO_2$ saturated rocks contain normative feldspars and/or orthopyroxene (enstatite or hypersthene) minerals, but lack either quartz (an indicator of $SiO_2$ oversaturation) or magnesium olivine (forsterite) or feldspathoids (indicators of $SiO_2$ undersaturation) as discussed below.

**$SiO_2$ undersaturation** occurs when, during CIPW normative calculations, $SiO_2$ is depleted before all the other oxides have been used to form normative minerals. In this case there is insufficient $SiO_2$ to make not only quartz, but also feldspars (such as albite, anorthite or orthoclase) or orthopyroxenes (enstatite or hypersthene). $SiO_2$ undersaturated rocks commonly contain feldspathoid or magnesium olivine (forsterite) minerals that cannot coexist with quartz because they crystallize from silica undersaturated melts.

As we shall see, IUGS rock classifications based on modal mineralogies strongly reflect the concept of $SiO_2$ saturation because the primary mineral indicators include quartz ($SiO_2$ oversaturated), feldspars ($SiO_2$ saturated) and feldspathoids ($SiO_2$ undersaturated). It is also worth noting that most ultrabasic rocks are also $SiO_2$ undersaturated whereas most acidic rocks are $SiO_2$ oversaturated.

*Relative abundance of aluminum oxide ($Al_2O_3$)*

Aluminum oxide is the second most abundant compound in Earth's crust. Igneous rocks are classified based upon the relative proportions of $Al_2O_3$ to CaO, $Na_2O$, and $K_2O$. The relative proportion of these oxides yields the following descriptive terms: peraluminous, metaluminous, subaluminous, and peralkaline (Table 7.7). As feldspars are the most abundant mineral group in Earth's crust, the relative abundance of $Al_2O_3$ as compared to $CaO + Na_2O + K_2O$ affects the

**Table 7.6** Silica saturation and key mineral indicators.

| Silica ($SiO_2$) saturation | Key mineral indicators |
|---|---|
| $SiO_2$ oversaturated | Contains quartz ± feldspars and/or Mg orthopyroxene. Excludes felspathoids and Mg rich olivine (forsterite). |
| $SiO_2$ saturated | Contains feldspars and/or Mg orthopyroxene only. Excludes quartz Mg-rich olivine and feldspathoids |
| $SiO_2$ undersaturated | Contains forsterite olivine, and feldspathoids. May also contain feldspars and/or orthopyroxene minerals. Excludes quartz. |

**Table 7.7** Aluminum oxide classification and associated modal minerals.

| $Al_2O_3$ abundance | $Al_2O_3$ vs. CaO, $Na_2O$, $K_2O$ | Common minerals |
|---|---|---|
| Peraluminous | $Al_2O_3 > CaO + Na_2O + K_2O$ | Muscovite, corundum, topaz, garnet, tourmaline, cordierite, andalusite, biotite |
| Metaluminous | $Al_2O_3 < CaO + Na_2O + K_2O$ and $> Na_2O + K_2O$ | Hornblende, epidote, melilite, biotite, pyroxene |
| Subaluminous | $Al_2O_3 = Na_2O + K_2O$ | Olivine, orthopyroxene, clinopyroxene |
| Peralkaline | $Al_2O_3 < Na_2O + K_2O$ | Aegerine, riebeckite, arfvedsonite, aenigmatite, astrophyllite, columbite, pyrochlore |

mineral assemblages that develop in igneous rocks (Shand 1951; Hyndman 1985). This classification is particularly useful for the discrimination of granitic rocks. **Peraluminous** rocks contain minerals with very high $Al_2O_3$ contents. **Peralkaline** rocks contain minerals with high $K_2O$ and/or $Na_2O$ contents. **Metaluminous** rocks contain mafic minerals with average aluminum contents. **Subaluminous** rocks contain mafic minerals with low aluminum concentrations.

## 7.7 IUGS IGNEOUS ROCK CLASSIFICATION

Many different igneous rock classifications exist that include hundreds of possible rock names. The most widely used rock classifications identify igneous rocks based on texture and: (1) modal minerals identified in the rock, (2) theoretical normative minerals calculated from chemical composition data from laboratory analyses, or (3) chemical composition of the rock based on laboratory analytical methods. One attempt to standardize igneous rock nomenclature was initiated in the 1960s by Albert Streckeisen on behalf of the International Union of Geological Sciences (IUGS). The IUGS recommends a classification system for both plutonic and volcanic rocks using essential mineral groups as endpoints in triangular- and diamond-shaped diagrams (Streckeisen 1976; LeBas and Streckeisen 1991; LeMaitre 2002). While the IUGS classification is generally accepted, it is not comprehensive for all igneous rocks and pre-existing rock nomenclature remains in use. The following discussion provides a summary of the IUGS classification system, highlighting benefits of a unified classification approach and its drawbacks.

Mole % are calculated by taking the weight % of a mineral and dividing by the mineral's molecular weight.

The **IUGS classification** is based upon the modal concentrations of five essential mineral groups abbreviated by the letters: Q, A, P, F, and M.

**Q** = Quartz, tridymite, cristobalite, and all $SiO_2$ group minerals

**A** = Alkali feldspar, including orthoclase, microcline, sanidine perthite, anorthoclase, and albite plagioclase with up to five mole % Anorthite ($An_0$–$An_5$).

**P** = Plagioclase($An_5$–$An_{100}$) and scapolite (altered plagioclase)

**F** = Feldspathoids, also known as foids. The term foid is derived from being "*Feldspath oid*" rich. Feldspathoids include the minerals nepheline, sodalite, cancrinite, leucite, analcite, nosean, hauyne, and kalsilite. In naming a rock, we use the major feldspathoid mineral as either an adjective or as part of the noun. For example, instead of naming a leucite rich syenite a "foid bearing syenite," the rock would be called a "leucite bearing syenite" or a nepheline rich rock would be a nepheline syenite.

**M** = Mafic and related minerals, including olivine, pyroxene, amphiboles, iron or magnesium rich micas, melilite, opaque minerals, garnet, epidote, calcite, allanite, zircon, apatite, sphene, titanite (Streckeisen 1976).

The QAPF modal classification applies to igneous rocks with >10% felsic minerals and <90% mafic mineral (M) content by volume; that is, for rocks with color indices (CI) of <90. The QAPF plot is a diamond-shaped diagram (Figure 7.23) on which the modal percentages of quartz or feldspathoid mineral groups, and the ratio of alkali to plagioclase feldspar groups are plotted. The upper QAP portion of the diagram distinguishes $SiO_2$ oversaturated rocks containing quartz. The lower FAP portion of the diagram consists of $SiO_2$ undersaturated rocks containing feldspathoids minerals but not quartz. Line A–P represents the line of $SiO_2$ saturation that separates the (upper triangle) oversaturated realm from the (lower inverted triangle) undersaturated realm. The QAPF mineral constituents are normalized so that the sum total of the four end member mineral constituents equals 100%. Remember that in all cases, either Q and/or F will be zero as quartz and feldspathoid minerals do not crystallize together as magma can't be both oversaturated and undersaturated with respect to $SiO_2$.

Let's consider a plutonic rock whose modal mineralogy consists of: 20% Quartz, 30% plagioclase, 25% K-feldspar, 12% biotite, 10% hornblende and 0% feldspathoid.

Since Q + A + P + F > 10%, we will normalize three nonzero end member minerals to exclude all components except Q, P, and A.

Q (20%) + P (30%) + A (25%) = 75% will be recalculated (normalized) to 100%.

Normalized 3 end member values: Q (20/75) = 27%, P (30/75) = 40% and A (25/75) = 33%

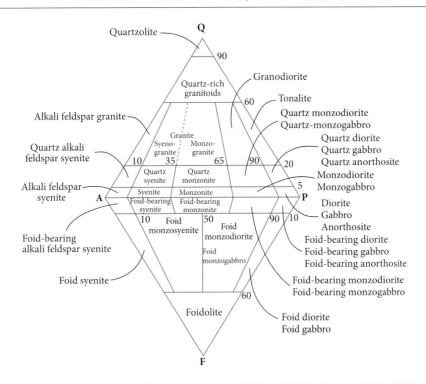

**Figure 7.23**    QAPF diagram for plutonic igneous rocks with >10% felsic minerals and <90% mafic minerals. Classification based upon IUGS classification. *Source*: Based on Streckeisen (1976) and LeMaitre (2002). © John Wiley & Sons.

Normalizing the two feldspar end members:

$$Plagioclase(P) = (30 / 55) = 55\%P$$
$$K\text{-feldspar}(A) = (25 / 55) = 45\%A$$

The plutonic rock would plot as a monzogranite on the QAPF IUGS diagram.

The QAPF diagram was originally designed for the classification of plutonic igneous rocks. Thereafter, it was adapted for volcanic crystalline rocks with variable degrees of success, as fine grain size can limit the ability to identify modal minerals.

### 7.7.1    IUGS plutonic rock classification

The IUGS classification diagram for plutonic igneous rocks with >10% felsic minerals (Q, A, P or F) and <90% mafic minerals is presented in Figure 7.23. Thirty-five different plutonic rocks are recognized within the plutonic QAPF classification.

Careful observation illustrates some problems with the IUGS system. Near the plagioclase (P) corner, quartz diorite and quartz gabbro occupy the same region; diorite, gabbro, and anorthosite also coexist in the same location. Similarly, feldspathoids-bearing gabbro, diorite, and anorthosite rocks also occur in the same foid regions. How do we discriminate among these rocks? For anorthosite, the answer is straightforward as these rocks contain more than 90% plagioclase. Distinguishing between gabbro and diorite as well as quartz gabbro and quartz diorite is a bit more complex. As indicated in Figure 7.23, gabbro and diorite each contain <5% quartz; while quartz gabbro and quartz diorite each contain 5–20% quartz. For both sets of rocks, three different factors distinguish quartz gabbro and gabbro from quartz diorite and diorite:

1   Gabbros/quartz gabbros contain 70–90% mafic minerals whereas diorites/quartz diorites contain 30–70% mafic minerals. This approach is based on visual identification of hand samples and is the method used in the IUGS system.

2   Gabbros/quartz gabbros are more calcic with plagioclase An contents >50. Diorites/quartz diorites are more sodic with plagioclase

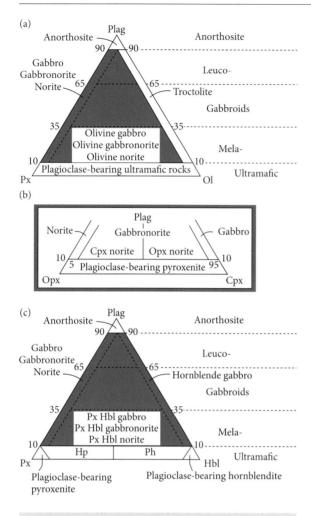

**Figure 7.24** Modal classification of gabbroic rocks based on proportions of three modal minerals that include plagioclase (Plag), olivine (Ol), orthopyroxene (Opx), clinopyroxene (Cpx) and hornblende (Hbl) Note the terms leuco- and mela- may be used as adjectives to denote light-colored and dark-colored rocks, respectively. (a) Gabbroic rocks classified with plagioclase, pyroxene, and olivine. (b) Gabbroic rocks classified using plagioclase, clinopyroxene, and orthopyroxene. (c) Gabbroic rocks classified on plagioclase, pyroxene, and hornblende. *Source*: Based on Streckeisen (1976) and LeMaitre (2002). © John Wiley & Sons.

An contents <50. Determination of An content requires microscopic analysis of thin sections in a laboratory or more advanced geochemical analysis.

3 Gabbros/quartz gabbros contain 45–52% SiO$_2$ and diorites/quartz diorites contain

53–66% SiO$_2$. Determination of the SiO$_2$ content requires detailed geochemical analysis of powdered rock specimens in a laboratory.

### 7.7.2 IUGS gabbroic rock classification

Gabbros are classified in more detail based on the modal mineral proportions of plagioclase (Pl), and the mafic minerals olivine (Ol), pyroxene (Px) and hornblende (Hb) as illustrated in Figure 7.24. In these triangular diagrams, mineral abundances are recalculated (normalized) so that the sum of the three mineral proportions equals 100%. At the top apex of the triangle, anorthosite rocks contain more than 90% recalculated plagioclase. Rocks that contain 10–90% recalculated plagioclase consist of a variety of gabbros and norites. The distinction between these rocks is based upon whether the pyroxene minerals are orthopyroxenes (norite) or clinopyroxenes (gabbro). Orthopyroxenes crystallize in the orthorhombic system and include the minerals enstatite and hypersthene. Clinopyroxenes crystallize in the monoclinic system and include the minerals augite and pigeonite. Ultramafic rocks containing >90% recalculated pyroxene minerals and less than 10% plagioclase are located near the base of the gabbroic rock classification triangle, and are discussed in the following section.

### 7.7.3 IUGS ultramafic rock classification

As stated previously, the QAP classification scheme is applied when felsic minerals compose >10% of the rock (Streckeisen 1973; LeBas and Streckeisen 1991; LeMaitre 2002). A different set of triangular rock discrimination plots (Figure 7.25) are used for ultramafic plutonic rocks containing >90% dark-colored minerals. The ultramafic family includes peridotites (>40% olivine), pyroxenites (pyroxene rich rock with <40% olivine) and hornblendites (hornblende rich rock with <40% olivine). The minerals are recalculated (normalized) so that the sum of the three mineral proportions of each triangular diagram equals 100%.

Igneous rock terminology continues to be cumbersome. Glazner et al. (2019) have proposed a modified classification that reduces the number of plutonic rock names, includes quantitative information about the modal mineral content, and denotes the major minerals not

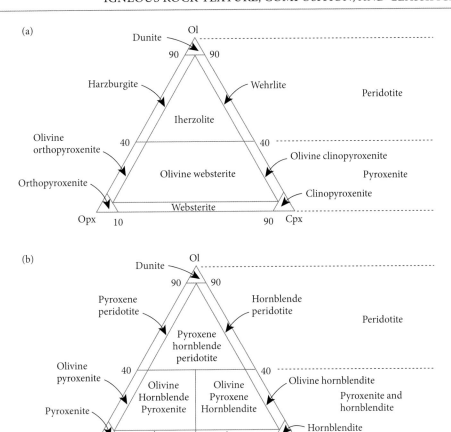

**Figure 7.25** Modal classification of ultramafic plutonic rocks based on the proportions of olivine (Ol), orthopyroxene (Opx), clinopyroxene (Cpx), pyroxene (Px) and hornblende (Hbl). (a) Plots of olivine, orthopyroxene, and clinopyroxene. (b) Plots of olivine, pyroxene, and hornblende for plutonic rocks that contain more than 10% essential hornblende. Note also that the terms pyroxenite and hornblendite may be used as generalized rock terms with preceding adjectives (e.g. olivine pyroxenite) for rocks enriched in either pyroxene or hornblende but containing less than 40% olivine. A more specific use for these rock names is shown at the lower two corners of Figure 7.25b in which pyroxenites are defined as containing more than 90% pyroxene and hornblendites are defined as containing more than 90% hornblende *Source*: Based on Streckeisen (1973) and LeMaitre (2002). © John Wiley & Sons.

included on the triangle or diamond apices. For example, a rock with 20% quartz, 20% K-Feldspar and 50% plagioclase with 10% mafic minerals biotite>hornblende would be named "hbl-bio 20, 20, 50 granodiorite," and plotted as a red dot on the QAPF diagram of Figure 7.26.

### 7.7.4   IUGS volcanic rock classification

The IUGS rock classification system also includes a QAPF diagram for volcanic rocks. (Figure 7.27) based upon the relative abundances

of mineral groups: quartz (Q), alkali feldspars (A), plagioclase (P) and feldspathoids (F). The IUGS system does not include separate diagrams for mafic or ultramafic volcanic rocks.

### 7.7.5   IUGS classification drawbacks

Together the QAPF and the mafic/ultramafic classification diagrams adequately discriminate most plutonic igneous rocks based on modal mineralogy. The IUGS modal classification

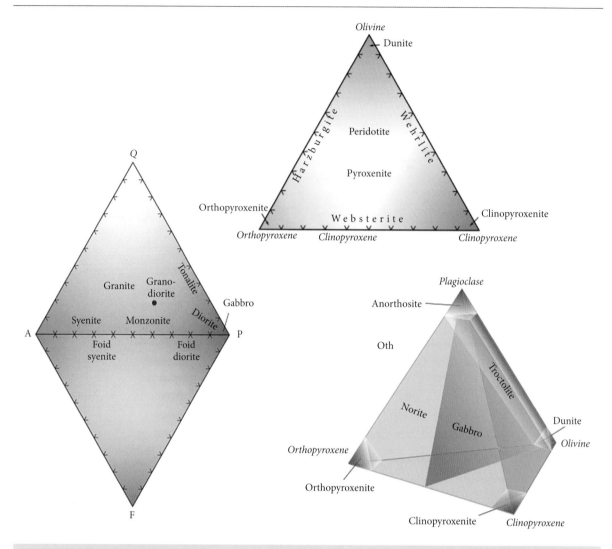

**Figure 7.26**   Simplified rock classification proposed by Glazner et al. (2019) for QAPF, mafic, and ultramafic plutonic rocks. The red circle in the QAPF diagram corresponds to a rock with 20% quartz, 20% K-Feldspar and 50% plagioclase with 10% mafic minerals biotite > hornblende and named "hbl-bio 20, 20, 50 granodiorite." *Source*: Courtesy of Allen Glazner.

can be utilized for QAPF volcanic rocks for which mineral identification is possible. The IUGS classification system does have its drawbacks. The IUGS system focuses on rocks whose primary minerals are in the QAPF or M categories, and does not emphasize rarer rocks such as carbonatites. Also, while the IUGS system is designed to enforce a uniformity in rock naming procedure, it cannot completely undo the wide variety of rock names already entrenched in the literature. The IUGS system is a modal system, which requires that three or four key minerals can be identified. Applying this system to volcanic rocks is problematic in that accurate identification of minerals in fine-grained rocks is not always possible. Furthermore, it completely ignores rocks with noncrystalline textures. As a result, the IUGS modal classification is less useful with volcanic rocks.

Other classification schemes that utilize chemical analyses can be used to distinguish and name rock types. Chemical composition of very fine grained rocks is determined in a geochemistry laboratory, typically by crushing and dissolving the rock specimen and using a spectrometer or other analytical techniques. One of the most common classification schemes

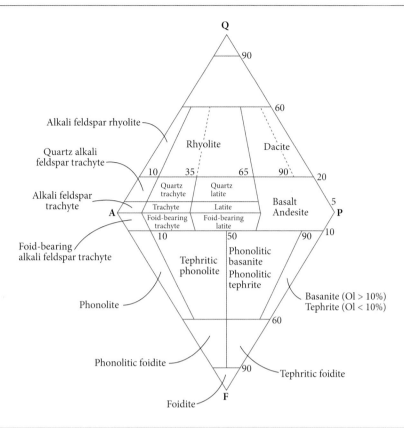

**Figure 7.27**   QAPF diagram for volcanic igneous rocks with >10% felsic minerals and <90% mafic minerals. Classification based upon IUGS classification *Source*: Based on Streckeisen (1976) and LeMaitre (2002). © John Wiley & Sons.

utilizes alkali-$SiO_2$ plots for the classification of very fine-grained igneous rocks. The chemical parameters used to classify fine-grained igneous rocks are $SiO_2$ and the alkali element compounds $Na_2O + K_2O$.

### 7.7.6   Total alkali-silica (TAS) classification

In addition to the IUGS system, several alkali-silica classification schemes have been developed for identifying volcanic rocks (LeMaitre et al. 1984; LeBas et al. 1986). These classification schemes are useful for volcanic rocks in which the mineral mode cannot be determined due to the presence of glass or cryptocrystalline texture. Where these conditions exist, chemical analyses present the only available option. The **Total Alkali to Silica (TAS)** classifications are suitable fine-grained or glassy volcanic rocks that are relatively unaltered and contain less than 2% $H_2O$ and 0.5% $CO_2$. In a TAS system, total alkalis

$(Na_2O + K_2O)$ are plotted on the ordinate (vertical y-axis) against $SiO_2$, which is plotted on the abscissa (horizontal x-axis). These rock fields are defined by total alkalis $(Na_2O + K_2O)$ that range from 0 to 16% and percent silica that ranges from 35 to 77%. The TAS classification identifies 16 different volcanic rock fields (Figure 7.28).

Igneous rock classification is based upon texture and composition. Despite the rather straightforward approach to identifying rocks, hundreds of igneous rock names have been proposed over the past 200 years. The IUGS system is one approach, designed to simplify the nomenclature in an attempt to standardize the nomenclature process. Computer programs have also been developed to assist in the IUGS rock identification process (Verma and Rivera-Gómez 2013). In Chapters 8 and 10 we will look further at why such a great diversity of igneous rocks occur on Earth as well as the tectonic settings in which they occur.

(a)

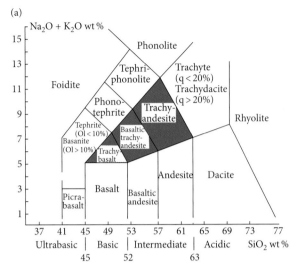

(b)

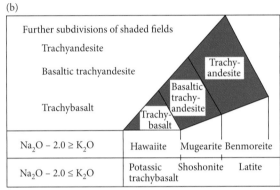

**Figure 7.28** (a) Alkali oxide versus silica (TAS) classification diagram for volcanic rocks. (b) Rocks falling in the shaded areas may be further subdivided based on relative concentrations of $Na_2O$ and $K_2O$. This classification utilizes IUGS rock terms *Source*: Modified from LeBas et al. (1986). © Oxford University Press.

## CONTENT ASSESSMENT

1  Describe two roles gases serve as magma migrates toward Earth's surface.

2  Describe the distinction between the terms mafic and basic.

3  Describe three factors that affect magma viscosity.

4  Identify the name of pyroclastic rocks that contain the following grain size diameters

| Rock name | <2 mm | 2–64 mm | >64 mm |
|-----------|-------|---------|--------|
|           | 30%   | 40%     | 30%    |
|           | 10%   | 80%     | 10%    |
|           | 60%   | 20%     | 20%    |

5  Describe how the seven major oxide compounds occurring in magma trend with respect to increases in silica ($SiO_2$)?

6  Identify minerals indicating silica oversaturation, saturation, and under saturation.

7  Identify key minerals that occur in peraluminous and peralkaline rocks.

8  Gabbro, diorite, and anorthosite occupy the same domain in the IUGS diagram. Describe the means by which they may be differentiated.

9  Provide IUGS names for plutonic rocks with weight percent mineral contents listed:

| Rock name | Quartz | Plagioclase | Alkali feldspar | Felds pathoid |
|-----------|--------|-------------|-----------------|---------------|
|           | 35     | 50          | 15              | 0             |
|           | 4      | 87          | 9               | 0             |
|           | 15     | 45          | 40              | 0             |
|           | 25     | 5           | 70              | 0             |
|           | 0      | 35          | 50              | 15            |

**10** Provide IUGS names for plutonic rocks with weight percent mineral contents listed:

| Rock name | Olivine | Plagioclase | Hornblende | Pyroxene |
|---|---|---|---|---|
| | 91 | 4 | 0 | 5 |
| | 35 | 5 | 35 | 25 |
| | 1 | 92 | 1 | 6 |
| | 45 | 27 | 23 | 5 |
| | 0 | 5 | 8 | 87 |

**11** Provide IUGS names for plutonic rocks with weight percent mineral contents listed:

| Rock name | Olivine | Hornblende | Augite (CPX) | Hypersthene (OPX) |
|---|---|---|---|---|
| | 45 | 5 | 5 | 45 |
| | 30 | 2 | 30 | 38 |
| | 31 | 52 | 11 | 6 |
| | 2 | 2 | 91 | 5 |
| | 7 | 3 | 50 | 40 |

**12** Identify the IUGS name of volcanic igneous rocks with mineral contents listed below:

| Rock name | Quartz | Plagioclase | Alkali feldspar | Feldspathoid |
|---|---|---|---|---|
| | 45 | 50 | 5 | 0 |
| | 7 | 87 | 6 | 0 |
| | 15 | 45 | 40 | 0 |
| | 25 | 5 | 70 | 0 |
| | 0 | 85 | 3 | 12 |

**13** Identify the IUGS name of volcanic igneous rocks with chemical compositions in weight percent listed below:

| Rock name | $SiO_2$ | $Na_2O$ | $K_2O$ |
|---|---|---|---|
| | 42 | 1 | 1 |
| | 49 | 6 | 1 |
| | 55 | 2 | 2 |
| | 56 | 1 | 8 |
| | 70 | 3 | 4 |
| | 75 | 5 | 4 |

**14** Basalt and andesite occupy the same domain on the IUGS diagram. Provide three means by which they may be distinguished.

## REFERENCES

Best, M.G. (2003). *Igneous and Metamorphic Petrology*, 2e. Oxford, UK: Blackwell Publishing 752 pp.

Brandeis, G., Jaupart, C., and Allegre, C.J. (1984). Nucleation, crystal growth and the thermal regime of cooling magmas. *Journal of Geophysical Research 89* (B12): 10161–10177.

Cann, J.R. (1970). New model for the structure of the ocean crust. *Nature 226*: 928–930.

Cann, J.R. (1971). Major element variations in ocean floor basalts. *Philosophical Transactions of the Royal Society of London, Series A 268*: 495–506.

Cashman, K.V., Sturtevant, B., Papale, P., and Navon, O. (2000). Magmatic fragmentation. In: *Encyclopedia of Volcanoes* (ed. H. Sigurdsson), 421–430. New York: Academic Press.

Cross, W., Iddings, J.P., Pirsson, L.V., and Washington, H.S. (1902). A quantitative chemicominerological classification and nomenclature of igneous rocks. *Journal of Geology 10*: 555–690.

Faure, G. (1977). *Principles of Isotope Geology*. New York: Wiley 464 pp.

Fisher, R.V. (1961). Proposed classification of volcaniclastic sediments and rocks. *Geological Society of America Bulletin 72*: 1409–1414.

Fisher, R.V. (1966). Rocks composed of volcanic fragments. *Earth Science Reviews 1*: 287–298.

Fisher, R.V. and Schmincke, H.U. (1984). *Pyroclastic Rocks*. New York: Springer-Verlag 472 pp.

Glazner, A. and Johnson, B.R. (2013). Late crystallization of K-feldspar and the paradox of megacrystic granites. *Contributions to Mineralogy and Petrology 166*: 777–799.

Glazner, A.F., Bartley, J.M., and Coleman, D.S. (2019). A more informative way to name plutonic rocks. *GSA Today 29*: 4–10.

Hess, P.C. (1989). *Origins of Igneous Rocks*. Cambridge, MA: Harvard University Press 336 pp.

Higgins, M.D. (2011). Textural coarsening in igneous rocks. *International Geology Review 53*: 354–376.

Hunter, R.H. (1996). Textural developments in cumulate rocks. In: *Layered Intrusions: Developments in Petrology* (ed. R.G. Cawthorn), 77–101. Amsterdam: Elsevier.

Hyndman, D.W. (1985). *Petrology of Igneous and Metamorphic Rocks*, 2e. New York: McGraw Hill 786 pp.

Jackson, M.D., Blundy, J., and Sparks, R.S.J. (2018). Chemical differentiation, cold storage and remobilization of magma in the Earth's crust. *Nature 564*: 405–409.

Kelsey, C.H. (1965). Calculation of the C.I.P.W. norm. *Mineralogical Magazine 34*: 276–282.

LeBas, M.J. and Streckeisen, A.L. (1991). The IUGS systematics of igneous rocks. *Journal of the Geological Society of London 148*: 825–833.

LeBas, M.J., LeMaitre, R.W., Streckeisen, A., and Zanettin, B. (1986). A chemical classification of volcanic rocks based on the total alkali-silica diagram. *Journal of Petrology* 27: 745–750.

LeMaitre, R.W. (1984). A proposal by the IUGS subcommission on the systematics of igneous rocks for a chemical classification of volcanic rocks based on the total alkali silica (TAS) diagram. *Australian Journal of Earth Science* 31: 243–255.

LeMaitre, R.W. (2002). *Igneous Rocks: A Classification and Glossary of Terms*, 2e. New York: Cambridge University Press 236 pp.

Lesher, C.E. and Spera, F.J. (2015). Thermodynamic and transport properties of silicate melts and magma. In: *The Encyclopedia of Volcanoes* (eds. H. Sigurdsson, B. Houghton, H. Rymer, et al.), 113–141. ISBN: 9780123859389.

Lofgren, G. (1980). Experimental studies on the dynamic crystallization of silicate melts. In: *Physics of Magmatic Processes* (ed. R.B. Hargraves), 487–551. Princeton, NJ: Princeton University Press.

Lofgren, G. (1983). Effect of heterogeneous nucleation on basaltic textures: a dynamic crystallization study. *Journal of Petrology* 24: 229–255.

MacKenzie, W.S., Donaldson, C.H., and Guilford, C. (1984). *Atlas of Igneous Rocks and Their Textures*. Essex, England: Wiley 148 pp.

McBirney, A.R. and Nicolas, A. (1997). The Skaegaard layers series: part II. Dynamic layering. *Journal of Petrology* 38: 569–580.

Mullen, E.D. (1983). Mn/$TiO_2$/$P_2O_5$: a minor element discriminant for basaltic rocks of oceanic environments and its implications for petrogenesis. *Earth and Planetary Science Letters* 62: 53–62.

Pearce, J.A. and Cann, J.R. (1973). Tectonic setting of basic volcanic rocks determined using trace chemical analyses. *Earth and Planetary Science Letters* 19: 290–300.

Pearce, T.H., Gorman, B.E., and Birkett, T.C. (1975). The $TiO_2$–$K_2O$–$P_2O_5$ diagram: a method of discriminating between oceanic and non-oceanic basalts. *Earth and Planetary Science Letters* 24: 419–426.

Pearce, T.H., Gorman, B.E., and Birkett, T.C. (1977). The relationship between major element chemistry and tectonic environment of basic and intermediate volcanic rocks. *Earth and Planetary Science Letters* 36: 121–132.

Saccani, E. (2015). A new method of discriminating different types of post-Archean ophiolitic basalts and their tectonic significance using Th-Nb- and Ce-Yb-Yb systematics. *Geoscience Frontiers* 6 (4): 481–501.

Schmid, R. (1981). Descriptive nomenclature and classification of pyroclastic deposits and fragments: recommendations of the IUGS Subcommision on the Systematics of Igneous Rocks. *Geology* 9: 41–43.

Shand, S.J. (1951). *Eruptive Rocks: Genesis, Composition, Classification and Relation to Ore Deposits*. New York: Hafner Publishing 488 pp.

Sparks, R.S.J. (1978). The dynamics of bubble formation and growth in magma: a review and analysis. *Journal of Volcanology and Geothermal Research* 3: 1–37.

Spera, F. (2000). Physical properties of magma. In: *Encyclopedia of Volcanoes* (ed. H. Sigurdsson), 171–190. New York: Academic Press.

Špillar, V. and Doleјš, D. (2013). Calculation of time-dependent nucleation and growth rates from quantitative textural data: inversion of crystal size distribution. *Journal of Petrology* 54: 913–931.

Streckeisen, A. (1973). Plutonic rocks: classification and nomenclature recommended by the IUGS Subcommission on the Systematics of Igneous Rocks. *Geotimes* 18 (10): 26–30.

Streckeisen, A. (1976). To each plutonic rock its proper name. *Earth Science Reviews* 12: 1–33.

Swanson, S.E. (1977). Relation of nucleation and crystal growth rate to the development of granitic textures. *American Mineralogist* 62: 966–978.

Verma, S.P. and Rivera-Gómez, M.A. (2013). Computer programs for the classification and nomenclature of igneous rocks. *Episodes* 36 (2): 115–124. http://tlaloc.ier.unam.mx/igrocs

Wallace, P. and Anderson, A.T. Jr. (2000). Volatiles in magma. In: *Encyclopedia of Volcanoes*, vol. 39 (ed. H. Sigurdson), 29–60. New York: Academic Press.

Wilson, L. (1980). Relationships between pressure, volatile content and ejecta velocity in three types of volcanic explosion. *Journal of Volcanology and Geothermal Research* 8: 297–313.

# Chapter 8

# Magma and intrusive structures

## 8.1  ROCK MELTING

The great variety of igneous rocks found in Earth's crust is largely attributable to **anataxis**, which refers to the partial melting of a parental source rock. Anatexis produces: (1) a liquid melt fraction enriched in lower temperature, incompatible constituents; and (2) a residual rock component enriched in a higher temperature (refractory), compatible elements and depleted in lower melting temperature components. The type of magma produced by partial melting and subsequent processes depends upon factors such as:

1  The composition, temperature, and depth of the source rock;
2  The percent partial melting of the source rock;
3  The source rock's previous melting history;
4  Diversification processes that change the composition of the magma after it leaves the source region.

Earth scientists are unable to penetrate Earth's deep interior to observe complex rock melting processes firsthand. Instead, rock melting has been experimentally modeled in research laboratories since the early twentieth century, providing insight into both magma genesis and the crystallization processes generating igneous rocks. In the simplest approach, rock melting is modeled as two idealized end member processes: (1) "equilibrium melting" and (2) "fractional melting."

*Earth Materials*, Second Edition. Kevin Hefferan and John O'Brien.
© 2022 John Wiley & Sons Ltd. Published 2022 by John Wiley & Sons Ltd.
Companion website: www.wiley.com/go/hefferan/earthmaterials2

### 8.1.1 Equilibrium melting

Equilibrium melting occurs in a closed system where chemicals are neither added nor removed from the plutonic environment. Equilibrium melting requires that the melt remains in contact with the residual rock throughout the melting process. As modeled in phase diagrams, equilibrium melting implies that the overall composition of the system remains the same while the composition of melts and solids evolves. In equilibrium melting of a rock:

1 The melt produced is enriched in low melting temperature, incompatible constituents. Enrichment is most pronounced with small degrees of partial melting (<10%) because the lowest temperature constituents preferentially enter the melt first.
2 Increased degrees of partial melting (>30%) dilute the enrichment of low temperature constituents in the melt; the melt becomes progressively less enriched in low melting temperature as higher temperature constituents enter the melt.
3 The solid refractory residue is enriched in high melting temperature, compatible constituents.
4 With continued partial melting, the solid residue becomes progressively more enriched in the refractory constituents.
5 If the rock is 100% melted, the melt produced will have the same composition as the original parent rock. This is because equilibrium melting infers that rock melting occurs in a closed system that does not allow material to be added or removed from the plutonic environment, and that crystals and melt remain in contact throughout the melting process.

### 8.1.2 Fractional (disequilibrium) melting

Fractional melting implies that solids and melt separate into isolated fractions that do not continue to react together during the melting process. In this idealized process, liquids and crystals do not remain in equilibrium because melts are separated from the refractory crystals before equilibrium reactions are complete. Fractional melting produces a melt that is more evolved than the parent source rock

from which it was derived. For example, in fractional melting of a troctolite containing plagioclase and olivine, the early melts are highly enriched in low melting temperature constituents such as more sodium-rich ($An_{0-50}$) plagioclase (albite–andesine) and iron-rich olivine (fayalite). The unmelted refractory residual solid is enriched in calcium-rich ($An_{50-100}$) plagioclase (anorthite–labradorite), and magnesium-rich olivine (forsterite). With continued rock melting, each succeeding melt will be less enriched in low melting temperature constituents than the initial melts because the source rocks have been progressively depleted of low temperature constituents. As a general rule, small degrees of partial melting of undepleted (previously unmelted) source rocks produce melts that are highly enriched in low temperature constituents. Larger degrees of melting of previously depleted source rocks produce melts that are significantly less enriched in low temperature constituents. Fractional melting can produce a daughter magma with mineral chemistry distinctly different from that of the original parent rock. Fractional melting is especially important in the generation of tholeiitic basalt by the partial melting of mantle peridotite at ocean spreading ridges.

While equilibrium melting and disequilibrium melting represent two idealized end member models, they provide useful starting points for the more complicated melting and subsequent processes that generate different magma compositions within Earth.

## 8.2 FACTORS IN ANATEXIS AND INITIAL MELT COMPOSITION

Why do solid rocks melt within Earth? Anatexis is initiated by some combination of increasing temperature, decreasing pressure or increasing volatile content, especially water vapor pressure ($P_{H2O}$). Let us briefly discuss these melting factors.

### 8.2.1 Increasing temperature

Increasing temperature is the most obvious cause for melting solid rock. Temperature increases with depth within Earth are represented by a sloping line called the **geothermal gradient**. The geothermal gradient is not uniform vertically or laterally within Earth. Earth's average geothermal gradient is ~25 °C/km for the upper 10 km, but decreases with depth. The geothermal gradient also varies based upon rock age and tectonic setting, ranging from 5 to 10 °C/km in old continental lithosphere to 30–50 °C/km at hot spots, ocean spreading ridges, and volcanic arcs. Elevated geothermal gradient sites constitute locations within Earth where mantle peridotite or crustal rocks may partially melt as a result of increasing temperature.

### 8.2.2   Decreasing pressure

Decompression melting, also known as **adiabatic melting**, results from a decrease in pressure associated with uplift. Pressure is related to rock depth whereby 10 km depth corresponds approximately to 3.3 kbars (or 3.3 km ~ 1 kbar). Adiabatic melting of solid rock can occur without any increase in temperature or volatile content. In volatile poor systems, for example those with low water vapor contents, melting temperatures are proportional to pressure. Assuming all other factors are constant, the higher the pressure, the higher the melting temperature. Conversely, as pressure decreases, melting temperature also decreases significantly resulting in intersection with the melting curve (Figure 8.1). Adiabatic melting dominates extensional tectonic environments in which the crust is stretched and thinned. Crustal thinning allows upwelling of the underlying mantle effectively decreasing the lithostatic stress. Uplift and decompression effectively reduce rock-melting temperatures. Adiabatic melting of mantle peridotite constitutes the primary means by which basaltic magmas are generated at ocean spreading ridges, back arc basins and continental rifts. Decompression also plays a contributing role in hot spot regions where warm rocks become less dense, resulting in uplift and decompression melting.

### 8.2.3   Volatile induced melting

Magma commonly contains dissolved fluids such as water vapor ($P_{H2O}$). Elevated volatile content under pressure, particularly $P_{H2O}$, may significantly lower melting temperatures (Figure 8.2). In addition to $H_2O$, other volatiles include compounds such as CO, $CO_2$,

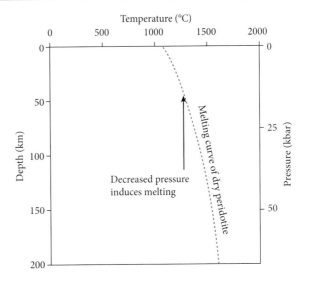

**Figure 8.1**   Pressure decrease accompanying uplift triggers melting as the upwelling mantle rock intersects the dry melting curve for dry (anhydrous) peridotite. As shown by the vertical black arrow, solid dry mantle peridotite exists at 120 km depth. Rock melting is initiated as crustal thinning uplifts the peridotite to 50 km depth. Note that no temperature changes occur during uplift and melting.

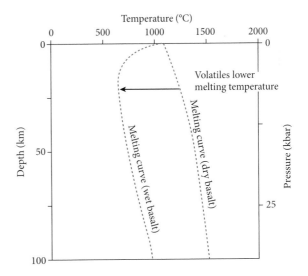

**Figure 8.2**   Note the lower temperature melting curve for wet basalt versus dry basalt. The addition of volatiles reduces the melting temperature of wet basalt by hundreds of °C. *Source*: Modified from Green and Ringwood (1967). John Wiley & Sons.

OH, $SO_2$, $H_2S$, $NH_3$, $K_2O$, $Na_2O$, HCl, HF, as well as the elements H, F, Cl, S, He and Ar. Volatiles serve as a **flux**, attacking bonds and reducing the melting temperature of a substance. Water vapor dissolved in magma under pressure tends to weaken Si–O bonds in silicate minerals. As silica bonds are weakened, progressively lower temperatures are necessary to melt solid rock. Therefore, an increase in $P_{H2O}$ can induce rock melting at a constant temperature and pressure.

The addition of relatively small amounts of water, even less than 1%, under high pressures, can dramatically lower rock melting temperatures. Elevated $P_{H2O}$ also allows rocks to melt over a wider temperature range, strongly counteracting the effects of lithostatic pressure. High volatile content also tends to decrease the amount of FeO and MgO incorporated into partial melts. As a result, magmas produced by $P_{H2O}$ induced partial melting ("wet" melting) tend to be less mafic than dry melts.

The addition of volatile compounds such as $H_2O$, $CO_2$, $K_2O$, and $Na_2O$ is the major cause for partial melting at convergent plate boundaries. As subducted lithosphere penetrates to depths of 80–150 km, temperature–pressure conditions cause hydrous minerals such as amphibole, mica and serpentine to become unstable. Dehydration of these minerals releases water vapor and other volatiles, raises volatile pressures, and induces melting in the overlying mantle wedge. The addition of volatile compounds may depress rock melting temperatures hundreds of °C in the overlying wedge above the subduction zone (Figure 8.2).

Partial melting changes the parent rock composition by enriching the melt in low temperature components and enriching the refractory residue in high temperature components. But how can we account for such a wide variety of igneous rock types?

### 8.2.4 Partial melting and melt composition

Anatexis is one means by which diverse suites of igneous rocks are generated. Anatexis generates a daughter magma derived from partial melting and leaves behind an unmelted, residual, parental rock. Both the melt and the residual rock have compositions that differ from that of the original parental source rock. Anatectic melts tend to be more enriched in low melting temperature constituents than the parent, refractory, residual rock. Partial melting removes incompatible elements, including large ion lithophile elements as well as $SiO_2$, $K_2O$, and $Na_2O$ from the parent rock and concentrates these low melting temperature constituents in the daughter magma. As a result of their concentration in the melt, the residual rock is depleted in incompatible elements and low melting temperature constituents. Rocks crystallizing from these melts tend to contain more felsic minerals, such as quartz, K feldspar, and Na plagioclase, than their parental source rock. The refractory residual rock tends to be enriched in high temperature constituents such as MgO, CaO, and compatible elements such as high field strength (HFS) elements. Common refractory minerals include olivine, pyroxene and calcium plagioclase. Despite these chemical changes, the anatectic melts and the residual rock remain genetically and chemically related to the original source rock. Thus, partial melting generates magmas with greater incompatible element and other low temperature constituent concentrations; in contrast, partial melting leaves behind a refractory residuum rock with higher concentrations of compatible elements and other high temperature constituents.

The percentage of partial melting plays a critical role in determining the degree of segregation between low melting temperature and high melting temperature constituents. As the percent partial melting increases, the degree of enrichment of incompatible elements decreases. Recall from Chapter 7 that light rare earth elements (LREE) are generally more incompatible with solid minerals than are heavy rare earth elements (HREE). This means that LREE (La–Sm) are preferentially incorporated into melts; the smaller the degree of partial melting, the higher the LREE enrichment in the melt. Figure 8.3 illustrates how LREE enrichment may be used to infer the degree of partial melting of a garnet bearing peridotite because garnet has a strong preference for HREE. At 1% partial melting, LREE are strongly enriched relative to HREE (Eu–Lu). However, the degree of LREE enrichment decreases progressively as the degree of partial melting increases progressively to 30%.

Following partial melting, the residual mantle rock is depleted in incompatible elements, such as the LREE and LIL's (K → Cs), com-

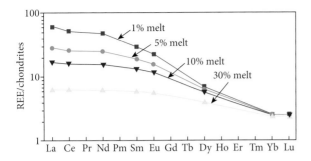

**Figure 8.3** Incompatible light rare elements (LREE) are progressively enriched with smaller percentages of partial melting. As the percent of partial melting increases, the rare Earth element pattern more closely approximates that of the parental material such as chondrites as shown by a flatter curve. Chondrites are thought to be chemically equivalent to the bulk composition of early Earth. *Source*: Courtesy of Stephen Nelson.

pared to the original mantle composition. Conversely, melts derived from mantle melting are enriched in incompatible elements compared to original mantle concentrations. In this way, we can infer the nature and history of the source region by looking at the pattern of REE's and LIL's in the resulting melts and the rocks that form from them.

Multiple partial melting episodes of source rocks can produce the wide range of igneous rocks observed in Earth's crust. Because the low melting temperature constituents are concentrated in the melt, partial melts are characterized by increases in $SiO_2$ and LIL content. In a very simplified model, with many local exceptions, partial melting of ultramafic peridotite generates a mafic (basalt/gabbro) melt; mafic rock partial melting produces an intermediate magma (andesite/diorite); and partial melting of an intermediate rock yields a felsic melt (rhyolite/granite).

In this way, it is possible to imagine how a mafic oceanic crust and a felsic-intermediate continental crust might have been extracted by multiple partial melting cycles. These cycles are dominated by decompression melting of a refractory mantle at ocean spreading ridges and flux melting of ocean lithosphere at convergent plate boundaries generating arc and continental crustal material. In addition to anatectic processes, the wide range of igneous

rocks in Earth's crust is also due to magma diversification processes that occur after the melt separates from the source region, as discussed below.

## 8.3 DIVERSIFICATION: DIFFERENTIATION, MIXING AND ASSIMILATION PROCESSES

In addition to partial melting of different source rocks, complex magma diversification processes are also responsible for producing Earth's wide array of magmas and igneous rocks. Diversification processes involve changes in bulk magma chemistry after its initial generation. Diversification processes include closed system differentiation processes as well as open system processes related to magma–country rock interaction or interaction with other magmas. Common closed system and open system diversification processes are discussed below.

### 8.3.1 Differentiation processes

**Differentiation** includes all closed system diversification processes in which the original parent melt evolves into one or more daughter melts with a different composition, without material being exchanged with an external source. Closed system differentiation processes include several fractional crystallization and separation processes in which early formed crystals segregate from the remaining melt. Norman Bowen (1928), an early proponent of fractional crystallization, suggested that fractionation may be accomplished "through the relative movement of crystals and liquid." Bowen is regarded as the "father of modern petrology," so let us consider a few of his many important contributions.

In the late nineteenth century, geologists such as Reginald Daly began investigating the origin, composition and crystallization processes involved in generating igneous rocks. Bowen, a Canadian protégé of Daly, began working as an igneous petrologist at the Carnegie Institute of Washington in 1910. Bowen analyzed rocks from the Palisades sill, a shallow igneous intrusion situated in New Jersey just west of New York City. The ~200 million year old Palisades Sill intrusion consists largely of mafic rocks including basalt, diabase (fine-grained gabbro) and gabbro. Bowen noted that the mineralogy of the sill varied vertically. Lower levels within the sill were enriched in olivine, while other levels were enriched in orthopyroxene, clinopyroxene, plagioclase, and even in K feldspar and quartz. Through field observations and laboratory rock melting experiments, Bowen theorized that a **crystal–liquid fractionation** process must have been involved that allowed early formed crystals to be sequentially segregated from the remaining melt. Removal of newly formed crystals caused the remaining melt chemistry to change creating progressive layers of different composition. Bowen melted igneous rock samples from the Palisades Sill in his Carnegie laboratory and developed a model based on the crystallization sequence of major igneous minerals. This model is referred to as **Bowen's reaction series** (Figure 8.4). Bowen recognized eight major mineral groups that constitute the bulk of most igneous rocks. These mineral groups are: olivine, pyroxene, plagioclase, amphibole, biotite, muscovite, K-feldspar, and quartz.

On the basis of his field and laboratory work, Bowen (1928) proposed a reaction series for cooling mafic magma. Bowen's reaction series consists of a discontinuous and continuous crystallization series. The **discontinuous reaction series** involves distinctly different mineral groups, namely olivine, pyroxene, amphibole, and biotite. In the discontinuous reaction series, high temperature refractory minerals such as Mg-rich olivine crystallize first, removing MgO from the remaining melt. Recall from our discussion of phase diagrams that as a melt cools to the peritectic, forsteritic olivine crystals react with the melt to produce the orthopyroxene mineral enstatite. This is a discontinuous melt–solid reaction that produces a new mineral with a different structure and composition. Such discontinuous reactions are possible if crystals are allowed to react (equilibrium conditions) with the melt. At still lower temperatures, discontinuous reactions convert pyroxene to amphibole and amphibole to biotite. These biotite is hardly a refractory mineral contribute to a variety of mafic rocks layers such as basalt, diabase, and gabbro with various proportions of olivine, orthopyroxene, clinopyroxene, amphibole, and biotite.

The **continuous reaction series** involves plagioclase which is characterized by a complete solid solution series within a single mineral group. The solid solution series exists between higher crystallization temperature Ca-rich plagioclase and progressively lower temperature, more Na-rich plagioclase. In equilibrium conditions, early formed Ca-rich plagioclase crystals react with the melt to produce increasingly Na-rich plagioclase crystals. Each of the minerals in the discontinuous and continuous reaction series form in equilibrium conditions and follow the phase relationships detailed in Chapter 3.

When discontinuous and continuous reactions do not proceed to completion under equilibrium conditions, chemical variations can be preserved within crystals recording zoned crystals and reaction rims. **Zoned crystals** display a systematic pattern of chemical variations from the periphery of the crystal towards its center, recording an incomplete continuous chemical reaction between the crystal and the surrounding melt. For example, incomplete solid solution reactions involving plagioclase produce zoned crystals in which a Ca-rich (anorthite) plagioclase core may be rimmed by Na plagioclase. **Reaction rims** display a new mineral along the crystal periphery that surrounds an anhedral, partially resorbed core composed of a different mineral. For example, hornblende rims commonly develop on augite cores. Reaction rims indicate incomplete discontinuous chemical reactions between crystals and melts.

Bowen's pioneering work on crystal–liquid reactions provided the foundation for understanding mineral crystallization processes and magma evolution. From an original mafic melt, Bowen inferred that a wide variety of rocks could crystallize through a sequential crystal–liquid separation process. Bowen outlined a fractional crystallization process whereby, as magma temperatures cooled below 1200 °C, a distinct silicate mineral crystallization sequence ensued beginning with MgO-rich olivine, followed by CaO plagioclase and FeO olivine generating ultramafic to mafic rocks such as peridotite, basalt and gabbro. With the removal of MgO, FeO, and CaO, the remaining melt is progressively enriched in $SiO_2$, $K_2O$, and $Na_2O$ and $H_2O$. With decreasing temperatures, increasingly more Na-rich plagioclase continues to crystallize along with hydrous minerals such as hornblende and biotite

(a)

(b)

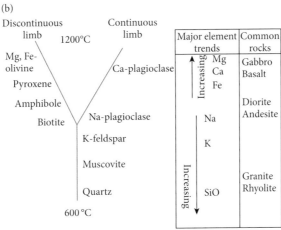

**Figure 8.4** (a) Bowen's reaction minerals consisting of the ferromagnesian minerals olivine, augite pyroxene, hornblende amphibole and biotite mica. Non ferromagnesian minerals include Ca- and Na-rich plagioclase, K-feldspar, muscovite mica and quartz. (b) Bowen's reaction series envisions minerals derived from a mafic magma that successively crystallize with decreasing temperatures. The transition from Ca- to Na-rich plagioclase represents a continuous reaction series within one mineral group. The olivine, pyroxene, amphibole, and biotite transition represents a discontinuous reaction series as these represent distinctly different ferromagnesian mineral groups that sequentially crystallize upon cooling.

producing intermediate rocks such as andesite and diorite. After most of the magma has crystallized, a $SiO_2$, $K_2O$ $Na_2O$-rich magma remains, from which the minerals muscovite, K-feldspar, and quartz crystallize at low temperatures. These three minerals are not part of a reaction series because they do not react with the magma, continuously or discontinuously, to produce minerals with a different composition. These three minerals form the silicic rocks such as granite and rhyolite.

Bowen suggested that fractional crystallization of mafic magma could explain the widespread occurrence of silicic rocks such as the immense granite and granodiorite bodies exposed on continents. It is now recognized that fractional crystallization in itself cannot generate vast volumes of silicic magma observed on Earth. Other processes must also be at play. Nevertheless, Bowen's reaction series remains a fundamental teaching tool useful in the visualization of magma crystallization sequences, crystal–melt interactions and the common association of major minerals in rock assemblages.

**Fractional crystallization** involves the effective separation of crystals and melt from an originally homogeneous magma. Models for fractional crystallization involve marginal accretion, gravitational separation, convective flow, and filter pressing. Each of these will be briefly discussed below.

The interior of a magma chamber is well insulated, retains heat and cools relatively slowly. The periphery of the magma chamber in contact with surrounding, cool country rock loses heat by conduction, resulting in relatively rapid crystallization. In addition, crystals in the wall rocks provide nucleation sites for crystal growth. Crystallization along the walls of the magma chamber in which crystals preferentially form and adhere to the edges results in **marginal accretion**. As additional layers of crystal accrete to the walls, earlier formed crystals are no longer in contact with the magma and have been effectively segregated from it.

Marginal accretion (Figure 8.5) may be subdivided based on the location within the pluton. Heat rises so that the upper margin of the magma chamber may cool relatively quickly. As a result, **roof accretion** results from early crystallization of minerals along the ceiling or roof due to preferential heat loss. **Sidewall**

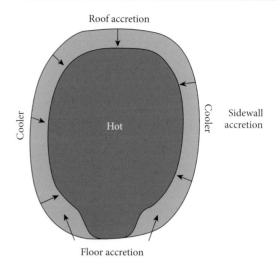

**Figure 8.5** Marginal accretion due to preferential cooling of the perimeter of the magma chamber produces sidewall, roof, and floor accretion of early formed crystals. *Source*: Courtesy of Stephen Nelson.

**accretion** develops as the magma chamber releases heat to the relatively cold country rock, generating crystals that adhere to the side margins of the magma chamber. **Floor accretion** occurs as crystals form along the base of the magma chamber. Each of these marginal accretion processes may effectively separate early formed crystals from the remaining melt, enriching it in lower-temperature constituents. Because they lose heat rapidly to the wall rocks, pluton margins adjacent to cool country rock commonly display a finer grained **chilled margins** compared to the coarser grained crystals of the pluton interior.

**Gravitational separation** includes fractionation processes that occur when crystals develop with significantly different densities than the surrounding magma (Figure 8.6). Gravitational separation includes crystal settling and crystal flotation processes.

**Crystal settling** occurs when higher density, ferromagnesian crystals settle to the base of a magma chamber relative to the lower density liquid magma. Crystal settling may result in discrete layers of crystal mush such that banded or layered cumulus crystals may precipitate via magmatic "sedimentation" on the pluton floor (Irvine et al. 1998). **Crystal flotation** can occur if early formed crystals, such as plagioclase, are less dense than the magma. As a result crystals

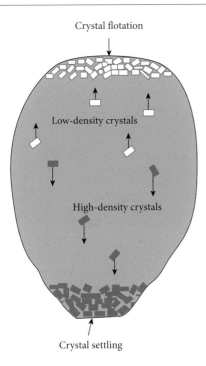

Crystal flotation

Low-density crystals

High-density crystals

Crystal settling

**Figure 8.6** Gravitational separation in a magma chamber involves: crystal settling of higher density crystals (commonly ferromagnesian minerals such as olivine and pyroxene); and crystal flotation of lower density (commonly non-ferromagnesian minerals such as plagioclase feldspars) crystals. *Source*: Courtesy of Stephen Nelson.

may float towards the roof of a magma chamber, effectively segregating them from the remaining melt (Bottinga and Weill 1970).

Both crystal settling and crystal flotation imply that individual crystals migrate through low viscosity magma that has a different density than the crystals. However, if the differences in density between crystals and magma are vanishingly small and high viscosity magma would impede crystal migration. Gravitational migration is possible for large clusters of early formed crystals with higher settling or flotation velocities in very high temperature, low viscosity magma. Massive Proterozoic anorthosite bodies, containing >90% plagioclase, likely owe their origin in part to crystal flotation processes in mafic magma plutons (Fram and Longhi 1992; Namur et al. 2011). Crystal migration is far less likely to occur in lower temperature, higher viscosity magmas.

**Convective flow segregation** processes occur whereby liquids and crystals are segregated due to factors such as velocity, density or temperature. Variations in flow velocity within low viscosity magma can result in sorting of crystals. The largest or densest crystals will accumulate in regions of high velocity while smaller or lower density crystals accumulate in flow paths of lower velocity. Higher density crystals may also migrate by gravitational flow towards the base of the magma chamber. In some cases, large masses of unconsolidated or partially consolidated crystals break off roofs and sidewalls and flow via density surge currents to the chamber floor. They may then settle at different velocities to produce graded bedding and other stratification features (Irvine et al. 1998).

In Bowen's fractional crystallization model, he alluded to variations in pluton heat loss as a contributing factor to crystal–liquid separation. The periphery of the pluton loses heat more readily, resulting in crystallization; the pluton interior retains heat and remains predominantly liquid. Thus, from an originally homogeneous magma, differential heat loss may generate thermally induced convective flow and separate crystals from melt. Bowen (1928) recognized the role of temperature variations inducing convective circulation in a magma chamber. Such convective currents can produce depositional structures usually associated with sedimentary processes such as graded bedding and cross bedding. Bowen also envisioned separation of crystals from liquid during the later stages of magma chamber cooling. In this case, cooling can be accompanied by deformation in which crystals are compacted and rotated while liquids experience expulsion. This variation on crystal–liquid fractionation involves a set of processes under the umbrella of "**filter pressing**" (Figure 8.7).

Crystal settling, crystal flotation, marginal (sidewall) accretion, convective flow, and filter pressing continue to be recognized as key crystal–liquid fractionation processes (Sleep 1988; Irvine et al. 1998; Sisson and Bacon 1999). Evidence for convective flow and crystal settling is perhaps best displayed in the Tertiary age Skaergaard intrusion of Greenland. This intrusion contains features such as cumulate layering, graded bedding, cross bedding, and slump structures, normally associated with sedimentary rocks.

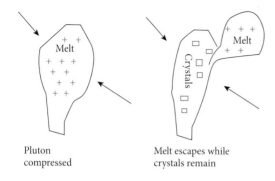

Pluton compressed

Melt escapes while crystals remain

**Figure 8.7** Filter pressing in which a magma chamber containing crystals and liquids is compressed, squeezing out more mobile liquid and leaving behind a crystal residue in the compressed chamber. *Source*: Kevin Hefferan. © John Wiley & Sons.

**Figure 8.8** Liquid immiscibility occurs with the cooling of chicken soup. At high temperatures, fat is dissolved in solution. With decreasing temperatures, higher crystallization temperature fat separates from the solution as a separate phase and rises to the top. *Source*: Kevin Hefferan. © John Wiley & Sons.

In addition to crystal–liquid fractionation, closed system differentiation may occur due to liquid fractionation. In liquid fractionation, one parent magma fractionates to produce two or more distinctly different daughter magmas. Liquid fractionation processes include liquid immiscibility and differential diffusion. **Differential diffusion** involves the preferential diffusion of select ions within the magma in response to compositional, thermal or density gradients as well as water content. This causes the parent magma to differentiate into daughter magmas with different compositions.

**Liquid immiscibility, also called liquid–liquid fractionation,** occurs when magma separates into two distinct immiscible liquid phases. Bowen (1928) and Daly (1933), early proponents of liquid immiscibility, suggested that a magma may segregate into two separate magmas of different composition upon cooling. The separation of fat from chicken soup as it cools is an excellent analog for liquid immiscibility (Figure 8.8). The fat is soluble or miscible in high temperature soup. However, fat becomes less soluble as the soup cools. Upon cooling, fat rises as a less dense, solid phase floating on the liquid soup. Liquid immiscibility may occur to a limited extent with segregation of granitic liquids from basaltic magmas, as well as segregation of sulfide-rich liquids from silicate magmas, and carbonate liquids from alkaline magmas. In laboratory studies, $CO_2$ exsolves as a separate fluid phase from alkalic magmas, generating a carbonate-rich fluid which crystallizes as a carbonatite. Liquid immiscibility has also been demonstrated in basaltic rocks from the Deccan

Traps (De 1974). Philpotts (1979) also documented liquid immiscibility in two basaltic glasses from the Triassic–Jurassic Rattlesnake Hill Basalt of Connecticut where two different melts separated to form immiscible droplets in each other that solidified as glasses, resembling to some degree the fat droplets in chicken soup analogy.

### 8.3.2 Open system diversification processes

Open system processes such as assimilation, magma mixing, and magma mingling involve chemical changes due to interaction of magma with the surrounding country rock or with other magmas entering the system.

**Assimilation** occurs whereby the surrounding wall rock (country rock) is intruded by, and chemically reacts with, the magma. Magma composition evolves when elements from the wall rock are incorporated into the magma and/or when elements from the magma are transferred into the wall rocks. Forceful injection of magma commonly fractures and dislodges the surrounding wall rock in a process called **stoping**. Country rock fragments fall (**stope**) into the magma as **xenoliths** (Figure 8.9). As foreign rock bodies, xenoliths produce changes in magma composition by partially melting and/or reacting with the magma. Xenoliths that remain as solid rock may display reaction rims, indicat-

**Figure 8.9** Mafic gneiss xenolith entrained within a Proterozoic granite west of Golden, Colorado. Note the black reaction rim around the periphery of the xenolith. *Source*: Photo by Kevin Hefferan. © John Wiley & Sons.

**Figure 8.10** Olivine xenocrysts in vesicular basalt from appropriately named Peridot, Arizona. *Source*: Photo by Kevin Hefferan. © John Wiley & Sons.

ing that chemical reactions occurred with the encasing magma.

**Xenocrysts** are foreign crystals not generated by crystallization of the surrounding magma (Figure 8.10). Chemical reactions between xenocrysts and magma may significantly alter the magma's chemical composition (Bailey et al. 1924; Daly 1933).

Xenolith and xenocryst assimilation provides a direct means to introduce new, externally derived chemical components that alter magma composition and resultant mineral assemblages upon cooling.

In active plutonic environments, magma injection is unlikely to be a singular event.

Rather, intermittent pulses of magma are injected into the magma chamber from either the same source or different source, essentially replenishing the magma supply over short or long periods of time. Magma replenishment produces complex relationships and constitutes an important diversification mechanism. Streck et al. (2008) document short term multiple dacitic magma injections in the Mt. St. Helens 2004–2005 eruptive cycles. Multiple cycles of dacitic magma injection created a series of normal zoning bands followed by crystal resorption, dissolution, magma replenishment, and renewed zoning. These events produced a series of normal zoning with anorthite (An) calcium-rich plagioclase cores and An poor, more sodium-rich rims, separated by dissolution surfaces (Figure 8.11).

**Magma mixing** involves thorough mixing of two or more magmas so that the individual magma components are no longer recognizable. For example, two magmas can mix to produce a magma of intermediate composition. **Magma mingling** occurs when two or more dissimilar magmas coexist, displaying contact relations but retaining their distinctive individual magma characteristics (Figure 8.12). Magma mingling occurs when the magmas are interjected but do not combine thoroughly. In the case where mafic and felsic magmas mingle, the rocks may display a "marble cake" appearance. The inability of two or more magmas to homogenize thoroughly may be due to significant differences in temperature, density, viscosity, or to insufficient convection. D'Lemos (1992) cites magma mingling in the development of cylindrical pipes. These pipes formed as granitic melts intruded into diorite magma in Guernsey, Channel Islands. The pipes are thought to have developed from a less dense granitic magma that buoyantly injected itself into denser, overlying diorite magma.

Magma mixing is particularly important at continental rifts and convergent plate boundaries. In continental rift settings, hot mafic magma generated by partial melting of the mantle rises through overlying intermediate to silicic crust. The hot mafic magma partially melts the overlying crust generating a second, more silicic magma source that may mingle or partially mix together. Magma mixing at convergent margins may be more complex, with various components derived by flux melting of the overlying crustal and mantle wedge above the subducting

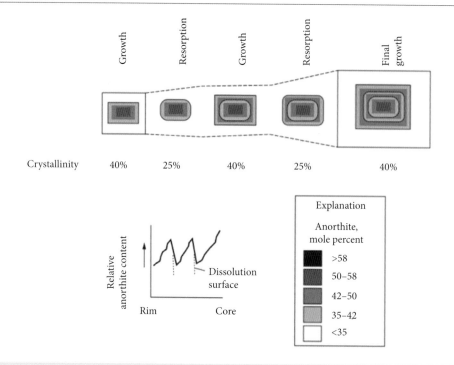

**Figure 8.11** In analyzing dacite rocks that formed during the 2004–2005 Mt. St. Helens eruptions, Streck et al. (2008) document multiple normal zoning cycles whereby plagioclase crystal cores are Ca rich (An > 42) compared to crystal rims (An < 42). Each normal zoning cycle is separated by a resorption cycle in which crystal rims experience resorption reducing the percent crystallinity and creating a dissolution surface. Streck et al. (2008) attribute these cycles to multiple dacitic magma recharge events. *Source*: U.S. Geological Survey, United States Department of the Interior.

plate, partial melting of the downgoing lithospheric slab, or from forearc basement and sediment incorporated into the magma. Dehydration reactions associated with the subducted slab release fluids increasing the volatile content and lowering the melting temperatures of minerals. The addition of volatiles also results in hydrous minerals such as amphiboles and micas, producing intermediate rocks such as andesite, dacite, and monzonite.

Magma mixing and mingling were discounted in the early twentieth century when the fractional crystallization hypotheses held sway. However, they are now recognized as significant processes by which of various compositions magmas may be generated by processes that may or may not be related to fractionation. Plutons, particularly those at convergent plate boundaries, experience multiple magma injections and elevated temperature and pressure conditions for millions of years. These long term conditions produce incremental mineral growth, chemical reaction stews and mineral replacement. For

example, Coombs and Gardner (2004) document pyroxene rims on olivine phenocrysts as well as olivine phenocrysts occurring within an andesitic groundmass in rocks formed in a convergent plate tectonic setting. Such disequilibrium assemblages suggest mixing of magmas of different compositions, producing unusual mineral assemblages and complex reactions. Magma mixing and crystal–liquid reactions may occur when the pluton is largely molten (<50% magma) as well as when rheological lockup begins where crystal volume exceeds the molten magma component. Solid state textural equilibration processes alter crystal boundaries following initial crystallization (Higgins 2011). Glazner and Johnson (2013) propose that large K-feldspar megacrysts (crystals larger than 5 cm in diameter) occurring in granite and granodiorite rocks likely formed as late stage crystal mineral replacement, recrystallization, and grain coarsening. Glazner and Johnson (2013) consider the combined effects of magma mixing, albite–orthoclase phase

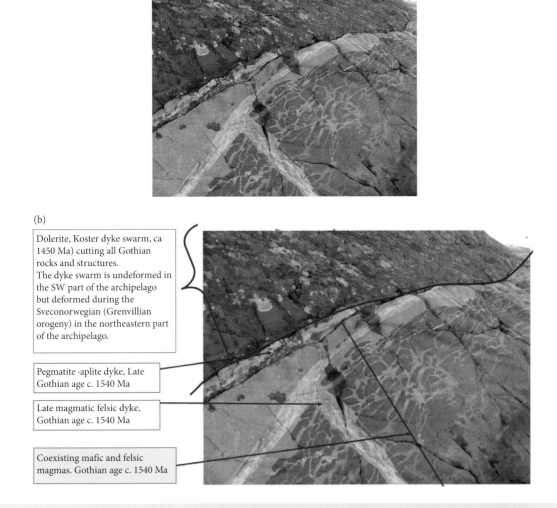

**Figure 8.12**    (a) Coexisting 1540 Ma mafic (gray) and felsic (tan) magmas produced magma mixing and magma mingling textures in Proterozoic rocks, Kosterhavet National Park on Nordkoster island, Koster Islands, Sweden. The 1540 Ma mafic–felsic mixed and mingled rocks are overlain by 1450 Ma diabase dikes (dark gray color). (b) Image interpretation provided by Thomas Eliasson. *Source*: Photo courtesy of Thomas Eliasson and the Geological Survey of Sweden https://www.flickr.com/people/geologicalsurveyofsweden/.

diagram chemical reactions and essentially greenschist facies metamorphic conditions (moderate pressures and 400–600 °C temperatures) to induce solid state K-feldspar mineral growth in felsic plutonic rocks.

## 8.4    MAGMA SERIES

Despite compositional variations resulting from diversification processes, rocks derived from a given parent magma display distinct geochemical signatures indicating a common origin. As an analogy, consider the role of DNA in assessing lineage. Despite the genetic changes that offspring experience through successive generations, a genetic DNA link to a common ancestor may still be established over thousands of years. Successive changes to magma composition through time bear a similar link to a common ancestor, so long as significant metamorphism has not altered rock chemistry.

Different processes have been suggested to explain the wide variety of magma and igneous rock compositions. Variations in magma and rock compositions may be due to:

1   Widely varying chemistry of the initial magma, related to source rock composition, melting history, and degree of melting;
2   Alterations to the chemical composition of the magma through differentiation;
3   Interactions between magmas such as magma mingling and mixing;
4   Assimilation reactions with the surrounding country rock; and
5   Late to post crystallization alteration processes that may include textural equilibration and metamorphism.

The net result is that magma chemistry can evolve over time resulting in a series of genetically related magmas and rocks of different compositions. These compositional trends may be broadly grouped into "magma series." A **magma series** consists of genetically related magmas that have changed composition from a common original parental magma (Gill 1981).

Magma series include the (subalkaline) calc-alkaline and tholeiite series and the alkaline (or alkali) series. Calc-alkaline magmas produce igneous rocks with moderate to high $Na_2O$, $K_2O$, and $SiO_2$ concentrations. Tholeiite magmas generate rocks with low $Na_2O$ and $K_2O$ but high FeO and CaO concentrations. The alkaline series crystallize rocks with high $Na_2O$ and $K_2O$ concentrations compared to $SiO_2$ and are generally undersaturated with respect to $SiO_2$.

In studying Greenland's mafic Skaergaard intrusion, Wager and Deer (1939) identified the calc-alkaline and tholeiitic series based on changes in iron concentration with progressive differentiation. These progressive changes in chemical composition were plotted on a triangular AFM (A = alkali, F = iron, M = magnesium) diagram. Tholeiitic suites display early iron enrichment with increased fractionation while the calc-alkaline suite does not display iron enrichment as illustrated in the AFM diagram (Figure 8.13).

When combined with Harker diagrams, it can be demonstrated that all the rocks in the calc-alkaline series can be generated by diversification of basaltic or basaltic andesite parent magma. Whether this is always the case is still a matter of contention.

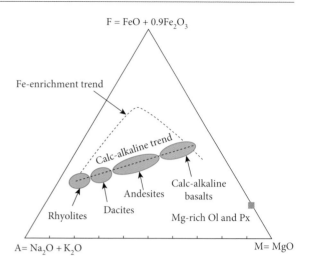

**Figure 8.13**   AFM diagram illustrating the calc-alkaline trend with progressive iron depletion and alkali and $SiO_2$ enrichment. Fractionation of a calc-alkaline basaltic magma results in progressive $SiO_2$ and alkali enrichment producing rocks such as andesites, dacites, and rhyolites. The tholeiite trend, displaying iron enrichment with fractionation is also illustrated.

### 8.4.1   Calc-alkaline magma series

Upon fractionation, **calc-alkaline** magmas record a progressive decrease in iron and magnesium with increasing $SiO_2$ and alkali concentrations (Figure 8.13). Available FeO and MgO are progressively removed from the melt through crystallization of olivine, pyroxenes, amphiboles, biotite and iron oxide minerals such as magnetite. The removal of ferromagnesian minerals enriches the melt in $Na_2O$, $K_2O$, and $SiO_2$. Thus, with increasing fractionation from magma, the calc-alkaline series exhibit increases in $SiO_2$, $Na_2O$ and $K_2O$ and decreases in CaO, MgO, and total iron (Bowen 1928; Miyashiro 1974; Grove and Kinzler 1986). The calc-alkaline series consists largely of andesite, dacite, and rhyolite as well as high alumina (16–20% $Al_2O_3$) basalt (Irvine and Barager 1971). Calc-alkaline magmas dominate convergent margin environments, such as the modern Circum Pacific region, wherein batholiths and composite volcanoes develop above subduction zones. The rocks in subduction zone volcanoes are known colloquially as the BADR series for the *b*asalts, *a*ndesites, *d*acites, and *r*hyolites that occur in them.

## 8.4.2 Tholeiite magma series

With increasing fractionation, **tholeiitic** magmas experience early iron enrichment at low to moderate $SiO_2$ concentrations (Figure 8.13). Early crystallization of forsterite olivine and Ca plagioclase depletes the melt of MgO and CaO causing the remaining melt to become enriched in FeO, as well as $SiO_2$ and alkalis (Miyashiro 1974; Grove and Kinzler 1986). Thereafter, FeO-rich minerals such as fayalitic olivine, and Fe-rich pyroxenes crystallize, depleting the melt in FeO while continuing to enrich it in $SiO_2$ and alkalis. Tholeiitic magmas generate large volumes of basalt with subtle compositional variations (Irvine and Barager 1971). Extreme fractionation of tholeiitic magmas can produce small amounts of more silicic magmas and rocks. Tholeiitic magmas and basalts dominate extensional environments such as ocean ridges and continental rifts. Tholeiites also occur over hotspots in intraplate settings and in the frontal portions of immature (relatively young) volcanic arcs characterized by a thin volcanic arc crust (Miyashiro 1974).

## 8.4.3 Alkaline magma series

**Alkaline magma** are highly enriched in $Na_2O$ and/or $K_2O$ and are less common than either calc-alkaline or tholeiitic magma. Alkaline magmas contain high concentrations of the alkali (lithium, sodium, potassium, rubidium, cesium, and francium) elements. Alkaline magmas are typically $SiO_2$ undersaturated and produce rocks that contain feldspathoid and/or forsteritic olivine minerals. Alkaline feldspathoid minerals include: nepheline, leucite, cancrinite, sodalite, and analcite (a zeolite mineral closely associated with feldspathoids). Less undersaturated alkaline magmas are approximately $SiO_2$ saturated, and generate rocks with significant amounts of alkali feldspars. Alkali feldspar minerals include sanidine, anorthoclase, perthitic K-feldspar, microcline, orthoclase and albite. Alkaline magmas never produce equilibrium quartz because of the lack of silica. Alkaline magmas are characterized by extremely diverse chemical compositions with $SiO_2$ contents ranging from 0 to 65%, with magnesium numbers (MgO/MgO + FeO) ranging from 0 to 0.88. Many alkaline magmas display geochemical signatures indicative of a mantle source. Most alkaline rocks contain feldspathoids, alkali feldspars and/or sodic amphibole or pyroxene minerals (Sorensen 1974). Sodic amphibole minerals that crystallize from alkaline magmas include riebeckite and richterite. Pyroxene minerals that occur in alkaline rocks include aegerine, aegerine–augite, and spodumene.

Alkaline rocks encompass a wide variety of compositions that include volcanic rocks such as alkali basalts, hawaiite, benmoreite, mugearite, trachyte, nephelinite, phonolite, lamprophyre, carbonatite, komatiite, kimberlite, and others. Plutonic alkaline rocks include syenite nepheline syenite, monzosyenite, monzogabbro, may nephelinite and others. Alkaline rocks occur in a number of environments which include stable cratonic interiors, continental rifts (Bailey 1974), ocean islands distant from spreading centers (Sorensen 1974) and subduction zones (Burke et al. 2003).

Both alkaline and tholeiitic basaltic magmas occur at hotspots, ocean islands and rift environments that derive magma from partial melting of mantle rocks. What is the relationship between these two magma series? Alkaline magmas and tholeiitic especially if the source rocks are relatively undepleted or even enriched in alkalis. Tholeiitic magmas are most likely generated by more extensive melting at lower temperature and pressure conditions of a previously depleted source rock. In addition, limited partial melting of tholeiitic source rock may yield alkali magmas, particularly under high pressure conditions in the lower crust/upper mantle (Frey et al. 1979; Frey et al. 1990; Naumann and Geist 1999).

## 8.4.4 Bimodal magmas

**Bimodal** magma suites are characterized by the voluminous occurrence of felsic and mafic (e.g. rhyolite–dacite and basalt) rocks. Bimodal volcanism is commonly associated with continental rifts. The mafic component is derived by partial melting of the mantle. The felsic component is derived by partial melting of the continental crust heated by rising mafic magmas. Bimodal volcanic activity is widespread within the Basin and Range province of the southwestern United States.

The different tectonic environments in which the major magma series commonly occur are summarized in Figure 8.14. Note that:

1   Calc-alkaline magmas occur almost exclusively at convergent margins;
2   Tholeiitic magmas are generated primarily at ocean spreading centers, but also occur in continental rifts, back arc basins, ocean islands, and hotspots;
    Alkaline magmas occur primarily in hotspot ocean island, and rift environments.
3   Bimodal volcanism most commonly occurs at continental rifts and continental hotspots.

Major, minor, and trace element concentrations provide critical information for understanding the origin and evolution of igneous rocks and igneous rock series. However, assessing the data for large numbers of rock samples can be overwhelming. One approach to solving this dilemma involves constructing variation diagrams that depict geochemical data graphically in a way that reveals patterns or trends. However, geochemical data input and the geochemical trends produced must be critically evaluated to determine if they are of geologic value. It is essential to observe rock relations in the field setting and analyze rock thin sections using a petrographic microscope, as well as evaluate geochemical relations. Researchers must also read scientific journal literature that critique variation diagrams to be aware of their strengths and their significant weaknesses (Chayes 1960, 1962, 1964; Butler 1979, Thomas and Aitchison 2005; Saccani 2015).

## 8.5   VARIATION DIAGRAMS

**Variation diagrams** provide a means to visually display rock chemistry. The hope is that chemical variations between samples can assist in establishing rock origins, histories, and relationships. Bivariate diagrams consist of two variables plotted in an x–y coordinate system. Trivariate diagrams contain three variables plotted in a triangular field. The variables may consist of major, minor or trace elements plotted as weight percent oxides, molar concentrations or ratios. Linear or curvilinear data trends in which data points fall near a "best fit" line may indicate a degree of relationship between rock samples. In well documented studies, linear data trends may be explained by progressive changes in magma composition related to anatexis of a common source and/or one or more diversification processes. On the other hand, random data suggest that the rocks are derived from different sources or have suffered severe secondary alteration associated with metamorphism such that their original chemical identity has been overprinted beyond recognition.

Alfred Harker (1909) devised Cartesian (x, y) coordinate **Harker diagrams** which can display data from hundreds of rock samples. Harker diagrams are bivariate diagrams in

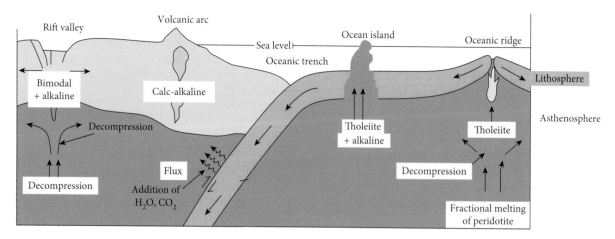

**Figure 8.14**   Generalized cross section illustrating relationships between anatexis, magma series, and tectonic environments. *Source*: Courtesy of Stephen Nelson.

which the vertical ordinate (y axis) displays the weight percent of major or minor oxide compounds such as FeO, MgO, CaO, $Na_2O$, $K_2O$ or $TiO_2$. The horizontal abscissa (x axis) displays weight percent $SiO_2$ content. The objective is to concisely display variations in major or minor oxide concentrations with respect to changes in $SiO_2$. Harker diagrams permit overall trends of major element variation to be deduced.

If Harker diagrams display consistent trends for geochemical data points, then the rocks may be genetically related; that is, the major element variations may reflect a liquid line of descent from a common source. For example as predicted by Bowen's reaction series for a mafic magma, Harker diagrams may show MgO, $Fe_2O_3$, and CaO decreasing with increasing $SiO_2$ due to the early crystallization of olivine, pyroxene, and Ca plagioclase from a basic magma. $Al_2O_3$ may initially increase but then decrease as plagioclase continues to crystallize, along with amphibole and biotite to form andesitic melts. With continued cooling and fractional crystallization, $Na_2O$ and $K_2O$ both increase with increasing $SiO_2$. However, $Na_2O$ will begin to decrease as Na plagioclase begins to crystallize forming dacitic melts. $K_2O$ increases linearly with increasing $SiO_2$ content until the remaining $K_2O$-rich melt crystallizes forming rhyolite.

Conversely, Harker diagrams may serve to distinguish two or more sets of rocks based upon significant variations in chemical composition. For example, Stone et al. (2017) show that in the Providence Mountains of San Bernadino County, California, the 18.8 Ma Peach Spring Tuff can be distinguished from most of the 17.7 Ma Wild Horse Mesa Tuff units based on variations in $TiO_2$ concentrations (Figure 8.15). The older Peach Spring Tuff contains abundant sphene, which accounts for the high $TiO_2$ concentrations, and was derived by a supervolcano explosion associated with the Silver Creek caldera near Oatman, Arizona (Ferguson et al. 2013).

**Spider diagrams** plot trace element concentrations on the abscissa (x axis) relative to a standard reference for each element on the ordinate (y axis). The standard reference varies depending upon the rock type analyzed and the tectonic relationship that the researchers wish to demonstrate. The standard reference may include chondrite meteorites, primitive mantle, normal mid ocean ridge basalt (N-MORB), enriched mid ocean ridge basalt (E-MORB), or an ocean island basalt (OIB). Flat trends indicate a close

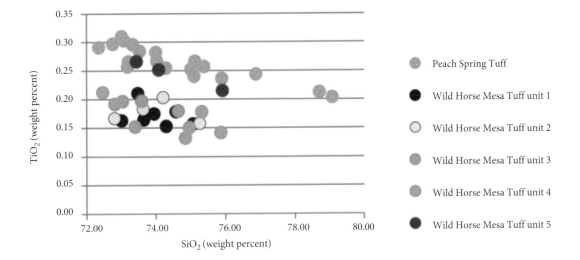

**Figure 8.15** This Harker diagram shows that a relatively high titanium content relative to $SiO_2$ helps distinguish the Peach Spring tuff unit from units 1–4 of the Wild Horse Mesa in Providence Mountains of California. The Peach Spring Tuff samples contain ~72–79% $SiO_2$ and 0.20–0.30% $TiO_2$; while the Wild Horse Mesa Tuff units 1–4 contain ~72–76% $SiO_2$ and 0.12–0.22% $TiO_2$. Wild Horse Mesa Tuff unit 5 more closely aligns with the geochemical signature of the Peach Spring Tuff which may indicate the need for further investigation. *Source*: U.S. Geological Survey, United States Department of the Interior.

chemical relationship to the standard reference. Spider diagrams are useful in demonstrating the enrichment or depletion of compatible versus incompatible elements, light rare earth element versus heavy rare earth element variation, or other data that may indicate diversification trends or magmatic source. The relative abundance of specific elements may also yield information regarding the fractionation of particular minerals. For example, depletion of Eu indicates plagioclase fractionation from the melt which effectively removes Eu from the magma. On the other hand, Eu enrichment would indicate assimilation of plagioclase components into the melt.

Figure 8.16 displays data comparing the Ricardo basalts to N-MORB, E-MORB,

OIB, and primitive mantle compositions. Note that the trace element concentrations of the Ricardo basalts are most similar to OIB. Note also that LREE and largest LIL's are enriched relative to HREE in a way highly dissimilar from N-MORB and somewhat dissimilar from E-MORB. Ricardo basalts are between 10 and 100 times more enriched in LIL's and REE's compared to the primitive mantle. It is likely that the parent melts for the Ricardo basalts were produced by small degrees of partial melting of a relatively undepleted mantle source associated with hot spot melting.

While we have provided a general introduction, detailed geochemical analyses require a thorough understanding of geochemistry and thermodynamics beyond the scope of this text-

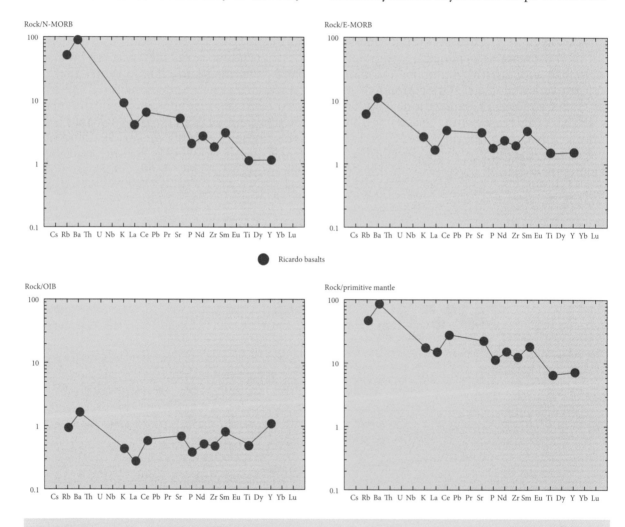

**Figure 8.16**   Spider diagrams of Ricardo basalts collected in the vicinity of the Garlock Fault in southeastern California (Anderson and Jessey 2005). The Ricardo basalt display trace element similarities to ocean island basalts (OIB). N-MORB = normal mid ocean ridge basalt; E-MORB = enriched mid ocean ridge basalt. *Source*: Anderson and Jessey (2005). © Geological Society of America.

book. Detailed geochemistry understanding and computer software can provide insight into the origin and history of igneous and metamorphic rocks. Computer programs have been developed to assist in the geochemical analysis of rock data such as Igpet (Carr and Gazel 2017) and GCDkit : GeoChemical Data ToolKIT (Janoušek et al. 2006). However, it must be emphasized that geochemical analyses must be conducted in concert with field and laboratory analysis of rock samples.

So far in this chapter we have focused upon the origin and diversification of magma and some geochemical considerations. In the following discussion, we will discuss plutonic rock bodies created by magma cooling at depth within Earth.

## 8.6   INTRUSIVE STRUCTURES

Plutonic structures form when magma intrudes pre-existing country rock within Earth. The critical distinctions in classifying plutonic structures include the size, shape, and orientation of the intrusion. Batholiths and stocks are irregularly shaped plutons. Plutons that have a more tabular shape, resembling table tops or walls, are classified as concordant or discordant depending upon the orientation of the intrusive contacts with respect to the surrounding country rock layers. The contacts of **concordant** intrusions are oriented parallel to the pre-existing layering in the surrounding country rock. The pre-existing geologic layering may be bedding, metamorphic foliation, volcanic layers or other structural forms. In the event that no geologic layering is discernible in the country rock, concordant intrusions are parallel to Earth's surface. The boundaries of **discordant** intrusions are oriented at a sharp angle to the country rock layering and cut across it. If no geologic layering is observable, discordant boundaries would be inclined at an angle rather than be parallel to Earth's surface. Intrusive structures we will discuss below include batholiths and stocks, sills, laccoliths, lopoliths, necks, and dikes (Figure 8.17).

### 8.6.1   Batholiths and stocks

**Batholiths** are defined as plutons of more or less irregular shape with surface exposures $\geq 100 \, km^2$. **Stocks** are plutons with surface exposures $\leq 100 \, km^2$. Batholiths and stocks can

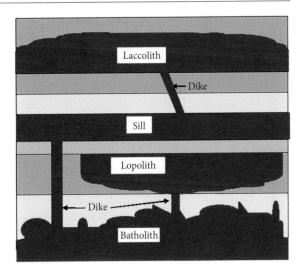

**Figure 8.17**   Cross-sectional view of plutonic structures.

form wherever magmatism occurs. Magmatism occurs primarily at convergent and divergent plate boundaries and hot spots. Divergent margin and hotspot stocks and batholiths are largely gabbroic in composition, although these are not generally exposed for direct field analysis. Convergent margin batholiths are generally intermediate to silicic in composition and produce rocks such as diorite, granodiorite, quartz diorite or granite. Convergent margin batholiths are well exposed in many localities where erosion has exposed rocks formed in magmatic arcs and orogenic belts such as in the Cordilleran (Rocky Mountain) system of North America. Convergent margin batholiths may be enriched in valuable metals such as molybdenum, tin, gold, and silver. Large silicic batholiths also form over hot spots in continental interiors as will be discussed further in Chapter 10.

Formerly, batholiths were considered to represent vast chambers of molten rock that existed at depth and cooled over thousands to a few million years (Buddington 1959). However, seismic refraction studies have failed to identify large molten bodies, even at fast spreading, mid ocean ridge systems (Detrick et al. 1990). Modern models suggest that batholiths may consist of <10% melt at any time and can develop over tens of millions of years as incremental, composite assemblage of many separate plutons (Pitcher 1979; Glazner et al. 2004; Coleman et al. 2004). Major batholiths do not represent a single intrusion but rather a diachronous set of individual plutons that occur within a spatial domain, particularly along convergent or divergent margins. For example, the Coastal Batholith in the Andes Mountains convergent margin is a composite batholith consisting of over 700 separate plutons (Pitcher 1979). Magmatic intrusions at convergent margins play a pivotal role in the growth of continents through geologic time. Batholith emplacement along convergent plate boundaries is the major mechanism by which continental crust has grown since the Archean, 2.5 billion years ago.

Stocks and batholiths can display concentric or tabular zoning patterns. Zoning may develop due to fractionation and/or contamination of the magma with the surrounding country rock (Holder 1979). Normal zoning within a pluton is characterized by a mafic rim and a silicic core. Vance (1961) suggested that normal zoning develops as cooling and crystallization progress sequentially from the pluton margins towards the inner core. Zoning may result in the segregation of mafic minerals such as pyroxene and amphibole from more felsic minerals such as quartz and feldspars. In some batholiths, distinct couplets form whereby a pinstripe pattern develops with alternating mafic and felsic mineral layers (Figure 8.18). The cause for couplet development is uncertain but may involve multiple dike like intrusions and/or fractionation processes.

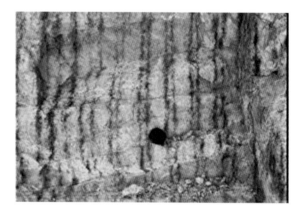

**Figure 8.18** White plagioclase-rich anorthosite with dark, vertical pyroxene mineral pinstripes displays rhythmic banding in a Montana batholith. *Source*: Photo by Kevin Hefferan. © John Wiley & Sons.

### 8.6.2 Concordant plutonic structures

**Concordant** plutonic structures, whose boundaries parallel country rock layers, include sills, laccoliths, and lopoliths. These concordant structures form when magma is injected into pre-existing planes of weakness between wall rock layers, prying them apart.

A **sill** is a tabular, concordant pluton that parallels country rock layers (Figure 8.19). Tabular plutons have shapes that resemble table tops or window sills. They have two mutually perpendicular long dimensions and a short dimension perpendicular to the other two. Sills develop through the injection of magma along a plane of weakness parallel to the layering in the country rock. Planes of weakness include bedding planes and metamorphic foliations. Sills commonly form from low viscosity magmas and are generally <50 m in thickness. However, sill thickness varies from thin tabular concordant plutons to much larger plutons such as the 300 m thick Palisades sill in New Jersey.

A **laccolith** (Gilbert 1877; Daly 1933) is a blister like concordant pluton characterized by a flat floor and domed roof (Figure 8.20). Laccoliths are similar to sills in that the magma is injected parallel to country rock layering. Laccoliths may begin as flat topped sill structures. The intrusion of additional magma produces sufficient force to bulge the sill roof upward creating a convex structure with a round dome top. Laccoliths were first described in the Henry Mountains of Utah by G.K. Gilbert in the 1870s. Prominent laccoliths include the Prospect Intrusion in New South Wales, Australia, and the Uwekahuna laccolith, Hawaii. Magma that produces laccoliths, sills, and other concordant structures generally rises from deeper sources via discordant dike feeder systems.

**Lopoliths** are dish shaped to funnel shaped, concordant plutons that resemble a wine glass in cross-sectional view (Figure 8.20). According to Daly (1933), F.F. Grout originally defined lopoliths in 1918 based on studies of the Proterozoic Duluth Gabbro complex of Minnesota. Other examples include Greenland's Skaergaard intrusion, South Africa's Bushveld complex, Finland's Koillismaa complex, Ontario's Sudbury intrusion and the Muskox intrusion in the Northwest Territories of Canada. Lopoliths can extend tens to hundreds of kilometers in length with thicknesses of up to several kilometers. The Bushveld

**Figure 8.19** Horizontal basaltic sill enveloped by rhyolite ignimbrites provides a stunning example of bimodal volcanism within the Yellowstone caldera. *Source*: Photo by Kevin Hefferan. © John Wiley & Sons.

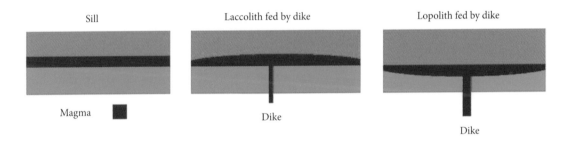

**Figure 8.20** Cartoon cross section of a horizontal, planar sill, a laccolith with a flat floor, and convex roof surface, and a lopolith with a convex floor and flat roof. Note that these concordant structures commonly receive magma from deeper batholiths via a dike conduit. *Source*: Kevin Hefferan (author).

complex has a length of 550 km and is 8 km thick. Many large lopoliths, such as the Muskox, consist of concentrically zoned mafic and ultramafic rocks. The wine glass lopolith shape can develop due to one or more of the following mechanisms: (1) meteorite impact and associated crustal melting as has been suggested for the Sudbury complex; (2) Normal faulting and crustal melting associated with rifting as in the Koillismaa complex; or (3) Sill like structure which receives upwelling magma from a conical feeder tube, which would correspond to the stem in the wine glass. Lopoliths, like many layered mafic–ultramafic intrusions, may be highly

enriched in valuable metals such as nickel, copper, chromium, platinum, and palladium.

**Veins** form as hot fluids flowing through fractures cool and crystallize. Veins may be concordant or discordant to the surrounding country rock. In a sense veins are small scale (millimeters to centimeters in diameter) sills or dikes. Veins can occur in great abundance resulting in **vein swarms**. Vein swarms may display random or preferred orientations. Preferred orientations of vein sets may be parallel (Figure 8.21), sub parallel, orthogonal, en echelon or sigmoidal. Orthogonal vein sets occur at right angles to one another. En echelon veins are parallel to sub parallel but offset. Sigmoidal veins are "S shaped" en echelon sets.

### 8.6.3 Discordant plutonic structures: necks, diatremes, and dikes

**Discordant** plutons are intrusive structures whose contacts cut across pre-existing country

**Figure 8.21** Swarm of parallel, concordant white felsic veins intruding dark gray volcaniclastic rocks near Golden, Colorado. These veins, concordant to the country rock, are essentially cm scale sills. Note pen in left for scale *Source*: Photo by Kevin Hefferan. © John Wiley & Sons.

rock layers. Discordant plutons include cylindrical necks, diatremes, and dikes. Many of these discordant structures serve as conduits that transport magma from deeper plutons to shallow intrusions or even surface eruptions.

**Necks** are cylindrical plutonic pipes exposed at the surface by subsequent erosion. They represent ancient conduit pipes that funneled magma upward to a volcano, that has long since been removed by erosion. Solidification of magma within the cylindrical pipe produces hard, crystalline rock within the volcano. The poorly consolidated overlying volcanic edifice and surrounding country rock erode at a faster rate over time. The hard, crystalline rocks of the cylindrical neck are erosion resistant. As a result, the cylindrical pipe is preserved as a vertical rock column. Notable examples include Ship Rock, New Mexico, and Devils Tower of Wyoming. Ship Rock has a central neck with radiating dike ridges (Figure 8.22). Necks are commonly called "volcanic" necks because they were the central vent of a volcano. This is actually a misnomer because necks represent plutonic cylindrical feeder pipes. Devils Tower contains a central neck structure with columnar joint pillars. The origin of Devils Tower has been suggested as possibly a pipe feeding a volcano, or a buried laccolith or stock that has been exposed by uplift and erosion (Figure 8.23). Native American legend has suggested that

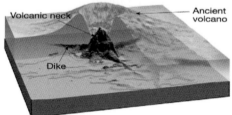
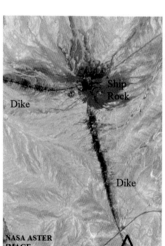

**Figure 8.22** Ship Rock contains radiating mafic dikes emanating from the central neck structure. Erosion of the soft Manco Shale country rock leaves behind vertical rock dike walls. *Source*: Photos by Kevin Hefferan. Images courtesy of NASA.

Devils tower neck

Columnar joints viewed at the base
of Devils Tower Neck

Did Devils Tower
Form by:

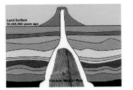

A. Dike feeding
a volcano?

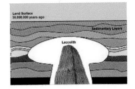

B. Remnant of a
buried laccolith?

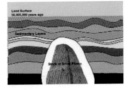

C. Remnant of a
buried stock?

**Figure 8.23**  Devil's Tower neck with columnar joints, Wyoming. Devils Tower origin may be related to dikes feeding a volcano, a laccolith or the remains of an ancient stock. *Source*: Kevin Hefferan photos. Illustrations courtesy of U.S. National Park Service. © The United States Department of the Interior

Devils Tower served as a protective shelter from a giant bear with powerful nails that clawed the sides of the tower creating the columnar columns.

**Diatremes** are carrot shaped, cylindrical pipes that can extend to depths of several kilometers. Diatremes develop via explosive intrusions that originate deep within the mantle. The explosiveness is due to the high volatile content which propels magma and xenolith fragments upward towards the surface. As will be discussed in Chapter 10, diatremes are extensively studied due to their association with ultramafic diamond bearing kimberlites. Volcanic structures such as tuff cones may have developed on Earth's surface above diatreme pipes. However, these surficial deposits are usually long since eroded, exposing deeper diatreme pipes.

**Dikes** are tabular intrusions that cross cut country rock layers. Because dike rocks are typically more resistant to weathering and erosion than surrounding country rock, dike exposures commonly form topographically elevated ridges. Dikes occur as individual structures or in dike sets or swarms, in which many dikes may show similar orientations, radial orientations, or random orientations. Figure 8.24 illustrates a granitic dike that cross cuts metamorphic schist to produce a stock into which four US Presidents heads have been carved at Mt. Rushmore, SD.

While most dikes exposed on land measure meters to tens of meters in thickness and hundreds of meters to a few km in length, exceptions exist. The 2.5 Ga Great Zimbabwe dike of southern Africa is ~700 km long and 8 km wide. Zimbabwe's Great dike is unusual for its length, ultramafic composition and rich ore deposits of platinum group metals and chromite.

**Dike swarms** consist of multiple dikes that can occur either in parallel, sub parallel, radiating, concentric or random orientations. **Radial dikes** are typically produced when vertical forces related to rising magma produce fractures that radiate outward from the central vent. The radial fractures are analogous to those produced in window glass when it is struck by a hard object.

**Figure 8.24** Granitic dike cross cutting metamorphic rock at Mt. Rushmore. *Source*: Photo by Kevin Hefferan. © John Wiley & Sons.

Radial dikes form when magma from the central vent is injected outward into the radial fractures. Ring dikes and cone sheets constitute networks of interrelated, discordant intrusions associated with concentric fracture sets (Figure 8.25). The concentric fracture sets commonly develop following the collapse of a volcanic summit, in subsiding structures such as calderas or due to meteorite impacts. **Ring dikes** are nearly vertical in cross section and circular in map view. Ring dikes occur in Tertiary volcanic suites on the Island of Mull, Scotland.

**Cone sheet** dikes are circular in map view and resemble ring dikes, except that cone sheet dikes converge at depth in cross-sectional view. Cone sheets possess an ice cream cone like structure. These structures develop as magma moves upward from a plutonic source. Pressures associated with the intrusion of magma produce fractures, which propagate upwards and outwards from a central point. Magma is injected into these concentrically nested fractures, generating the cone sheet structures. Schirnick et al. (1999) report ~500 cone sheets,

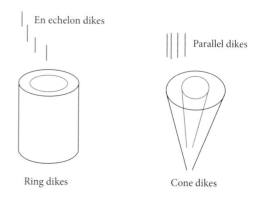

**Figure 8.25** Three-dimensional diagrams of en echelon, parallel, ring, and cone dikes.

radial dikes, and hypabyssal plutonic structures within the 20 km wide Tertiary age Tejeda Caldera Complex in the Canary Islands.

**En echelon dikes** consist of parallel, offset dikes that form in response to shear.

Parallel dike sets form perpendicular to extension and are particularly common in rift environments (Figure 8.25). **Parallel dikes**

**Figure 8.26** Sheeted dikes at Giant's Causeway, located on the Irish Sea coastline. The sheeted dikes contain parallel colonnade dikes overlain by radiating (entablature) dike orientations. *Source*: Photo by Kevin Hefferan. © John Wiley & Sons.

commonly develop within sheeted dike complexes, discussed below.

Steeply inclined **sheeted dikes** occur in ocean crust (Layer 2B) as mid ocean ridge feeder conduits convey magma to the overlying pillow basalt (Layer 2A). Sheeted polygonal mafic dikes form by the cooling and contraction of magma as it is injected into extensional fractures in oceanic rift valleys. Repeated injection into extension fractures occurs over millions of years as the ocean lithosphere diverges from the rift axis. As a result, new magma injections intrude fractures parallel to previous dikes producing massive, parallel, wall like to flower like intrusions called sheeted dikes. Hot magma injections into pre-existing dikes result in chilled margins along the periphery of the newly forming dike. Thus, sheeted dikes commonly contain finer grained chilled margins due to more rapid heat loss and crystallization, where younger dikes are in contact with older dikes. In addition to ocean floor divergent mar-

gins, sheeted dikes also occur in continental rift settings, intraplate hot spots (e.g. Yellowstone) and flood basalt provinces. Outstanding sheeted dike structures are exposed along the shores of the Irish Sea both in southern Scotland and along the north shores of Ireland. Hexagonal polygonal columns formed 50–60 million years ago as magma contracted upon cooling to form parallel colonnade structures and flowery, radiating entablature layers (Figure 8.26). Legends suggest the dikes are the remnants of an ancient causeway bridging the Scottish island of Staffa and the north shores of Ireland for the rovings of the giant Finn MacCool across the Irish Sea. Sheeted dike formation is discussed in more detail in the discussion of oceanic ridge volcanism and ophiolite sequences in Chapter 10.

In this chapter we have focused on the diversity of magma and intrusive rock structures. In Chapter 9, we will concentrate on volcanic structures and eruptions.

## CONTENT ASSESSMENT

1   Describe three processes that induce rock melting and relate them to geologic plate boundaries.

2   Describe a process by which the melting of a gabbro can produce a diorite.

3   Contrast the chemical crystallization trends of tholeiitic and calc-alkaline magmas on AFM diagrams.

4   Describe conditions in which normal zoning may occur.

5   Describe conditions by which reverse zoning may occur.

6   Describe how the concentration of incompatible elements in a magma varies with degrees of partial melting.

7   Describe how a sill differs from a laccolith.

8   Describe why dikes are fundamentally important in transporting magma to volcanoes.

## REFERENCES

Anderson, C.J., and Jessey, D.R. (2005) Geochemical analysis of the Ricardo volcanics, Southern El Paso Mountains, California. Geological Society of America Abstracts with Programs, 37, 47. http://gsa.confex.com/gsa/2005CD/finalprogram/abstract_84281.htm.

Bailey, D.L. (1974). Continental rifting and alkaline mamatism. In: *The Alkaline Rocks* (ed. H. Sorenson), 148–159. New York: Wiley.

Bailey, E.B., Clough, C.T., Wright, W.B. et al. (1924). *Tertiary and Post-Tertiary Geology of Mull, Loch Aline and Oban*. Memoirs of the Geological Survey of Scotland 445 pp.

Bottinga, Y. and Weill, D.F. (1970). Densities of liquid silicate systems calculated from partial molar volumes of oxide components. *American Journal of Science* 269: 169–182.

Bowen, N.L. (1928). *The Evolution of Igneous Rocks*. Princeton, NJ: Princeton University Press 332 pp.

Buddington, A.F. (1959). Granite emplacement with special reference to North America. *Geological Society of America Bulletin* 70: 671–747.

Burke, K., Ashwal, L.D., and Webb, S.J. (2003). New way to map old sutures using deformed alkaline rocks and carbonatites. *Geology* 31: 391–394.

Butler, J.C. (1979). Trends in ternary petrologic variation diagrams – fact or fantasy. *American Mineralogist* 64: 1115–1121.

Carr, M.J. and Gazel, E. (2017). Igpet software for modeling igneous processes: examples of application using the open educational version. *Mineralogy and Petrolrology*

111: 283–289. https://www.rockware.com/downloads/documentation/igpet/igpetpaperprintversion.pdf.

Chayes, F. (1960). On correlation between variables of constant sum. *Journal of Geophysical Research* 65: 4185–4193.

Chayes, F. (1962). Numerical correlation and petrographic variation. *Journal of Geology* 70 (4): 440–452.

Chayes, F. (1964). Variance–covariance relations in some published Harker diagrams of volcanic suites. *Journal of Petrology* 5 (2): 219–237.

Coleman, D.S., Gray, W., and Glazner, A.F. (2004). Rethinking the emplacement and evolution of zoned plutons: geochronologic evidence for incremental assembly of the Tuolumne Intrusive Suite, California. *Geology* 32: 433–436.

Coombs, M.L. and Gardner, J.E. (2004). Reaction rim growth on olivine in silicic melts: implications for magma mixing. *American Mineralogist* 89: 748–759.

Daly, R.A. (1933). *Igneous Rocks and the Depth of the Earth*. New York: McGraw Hill 598 pp.

De, A. (1974). Silicate liquid immiscibility in the Deccan traps and its petrogenetic significance. *Geological Society of America Bulletin* 85: 471–474.

Detrick, R.S., Mutter, J.C., Buhl, P., and Kim, I.I. (1990). No evidence from multichannel reflection data for a crustal magma chamber in the MARK area on the Mid Atlantic Ridge. *Nature* 347: 61–64.

D'Lemos, R. (1992). Magma mingling and melt modification between granitic pipes and host diorite, Guernsey, Channel Islands. *Journal of the Geological Society of London* 149: 709–720.

Ferguson, C.A., McIntosh, W.C., and Miller, C.F. (2013). Silver Creek caldera – the tectonically dismembered source of the Peach Spring Tuff. *Geology* 41 (1): 3–6.

Fram, M.S. and Longhi, J. (1992). Phase equilibria of dikes associated with proterozoic anorthosite complexes. *American Mineralogist* 77: 605–616.

Frey, F.A., Green, D.H., and Roy, S.D. (1979). Integrated models of basalt petrogenesis: a study of quartz tholeiites to olivine melilitites from southeastern Australia utilizing geochemical and experimental petrological data. *Journal of Petrology* 19: 463–513.

Frey, F.A., Wise, W.S., Garcia, M.O. et al. (1990). Evolution of Mauna Kea volcano, Hawaii: the transition from shield building to the alkalic cap stage. *Journal of Geophysical Research* 95: 1271–1300.

Gilbert, G.K. (1877). *Report on the Geology of the Henry Mountains*. Washington, DC: Department of the Interior, US Geographical and Geological Survey of the Rocky Mountain Region 170 pp.

Gill, J.B. (1981). *Orogenic Andesites and Plate Tectonics*. New York: Springer-Verlag 390 pp.

Glazner, A. and Johnson, B.R. (2013). Late crystallization of K-feldspar and the paradox of megacrystic granites. *Contributions to Mineralogy and Petrology* 166: 777–799.

Glazner, A.F., Bartley, M.M., Coleman, D.S. et al. (2004). Arc plutons assembled over millions of years by amalgamation from small magma chambers? *GSA Today* 14: 4–11.

Green, D.H. and Ringwood, A.J. (1967). An experimental investigation of the gabbro to eclogite transformation and its petrologic application. *Geochemica et Cosmochimica Acta* 31: 767–833.

Grove, T.L. and Kinzler, R.J. (1986). Petrogenesis of andesites. *Annual Review of Earth and Planetary Sciences* 14: 417–454.

Harker, A. (1909). *The Natural History of Igneous Rocks*. New York: McMillan Publishers 384 pp.

Higgins, M.D. (2011). Textural coarsening in igneous rocks. *International Geology Review* 53: 354–376.

Holder, M.T. (1979). An emplacement mechanism for post-tectonic granites and its implications for their geochemical features. In: *Origin of Granite Batholiths: Geochemical Evidence* (eds. M.P. Atherton and J. Tarney), 116–128. Orpington, Kent, UK: Shiva Publishing.

Irvine, T.N. and Barager, W.R.A. (1971). A guide to the chemical classification of the common volcanic rocks. *Canadian Journal of Earth Sciences* 8: 523–548.

Irvine, T.N., Anderson, J.C., and Brooks, C.K. (1998). Included blocks (and blocks within blocks) in the Skaergaard intrusion: geologic relations and the origins of rhythmic modally graded layers. *Geological Society of America Bulletin* 110: 1398–1447.

Janoušek, V., Farrow, C.M., and Erban, V. (2006). Interpretation of whole-rock geochemical data in igneous geochemistry: introducing Geochemical Data Toolkit (GCDkit). *Journal of Petrology* 47 (6): 1255–1259. http://www.gcdkit.org/.

Miyashiro, A. (1974). Volcanic rock series in island arcs and active continental margins. *American Journal of Science* 274: 321–355.

Namur, O., Charlier, B., Pirarc, C. et al. (2011). Anorthosite formation by plagioclase flotation in ferrobasalt and implications for the lunar crust. *Geochimica et Cosmochimica Acta* 75: 4998–5018.

Naumann, T.R. and Geist, D.J. (1999). Generation alkalic basalt by crystal fractionation of tholeiitic magma. *Geology* 27: 423–426.

Philpotts, A.R. (1979) Silicate liquid immiscibility in tholeiitic basalts. Journal of Petrology, 20, 99–118.

Pitcher, W.S. (1979). The nature, ascent and emplacement of granite magmas. *Geological Society of London Journal* 136: 627–662.

Saccani, E. (2015). A new method of discriminating different types of post-Archean ophiolitic basalts and their tectonic significance using Th–Nb-and Ce–Yb–Yb systematics. *Geoscience Frontiers* 6 (4): 481–501.

Schirnick, C., van den Bogaard, P., and Schmincke, H.U. (1999). Cone sheet formation and intrusive growth of an oceanic island – the Miocene Tejeda complex on Fran Caniara (Canary Islands). *Geology* 27: 207–210.

Sisson, T.W. and Bacon, C.R. (1999). Gas-driven filter pressing in magmas. *Geology* 27: 613–616.

Sleep, N.H. (1988). Tapping of melt by veins and dikes. *Journal of Geophysical Research* 93 (10): 10255–10272.

Sorensen, H. (1974). *The Alkaline Rocks*. New York: Wiley 622 pp.

Stone, P., Miller, D.M., Stevens, C.H. et al. (2017). *Geologic map of the Providence Mountains in parts of the Fountain Peak and adjacent 7.5′ quadrangles, San Bernardino County, California: U.S.*, vol. 3376. Geological Survey Scientific Investigations Map 52 pp doi:https://doi.org/10.3133/sim3376.

Streck, M.J., Broderick, C.A., Thornber, C.R. et al. (2008). Plagioclase populations and zoning in dacite of the 2004–2005 Mount St. Helens eruption: constraints for magma origin and dynamics. In: *A Volcano Rekindled: The Renewed Eruption of Mount St. Helens, 2004–2006* (eds. D.R. Sherrod, W.E. Scott and P.H. Stauffer), 791–808. U.S. Geological Survey Professional Paper 1750.

Thomas, C.W. and Aitchison, J. (2005). Compositional Data Analysis of Geological Variability and Process: A Case Study. *Mathematical Geology* 37: 753–772. https://doi.org/10.1007/s11004-005-7378-4.

Vance, J.A. (1961). Zoned granite intrusions – an alternative hypothesis of origin. *Geological Society of America Bulletin* 72: 1723–1728.

Wager, L.R. and Deer, W.A. (1939). Geological investigations in East Greenland. III. The petrology of the Skaergaard intrusion, Kangerdlugssuaq, East Greenland. *Medd Gronland* 105 (4): 1–352.

# Chapter 9

# Volcanic features and landforms

## 9.1   VOLCANOES

A **volcano** is a naturally occurring landform produced where magma rises up to Earth's surface and erupts as lava. Earth's volcanoes are preferentially located at convergent and divergent boundaries as well as intraplate hotspots. Volcanic activity vividly displays the dynamic nature of our hot, turbulent planet, and profoundly impacts Earth in many ways. Volcanic eruptions generate new land area, valuable mineral deposits, and arable soil. Volcanic eruptions can also kill tens of thousands of people and dramatically alter Earth's climate, causing major extinctions. In fact, nearly all of Earth's mass extinction events correspond with major volcanic events. In this chapter, we will describe common volcanic landforms such as craters, vents, fissures, and calderas as well as major types of volcanoes

and their deposits. We will also address why eruptive styles vary so much for different volcanoes and how volcanoes impact Earth's inhabitants.

### 9.1.1   Craters, central vents, and fissure vents

A **crater** is a nearly circular depression produced by the removal of rock during volcanic eruptions. Lava commonly erupts from a **central vent** located in a crater near the summit of the volcano. From an analysis of 5564 volcanic eruptions, most of which were above sea level, Simkin et al. (1981) suggested that nearly half of all volcanic eruptions occur from central vents. However, their analysis may be skewed because most volcanism likely occurs unobserved deep on the ocean floor along ocean spreading ridge fissures. **Fissures** are planar fractures that commonly develop due to

*Earth Materials*, Second Edition. Kevin Hefferan and John O'Brien.
© 2022 John Wiley & Sons Ltd. Published 2022 by John Wiley & Sons Ltd.
Companion website: www.wiley.com/go/hefferan/earthmaterials2

extension at (1) ocean spreading ridge systems such as the Mid Atlantic Ridge and East Pacific Rise; (2) continental rifts such as the East African Rift System or the Basin and Range of the southwestern USA, and (3) hotspots such as Hawaii (Figure 9.1). In addition to central vents and fissures, lava also commonly erupts from the sides or base of a volcano producing **flank eruptions**.

Central vents, fissures, and flank eruptions may occur alone or in combination with one another in complex magmatic plumbing systems. These plumbing systems may channel magma to rise and flow along shallow (hypabyssal) dike and sill systems, interacting with wall rocks, gases, and groundwater, before erupting on the surface as lava. North of Flagstaff (Arizona), SP Volcano contains a beautiful central crater vent and a vivid flank eruption flow at the base that extends northward over 6 km (Figure 9.2). Hawaii's Mauna Loa and Kīlauea magma plumbing systems provide conduits for lava to produce fissures, isolated central vents, and small flank eruptions. Low discharge flank eruptions produce small **parasitic volcanoes** on larger volcanoes and may show alignment over the linear fissures. Mauna Kea, an immense volcano located on the big island of Hawaii, contains over 100 parasitic volcanoes (Figure 9.3). Volcanic eruptions can evolve from one emission style to another. For example, the 1983 eruption at Kīlauea began as a fissure eruption and evolved into a central vent eruption over several months. After a period of quiescence, Kīlauea's caldera erupted again on 19 March 2008, marking the first explosive activity at the central vent summit since 1924.

**Figure 9.1** "Curtain of Fire" fissure eruption on Mauna Loa produces an elongated spatter rampart dike upon solidification of lava. *Source*: Photo courtesy of D.A. Clague, taken in March 1984, with permission of US Geological Survey. © U.S. Department of the Interior.

### 9.1.2 Calderas

Calderas are large, generally circular to oval depressions caused by subsidence of Earth's surface. Subsidence is generally initiated by the rapid loss of magma stored in the shallow subsurface. Magma loss may be due to: (1) subsurface withdrawal of magma from a shallow chamber as magma migrates to another location, or (2) cataclysmic volcanic eruptions that empty the shallow magma chamber.

Hawaii's Kīlauea caldera (Figure 9.4) is experiencing a very active phase of magma level fluctuation that affects the East Rift fissure zone, the central vent crater and the creation of lava lakes. Lave lake levels are erratic and intricately connected to magma reservoir pressure, which in turn may be influenced by the East Rift fissure zone and the magma supply rate. Kīlauea caldera's 19 March 2008 eruption created the new "Overlook pit crater" within Halema'uma'u crater. Sporadic lava lake activity began in 2008–2009 and more continuous lava lake level rise began in February 2010, before draining briefly in March 2011 due to eruptive activity on the East Rift Zone. Lava lake levels rose again beginning in 2012, reaching the Overlook crater rim and flowing on the Halema'uma'u floor in 2015, 2016, and 2018. By 2018, the lava lake was approximately 280 × 200 m (~42 000 m²), one of the two largest lava lakes on Earth; the other being the Nyiragongo Volcano lava lake in the Democratic Republic of the Congo (Patrick et al. 2021). Kīlauea caldera experienced a renewed volcanic eruptive cycle that began 3 May 2018 and continues to the time of this textbook printing.

While Hawaii's Kīlauea caldera predominantly represents a magma withdrawal caldera, many great calderas are generated by massive explosive volcanic eruptions such as Yellowstone (USA), Mt. Toba (Sumatra), Lake Taupo (New Zealand) and Mt. Mazama (USA). Mt. Mazama in Oregon's Cascade Range (USA) erupted in a cataclysmic event ~5677 BCE ejecting an estimated 1000 km³ of rock material (Zdanowicz et al. 1999). Oregon's Crater Lake occupies the former site of Mt. Mazama; this caldera averages 8 km in diameter and is 1.6 km deep.

Explosive caldera eruptions also occur in the ocean, despite the high hydrostatic pressure environment. Calderas have been observed in submarine volcanoes associated

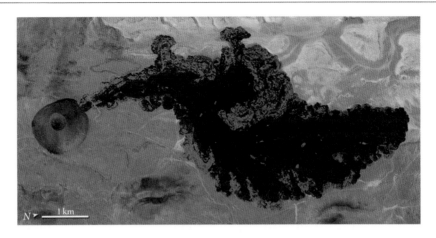

**Figure 9.2**   A central summit crater vent and flank eruption basalt lava flow (4 000–71 000 years old) are clearly visible on SP Volcano in the San Francisco Volcanic Field. *Source*: NASA. NASA Earth Observatory image from April 17, 2010 by Jesse Allen and Robert Simmon using Advanced Land Imager on NASA's Earth Obwserving-1 satellite.

**Figure 9.3**   Small parasitic volcano cones at the Mauna Kea summit. Subsurface fissures channel magma to the small parasitic eruptive centers. *Source*: Photo by Kevin Hefferan. © John Wiley & Sons.

with subduction zones. McCauley caldera, located near the Kermadec volcanic arc offshore of New Zealand, is 6–10 km across and over 1 km deep. Since May of 2018, over 7000 earthquakes have occurred on Mayotte, the easternmost island of the Comoros archipelago, situated in the Indian ocean between Africa and Madagascar. Dead fish, strong hydrogen sulfide odors and seismicity in a generally aseismic region revealed a new submarine volcano 50 km east of the island. Since 2018,

researchers have documented that over 1.3 km³ of magma drained from a 10 to 15 km diameter magma reservoir 25–55 km depth, below the Moho. The Mayotte submarine eruption is the largest documented underwater volcanic eruption by volume ever observed and continues into 2021 (Cesca et al. 2020; Tripathy-Lang 2020).

## 9.2   VOLCANO CLASSIFCATION, FEATURES, AND SETTINGS

Volcanoes are classified based on their eruptive history as well as their chemical and physical characteristics. Eruptive history refers to the previous record of volcanic eruptions and likelihood for recurrence. **Active volcanoes** have erupted within the past several hundred years and are likely to do so again. **Dormant volcanoes** are potentially active, but have not erupted in thousands of years; thus, they are "sleeping." **Extinct volcanoes** have not erupted for tens of thousands of years and volcanic activity is not likely to recur.

The shape and size of volcanic landforms is directly related to the frequency and duration of eruptions, the underlying magma type and the volume available. Explosiveness and lava flow characteristics depend upon the composition, temperature, volatile content, and viscosity of the underlying magma. Viscosity increases with increasing silica and decreasing temperature so that lower tempera-

(a)

(b)

**Figure 9.4**   (a) Kīlauea caldera rim, Halema'uma'u crater, and Overlook pit crater lava lake are visible on 6 March 2015. *Source*: US Geological Survey. © U.S. Department of the Interior. (b) US Geological Survey scientist at Hawaii Volcano Observatory monitoring the elevation of the Halema'uma'u lava lake using a laser rangefinder on 12 January 2021. The lava lake measures 196 m (643 ft) deep. *Source*: U.S. Geological Survey photo by M. Patrick. https://www.usgs.gov/media/images/increased-surface-activity-halema-uma-u-2. © U.S. Department of the Interior.

ture, silica rich magmas are more viscous. Baker et al. (2004) suggest using food analogies when discussing magma viscosity such that at 25 °C and 1 bar pressure (a typical kitchen environment), smooth peanut butter has a viscosity of $\approx 10^8$ Pa/S, nearly equal to that of rhyolite melt with 2% dissolved water vapor by weight at 800 °C. In contrast, ketchup has a viscosity of $\approx 10^2$ Pa/S, nearly equal to that of anhydrous (volatile poor) tholeiitic basalt at 1200 °C.

Viscosity plays a critical role in the development of volcanic structures, the aerial distribution and thickness of lava, and the explosiveness of eruptions. Low viscosity mafic lava moves great distances from the source vent, producing aerially extensive, non-explosive flows. In some instances, flood basalts flow hundreds to thousands of kilometers from the vent. As a result, basaltic lava flows commonly generate extensive plateaus and broad, gently sloping, convex shaped (shield) volcanoes. High viscosity felsic lavas do not flow readily. As a result, viscous felsic lava accumulates locally, producing thick deposits around the vent. If lava erupts from a central crater, a volcanic dome develops. Viscous lava also entraps exsolved gases to produce explosive eruptions. High viscosity, volatile rich volcanic eruptions can be devastatingly destructive as discussed later in this chapter.

Let us now investigate the major volcanic landforms one might encounter along islands and mountains of the Pacific rim, eastern Indian Ocean, Iceland, Mediterranean Sea, East Africa or in western North America. These locations provide stunning examples of Earth's vitality and gargantuan capacity for sudden, catastrophic change.

### 9.2.1   Flood basalts

**Flood basalts** erupt as massive outpourings of low viscosity basaltic lava that can envelop hundreds of thousands of square kilometers generating **large igneous provinces (LIPs)**. Large igneous flood basalt provinces form above mantle hotspots in the marine setting as **oceanic flood plateaus** or on land as **continental flood plateaus**. Oceanic flood basalts include the Ontong Plateau deposits in the western Pacific Ocean basin and the Kerguelen Plateau in the Indian Ocean. Spectacular examples of continental flood basalts include India's Deccan Traps, the Siberian flood basalts, East Africa's Karoo flood basalts, and North America's Columbia River/Snake River Plain flood basalts. The largest flood basalt of all – the Central Atlantic Magmatic Province (CAMP) – includes both marine and continental deposits distributed throughout the Atlantic Ocean margins that record the breakup of the Pangea supercontinent.

Large igneous province flood basalts are somewhat unique among volcanic features in that no modern examples occur, which prevents the direct observation of eruptive styles and their impact upon Earth's climate and inhabitants. The Columbia River/Snake River Plain flood basalts are the youngest large igneous province flood basalt deposits on Earth, having erupted within the past 17 Ma (Figure 9.5). In addition to LIPs, smaller flood basalt deposits can form in rift basins, hotspots, and ocean basins.

Flood basalts typically release lava through multiple fissures. Multiple long fissures and low lava viscosity together with large volumes of magma promote widespread flooding as opposed to localized accumulation around central vents. Where fluid lavas flow into basins, deep lava lakes may form to produce unusually thick accumulations. Flood basalts may erupt repeatedly over millions of years to produce stacked sequences of flow deposits, hundreds to thousands of meters thick. The tops of each individual flow may show vessiculation as gas bubbles migrate toward the top of flows. Thick flows display columnar jointing, formed as the lava cools and contracts. The joint sets create polygonal columns less than 1 m in diameter, but meters to hundreds of meters in height where lava lakes once existed. The relatively flat surfaces produced as fluid lavas accumulate in low areas generate immense **lava plateaus**. Figure 9.6 illustrates columnar joints exposed at Giant's Causeway in Ireland, an ancient lava flow which formed from a fissure eruption in the Irish Sea 50–60 Ma.

### 9.2.2 Ocean ridge fissure eruptions

Ocean ridges consist of a 65 000 km long global network of submarine rift mountains characterized by horizontal extension and basaltic volcanism. The crests of ocean spreading ridges are generally 2–3 km below sea level. However, Iceland, the Galapagos Islands and the Azores Islands represent rare examples of ocean ridges exposed above sea level due to unusually large volumes of erupted lava; these ocean islands represent regions where hotspots locally underlie the spreading ocean ridge system.

Ocean ridge eruptions are generally not explosive due to basaltic magma's low volatile content and low viscosity, and the relatively high hydrostatic pressures associated with the water depth. All of these factors inhibit gas exsolution and reduce explosivity. Tensional stresses associated with rifting result in eruptions from elongated fissures sub parallel to the ridge axis. As basaltic magma rises through ocean ridge fissure vents, interaction with cold sea water produces a variety of remarkable features which include pillow basalts, hyaloclastites, black smokers, and mineralized metal sulfide mound deposits.

As hot (1100–1300 °C) basaltic magma rises upward and reacts with cold seawater, spheroidal **pillow lavas** develop (Figure 9.7). The outer shell of the pillow is quenched instantaneously producing a glassy rind. The interior of the pillow cools more slowly. After the outer glassy rind forms, radial cooling joints develop within the interior of the pillow. Repeated fracturing of the lava flow and the inflation of larger pillows provides conduits from which interior lava escapes to produce additional pillows. These slide downslope and accumulate one atop the other. A series of pillows form, flow, and subsequently breach, generating cascading pillow piles parading downslope. Individual pillows are generally less than 1 m in diameter and occur in pods. The cumulative thickness of pillow basalt pods ranges from meters to hundreds of meters. Pillow lavas are widespread along ocean ridge system, but also form in continental lakes and volcanic arc settings or anywhere subaqueous basalt eruptions occur.

In addition to pillow basalts, ocean ridges are sites of hydrothermal (hot water) activity produced as heated seawater flows through the hot, juvenile crust produced by sea floor spreading. Hydrothermal activity produces **black smoker** vents which emit plumes of dark, hot (350–400 °C) metal and sulfide rich solutions. These solutions precipitate black sulfide rich mounds, which can grow to form chimney like tower structures (Figure 9.8). Black smoker towers and sulfide mounds were first discovered along the axis of the East Pacific Rise (Francheteau et al. 1979). The black color is due to the suspended fine grained sulfides. Black smoker chimneys observed along the East Pacific Rise and Mid Atlantic Ridge grow at the rate of up to 6 m/yr, with the largest

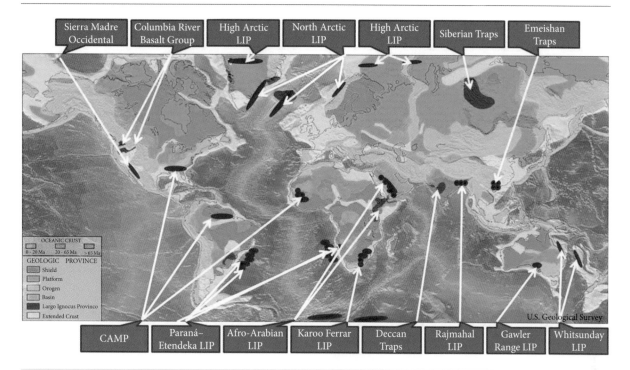

**Figure 9.5** Flood basalts occur in both continental and oceanic environments in which low viscosity basaltic lava travels up to thousands of kilometers from the vent source. *Source*: Image courtesy of Williamborg and Wikimedia Commons CC BY-SA 3.0. Image created by Williamborg through adaptation of a U.S. Geological Survey image. https://creativecommons.org/licenses/by-sa/3.0, via Wikimedia Commons. https://upload.wikimedia.org/wikipedia/commons/0/02/Flood_Basalt_Map.jpg.

chimneys ~50 m high and up to 12 m in diameter. Black smoker towers and associated mounds are enriched in metals such as cobalt, nickel, copper, zinc, and silver. This mineralization process is the source for economically viable volcanogenic massive sulfide deposits found in modern and ancient plate margins.

Ocean ridges also produce **white smokers** which emit lower temperature (100–300 °C) solutions that precipitate light colored minerals such as calcite, gypsum, barite, and quartz. In December of 2000, calcite towers 20–30 m high were viewed by geologists in the Alvin Submersible along a previously unknown volcanic vent field, located 9 miles west of the Mid Atlantic Ridge. These calcite towers, dubbed "Lost City Vent Field," formed by hydrothermal reactions between ocean mantle harzburgites and cold seawater over the past 30000 years. As a byproduct of hydrothermal alteration, harzburgite alters to serpentinite, and calcium carbonate, silica, and sulfate minerals precipitate generating white colored rock columns (Fruh-Green et al. 2003).

Amazingly, many life forms such as crabs, clams, tube worms, mussels, and long necked barnacles, dwell in the vicinity of ocean ridge vent systems. These organic communities do not depend on sunlight and photosynthesis. Instead, chemosynthetic bacteria fix energy from the hydrothermal vents into organic molecules and marine life forms feed on the chemical nutrients fixed by the bacteria. The diversity of life forms was first observed by researchers aboard the Alvin submersible along the East Pacific Rise from 1977 to 1979 (Francheteau et al. 1979). The ocean ridge environment is now considered as one of the possible environments in which life first developed on Earth.

### 9.2.3 Shield volcanoes

**Shield volcanoes** are broad sloping edifices that cover hundreds to thousands of square kilometers with convex shapes that resemble the defensive shields of ancient warriors. Of the conical volcanic landforms, shield volca-

(a)

(b)

**Figure 9.6** (a) 12 m high "Pipe Organ" basalt columns, Giants Causeway, North coast of Ireland. (b) Most columns are six sided (hexagonal) but four, five and eight sided columns also exist. Note also the radial cooling fractures in each of the columns. Pen for scale. *Source*: Photos by Kevin Hefferan. © John Wiley & Sons.

noes encompass the greatest volume. Shield volcanoes are produced by hot, low viscosity mafic magma that rises and flows great dis-

**Figure 9.7** Glassy rinds and stretch marks on billowed pillow basalts that erupted on the south Pacific seafloor. *Source*: Photo courtesy of NOAA.

tances from the vent. The slopes of shield volcanoes are not steep, generally 2°–10°. Shield volcanoes occur over hotspots that emit large volumes of basaltic lava principally from central vents, fissure rifts, and flank eruptions.

In the Pacific Ocean, shield volcano growth over the past 80 Ma produced the Emperor Seamount and Hawaiian Island hotspot chain. The Big Island of Hawaii is an amalgamation of five shield volcanoes: Kīlauea, Mauna Loa, Mauna Kea, Hualalai, and Kohala (Figure 9.9). Mauna Loa and Mauna Kea are immense shield volcanoes over 4 km in elevation. Mauna Kea is 4.2 km (13 796 ft) above sea level and over 11 km (33 000 ft) above the ocean floor; its relief exceeds that of Mount Everest above the plains of India. While these shield volcanoes are dominated by basaltic lava flows, ash eruptions have occurred as recently as the summer of 2018 (Figure 9.10).

The Hawaiian eruptions that produce shield volcanoes, also generate fiery basaltic **lava fountains** up to 100 m high (Figure 9.11). Airborne blobs of liquid lava emitted by fountains are referred to as **spatter**. Spatter that solidifies at the base of lava fountains is named **welded spatter** or **agglutinate deposits**. **Spatter cones** (Figure 9.12), typically less than 20 m in height, form where welded spatter accumulate around a central vent. Lava fountains that erupt via fissures (Figure 9.1) produce planar ridges called **spatter ramparts** on either side of the fissure (Figure 9.13). These features are well displayed on the shield

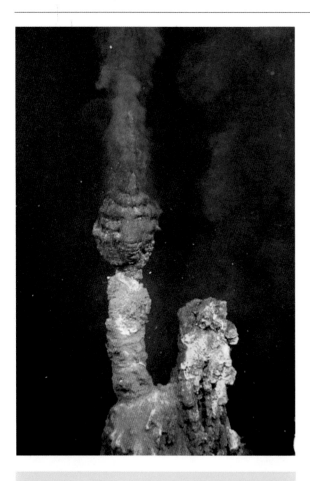

**Figure 9.8** Black smoker from hydrothermal vent fields along the Kermadec Arc *Source*: Image courtesy of New Zealand American Submarine Ring of Fire 2007 Exploration, NOAA Vents Program, the Institute of Geological & Nuclear Sciences and NOAA-OE. https://oceanexplorer.noaa.gov/explorations/07fire/logs/july31/media/brothers_blacksmoker.html.

volcanoes of Hawaii, where they can be produced in a matter of days or weeks. Other volcanic features common, but not limited, to Hawaii are described below.

Gas rich lava propelled high into the air may also cool, solidify, and fall to the ground as vesicular clots of scoria. Airborne mafic lava droplets quenched in flight may form black, glassy, streamlined particles called **Pele's tears** (Figure 9.14), named in honor of the Hawaiian goddess of fire. As small lava droplets are propelled through the air, some are stretched and elongated into golden, acicular strands called Pele's hair (Figure 9.15).

Newly formed deposits display the tears attached to the hair follicles like a lead weight attached to a fishing line. Pele's hair particles have diameters of ~0.5 mm, lengths up to 2 m and can remain airborne over distances of tens of kilometers.

Low viscosity lava can erupt as massive sheet flows that inundate a landscape, or as channelized flows within lava levees or lava tubes. **Lava levees** are elevated lateral banks that contain the lava flow within a stream like channel. **Lava tubes** (Figure 9.16) are shallow subterranean tunnels channeling lava beneath thin, solidified basaltic roofs. In some tubes, skylights (Figure 9.17) develop as a portion of the tunnel roof collapses, allowing the cautious observer a window to peer into the flowing lava tube. When the lava source is depleted, continued flow of lava can empty the tunnel to form an evacuated lava tube. An evacuated lava tube may have lava stalactites hanging from the ceiling, lava drips, and flow lines occurring on the walls and gas vesicles occurring around the tube periphery. Lava tubes range from less than 1 m to tens of meters in diameter and may extend several tens of kilometers in length. Lava tubes several kilometers long occur in the basaltic pahoehoe flows of Hawaii, Newberry Crater, Oregon, Italy, Japan, the Canary Islands, Tenerife Island (Spain), and in New South Wales, Australia.

Lava flows, lava tubes, and the role of gases are all interrelated and on full display in Hawaii. U.S. Geological Survey Hawaii Volcano observatory geophysicist Jim Kauahikaua describes the development of pahoehoe as follows: "When lava first comes out at the vent, it is highly charged with gas, so much so that the lava is more than 85% bubbles. The lava is more like foam, and the pahoehoe that forms when the lava cools is very "shelly"; so-called because walking on it is like walking on large egg shells. The lava then tends to form channels that carry the fluid lava away from the vent, thereby beginning a lava flow. It may take a while for a channel to develop or it may happen almost immediately. In either case, a roof eventually forms over the channel, making a lava tube. A tube insulates the lava inside so that it can stay hot and fluid as it flows away from the vent. The lavas are far less bubbly at this stage than when they first came out at the vent, because a lot of gas has escaped into the air, but they can still be more than 30% bubbles. Sometimes we can look into skylights (openings) in the roof of a tube and

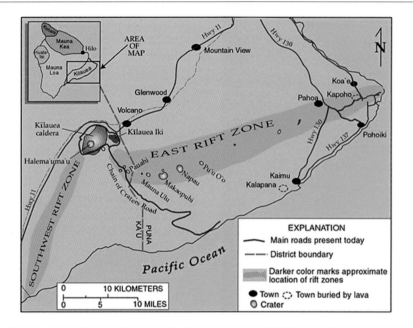

Figure 9.9 Shield volcano map of the Big Island of Hawaii. *Source*: Courtesy of the US Geological Survey. © U.S. Department of the Interior.

Figure 9.10 On 6 June 2018, an explosion within Halema'uma'u crater blasted an ash plume over 3 km high The explosion created a 5.6 magnitude earthquake shaking much of Hawaii. This event punctuated a renewed eruptive phase in Hawaii's long lived hotspot trail. *Source*: US Geological Survey and US National Park Service. © U.S. Department of the Interior.

Figure 9.11 A 8 June 2018 eruption produced 70 m high lava fountains spouting from Hawaii's East Rift Zone Fissure. *Source*: US Geological Survey. © U.S. Department of the Interior.

see large bubbles breaking on the lava surface. Whenever the advancing lava flow encounters large, flat areas, a different kind of tube can develop. The lava slows down, rapidly forms a crust, and thickens by continued injection of lava into the cooled shell, thereby forming a tumulus.

The lava flows also spread out creating broad areas covered by large **tumuli**, which are hill- or plateau-shaped pahoehoe features that crack and bulge upward as the flow thickens. These can continue to grow as long as lava is being injected into them. These tumuli have grown over 10 m thick after starting as a 30 cm thick advancing lava flow. Probably the largest known tumulus in Hawaii is

**Figure 9.12** Approximately 5 m tall spatter cone constructed by a lava fountain eruption on the Pu'u 'O'o spatter and cinder cone complex, Hawaii. *Source*: Photo by T.N. Mattox on 3 March 1992 caldera. Photo courtesy of US Geological Survey. © U.S. Department of the Interior.

**Figure 9.14** Pele's tears collected downwind from Kīlauea Volcano, Hawaii. Note the U.S. dime for scale in lower right caldera. *Source*: US Geological Survey. © U.S. Department of the Interior.

**Figure 9.13** Approximately 5 m high, 100 m long spatter rampart in Kīlauea from a 1980s fissure eruption. *Source*: Photo by Kevin Hefferan. © John Wiley & Sons.

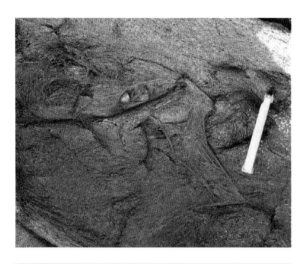

**Figure 9.15** Pele's blonde hair created by windblown streamlining of silica glass. Metal scale is 10 cm in length. *Source*: Photo courtesy of Neil Heywood.

the so-called Uwekahuna laccolith exposed in the northwest wall of Kīlauea Crater. A laccolith is formed when magma rising toward the Earth's surface stalls and solidifies at a shallow depth rather than erupting. However, our current knowledge suggests that this feature is a tumulus that once stood over 20 m (60 ft) high and was subsequently buried by younger lava flows. Because the lava actually lifts the cooling and thickening upper crust of tumuli, it is under pressure, and some of the gases become re-dissolved, lowering the volume of bubbles in the lava. We see

this in a very characteristic glassy, steel-bluish variety of pahoehoe that only comes out of the bases of tumuli several weeks after they start to form." (Courtesy of Jim Kauahikaua and the U.S. Geological Survey https://www.usgs.gov/center-news/volcano-watch-how-does-pahoehoe-flow).

**Pahoehoe** lava consists of low viscosity, basaltic lava which produces well defined flows with a billowing, rippled or ropey surface (Figure 9.18). Rapid cooling produces a glassy rind which becomes convoluted due

**Figure 9.16** Thurston Lave Tube in Hawaii formed during fissure eruptions in the 1400s. *Source*: Photo by Kevin Hefferan. © John Wiley & Sons.

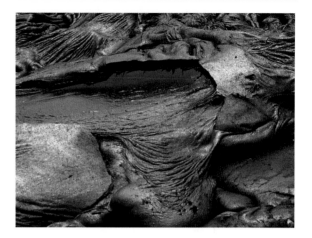

**Figure 9.18** Ropey, billowy pahoehoe flow and a skylight in Hawaii. *Source*: US Geological Survey. © U.S. Department of the Interior.

**Figure 9.17** Skylights into lava tube system in Hawaii caldera. *Source*: US Geological Survey. © U.S. Department of the Interior.

to the continued lava flow and expansion of gas beneath the thin solid glassy rind. Entrapment of gas bubbles beneath the solid, glassy rind typically produces a vesicular zone just beneath it. Pahoehoe lava is well developed on shield volcanoes but can occur anywhere low viscosity lava flows down moderate slopes. Gas expansion results in dramatic inflation and thickening of pahoehoe sheets upon cooling. Modern pahoehoe flows (<1 km³) in Hawaii and Iceland (10–20 km³) inflate and thicken by at least an order of magnitude through internal gas inflation. Pahoehoe inflation and thickening has also been proposed for ancient flood basalt prov-

inces, based upon vertical distribution of vesicles and other pahoehoe features observed in the Columbia River flood basalts (Self et al. 1996).

With continued flow pahoehoe lava can evolve into blocky **aa** lava flows with angular, jagged, fractured surfaces (Figure 9.19). The angular blocks that cover the surface of aa lava flows are described as spiny, rubbly or clinkery. Typical aa flows are 2–8 m thick. When aa lavas flow into a subaqueous environment, water seeps into fractures. The injection of cold water into hot fractures at shallow depths produces steam, yielding an explosive reaction and the ejection of airborne pyroclastic fragments. Explosive volcanic eruptions are limited to shallow depths. At greater depths, higher pressures impede volatile expansion limiting explosive eruptions (Francis 1993).

Pahoehoe and aa lavas may be generated from the same volcanic source. Ropey pahoehoe lavas can evolve into blocky aa lavas by cooling, loss of volatiles from solution or gas bubble escape, or change in topographic slope. However, whereas pahoehoe flows are restricted to basaltic lava, higher viscosity aa lavas may be basaltic to andesitic in composition. High viscosity results from higher molecular linkage due to higher $SiO_2$ content, lower dissolved gas content, lower temperatures or higher shear stresses acting on the flow surface (shear stresses increase with topographic slope angle). Spectacular transitional pahoehoe to aa basaltic flows occur in Hawaii

**Figure 9.19** An aa flow erupted on Kīlauea Volcano's East Rift Zone on 1 June 2018. Note that the interior of a lava flow is incandescently hot while surface cooling creates a dark rubbly solid surface. It will require ~130 days for this 4.5 m thick flow to cool to completely solidify at a temperature of ~ 200 °C. *Source*: Photo by A. Lerner and courtesy of the U.S. Geological Survey. © U.S. Department of the Interior.

where lava flows over steep crater walls and experience increased shear stress. At the base of the crater, aa lavas revert to pahoehoe as they flow along the relatively flat crater floor. At Pisgah Crater in California, pahoehoe flows change to aa because of cooling, dissolved gas loss, and increased slope gradients. Several factors (temperature, viscosity, volatile content, topographic slope) play critical roles in determining lava flow characteristics.

While not everyone can visit an active lava flow, laboratory experiments can yield important insights. Since 2010, Jeff Karson and Robert Wysocki have created the Lava Project at Syracuse University, melting 1.2 billion year old Keweenaw basalt from Wisconsin in a furnace at temperatures exceeding 1100 °C, creating the first lava flows in New York State in 200 million years. The controlled lava flow experiments utilize different flow rates, temperatures, slopes, and interaction with other materials such as snow and sediment to mimic conditions that occur in Hawaii, New Zealand or Iceland (https://www.syracuse.edu/stories/lava-makers).

Hawaii's 2018 volcanic eruption highlighted the importance of gas bubbles, as gas bubbles constituted more than 50% by volume in some rock cores collected after the flows solidified. Atsuko Namiki and Janine Birnbaum noticed the effect of gas bubbles on Hawaiian lava flows and sought to model lava behavior using corn syrup, citric acid, and baking soda. They created three types of flow: no bubble flows, bubbly corn syrup flows, and bubbly corn syrup flows with suspended particles. Cameras and laser sensors recorded the corn syrup flows down a meter long plastic plank and observed different flow characteristics. Corn syrup with no bubbles recorded the highest velocities while the bubbly corn syrup with suspended particles recorded the lowest velocities. Bubbly corn syrup also experienced gravitational separation with bubbles rising to the top and creating pahoehoe flows while the corn syrup liquid in the center continued flowing producing the characteristic ropey texture (Gasparini 2020). You can observe a video of their corn syrup flow at https://eos.org/articles/corn-syrup-reveals-how-bubbles-affect-lavas-flow&utm_campaign=ealert.

Compare the corn syrup flows with drone footage of the June 2018 Hawaii volcano flows at https://www.youtube.com/watch?v=F11GTK8FZyI.

Laboratory experiments provide useful tools for students and researchers to understand lava flow behavior; they also can recreate insightful predictive scenarios to help protect lives and preserved anthropogenic features caught in the path of oncoming lava flows.

### 9.2.4 Pyroclastic cone volcanoes

Whereas shield volcanoes encompass hundreds of square kilometers and attain heights of over 4 km above sea level, pyroclastic cones are modest in scale and can develop within a period of years. Pyroclastic cones are relatively small, usually encompassing areas of less than 20 km² and with heights typically less than 500 m. **Pyroclastic cones** are steep sided (~30°–35°) conical features composed of tephra. **tephra** consist of volcanic rock fragments of various sizes and compositions emitted during explosive eruptions. Common ash to bomb sized fragments include basalt, scoria, and andesite, with lesser amounts of other fragments. Pyroclastic cones develop from tephra emitted from vents usually in central, bowl shaped craters. Pyroclastic cones

## Box 9.1   Hawaii 2018 eruption

Since 1983, fissure eruptions along Hawaii's East Rift Zone (ERZ) have produced brilliant lava fountains, incandescent pahoehoe, and aa fissure flows and the creation of the Pu'u 'Ō'ō cinder cone. Beginning in 2008 within Kīlauea caldera, the Halema'uma'u crater began filling with lava which reached the caldera floor at the summit. On 30 April 2018, magma flow generated a series of powerful earthquakes that opened new conduits and resulted in draining lava from Halema'uma'u crater, collapse of Pu'u 'Ō'ō cone, and redirection of magma flow eastward toward Kapoho (Figure B9.1). On 4 May 2018, days after the collapse of Pu'u 'Ō'ō, the lava lake level in Halema'uma'u diminished and Kīlauea summit caldera subsided at a high rate (Figure B9.2).

The lava lake surface disappeared from view on 10 May 2018, dropping to a depth more than 325 m below the Halema'uma'u crater floor. Magma withdrawal from Halema'uma'u crater lowered

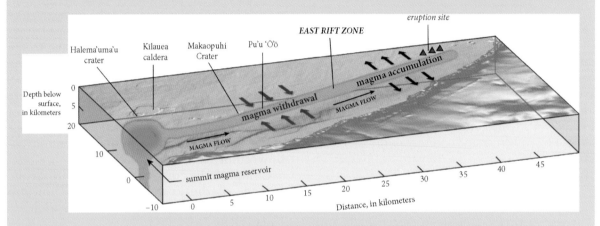

**Figure B9.1**   USGS block diagram illustrating magma flow from the summit caldera to the fissures of the East Rift Zone (ERZ). Blue arrows = contraction across the upper and middle rift zone, black arrows = expansion in ERZ. *Source*: Courtesy of US Geological Survey. © U.S. Department of the Interior.

**Figure B9.2**   Kīlauea summit lava lake has been active from April 2018 to 2021. View of west rim of Halemaʻumaʻu crater on 7 April 2021 as lava continues to erupt from the west vent, where a diffuse gas plume is visible in the lower left. The active west part of the lava lake (lower center) is a lighter gray color, compared to the darker appearance of the solidified surface crust to the east. *Source*: Courtesy US Geological Survey. https://www.nps.gov/havo/planyourvisit/upload/PrelimSum_LERZ-Summit_2018_508.pdf. © U.S. Department of the Interior.

## Box 9.1   *Continued*

the magma level below the water table. As a result of interaction of magma and water, explosive ash and steam eruptions occurred at Kīlauea's summit for the first time since 1924. Magma withdrawal and explosions produced earthquakes, faulting, subsidence, rock falls, and enlargement of the caldera complex (Figures B9.3 and B9.4). The shallow magma reservoir beneath the east margin of Halema'uma'u continued to drain and redirect flow of lava eastward along the ERZ. Throughout the spring of 2018, over 20 ERZ fissure eruptions destroyed homes from Leilani Estates eastward to Kapoho and threatened the Puna Geothermal Venture (PGV) Energy facility (Figure B9.5). According to the U.S. Geological Survey, the 2018 eruption resulted in 35 km² of land being inundated by lava 10–25 m thick, 875 acres of

**Figure B9.4**   Halema'uma'u crater experienced loss of its lava lake, summit deflation and aerial expansion from April to August 2018. The parking lot for the National Park's Halema'uma'u Overlook (closed since 2008 due to volcanic hazards) is visible to the left of the crater.

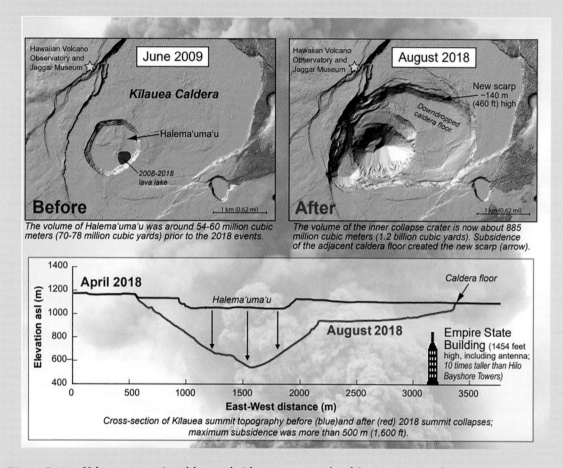

**Figure B9.3**   Kīlauea summit caldera subsidence as a result of Spring 2018 volcanism. *Source*: Courtesy of Mark Wasser, David Benitez of the US National Park Service. © U.S. Department of the Interior.

*Continued*

**Box 9.1** *Continued*

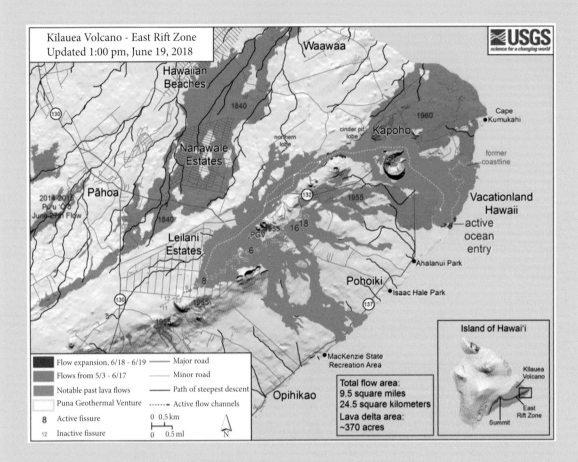

**Figure B9.5** 9 August 2018 map denoting volcanic activity along the East Rift Zone. *Source*: US Geological Survey. © U.S. Department of the Interior.

new land were created by ocean lava flow entries, 716 dwellings were destroyed by lava, ~1 billion m³ of lava erupted, and 60,000 earthquakes occurred between 30 April–4 August 2018 (4,400 were M3 and higher; largest: M6.9 on 4 May). Hawaii's 2018 also affected surface water bodies. Green Lake, Hawaii's largest and deepest (60 m) freshwater lake and a popular swimming spot, completely evaporated when lava flowed into the lake basin. Similarly, Kapoho Bay and its famous tide pools were also filled by lava flows. The interaction of lava and surface water produces volcanic fog, called **vog**, and also generates laze. **Laze** is a poisonous volcanic haze containing hydrochloric acid and other toxic volcanic gases that can be lethal. Hawaii is certainly alive; luckily, so are its inhabitants. Although some injuries were noted, no human loss of life occurred which, as we will discuss later in this chapter, is typical of Hawaiian style volcanic eruptions. U.S. Geological Survey volcanologists utilized global positioning satellites (GPS), tiltmeters, seismometers, drones, co-spectrometers, geochemical analyses, satellite radar (InSAR) and gas and thermal sensor data to monitor magma withdrawal from Kīlauea summit and flow to the ERZ. The magmatic plumbing system of this dynamic hotspot continues to amaze volcanologists as scientists track magma flow and also document the reactivation of an older, cooler (1070 °C), more evolved magma source beneath the East Rift Zone. The 2018 volcanic episode subsided by August 2018. However, Kīlauea crater began a new phase of lava lake eruptive activity in December 2020. A 28 December 2020 US Geological Survey video of Kīlauea crater can be viewed at https://www.youtube.com/watch?v=fiK576HNXeM.

Kīlauea continues to be the most active volcano on Earth.

include: (1) **Scoria cones** composed predominantly of vesicular basaltic material; (2) **Cinder cones** consisting of ash, lapilli, and bomb-sized particles of various compositions that accumulate as circular to oval shaped conical volcanoes.

Pyroclastic cones occur in a wide variety of settings including continental rifts, convergent plate boundaries, divergent plate boundaries, and over hotspots. They can occur as discrete volcanic features on basaltic lava plains or as parasitic features on more massive shield or composite volcanoes. Mauna Kea shield volcano in Hawaii contains over 100 small satellite pyroclastic cones (Figure 9.20).

The Late Miocene to Holocene San Francisco volcanic field in Arizona, which encompasses an area of approximately 500 km², provides an excellent example of a volcanic field. San Francisco volcanic field includes approximately 600 cinder and scoria cones, the most famous of which is Sunset Crater. Sunset Crater is among the youngest volcanic feature in the conterminous United States, having erupted ~1000 years ago (Smiley 1958). San Francisco Peaks (Figure 9.21), the largest feature in this volcanic field, is a composite volcano that bears a striking resemblance to post-1980 Mt. St. Helens in having an amphitheater shaped crater along its northeastern flank. Observing the

**Figure 9.21** Cinder cones of the San Francisco volcanic field in the foreground with the San Francisco Peaks composite volcano in the background. *Source*: Photo by Kevin Hefferan. © John Wiley & Sons.

1980 Mt. St. Helens eruption, volcanologists recognized that the San Francisco Peaks amphitheater likely formed by a similar lateral eruption blast that had hitherto not been recognized.

### 9.2.5   Composite volcanoes

**Composite volcanoes** are majestic cone shaped mountains encompassing tens to hundreds of square kilometers in area with slopes ranging from 10° to 30°. Composite volcanoes consist of alternating layers of pyroclastic debris and lava flows that commonly develop on the overlying plate at convergent plate boundaries. These volcanoes are a composite of many different eruptive phases and rock types, that generate stratified (hence the alternative name "**stratovolcanoes**") layers. Although composite volcanoes such as Mt. Rainier are large and majestic, many exceeding 4 km in elevation, their volume pales in comparison to immense shield volcanoes (Figure 9.22).

Composite volcanoes contain a wide variety of igneous rocks whose composition ranges from basalt to rhyolite, but are generally dominated by andesite. Andesitic lava typically has higher viscosity and greater yield strength than basalt. As a result, andesitic lava flows are thicker, slower, and more likely to build up around the vent source than more fluid basaltic flows in shield volcanoes. Andesitic lava's high viscosity generates block lava flows, which are similar to, but coarser, than aa

**Figure 9.20** On the Big Island of Hawaii: Mauna Loa (background), Mauna Kea (foreground) and cinder cone to the right. Note the dramatic difference in scale between goosebump like cinder cones and massive shield volcanoes. *Source*: Photo by Kevin Hefferan. © John Wiley & Sons.

## Box 9.2   (GD): Parícutin

You probably live in an area entirely devoid of volcanic activity. Could you imagine steam rising from a newly formed crack in your yard, which over the period of months and years develops into a volcano? This is exactly what occurred in Mexico, shocking local farmers and geologists throughout the world. The Parícutin cinder cone volcano unexpectedly formed in a cultivated cornfield in February of 1943. Within a few years a level farmland tract catapulted into a volcanic cone rising over 400 m above the surrounding terrain and covering approximately 15 km². The fiery lava and ash eruption (Figure B9.6) continued to erupt until 1952. Parícutin is a monogenetic volcano; that is, lava and ash erupted from a singular eruptive event over a limited time frame (1943–1952). The nine-year eruption of Parícutin volcano documented an active cinder cone eruption and provided insight into the development of cinder cone volcanoes throughout the world. Having observed a modern day cinder cone develop, geologists use Parícutin as a model in the analysis of older cones.

**Figure B9.6**  The volcano of Parícutin soon after its birth in 1943. *Source*: Photo by K. Segerstrom, U.S. Geological Survey, image in public domain. (K. Segerstrom photo courtesy of US Geological Survey). © U.S. Department of the Interior.

lava. **Block lava** consists of smooth sided blocks up to several meters in diameter that tumble downslope. Block lavas lithify as volcanic breccia deposits (Francis 1993). Block lavas occur in a wide variety of settings including convergent margins and hotspots (Figure 9.23).

Because they erupt a wide variety of lava compositions, composite volcanoes consist of a variety of aphanitic, porphyritic-aphanitic,

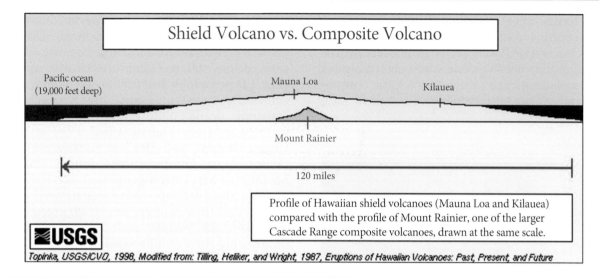

**Figure 9.22** Relative scale of composite volcanoes versus shield volcanoes. *Source*: Photo courtesy of US Geological Survey. © U.S. Department of the Interior.

pyroclastic, and glassy rocks. The lava flows are dominated by andesitic rocks, but basaltic, dacitic, and even rhyolitic rocks also occur. The pyroclastic rocks are dominated by andesitic, dacitic, and rhyolitic tuffs and lapilli tuffs, with their higher viscosity and dissolved gas contents. Dacite and rhyolite magmas are even more viscous than andesitic magma, resulting in even slower rates of movement. The slow rate of magma movement can produce spines and domes in which viscous magma solidifies within the composite volcano crater, essentially acting as a plug inhibiting the release of lava.

**Spines** develop when a solidified magma plug is pushed up through the conduit, forming a vertical column on Earth's surface. **Plugs** consist of magma that solidifies within the "throat" or conduit of the volcano. The term "plug" is entirely appropriate because it performs the same role as a stopper in a bathtub, preventing the release of fluid. While plugs may be referred to as volcanic plugs, the rocks of which they are composed actually formed as shallow (hypabyssal) plutonic features. When the hot, plug is extruded onto the surface it becomes a spine. Following the 1902 eruption of Mt. Pelée in the Caribbean, a 45 m high vertical spine was exposed. The Mt. Pelée spine was not a long lasting feature, crumbling within a few years. From 2004–2006, Mt. St. Helens developed a short lived spine within its resurgent dome (Figure 9.24).

**Figure 9.23** Block lava composed of trachyandesite (an extrusive rock, intermediate in composition between trachyte and andesite) at Craters of the Moon National Monument, Idaho. *Source*: US Geological Survey. © U.S. Department of the Interior.

**Domes** are steep sided, blister like forms that occur within volcanic craters (Figure 9.25). Domes are inflated areas generated by the accumulation of viscous, often glassy, blocky lava of dacitic, rhyolitic or trachytic composition, on the surface or in the shallow sub-

**Figure 9.24**  Fin shaped spine at Mount St. Helens, February 2005. *Source*: Photo courtesy of Steve Schilling and the US Geological Survey. © U.S. Department of the Interior.

**Figure 9.25**  Mount St. Helens' lava dome as viewed on 22 August 1981 from 800 m away within the crater. On 22 August 1981, the dome measured 163 m high and 400 m wide, approximately the height of a 44 story building and wider than four football fields. By 1986, the dome had risen to a height of 230 m (66 story building) and a width of 800 m which is equivalent to nine football fields. *Source*: Lyn Topinka photo, courtesy of the US Geological Survey. © U.S. Department of the Interior.

surface. Domes also contain large, viscous bombs intermixed with tephra. Dome growth requires monitoring because volcanic dome collapse generates deadly pyroclastic flows which, but for early evacuations, can result in massive loss of life. Recent catastrophic dome collapse eruptions occurred in 1980 at Mt. St. Helens (USA), 1991 at Mt. Unzen (Japan), and 1997 at Soufriere Hills (Montserrat). The June 1991 dome collapse on Japan's Mt. Unzen generated a pyroclastic flow (nuée ardente) that raged downslope. When the flow unexpectedly changed its direction, 38 people including geologists monitoring the event perished. In June 2018, Guatemala's Fuego volcano erupted killing more than 100 people in pyroclastic flows. Pyroclastic flows and lahars represent the two greatest recurring hazards from composite volcanoes.

Composite volcanoes are widespread in the Pacific "ring of fire" and the eastern Indian Ocean, where they occur at:

1   Continent–ocean convergent margins, where **continental margin volcanic arcs** develop on the overlying continental lithosphere. Partial melting of a thick overlying wedge containing continental lithosphere and mantle produces the calc alkaline series that include basaltic, andesitic, dacitic, and rhyolitic rocks. Continental margin arcs occur along the west coast of South America, Central America, the Pacific Northwest, and Alaska in North America.
2   Ocean–ocean convergent margins, where **island volcanic arcs** develop above the overlying ocean plate. Partial melting of a thin overlying wedge containing ocean lithosphere and mantle produces calc alkaline basalts and andesites, which are the predominant volcanic rocks that develop in these settings. However, tholeiitic basalts, boninites, and other rocks also occur in young island arc systems. Island arc volcanoes occur along the Aleutian Range of the Northern Pacific, throughout the western Pacific, the eastern Indian Ocean region and in the Caribbean Sea.

Although intimately associated with convergent plate boundaries, explosive volcanoes also occur in the ocean ridge setting of Iceland, which has an unusually thick accumulation of oceanic crustal rock, created by long lived hotspot activity at a divergent plate margin. The March–May 2010 of Iceland's Eyjafjallajökull volcano produced a massive ash cloud that disrupted European and transatlantic air travel for over a month (Tarasewicz et al. 2012).

Stratovolcanoes eject enormous amounts of pyroclastic debris into the atmosphere and onto Earth's surface. By far the most voluminous pyroclastic rocks produced in continental settings are those generated by highly explosive volcanic eruptions of silicic composition. Why are eruptions of silicic magma so explosive? Silicic magmas can dissolve large amounts of water vapor deep below Earth's surface under high confining pressures. As silicic magma rises toward the surface, the water vapor begins to separate (exsolve) as a separate phase from the magma as it crosses the level of exsolution. Exsolution of gas generates expanding gas bubbles and lowers the $P_{H2O}$ of the magma. Decreased water vapor allows silica tetrahedra to rapidly link, dramatically increasing magma viscosity. This increase in magma viscosity severely inhibits the formation of crystals and creates a mass of hot, highly vesicular, low density, semi-solid glassy material. As this viscous, glassy mass continues to ascend toward the surface, vesicles expand, and increase in number to produce zones of frothy, glassy magma that are essentially hot pumice. Each expanding gas bubble is surrounded by glassy, semi-solid bubble walls which thin as the bubbles expand in much the same way that the walls of a balloon or a piece of bubble gum thin as they expand. If the confining pressure on this material is decreased slowly as the magma rises or suddenly, as when a vent opens to the surface, the material will cross the **level** (or **surface**) of **fragmentation**. Fragmentation occurs where the outward pressure of the expanding gas bubbles ruptures the thinned bubble walls generating glassy pumice fragments called **shards**. Once the gas bubble wall material has been shattered, a mixture of shards and hot gases is created. This mixture, characterized by extremely low density and viscosity, accelerates rapidly upward toward the surface through any available opening. Gas ascension triggers a chain reaction in which large volumes of frothy pumice fragments accelerate upward and are hurtled out of a vent at great velocities to great heights producing cataclysmic explosive eruptions.

Explosive volcanic eruptions produce a **vertical plume** or **eruption column** that can be (Figure 9.26) divided into three distinct parts (Sparks and Wilson 1976; Sparks 1986):

1 A lower **gas thrust region** that consists of material thrust from the vent by expanding gases at velocities ranging from 100 to 600 m/s.
2 An upper **convective thrust region** produced by the convective rise of heated atmospheric gases and fragments.
3 An **umbrella region**, similar to the mushroom shaped heat cloud produced by thermonuclear bomb explosions column, begins to spread laterally as the result of temperature inversions in the atmosphere.

The vertical plume consists largely of expanding gases and glassy (vitric constituents) pumice fragments and shards created when the magma crosses the fragmentation surface. The vertical plume may also contain subordinate rock fragments (lithic constituents) ripped from vent walls and fragments of any crystals (crystal constituents) that formed in the magma below the level of exsolution. The buoyancy of the plume is largely dependent upon the degree of mixing and heating that occurs with the surrounding atmosphere (Sparks 1986).

Two modes of eruptive column behavior, convective thrust, and gravitational collapse, result in two kinds of deposits: pyroclastic fall (air fall) deposits and pyroclastic flow deposits.

1 Convective thrust occurs when the eruptive plume exhibits buoyancy. Buoyancy is due to low particle concentration within the hot plume as well as heating and mixing with surrounding atmospheric air. Convective thrust produces a vertical plume that equilibrates forming a mushroom shaped region from which pyroclastic particles ultimately descend as air fall or pyroclastic fall deposits.
2 Gravitational collapse of the plume occurs due to negative buoyancy resulting from high particle concentrations within the dispersion and/or insufficient heating and mixing with the surrounding atmospheric

Umbrella Region

Convective Thrust Region

Gas Thrust Region

**Figure 9.26** 22 June 1980 explosive eruption of Mt. St. Helens sent pumice and ash 18 km into the air. *Source*: Photo courtesy of Mike Doukas, with permission of the US Geological Survey. https://commons.wikimedia.org/wiki/File:MSH80_st_helens_eruption_plume_07-22-80.jpg. © U.S. Department of the Interior.

air. Gravitational collapse of a plume due to higher density than the surrounding air is one means by which pyroclastic flows are initiated (Sparks and Wilson 1976).

Let's consider air fall deposits and pyroclastic flow deposits in greater detail.

**Pyroclastic fall (air fall) deposits** are produced by airborne pyroclasts propelled upward in an eruption column (Figure 9.26). Pyroclasts in the eruptive column (tephra) may stay aloft for considerable periods of time and be dispersed over great distances. The relatively small 1980 eruption of Mt. St. Helens resulted in thin ash fall tuff deposits 2,000 km downwind from the volcano. The ejected ash eventually spreads outward and downwind and begins to settle downward to accumulate as sorted, stratified air fall deposits on Earth's surface.

The densest air fall particles, blocks, and bombs, are deposited first, followed by increasingly finer grained particles (Wright et al. 1980). Because the largest pyroclasts (bombs and pyroclastss) tend to settle first and closest to the vent and the smaller particles tend to settle more slowly and farthest from the vent, pyroclast size in air fall deposits generally decreases upward within the deposit and with distance from the source. Some proximal solidified air fall deposits grade upward from coarse breccias near the base through lapilli tuffs into fine tuffs near the top. Distal ash sized particles are deposited, compacted, and cemented together as **ash fall tuffs**. Ash fall tuffs (Figure 9.27) are generally lightweight, porous, poorly cemented, and rather soft rocks (Sparks and Wilson 1976). The internal "blanket" layering, sorting, and fining upward graded bedding of air fall deposits allows them to be distinguished from pyroclastic flow deposits. Fining upward means that the coarsest grains are deposited at the base and finer sized grains are deposited toward the top of a layer.

Many pyroclastic rocks form from **pyroclastic flows**, consisting of turbulent mixtures of hot rock fragments and gases (Figure 9.28).

**Figure 9.27** Loosely welded tuff from, of all places, Kīlauea Crater, Hawaii. Although Hawaiian volcanoes generally erupt liquid basaltic lava, explosive ash eruptions do occur, as in 1924, 2007 and 2018. *Source*: Photo by Kevin Hefferan. © John Wiley & Sons.

Unlike air fall deposits, **pyroclastic flow deposits** generally lack stratification and are poorly sorted because chaotic mixtures of clast sizes are deposited rapidly. Some pyroclastic flows display inverse grading where finer grained particles are overlain by coarser grained deposits. Pyroclastic flows are commonly lobate in map (plan) view and lens like in cross section.

Three major types of pyroclastic flows are recognized:

1  **Pyroclastic surges** are very low density, extremely hot gaseous flows containing ash to lapilli size particles that travel in excess of 100 km/h. Low densities allow pyroclastic surges to defy gravity and climb upwards from valleys enveloping higher slopes and ridges. In volcanically active areas, warning signs advise people to climb to higher elevations. In the case of pyroclastic surges, such upward mobility may be in vain. In 1995, pyroclastic surges inundated the Caribbean island of Montserrat turning the verdant, green island into an eerie monochromatic, gray landscape resembling a lunar surface (Figure 9.29). Surges are generated by plume collapse or laterally-directed blasts.

**Figure 9.28** Pyroclastic ash cloud and pyroclastic flow generated by dome collapse; Mayon Volcano, Philippines, 23 September 1984. Ash cloud column extended 15 km above sea level. *Source*: C.G. Newhall photo, courtesy of US Geological Survey. © U.S. Department of the Interior.

**Figure 9.29**   Pyroclastic surges buried Plymouth on the Island of Montserrat following Soufriere Hills dome collapse on 25 June 1997. *Source*: Photo by Kevin Hefferan. © John Wiley & Sons.

**Figure 9.30**   Montserrat Lahar deposits from 1995, with building size blocks entrained in a muddy matrix. *Source*: Photo by Kevin Hefferan.

2   Pumice flows are low to moderate density, hot vessiculated flows. Siliceous pumice flows produce light colored, vesicular **ignimbrites**. Andesitic to basaltic flows produce vesicular **scoria flows**. Massive explosive eruptions, such as the 1912 Katmai eruption in Alaska, produce immense ignimbrites. Thick ignimbrites contain features such as columnar joints which indicate relatively slow cooling. Ignimbrites are generated by the collapse of vertical ash plumes.

3   **Nuées ardentes** (French for "fiery clouds"), are fluidized mixtures of hot, incandescent rock fragments, and gases that flow along the surface as a glowing cloud of billowing pyroclastic debris. Searing gases generated by bubble fragmentation and the incorporation of atmospheric gases provide buoyant support for the mixture to behave as a fluid. Pyroclastic flows have temperatures that can approach 1000°C, move at top speeds exceeding 150 km/h, over distances of as much as 100 km, and can form pyroclastic deposits up to 1 km thick. Pyroclastic flows can kill tens of thousands of people within minutes. High density, vesicle poor **block and ash** pyroclastic flow deposits are generated by nuée ardentes (Wright et al. 1980; Francis 1993). Nuées ardentes are produced by the collapse of lava domes.

**Lahars** are volcanic debris flows up to tens of meters thick with the consistency of wet cement. Lahars are dense slurries of ash to bomb sized blocks that are easily remobilized in the presence of water. Lahars are particularly dangerous in steep terrain, such as the slopes of composite volcanoes, and can travel tens of kilometers with velocities approaching 100 km/h. Lahars can bury thousands of people in a matter of minutes. Lahars form at three different time stages:

1   Synchronous with volcanism (syn-eruption lahars) and accompanying active pyroclastic flow (Figure 9.30).
2   Soon after volcanism has ceased (early post-eruption lahars) where recent pyroclastic deposits are remobilized. Factors that induce remobilization include ground shaking, slope instability, heavy precipitation and/or melting of ice and snow;
3   On the slopes of dormant or inactive volcanoes (late post-eruption lahars) where old pyroclastic deposits are remobilized.

Syn-eruption lahars are commonly generated by combining pyroclastic flow and/or airfall deposits with glacial meltwater or precipitation. The 1991 Mt. Pinatubo (Philippines) eruption involved both syn-eruption and post-eruption lahars. Mt. Pinatubo's eruption occurred during a monsoon, so the precipitation mixed with ash fall/flow depos-

## Box 9.3   A tale of two cities: past and future?

Major metropolitan cities, towns, and villages throughout the world thrive in the shadow of composite volcanoes. Composite volcanoes pose substantial hazards, especially from pyroclastic flows and lahars. Let us consider two regions built in similar geologic environments. The Nevado del Ruiz eruption in Colombia was a particularly gruesome disaster, with a massive loss of life which was preventable. Nevado del Ruiz is a 5389 m high, snow-capped composite volcano in the Andes volcanic chain. The village of Armero, 75 km downstream from the volcano, was built directly on top of an 1845 lahar that killed over 1000 people. However, fertile soils in the valleys surrounding Nevado del Ruiz enticed people to rebuild on these deposits.

Throughout 1985, minor earthquakes and steam explosions rocked Nevado del Ruiz, prompting a scientific study of the largest volcano in Colombia. The National Bureau of Geology and Mines produced a report and hazard assessment map on 7 October 1985 that indicated lahars could severely impact Armero and surrounding regions downslope from volcano. Fearing public unrest and costs, government officials discounted the report, ignoring its findings as well as the hazard assessment map (Figure B9.7). On 13 November 1985 an eruption occurred which melted the glacial ice atop Nevado del Ruiz. A combination of glacial meltwater and pyroclastic debris released by the eruption combined to produce multiple lahars that flowed down tributary valleys and joined to form several massive lahars. Within minutes, one of the lahars raged down a valley burying 23 000 people under approximately 10 m of debris (Figure B9.8).

Several thousand kilometers to the north, the Cascade composite volcano chain occurs on the west coast of North America. The Cascade volcanoes were produced by the same subduction processes as

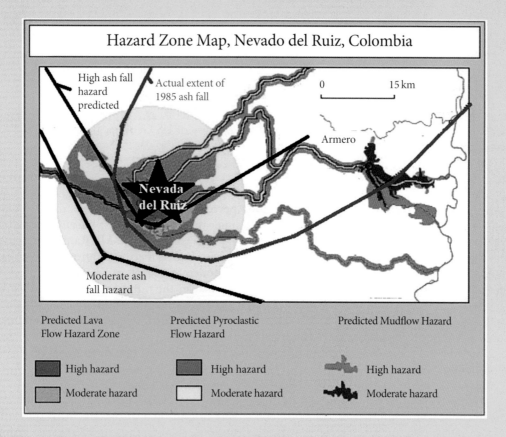

**Figure B9.7**   Hazard assessment map of Colombia's Nevado (Navada) del Ruiz volcano. *Source*: Courtesy of the US Geological Survey. © U.S. Department of the Interior.

*Continued*

**Box 9.3** *Continued*

**Figure B9.8** Armero buried by a lahar. *Source*: Map courtesy of US Geological Survey. © U.S. Department of the Interior.

Nevado del Ruiz and contain similar characteristics: steep, majestic, glacier covered volcanoes capable of producing immense pyroclastic flows and lahars. Approximately 5600 years ago, the Osceola volcanic mudflow (lahar) flowed northwest from Mt. Rainier to the Pacific coast burying everything in its path. Land use has changed substantially since the Osceola roared down Mt. Rainier's slopes. The Osceola mudflow now underlies the rapidly growing Tacoma-Seattle metropolitan area. Note that the recurrence interval for lahars in this region suggests that moderate size lahars are expected an overage of once every 100–500 years (Figure B9.9). One should consider the ramifications of urban development in the shadows of Mt. Rainier and on a geologically young volcanic mudflow. A lahar warning system has been set up for Mt. Rainier. If a lahar begins to flow down valley on the mountain, pressure sensors will be activated that set off an alarm system at sites further down valley, so that people will be warned to move to higher ground. If warning procedures are effective, massive loss of life should not occur.

Volcanic eruptions in marine settings may induce awesome tsunamis that overwhelm populations scattered thousands of kilometers from the eruptive center. The Banda Aceh tsunami of December 2004, which killed over 200 000 people, was caused by a displacement of the seafloor related to a major earthquake. However, volcanic eruptions, such as the 1883 Krakatoa eruption, produce displacements of the sea floor capable of generating massive tsunamis. Krakatoa is located between the islands of Java and Sumatra in Indonesia. Prior to the 1883 eruption, the island of Krakatoa consisted of three contiguous stratovolcanoes (Perboewatan, Danan, and Rakata) situated in an ancient caldera. Beginning in May 1883, ships in the eastern Indian Ocean noted ash

**Box 9.3**   *Continued*

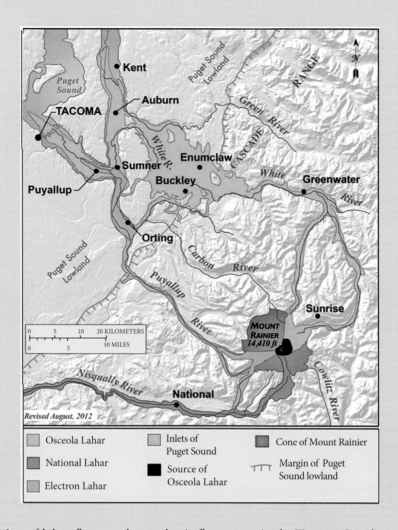

**Figure B9.9**   Map of lahar flows and pyroclastic flow zones in the Tacoma-Seattle region. *Source*: Map courtesy of US Geological Survey.

plumes ejected to heights greater than 10 km. On August 26, a series of cataclysmic eruptions occurred that produced violent pyroclastic plumes up to 25 km high. These eruptions ultimately generated a caldera collapse event that generated a series of catastrophic tsunamis that resulted in the deaths of over 30 000 people. An additional 5000 people were killed by the hot pyroclastic flows and airborne debris. On 22 December 2018 Anak Krakatau (Son of Krakatoa) erupted and the southwest portion of the volcano collapsed into the ocean producing a submarine disturbance that triggered a tsunami killing over 430 people in Java. Anak Krakatau, Krakatoa, and Banda Aceh are all located along Indonesia's Java-Sumatra subduction zone: history continues to repeat itself in a devastating web of subduction, earthquakes, volcanism, and mass wasting.

its and other accumulated pyroclastic debris producing deadly syn-eruption lahars. Sadly, the lahars did not cease with the eruption activity. The immense pyroclastic deposits infilled stream channels and were easily remo-

bilized with subsequent precipitation events as cool, post-eruption lahars. Fortunately, because two U.S. military bases were near the eruption, USGS volcanologists actively monitored Mt. Pinatubo with the Philippine

Institute of Volcanology and Seismology. Their combined efforts kept the human death toll to less than 500 people.

### 9.2.6 Rhyolite caldera complexes

**Rhyolite caldera complexes** are massive volcanic features that lack the typical highly elevated landform associated with most volcanoes. Rhyolite caldera complexes erupt with such unimaginable violence that the entire volcanic structure, rather than just the top of the volcano, collapses producing large caldera depressions. Because of the grand magnitude of these explosive eruptions, rhyolite caldera complexes have been labeled "**supervolcanoes**" As low depression calderas are the signature aspect, rhyolite caldera complexes are also known as "inverse volcanoes." Following a major eruptive pulse, rhyolite caldera complexes may continue to emit smaller scale eruptions over long time intervals and experience regional uplift, referred to as "resurgence." Rhyolite caldera complexes occur above continental rifts (Long Valley Caldera of California and Valles Caldera, New Mexico, USA), hotspots (Yellowstone,

USA) and subduction zones (Lake Toba, Indonesia and Lake Taupo, New Zealand).

Lake Toba, on the island of Sumatra, is the largest of Earth's rhyolite caldera complex supervolcanoes. Lake Toba is situated in a caldera which measures 100 km by 30 km. The Toba complex has been volcanically active for the past 1.2 Ma, releasing ~3400 km³ of magma. Approximately 74 000 years ago, 2800 km³ of rhyolitic debris erupted in an immense explosion covering an area greater than 20 000 km². Ash was deposited up to 3100 km from the vent source (Rose and Chesner 1987; Chesner and Rose 1990). New Zealand's Lake Taupo is located within a rhyolite caldera complex system created by a cataclysmic eruption 26 500 years ago in which 800 km³ of tephra erupted. Subsequent major eruptions have occurred as recently as 181 AD. Lake Toba and Lake Taupo rhyolite caldera complexes overlie active subduction zones and will certainly produce devastating eruptions in the future.

**Yellowstone** is the best known intraplate rhyolite caldera complex. Yellowstone is among the largest resurgent calderas on Earth at ~45 km × 75 km in diameter. Yellowstone has

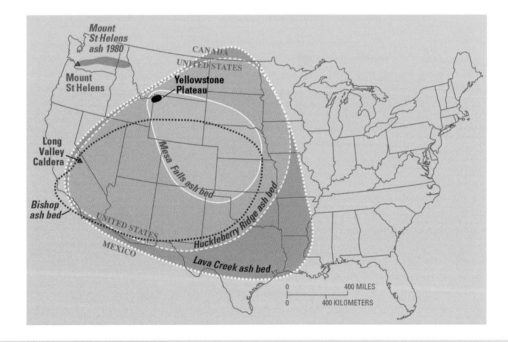

**Figure 9.31** Yellowstone's Quaternary eruptions include three of the four largest North American rhyolite caldera volcanic eruptions, distributing ash throughout the region west of the Mississippi. The fourth is the Bishop tuff which erupted from Long Valley Caldera 700 Ka. *Source*: Courtesy of US Geological Survey. © U.S. Department of the Interior.

experienced three major cataclysmic eruptions within the Quaternary Period (Figure 9.31); these explosive eruptions occurred 2 Ma, 1.2 Ma and 600 000 years ago. Currently, Yellowstone volcanic activity is limited to spectacular geysers and thermal springs, which we will discuss shortly. At the present time, we can enjoy the beautiful hydrothermal features within Yellowstone National Park without any reasonable fear for our safety. Yellowstone's eruptive past foretells the future as the underlying magma chamber will again demonstrate its vitality. Rhyolite caldera complexes explosively erupt infrequently, even by geologic standards. However, these eruptions rank among the largest explosive volcanic events in Earth history.

### 9.2.7   Phreatomagmatic and phreatic eruptions

Phreatomagmatic and phreatic eruptions are steam driven explosions produced when magma or lava heats groundwater, converting it to steam. As a result of steam generation within rock fractures, violent explosions occur in which steam, hot water and/or pyroclastic debris are hurled into the air. **Phreatic** activity refers to the eruption of heated ground water unaccompanied by the eruption of magma or lava. **Phreatomagmatic** eruptions involve the eruption of both steam and pyroclasts derived from magma or lava.

**Phreatomagmatic eruptions** involve both magma and heated groundwater. Current phreatomagmatic eruptions in Hawaii are attributed to magma levels dropping below the water table. As a result of the interaction of magma and water near the water table, liquid water vaporizes and serves as a propellant fuel for the explosive eruption of tephra (Figure 9.32). Phreatomagmatic eruptions have occurred at Hawaii's Halema'uma'u Crater in 1924, 2008, 2018 and 2020–2021. Volcanologists monitored Kīlauea caldera from May to August 2018 as the lava lake

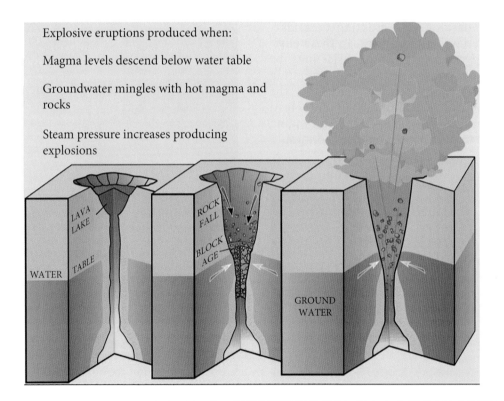

**Figure 9.32**   Phreatomagmatic volcanic activity is particularly common in places such as Hawaii where both magma and water are in great supply. This scenario occurred in May 2018 in Kīlauea Caldera when the lava lake drained below the water table triggering explosive ash eruptions occurred. *Source*: Courtesy of the US Geological Survey. © U.S. Department of the Interior.

levels dropped below the water table and scientists anticipated violent, punctuated steam and ash eruptions. These eruptions document the dynamic interaction between lava and groundwater levels. When magma levels drop below the groundwater table, liquid water vaporizes and serves as a propellent fuel for the explosive eruption of tephra. Volcanic features produced by phreatomagmatic eruptions include tuff rings, tuff cones and maars. Tuff rings and tuff cones are both positive relief features that form where magma interacts with shallow aquifers, wetlands, coastal environments or bodies of surface water. Maars are low relief depressions caused by explosive eruptions in water. These phreatomagmatic structures are ephemeral features as they erode readily and are not well preserved in the geologic record.

Tuff rings and tuff cones are ephemeral volcanic features created by the shallow aqueous eruption of basaltic lava. **Tuff rings** are gently sloping, circular, phreatomagmatic structures thought to have erupted in a high water to magma environment. Tuff rings commonly develop by basaltic lava erupting in a lake, beach or wetland environment. In these environments, pyroclastic flow material may be reworked by wave action, redistributing basaltic tephra around the vent. On the Hawaiian island of O'ahu, Diamond Head tuff ring (Figure 9.33), has a relief ranging from 100 m to over 230 m high and contains a central crater 1 km in diameter (Francis 1993; Sohn 1996).

**Tuff cones** display steep sided, circular volcanic cone structures (Figure 9.34). Tuff cones occur in association with tuff rings; however, tuff cones may have a higher magma to water component and greater pyroclastic ash fall deposition resulting in the conical form (Sohn 1996). Okmok Volcano, a 10 km diameter Holocene caldera system in the central Aleutians, began erupting 12 July 2008. Through August 2018, hundreds of million cubic meters of tephra and lahars covered Umnak Island. Within the caldera, magma interacting with groundwater and surface water bodies produced phreatomagmatic eruptions. Theses eruptions formed new crater lakes and a 200 m high tuff cone. The Okmok

**Figure 9.34** A lake in the bottom of a tuff cone crater within the caldera at Okmok volcano. *Source*: Photo by Janet Schaefer, Image courtesy of the Alaska Volcano Observatory/Alaska Division of Geological & Geophysical Surveys. https://avo.alaska.edu/images/image.php?id=98431. © U.S. Department of the Interior.

**Figure 9.33** Diamond Head Tuff Ring. *Source*: Photo by Steve Dutch. http://www.uwgb.edu/DutchS/EarthSC202Slides/volcslid.htm.

eruption was the first phreatomagmatic volcanic eruption in the United States since the 1977 Ukinrek Maar eruption.

**Maars** are low relief volcanic craters that form by shallow explosive phreatomagmatic eruptions (Figure 9.35). Commonly, the volcanic crater fills with water to create either a freshwater or saline lake. As an example, Lake Becharof, Alaska's second largest freshwater lake, occupies Ukinrek Maar in Alaska. Ukinrek Maar, which is approximately 100 m deep and 300 m in diameter, formed due to explosive eruptions triggered by the injection of basaltic magma into water saturated glacial till during a ten day eruption in April 1977. Tuff cones, tuff rings and maars are common landforms produced by phreatomagmatic eruptions. Next we'll address landforms produced by phreatic eruptions.

The eruption of heated water and steam without magma, characterizes phreatic eruptions. **Phreatic eruptions** produce hot springs, geysers, and fumaroles (**solfataras**). These features form when groundwater percolates downward toward a high temperature magma reservoir. As the water is heated, it expands, becomes less dense and rises along fractures allowing for the release of steam and hot water. Phreatic eruptions produce siliceous sinter deposits, tufa limestone, sulfides, and other minerals associated with hydrothermal deposits. Three types of phreatic eruptions occur:

1   **Hot springs** contain groundwater heated by proximity to magma (Figure 9.36).
2   **Geysers** are eruptive hot springs that eject fountains of heated water periodically (Figure 9.37) Geysers erupt as groundwater is heated to temperatures above 100 °C. Because the boiling point of water is higher under pressure, superheated water rises, converts to steam and is ejected explosively as a geyser.
3   **Fumaroles** (Figure 9.38) emit mixtures of steam and other gases such as hydrogen sulfide. Sublimation of the sulfide gases to solid form produces economically viable sulfur deposits used in making agricultural fertilizers and other uses in the chemical industry.

**Figure 9.35**   In April of 1977, Ukinrek volcano experienced a 10 day phreatomagmatic eruption allowing geologists to observe the creation of a maar. (a) Map view of the 300 m wide maar crater. (b) April 1977 Ukinrek volcano phreatomagmatic eruptive ash plume. (c) Cross section view of maar wall with approximately 15 m of tephra and ash flow deposits. *Source*: Photo courtesy of the United States Geological Survey. © U.S. Department of the Interior.

Hot springs, geysers, and fumaroles develop in areas of unusually high geothermal gradients and are important energy sources where hot rocks and/or magmas lie close to the surface.

Volcanic activity beneath glaciers is also significant in alpine volcanoes, glaciated continents and islands. Sub-glacial volcanic eruptions can result in the release of glacial meltwater floods called **jökulhlaups** (Icelandic term pronounced Yo-kul-hloips). Jökulhlaups are produced by the sudden

**Figure 9.36**   Hot spring pools precipitate opaline (silica) sinter deposits within Yellowstone National Park, Wyoming, USA. *Source*: Photo by Kevin Hefferan. © John Wiley & Sons.

**Figure 9.38**   (a) This fumarole at Kīlauea volcano is releasing sulfur gases which, upon cooling, sublimate (gas to solid) around the vent as sulfaterra. Sulfaterra is sulfur rich soil and rock. (b) Sulfur crystals sublimate from gases released at Kīlauea. *Source*: Photos by Kevin Hefferan. © John Wiley & Sons.

**Figure 9.37**   Beehive geyser at Yellowstone. *Source*: Photo by Kevin Hefferan. © John Wiley & Sons.

flood burst of glacial lake water or water contained within a glacier. For example, Iceland's Vatnajokull ice cap overlies seven active volcanoes including Grímsvötn. On 30 September 1996 scientists detected seismic activity associated with a sub-glacial eruption of the Grímsvötn caldera. On 5 November, vertical fractures on the glacier released a large jökulhlaup that produced peak flow of $50\,000\,m^3/s$ and lasted for several days, destroying roads and bridges.

Similarly, the 2010 eruption of Iceland's Eyjafjallajökull volcano also created a jökulhlaup with a peak flow of $3000\,m^3/s$ on 14th April 2010. Not only did the 2010 sub-glacial eruption of Eyjafjallajökull generate a jökulhlaup, but the cold water from the melting ice instantaneously chilled the lava, causing lava to fragment into small silica glass shards and ash. Under great pressure, the shards and ash were ejected in an eruptive plume high into the upper atmosphere. Eyjafjallajökull's ash plume had an immense impact upon airline traffic in the Western Hemisphere for several weeks, resulting in the cancelation of over $100\,000$ flights. Volcanic plumes pose a particular hazard to aircraft as ash has a melting point of ~1100 °C, which is lower than the operating temperature of jet engines (1400 °C). Upon entering the aircraft engines, volcanic ash can melt and fuse into a solid glass coating within the engines, triggering aircraft engine temperature sensors to an auto shutdown mode. Ash also blocks engine cooling holes, abrades turbine components and diminishes fuel flow. In 1982, British Airways Flight 9 flew through Mount Galunggung (Indonesia) ash cloud and lost power in all four engines resulting in 7 km of rapid descent before the engines could be restarted. In 1989, KLM Flight 867 also temporarily lost power in all four engines after flying through Alaska's Mt. Redoubt eruption cloud. Flight 867 rapidly descended over 4 km and was forced to make an emergency landing at Alaska's Anchorage Airport. Mind you, these were relatively small eruption events for which the airline industry was largely unprepared, although pilot crews admirably saved the aircrafts and all aboard. A historic eruption could have a much greater impact upon air travel and our lives.

So far in this chapter, we have classified volcanic features largely based on landform size, shape, and physical characteristics of the volcanic rock material. These factors are related to: magma budget, silica content, magma viscosity and the presence/absence of water. Volcanoes are also classified by their explosiveness.

## 9.3 CLASSIFIYING VOLCANIC EXPLOSIVITY

Perhaps most importantly, volcanoes can be classified based on explosivity, the volume of pyroclastic debris and the height of the eruption column during eruptive episodes. These factors are related to the nature of the vents, the viscosity of the lava, the involvement of water, and the type of explosive activity. Volcanoes vary from being "quiet" to cataclysmic in eruptive style, as summarized in Table 9.1 and discussed below.

### 9.3.1 Quiescent eruptions

Magma producing quiet volcanic eruptions is characterized by low silica content, low viscosity, low volatile content and high temperature. As discussed in Chapter 7, silica poor magmas ($<52\%$ $SiO_2$) generally contain relatively small amounts of dissolved water vapor ($<1\%$) at moderate to high pressures. In such magmas, the dissolved water vapor exsolves as the magma approaches the surface and crosses the level of exsolution. The dissolved gases exsolve or bubble out of solution and escape upward through the magma. Similar processes occur with the release of carbon dioxide on opening a container of any carbonated beverage. Following the release of the confining cork, you can watch the $CO_2$ bubbles exsolve from a carbonated beverage and rise toward the surface. Pour the drink and observe the gaseous, vesicular "head" that floats above the liquid. In a similar process, gas bubbles released by the exsolution of magmatic gases generate vesicular textures on the surface of lava flows. Gas bubbles buoyantly ascend at a faster rate than the magma. As result, basaltic eruptions, except for small lava fountains, are typically non-explosive. The low percentage of gas bubbles do not lower the melt density enough to propel the

**Table 9.1** Common volcano explosivity classification.

| Eruption name | Description | Examples |
|---|---|---|
| *Quiescent eruptions* | | |
| Hawaiian | Eruptions begin as fissures, evolving to central vent flows and the generation of large shield volcanoes, fiery basaltic lava fountain eruptions, quiet lava flows and cinder cones | Kilauea, Hawii |
| Icelandic | Persistent fissure eruption of low viscosity basaltic lava flows. Prolonged quiet eruptions may generate lava plateaus and flood basalts | Laki, Iceland, 1783 $12 \, km^3$ of lava |
| Surtseyan (phreatomagmatic) | Explosive, steam-blast eruptions with lava flows and pyroclastic debris. Surtseyan eruptions are named after the volcanic island of Surtsey, which rose above sea level on 14 November 1963. Within two months Surtsey, a newly created island south of Iceland was 1.3 km long and 174 m high (Decker and Decker 2006). Surtsey continued to erupt until 1967 | Surtsey, Iceland, 1963 |
| Strombolian | Periodic bursts ("burps") of moderately explosive eruptions | |
| *Explosive eruptions* | | |
| Vulcanian | Explosive eruptions of basaltic to rhyolitic viscous lava and large volumes of volcanic ash plumes (<25 km high) and pyroclastic debris | Vulcano, Italy |
| Vesuvian | Violent eruptions of volcanic debris ejected, scattering ash over thousands of square kilometers | Mt. Vesuvius, Italy |
| Plinian | Tephra eruptions emit immense ash clouds >11 km in height into the stratosphere | Krakatoa, 1883 |
| Ultraplinian | Violent tephra eruptions of volumes $>1 \, km^3$ and ash cloud heights 25–55 km | Mt. Taupo, New Zealand, 181 AD |

*Source*: After Walker (1973).

basaltic magma upward from the vent at high velocities. The release of small amounts of gases in a fluid magma prevents truly explosive eruptions. Upon reaching Earth's surface, low viscosity lava produces gas charged fountains, rivers of lava (lava flows) and lava lakes, all of which attest to the low viscosity of basic lavas. The low resistance to flow also allows transport of lava great distances. Long distance transport is particularly apparent in flood basalts, where lavas flow hundreds to thousands of kilometers. Quiescent eruptions include Hawaiian, Icelandic and, to a lesser degree, Strombolian eruptions where ash cloud heights may reach 10 km. Let's consider some different types of quiescent eruptions, beginning with Hawaiian events.

**Hawaiian eruptions** are among the mildest form of volcanic activity, erupting hot, low viscosity basaltic lava from central vents or fissures. Despite their quiescent nature, even these eruptions can generate gas propelled ejection of lava up to 1 km into the air. The low viscosity and the low yield strength of the magma allows for the exsolution and vesiculation of volatiles without explosively fragmenting the magma (Francis 1993). As mentioned earlier in this chapter, Hawaiian eruptions typically begin with the eruption of gas bubble rich lava fountains, followed by the emission of fluid lava flows from central vents with smaller fissures or flank eruptions. The end result is the construction of massive shield volcanoes or flood basalts over large areas.

**Icelandic eruptions** are named after the island nation of Iceland which is located above the Mid Atlantic Reykjanes ridge. Beneath the ridge, extension of Earth's lithosphere creates parallel or en echelon sets of fractures which facilitate fissure eruptions. Fissure eruptions produce linear "curtains of fire," Ocean ridges, flood basalts and basalt plateaus may be generated by long lived Icelandic fissure eruptions.

**Surtseyan or Phreatomagmatic** eruptions occur when basaltic lavas are erupted into a lake or shallow portion of the ocean producing tuff rings, cones, and maars. Water is converted to stream by the magma and magma is converted to pyroclasts by rapid cooling as both rise explosively to create mafic tuff cones or tuff rings. Surtseyan eruptions are named for the phreatomagmatic eruption that produced the island of Surtsey, just south of Iceland in 1964. Beginning as a submarine eruption, sufficient hydromagmatic debris accumulated on the ocean floor to build the island above sea level as a small mafic tuff cone. Phreatomagmatic eruptions may also occur at "within plate" locations, such as the Hawaiian Islands, where groundwater and magma interact to generate explosive stream and pyroclastic eruptions. Explosive phreatomagmatic eruptions occurred at Kīlauea in 1790, in 1924, 2007–2018, and 2020–2021.

**Strombolian eruptions** are punctuated by mildly explosive, thunderous bursts of pasty, viscous basaltic to andesitic lava ejected tens to hundreds to thousands of meters into the air. Unlike Hawaiian eruptions, the airborne material in Strombolian eruptions returns to the Earth as solid scoria rather than spatter (Macdonald 1972; Walker 1973). Strombolian eruptions commonly produce pyroclastic scoria cones. In rare cases, large composite volcanoes may also develop over sustained periods of Strombolian activity as in Stromboli volcano, the namesake for this style of eruption. Strombolian volcanoes also include Mt. Etna (Italy), Mt. Erebus (Antarctica) and Paricutin Volcano in Mexico.

## 9.3.2  Explosive eruptions

Explosive volcanic eruptions include the Vulcanian, Peléean, Sub-Plinian, Plinian and Ultraplinian eruptions. These explosive eruptions generate ash clouds between 10 and 55 km in height. **Vulcanian** eruptions consist of a series of short, explosive bursts, generated by the shattering of solid rock plugs of andesite to dacite composition. The early, explosive bursts are considered a throat clearing phase in which the magma conduit is evacuated of viscous plugs and constraining debris. Plug shattering is followed by an extensive period of more sustained but less explosive eruptions. Vulcanian eruptions are named for the island of **Vulcano**, located off the coast of Italy. Vulcanian eruptions may emit pyroclastic bombs as well as dense clouds of ash laden gas from a central crater. Vulcanian activity consists of vertical plumes of fragmented rock particles propelled by exsolved gases up to 25 km into the air. Vulcanian eruptions are also referred to as Peléean eruptions.

The most explosive eruptions are termed **Plinian** eruptions. Plinian eruptions involve rapidly ascending andesitic to rhyolitic magma and exsolved gases moving upward at velocities of several hundred meters per second. Instead of propelling ahead of the magma, the entrained volatile bubbles rise within the magma stream resulting in sustained jets of violently propelled lava. As a result, the gases are violently discharged creating mushroom shaped plumes of ash. Plinian eruptions occur in highly viscous silicic magmas. Plinian eruptions may be subdivided based on ash cloud height and tephra volume. Plinian eruptions emit immense clouds of tephra over 11 km in height into the stratosphere and eject tephra in volumes $>1\,km^3$. **Ultraplinian** eruptions are the most explosive emitting ash clouds up to 55 km and tephra volumes $>10\,km^3$.

**Plinian eruptions** are also referred to as or **Vesuvian** or **Krakatoan** eruptions, named after the eruptions of Mt. Vesuvius (Italy) and Krakatoa volcano (Indonesia). Plinian eruptions are named for Pliny the Younger, who recorded the eruption of Mt. Vesuvius in Italy on 24 August, 79 AD. Mt. Vesuvius erupted in a cataclysmic display of fury that buried the towns of Pompeii and Herculaneum along the Bay of Naples. Although this eruption occurred almost 2000 years ago, two letters written by Pliny the Younger, to a historian named Tacitus, accurately recorded the dramatic chain of events. The eruption involved the ejection of several cubic kilometers of tephra into the air, typical of what is now referred to as Plinian eruptions. In addition to the tephra, incandescent, ground enveloping nuée ardentes surges swept down the countryside with velocities up to 100 km/h. Together the ash fall and pyroclastic surges completely buried the towns of Pompeii and Herculaneum respectively, both located within 10 km of Mt. Vesuvius. No reliable estimates exist on the number of people killed (Francis 1993).

Recent Plinian eruptions include: the 1815 Tambora eruption, the 1883 Krakatoa eruption (Indonesia), the 1912 Katmai Eruption (Alaska), the 1980 Mt. St. Helens eruption, and the 1991 eruption of Mt. Pinatubo (Philippines). Ultraplinian eruptions include the 5677 BC. Mazama eruption that created Crater Lake Caldera (USA), the 74 000 year-old Toba eruption in Sumatra, the Taupo (New Zealand) eruption and the 2 million year old Yellowstone Lava Creek eruption. Ultraplinian eruptions are climate changing, cataclysmic events.

Volcanic activity has dramatically enhanced our planet on a global scale. Most of Earth's water was derived from volcanic degassing of water from the Earth's interior. Volcanic activity has generated valuable metal deposits, as discussed in Chapter 19. Volcanic activity has also created vast tracts of arable land noted for their rich productive soils. Volcanic activity may also be the source for early life forms on Earth in hydrothermal vents at ocean ridges or hotspots.

On the other hand, volcanic activity presents significant hazards to our world.

Volcanic activity, particularly from continental volcanoes located at convergent plate boundaries or over continental hotspots, can result in widespread devastation and loss of life in the hundreds to tens of thousands. Composite volcano eruptions at convergent margins generate localized (tens of kilometers) pyroclastic flows and lahars that can bury tens of thousands of people in tens of meters of debris. Twentieth century pyroclastic flows and lahars at localities such as Nevado del Ruiz, Mt. Pelée and Mt. Pinatubo vividly display the horror that can devastate entire cities.

Plinian and Ultraplinian eruptions, as well as flood basalt eruptions, can dramatically alter Earth's climate. The 5 April 1815 eruption of Mount Tambora (Indonesia) killed 92 000 and caused disastrous crop failures and summer snowfalls as far away as North America. Presumably, cataclysmic eruptions such as occurred at Toba and Yellowstone would produce a global impact significantly longer than a few years. Exactly how many years? That is a good scientific question without a suitable answer.

Volcanic eruptions affect climate in a variety of interrelated processes. Volcanic dust ejected into the atmosphere causes short term cooling, on the order of days to months. The cooling is due to the blockage of solar radiation, preventing the Sun's warm infrared rays from penetrating to the Earth's surface. The amount and duration of cooling depends upon the volume of dust suspended in the atmosphere. Large volumes can block out sun's rays and cause a volcano winter scenario that would drastically reduce photosynthetic activity, undermining the food web both in the terrestrial and marine environments.

Volcanic eruptions release volatiles such as sulfur oxide, sulfur dioxide, nitrous oxides, carbon oxides and halogens into the atmosphere and oceans. Sulfur, nitrous, and carbon oxides, as well as fluorine and chlorine gases, react with water to form acidic solutions in the atmosphere and in bodies of water. The reflective properties of the sulfuric acid haze result in decreased penetration of infrared rays on the order of one to five years, depending upon the amount and duration of haze in the atmosphere. Halogens cause depletion of ozone and increased penetration of harmful ultraviolet light. Climate change may also occur in response to the release of greenhouse gases, such as carbon dioxide, during a volcanic eruption. Climate warming associated with greenhouse gas buildup results in ocean warming and increasingly anoxic conditions.

Massive flood basalt eruptions appear to coincide with massive extinction events in the geologic record. Rampino et al. (2017) suggest that explosive reactions between magma and coal during the Siberian flood basalt eruptions released large amounts of $CO_2$ and $CH_4$ into the atmosphere, causing severe global warming and the end-Permian mass extinction. It is likely not a coincidence that all major extinctions on Earth coincide with massive flood basalt eruptions. The "Big Five"

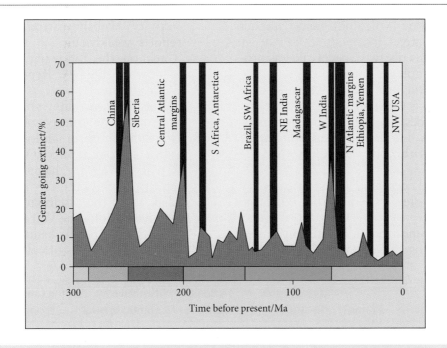

**Figure 9.39** Mass extinctions (gray) and flood basalts (red) over the past 300 million years. *Source*: Courtesy of Wiley Publishers. © John Wiley & Sons.

mass extinctions include the Ordovician–Silurian (444 Ma), Devonian–Carboniferous (360 Ma), Late Permian (251 Ma), Late Triassic (200 Ma), and end-Cretaceous (65.6 Ma). While evidence for flood basalt activity at the Ordovician–Silurian extinction event has been lacking, Jones et al. (2017) discovered high mercury concentrations from Nevada (USA) and south China that are considered to have originated from massive basalt eruptions. Mass extinctions are also recognized at the Cambrian–Ordovician boundary (488 Ma) and the Guadalupian extinction during the Permian (260 Ma), among others listed in Figure 9.39. Massive flood basalts are unlikely to emit massive ash clouds but the shear volume of lava and gases emitted is staggering. No humans have ever witnessed the fierce destructive nature of flood basalt or rhyolite caldera complex eruptions. The consequences to life on Earth from such massive eruptions are beyond our ability to appreciate fully. Mass extinctions likely involve a number of different factors – volcanism, meteorite impact, glaciation, orogenic mountain building – that when coupled together produce conditions incompatible with life on Earth. Of these factors, available evidence suggests volcanic eruptions may be among the chief culprits responsible for the sudden demise of life forms on Earth.

On the other hand, volcanic eruptions may hold the key to the development of Earth's life through chemosynthetic activity at ocean spreading ridges. As demonstrated in this chapter, Earth's dynamic volcanism rocks our planet and impacts our world beyond human scope. Our next chapter will expand upon the relationship of magmatic activity, plate tectonics and igneous rock associations.

## CONTENT ASSESSMENT

1  Name and describe three major types of volcanoes and provide an example of each.
2  Describe the eruptive deposits and impacts at Pompeii in 79 AD.
3  Describe volcanic events and impacts in Nevado del Ruiz in 1985.
4  How did events in Alaska's Redoubt Volcano in 1989 and Iceland in 2010 affect aviation?
5  Name and describe three types of pyroclastic material.
6  How do volcanic gases impact health?

7 Explain how volcanic gases impact global climate.

8 Name and describe two types of calderas providing examples of each.

9 Describe and provide an example of a large igneous province.

10 Name and describe volcanic deposits associated with sea floor volcanism. https://www.youtube.com/watch?v=XHe7-QHETfg.

11 Name and describe three means by which geologists monitor volcanic activity.

12 Go to the Smithsonian Institution/United States Geological Survey Weekly Volcanic Activity Website http://volcano.si.edu/reports_weekly.cfm.

Click on the following box and drop down links: Reports, Current Eruptions

Determine how many volcanoes are currently in an eruptive phase.

13 Go to the Smithsonian Institution/United States Geological Survey Weekly Volcanic Activity Website http://volcano.si.edu/reports_weekly.cfm.

Click on the following box links: Database, Holocene Volcanoes

Identify one specific example for each of the following types of volcanoes.

| Volcano type | Name | Geographic location | Tectonic setting | Eruptive characteristics | Example of recent activity |
|---|---|---|---|---|---|
| Shield | | | | | |
| Stratovolcano | | | | | |
| Fissure eruption | | | | | |
| Caldera | | | | | |
| Submarine | | | | | |
| Ocean ridges | | | | | |

14 Observe the Syracuse University Lava Flow at the website below and note different volcanic features that develop. https://www.youtube.com/watch?v=Fcz3vBdI7Nc.

15 To watch a seven minute summary of the Mt. St. Helens 1980 eruption refer to the US Geological Survey video at http://www.youtube.com/watch?v=xP2dreOI8gI.

16 Interested in Kīlauea's activity as of January 2020. Watch a 45 minute presentation by US Geological Survey geologist Carolyn Parcheta at https://www.youtube.com/watch?v=85WXeY4U56g

17 The University of Hawaii at Hilo Center for the Study of Active Volcanoes (CSAV) has an excellent website detailing videos and aspects of volcanism at https://hilo.hawaii.edu/csav/

18 Describe the aa lava flow exhibited in this U.S.G.S video at:

https://www.usgs.gov/media/videos/k-lauea-east-rift-zone-eruption-may-6-2018-0

19 What new events are occurring in Hawaii? Refer to the following U.S. Geological Website: https://www.usgs.gov/volcanoes/kilauea/current-eruption?qt-science_support_page_related_con=0#qt-science_support_page_related_con

20 Comment on current volcano monitoring tools after reviewing material at: https://www.usgs.gov/natural-hazards/volcano-hazards/monitoring https://phys.org/news/2020-08-iceland-volcano-eruption.html

## REFERENCES

Baker, D., Dalpe, C., and Poirier, G. (2004). The viscosities of food as analogs for silicate melts. *Journal of Geoscience Education* 52: 363–367.

Cesca, S., Letort, J., Razafindrakoto, H.N.T. et al. (2020). Drainage of a deep magma reservoir near Mayotte inferred from seismicity and deformation. *Nature Geoscience*: 1387–1393. https://doi.org/10.1038/s41561-019-0505-5.

Chesner, C.A. and Rose, W.I. (1990). Stratigraphy of the Toba Tuffs and the evolution of the Toba Caldera Complex, Sumatra, Indonesia. *Bulletin of Volcanology* 53: 343–356.

Decker, R. and Decker, B. (2006). *Volcanoes*, 4e. New York: W.H. Freeman Publishing 326 pp.

Francheteau, J., Needham, H.D., Choukroune, P. et al. (1979). Massive deep-sea sulfide ore deposits discovered by submersible on the East Pacific Rise: Project RITA, 21°N. *Nature 277*: 523–528.

Francis, P. (1993). *Volcanoes: A Planetary Perspective.* New York, OX: Clarendon (Oxford University Press) 443 pp.

Fruh-Green, G.L., Kelley, D.S., Bernasconi, S.M. et al. (2003). 30,000 years of hydrothermal activity at the Lost City Vent Field. *Science 301*: 495–498.

Gasparini, A. (2020). Corn syrup reveals how bubbles affect lava's flow. *Eos 101* https://doi.org/10.1029/2020EO152868.

Jones, D.S., Martini, A.M., Fike, D.A., and Kaiho, K. (2017). A volcanic trigger for the Late Ordovician mass extinction? Mercury data from south China and Laurentia. *Geology 45*: 631–634.

Macdonald, G.A. (1972). *Volcanoes.* New Jersey: Englewood Cliffs, Prentice-Hall. Inc 510 p.

Patrick, M., Orr, T., Swanson, D. et al. (2021). Kīlauea's 2008–2018 summit lava lake—chronology and eruption insights, chap. A. In: *The 2008–2018 Summit Lava Lake at Kīlauea Volcano, Hawaii*, Professional Paper, vol. *1867* (eds. M. Patrick, T. Orr, D. Swanson and B. Houghton). U.S. Geological Survey 50 pp, doi:https://doi.org/10.3133/pp1867A.

Rampino, M.R., Rodriguez, S., Baransky, E., and Cai, Y. (2017). Global nickel anomaly links Siberian Traps eruption and the Latest Permian mass extinction. *Scientific Reports 7*: 12416. https://doi.org/10.1038/s41598-017-12759-9.

Rose, W.I. and Chesner, C.A. (1987). Dispersal of ash in the great Toba eruption, 75 Ka. *Geology 15*: 913–917.

Self, S., Thordarson, T., Keszthelyi, L. et al. (1996). A new model for the emplacement of Columbia River basalts as large, inflated pahoehoe lava flow fields. *Geophysical Research Letters 23* (19): 2689–2692.

Simkin, T., Siebert, L., McClelland, L. et al. (1981). *Volcanoes of the World.* Washington: Smithsonian Institution 232 pp.

Smiley, T.L. (1958). The geology and dating of Sunset Crater, Flagstaff, Arizona. In: *Guidebook of the Black Mesa Basin, Northeastern Arizona, New Mexico. 9th Field Conference* (eds. R.Y. Anderson and J.W. Harshbarger), 186–190. Geological Society Guidebook.

Sohn, K. (1996). Hydrovolcanic processes forming basaltic tuff rings and cones on Cheju Island, Korea. *Geological Society of America Bulletin 108*: 1199–1211.

Sparks, R.S.J. (1986). The dimensions and dynamics of volcanic eruption columns. *Bulletin of Volcanology 48*: 3–15.

Sparks, R.S.J. and Wilson, L. (1976). A model for the formation of ignimbrite by gravitational column collapse. *Journal of the Geological Society of London 132*: 441–451.

Tarasewicz, J., Brandsdóttir, B., White, R.S. et al. (2012). Using microearthquakes to track repeated magma intrusions beneath the Eyjafjallajökull stratovolcano, Iceland. *Journal of Geophysical Research 117*: B00C06. 13 pp.

Tripathy-Lang, A. (2020). New Volcano, Old Caldera. *Eos 101* https://doi.org/10.1029/2020EO152781.

Walker, G.P.L. (1973). Explosive volcanic eruptions – a new classification scheme. *Geologie Rundschau 62*: 431–446.

Wright, J.V., Smith, A.L., and Self, S. (1980). A working terminology of pyroclastic deposits. *Journal of Volcanology and Geothermal Research 8*: 315–336.

Zdanowicz, C.M., Zielinski, G.A., and Germani, M.S. (1999). Mount Mazama eruption; calendrical age verified and atmospheric impact assessed. *Geology 27*: 621–624.

# Chapter 10

# Igneous rock associations

## 10.1 PETROTECTONIC ASSOCIATIONS

Petrotectonic associations are suites of rocks that form in response to similar geologic conditions. Igneous rock associations most commonly occur at divergent plate boundaries, convergent plate boundaries, and hot spots (Figure 10.1a). Approximately 40–50 hot spots exist both at lithosphere plate boundaries (e.g. Iceland) and in intraplate settings (e.g. Hawaii). Fisher and Schmincke (1984) estimate the percent of magma generated at modern divergent, convergent and hot spot regions as 62%, 26%, and 12%, respectively.

While plate tectonic activity plays a critical role in the development of petrotectonic associations, it is not the sole determining factor. For example, the earliest onset of modern plate tectonics continues to be debated, with some researchers (Kusky et al. 2001; Parman et al. 2001) favoring Archean (>2.5 Ga) onset and others (Hamilton 1998, 2003; Stern 2005, 2008; Ernst 2007; Condie and Kröner 2008) proposing Proterozoic initiation of deep subduction ~1 billion years ago (1 Ga). If the latter is true, then ~78% of Earth's igneous activity occurred under conditions that predate the onset of modern plate tectonic activity given that Earth formed 4.56 Ga. In addition to questions regarding magmatism at Precambrian plate tectonic boundaries, Phanerozoic intraplate magmatism may or may not be influenced by lithospheric plate boundaries (Hawkesworth et al. 1993; Dalziel et al. 2000). So while the plate

*Earth Materials*, Second Edition. Kevin Hefferan and John O'Brien.
© 2022 John Wiley & Sons Ltd. Published 2022 by John Wiley & Sons Ltd.
Companion website: www.wiley.com/go/hefferan/earthmaterials2

(a)

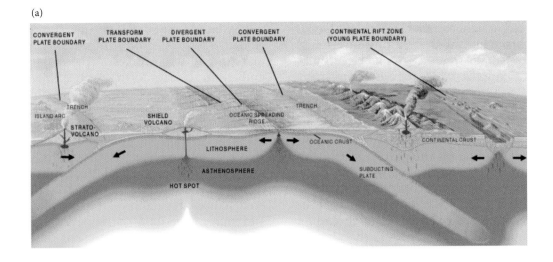

(b)

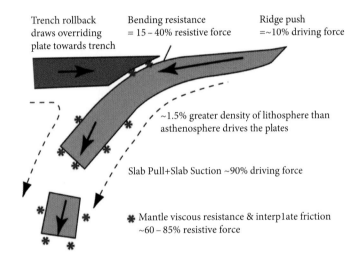

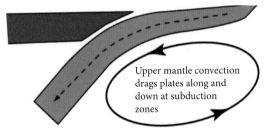

**Figure 10.1** (a) Major tectonic environments where igneous rocks occur. *Source*: Courtesy of the U.S. Geological Survey and the U.S. National Park Service, United States Department of the Interior. (b) Slab pull and slab suction provide ~90% of force driving plate motion, while ridge push at divergent margins provides the remaining 10%. Mantle convection is not considered to be a significant driving force in plate motion. *Source*: Courtesy of Robert Stern.

tectonic paradigm is very useful, it does not address all igneous rock assemblages produced throughout Earth's tumultuous history. Nevertheless, modern plate tectonics has operated at least since the Neoproterozoic with plate motion driven by the lithosphere itself. Sinking of the lithosphere in subduction zones is estimated to provide 90% of plate motion force in two ways: (i) slab pull by directly pulling on the plate and (ii) slab suction by entraining the surrounding mantle to descend with it. Approximately 10% of the driving force is estimated to be provided by ridge push at the ocean spreading ridge (Figure 10.1b). Early ideas that mantle convection served as the primary driving force in lithospheric plate motion are considered obsolete (Conrad and Lithgow-Bertelloni 2002, 2004; Stern 2007). In the following sections we will address major igneous petrotectonic associations, beginning with divergent plate boundaries.

## 10.2 DIVERGENT PLATE BOUNDARIES

Decompression of the asthenosphere in response to lithospheric extension results in partial melting of mantle peridotite at divergent margins. Less dense mafic melts rise and solidify to produce oceanic crust, while the more dense refractory residues cool below a critical temperature to form the thickening mantle layer of ocean lithosphere. Ocean lithosphere is created primarily at spreading ridges such as the Mid-Atlantic Ridge, East Pacific Rise, and the Indian and Antarctic ocean ridge systems. Ocean lithosphere is also generated in back arc basin spreading ridges (e.g. Marianas Trough) and ocean hotspots (e.g. Hawaii and ocean flood basalts). In all cases, anatexis (partial melting) of the ultramafic mantle is the primary magma source for ocean lithosphere.

Ocean lithosphere contains four distinct layers as indicated in Figure 10.2. Layer 1 contains well stratified marine pelagic sediments and sedimentary rocks that accumulate on the ocean floor. Layer 2 consists of two basaltic rock layers: an upper layer (2A) contains pillow basalts that develop when mafic lavas flow onto the ocean floor, rapidly cool in the aqueous environment and solidify in

spheroidal pillow basalt masses. Beneath the pillow basalt, mafic magma injects into extensional fractures producing steeply inclined diabase and basalt dikes as the magma cools and contracts to form layer 2B. Repeated horizontal extension and magma intrusions generate dike sets arranged parallel to one another in a sheeted dike complex. Beneath the sheeted dike layer, mafic magma cools slowly allowing phaneritic crystals to nucleate and grow in layer 3. Layer 3 contains massive (isotropic) gabbro in the upper section, layered (cumulate) gabbro in a middle section, and increasing amounts of layered (cumulate) peridotite toward the bottom of the section, marking the base of ocean crust. The Mohorovičić discontinuity (Moho) separates cumulate rocks in layer 3 from noncumulate, metamorphosed rocks in layer 4. The Moho marks the rock boundary between the ocean crust, which is generally 5–7 km thick, and the underlying mantle. Layer 4 is composed of depleted mantle peridotite refractory residue (e.g. harzburgite, dunite). Layer 4 mantle peridotite is marked by high temperature, solid state strain fabric (metamorphosed) and represents the lowest layer of the oceanic lithosphere.

Layers 3 and 4 are typically unexposed on ocean floors because they are commonly overlain by layers 1–2. In some locations such as the Gakkel ridge in the Arctic Ocean and the Southwest Indian Ocean ridge, these deep layers are exposed on the ocean floor in ultraslow spreading (<2 cm/year) centers where ocean crust is anomalously thin (1–4 km thick). In other spreading centers, low angle detachment faults have tectonically exhumed (uplifted) deep layers toward the ocean floor. Layers 3 and 4 mantle exposures can also be exposed in the walls of transform faults due to transtensional or transpressional faults due to oblique slip between the spreading ridge and transform zone (Maia et al. 2016).

Ancient slices of thin ocean lithosphere are also preserved in alpine orogenic belts as ophiolite sequences. The anomalously thin ocean crust observed at ultraslow spreading ridges solves an enigma as to why many ophiolites do not conform to the normal ocean crustal thickness of 5–7 km (Snow and Edmonds 2007). Let us now consider

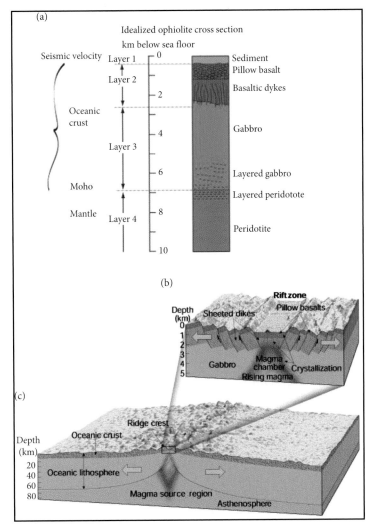

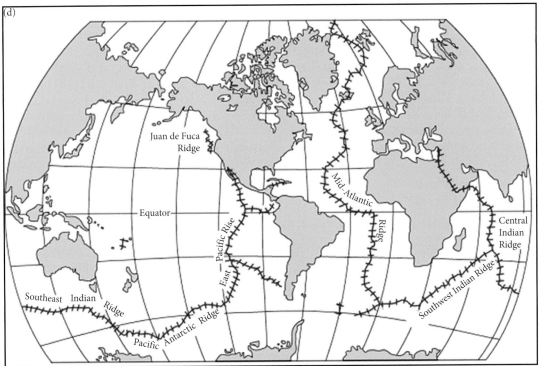

**Figure 10.2** (a) Cross section of ocean lithosphere. (b) Block diagram of mid ocean ridge divergent margins. (c) Ocean ridges are primary sites for the generation of ocean lithosphere. (d) Global array of divergent margins. *Source*: Ocean Drilling Project and the U.S. Geological Survey, United States Department of the Interior.

petrotectonic assemblages that form at ocean ridge spreading centers.

### 10.2.1 Mid ocean ridge basalts (MORB)

At ocean spreading centers and some back arc basin spreading centers, partial melting of lherzolite (peridotite) generates voluminous, geochemically distinct, mid-ocean ridge basalts (MORB) and gabbros that contain minerals such as plagioclase, augite, hypersthene, pigeonite, diopside, and olivine. Typical major and minor element concentrations are indicated in Table 10.1. MORB are low $SiO_2$ (45–52%), low K (<1% $K_2O$), tholeiites with high MgO (~7–10%), $Al_2O_3$ (15–16%) and compatible element concentrations (Ni, Cr ~100–500 ppm). MORB develop from partial melting of a depleted mantle source, as indicated by low $^{87}Sr/^{86}Sr$ ratios (0.702–0.704), low volatile incompatible element concentrations, and high compatible element concentrations (Cann 1971). "Depleted source" refers to mantle lherzolite that has undergone previous melt cycles which largely removed mobile incompatible elements.

MORB are subdivided into normal MORB (N-MORB) and enriched MORB (E-MORB) based upon minor and trace element abundances (Table 10.1 and Figure 10.3). N-MORB are strongly depleted in highly incompatible elements such as large ion lithophile elements (LILs such as Cs, Rb, Ba), high field strength elements (HFS such as Nb, Ta) and light rare Earth elements (LREEs such as La, Ce, Pr, Nd, and Sm). These geochemical characteristics suggest that N-MORB magma represent 20–30% partial melting of a well-mixed, depleted mantle source (Frey and Haskins 1964; Gast 1968).

Although the major element and heavy rare Earth elements (HREEs ranging from Eu-Lu) concentrations are comparable, E-MORB have higher incompatible element (LREE, HFS, LIL) concentrations relative to N-MORB. Specifically, E-MORB are defined by having chondrite normalized La/Sm ratios >1. La (a LREE) may occur in concentrations of 1–5 ppm in N-MORB but up to 50–100 ppm in E-MORB.

How can we account for chemical variations between N-MORB and E-MORB? Several different hypotheses have been proposed. First, E-MORB may represent smaller degrees (~10–15%) of partial melting of residual mantle rock so that the incompatible elements are more highly concentrated in E-MORB magmas. Second, E-MORB could be tapping a mantle plume source that has not been previously melted. Third, E-MORB could represent magma enriched from magma mixing, assimilation or partial melts derived from subducted ocean lithosphere. For example, Eiler et al. (2000), based on a study of 28 basalt samples from the Atlantic, Pacific and Indian ridges, propose that E-MORB include a component of partially melted oceanic lithosphere that has been recycled into the upper mantle from ancient subduction zones.

Within the classic N- and E-MORB classification, subgroups are developing. Pearce (2008) proposes a plume-type MORB (P-MORB) that displays higher enrichment in incompatible element and REE concentrations than E-MORB, comparable to those of ocean island basalts (OIB). However, whereas OIB have an alkaline signature with Nb/Y > 1, P-MORB have a sub-alkaline chemistry with Nb/Y < 1 (Pearce 1982, 2008). Saccani (2015) proposes a MORB with a garnet signature

**Table 10.1** Trace element abundances for N-MORB and E-MORB in parts per million (ppm).

| | LIL | | | | | HFS | | LREE | | | | | HREE | | | | |
|---|---|---|---|---|---|---|---|---|---|---|---|---|---|---|---|---|---|
| | Cs | Rb | Ba | Th | U | Nb | Ta | La | Ce | Pr | Nd | Sm | Zr | Eu | Gd | Yb | Lu |
| N-MORB | 0.007 | 0.56 | 6.3 | 0.12 | 0.47 | 2.33 | 0.132 | 2.5 | 7.5 | 1.32 | 7.3 | 2.63 | 74 | 1.02 | 3.68 | 3.05 | 0.455 |
| E-MORB | 0.063 | 5.04 | 57 | 0.6 | 0.18 | 8.3 | 0.47 | 6.3 | 15 | 2.05 | 9 | 2.6 | 73 | 0.91 | 2.97 | 2.37 | 0.354 |

*Source*: Based on Best (2003) and Sun and McDonough (1989). © John Wiley & Sons.
E-MORB, enriched mid-ocean ridge basalt; HFS, high field strength; HREE, heavy rare Earth elements; LIL, large Earth elements; LREE, light rare Earth elements; N-MORB, normal mid ocean ridge basalt.

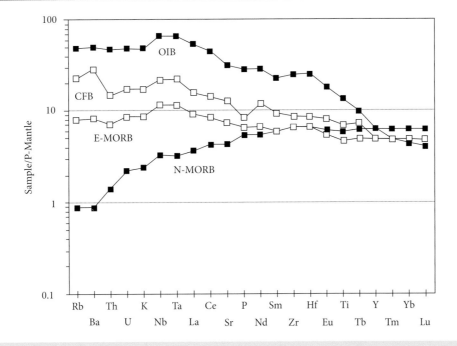

**Figure 10.3**   A spider diagram illustrates minor and trace element variations for normal mid-ocean ridge basalts (N-MORB), enriched mid ocean ridge basalts (E-MORB), continental flood basalts (CFB) and ocean island basalts (OIB) compared to primitive mantle (P-Mantle) values used by Sun and McDonough (1989). Note that the N-MORB are depleted in LILS (Rb, Ba, K), HFS (Th, U, Nb, Ta), and LREE (La, Ce) indicating partial melting of a previously depleted mantle source. Rock samples collected from Ocean Drilling Program Hole 899B located at the Iberia Abyssal Plain (Seifert and Brunotte 1996) in the eastern Atlantic Ocean *Source*: Seifert and Brunotte (1996). Licenced under CC BY-4.0.

(G-MORB) which formed during early stages of continental rifting that produce young ocean basins. G-MORB display significant depletion in HREE (Tm, Yb, Lu) and are more enriched in medium REE (Sm, Eu, and Gd), LREE (Ce), and Th than N-MORB. Saccani (2015) interprets the HREE/MREE depletion as an indicator of a garnet-enriched mantle source.

While MORB is the dominant volcanic rock type at ocean spreading ridges, other rock types occur in varying proportions. Ocean ridges also produce high aluminum basalts, which contain $Al_2O_3$ concentrations ≥16%. Andesite, icelandite, ferrobasalt, trachyte, hawaiite, mugearite, trachybasalt, trachyandesite, dacite (plagiogranite), and rhyolite can occur as minor components at ocean ridges as well as in "leaky" transforms, continental rifts, and ocean islands. The andesitic to rhyolitic volcanic rocks at ocean ridges have higher $TiO_2$ (>1.3%) concentrations compared to the more common convergent margin varieties and are always subordinate to basalt (Gill 1981).

Divergent margins generate the bulk of ocean floor rocks which represent 71% of Earth's area. As a result, mid-ocean ridge basalt and underlying gabbro and peridotite are widespread in our relatively young ocean basins, almost all of which are less than 200 million years old. Granot (2016) proposes based on magnetic anomaly studies that 340 million year old ocean crust exists in the Eastern Mediterranean Sea, perhaps a remnant of ancient Tethys Ocean prior to the breakup of Pangea. What happens to all the other old ocean lithosphere? For more than a billion years (Ga), ocean lithosphere has been subducted and recycled at convergent margins.

## 10.3 CONVERGENT PLATE BOUNDARIES

While divergent plate boundaries are dominated by MORB, chemically diverse igneous assemblages occur in the convergent margins widely distributed in the Pacific Ocean, eastern Indian Ocean and the Caribbean and Scotia Seas (Figure 10.4).

## Box 10.1 Basalt discrimination using trace elements

Various attempts have been made to discriminate basalt types based on geochemical variation diagrams. Analyzing over 2000 ophiolite basalt samples and 560 modern basalt samples from known tectonic settings, Emilio Saccani (2015) has proposed a simple N-MORB normalized $Th_N$ vs. $Nb_N$ diagram to distinguish 10 different types of basalt using absolute measures plotted in a binary diagram as opposed to element ratios or ternary diagrams (Box 10.1, Figure B10.1). The $Th_N$ vs. $Nb_N$

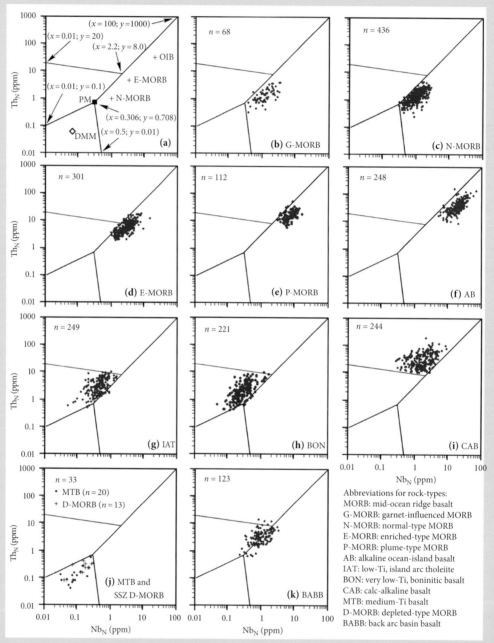

**Figure B10.1.** Over 2000 Proterozoic and younger ophiolite rocks were plotted on $Th_N$ vs. $Nb_N$ diagram to distinguish different types of basalts generated in subduction and non-subduction related environments. (a) Primitive mantle (PM), depleted mantle (DMM), normal mid ocean ridge basalt (N-MORB), enriched mid ocean ridge basalt (E-MORB) and ocean island basalt (OIB). (b) Garnet influenced mid ocean ridge basalt (G-MORB). (c) Normal mid ocean ridge basalt (N-MORB). (d) Enriched mid ocean ridge basalt (EMOR). (e) Plume type mid ocean ridge basalt (P-MORB). (f) Alkaline ocean ridge basalt (AB). (g) Ocean island basalt (OIB). (h) Very low titanium boninitic basalt (B). (i) Calc-alkaline basalt (CAB). (j) Medium titanium basalt (MTB) and supersubduction zone depleted mid ocean ridge basalt (SSZ D-MOR). (k) Back arc basin basalt (BABB). *Source:* Images courtesy of Emilio Saccani.

## Box 10.1 *Continued*

diagram distinguishes basalts generated in oceanic, subduction unrelated settings, rifted margins, and ocean continent transition zones from subduction related basalts. Additionally subduction related basalts can be subdivided into intraoceanic island arc, island arc tholeiites, boninites, and medium Ti basalt classifications (Box 10.1, Figure B10.2). Emilio Saccani (2015) developed a chondrite normalized $(Ce/Yb)_N$ vs. $(Dy/Yb)_N$ diagram to discriminate MORBs with a garnet signature (G-MORB) that has been recognized in alpine-type continental rifts and ocean-continent transition zones (Box 10.1, Figure B10.3). These newly developed discrimination diagrams provide powerful tools, when used in conjunction with detailed field mapping and thin section analysis, to decipher the geologic origin of ancient rocks exposed in orogenic belts, particularly ophiolites.

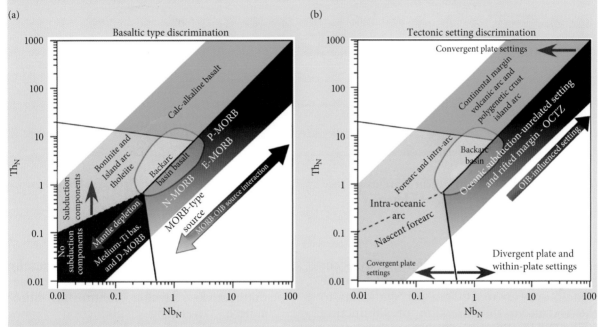

**Figure B10.2.** $Th_N$ vs. $Nb_N$ diagrams are used to identify basalt-type and tectonic settings. *Source*: Images courtesy of Emilio Saccani.

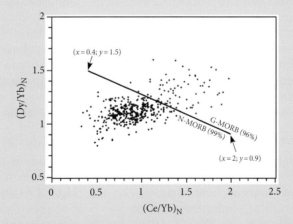

**Figure B10.3.** A chondrite normalized $(Ce/Yb)_N$ vs. $(Dy/Yb)_N$ diagram is used to discriminate MORBs with a garnet signature (G-MORB) that have been recognized in Alpine-type continental rifts and ocean continent transition zones. *Source*: Images courtesy of Emilio Saccani.

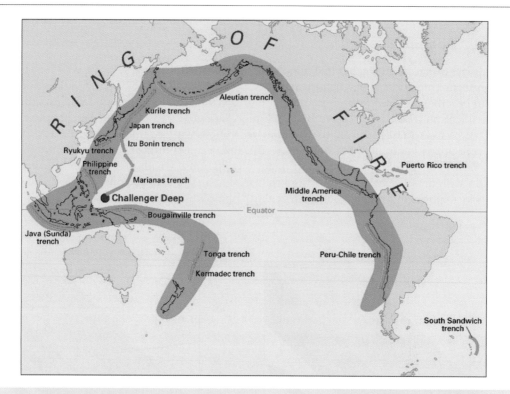

**Figure 10.4**   Earth's convergent margins marked by ocean trenches. *Source*: U.S. Geological Survey, United States Department of the Interior.

Convergent margin magmatism may occur for thousands of kilometers parallel to the trench, and up to 500 km perpendicular to the trench in the direction of subduction (Gill 1981). Plutonic rocks at convergent margins include diorite, granodiorite, quartz diorite, granite, gabbro, tonalite and trondhjemite. This plutonic suite of rocks occurs in batholiths above subduction zones and provides magma to overlying volcanic arcs. The spectrum of possible volcanic rock types varies widely. Youthful island arc environments are dominated by arc basalts and basaltic andesites. In contrast, mature continental arc systems are comprised largely of andesite, with lesser amounts of basalt, dacite, rhyodacite and rhyolite.

As opposed to relatively simple decompression melting of the mantle at divergent margins, convergent margin magmatism is affected by more variables, each of which can change magma composition (Figure 10.5). These variables include:

1   Composition and thickness of the overlying converging plate. Thinner ocean lithosphere in the overlying plate generally produces metaluminous, mafic to intermediate rocks. Thicker continental lithosphere overlying subduction zones commonly yields peraluminous, potassic, intermediate to silicic rocks.

2   Composition of rock material experiencing anatexis. Earth materials experiencing partial melting may include overlying ultramafic mantle wedge, mafic to silicic forearc basement, subducted mafic-ultramafic ocean lithosphere, and marine sedimentary material. The relative proportion of each of these components affects the composition of plutonic and volcanic rocks generated in the arc system.

3   Flux melting in which volatile-rich minerals such as micas, amphiboles, serpentine, talc, carbonates, clays, and brucite release $H_2O$, $CO_2$ or other volatile vapors. These volatiles flux, or lower the melting temperature of, the mantle peridotite and eclogite rocks overlying the subducted slab. Eclogites are high pressure rocks derived from metamorphism of basalt and gabbro in the subducted slab.

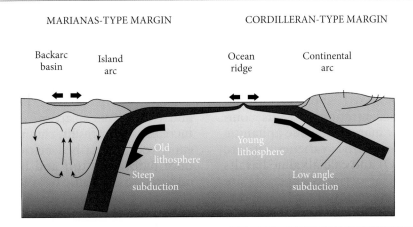

MARIANAS-TYPE MARGIN                    CORDILLERAN-TYPE MARGIN

**Figure 10.5** The steeply dipping Marianas-type island arc subduction and the shallowly dipping Cordilleran continental arc subduction models are illustrated. Note the thick, deep, mantle wedge overlying the Marianas-type margin and the thin mantle wedge in the Cordilleran-type model. *Source*: Kevin Hefferan. © John Wiley & Sons.

4   Diversification processes such as fractionation, assimilation, and magma mixing as well as metamorphic reactions strongly alter the composition of magmas generated in the overlying wedge of the arc system.

5   Dip angle of the subduction zone where old, cold, dense lithosphere favors steep subduction and young, warm, buoyant lithosphere produces shallow subduction zones (Figure 10.5). Steeply inclined subduction zones allow for the melting of thick wedge shaped mantle wedges in the overlying plate. Shallowly dipping subduction zones allow only a thin wedge of mantle to intervene above the subducting plate, minimizing overlying mantle wedge input. The negative buoyancy of old, cold, dense ocean lithosphere is the key force driving deep lithosphere subduction in modern plate tectonics over the past 1 Ga. This negative buoyancy may not have been present in the hot, buoyant Archean ocean lithosphere such that deep subduction may not have been possible before 2.5 Ga (Davies 1992; Ernst 2007; Stern 2008). This may explain the origin of some unique Archean rock assemblages as well as the virtual absence of Archean blueschists discussed later in this text.

While magma composition is highly variable based on the factors described above, Phanerozoic convergent margins are dominated by the calc-alkaline suite of rocks whose compositions are enriched in $SiO_2$,

alkalis ($Na_2O$ and $K_2O$), LIL, LREE elements and volatiles and relatively depleted in FeO, MgO, HFS, and HREE concentrations (Miyashiro 1974; Hawkesworth et al. 1993; Pearce and Peate 1995). The presence of hydrous minerals such as hornblende and biotite indicates that arc magmas contain >3% $H_2O$. Volatiles play an important role in subduction zone flux melting.

The calc-alkaline association (Figure 10.6) of basalt, andesite, dacite, and rhyolite (BADR) is the signature volcanic rock suite of convergent margins and constitutes one of the most voluminous rock assemblages on Earth, second only to MORB (Perfit et al. 1980; Grove and Kinzler 1986). Harker diagram plots of major elements generally indicate a liquid line of descent from a common source, such that BADR rocks are derived from a common parent magma of basaltic composition. Arc tholeiite and calc-alkaline basaltic magmas are derived by partial melting of Earth's mantle at subduction zones; basaltic magmas evolve by a combination of crystal fractionation, assimilation, and magma mixing processes to more silicic andesite, dacite, and rhyolite arc magma suites (Defant and Drummond 1990).

Andesite, named for South America's Andes Mountains, parallel to South America's Peru-Chile trench, is by far the most common calc-alkaline volcanic rock forming at convergent margins (Figure 10.4). The more silicic (dacite, rhyolite) members of the BADR group represent more highly fractionated daughter products. We will discuss each of these below.

The major rock types in volcanic arc systems can be distinguished based upon major element concentrations such as $SiO_2$, $K_2O$, and $Al_2O_3$ content. Basalts contain 45–52% $SiO_2$ and can be subdivided into a number of different varieties based upon major and minor element concentrations. Basalts common in convergent margins include aphanitic and aphanitic-porphyritic varieties of arc tholeiites (low $K_2O$) and calc-alkaline basalts (moderate $K_2O$). Plagioclase phenocrysts are common. The arc tholeiites differ from other tholeiitic basalts (MORB and ocean islands) in containing higher concentrations of $Al_2O_3$, typically in concentrations greater than 16 wt. %. As a result, arc tholeiites are also referred to as high aluminum basalts. The calc-alkaline basalts differ from tholeiites by having higher alkali (notably $K_2O$) concentrations and not displaying iron enrichment typical of tholeiitic fractionation trends (Figure 10.6). As discussed in Chapters 7 and 9, magma viscosity and explosiveness are proportional to $SiO_2$ increases. As a result, the more siliceous volcanic rocks described below commonly produce pyroclastic tuff and breccia deposits in addition to aphanitic to aphanitic-porphyritic crystalline textures.

Andesites are volcanic rocks containing 52–63% $SiO_2$. Andesites can be subdivided based upon the range of $SiO_2$: basaltic andesites, common in youthful island arc

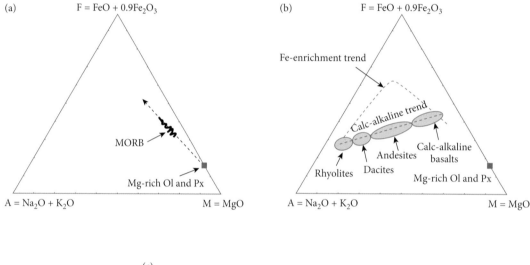

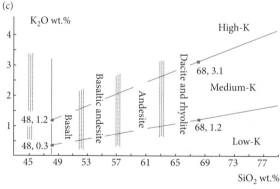

**Figure 10.6** (a) Tholeiitic mid ocean ridge basalts (MORB) display iron enrichment due to the early crystallization of magnesium-rich olivine and pyroxenes; (b) The calc-alkaline suite does not display significant iron enrichment but displays alkali enrichment with progressive crystallization. (c) Three volcanic rock suites are recognized based on percent $Si_2O$ and $KO_2$. Low K assemblages consist of tholeiitic basalt. Medium K assemblages contain calc-alkaline assemblages. The high K suite consists of high K calc-alkaline rocks and trachyandesite rocks referred to as shoshonites *Source*: Based on Gill (1981) and Le Maitre et al. (1989, 2002).

systems, contain >52–57% $SiO_2$ while more silicic andesites, common in mature continental arc systems, contain >57–63% $SiO_2$. Andesites commonly occur as gray, porphyritic-aphanitic volcanic rocks with phenocrysts of plagioclase, hornblende or biotite. Plagioclase phenocrysts are most common and often display euhedral, zoned crystals. Hornblende phenocrysts are also common and may display reaction rims. Pyroxene (principally augite, hypersthene or pigeonite) phenocrysts less commonly occur. Quartz, K feldspar, olivine and feldspathoid phenocrysts are rare. Interestingly, the bulk composition of andesite (and its plutonic equivalent diorite) approximates that of terrestrial crust, suggesting that subduction zone processes have played a significant role in the development of continental crust (Hawkesworth and Kemp 2006). The generation of voluminous andesite is favored by subduction angles greater than ~25°, anatexis of thick (> ~25 km) continental hanging wall plate and partial melting of subducted slabs at depths of 70–200 km (Gill 1981). More silica-rich volcanic rocks such as dacites and rhyolites form in similar settings and involve partial melting of mafic descending slabs, magma mixing, fractionation, and assimilation of the overlying continental crustal slab (Defant and Drummond 1990).

Dacites are quartz phyric volcanic rocks, intermediate between andesite and rhyolite with 63–68% $SiO_2$ (Gill 1981). Dacites are rich in plagioclase and are the volcanic equivalent of granodiorites, in which alkali feldspars are subordinate to plagioclase. When present, phenocrysts are commonly subhedral to euhedral, zoned, and generally consist of oligoclase to labradorite plagioclase or sanidine. Minor minerals commonly include biotite, hornblende, augite, hypersthene, and enstatite.

Trachyandesites (also known as latites and shoshonites) are composed of ~66–69% $SiO_2$ and commonly contain phenocrysts of andesine to oligoclase plagioclase feldspar amidst a groundmass of orthoclase and augite. Trachyandesites are monzonite (plutonic) equivalents and are less oversaturated with $SiO_2$ than the other rock types discussed here; as a result, trachyandesites tend not to be quartz phyric.

Rhyolites (≥69% $SiO_2$) and rhyodacites (~68–73% $SiO_2$) are associated with explosive silicic eruptions that produce fragmental, glassy,

and also occur with aphanitic to aphanitic-porphyritic textures. Rhyodacite is a rock term, not recognized by the IUGS system, for transitional volcanic rocks that bridge the dacite/rhyolite boundary. These rocks can occur as glasses (obsidian or pumice), pyroclastic tuffs, and breccias, or as aphanitic to aphanitic-porphyritic crystalline rocks. Common phenocrysts include alkali feldspar or quartz, with minor concentrations of hornblende and biotite.

In addition to variations in $SiO_2$, arc rocks display significant variation in $K_2O$ concentrations, ranging from low (tholeiitic), medium (calc-alkaline), and high $K_2O$ (calc-alkaline to shoshonite) rock suites (Gill 1981). The progression from tholeiite to calc-alkaline to shoshonite (trachyandesite) reflects increasing $K_2O$ and $K_2O/Na_2O$ and decreasing iron enrichment (Jakes and White 1972; Miyashiro 1974). $K_2O$ content in convergent margin volcanic suites broadly correlates with the thickness of the overlying slab in convergent margin systems (Figure 10.6c). Low K tholeiites dominate with overlying slab thicknesses ranging from ~0 to 20 km; medium to high K calc-alkaline andesites are associated with overlying slab thicknesses of ~20–40 km; high K shoshonites commonly develop where the overlying slab is >40 km thick (Gill 1981).

Three major types of convergent margins occur: (i) ocean–ocean convergence generating youthful island arc volcanic complexes; (ii) ocean-continent convergence generating mature continental arc complexes, and (iii) continent–continent convergent margins marked by the cessation of subduction and consequent continental collision. Many of the same rock types can be found in all three environments; however, continental systems contain a greater proportion of silicic calc-alkaline rocks enriched in quartz and K feldspar; in contrast, island arcs contain a greater proportion of mafic to intermediate rocks as described below

### 10.3.1 Island arcs

Ocean lithosphere is subducted beneath an overlying plate composed of oceanic lithosphere (Figure 10.7a) to produce island arc chains in the eastern Indian Ocean, the Caribbean and Scotia Seas and the Western Pacific Ocean, notably the Marianas Islands.

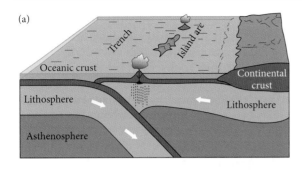

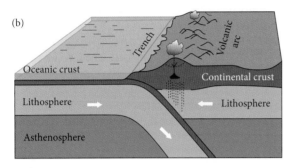

**Figure 10.7** (a) Ocean–ocean convergence producing island arc volcanoes and back arc basins. (b) Ocean-continent convergence producing continental volcanic arcs *Source*: U.S. Geological Survey, United States Department of the Interior.

Island arcs develop on the overlying ocean lithosphere plate, above the subduction zone. Island arc volcanoes are underlain by intermediate to mafic plutonic suites dominated by diorite, quartz diorite, granodiorite, tonalite, and even gabbro. Diorites contain <5% quartz and quartz diorites contain 5–20% quartz; both of these rocks are enriched in plagioclase and hornblende, with lesser amounts of pyroxene and biotite. Hornblende, and to a lesser degree pyroxene and biotite, impart dark (mafic) colors while the plagioclase tends to result in lighter (felsic) hues; together, these mineral suites tend to occur in approximately equal concentrations, producing a speckled light and dark coloration. Diorites and quartz diorites commonly occur in batholiths that overlie ocean subduction beneath youthful volcanic arcs. Granodiorites, which represent the plutonic equivalent of dacites and rhyodacites, contain >20% quartz and more plagioclase than K feldspar. Island arc granodiorites are generally metaluminous, containing hornblende, biotite, and minor amounts of muscovite. Island arc plutons can also contain tonalites and trondhjemites, which are plutonic rocks enriched in plagioclase feldspar and quartz. Tonalites, first described from Monte Adamello near Tonale in the Eastern Alps, contain Ca plagioclase and quartz with minor amounts of K feldspar, biotite, and hornblende. Trondhjemites, also known as plagiogranites, are granodioritic rocks in which Na plagioclase represents half to two-thirds of the total feldspar component.

In addition to the voluminous calc-alkaline rock suite dominated by andesites and basaltic andesites, young island arc systems also produce low K arc tholeiite basalts as well as relatively rare rocks named boninites and adakites. Low K arc tholeiites occur on the oceanward side of the volcanic arc, nearest the trench. Tholeiitic magmas commonly form at subduction zones where the overlying plate is relatively thin. Major element concentrations of the tholeiitic island arc basalts are very similar to MORB, as indicated by their relatively low $K_2O$ concentrations and iron enrichment, suggesting a similar depleted mantle source. The magma is derived by flux melting of the ultramafic mantle wedge overlying the subducted slab as well as melting of the subducted (high pressure eclogite due to metamorphism of basalt and gabbro) slab itself. Island arc tholeiite basalts can be distinguished from MORB by greater concentrations of K and other large ion lithophile elements (LILs such as Ba, Rb, Sr, Cs, Rb) and lower concentrations of high field strength elements (HFS such as Hf, Ta, Ti, V) elements (Perfit et al. 1980; Hawkesworth et al. 1993; Pearce and Peate 1995). Tholeiitic island arc magmas commonly produce basalts, basaltic andesites, and andesites in the volcanic arc and diorite, tonalite (plagiogranite) or lesser granodiorite plutons in the underlying magmatic arc.

Boninites, named for the Bonin Islands in the Western Pacific Ocean, are high Mg (MgO/MgO + total FeO > 0.7) intermediate volcanic rocks that contain a $SiO_2$ saturated (52–68% $SiO_2$) groundmass. These rare rocks contain phenocrysts of orthopyroxene, and notably lack plagioclase phenocrysts (Bloomer and Hawkins 1987). Boninites are enriched in Cr (300–900 ppm), Ni (100–450 ppm), volatile elements, light rare Earth elements as well as Zr, Ba, and Sr. Boninites are depleted in heavy rare Earth elements (HREE) and high field strength (HFS) elements. These unusual rocks occur proximal to the trench and bear the geochemical signature of primitive mantle derived magmas produced early in the subduction cycle (Hawkins et al. 1984; Bloomer and Hawkins 1987; Pearce and Peate 1995). Thus, boninites are a product of subduction related melting below the forearc of youthful island arc systems. Van der Laan et al. (1989) suggest that boninites are produced by high temperature, low pressure remelting of previously subducted ocean lithosphere (in Wyman 1999). Interestingly, boninites can be associated with rare ultramafic komatiites that we will discuss later in this chapter.

Adakites are silica saturated (>56% $SiO_2$) rocks with high Sr/Y and La/Yb ratios (LREE enriched relative to HREE) and low HFS (such as Nb, Ta) concentrations. Adakites, named for Adak Island of the Aleutian Island chain, have been thought to be derived by slab melting of eclogite and/or garnet amphibolite from the descending ocean lithosphere (Kay 1978; Defant and Drummond 1990;

Stern and Killian 1996; Reay and Parkinson 1997). While it was initially believed that adakites only form only where young (<25 Ma), thin, hot ocean lithosphere is subducted beneath island arc lithosphere, adakites are now known to form at continent-continent collision sites as a result of shallow slab subduction of continental lithosphere (Chung et al. 2003). Early plate tectonic models suggested that continental lithosphere did not undergo subduction. Recent studies document that indeed subduction of continental subduction to depths exceeding 120 km does occur, producing coesite and diamond bearing garnet peridotites and eclogites. Shallow slab subduction and lithosphere recycling at subduction zones may play a significant role in the development of adakites as well as their plutonic equivalents trondhjemites and tonalites. Research continues to determine possible relationships of adakite formation with Archean trondhjemite-tonalite and granodiorite (TTG) associations and the evolution of continental crust (Drummond and Defant 1990; Castillo 2006; Gomez-Tuena et al. 2007).

---

## Box 10.2 Tonalite, trondhjemite and granodiorite association (TTG)

Plutons containing tonalite, trondhjemite, and granodiorite (TTG) are found in subduction zone environments ranging from the Archean to the Recent. However, Archean (>2.5 Ga) subduction zone plutonic rocks consist dominantly of tonalite, trondhjemite, and granodiorite. Trondhjemites were named by V.M. Goldschmidt in 1916 for holocrystalline, leucocratic Norwegian rocks enriched in Na-plagioclase and quartz and depleted in biotite and potassium feldspar (Barker 1979). Trondhjemite are similar to tonalites but contain greater concentrations of Na plagioclase (oligoclase to albite) and more variable K feldspar concentrations (Box 10.2, Figure B10.4a). Tonalites and trondhjemites are also known as plagiogranite. **Charnockites**, an orthopyroxene bearing suite of rocks of generally granitic composition, also occur with the TTG association.

TTG associations occur in Archean rocks such as the Pilbara craton of Australia, and the Beartooth and Big Horn Mountains of Wyoming. In contrast, Proterozoic and younger convergent margin granitic rocks consist predominantly of granite and granodiorite, with TTG associations representing a very small component in ocean–ocean convergence. What conditions changed at ~2.5 Ga? Shallow dipping subduction zones "pinch out" the overlying mantle wedge such that the wedge component plays a relatively minor role in magma genesis. Archean subduction involved higher geothermal gradients, shallow subduction, and melting of the down going ocean lithosphere, with minimal input from the overlying lithosphere wedge. Shallow subduction of oceanic lithosphere at unusually low angles (Figure B10.4b) has been proposed as a model for the growth of Archean continental crust through the generation of TTG plutonic rocks and their volcanic equivalents – adakites (Martin 1986; Smithies et al. 2003).

*Continued*

**Box 10.2** *Continued*

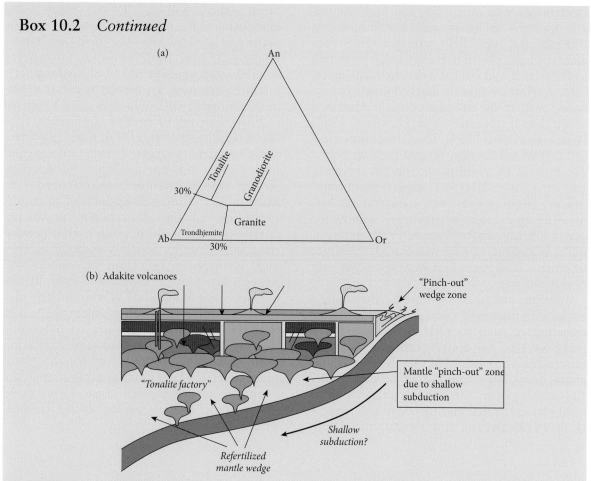

**Figure B10.4** (a) Classification of some granitoid rocks enriched in plagioclase and > 20% quartz. Trondhjemite is a light colored tonalite containing Na-rich oligoclase-albite. An = anorthite (Ca plagioclase), Ab = albite (Na plagioclase) and Or = orthoclase (K feldspar). *Source*: Based on Barker (1979) and Best (2003). © John Wiley & Sons. (b) Shallow subduction of Archean ocean lithosphere may have produced tonalite plutons and adakite volcanic rocks. *Source*: Geological Survey of Canada, Natural Resources of Canada 2009.

Insofar as the overlying arc lithosphere is relatively thin in immature island arc systems, young volcanic arcs are dominated by basalts and basaltic andesites with rare boninites and adakites. Prolonged subduction in island arc systems generates increasingly thicker arc lithosphere. As island arc lithosphere thickens, andesites and dacites predominate as the $Si_2O$ and $K_2O$ contents of all rocks increase with the development of continental-type arc lithosphere (Miyashiro 1974). In nearly all convergent margins, the calc-alkaline association is generated by fractional crystallization of basaltic magma derived by partial melting of overlying mantle peridotite, fluxed by fluids released from the dehydrated subducted oceanic lithosphere slab. The continued removal of crystals from melt leads to continuous variation in the residual liquid (liquid line of descent) generating basalt, andesite, dacite, and rhyolite. In addition to fractionation, open system diversification processes such as assimilation and magma mixing alter magma chemistry (Grove and Kinzler 1986) and influence the chemical composition of magmas generated in the mantle wedge overlying the subducted slab. In addition to magmatism within the island arc complex, igneous activity also occurs behind the island arc in back arc basins.

### 10.3.2 Back arc basins

Although compressional forces dominate island arc settings, lithospheric extension can occur in the overlying plate, behind the arc, resulting in the development of **back arc basins** (Figure 10.7a). How does back arc extension occur? "Trench Pull" forces move the volcanic arc toward the subduction zone resulting in lithospheric extension behind the volcanic arc (Chase 1978). Extension is manifested as normal faults and back arc spreading. The western and northern Pacific Ocean (Figure 10.8) provide excellent examples of back arc basins which include the Sea

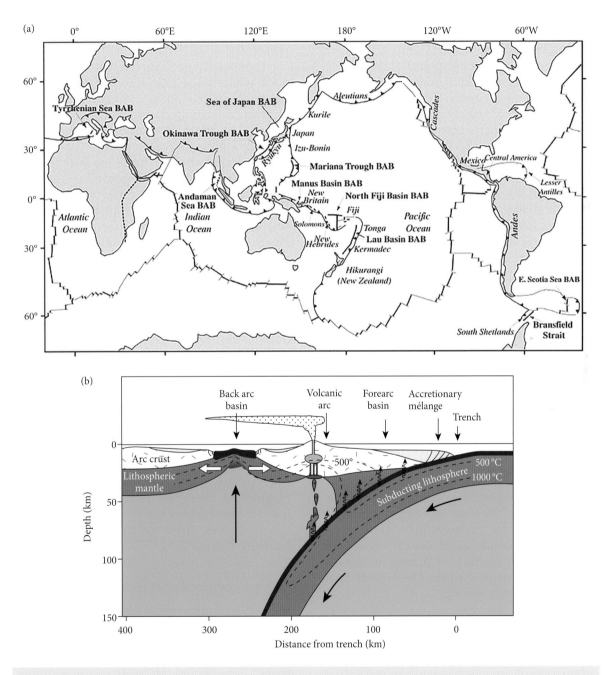

Figure 10.8 (a) Modern back arc basins are concentrated in the Western Pacific Ocean. (b) Back arc basins form by extension within the arc crust. *Source*: Back-arc basin, Wikipedia. Licensed under CC BY-SA 3.0.

of Japan, the Bering Sea, the Lau Basin-Havre Trough, Manus Basin and the Marianas Trough.

Back arc basins erupt a diverse suite of volcanic rocks including basalt, basaltic andesite, andesite, and dacite, but tholeiitic and alkalic basalts commonly dominate. As suggested in our earlier discussion of island arc tholeiitic basalts, the relative proportions of andesitic versus basaltic magmas is related to the nature and thickness of the lithospheric wedge above the subduction zone (Gill 1981). In some island arc settings (e.g. Kermadec, Marianas, Scotia, and Vanuatu arcs), basalts dominate over andesites in the back arc and the volcanic arc regions. Fryer et al. (1981), on the basis of trace element chemistry, identified a distinctive group of rocks known as arc basin basalts (BABB). BABB are tholeiitic, with geochemical similarities to both MORB and arc tholeiite trends. Relative to MORB, BABB display greater enrichment of $H_2O$, alkali elements, and large ion lithophile elements. BABB are slightly depleted in Ti, Y, and Nb and display flat REE patterns 5–20 times that of chondrites (Sinton et al. 2003). Back arc basin basalts may show relative enrichment in volatile elements, Th, and LREE, which suggests the involvement of subduction related fluids in magma genesis (Pearce and Peate 1995).

Why do back arc basins produce a wide array of rock types that range from near MORB to calc-alkaline compositions? Extension in the back arc results in partial melting of mantle peridotite producing MORB-like rocks. Calc alkaline magma sources include the mantle wedge situated above the downgoing slab, recycled subducted lithospheric slab, and subducted marine sediment. Thus, back arc basin basalts are produced by a combination of partial melting of lherzolite upper mantle wedge that has been fluxed by volatiles released by the subducted ocean lithosphere, decompression melting of mantle peridotite at back arc spreading ridges (Fretzdorff et al. 2002) and magma mixing.

### 10.3.3 Continental margin arcs

Mature convergent margins, involving the subduction of ocean lithosphere beneath thick continental lithosphere, occur along the Eastern Pacific margin and extend from the Cascades southward to the Andes Mountains (Figure 10.7b). Ascending hydrous fluids from the subducted ocean slab chemically react with the overlying wedge composed of mantle and thick continental lithosphere. These magmas produce continental arc plutons that are more silicic than island arc plutons as a result of the thick overlying continental lithosphere through which subduction zone fluids must penetrate. Extensive assimilation and fractionation within the overlying continental slab result in $K_2O$ and $SiO_2$ enrichment within plutons.

Ocean–continent convergent margins produce voluminous granodiorite, diorite, granite, and tonalite plutons. Large granodiorite plutons commonly dominate ocean-continent convergent plate boundaries. For example, the Sierra Nevada consists of 25–30 km thick upper crustal rocks containing hundreds to thousands of individual granodioritic, dioritic, and tonalitic plutons. The Sierra Nevada composite batholith developed due to eastward dipping subduction beneath western North America. Arc magmatism continued from 220 to 80 Ma (Fliedner et al. 2000; Ducea 2001).

Magma from these intermediate-silicic plutons erupts onto Earth's surface producing composite volcanoes. Together with voluminous andesites, rocks such as dacites, rhyodacites, rhyolites, and latites display, that aphanitic-porphyritic to pyroclastic textures occur in composite volcanoes such as Mt. St. Helens and Crater Lake in the Cascade Range (U.S.). Highly alkalic shoshonites also erupt over thick continental lithosphere. **Shoshonites** are dark colored, K-rich trachyandesites, commonly containing olivine and augite phenocrysts with a groundmass of labradorite plagioclase, alkali feldspar, olivine, augite, and leucite. Shoshonites occur in thickened lithosphere farthest from the trench region and in back arc basins.

### 10.3.4 Continental collision zones

In long lived convergent margins, the termination of ocean lithosphere subduction marks the transition from mature arc-continent convergence to a final collision of two continental blocks. The Alpine-Himalayan orogenic system is the modern classic model of continent-continent collision following tens of millions of years of ocean lithosphere subduction. Ancient examples include the complete subduction of multiple ocean basins 250 million years ago, resulting in continent-continent collision between Europe and Asia producing the Ural Mountains and of the Americas, Africa and Europe to create the supercontinent Pangea.

In continent-continent collisions, the lower plate continental crust traditionally been viewed as excessively buoyant such that deep subduction to great depths could not occur. In this plate tectonic model, the lower plate continental crustal slab breaks off (delaminates) and underplates the overlying continental crustal plate producing a doubly thick crustal mass. However, Himalayan research involving the collision of India and Asia over the past 90 Ma documents continental crust subduction to depths of over 200 km. Ultra high pressure (UHP) indicators such as eclogite, majoritic garnet peridotite, coesite, and microdiamond bearing rocks in the Dabie-Sulu UHP metamorphic belt of China indicate that deep continental crusts subduction indeed occurs, altering the traditional view of collisional tectonism (Ye et al. 2000; Liou et al. 2009).

Magmas generated at continental collision zones are diverse. Melting at the base of doubly thick collisional zones produces $Al_2O_3$-, $K_2O$-, and $SiO_2$-rich igneous rocks such as rhyolites, rhyodacites, and shoshonites and plutonic rocks of increasingly granitic composition. These magmas are the result of relatively flat subduction ($<\sim25°$), thick continental lithosphere ($>25$ km), higher degrees of partial melting of continental lithosphere and/or arc basement and the diminished role of ocean lithosphere subduction. Alkaline basalts also occur in continent-continent collisions as a result of upwelling mantle melts. Rhyolites and rhyodacites are characterized by high viscosity which retards lava flow, resulting in thick accumulations of limited aerial extent. Common minerals include quartz, K feldspar, biotite, plagioclase, anorthoclase, and magnetite. Shoshonitic magmas are generated furthest from the trench, wherein melts assimilate $K_2O$ and $Na_2O$ as they rise through thick slabs of overlying continental lithosphere. Volcanic eruptions of rhyolite to shoshonite lavas can erupt explosively, generating voluminous pyroclastic tuffs and breccias, or produce lava flows that solidify to produce glassy and/or aphanitic-porphyritic textures.

The plutonic equivalent of rhyodacite and rhyolite are granite or granitoid rocks. The term "granitic" or "granitoid" is loosely used for silica oversaturated plutonic rocks that contain essential potassium feldspars and quartz. Granitoids include the IUGS fields of quartzolite, quartz-rich granitoid, alkali feldspar granite, granite, quartz granitoid, granodiorite, and tonalite. Granitoid rocks that form at mature convergent margins tend to be peraluminous to metaluminous, containing hornblende, biotite, and/or muscovite. Although granites of variable composition occur, S-type and I-type granites tend to predominate. I-type magmas form by partial melting of mafic to intermediate igneous rocks in or above the subduction zone at ocean–ocean or ocean-continent convergent margins. Peraluminous, K-rich, S-type granites, and granodiorites are particularly common at continent-continent collisions. The peraluminous sedimentary component is derived from phyllosilicate minerals in graywackes and mudstones of the continental crust and accretionary wedge. These sedimentary materials melt to produce two mica granites containing biotite and muscovite.

## Box 10.3 Granite classification

Strictly speaking, the term "granite" is restricted to plutonic rocks containing 20–60% quartz and 35–90% alkali feldspars. Thus, the two essential mineral groups in granite are quartz and feldspars. Other minor minerals include hornblende, biotite, and muscovite. Accessory minerals include magnetite, rutile, tourmaline, sphene, apatite, molybdenite, gold, silver, and cassiterite. Granite plutons are genetically associated with Precambrian cratons (Pitcher 1982). Phanerozoic plutonic belts are found along continent-ocean subduction zones or at continent-continent suture zones. Within orogenic

*Continued*

## Box 10.3 *Continued*

settings, granites may be emplaced synchronous (syn-kinematic) with convergence, as late stage collisional plutons, or as post-kinematic intrusions.

Granites have been subdivided by a number of methods, one of which attempts to infer source rock origin. On the basis of primarily chemical and mineralogic characteristics, Chappell and White (1974) and others recognize four distinct types of granite (M, I, S, and A types) based upon the nature of the inferred parental source rock. I, S, and M types are orogenic granites associated with subduction, whereas A-type granites are anorogenic in origin.

- **M-type granites** (Pitcher 1982) are derived from mantle derived parental magmas, as indicated in low $Sr^{87}/Sr^{86}$ ratios (<0.704). M-type granites are associated with calc-alkaline tonalites, quartz diorites, and gabbroic rocks. In addition to quartz and feldspars, hornblende, clinopyroxene, biotite, and magnetite are among the major minerals. M-type granites develop in island arc settings. Copper and gold mineralization is associated with M-type granites.
- **I-type Granites** (Chappell and White 1974) are generated by melting of an igneous protolith from either the downgoing oceanic lithosphere or the overlying mantle wedge. I-type granites contain a broad range of rock compositions. More mafic varieties are enriched in hornblende, biotite, magnetite, and sphene. I-type granites are enriched in $Na_2O$ and $Ca_2O$ and contain lower $Al_2O_3$ concentrations. I-type granites have $Sr^{87}/Sr^{86}$ ratios less than 0.708, usually in the range of 0.704–0.706, indicating magma derived from a mantle source. Porphyry copper, tungsten, and molybdenum deposits are associated with I-type granites. I-type granites are prevalent along the Mesozoic-Cenozoic Andes Mountains (Chappell and White 1974; Beckinsale 1979; Chappell and Stephens 1988).
- **S-type Granites** (Chappell and White 1974) are produced by the melting of sedimentary crustal rocks in collision zones. Unlike I-type granites, S-type granites are restricted to high $SiO_2$ types. S-type granites are also known as two mica granites in that they commonly contain both muscovite and biotite, reflecting the peraluminous content of the sedimentary source rock rich in phyllosilicate minerals. Hornblende is conspicuous by its absence. Other minerals include monazite and aluminosilicate minerals such as garnet, sillimanite, and cordierite. S-type granites are depleted in $Na_2O$ but enriched in $Al_2O_3$ (peraluminous). S-type granites have $Sr^{87}/Sr^{86}$ ratios >0.708, indicating that source rocks had experienced an earlier sedimentary cycle. Tin deposits are associated with S-type granites (Chappell and White 1974; Beckinsale 1979).
- **A-type Granites** (Loiselle and Wones 1979) are anorogenic rocks produced by activities that do not involve the subduction and collision of lithospheric plates. Anorogenic environments include stable cratons, continental rifts, ocean islands and inactive, post-collisional continental margins. Associated with A-type granites are alkali rich, relatively anhydrous rocks that can include alkali granite, syenite, alkali syenite, and quartz syenite. A-type granites are enriched in alkaline elements with high K/Na and (K+Na)/Al ratios as well as high Fe/Mg, F, HFS elements (Zr, Nb), and REE concentrations (Nb, Ga, Y, Ce); A-type granites are depleted in Mg, Ca, Al, Cr, Sr, Ni and so on and have low water contents and high Ga/Al ratios (Collins et al. 1982; Whalen et al. 1987). Relative to I-, S-, and M-type granites, A-type granites are more enriched in large ion lithophile elements and depleted in refractory elements (Creaser et al. 1991). A-type granites are peralkaline and commonly contain biotite, alkali pyroxenes, alkali amphiboles, and magnetite (Collins et al. 1982).

Anorogenic granite, alkali granite, and syenite were particularly common 1.1–1.4 Ga following the assembly of the Mid Proterozoic Columbia Supercontinent. These A type, granitic intrusions were widespread in North America, extending from Mexico to the Lake Superior region. Significant volumes of anorogenic granites occur in Precambrian cratons throughout the world. These Mid Proterozoic granitoid rocks were remarkably similar in age, composition, and appearance, displaying rapakivi texture.

A number of models have been proposed for the origin of A-type granites. One model proposes the overthickening of continental lithosphere such that the upper mantle and base of the crust partially melt generating silicic magma that subsequently rises and cools at shallower depths to form anorogenic

## Box 10.3  *Continued*

granite. Other "residual source models" propose that A-type granites, such as the Pikes Peak Batholith in Colorado (USA), develop from the partial melting residual silicic granulites that had previously generated I-type granites (Barker et al. 1975; Collins et al. 1982). Alternative models suggest that A-type granites are derived by melting quartz diorite, tonalite or granodiorite parent rocks (Anderson 1983). The table below summarizes the major features of A-, S-, M-, and I-type granitoids.

**Table B10.1**  The major features of M-, I-, S-, and A-type granitoids.

|  | A | S | I | M |
|---|---|---|---|---|
| $SiO_2$ | 60–80% | 65–79% | 53–76% | 54–73% |
| $Na_2O$ | >2.8% | Low, <3.2% | High, >3.2% | Low, <3.2% |
| $K_2O/Na_2O$ | High | High | Low | Very low |
| $Sr^{87}/Sr^{86}$ | 0.703–0.712 | >0.706 | <0.706 | <0.704 |

Convergent margins contain a diverse range of rock types. In addition to intermediate and silicic igneous rocks, mafic, and ultramafic assemblages also occur at convergent margins due to magmatic processes and/or tectonic displacement. Let us first consider tectonically emplaced Alpine orogenic complexes and then we'll briefly discuss mafic and ultramafic intrusive complexes.

### 10.3.5  Alpine orogenic complexes

**Alpine orogenic complexes** are fault bounded, deformed rock sequences that mark the site of present or former convergent margins. Unlike the intermediate to silicic igneous rocks that develop in situ (in place) as a result of subduction induced magmatism, alpine orogenic complexes have been transported far from their site of origin by thrust faulting and shearing. Because such tectonism can be intense, these complexes are commonly dismembered into fault blocks and jumbled together in a haphazard fashion such that their original layering may be disrupted. Alpine orogenic bodies contain disrupted pelagic sediment layers, basalt, cumulate mafic, and ultramafic layers as well as tectonized mantle slices of ocean lithosphere and calc-alkaline intrusive and volcanic assemblages. Alpine orogenic bodies are commonly associated with tectonic mélanges.

A tectonic **mélange** (from French word for mixture) is an intensely sheared, heterogeneous rock assemblage embedded within a highly deformed mud matrix. Mélanges form at subduction zones where rocks and tectonic blocks are sliced from the downgoing oceanic lithosphere and often mixed with rocks formed in forearc settings. The diverse suite of rocks may include: (i) deformed and altered mid ocean ridge, ocean island, and ocean plateau basalts; (ii) limestone, chert, and other marine sedimentary rocks; and (iii) slices of eclogite, peridotite and blueschist from subducted oceanic or forearc lithosphere. Blueschists are high pressure/low temperature metamorphic rocks characteristic of subduction zones.

**Ophiolites** constitute one type of alpine deposit in which oceanic or volcanic arc lithosphere is preserved in orogenic belts. The term ophiolite was first proposed by Steinmann (1905) for serpentinized rocks in the Alps. Over the next several decades, Steinmann recognized a suite of rocks that, thereafter, became known as the "Steinmann trinity." These three rock types consist of: pelagic chert, serpentinite (hydrothermally altered peridotite), and spilites (altered pillow basalts). As the term is currently used, ophiolites are thought to represent coherent slices of oceanic lithosphere, volcanic arc basement, and/or fore arc or back arc basin lithosphere "obducted," or thrust, onto the edge of continents above subduction zones.

Ophiolites may form at continental rifts, ocean spreading centers, hotspot plumes or at convergent plate boundaries (Dilek and Furnes 2011; Saccani 2015). While it is intuitively obvious that ophiolites may originate by the generation of ocean lithosphere at emergent continental rifts and ocean spreading ridges, petrologists recognize that many ophiolites represent oceanic fragments produced in volcanic arc and suprasubduction zone (SSZ) settings. Suprasubduction zone ophiolites develop due to extension in the plate overlying the subduction zones. SSZ ophiolites may develop in back-arc basins, within the volcanic arc or in fore arc basins. Researchers use immobile elements as petrogenetic indicators to determine ophiolite sites of origin. Using geochemical data from over 2000 ophiolitic basalts of uncertain origin from around the world and 500 basalt samples from modern tectonic locations, Saccani (2015) created ophiolite discrimination diagrams, using Th–Nb and Ce–Dy–Yb systematics, that differentiate between oceanic, nonsubduction-related environments such as rifts, hotspots, and mid ocean ridge settings, versus subduction related ophiolites that form in continental arc, island arc suprasubduction zone environments. Strong enrichment in Th relative to Nb occurs with crustal input during subduction or crustal contamination. Interestingly, back-arc basin ophiolites show both subduction and nonsubduction-related geochemical patterns in different locations on Earth, depending upon the degree of subduction zone recycling affecting that particular tectonic site. Continental arc ophiolites contain the highest Th-Nb ratios whereas intra-ocean settings record the lowest Th-Nb ratios (Box 10.1). Both the origin of ophiolites and their means of emplacement on the edges of continents remain areas of intense research.

Complete ophiolite sequences display a stratigraphic sequence similar to that of ocean lithosphere (Figure 10.2a). The stratigraphy of an idealized ophiolite sequence was defined by the first Penrose Conference in 1972 as follows:

- Layer 1: pelagic, marine sedimentary rock such as ribbon chert, thin shale beds, and limestone derived from the lithification of siliceous ooze, clay, and calcareous ooze sediments, respectively.
- Layer 2A: mafic volcanic complex, which may contain pillow basalt
- Layer 2B: mafic sheeted dike complex
- Layer 3: cumulate gabbroic complex with basal cumulate peridotites and pyroxenites
- Layer 4: tectonized ultramafic complex consists of variably metamorphosed harzburgite and dunite. Podiform chromite deposits occur with dunite bodies. The tectonized ultramafic complex overlies a metamorphic basal sole thrust.

While the idealized stratigraphic layering of ophiolites mimics ocean lithosphere layering, the complete idealized stratigraphic sequence is rarely preserved. Most ophiolites are found in tectonic slices such that the complete stratigraphy is rarely preserved. Tectonically disrupted ophiolites are referred to as partial, or dismembered ophiolites. Ophiolites occur throughout the world (Figure 10.9) and mark the former location of ocean lithosphere subduction. Excellent examples of ophiolites include the following localities: Oman, Troodos (Cyprus), Coast Range, Newfoundland and Morocco.

Ophiolites are important in providing: (i) Valuable ore deposits containing Cu, Ag, Au, Zn, Ni, Co, Cr, and other metals; and (ii) Evidence documenting oceanic basin development and ocean lithosphere subduction dating from the Precambrian to the present. Well documented ophiolites less than 1 Ga occur throughout the world. Archean examples, dating as far back as 3.8 Ga, are highly controversial and may (Furnes et al. 2007) or may not (Nutman and Friend 2007; Hamilton 2007) represent true ophiolites.

In addition to their occurrence in alpine orogenic complexes and ophiolites, mafic, and ultramafic magmas intrude convergent margin assemblages to form the concentrically zoned or layered plutons discussed below.

## 10.3.6 Alaska-type (zoned) intrusions

**Alaska-type intrusions** consist of concentrically layered (zoned) plutons formed in convergent margin settings. Alaska-type intrusions are commonly several kilometers in diameter, exhibit a dunite core and pyroxenite shell, surrounded by massive gabbro. Late granitic zones may also occur around the perimeter of the intrusive structure. In contrast to tectonically emplaced alpine suites, Alaska-type plutons form *in situ* (in place) by intrusion of magma into the surrounding country rock. Originally recognized in Alaska and in the Ural Mountains, these plutonic bodies have since been identified in many other localities throughout the world. Buddington and Chapin (1929) noted the concentric layers in the Blashke Island complex of Alaska (Figure 10.10). Alaska-type intrusions are economically important as sources of metals, particularly platinum group elements (PGE).

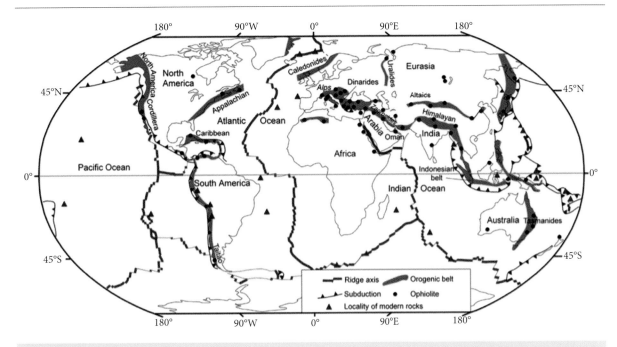

**Figure 10.9** Major ophiolite, orogenic, and ocean basin sampling locations throughout the world. *Source*: Courtesy of Emilio Saccani.

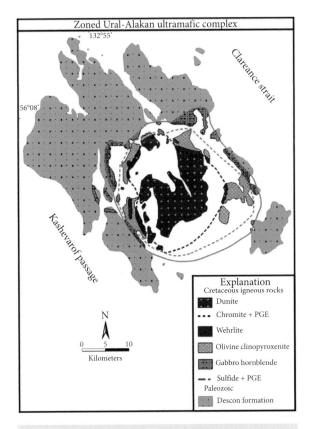

**Figure 10.10** Zoned intrusion from the Blashke Islands complex, southeast Alaska. PGE, platinum group elements. *Source*: Modified from Kennedy and Walton (1946). © John Wiley & Sons.

Irvine (1959) conducted extensive work on the Cretaceous age Duke Island (Alaska) Complex. The Duke Island Complex consists of a 3 km diameter plutonic body at Judd Harbor and a 5 km diameter intrusion at Hall Cove. These two intrusions, which are likely continuous at depth, contain a dunite core surrounded by successive rings of peridotite, olivine pyroxenite, hornblende-magnetite pyroxenite and gabbro; all of which are cut by late granitic rocks. Irvine (1959) noted layers that exhibited graded beds within ultramafic rocks at Duke Island, in which crystal diameters are coarse grained at the base and fine upward within a layer. Duke Island remains an active exploration site with economic deposits of Cu, Ni, Pt, and Pd.

Alaska-type ultramafic-mafic plutons commonly occur as post-orogenic intrusions in volcanic arc or accretionary mélange terrains. A number of different processes have been suggested for their formation; these include: fractionation of ultramafic or mafic parent magma from the upper mantle, magma mixing at convergent plate boundaries or magmas from deep mantle plumes (Taylor Jr 1967; Tistl et al. 1994; Sha 1995; Ishiwatari and Ichiyama 2004).

While divergent and convergent margins produce the bulk of Earth's magmatism, igneous rocks also develop within lithospheric plates without any direct link to plate boundary processes as discussed below.

## 10.4 INTRAPLATE IGNEOUS ACTIVITY

Intraplate igneous activity refers to magma generation and volcanic eruptions that produce igneous rock suites within lithospheric plates, rather than at plate boundaries. Intraplate magmatism may be initiated by oceanic or continental hotspot activity, continental rifts, overthickened continental lithosphere or mantle derived melts.

Large igneous provinces (LIP), encompassing volumes $>10^6$ km$^3$ (Mahoney and Coffin 1997), are the greatest manifestation of intraplate igneous activity on Earth. LIPs erupt huge magma volumes of incredibly short duration (1–5 million years); a large proportion (>75%) of the total igneous volume is emplaced within a few hundred thousands of years (Bryan and Ernst 2008; Bryan and Ferrari 2013; Self et al. 2014). Most LIP (Figure 10.11) are basaltic in composition marked by pahoehoe lava flows. However silicic examples, known as SLIP (silicic large igneous provinces), also occur. The most widespread Phanerozoic intraplate magmatic features consist of massive tholeiitic flood basalts. These massive volcanic landforms form as both oceanic flood basalts (OFB) and continental flood basalts (CFB). In the following section, we will first consider oceanic intraplate magmatism and then later discuss continental intraplate assemblages.

### 10.4.1 Oceanic intraplate magmatism

Ocean islands and ocean plateaus form above mantle hot spots that erupt anomalously high volumes of tholeiitic and alkalic basaltic lava onto the ocean floor. Ocean islands are volcanic landforms that rise upward above sea level. Seamounts are volcanically produced peaks below sea level. Oceanic plateaus are broad, flat topped areas that result from massive outpourings of lava flowing laterally from source vents. Oceanic plateaus cover large areas of the ocean floor ranging up to 10 million km$^2$.

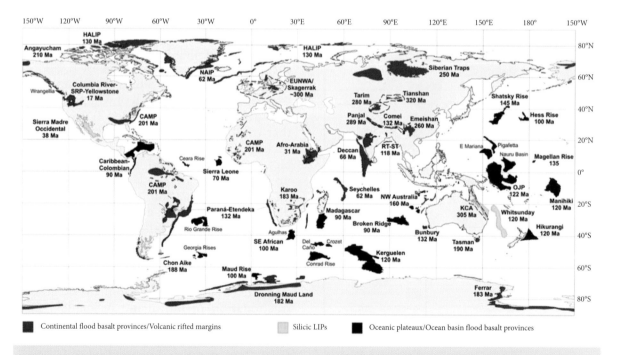

**Figure 10.11** Earth's large igneous provinces (LIP). CAMP, Central Atlantic Magmatic Province; NAIP, North Atlantic Igneous Province; HALIP, High Arctic large igneous province, KCA, Kennedy-Connors-Auburn, RT-ST, Rajmahal Traps–Sylhet Traps, EUNWA – European, northwest Africa.
*Source*: Courtesy of the Geological Society of America.

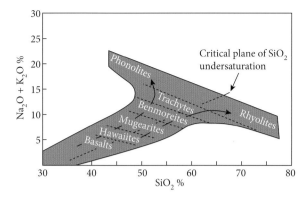

**Figure 10.12** The fractionation sequence occurring at ocean islands produces a diverse suite of volcanic rocks with variable alkali and silica concentrations. *Source*: Courtesy of Stephen Nelson.

### Ocean island basalts (OIB)

Ocean island basalts are a geochemically distinct suite of rocks distinctly different from MORB (Figure 10.12). In contrast to MORB, OIB are more alkalic and are less depleted; in fact, they may be somewhat enriched with respect to incompatible elements such as K, Rb, U, Th, and the LREE (Hofmann and White 1982; Hofmann 1997). The different geochemical signatures have been interpreted to represent different mantle source areas. While MORB were considered to represent partial melts of an upper, depleted (previously melted) mantle, OIB were considered to perhaps represent partial melts from a deeper, undepleted mantle source.

OIB display large variations in Sr, Nd, and Pb and other isotopic ratios, suggesting the role of multiple sources and processes. Various hypotheses have been proposed to account for OIB chemistry including:

1  Small degrees of melting of a primitive mantle source;
2  Melting of a mantle source enriched in alkali elements;
3  Incorporation of subducted oceanic crust in the source region; or
4  Entrainment of subducted sedimentary rocks in the source region (Hofmann and White 1982; Hofmann 1997; Kogiso et al. 1998; Sobolev et al. 2005).

The isotopic signatures of many OIB indicate that magmas were derived from nonprimitive sources of variable mantle composition. For example, Rb/Sr, Nd/Sm ratios are lower than primitive mantle ratios while U/Pb, Th/Pb, and U/Th ratios are higher than primitive mantle sources (Hofmann and White 1982). In fact, most OIB display isotopic ratios indicative of an enriched mantle source, particularly their elevated incompatible element and NiO concentrations. Why is the mantle composition so variable? One explanation involves mantle enrichment due to the incorporation of a recycled oceanic lithosphere derived from ancient subduction zones (Hofmann and White 1982; Hofmann 1997; Turner et al. 2007). Hirschmann et al. (2003) and Kogiso et al. (2003) propose that undersaturated, nepheline normative OIB magmas may be derived by partial melting of a garnet pyroxenite, which itself is derived by the mixing of subducted MORB basalt and mantle peridotite. Thus, at least some ocean island basalt hot spots tap mantle melts enriched by ocean lithosphere subducted up to 2.5 Ga (Turner et al. 2007). Let us consider the best known OIB location: the Hawaiian Islands.

From the ocean floor up, the Hawaiian Islands constitute the highest mountains on Earth with a relief greater than 9.5 km (~31 000 ft) high; Mt. Everest in contrast has an elevation just over 9 km (~29 000 ft.). The Hawaiian hotspot has been active for over 80 million years, generating a chain of seamounts and islands extending for a distance of 5600 km. The Hawaiian Islands are dominated by olivine tholeiites; tholeiitic basalts comprise ~99% of the exposed Hawaiian volcanic rocks with alkalic basalts contributing only a small fraction. Early eruptions of alkali basalts are followed by extensive tholeiitic basalts that generate massive shield volcanoes. Late stage Hawaiian volcanism reverts to alkali basalts (Hawaiites to benmoreites), perhaps indicating a lower degree of melting as the island moves away from the hot spot magma source and temperatures decline. The dominant tholeiitic iron enrichment trend likely results from fractional crystallization of early formed magnesium-rich olivine and pyroxene. As magmas become more enriched in iron, alkali-enriched hawaiite forms by fractionation of tholeiitic basalt. The addition of magnetite to the crystallizing assemblage causes magmas to become progressively less enriched in iron. Continued crystallization produces an iron depletion and alkali enrichment trend in which magmas evolve successively from hawaiite to mugearite and benmoreite. Depending upon the degree of silica saturation, the final minerals to crystallize may include feldspathoids in silica undersaturated phonolite, K feldspar in silica saturated trachyte or quartz in $SiO_2$ oversaturated rhyolite (Figure 10.12).

Hawaiian basaltic magmas are thought to form by partial melting of a heterogeneous mantle plume source, perhaps composed of primitive garnet lherzolite, enriched by a subducted ocean lithosphere component. Using geochemical data such as Ni, Mg, Pb, O, Hf, and Os isotopic ratios, researchers (Blichert-Toft et al. 1999; Sobolev et al. 2005) propose that the Hawaiian magma source is generated from a mantle enriched from the recycling of eclogite (subducted ocean lithosphere) which magmatically mixes with mantle peridotite. Using geochemical constraints such as high NiO concentrations, Sobolev et al. (2005) estimate that recycled (subducted) ocean crust accounts for up to 30% of the Hawaiian source material. These important studies suggest that the Hawaiian intraplate magmatism may be related to ancient subduction processes. While OIB volcanism at locations such as the Hawaiian Islands is impressive, humans have yet to witness the immense volcanism necessary that erupts gargantuan ocean plateaus.

### Oceanic flood basalt (OFB) plateaus

The Ontong-Java Plateau, located near the Solomon Islands in the Western Pacific Ocean, is the largest OFP on Earth. As summarized by Fitton and Godard (2004), the dates for Ontong-Java eruptions are somewhat enigmatic. The Ontong-Java flood basalts erupted either in a single massive flood eruption (~122 Ma) or in a series of eruptions spread over 10 million years with the initial massive outpouring occurring ~122 Ma. The Ontong-Java Plateau encompasses a surface area of 2 million $km^2$ and a volume of 60 million $km^3$. On the basis of 10 Ocean Drilling Project (ODP) rock core analyses, the Ontong-Java Plateau region is thought to consist largely of a relatively homogeneous low K tholeiite which erupted as massive sheet flows and pillow basalts, accompanied by minor volcaniclastic and vitric tuff deposits (Coffin and

Eldholm 1994; Fitton and Godard 2004). On the basis of geochemical data (e.g. enrichment in incompatible elements Zr-Lu) Tejada et al. (2004) suggest that the Ontong-Java basalt magma was derived by 30% melting of a primitive, enriched, high magnesium (15–20 wt. % MgO) mantle source.

The origin of oceanic intraplate magmatism continues to be an active area of research. At least in some cases, intraplate magmatism may be partially derived by deep convection cells involving ancient subducted ocean lithosphere. For example, Seno and Rehman (2011) propose that subduction of the Indian continental crust beneath Hindu Kush and Burma in the Himalayas traveled over the Reunion and Kerguelen hotspots from 100 to 126 Ma and chemically interacted with the upwelling hot spot plumes. In other cases, intraplate magmatism may be driven by mantle plumes unrelated to plate boundary activities. Hot spots tap magmas from different depths such as the lower crust, upper mantle or mantle/core boundary. Geochemical analyses of basalt and seismic tomography continue to provide insight in our attempts to understand these perplexing igneous processes. Our discussion of oceanic intraplate magmatism centered on varieties of basalt, which dominate these settings. In sharp contrast, continental intraplate magmatism produces a wider range of igneous rock types, discussed in the section that follows.

### 10.4.2 Continental intraplate igneous activity

Continental intraplate magmatism and volcanism produce continental flood basalts, continental rift assemblages, bimodal volcanism, layered mafic, and ultramafic intrusions, ultramafic suites that include komatiites and kimberlites, and an unusual array of alkaline rocks and anorogenic granitoid rocks.

### Continental flood basalts

Examples of huge outpourings of continental flood basalts (CFB) include the: Siberian flood basalts of Russia, Central Atlantic Magmatic Province (CAMP), Deccan Traps of India, Columbia River, Snake River plain (SRP), Karroo basalts of Africa, and the Keweenaw flood basalts of the United States. The three largest flood basalt events – Siberian, Deccan Traps, and Central Atlantic Magmatic

Province – correspond with among the largest extinction events in Earth's history in the Permian, Triassic, and Cretaceous Periods (Renne 2002).

Although less common, silicic large igneous provinces (SLIP) also occur in association with continental breakup, intraplate magmatism, and back arc basin magmatism. SLIP are silicic dominated provinces containing rhyolite caldera complexes and ignimbrites. SLIP occur notably in the Whitsunday volcanic province of eastern Australia and the Chon Aike province of South America (Bryan et al. 2002). Below we briefly describe several well-known continental flood basalt provinces.

The Permian **Siberian flood basalts** (Figure 10.11) consist predominantly of tholeiitic basaltic lava flows tens to a few hundreds of meters thick with minor trachyandesites, nephelinites, picrites, volcanic agglomerates, and tuffs (Zolotukhin and Al'Mukhamedov 1988; Fedorenko et al. 1996). The 251 Ma Siberian flood basalts were already recognized as one of the greatest known outpourings of lava when in 2002, the Western Siberian Basin flood basalt province was discovered which effectively doubled the aerial extent of the Siberian traps to approximately 3 900 000 km$^2$ (Reichow et al. 2002). It is analogous to burying half of the contiguous United States in lava. In the Maymecha-Kotuy region of Russia, Kamo et al. (2003) suggest that the entire 6.5 km thick basalt sequence erupted within 1 million years, based upon U-Pb dates obtained from the base (251.7 ± 0.4 Ma) and top (251.1 ± 0.3 Ma) of the basalt sequence. The volume of magma erupted in such a short time span boggles the mind.

The **Central Atlantic Magmatic Province (CAMP)** formed during the Triassic-Jurassic breakup of the Pangea supercontinent, which produced rift basins and flood basalts in North America, South America, Europe, and Africa (Figure 10.11). These once contiguous tholeiitic basalts are now widely distributed across the entire Atlantic Ocean realm, encompassing a total area of more than 7 million km$^2$. The Ar$^{40}$/Ar$^{39}$ ages indicate that the CAMP basalts erupted between 191 and 205 Ma, with a peak age of 200 Ma. CAMP rocks consist of tholeiitic to andesitic basalts, with rare alkaline and silicic rocks. CAMP tholeiites have low TiO$_2$ concentrations, negative mantle normalized niobium anomalies and moderate to strongly enriched rare Earth element patterns. These geochemical patterns indicate an anomalously hot mantle plume which resulted in the partial melting of the overlying lithosphere (Marzoli et al. 1999).

The Cretaceous **Deccan traps** (Figure 10.11) encompass an area of 500 000 km$^2$ in southwestern India. Over 1 000 000 km$^3$ of flood basalt erupted in southwestern India between 65 and 69 Ma (Courtillot et al. 1988). Individual lava flows generally vary from 10 to 50 m in thickness with total flow thicknesses varying from less than 100 m to more than 2 km (Ghose 1976; Sano et al. 2001). The flood basalts and related dike swarms are interpreted to result from rifting as the Indian plate migrated over a mantle plume (Muller et al. 1993). The Deccan traps are dominated by tholeiitic basalts with minor amounts of alkalic basalts. Geochemical studies suggest that the Deccan basalts originated by fractional crystallization of shallow magma chambers (~100 kPa, 1150–1170 °C). The basaltic magma experienced variable degrees of contamination as it ascended and assimilated granitic crustal rocks (Mahoney et al. 2000; Sano et al. 2001).

Although relatively small compared to the flood basalt provinces listed above, the Cenozoic **Columbia River flood basalts** (Figure 10.11) are among the most studied CFB on Earth. The Columbia River flows consist largely of quartz tholeiites and basaltic andesite, with 47–56 wt. % silica (Swanson and Wright 1980; Reidell 1983). Columbia River basalts crop out in the U.S. states of Washington, Oregon and Idaho, encompassing an area of approximately 163 700 ± 5000 km$^2$ (Tolan et al. 1989). The total volume of erupted lava has been estimated to be approximately 175 000 ± 31 000 km$^3$ (Tolan et al. 1989). The Columbia River basalt Group has been subdivided into five formations: the Imnaha Basalt, Grande Ronde and coeval Picture Gorge, Wanapum Basalt, and Saddle Mountain Basalt. The Grande Ronde Basalt, which erupted 15.5–17 Ma, comprises approximately 87% of the total volume of the Columbia River basalt (Swanson and Wright 1981). Over 300 individual lava flows erupted from northwest trending fractures between 6 and 17 Ma, making this the youngest continental flood basalt province on Earth. Individual flows traveled as much as 550 km, erupting in north-central Idaho and flowing to the Pacific Ocean (Hooper 1982). Unlike most other flood basalt provinces, the Columbia River basalts lack early picritic basalt eruptions and interbedded silicic lavas and have less than 5% phenocrysts (Durand and Sen 2004). The low concentrations of phenocrysts are thought to be related to either rapid ascent of magma (McDougall 1976) or to a high water content of ~4.4%, which effectively lowered the melting temperature and inhibited the early development of large crystals (Lange 2002).

Various hypotheses have been proposed for the origin of the Columbia River basalt flows. One set of hypotheses propose that the basalts crystallized from primary magma (Swanson and Wright 1981). The relatively low $Sr^{87}/Sr^{86}$ (0.7043–0.7049) ratios indicate a mantle source (McDougall 1976). However, the high $SiO_2$ concentrations, high total FeO (9.5–17.5 wt. %) and low MgO concentrations (3–8 wt. %) suggest that the parental magma was not primary (Hooper 1982; Lange 2002). A second set of hypotheses assert that the basalts are the product of diversification processes (McDougall 1976; Reidell 1983). Viable diversification models suggest that partial melting of pyroxenite (Reidell 1983) or eclogite (Takahashi et al. 1998) parental rock, was followed by the injection of separate magmatic pulses, subsequent magma mixing and assimilation of crustal rock. Trace element data suggest that the Columbia River basalts were not derived by fractionation of a single magmatic pulse. Thus, a unified model suggests that the Columbia River basalts were created by multiple pulses of heterogeneous mantle derived magmas, contaminated by continental crust during magma ascent and magma mixing (Hooper 1982; Reidell 1983). The tectonic origin of the Columbia intraplate magmatism has been the subject of debate. Possible tectonic causes include: heating associated with the subduction of the Juan de Fuca ridge, back arc spreading, and association with the Yellowstone hotspot.

### 10.4.3 Continental rifts

**Continental rifts** produce a wide array of rocks that include alkalic basalt as well as alkaline and silicic rocks. Alkaline rocks include phonolite, trachyte and lamproite. Silicic rocks include rhyolite and rhyodacite, which occur in lava domes or as pyroclastic flow and ash fall deposits. Plutonic rocks vary from syenite, alkali granite to gabbroic rocks.

Continental rifting occurs in regions such as the East African rift basin, Lake Baikal (Russia), the Basin and Range and the Rio Grande rift system (USA). The East African rift system (Figure 10.13) erupts abundant alkali basalt as well as phonolite, trachyte, rhyolite, and carbonatite lava. Ancient continental rifts include the Permian age (~250 Ma) Rhine Graben (Germany) and Oslo Graben (Norway, Triassic (~200 Ma) rift basins of the Atlantic Ocean basin and the 1.1 Ga Keweenaw rift of the Lake Superior Basin (USA). Continental rift zones can contain important hydrocarbon reservoirs because of the rapid deposition of organic-rich sediments. Volcanic flows also can provide valuable metallic ore deposits such as nickel and copper.

**Carbonatites** are shallow intrusive to volcanic rocks that contain >50% primary carbonate minerals that include calcite, dolomite, magnesite, natrolite, trona, and ankerite. Barite, fluorite, apatite, magnetite, and pyroxene may occur as minor minerals as well as silica undersaturated minerals such as olivine, nepheline, and sodalite. The origin of carbonatite was a contentious issue prior to the 1960 eruption of Oldoinyo L'Engai volcano in the East African Rift Valley of Tanzania. Oldoinyo L'Engai erupted unusually low viscosity, pahoehoe, natrocarbonatite lavas at temperatures of ~500°C. Natrocarbonatites are a carbonate-rich lava enriched in the minerals nyerereite ($Na_2Ca(CO_3)_2$) and gregoryite (($Na_2 K_2 Ca)CO_3$).

Carbonatites form in stocks, dikes, and cylindrical structures primarily at continental rifts (Dawson 1962). Carbonatites crystallize from alkaline magmas enriched in incompatible rare Earth elements (REE such as Ba, Cs, and Rb) and depleted in compatible elements (such as Hf, Zr, and Ti) suggesting that these magmas formed by small degrees of partial melting. Carbonatite rocks are distinct from marbles and sedimentary carbonate rocks by the occurrence of igneous minerals. Carbonatites have the highest concentrations of REE among igneous rocks and are the primary source of niobium and light REE such as La, Ce, Pr, and Nd. Over 500 carbonatite

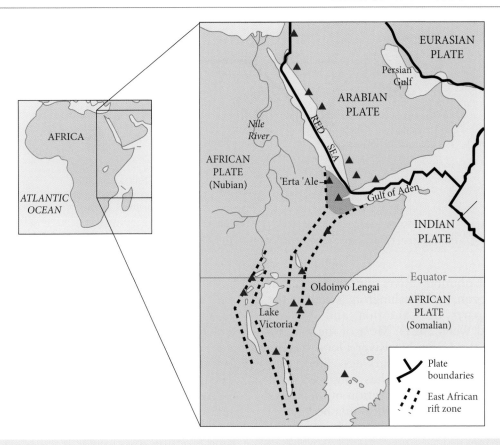

**Figure 10.13**  East African rift system represents the third leg to the Gulf of Aden and Red Sea rift chain. *Source*: U.S. Geological Survey, United States Department of the Interior.

deposits are known worldwide, four of which are currently being mined including the Mountain Pass deposit in California, USA, and the Bayan Obo, Maoniuping, and Dalucao deposits in China (Verplanck et al. 2016).

What is the driving force behind the lithospheric extension that leads to the development of continental rifts? Various hypotheses have been proposed which include (Figure 10.14):

1  Upwelling of hot plumes generated by the return convective loop of downgoing oceanic lithosphere;

2  Partial melting at great depths of over-thickened continental lithosphere following supercontinent assembly;

3  Subduction of ocean spreading ridges resulting in shallow sub-lithospheric melting producing back arc basin-type extension within continental lithosphere.

### 10.4.4  Bimodal volcanism

The widespread occurrence of basalt and rhyolite, without significant andesite is referred to as **bimodal volcanism**. Bimodal volcanism occurs at continental rifts and hotspots underlying continental lithosphere. Partial melting of the mantle generates basaltic magma. The rising basaltic magma partially melts continental crust resulting in the dual occurrence of basalt and rhyolite. A classic example occurs in Yellowstone National Park in Wyoming (USA). Yellowstone's magmatic source is related to a mantle hotspot which has been active for over 50 million years. Recent seismic tomography indicates that the Yellowstone hotspot is a high temperature, low density conduit that extends through the lower mantle and is sourced at the core–mantle boundary (Nelson and Grand 2018; Steinberger et al. 2019). The Yellowstone hotspot (Figure 10.11) has provided the source material for rhyolites, adakites, and the

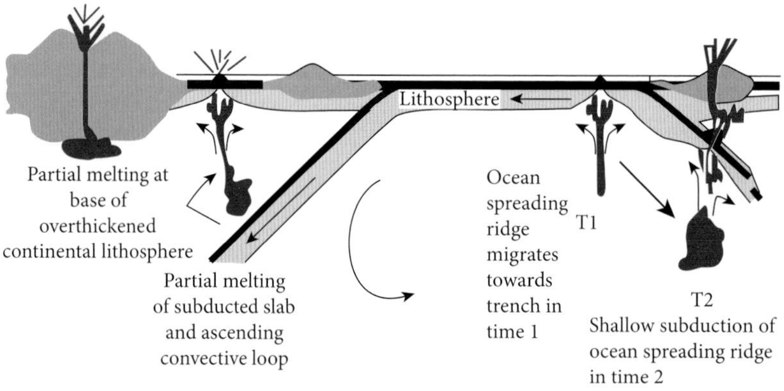

**Figure 10.14** Possible tectonic causes for continental rifts.

immense Columbia River flood basalts in Idaho, Oregon, and Washington as well as the northern parts of California and Nevada (Camp and Wells 2021). Most of the magma producing the Columbia River Flood basalts erupted 15–17 million years ago. Since that time, as North America has migrated in a southwest direction, the position of the active hotspot has migrated ~800 km in a northeasterly direction to its present location at Yellowstone.

Christiansen (2001) recognized three immense rhyolitic lava deposits at Yellowstone: The 2.1 million year old Huckleberry Ridge Tuff, the 1.3 million year old Mesa Falls Tuff and the 640 000 year old Lava Creek Tuff. Together, these three tuff deposits constitute the Yellowstone Group. The 2.1 Ma Huckleberry Ridge eruption dispersed 2450 km$^3$ rhyolite deposits over an area of 15 500 km$^2$ and produced a caldera over 75 km long. The 1.3 Ma Mesa Falls eruption produced tuff deposits largely within the Huckleberry Ridge caldera. While the Mesa Falls eruptive deposits were restricted to the pre-existing caldera, a new 16 km caldera developed along the northwest end of the Huckleberry Ridge Caldera. The youngest Lava Creek cycle of eruptive activity began around 1.2 Ma and continued for approximately 600 000 years. The Lava Creek eruption produced a large caldera and scattered rhyolitic deposits over an area of 7500 km$^2$.

The Yellowstone Caldera is a composite caldera generated by three separate rhyolitic eruptive events. In the intervening time between each of these rhyolitic eruptions, basaltic lava also erupted (Figure 10.15). The basalt eruptions appear to be independent of the rhyolite eruptive cycles. The rhyolite and basalt eruptions represent two distinctly different magmatic sources. The rhyolitic magma is derived from the successive emplacement of granitic batholiths within the crust. The basaltic magma is generated by partial melting of the peridotite-rich upper mantle. The Yellowstone Caldera possesses two ring fracture zones within this composite caldera structure. Ring fractures are circular fracture sets generated by ground subsidence following the release of magma from a shallow pluton.

Approximately 40 rhyolite eruptions have occurred in the past 640 000 years, since the last of the three cataclysmic Quaternary eruptions at Yellowstone. No lava has erupted in Yellowstone over the past 70 000 years. Two resurgent domes are currently being constructed within the Yellowstone caldera and the ground surface is slowly being inflated, with uplift as much as 1 m since the 1920s. While the eruption of lava at Yellowstone is not anticipated in the next few thousand years, the area is presently experiencing uplift, perhaps the early warning signs of a new eruptive phase (Christiansen 2001). Yellowstone has been the subject of a movie entitled "Supervolcano," which is entirely appropriate: eruptions there were among the largest of the Quaternary Period. The magma erupted from Yellowstone 2.1 million years ago was approximately 6000 times greater than the volume released in the 1980 eruption of Mount St. Helens. The smallest of Yellowstone's three Quaternary eruptive events released five times more debris than the massive 1815 Tambora, Indonesia eruption. It is estimated that

**Figure 10.15**  Dual columnar basalt flows separated above and below by massive rhyolite ignimbrite deposits in the Yellowstone Caldera, USA. *Source*: Photo by Kevin Hefferan. © John Wiley & Sons.

25 000 km³ of magma are contained within the 7 km deep Yellowstone batholith. Should a portion of that magma erupt from the Yellowstone Caldera, North America would experience a devastating eruption unlike any other witnessed in human history.

### 10.4.5 Layered mafic-ultramafic intrusions

**Layered mafic-ultramafic intrusions** develop by partial melting of peridotite and a high pressure gabbroic rock called eclogite from the lower crust and mantle. Gravitational separation is likely to play a key role in the accumulation of crystal layering. Layered mafic-ultramafic intrusions include shallow tabular sills and dikes as well as funnel shaped lopoliths. Igneous layers may occur as flat, planar structures or display features commonly associated with sedimentation such as cross bedding, graded bedding, channeling or slump structures. Layers occur on the scale of meters, centimeters or as microscopic cryptic lenses. Cryptic (hidden) layering also occurs as revealed only by subtle changes in chemical composition.

Layered intrusions are commonly anorogenic bodies injected into stable continental cratons at moderate depths. These intrusions commonly contain layers of rocks such as norite, gabbro, anorthosite, pyroxenite, dunite, troctolite, harzburgite, and lherzolite. Minor silicic rocks such as granite can also occur. Common major minerals include olivine, orthopyroxene (enstatite, bronzite, hypersthene), clinopyroxene (augite, ferroaugite, pigeonite), and plagioclase.

Most of the world's valuable platinum group elements (PGE) occur as *"reefs"* in layered mafic-ultramafic intrusions. PGE reefs are stratified PGE-enriched ore deposits in mafic to ultramafic layered intrusions. These igneous reefs consist of thin layers of silicate rocks that contain concentrated sulfides enriched in PGE (platinum, palladium, rhenium, Ru, Ir, Os), chromium, and nickel metallic ore deposits. The enriched ore zone is commonly centimeters to several meters thick and contained in intrusions that are commonly hundreds to thousands of meters thick. Although the ore zone itself is relatively thin, reef mineralization can extend tens to hundreds of kilometers along strike.

In addition to the reef ore deposits, PGE sulfide mineralization also occurs along the margins of the layered mafic-ultramafic intrusions, where hot magma is in contact with cold country rock. Contact-type PGE mineralization consists of disseminated to massive concentrations of iron, copper, nickel, and PGE-enriched sulfide mineral zones tens to hundreds of meters thick. Reef-type and contact-type deposits are the primary source of the world's platinum and rhodium, particularly from South Africa's Bushveld Complex. Reef-type PGE deposits contain platinum, palladium, and rhodium; copper, nickel, ruthenium, iridium, osmium and gold will be recovered as by products. PGE reef ores are mined in the Bushveld Complex, the Stillwater Complex, and Zimbabwe's Great Dyke. PGE-enriched contact-type polymetallic deposits contain copper, nickel, PGE, and by product gold. Contact-type PGE deposits are mined primarily in the Bushveld Complex (IIjina and Lee 2005; Zientek 2012) (Figure 10.16).

Various hypotheses put forth for the formation of layered mafic-ultramafic intrusion ore deposits include:

1 gravitational settling of denser crystals that precipitate from magma and descend onto the magma chamber floor;
2 hydrodynamic deposition and sorting of crystals from convecting silicate magma currents;
3 mass wasting slumping of crystal masses in the magma chamber;
4 *in situ* crystallization in which crystals grow from the pluton margins toward the center, or upward from the pluton floor as magma cools. *In situ* crystallization invokes immiscible sulfide droplets enriched in metallic elements crystallizing directly on the magma chamber floor supplemented by some crystals being deposited by convecting silicate magma currents (Chistyakova et al. 2019).

5 crystal compaction and secondary aging processes. Crystal aging is an equilibration process whereby small crystals are consumed by larger crystals producing segregate layers. Chemical diffusion and dissolution in the aging process results in lower density (e.g. plagioclase) material to move upward and while denser metallic elements react at different aging rates producing plagioclase, olivine, pyroxene, and chromite layers. Repeated cycles of crystal aging may produce couplets (Figure 10.17) observed in the Stillwater Complex (Boudreau 2010).

6 Metallic ore enrichment by secondary remobilization and volatile fluxing by halogen rich (e.g. chlorine) fluids derived from the assimilation of crustal rock (Boudreau et al. 1997; Boudreau 2016).

Three of the largest layered intrusions on Earth are the Stillwater Complex in Montana (U.S.), the Bushveld Complex in South Africa

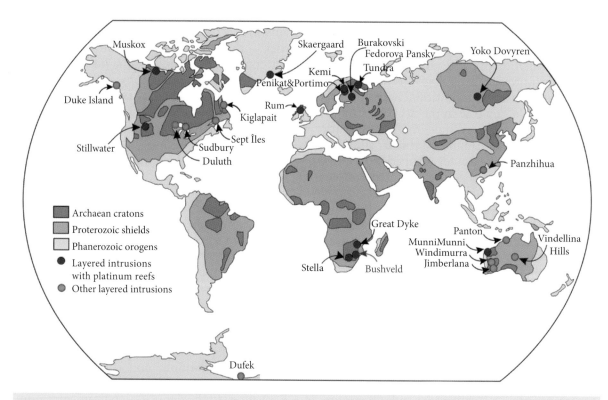

**Figure 10.16** Map indicating the location of layered mafic-ultramafic intrusions. *Source*: Courtesy of Sonya Chistyokova.

**Figure 10.17** Close up of rhythmic layers within a channel structure in the Stillwater Complex. *Source*: Photo by Kevin Hefferan. © John Wiley & Sons.

and the Skaergaard Intrusion in Greenland. Other significant layered intrusions include the Muskox intrusion of Northwest Territories (Canada), Keweenaw, and Duluth intrusion of Minnesota (U.S.) and the Great Dyke of Zimbabwe, Africa. Nearly all of the massive layered igneous intrusions are associated in time and space with large igneous provinces. Fractionation patterns in the Skaergaard Intrusion indicate magma solidified in a closed system. The intrusion crystallized from the pluton margins inward toward the center. Changes in mineral composition from margins to pluton center conform to the characteristic mafic magma crystallization pattern (McBirney 2007). Most of the other large layered igneous intrusions, such as the Bushveld Complex, the Stillwater Complex, and the Great Dyke show evidence for open system behavior with repeated injections of magma (Zientek 2012).

The 2.7 Ga **Stillwater complex** is a large, layered mafic-ultramafic igneous intrusion in the Beartooth Mountains of southwestern Montana. The Stillwater complex is exposed along a northwesterly strike for a distance of 48 km, with observable thicknesses up to 6 km. The Stillwater complex, which formed by multiple injections of mafic-ultramafic magma intruded metasedimentary rocks, is the finest exposed layered intrusion in North America and contains economic deposits of platinum group metals as well as chromium, copper, and nickel sulfides (McCallum et al. 1980, 1999; Premo et al. 1990).

The Stillwater complex consists of three main units which include a lowermost basal zone, an ultramafic zone and an upper banded zone. The basal zone consists of norite-, harzburgite-, and bronzite-rich orthopyroxenite layers. The ultramafic zone consists of dunite-, harzburgite-, bronzite-rich orthopyroxenite, and chromite-rich peridotite layers. The basal and ultramafic zones contain Cu, Cr, and Ni sulfide ore deposits. The upper banded zone consists largely of repetitive layers of alternating norite, gabbro, anorthosite, and troctolite and is enriched in Cu, Ni, and PGE ore deposits (McCallum et al. 1980, 1999; Todd et al. 1982). Chlorine-rich magmatic fluids played a key role in leaching background metal deposits within the intrusion and concentrating these metals in discrete enriched layers called reefs within the banded zone (Boudreau et al. 1986, 1997; Meurer et al. 1999). Crystal compaction and late stage crystal aging processes enhanced mineral segregation and layering (Boudreau 2010, 2016).

South Africa's 2.06 Ga **Bushveld complex**, a massive lopolith or funnel structure, is the world's largest layered igneous intrusion. Extending over 400 km in length, up to 8 km thick and underlying an area of 60 000 km², this complex contains a layered sequence of mafic and ultramafic rocks, capped locally by granite.

The Bushveld Complex consists of four main zones (Daly 1928; Vermaak 1976; Cawthorn 1999). These include, from top to bottom, the following: (i) an upper zone consists of granophyre, gabbro, and norite; (ii) a main zone contains gabbro and anorthosite; (iii) a critical zone consists of anorthosite, norite, and pyroxenite; and (iv) a basal zone consists of orthopyroxenite, harzburgite, dunite, and peridotite. A chromite horizon occurs at the top of the basal series.

The Bushveld Complex hosts the largest reserves of vanadium, chromium, and platinum group metals in the world. PGE are concentrated within the Merensky Reef within the Critical Zone. Anorogenic granitic rocks capping the complex contain tin, fluorine, and molybdenum. The Bushveld complex layering formed through differentiation processes accompanied by a series of magmatic injections, resulting in a massive laccolith or domal structure. As in the Stillwater intrusion described above, chlorine-rich magmatic fluids are thought to have played a role in concentrating PGE in the Bushveld Complex (Boudreau et al. 1986).

Whereas most layered mafic-ultramafic intrusions are Precambrian in age, Greenland's 55 Ma **Skaergaard Intrusion** is the youngest of the great PGE enriched intrusions. The Skaergaard lopolith intrusion crops out along Greenland's eastern shores and offers exceptionally good exposures of layering formed by differentiation and convective current structures. The Skaergaard Intrusion, with a volume of $500 km^3$, is heralded as the finest example on Earth of closed system fractional crystallization, displaying layered sequences of euhedral to subhedral crystals as well as distinctive structures usually associated with sedimentary beds. These structures include cross bedding, graded bedding, and slump structures (Wager and Deer 1939; Irvine 1982; Irvine et al. 1998).

Zoned and layered mafic-ultramafic intrusive complexes provide rare but massive examples of magma diversification yielding segregated mineral zones and valuable metallic ore deposits.

### 10.4.6 Anorthosites, komatiites, and kimberlites

Plagioclase-rich (>90%) **anorthosites** are enigmatic plutonic rocks that range in age from Archean to Recent. In addition to their widespread occurrence on Earth dating back to the Archean, anorthosites are also the oldest known rocks (~4.3 billion years old) on the moon. Archean anorthosites (~3.2–2.8 billion years old) commonly contain equant, euhedral calcium-rich plagioclase ($An_{80-90}$) crystals. Archean anorthosites are associated with continental greenstone belts that may delineate ancient convergent plate boundaries (Polat et al. 2018). Proterozoic anorthosites (~1.8–0.9 billion years old) are less enriched in calcium ($An_{40-60}$), generally corresponding to labradorite plagioclase and andesite composition, and are associated with granite complexes formed in the supercontinents Columbia (Nuna) and Rodinia. Proterozoic anorthosites widely occur in the southwestern United States, Appalachian Mountains, eastern Canada, Scandinavia, and eastern Europe. Supercontinent reconstructions superimpose these disparate localities into one continuous belt during the Proterozoic. Bybee et al.

(2014), on the basis of association with high pressure orthopyroxene minerals in Canada and Norway, suggest these Proterozoic anorthosites were derived from long lived (~100 million year old) ultramafic magmas that were derived from melting at Andean-type convergent plate boundaries. Other Proterozoic and Phanerozoic anorthosites occur with layered mafic intrusions such as the Bushveld (South Africa) and Stillwater (Montana, USA) complexes, as well as along some ocean spreading ridges, ophiolites, and as xenolith fragments in granites, basalts, and even kimberlites. These Anorthosites are likely derived from lower crustal to mantle mafic magmas in which plagioclase experiences crystal flotation segregation resulting in this low density feldspar group into concentrate along the roofs of magma chambers. The origin of anorthosites has long been debated and dates back to the work of Norman Bowen (1917) who published a paper on the problem of the anorthosites. Bowen (1917) demonstrated that plagioclase normally crystallizes with pyroxene and/or olivine. Insofar as plagioclase and pyroxene have different settling velocities in basaltic magma, Bowen (1917) predicted layers of pyroxenitic and feldspathic cumulates. However, in massif-type anorthosites, cumulate layers enriched in pyroxene or olivine are rare or lacking. In layered intrusions, the olivine and pyroxene cumulates are present. However, the anorthosites commonly have an intrusive relationships with their mafic–ultramafic host cumulates (Maier et al. 2021).

Research continues on the origin and tectonic significance of these anorthosite rocks which widely occur on Earth's continents as large massifs associated with A-type granites, and in layered intrusions with mafic and ultramafic rocks. Eastern Canada's anorthosite complex encompasses an area of $155 000 km^2$. South Africa's Bushveld complex exposes over $55 000 km^2$ of anorthosite and New York's Adirondack Mountains contain $4000 km^2$ of anorthosite. Maier et al. (2021) suggest that in the Bushveld complex, the anorthosites may have been derived from reactive flow of acidic, alkali, and silica undersaturated magmas altering older mafic and ultramafic layers. In essence, anorthosite recrystallized due to reactive fluid percolation that preferentially leached mafic minerals from mafic and ultramafic rocks such as norite and created plagioclase-rich anorthosite in massive secondary reactions.

Another group of curious igneous rocks, komatiites and kimberlites, are relatively rare ultramafic, alkaline deposits that reflect unusual conditions as discussed below.

**Komatiites** are ultramafic volcanic rocks that occur almost exclusively in Precambrian greenstone belts over 2.6 Ga. Greenstone belts are metamorphosed assemblages of green colored rocks that contain layers of ultramafic and mafic rocks overlain by silicic rocks and sediments. Komatiites, named after the 3.5 Ga Komatii region of Barberton, South Africa, are high magnesium (>18% MgO), olivine-rich volcanic rocks, depleted in titanium and light rare Earth elements. The high magnesium content and light rare Earth element depletion indicate a previously depleted mantle source (Sun and Nesbitt 1978

in Walter (1998)). Komatiite flows, first recognized in the Barberton region in 1969, commonly contain **spinifex** texture (Figure 10.18). Spinifex texture consists of needle-like, acicular olivine, pyroxene (augite and/or pigeonite), and chromite phenocrysts in a glassy groundmass (Viljoen and Viljoen 1969; Arndt 1994). Spinifex texture commonly occurs in the upper parts of komatiite flows or in the chilled margins of sills and dikes where rapid quenching produced skeletal, acicular crystals (Arndt and Nesbitt 1982). In addition to spinifex texture, circular varioles, radiating spherulites, and tree-like dendritic textures also occur. These textures are attributed to rapid undercooling or quenching of extremely hot lavas (Fowler et al. 2002). Nearly all komatiites erupted during the Archean Eon (>2.5 Ga) when the early Earth was much hotter.

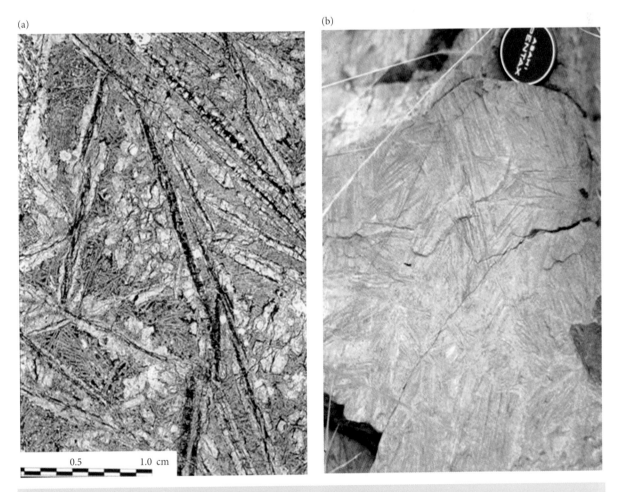

(a)

(b)

0.5   1.0 cm

**Figure 10.18**   Komatiite (a) Photomicrograph and (b) field photo of spinifex texture. *Source*: Photos courtesy of Maarten de Wit.

Interestingly, the only known Phanerozoic (<544 Ma) komatiites occur on Gorgona Island, Colombia, where 88 Ma komatiites erupted as >1500 °C ultramafic lava flows. Gorgona Island, located 80 km west of Colombia in the Pacific Ocean, is composed largely of gabbro and peridotite (Echeverria 1980; Aitken and Echeverria 1984). Gorgona Island is also notable for the rare occurrence of ultramafic pyroclastic tuffs which record explosive volcanism (Echeverria and Aitken 1986).

Why are komatiites so rare in the last 2.6 Ga? Archean komatiites may indicate elevated liquidus temperatures of 1575–1800 °C (1 atm pressure) in the 2.7 Ga upper mantle (Arndt 1976; Green et al. 1975; Wei et al. 1990; Herzberg 1992, 1993 in de Wit 1998). The virtual absence of Phanerozoic komatiites may be attributed to lower upper mantle temperatures which precludes the extensive mantle melting required to produce ultramafic melts. Hypotheses for the origin of komatiites include:

1   Melting in the hydrated mantle wedge above Precambrian subduction zones (Allegre 1982; Grove et al. 1997; Parman et al. 2001);
2   Deep mantle plume hotspot which led to large degrees of partial melting producing oceanic plateaus (Storey et al. 1991);
3   Partial melting (10–30%) of a garnet peridotite at pressure of 8–10 GPa (Walter 1998).

Komatiites, like layered gabbroic intrusions, are associated with valuable metallic ore deposits such as nickel, copper, and platinum metals. For example, komatiite metallic ore deposits occur in the 2.7 Ga Yilgarn Craton of Western Australia, the 3.5 Ga South African Barberton region and the 2.7 Ga Canadian Shield. The nickel sulfide ore deposits are thought to have originated in ultramafic lava tubes that concentrated high density metals in channel beds.

Ultramafic volcanic rocks such as komatiites and kimberlites, discussed below, form almost exclusively (with Gorgona Island being the only known exception) in continental settings. In most cases, these mantle derived assemblages occur in stable Precambrian shields.

**Kimberlites** are magnesium-rich, ultramafic volcanic rocks that rapidly rise to Earth's surface from the crust and mantle. Kimberlites consist of a volcanic component and an underlying diatreme pipe structure that served as a conduit extending upward from the mantle. Diatremes vary greatly in surface area, ranging from a few m² to km². Most diatremes taper downward, resembling an inverted cone in cross section view. Kimberlite pipes occur with other plutonic structures such as dikes and sills. Kimberlites, which originate at temperatures of 1200–1400 °C and depths exceeding 150 km, rise explosively through thick continental lithosphere.

Volatile enriched, very low viscosity mantle melts rocket upward toward Earth's surface at velocities ~15–72 km/h. (Sparks et al. 2006). The magma is propelled upward by either the degassing of $CO_2$-enriched magma and/or by phreatomagmatic processes. Phreatomagmatic processes require a water source to interact with the kimberlite magma.

The high volatile content serves two primary purposes in that it: (i) lowers the melting temperature preventing crystallization and (ii) provides the propellant "jet fuel" to accelerate kimberlite magma to Earth's surface. Sparks et al. (2006) suggest that kimberlite eruptions generate up to 10 000 m³ of pyroclastic debris over hours to months, producing Plinian ash plumes up to 35 km high. Strangely, no ultramafic lavas have been documented with kimberlite deposits. This is probably due to their low preservation potential and the extremely high volatile (up to 20%) content of kimberlite magma which can produce 70% vessiculation in the erupting lava (Sparks et al. 2006).

Kimberlite eruptions form maar craters that are largely filled with pyroclastic debris (Dawson 1980; Mitchell 1986; Sparks et al. 2006). Due to the association of high temperatures, pressure, volatile content and velocity, kimberlites commonly exhibit extensive hydrothermal alteration and are intensely fractured (Dawson 1980). Altered olivine and phlogopite phenocrysts occur within a fine groundmass of serpentine, calcite, and olivine. Olivine constitutes the major original mineral in the vast majority of kimberlites. However, in many samples olivine is completely replaced by serpentine, mica or clay minerals (Skinner 1989). Kimberlites also contain the high pressure minerals pyrope garnet, jadeite pyroxene and diamond which are stable at mantle depths >150 km.

Kimberlites were first discovered in the Kimberly region of South Africa where they are intimately associated with diamonds. Although best known from South Africa, kimberlites crop out in continental lithosphere throughout the world, commonly occurring with carbonatites and alkaline igneous rocks. Carbonatites, igneous rocks enriched in carbonate minerals such as calcite, dolomite or ankerite, are important $CO_2$ energy sources propelling kimberlites up from mantle depths. Kimberlites are also associated with reactivated shear zones and fracture zones (White et al. 1995; Vearncombe and Vearncombe 2002). Kimberlites intrude Early Proterozoic to Archean age cratons (2–4 Ga); intrusion ages vary and kimberlites as young as Tertiary age (~50 Ma) are known (Dawson 1980).

## 10.4.7 Lamprophyres and lamproites

Alkalic rocks that occur in stable continental lithosphere include lamprophyres, and lamproites. These $SiO_2$ undersaturated rocks typically occur in shallow (hypabyssal), volatile-rich dikes and may be associated with kimberlites. **Lamprophyres** are magnesium-rich, volatile-rich, porphyritic rocks containing mafic phenocrysts such as olivine, biotite, phlogopite, amphibole, clinopyroxene, and melilite.

Lamprophyres are associated with kimberlites but also occur as dikes intruding granodiorite plutons at convergent margin settings. **Lamproites** are K-rich, peralkaline rocks containing minerals such as leucite, sanidine, phlogopite, richterite, diopside, and olivine. Lamproites are enriched in Ba (>5000 ppm), La (>200 ppm), and Zr (>500 ppm). Some lamproites are diamond bearing, such as the Argyll lamproite in Western Australia.

In Chapters 7–10 we have attempted to present a logical approach to the description, classification, and origin of igneous rocks and landforms. We have also tried to demonstrate the tectonic relations in an understandable framework. Hopefully we have been somewhat successful in helping you understand igneous processes. In succeeding chapters, we will investigate how rocks are altered in two ways: (i) by weathering and erosion at Earth's surface and (ii) though the effects of high temperatures, pressures, and hot fluids via metamorphic reactions. In all of these reactions, water plays a critical role in altering and mobilizing elements within Earth's crust.

## CONTENT ASSESSMENT

1  Explain why MORB have a relatively narrow range of chemical variation.
2  Why do convergent margins contain a diverse suite of igneous rocks.
3  Why do back arc basins produce a wide array of rock types?
4  Compare the composition and origin of boninites to shoshonites.
5  How can M type granites be distinguished from I type granites?
6  How are A type granites distinguished from S type granites?
7  On what basis are ocean island basalts distinguished from MORB?
8  Why are ophiolites considered analogues for ocean lithosphere?
9  How do subduction related ophiolites differ from nonsubduction related ophiolites.
10  Describe conditions that produce bimodal volcanism.
11  Explain the relative absence of Phanerozoic komatiites.
12  Watch the following video and explain the driving forces that result in lithospheric plate motion. https://www.iris.edu/hq/inclass/animation/what_are_the_forces_that_drive_plate_tectonics#:~:text=Heat%20and%20gravity%20are%20fundamental%20to%20the%20process&text=The%20energy%20source%20for%20plate,convection%20could%20drive%20plate%20motions.

## REFERENCES

Aitken, B.G. and Echeverria, L.M. (1984). Petrology and geochemistry of komatiites and tholeiites from Gorgona Island, Columbia. *Contributions to Mineralogy and Petrology* 86: 94–105.

Allegre, C.J. (1982). Genesis of Archaean komatiites in a wet ultramafic subducted plate. In: *Komatiites* (eds. N.T. Arndt and E.G. Nisbet), 495–500. Berlin: Springer-Verlag.

Anderson, J.L. (1983). Proterozoic anorogenic granite plutonism of North America. In: *Proterozoic Geology* (eds. L.G. Medaris, C.W. Byers, D.M. Michelson and W.C. Shanks), 133–154. Geological Society of America Memoir 161.

Arndt, N.T. (1976). Ultramafic lavas in Munro Township: economic and tectonic implications. In: *Metallogeny and Plate Tectonics* (ed. D.F. Strong), 617–658. Ottawa: Geological Association of Canada.

Arndt, N.T. (1994). Komatiites. In: *Archean Crustal Evolution* (ed. K.C. Condie), 11–44. Amsterdam: Elsevier Publishers.

Arndt, N.T. and Nesbitt, R.W. (1982). Geochemistry of Munro Township basalts. In: *Komatiites* (eds. N.T. Arndt and E.G. Nisbet), 309–330. London: Allen and Unwin Publishers.

Barker, F. (1979). Trondhjemite: definition, environment and hypothesis of origin. In: *Trondhjemites, Dacites and Related Rocks* (ed. F. Barker), 1–11. New York: Elsevier Publishers.

Barker, F., Wones, D.R., Sharp, W.N., and Desborough, G.A. (1975). The Pikes Peak Batholith, Colorado, Front Range, and a model for the origin of the gabbro–anorthosite–syenite–potassic granite suite. *Precambrian Research 2*: 97–160.

Beckinsale, R.D. (1979). Granite magmatism in the tin belt of Southeast Asia. In: *Origin of Granite Batholiths, Geochemical Evidence* (eds. M.P. Atherton and J. Tarney), 34–44. Orpington, Kent: Shiva Publishing Ltd.

Best, M.G. (2003). *Igneous and Metamorphic Petrology*, 2e, 752 pp. Oxford: Blackwell Publishing.

Blichert-Toft, J., Frey, F.A., and Albarede, F. (1999). Hf isotope evidence for pelagic sediments in the source of Hawaiian basalts. *Science 285*: 879–882.

Bloomer, S.H. and Hawkins, J.W. (1987). Petrology and geochemistry of boninite series volcanic rocks from the Marianas Trench. *Contributions to Mineralogy and Petrology 97*: 361–377.

Boudreau, A. (2010). The evolution of texture and layering in layered intrusions. *International Geology Review*: 1–24.

Boudreau, A. (2016). The Stillwater Complex, Montana – overview and the significance of volatiles. *Mineralogical Magazine 80*: 585–637.

Boudreau, A.E., Mathez, E.A., and McCallum, I.S. (1986). Halogen geochemistry of the Stillwater and Bushveld Complex: evidence for transport of the platinum-group elements by Cl-rich fluids. *Journal of Petrology 27*: 967–986.

Boudreau, A.E., Stewart, M.A., and Spivack, A.J. (1997). Stable Cl isotopes and origin of high-Cl magmas of the Stillwater Complex, Montana. *Geology 25*: 791–794.

Bowen, N.L. (1917). The problem of the anorthosites. *Journal of Geology 25*: 209–243.

Bryan, S.E. and Ernst, R.E. (2008). Revised definition of large igneous provinces (LIPs). *Earth-Science Reviews 86*: 175–202.

Bryan, S.E. and Ferrari, L. (2013). Large igneous provinces and silicic large igneous provinces: progress in our understanding over the past 25 years. *Geological Society of America Bulletin 125*: 1053–1078.

Bryan, S.E., Riley, T.R., Jerram, D.A. et al. (2002). Silicic volcanism: an undervalued component of large igneous provinces and volcanic rifted margins. In: *Volcanic Rifted Margins* (eds. M.A. Menzies, S.L. Klemperer, C.J. Ebinger and J. Baker), 99–120. Geological Society of America Special Paper 362.

Buddington, A.F. and Chapin, T. (1929). Geology and mineral deposits of Southeastern Alaska. US Geological Survey Bulletin No. 800, 398 pp.

Bybee, G.M., Ashwal, L.D., Shirey, S.B. et al. (2014). Pyroxene megacrysts in Proterozoic anorthosites: implications for tectonic setting, magma source and magmatic processes at the MOHO. *Earth and Planetary Science Letters 389*: 74–85.

Camp, V.E. and Wells, R.E. (2021). The case for a long-lived and robust Yellowstone hotspot. *GSA Today 31* https://doi.org/10.1130/GSATG477A.1.CC-BY-NC.

Cann, J.R. (1971). Major element variations in ocean floor basalts. *Philosophical Transactions of the Royal Society of London, Series A 268*: 495–506.

Castillo, P.R. (2006). An overview of Adakite petrogenesis. *Chinese Science Bulletin 51* (3): 257–268.

Cawthorn, R.G. (1999). Geological models for platinum-group metal mineralization in the Bushveld Complex. *South African Journal of Sciences 95*: 490–498.

Chappell, B.W. and Stephens, W.E. (1988). Origin of intracrustal (T-type) granite magmas. *Royal Society of Edinburgh Transactions 79*: 71–86.

Chappell, B.W. and White, A.J.R. (1974). Two contrasting granite types. *Pacific Geology 8*: 173–174.

Chase, C.G. (1978). Extension behind island-arcs and motions relative to hot spots. *Journal of Geophysical Research 83*: 5385–5387.

Chistyakova, S., Latypov, R., Hunt, E.J., and Barnes, S. (2019). Merensky-type platinum deposits and a reappraisal of magma chamber paradigms. *Nature Scientific Reports 9*: 8807. https://doi.org/10.1038/s41598-019-45288-8.

Christiansen, R.L. (2001). The quaternary and pliocene yellowstone plateau volcanic field of Wyoming, Idaho, and Montana. US Geological Survey Professional Paper No. 729-G, 145 pp.

Chung, S.L., Liu, D., Ji, J. et al. (2003). Adakites from continental collision zones: melting of thickened lower crust beneath southern Tibet. *Geology 31*: 1021–1024.

Coffin, M.F. and Eldholm, O. (1994). Large igneous provinces: crustal structure, dimensions, and external consequences. *Reviews of Geophysics 32* (1): 1–36.

Collins, W.J., Beans, S.D., White, A.J.R., and Chappell, B.W. (1982). Nature and origin of A-type granites with particular reference to southeastern Australia. *Contributions to Mineralogy and Petrology 80*: 189–200.

Condie, K. and Kröner, A. (2008). When did plate tectonics begin? Evidence from the geologic record. In: *When Did Plate Tectonics Begin on Planet Earth?* (eds. K.C. Condie and V. Pease), 281–294. Geological Society of America Special Paper 440.

Conrad, C.P. and Lithgow-Bertelloni, C. (2002). How mantle slabs drive plate tectonics. *Science 298*: 207–209. https://doi.org/10.1126/science.1074161.

Conrad, C.P. and Lithgow-Bertelloni, C. (2004). The temporal evolution of plate driving forces: importance of "slab suction" versus "slab pull" during the Cenozoic. *Journal of Geophysical Research 109*: B10407. https://doi.org/10.1029/2004JB002991. https://agupubs.onlinelibrary.wiley.com/doi/10.1029/2004JB002991.

Courtillot, V., Feraud, G., Maluski, H. et al. (1988). Deccan flood basalts and the Cretaceous/Tertiary boundary. *Nature 333*: 843–845.

Creaser, R.A., Price, R.C., and Wormald, R.J. (1991). A-type granites revisited: assessment of a residual source model. *Geology 19*: 163–166.

Dalziel, I.W.D., Lawver, L.A., and Murphy, J.B. (2000). Plumes, orogenesis and supercontinental fragmentation. *Earth and Planetary Sciences 178*: 1–11.

Daly, R.A. (1928). Bushveld igneous complex of the Transvaal. *Geological Society of America Bulletin 39*: 703–768.

Davies, G.F. (1992). On the emergence of plate tectonics. *Geology 20*: 963–966.

Dawson, J.B. (1962). Sodium carbonate lavas from Oldoinyou Lengai, Tanganyika. *Nature 195*: 1065–1066.

Dawson, J.B. (1980). *Kimberlites and their Xenoliths*, 250 pp. Berlin: Springer-Verlag Publishers.

Defant, M.J. and Drummond, M.S. (1990). Derivation of some modern arc magmas by melting of young subducted lithosphere. *Nature 347*: 662–665.

Dilek, Y. and Furnes, H. (2011). Ophiolite genesis and global tectonics: geochemical and tectonic fingerprinting of ancient oceanic lithosphere. *Geological Society of America Bulletin 123*: 387–411. https://doi.org/10.1130/B30446.1.

Drummond, M. and Defant, M. (1990). A model for trondjhemite-tonalite-dacite genesis and crystal growth via slab melting: Archean to modern comparisons. *Journal of Geophysical Research 95*: 21505–21521.

Ducea, M. (2001). The California arc: thick granitic batholiths, eclogitic residues, lithospheric-scale thrusting, and magmatic flare-ups. *GSA Today 11* (11): 4–10.

Durand, S.R. and Sen, G. (2004). Pre eruption history of the Grande Ronde Formation lavas, Columbia River Basalt Group, American Northwest: evidence from phenocrysts. *Geology 32*: 293–296.

Echeverria, L. (1980). Tertiary or Mesozoic komatiites from Gorgona Island, Columbia: field relations and geochemistry. *Contributions to Mineralogy and Petrology 73*: 253–266.

Echeverria, L. and Aitken, B.G. (1986). Pyroclastic rocks: another manifestation of ultramafic volcanism on Gorgona Island. *Columbia. Contributions to Mineralogy and Petrology 92*: 428–436.

Eiler, J.M., Schiano, P., Kitchen, N., and Stolper, E.M. (2000). Oxygen-isotope evidence for recycled crust in the sources of mid-ocean-ridge basalts. *Nature 403*: 530–534.

Ernst, W.G. (2007). Speculations on evolution of the terrestrial lithosphere–asthenosphere system – plumes and plates. *Gondwana Research 11*: 38–49.

Fedorenko, A., Lightfoot, P.C., Naldrett, A.J. et al. (1996). Petrogenesis of the Siberian flood basalt sequence at Noril'sk. *International Geological Review 38*: 99–135.

Fisher, R.V. and Schmincke, H.U. (1984). *Pyroclastic Rocks*. New York: Springer-Verlag 472 pp.

Fitton, J.G. and Godard, M. (2004). Origin and evolution of magmas on the Ontong Java Plateau. In: *Origin and Evolution of the Ontong Java Plateau* (eds. J.G. Fitton, J.J. Mahoney, P.J. Wallace and A.D. Saunders), 151–178. Geological Society of London Special Publications No. 229.

Fliedner, M.M., Klemperer, S.L., and Christensen, N.I. (2000). Three-dimensional seismic model of the Sierra Nevada arc, California, and its implications for crustal and upper mantle compositions. *Journal of Geophysical Research 105*: 10,899–10,921.

Fowler, A.D., Berger, B., Shore, M. et al. (2002). Supercooled rocks: development and significance of varioles, spherulites, dendrites and spinifex in Archaean volcanic rocks, Abitibi Greenstone belt, Canada. *Precambrian Research 115*: 311–328.

Fretzdorff, S., Livermore, R.A., Devey, C.W. et al. (2002). Petrogenesis of the back-arc East Scotia Ridge, South Atlantic Ocean. *Journal of Petrology 43*: 1435–1467.

Frey, F.A. and Haskins, L.A. (1964). Rare earths in oceanic basalts. *Journal of Geophysical Research 69*: 775–779.

Fryer, P., Sinton, J.M., and Philpotts, J.A. (1981). Basaltic glasses from the Marina Trough. In: *Initial Reports of the Deep Sea Drilling Project*, vol. 60 (eds. D.M. Hussong, S. Uyeda, S.R. Knapp, et al.), 601–609. Washington, DC: US Government Printing Office.

Furnes, H., de Wit, M., Staudigel, H. et al. (2007). A vestige of Earth's oldest ophiotlite. *Science 315*: 1704–1707.

Gast, P.W. (1968). Trace element fractionation and the origin of tholeiitic and alkaline magma types. *Geochemica et Cosmochimica Acta 32*: 1057–1086.

Ghose, N.C. (1976). Composition and origin of the Deccan basalts. *Lithos 9*: 65–73.

Gill, J.B. (1981). *Orogenic Andesites and Plate Tectonics*, 390 pp. New York: Springer-Verlag.

Gomez-Tuena, A., Langmuir, C.H., Goldstein, S.L. et al. (2007). Geochemical evidence for slab-melting in the TransMexican volcanic belt. *Journal of Petrology 48*: 537–562.

Granot, R. (2016). Palaeozoic Ocean crust preserved beneath the eastern Mediterranean. *Nature Geoscience 9*: 701–705.

Green, D.H., Nicholls, I.A., Viljoen, R., and Viljoen, M. (1975). Experimental demonstration of the existence of peridotitic liquids in earliest Archean magmatism. *Geology 3*: 11–14.

Grove, T.L. and Kinzler, R.J. (1986). Petrogenesis of andesites. *Annual Review of Earth and Planetary Sciences* 14: 417–454.

Grove, T.L., de Wit, M.J., and Dann, J.C. (1997). Komatiites from the Komatii type section, Barberton, South Africa. In: *Greenstone Belts* (eds. M.J. de Wit and L.D. Ashwal), 436–450. Oxford: Oxford University Press.

Hamilton, W. (1998). Archean magmatism and deformation were not products of plate tectonics. *Precambrian Research* 91: 143–179.

Hamilton, W. (2003). An alternative Earth. *GSA Today* 13 (11): 4–12.

Hamilton, W. (2007). Comment on a vestige of Earth's oldest ophiolite. *Science 318*: 746.

Hawkesworth, C.J. and Kemp, A.I.S. (2006). Evolution of the continental crust. *Nature 443*: 811–817.

Hawkesworth, C.J., Gallagher, K., Hergt, J.M., and Keynes, M. (1993). Mantle and slab contributions in arc magmas. *Annual Review of Earth and Planetary Sciences 21*: 175–204.

Hawkins, J., Bloomer, S.H., Evans, C.A., and Melchior, J.T. (1984). Evolution of intra-oceanic arc-trench systems. *Tectonophysics 102*: 175–205.

Herzberg, C.T. (1992). Depth and degree of melting of komatiites. *Journal of Geophysical Research* 97: 4521–4540.

Herzberg, C.T. (1993). Lithosphere peridotites of the Kaapvaal craton. *Earth and Planetary Science Letters 120*: 13–29.

Hirschmann, M.C., Kogiso, T., Baker, M.B., and Stolper, E.M. (2003). Alkalic magmas generated by partial melting of garnet pyroxenite. *Geology 31*: 481–484.

Hofmann, A.W. (1997). Mantle geochemistry: the message from oceanic volcanism. *Nature 385*: 219–229.

Hofmann, A.W. and White, W.M. (1982). Mantle plumes from ancient ocean crust. *Earth and Planetary Science Letters 57*: 421–436.

Hooper, P.R. (1982). The Columbia River basalts. *Science 215*: 1463–1468.

Iljina, M.J. and Lee, C.A. (2005). PGE deposits in the marginal series of layered intrusions. In: *Exploration for Platinum Group Element Deposits*, vol. 35 (ed. J.E. Mungall), 75–96. Mineralogical Association of Canada Short Course Series.

Irvine, T.N. (1959). The ultramafic complex and related rocks of Duke Island, Southeastern Alaska. PhD dissertation, California Institute of Technology, Pasadena, CA, 337 pp. http://etd.caltech.edu/etd available/etd-03102006-161603/unrestricted/Irvine_tn_1959.pdf.

Irvine, T.N. (1982). Terminology of layered intrusions. *Journal of Petrology* 23: 127–162.

Irvine, T.N., Anderson, J.C., and Brooks, C.K. (1998). Included blocks (and blocks within blocks) in the Skaergaard intrusion: geologic relations and the origins of rhythmic modally graded layers. *Geological Society of America Bulletin 110*: 1398–1447.

Ishiwatari, A. and Ichiyama, Y. (2004). Alaskan-type plutons and ultramafic lavas in Far East Russia, Northeast China and Japan. *International Geology Review 46*: 316–331.

Jakes, P. and White, A.J.R. (1972). Major and trace element abundances in volcanic rocks of orogenic areas. *Bulletin of the Geological Society of America 83*: 29–40.

Kamo, S.L., Davis, D.W., Trofimov, V.R. et al. (2003). Rapid eruption of Siberian flood-volcanic rocks and evidence for coincidence with the Permian–Triassic boundary and mass extinction at 251 Ma. *Earth and Planetary Science Letters 214* (1–2): 75–91.

Kay, R.W. (1978). Aleutian magnesian andesites: melts from subducted Pacific Ocean crust. *Journal of Volcanology and Geothermal Research 4*: 117–132.

Kennedy, G.C. and Walton, W.S. (1946). Geology and associated mineral deposits of some ultrabasic rock bodies in southeastern Alaska. *United States Geological Survey Bulletin 947-D*: 65–84.

Kogiso, T., Hirose, K., and Takahashi, E. (1998). Melting experiments on homogeneous mixtures of peridotite and basalt: application to the genesis of ocean island basalts. *Earth and Planetary Science Letters 162*: 45–61.

Kogiso, T., Hirschmann, M.M., and Frost, D.J. (2003). High pressure partial melting of garnet pyroxenite: possible mafic lithologies in the source of ocean island basalts. *Earth and Planetary Science Letters 216*: 603–617.

Kusky, T.M., Li, J.H., and Tucker, R.D. (2001). The Archean Dongwanzi ophiolite complex, North China craton: 2.505-billion-year-old oceanic crust and mantle. *Science 292*: 1142–1145.

Lange, R. (2002). Constraints on the preeruptive volatile concentrations in the Columbia River flood basalts. *Geology 30*: 179–182.

Le Maitre, R.W., Bateman, P., Dudek, A. et al. (1989). A Classification of Igneous Rocks and Glossary of Terms: Recommendations of the International Union of Geological Sciences Subcommission on the Systematics of Igneous Rocks. In: Oxford, UK: Blackwell Scientific Publications.

Le Maitre, R., Streckeisen, A., Zanettin, B. et al. (eds.) (2002). *Igneous Rocks: A Classification and Glossary of Terms: Recommendations of the International Union of Geological Sciences Subcommission on the Systematics of Igneous Rocks*, 2e. Cambridge: Cambridge University Press, 236 pp doi:10.1017/CBO9780511535581.

Liou, J.G., Ernst, W.G., Zhang, R.Y. et al. (2009). Ultrahigh-pressure minerals and metamorphic terranes – the view from China. *Journal of Asian Earth Sciences 35*: 199–231.

Loiselle, M.C. and Wones, D.R. (1979). Characteristics and origin of anorogenic granites. *Geological Society of America Abstracts with Programs 11*: 468.

Mahoney, J.J., and Coffin, M.F. (1997). Large igneous provinces: continental, oceanic, and planetary flood volcanism. American Geophysical Union Monograph No. 100, 438 pp.

Mahoney, J.J., Sheth, H.C., Chandrasekharam, D., and Peng, Z.X. (2000). Geochemistry of flood basalts of the Toranmal section, northern Deccan Traps, India: implications for regional Deccan stratigraphy. *Journal of Petrology* 41: 1099–1120.

Maia, M., Sishel, S., Briais, A. et al. (2016). Extreme mantle uplift and exhumation along a transpressive transform fault. *Nature Geoscience* 9: 619–623.

Maier, W.D., Barnes, S.J., Muir, D. et al. (2021). Formation of Bushveld anorthosite by reactive porous flow. *Contributions to Mineralogy and Petrology* 176: 3. https://doi.org/10.1007/s00410-020-01760-7, https://link.springer.com/article/10.1007/s00410-020-01760-7

Martin, H. (1986). Effect of steeper Archean geothermal gradient on geochemistry of subduction-zone magmas. *Geology* 14: 753–756.

Marzoli, A., Renne, P.R., Piccirillo, E.M. et al. (1999). Extensive 200 million-year-old continental flood basalts of the Central Atlantic Magmatic Province. *Science* 284: 616–618.

McBirney, A.R. (2007). *Igneous Petrology*, 3e, 550 pp. Boston, MA: Jones and Bartlett Publishers.

McCallum, I.S., Raedeke, L.D., and Mathez, E.A. (1980). Investigations in the Stillwater Complex, part I. Stratigraphy and structure of the Banded zone. *American Journal of Science 280-A*: 59–87.

McCallum, I.S., Thurber, M.W., O'Brien, H.E., and Nelson, B.K. (1999). Lead isotopes in sulfides from the Stillwater Complex, Montana: evidence for subsolidus remobilization. *Contributions to Mineralogy and Petrology* 137: 206–219.

McDougall, I. (1976). Geochemistry and origin of basalt of the Columbia River group, Oregon and Washington. *Geological Society of America Bulletin 87*: 777–792.

Meurer, W.P., Willmore, C.C., and Boudreau, A.E. (1999). Metal redistribution during fluid exsolution and migration in the Middle Banded series of the Stillwater Complex, Montana. *Lithos* 47: 143–156.

Mitchell, R.H. (1986). *Kimberlites: Mineralogy, Geochemistry and Petrology*, 442 pp. New York: Plenum Press.

Miyashiro, A. (1974). Volcanic rock series in island arcs and active continental margins. *American Journal of Science* 274: 321–355.

Muller, R.D., Royer, J.Y., and Lawyer, L.A. (1993). Revised plate motions relative to the hotspots from combined Atlantic and Indian Ocean hotspot tracks. *Geology* 21: 275–278.

Nelson, P.L. and Grand, S.P. (2018). Lower-mantle plume beneath the Yellowstone hotspot revealed by core waves. *Nature Geoscience* 11 (4): 280–284. https://doi.org/10.1038/s41561-018-0075-y.

Nutman, A.P. and Friend, C.R.L. (2007). Comment on a vestige of Earth's oldest ophiolite. *Science* 318: 746.

Parman, S.W., Grove, T.L., and Dann, J.C. (2001). The production of Barberton komatiites in an Archean

subduction zone. *Geophysical Research Letters* 28: 2513–2516.

Pearce, J. (1982). Trace element characteristics of lavas from destructive plate boundaries. In: *Orogenic Andesites and Related Rocks* (ed. R.S. Thorpe), 528–548. Chichester, England: Wiley.

Pearce, J. (2008). Geochemical fingerprinting of oceanic basalts with applications to ophiolite classification and the search for Archean oceanic crust. *Lithos 100*: 14–28.

Pearce, J.A. and Peate, D.W. (1995). Tectonic implications of the composition of volcanic arc magmas. *Annual Review of Earth and Planetary Sciences 23*: 251–285.

Perfit, M.R., Gust, D.A., Bence, A.E. et al. (1980). Chemical characteristics of island-arc basalts: implications for mantle sources. *Chemical Geology 30*: 227–256.

Pitcher, W.S. (1982). Granite type and tectonic environment. In: *Mountain Building Processes* (ed. K. Hsu), 19–40. London: Academic Press.

Polat, A., Longstaffe, F.J., and Frei, R. (2018). An overview of anorthosite-bearing layered intrusions in the Archaean craton of southern West Greenland and the Superior Province of Canada: implications for Archaean tectonics and the origin of megacrystic plagioclase. *Geodynamica Acta 30*: 84–99. Open Access https://www.tandfonline.com/doi/full/10.1080/09853111.2018.1427408.

Premo, W.R., Helz, R.T., Zientek, M.L., and Langston, R.B. (1990). U-Pb and Sm-Nd ages for the Stillwater Complex and its associated sills and dikes, Beartooth Mountains, Montana: identification of a parent magma? *Geology 18*: 1065–1068.

Reay, A. and Parkinson, D. (1997). Adakites from Solander Island, New Zealand. *New Zealand Journal of Geology and Geophysics 40*: 121–126.

Reichow, M.K., Saunders, A.D., White, R.V. et al. (2002). Ar40/Ar39 dates from the West Siberian Basin: Siberian flood basalt province doubled. *Science 296*: 1846–1849.

Reidell, S.P. (1983). Stratigraphy and petrogenesis of the Grande Rhonde Basalt from the deep canyon country of Washington, Oregon and Idaho. *Geological Society of America Bulletin 94*: 519–542.

Renne, P.R. (2002). Flood basalts – bigger and badder. *Science 296*: 1812–1813.

Saccani, E. (2015). A new method of discriminating different types of post-Archean ophiolitic basalts and their tectonic significance using Th-Nb and Ce-Dy-Yb systematics. *Geoscience Frontiers 6*: 481–501. https://doi.org/10.1016/j.gsf.2014.03.006.

Sano, T., Fujii, T., Deshmukh, S.S. et al. (2001). Differentiation processes of Deccan trap basalts: contributions from geochemistry and experimental petrology. *Journal of Petrology 42*: 2175–2195.

Seifert, K. and Brunotte, D. (1996). Geochemistry of weathered mid-ocean ridge basalt and diabase clasts from Hole 899B in the Iberia Abyssal Plain.

*Proceedings of the Ocean Drilling Program Scientific Results 149*: 497–515. http://www-odp.tamu.edu/publications/149_SR/chap_29/chap_29.htm.

Self, S., Schmidt, A., and Mather, T.A. (2014). Emplacement characteristics, time scales, and volcanic gas release rates of continental flood basalt eruptions on Earth. In: *Volcanism, Impacts, and Mass Extinctions: Causes and Effects* (eds. G. Keller and A.C. Kerr), 319–337. Geological Society of America Special Paper 505.

Seno, T. and Rehman, H.U. (2011). When and why the continental crust is subducted: examples of Hindu Kush and Burma. *Gondwana Research 19*: 327–333.

Sha, L.-K. (1995). Genesis of zoned hydrous ultramafic/mafic-silicic intrusive complexes: an MHFC hypothesis. *Earth Science Reviews 39*: 59–90.

Sinton, J.M., Ford, L.L., Chappell, B., and McCulloch, M.T. (2003). Magma genesis and mantle heterogeneity in the Manus back arc basin, Papau New Guinea. *Journal of Petrology 44*: 159–195.

Skinner, E.M.W. (1989). Contrasting group-1 and group-2 kimberlite petrology: towards a genetic model for kimberlites. In: *Proceedings of the 4th International Kimberlite Conference, Perth, Australia* (eds. J. Ross, A.L. Jaques, J. Ferguson and D.H. Green), 528–544. Geological Society of Australia Special Publication No. 14.

Smithies, R.H., Champion, D.C., and Cassidy, K.F. (2003). Formation of Earth's early Archean continental crust. *Precambrian Research 127*: 89–111.

Snow, J.E. and Edmonds, H.E. (2007). Ultraslow-spreading ridges rapid paradigm changes. *Oceanography 20* (1): 90–101.

Sobolev, A.V., Hofmann, A.W., Sobolev, S.V., and Nikogosian, I.K. (2005). An olivine-free mantle source of Hawaiian shield basalts. *Nature 434*: 590–597.

Sparks, R.S.J., Baker, L., Brown, R.J. et al. (2006). Dynamical constraints on kimberlite volcanism. *Journal of Volcanology and Geothermal Research 155*: 18–48.

Steinberger, B., Nelson, P.L., Grand, S.P., and Wang, W. (2019). Yellowstone plume conduit tilt caused by large-scale mantle flow. *Geochemistry, Geophysics, Geosystems 20* https://doi.org/10.1029/2019GC008490.

Steinmann, G. (1905). Geologische Beobachtungen in den Alpen. II. Die Schardtsche Ueberfaltungstheorie und die geologische Bedeutung der Tiefseeabsatze und der ophiolitischen Massengesteine. *Berichte der Naturfoschenden Gesellschaft zu Freiberg, iB 16*: 18–67.

Stern, R.J. (2005). Evidence from ophiolites, blueschists and ultrahigh-pressure metamorphic terranes that the modern episode of subduction tectonics began in Neoproterozoic time. *Geology 33*: 557–560.

Stern, R.J. (2007). When did plate tectonics begin: theoretical and empirical considerations. *Chinese Science Bulletin 52* (5): 578–591.

Stern, R.J. (2008). Modern-style plate tectonics began in Neoproterozoic time: an alternative interpretation of Earth's tectonic history. In: *When Did Plate Tectonics Begin on Planet Earth?* (eds. K.C. Condie and V. Pease), 265–280. Geological Society of America Special Paper 440.

Stern, C.R. and Killian, R. (1996). Role of the subducted slab, mantle wedge and continental crust in the generation of adakites from the Andean Austral volcanic zone. *Contributions to Mineralogy and Petrology 123*: 263–281.

Storey, M., Mahoney, J.J., Kroenke, L.W., and Saunders, A.D. (1991). Are oceanic plateaus the site of komatiite formation? *Geology 19*: 376–379.

Sun, S. and McDonough, W.F. (1989). Chemical and isotopic systematics of oceanic basalts: implications for mantle composition and processes. In: *Magmatism in Ocean Basins* (eds. A.D. Saunders and M.J. Norry), 313–345. Boston, MA: Blackwell Scientific.

Sun, S.S. and Nesbitt, R.W. (1978). Petrogenesis of Archaean ultrabasic and basic volcanics: evidence from rare earth elements. *Contributions to Mineralogy and Petrology 65*: 301–325.

Swanson, D.A. and Wright, T.L. (1980). The regional approach to studying the Columbia River Basalt Group. *Geological Society of India Memoir 3*: 58–80.

Swanson, D.A. and Wright, T.L. (1981). Guide to geologic field trip between Lewiston, Idaho, and Kimberly, Oregon, emphasizing the Columbia River Basalt Group. In: *Guides to Some Volcanic Terranes in Washington, Idaho, Oregon, and Northern California*, 189 pp (eds. D.A. Johnston and N.J. Donnelly). US Geological Survey Circular No. 838.

Takahashi, E., Nakajima, K., and Wright, T.L. (1998). Origin of the Columbia River basalts: melting model of a heterogeneous plume head. *Earth and Planetary Science Letters 162*: 63–80.

Taylor, H.P. Jr. (1967). The zoned ultramafic complexes of southeastern Alaska. In: *Ultramafic and Related Rocks* (ed. P.J. Wyllie), 96–116. New York: Wiley.

Tejada, M.L.G., Mahoney, J.J., Castillo, P.R. et al. (2004). Pin-pricking the elephant: evidence on the origin of the Ontong Java Plateau from Pb-Sr-Hf-Nd isotopic characterizations of ODP Leg 192 basalts. In: *Origin and Evolution of the Ontong Java Plateau* (eds. J.G. Fitton, J.J. Mahoney, P.J. Wallace and A.D. Saunders), 133–150. Geological Society of London Special Publications No. 229.

Tistl, M., Burgath, K.P., Hohndorf, A. et al. (1994). Origin and emplacement of tertiary ultramafic complexes in Northwest Columbia: evidence from geochemistry and K-Ar, Sm-Nd and Rb-Sr isotopes. *Earth and Planetary Science Letters 126*: 41–59.

Todd, S.G., Keith, D.W., Le Roy, L.W. et al. (1982). The J-M platinum-palladium reef of the Stillwater Complex, Montana: I. Stratigraphy and petrology. *Economic Geology 77*: 1454–1480.

Tolan, T.L., Reidel, S.P., Beeson, M.H. et al. (1989). Revisions to the estimates of the aerial extent and volume of the Columbia River Basalt Group. In: *Volcanism and Tectonism in the Columbia River Flood-Basalt Province* (eds. S.P. Reidel and P.R. Hooper), 1–20. Geological Society of America Special Paper 239.

Turner, S., Tonarini, S., Bindeman, I. et al. (2007). Boron and oxygen isotope evidence for recycling subducted components over the past 2.5 Ga. *Nature* 447: 702–705.

Van der Laan, S.R., Flower, M.F.J., and Van Groos, A.F.K. (1989). Experimental evidence for the origin of boninites: near-liquidus phase relations to 7.5 kbar. In: *Boninites and Related Rocks* (ed. A.J. Crawford), 112–147. London: Unwin Hyman.

Vearncombe, S. and Vearncombe, J.R. (2002). Tectonic controls on kimberlite location, southern Africa. *Journal of Structural Geology* 24: 1619–1625.

Vermaak, C.F. (1976). The Merensky Reef – thoughts on its environment and genesis. *Economic Geology* 71: 1270–1298.

Verplanck, P.L., Mariano, A.N., and Mariano, A. Jr. (2016). Rare Earth element ore geology of carbonatites. In: *Reviews in Economic Geology*, vol. *18* (eds. P.L. Verplanck and M.W. Hitzman), 5–32. Society of Economic Geologists https://doi.org/10.5382/REV.18.

Viljoen, M.J. and Viljoen, R.P. (1969). Archean vulcanicity and continental evolution in the Barberton Region, Transvaal. In: *African Magmatism and Tectonics* (eds. T.N. Clifford and I.G. Gass), 27–39. Edinburgh: Oliver and Boyd.

Wager, L.R. and Deer, W.A. (1939). Geological investigations in East Greenland. III. The petrology of the Skaergaard intrusion, Kangerdlugssuaq, East Greenland. *Medd Gronland* 105 (4): 1–352.

Walter, M.J. (1998). Melting of garnet peridotite and the origin of komatiite and depleted lithosphere. *Journal of Petrology* 39: 29–60.

Wei, J.F., Tronnes, R.G., and Scarfe, C.M. (1990). Phase relations of alumina-undepleted and alumina depleted komatiites at pressures of 4–12 GPa. *Journal of Geophysical Research* 95: 15,817–15,828.

Whalen, J.B., Currie, K.L., and Chappell, B.W. (1987). A-type granites: geochemical characteristics, discrimination and petrogenesis. *Contributions to Mineralogy and Petrology* 95: 407–419.

White, S.H., de Boorder, H., and Smith, C.B. (1995). Structural controls of kimberlite and lamproite emplacement. *Journal of Geochemical Exploration* 51: 245–264.

de Wit, M.J. (1998). On Archean granites, greenstones, cratons and tectonics: does the evidence demand a verdict? *Precambrian Research* 91: 181–226.

Wyman, D.A. (1999). Paleoproterozoic boninites in an ophiolite-like setting, trans-Hudson orogen, Canada. *Geology* 27: 455–458.

Ye, K., Cong, B., and Ye, D. (2000). The possible subduction of continental material to depths greater than 200 km. *Nature* 407: 734–736.

Zientek, M.L. (2012). Magmatic ore deposits in layered intrusions—descriptive model for reef-type PGE and contact-type Cu-Ni-PGE deposits: U.S. *Geological Survey Open-File Report 2012–1010* 48 p.

Zolotukhin, V.V. and Al'Mukhamedov, A.I. (1988). Traps of the Siberian Platform. In: *Continental Flood Basalts* (ed. J.D. Macdougall), 273–310. New York: Kluwer Academic Publishers.

# Chapter 11

# Weathering, sediment production, and soils

## 11.1 WEATHERING

**Weathering** is the in place (*in situ*) breakdown of rock materials at or near Earth's surface. Weathering processes are fundamentally important in the generation of the soils, sediments, and sedimentary rocks that cover more than 80% of Earth's surface. Most sediment originates as solid detrital particles and dissolved solids produced during weathering. These materials are subsequently eroded and dispersed by water, wind, glaciers, and mass flows across Earth's surface to be deposited as detrital and biochemical sediments.

Weathering is also the dominant process in the production of soils, upon which so many essential human activities depend. Soils and sediments provide critical wildlife habitats in both terrestrial and aquatic environments serve as aquifers and aquitards critical for the storage and transmission of water and contain critical supplies of coal, petroleum, natural gas, and ore deposits. In addition, they are widely used as raw materials in the construction of roads, dams, buildings, and other structures. Soils are essential to agriculture and the production of forest products. Weathering processes determine soil texture, soil composition, soil nutrient content, and water retention properties, and therefore the types of crops that can be successfully grown in a given area. In short, we could not have survived on Earth without the sediment and soil-producing processes involved in weathering.

Soils and sediments are directly impacted by surface contaminants. Knowledge of the factors that permit soils and sediment to transmit pollutant plumes away from the initial site of contamination is especially critical, as is knowledge of how to construct containment barriers

*Earth Materials*, Second Edition. Kevin Hefferan and John O'Brien.
© 2022 John Wiley & Sons Ltd. Published 2022 by John Wiley & Sons Ltd.
Companion website: www.wiley.com/go/hefferan/earthmaterials2

or remediate near surface contamination. Soils present many challenges to engineers who must be able to evaluate their behavior before construction begins or risk the problems associated with ground failure.

Furthermore, because many types of soils are strongly dependent on climate and organic activity, ancient soils offer major clues to the emergence and evolution of life on continents, major long-term changes in global and local climates, and the long-term evolution of Earth's atmosphere.

For all of these reasons, this chapter is devoted to developing a deeper understanding of weathering, sediment production, and soils.

Weathering involves an interactive set of physical, chemical, and biological processes that result in the in situ breakdown of rock material at or near Earth's surface. Weathering may occur in the original source area where bedrock is exposed, or in rock materials that have been eroded, transported, and deposited thousands of kilometers away from their original source area.

Weathering and erosion are two important, but different, sets of processes. The distinguishing factor between weathering and erosion is that **weathering processes** involve the breakdown of rock material in a particular location, whereas **erosion processes** involve the removal of rock material from a geographic location which initiates its transportation to another location. Weathering and erosion are intimately related because weathering generally breaks rock materials down into smaller detrital, organic or dissolved constituents whose small size (unless they are cohesive) makes them more easily removed by erosion and dispersed by transportation. As will be seen in the discussions that follow, different rock materials weather at different rates, a process called **differential weathering**. Less resistant rocks that break down more rapidly tend to be eroded more rapidly. More resistant rocks that weather more slowly tend to erode more slowly. Differential weathering, combined with differences in rock durability, tends to produce differential erosion as the products of weathering are removed at different rates. Figure 11.1 illustrates differential erosion between durable, cliff-forming sandstones, and more easily eroded mudrocks in southern Utah.

Major weathering processes may be subdivided into disintegration and decomposition processes.

**Figure 11.1** Differential weathering between durable cliff and pillar-forming sandstone and less durable mudrocks in Utah. *Source*: Photo by Kevin Hefferan. © John Wiley & Sons.

**Disintegration** is the breakdown of larger, more coherent rock bodies into smaller fragments of the same composition. As discussed below, disintegration may involve physical (mechanical) and/or biological processes. Disintegration generates an increased number of smaller rock or mineral fragments of the original material that is being disintegrated. When such fragments are eventually transported and deposited as sediment, they may allow us to recognize the kinds of rock material originally exposed in the source area from which they were derived and to infer their dispersal pathways (Chapter 13). This may allow us to pinpoint the source or **provenance** of a particular sediment as being from a particular place where such rocks are (or were) exposed.

**Decomposition** is any breakdown of rock materials during weathering that involves changes in rock composition. Decomposition generally alters a rock's mineralogy so that minerals stable at higher subsurface temperatures or pressures are altered to minerals stable at the temperatures and pressures near Earth's surface. Decomposition involves both inorganic and organic chemical processes and is strongly dependent on the availability and chemistry of water, which plays a significant role in decomposition processes.

In the sections that follow, we will examine how the types of weathering processes and the degree of weathering that occur at a particular place vary greatly depending upon factors such as climate, rock type, slope, and time.

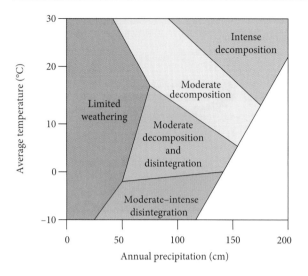

**Figure 11.2** The relative roles of mechanical disintegration and chemical decomposition as a function of average annual temperature and rainfall.

**Figure 11.3** Bryce Canyon, Utah showing pillars and windows formed by differential erosion, controlled by near vertical and horizontal joint sets.

Disintegration is more prevalent in cold and dry climates; decomposition processes dominate in warmer and wetter climates (Figure 11.2). Steep slopes favor short-term, incomplete decomposition as rock materials are removed rapidly by erosion before decomposition is complete, whereas gentle slopes favor longer term, more thorough weathering.

### 11.1.1 Disintegration

Disintegration includes any mechanical or organic weathering process that breaks rocks into smaller pieces of the same material. Several common disintegration processes are discussed in the paragraphs that follow.

*Joint formation*

**Joints** are fractures in rocks relative to which little or no tangential (fracture parallel) movement has taken place. Joints originate in response to stresses of various kinds and can have rather complicated histories. Most of the joints that act as conduits for fluids during weathering are produced by the decrease in confining pressure that occurs as formerly buried rocks approach the surface. Figure 11.3 illustrates joints in Eocene sandstones from Bryce Canyon, Utah. Stratification is not significantly offset parallel to the prominent,

near vertical joints. Weathering is clearly accelerated along the joints, which in turn influence the differential weathering and erosion responsible or the development of spires (hoodoos) bounded by joint surfaces and capped by resistant rock layers. More detailed information concerning the relationships between joints and landscape evolution can be found in DiPietro (2018), Willigoose (2018), Ritter et al. (2006), Hugget (2002), and Burbank and Anderson (2000).

Rocks can form at or be buried to depths of many kilometers. The pressure exerted on such rocks by overlying rocks is referred to as **lithostatic pressure** or **confining pressure**. Lithostatic pressure is an isotropic confining stress (force per unit area) that results from the weight of the overlying rocks that push inward equally in all directions. As rocks move closer to Earth's surface, lithostatic pressure diminishes with decreasing burial depth. This decrease in lithostatic pressure or load is called **unloading** or **decompression**. Unloading results from erosion and/or faulting that remove overlying rocks. As buried rocks experience unloading they tend to expand, and when they expand by more than 1 or 2%, they tend to fracture, resulting in joints. Joints may have many different orientations. They commonly occur along pre-existing weaknesses in rocks, many of which originated from earlier tectonic stresses that affected the rocks during burial. As decompression and expansion proceed, such fractures propagate and become more numerous. As a result, formerly intact rock is progressively fractured into smaller pieces. An interesting example occurs in quarries where workers excavate intact rock; within a few days, many fractures develop in the unexcavated rock on the quarry floor.

**Sheet joints** are rock fractures that open subparallel to Earth's surface and develop to depths that may exceed 100 m. They tend to form under upwardly convex surfaces such as domes and ridges in mechanically homogeneous rocks like granites. In such rocks, maximum tensile stress is roughly perpendicular to the convex surfaces. The term **exfoliation** is used to describe sheet joints that resemble the curved surface of an onion. Other joints open at high angles to the surface (Figure 11.4).

Fracture propagation is strongly abetted in some climatic settings by a second set of processes called frost action.

### Frost action

**Frost action (shattering)** occurs when preexisting fractures and weak surfaces are enlarged by the expansion of water as it freezes (Figure 11.5). Water penetrates even small fractures or the boundaries between crystals. When it freezes, water expands by nearly 10%, creating tensile stresses that cause the crystals to separate or the small fractures to become enlarged. When the ice melts, the enlarged fracture can hold more water. When the water freezes again, the

**Figure 11.4**  Joints in anorthosite bedrock, Saranac Lake, New York. *Source*: Courtesy of Anita O'Brien.

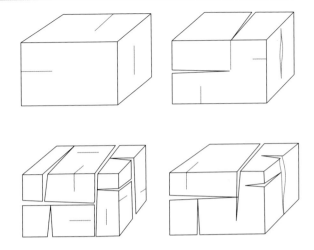

**Figure 11.5**  Sequential diagram (clockwise from top left) showing disintegration as visible fractures are enlarged and lengthened during frost action (shattering) and microfractures are enlarged to become visible fractures.

fracture grows again. Repeated freezing and thawing lead to the formation of many new and enlarged, intersecting fractures that facilitate the progressive disintegration of the original rock into smaller pieces of the same material. **Frost wedging** occurs along fractures oriented steeply to Earth's surface. **Frost heaving** develops along surfaces subparallel to Earth's surface when water freezes along less permeable surfaces, bedding planes, and/or sheet fractures, often lifting soil particles several centimeters above the surface.

Although joint formation and frost action dominate disintegration globally, several other processes are locally important. **Crystal growth** (other than ice crystals) may pry rock material apart as the crystal grows in a fracture or pore space. Rock expansion due to the growth of evaporate minerals such as gypsum is particularly important in desert environments and can be especially significant in the development of some arid climate soils, as discussed later in this chapter. **Slaking** occurs when minerals such as clays and micas expand when wetted. Significant mineral expansion generates stresses that cause adjacent minerals to disintegrate. Some workers have suggested that **thermal volume changes (insolation)** that result from daily or seasonal changes in rock temperature may cause significant amounts of disintegration. Most solids, including rocks, wood, and concrete structures, such as highways, bridges, and sidewalks, expand with increasing temperature and contract as temperatures decrease. Thermal expansion and contraction could induce stresses that would cause spalling of small particles of rock off the surface of a rock body. But the coefficient of thermal expansion (the rate at which volume changes with temperature change) in rocks and minerals is very small and likely insufficient to produce much spalling, even from the surface of rock bodies.

### Biological processes in disintegration

**Root growth** is important in disintegration. Rocks are pried apart and fractures enlarged as root systems, which grow along fractures, expand during growth (Figure 11.6).

A host of **animal activities** in which organisms crack, drill, bore, burrow, mix, and feed on rock material, causing it to be broken down into smaller pieces, are also significant in disintegration of surficial rock materials. But these activities pale compared to **human activities**. Every time we move surficial material to farm or plant crops, or quarry material to build houses, skyscrapers, and shopping centers, to construct dams, canals, bridges, tunnels and highways, and to strip or otherwise mine near surface resources, humans disintegrate huge volumes of surficial material. By changing the way such surface materials are organized, we create multiple geohazards and change the ways in which the materials undergo further weathering.

**Figure 11.6**  Biological weathering by tree root growth in Silurian dolostone, Door County, Wisconsin. *Source*: Kevin Hefferan. © John Wiley & Sons.

### Disintegration and decomposition

Rock disintegration processes enhance chemical decomposition by increasing the surface area of the resulting rock fragments (Figure 11.7). This increase in surface area enlarges the surface over which chemical reactions can occur when rock surfaces are in contact with a solvent such as soil water or groundwater. This increase in chemically reactive surface area accelerates the rate at which decomposition reactions occur, so that a positive interaction between disintegration and decomposition results.

A closely related example of the interplay between disintegration and decomposition is shown by **spheroidal weathering** in which massive, well-jointed rocks such as granite, gabbro, and basalt weather into spheroidal forms (Figure 11.8). Spheroidal weathering begins with the formation of spaced rectangular joint sets that split the bedrock into multiple blocks or parallelepipeds, similar in shape to sugar cubes. This increases both the surface area of the rock and access for the infiltration of groundwater. As with sugar cubes, dissolution proceeds most rapidly at the corners of the blocks where three chemically active faces intersect. Dissolution is slower along the edges of the blocks where two faces intersect and slower still in the centers of the individual faces. As the corners and edges are decomposed and eroded, blocks become rounded. Decomposition proceeds from the surface inward, so that more decomposed outer layers peel off before less decomposed inner layers, producing the forms shown in Figure 11.8.

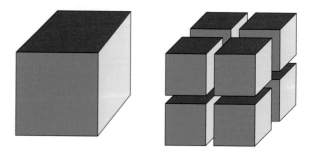

**Figure 11.7**  The increase in surface area resulting from disintegration of rock material: an initial block of $1\,m^3$ divided by three fractures produces eight cubes and doubles the surface area from $6\,m^2$ ($6 \cdot 1\,m^2 = 6\,m^2$) to $12\,m^2$ ($8 \cdot 6$ faces = $48 \cdot 0.25\,m^2 = 12\,m^2$).

**Figure 11.8** Spheroidal weathering of basaltic rock at Table Mountain, Golden, Colorado. *Source*: Photo by Kevin Hefferan. © John Wiley & Sons.

### 11.1.2 Decomposition

Decomposition processes produce compositional changes in rocks that are caused by chemical and biological (biochemical) reactions between minerals and pore fluids, such as water (aqueous solutions) and air (soil gases), and organisms and organic compounds. Of these, the most significant agent in decomposition is downward percolating water. Every time it rains, water percolates through joints and between particles in the soil, causing additional increments of decomposition to occur. These change the composition of the original minerals and the composition of the pore fluids and often involve the formation of new minerals that can become significant components of soils over time. Biochemical decomposition includes a complex set of interactive processes that depend upon:

1  Bedrock composition and the ease with which different minerals decompose.
2  Climatic factors such as temperature range, rainfall, and groundwater levels.
3  Organic factors such as vegetative cover, soil communities, and bacterial activity.
4  Soil water geochemical factors such as ionic concentrations, acidity–alkalinity, and oxidation reduction potential.
5  Topographic factors such as slope.

The following section concerns some of the major processes of decomposition summarized in Table 11.1.

*Dissolution*

**Dissolution** occurs when a mineral or other soil component is wholly or partially dissolved during chemical decomposition. The dissolution of the mineral clearly changes the chemical composition of both the original rock from which the mineral has been removed and the solution to which the dissolved solids have been added. Two common examples, the dissolution of halite (NaCl) and the dissolution of calcite ($CaCO_3$) are shown below. In the first example, halite reacts with polarized water molecules that separate sodium and chloride into ions that are dissolved in the water and surrounded by hydration sheaths.

$$1 \quad NaCl + H_2O \rightarrow Na^{+1}_{(aq)} + Cl^{-1}_{(aq)} + H_2C$$

In the second example, carbon dioxide gas dissolves in water to produce dissolved carbonic acid. During dissolution, the mildly acidic carbonic acid reacts with the calcite to produce dissolved calcium ions and dissolved bicarbonate ions in a process called **carbonation**.

$$2 \quad H_2O + CO_2 \rightarrow H_2CO_{3(aq}$$

$$3 \quad H_2CO_{3(aq)} + CaCO_3$$
$$\rightarrow Ca^{+2}_{(aq)} + 2\left(HCO_3\right)^{-1}_{(aq)}$$

In areas where soluble rocks such as limestones, dolostones, or evaporites are common, long-term dissolution may lead to the formation of caves, sinkholes, and other features characteristic of karst topography. Karst development, named after the Karst region in Slovenia, poses significant environmental issues related to surface collapse events and aquifer flow complexity (Box 11.1). Karst features are widespread in the United States (Figure 11.9) and throughout the world.

*Ion exchange*

**Ion exchange** occurs when ions are directly exchanged between a mineral and a solution. In the first example below, a hydrogen ion in aqueous solution is exchanged with a potassium ion in the potash feldspar ($KAlSi_3O_8$) with which the solution is in contact. Repetition of such exchanges enriches soil water in dissolved potassium,

**Table 11.1** Major processes and products of decomposition.

| Decomposition process | Examples | Decomposition products |
|---|---|---|
| Dissolution | $NaCl + H_2O \rightarrow Na^{+1}_{(aq)} + Cl^{-1}_{(aq)} + H_2O$ <br> $H_2CO_{3(aq)} + CaCO_3 \rightarrow Ca^{+2}_{(aq)} + 2(HCO_3)^{-1}_{(aq)}$ | Dissolved solids including $Ca^{+2}$, $Mg^{+2}$, $Na^{+1}$, $K^{+1}$, $H_4SiO_4$, $CO_3^{-2}$, $SO_4^{-2}$ |
| Ion exchange | $KAlSi_3O_8 + H^{+1}_{(AQ)} \rightarrow HAlSiO_3 + K^{+1}_{(AQ)}$ <br> $NaAlSi_3O_8 + H^{+1}_{(aq)} \rightarrow HAlSi_3O_8 + Na^{+1}_{(aq)}$ | Dissolved solids including $Na^{+1}$ and $K^{+1}$ |
| Hydrolysis | $2KAlSi_3O_8 + 2H^{+1}_{(aq)} + 9H_2O$ <br> $\rightarrow Al_2Si_2O_5(OH)_4 + H_4SiO_{4(aq)} + 2K^{+1}_{(aq)}$ | Clay minerals such as kaolinite, illite, smectite |
| | $Mn_2SiO_4 + 4H_2O \rightarrow 2Mn(OH)_2 + H_4SiO_{4(aq)}$ | Other hydroxides |
| Hydration | $CaSO_4 + 2H_2O \rightarrow CaSO_4 \cdot 2H_2O$ <br> $Fe_2O_3 + H_2O \rightarrow 2FeO \cdot OH$ | Hydrated and hydrous oxide minerals |
| Oxidation | $2Fe^{+2}_2SiO_4 + 4H_2O + O_2 \rightarrow 2Fe^{+3}_2O_3 + 2H_4SiO_{4(aq)}$ <br> $4Fe^{+2}S^{-2}_2 + 15O_2 + 8H_2O$ <br> $\leftrightarrow 2Fe^{+3}_2O_3 + 8S^{+6}O^{-2}_{4(aq)} + 16H^{+1}_{(aq)}$ | Oxide mineral and dissolved solids, e.g., $SO_4^{-2}$, $H_4SiO_4$ |
| Chelation | | Dissolved metals contained in organic ring complexes (chelates) |

## Box 11.1  Karst development and its implications

The effects of dissolution are well illustrated by the formation of caves, sinkholes (dolines), and other karst features. Atmospheric and soil carbon dioxide, produced largely by organic respiration and bacterial decomposition, combine with water to produce carbonic acid, which progressively dissolves carbonate or evaporite minerals (Figure B11.1a). Karst features are particularly common in warm, humid climates where abundant water and biotic activity combined with high temperatures favor dissolution. As acidic groundwater flows along joints, faults, and bedding planes, dissolution gradually enlarges them. This produces a positive feedback loop in which enlarged conduits permit more groundwater flow, which produces more dissolution. Dissolution occurs during relatively rapid flow in the vadose zone above the water table, in the zone near the fluctuating top of the water table and possibly during slower flow in the phreatic zone below the water table.

One result of such large-scale dissolution is the formation of networks of large cavities in the form of **caves** that frequently contain underground streams that enter the subsurface down dissolution features and emerge as cave springs (Figure B11.1b(i)). Sinkholes (dolines) are roughly circular to ovoid depressions (Figure B11.1b(ii)) that form by (1) gradual loss of surface soluble rocks by accelerated dissolution; (2) gradual surface sinking of less soluble overburden (soil, sediment) by infiltration into cavities produced by dissolution of subsurface rock layers; or (3) sudden collapse of the surface by the collapse of rock or soil into an underlying cavity. The features and processes associated with karst development present many natural hazards that can be compounded by human activities. Natural karst hazards include severe flooding of sinkholes and karst valleys, continued dissolution, and subsidence and/or sudden collapse of the surface during sinkhole formation. Many human activities accelerate subsidence and collapse. Diversion of water into underground systems can accelerate dissolution, leading to collapse of bedrock or soil into underground cavities. Overextraction of water from karst aquifers lowers the water table to the point where loss of water pressure triggers the collapse of surface materials into underlying cavities. In situations where collapse is imminent, additional surface loading during construction projects may trigger subsidence or collapse. Building houses, roads, and commercial structures in karst regions presents real challenges. In addition, surface pollutants that rapidly enter groundwater in

**Box 11.1** *Continued*

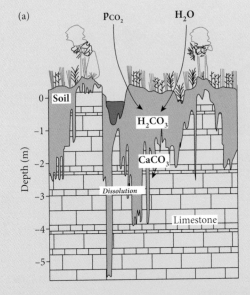

(a)

$$CaCO_3 + H_2O + CO_2 \rightarrow Ca^{2+} + 2HCO_3$$

- Bacterial and root respiration in the soil increase $P_{CO_2}$

- As $P_{CO_2}$ increases, so do dissolution rates

Tropical latitudes = high evapotranspiration + well-developed soil + high atm. $P_{CO_2}$

= KARST

(Yucatan, Caribbean, Florida, South China...)

(b) (i)

(ii)

**Figure B11.1** (a) Dissolution of carbonates by carbonated acidic groundwater. (b) (i) Cave spring with an underground river emerging from a cave produced by subsurface dissolution, Virginia. (ii) Sinkhole, Winter Park, Florida produced by the sudden collapse of the surface into a subsurface cavity, May, 1981. *Source*: Courtesy of the US Geological Survey.

karst areas due to high recharge rates are quickly dispersed, thus polluting wells, springs, and underground rivers over wide areas. Extreme care must be used when disposing of any hazardous materials in karst areas.

and the addition of hydrogen ions weakens the structure of the potash feldspar as it is progressively decomposed. Soil waters in warm, humid climates have a low pH, are rich in hydrogen ions, and are acidic. Such solutions decompose feldspars faster than more alkaline solutions with higher pH and fewer hydrogen ions, so that feldspar decomposition is accelerated in areas with warm, wet climates.

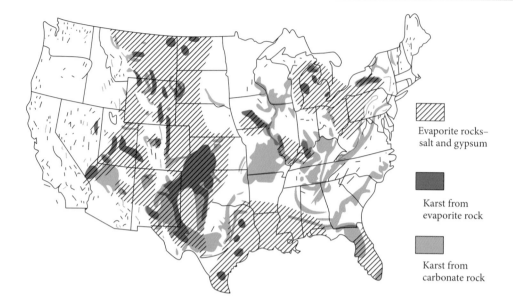

**Figure 11.9** Distribution of karst dissolution features in the United States. *Source*: Davies and LeGrand (1972); with permission of the US Geological Survey.

1 $KAlSi_3O_8 + H^{+1}_{(aq)} \rightarrow HAlSiO_3 + K^{+1}_{(aq)}$

The second example shows a similar ion exchange involving albite ($NaAlSi_3O_8$), the sodium plagioclase feldspar. Since feldspars are the most abundant mineral group in Earth's crust, ion exchange processes are an important part of chemical decomposition.

2 $NaAlSi_3O_8 + H^{+1}_{(aq)} \rightarrow HAlSi_3O_8 + Na^{+1}_{(aq)}$

The third example involves a reversible ion exchange between the potassium clay mineral illite $[KAl_2AlSi_3O_{10}(OH)_4]$ and dissolved hydrogen ion. Removal of potassium from illite has the potential to convert the illite into the clay called kaolinite that commonly dominates the clay mineral assemblages in acidic soils generated in warm, humid climates.

3 $KAl_2AlSi_3O_{10}(OH)_4 + H^{+1}_{(aq)}$
$\leftrightarrow HAl_2AlSi_3O_{10}(OH)_4 + K^{+1}_{(aq)}$

*Hydrolysis*

In weathering, **hydrolysis** is a chemical reaction between a mineral and water in which dissolved hydrogen ions and/or hydroxyl ions are added to form one or more new minerals. In most hydrolysis reactions, the original mineral is a silicate mineral and the new mineral is a hydroxide or clay mineral. Feldspars, the most abundant group of silicate minerals in Earth's crust, decompose by hydrolysis into clays. As a result, clay minerals are the most abundant group of new minerals produced during chemical decomposition. Because clay mineral crystals tend to be very small (<4 μm), clay minerals are the most abundant constituents of the mud fraction of detrital sediments in soils that are later dispersed by erosion and transportation and eventually deposited in surface environments.

In the first example below, potassium feldspar ($KAlSi_3O_8$) reacts with dissolved hydrogen ions and water to produce the clay mineral kaolinite $[Al_2Si_2O_5(OH)_4]$ plus dissolved orthosilicic acid (commonly referred to as dissolved silica) and dissolved potassium ions.

1 $2KAlSi_3O_8 + 2H^{+1}_{(aq)} + 9H_2O$
$\rightarrow Al_2Si_2O_5(OH)_4 + H_4SiO_{4(aq)} + 2K^{+1}_{(aq)}$

The second reaction is similar but involves the sodium plagioclase feldspar albite ($NaAlSi_3O_8$) reacting with hydrogen ions and water to form kaolinite $[Al_2Si_2O_5(OH)_4]$ plus orthosilicic acid and dissolved sodium ions. In both cases, the reactions may continue over long periods of time so that feldspar crystals are progressively converted into clay minerals by hydrolysis.

$$2 \quad 2NaAlSi_3O_8 + 2H^{+1}_{(aq)} + 9H_2O$$
$$\rightarrow Al_2Si_2O_5(OH)_4 + H_4SiO_{4(aq)} + 2Na^{+1}_{(aq)}$$

The third example generally occurs during the weathering of manganese-bearing minerals such as ferromagnesian silicates in which manganese ($Mn^{+2}$) substitutes for ferrous iron ($Fe^{+2}$). In this reaction, manganese-bearing olivine ($Mn_2SiO_4$) reacts with water to produce the hydroxide mineral pyrolusite [$Mn(OH)_2$] plus dissolved orthosilicic acid.

$$3 \quad Mn_2SiO_4 + 4H_2O$$
$$\rightarrow 2Mn(OH)_2 + H_4SiO_{4(aq)}$$

*Hydration and dehydration*

**Hydration** involves the addition of water to a crystal structure during the reaction between a mineral and the aqueous solution. In the first example below, the anhydrous calcium sulfate mineral anhydrite is converted into the hydrated calcium sulfate mineral gypsum by the addition of water. The reversal of this reaction is called **dehydration** and involves the conversion of hydrated gypsum into anhydrous anhydrite with loss of water from the crystal structure.

$$1 \quad CaSO_4 + 2H_2O \rightarrow CaSO_4 \cdot 2H_2O$$

The second example illustrates the conversion of the anhydrous iron oxide mineral hematite into the iron oxyhydroxide mineral goethite by the addition of water. This reaction is also reversible with goethite being converted to hematite by dehydration.

$$2 \quad Fe_2O_3 + H_2O \rightarrow 2(FeO \cdot OH)$$

*Oxidation*

**Oxidation** is a chemical reaction in which one or more electrons are transferred from a cation in the mineral to an anion, increasing the valence of the cation. Oxygen is strongly electronegative and therefore tends to capture electrons. Because oxygen is abundant in many weathering environments, the majority of oxidation reactions involve the transfer of electrons from a cation to oxygen as they react chemically to produce an oxide mineral. The production of an oxide mineral is not required by oxidation reactions. Only the loss of an electron is required. But because oxygen is the most abundant electronegative element on Earth, it dominates oxidation reactions.

In the first example below, ferrous ($Fe^{+2}$) iron-bearing olivine ($Fe^{+2}_2SiO_4$) combines with dissolved oxygen to form the ferric ($Fe^{+3}$) iron mineral hematite ($Fe^{+3}_2O_3$) plus dissolved orthosilicic acid. The reaction involves the loss of an electron from the iron to the oxygen, that is, oxidation, which increases the valence state of the iron from +2 to +3.

$$1 \quad 2Fe^{+2}_2SiO_4 + 4H_2O + O_2$$
$$\rightarrow 2Fe^{+3}_2O_3 + 2H_4SiO_{4(aq)}$$

The second example is similar but involves the conversion of the manganese silicate mineral rhodonite ($Mn^{+2}_2SiO_3$) by reaction with water and dissolved oxygen into the manganese oxide mineral *manganite* ($Mn^{+3}O_2$) plus dissolved orthosilicic acid.

$$2 \quad 2Mn^{+2}_2SiO_3 + 4H_2O + O_2$$
$$\rightarrow 2Mn^{+3}O_2 + 2H_4SiO_{4(aq)}$$

These two oxidation equations are simple, but representative, examples of the many oxidation reactions involved in the decomposition of ferromagnesian silicate minerals.

The third example illustrates the oxidation of the iron sulfide mineral pyrite ($Fe^{+2}S_2$) to the iron oxide mineral hematite ($Fe^{+3}_2O_3$). The oxidation (electron loss) of the iron is accomplished when the ferrous ($Fe^{+2}$) sulfide mineral pyrite reacts with water and dissolved oxygen to produce the ferric ($Fe^{+3}$) oxide mineral hematite plus dissolved sulfate ions and dissolved hydrogen ions. In the reaction below, the sulfur in the sulfide (+2) pyrite gains electrons to become sulfur (+6) in the sulfate ($SO_4^{-2}$) anion.

$$3 \quad 4Fe^{+2}S^{-2}_2 + 15O_2 + 8H_2O$$
$$\leftrightarrow 2Fe^{+3}_2O_3 + 8S^{+6}O_4^{-2}{}_{(aq)} + 16H^{+1}_{(aq)}$$

Under reducing conditions (low oxidation–reduction potential), chemical reaction number three is easily reversible. When it is reversed, the iron ions gain electrons so that the valence state is reduced from ferric iron ($Fe^{+3}$) to ferrous iron ($Fe^{+2}$). This occurs as hematite combines with sulfate and hydrogen ions and is converted to pyrite with the release of oxygen and water. Such reactions that involve the loss of electrons are called **reducing reactions** and the general name for processes involving the loss of electrons is **reduction**. Oxidation–reduction reactions are discussed further in Chapter 14.

*Organic decomposition and chelation*

Organic activity plays a major role in chemical decomposition processes. **Chelates** are organic hydrocarbon ring complexes produced directly by lichen and indirectly by the decay of humus. These highly soluble organic molecules tend to bind metallic elements such as $Ca^{+2}$, $Mg^{+2}$, $Al^{+3}$, $Fe^{+2}$, $Fe^{+3}$, $K^{+1}$, and $Na^{+1}$, effectively removing them from solution. This process is called **chelation**. Chelation often involves the exchange of hydrogen ($H^{+1}$) ions from the chelating agent to the solution and metal ions from the solution to the chelate. The increase in hydrogen ions decreases the pH of the solution, making it more acidic, while the decrease in metal ions in solution makes the metals in the remaining minerals more soluble. Both processes tend to increase the decomposition of metal-bearing minerals significantly. Chelates are also very important to plants as they provide nutrient metals to plants in a form that makes the metals readily available for absorption.

Organic activity also indirectly affects decomposition rates. Respiration produces carbon dioxide ($CO_2$) as a by-product. When $CO_2$ combines with water ($H_2O$), carbonic acid ($H_2CO_3$) forms (equation 11.1).

$$CO_2 + H_2O = H_2CO_3 \quad \text{(equation 11.1)}$$

$$H_2CO_3 = \left(HCO_3\right)^{-1} + H^{+1} \quad \text{(equation 11.2)}$$

$$\left(HCO_3\right)^{-1} = \left(CO_3\right)^{-2} + H^{+1} \quad \text{(equation 11.3)}$$

When carbonic acid dissociates, a hydrogen ion is released during the formation of a bicarbonate [$(HCO_3)^{-1}$] ion (equation 11.2 above). In certain situations, the bicarbonate ion dissociates into carbonate [$(CO_3)^{-1}$] ion (equation 11.3 above) and an additional hydrogen ($H^{+1}$) ion is produced. The release of hydrogen ions lowers the pH of the soil water and

increases its acidity. Thus, significant increases in soil $CO_2$ content ultimately generate significant increases in dissolved hydrogen ($H^{+1}$) ions making soil waters significantly more acidic. This tends to accelerate chemical decomposition of most common rock-forming minerals.

## 11.2 DISSOLVED SOLIDS

As noted above, when water infiltrates into fractures and spaces between grains, it dissolves constituents from the rocks and minerals with which it is in contact. An essential by-product of the decomposition reactions discussed in the previous section is an abundance of dissolved solids in soil and groundwater. Some of these dissolved solids are reprecipitated as new minerals in soils and others are carried underground and precipitated as mineral cements during sedimentary rock diagenesis, as discussed in Chapters 13 and 14. A large proportion of the dissolved solids in groundwater are discharged by springs into surface waters (Box 11.2). Some dissolved solids reside in surface water for a considerable period of time, especially in lakes. However, most of the dissolved load in surface waters eventually flows into the oceans via surface runoff where it is joined by smaller amounts of dissolved solids discharged directly into the ocean by submarine springs and submarine rock alteration processes to make the sea "salty" (rich in total dissolved solids). In this way, dissolved solids are widely dispersed through groundwater, surface water, and the ocean. Over time, many of these dissolved solids are removed from solution to form solid biochemical sediments that accumulate on Earth's surface, as detailed in Chapter 14, and hydrothermal metamorphic rocks on the sea floor, as discussed in Chapter 18.

In the sections that follow, we focus on the detrital sediments and soils that are generated by weathering processes.

## 11.3 DETRITAL SEDIMENTS

The solid, inorganic components of residual soils are **detrital sediments**. These detrital sediments are either **resistates**, which are residual

## Box 11.2   People and Earth materials: mineral water

Mineral water sales worldwide exceed $10^{14}$ l/yr and are rising rapidly. But just what is mineral water and where does it come from? Natural mineral water is obtained from springs or from wells drilled into the aquifer that supplies the spring. The "minerals" in mineral water are dissolved solids in concentrations that exceed 250 parts per million (250 ppm). Where do these dissolved solids come from? Natural rainwater has less than 10 ppm dissolved solids. Once it infiltrates into the ground, it finds itself in contact with rocks and minerals with varying degrees of solubility. As decomposition and related chemical reactions occur, natural waters leach constituents from the rocks and minerals and their total dissolved solid concentrations increase to where they become mineral waters. Mineral water taste depends largely on what "minerals" are dissolved, and an analysis of a mineral water permits one to infer the types of rocks and minerals that the groundwater was in contact with prior to being discharged as a spring. For example, hydrogen ($H^{+1}$) ions are sensed as sour, sodium ($Na^{+1}$) ions as salty and certain organic sulfate and chloride substances as bitter. In addition, natural mineral waters are carbonated to different degrees, which depend primarily on the amount of dissolved carbon dioxide gas they contain; thus the familiar choice between mineral water with and without gas. To capture the "spritz," naturally carbonated waters are usually recovered from a well and bottled under pressure so that the dissolved carbon dioxide is not lost. Given the commercial importance of the spring and mineral water industries, great care is taken to prevent the infiltration of pollutants into the groundwater that supplies these waters. It should be noted that many carbonated mineral waters are artificially carbonated by passing carbon dioxide gas through them under pressure.

Mineral water is also important in the manufacture of beer. The English city of Burton-upon-Trent, home of the Bass Brewery and several others, is famed for its hard, calcium-rich mineral waters, which are believed to be ideal for brewing fine ales. These waters, obtained from wells, owe their properties primarily to the dissolution of gypsum beds that underlie the valley of the Trent River. Breweries all over the world "burtonize" their own water sources by the addition of soluble gypsum in their quest for a better brew.

mineral and rock fragments of the original parent rock that have survived (resisted) decomposition, or **new minerals** generated by decomposition processes during weathering. As discussed in the previous section, new minerals are produced by processes such as hydrolysis (e.g., clay minerals), oxidation (e.g., hematite, goethite, pyrolusite), hydration (e.g., gypsum), and carbonation (e.g., calcite).

### 11.3.1   Resistates and chemical stability

The population of resistate (residual) rock and mineral fragments that occur in a particular source area depends on several factors. One important factor is the mineral composition of the source bedrock. One cannot find a resistate rock or mineral fragment in a residual soil that was not in the original bedrock. When such resistate fragments are eroded, transported, and deposited, they carry with them information about the rock types and minerals that existed in the source area at the time they were produced. These in turn may offer vital clues to the tectonic setting in which deposition likely occurred and help answer a variety of other questions, as detailed in Chapter 13.

Resistate rock and mineral fragments from the parent rock occur in residual soils only if they survive decomposition. Whether such rock and mineral fragments survive decomposition depends primarily on:

1   The resistance of each mineral to decomposition, based upon its chemical stability and the geochemical environment in the soil.
2   The rate of decomposition, which depends primarily on climatic factors such as

precipitation, temperature, vegetation, and organic activity.

3 The duration of decomposition that depends primarily on erosion rates, which in turn depend on relief (slope), vegetative cover, and rainfall.

Minerals that strongly resist decomposition are said to be chemically stable; minerals easily decomposed are said to be chemically unstable. As discussed in Chapter 2, ionically bonded substances with weak bonds are especially susceptible to dissolution, whereas strongly bonded covalent minerals are much more resistant to dissolution.

### Chemical stability of minerals

The **chemical stability** of any mineral – its resistance to decomposition – depends on the details of climate and soil geochemistry, but some useful generalizations can be made. Goldich (1938) essentially inverted Bowen's reaction series (Chapter 8) and applied it to an entirely separate set of geological processes that involve mineral stability during weathering. **Goldich's rule** states that the susceptibility of common igneous minerals is inversely proportional to their crystallization temperatures as summarized in Bowen's reaction series (Table 11.2). Minerals that crystallize at high temperatures, such as olivines, pyroxenes, and calcium-rich plagioclases, are chemically unstable in the low temperature and low pressure environment of Earth's surface. Because of this, they are far more susceptible to decomposition on Earth's surface than minerals that crystallize at lower temperatures, such as potassium feldspars, muscovite, and quartz. As a result, the low temperature minerals are preserved more commonly as resistate minerals.

Of course, source rocks contain many other igneous, metamorphic, sedimentary, and hydrothermal minerals, and they too possess different chemical stabilities. Chemically unstable minerals, such as halite, calcite, olivine and pyroxenes, tend to become relatively depleted as they are decomposed and removed from the resistate population. In contrast, minerals stable at low temperatures and pressures, such as quartz, clays, and iron oxides, tend to become relatively enriched in the resistate population. The relative chemical stability or susceptibility to chemical decomposition of common minerals is generally known and is shown by selected examples on the left side of Table 11.2. Among the rarer heavy minerals (specific gravity >2.8) that are least susceptible to decomposition are rutile, tourmaline, and zircon, and these too are likely to survive decomposition. The tendency of chemically stable minerals to survive decomposition means that they have the greatest potential to survive weathering, to be dispersed from the source area and to occur in sediments deposited elsewhere on Earth's surface. The high chemical stability of quartz is one of the reasons why it is the most abundant mineral in most sandstones and gravelstones (conglomerates and breccias), even though feldspars are more common than quartz in primary source rocks. The high chemical stability of clay minerals helps to explain why they are the major constituents of mudrocks such as shales.

### Rate and duration of decomposition

Rapid rates of decomposition accelerate depletion of chemically unstable minerals in the resistate population while simultaneously accelerating the relative enrichment of chemically stable constituents that resist decomposition. As noted earlier, heat and rainfall increase decomposition rates because heat and water catalyse decomposition reactions. Thus, rates of decomposition are highest in areas with warm, humid climates (see Figure 11.2; Table 11.3).

The duration of weathering in the source area is extremely important in determining the survival rates of residual rock and mineral fragments and depends primarily on rates of erosion. If erosion rates are high, rock material will be removed and dispersed from the source area before significant amounts of decomposition have occurred, so that a higher proportion of chemically unstable residual detritus will be dispersed into areas of deposition. If erosion rates are low, rock material will stay in the source area and will decompose for a longer period of time, which results in only the most resistant residual detritus being delivered to areas of deposition (Table 11.3).

**Table 11.2**  Chemical stability of major minerals under average weathering conditions.

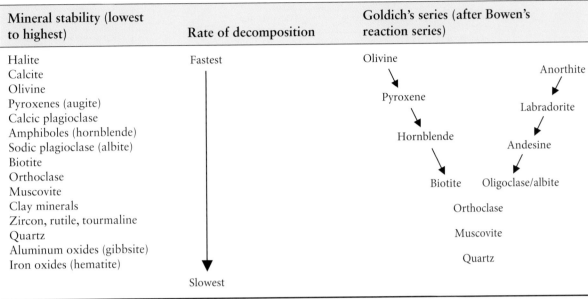

| Mineral stability (lowest to highest) | Rate of decomposition | Goldich's series (after Bowen's reaction series) |
|---|---|---|
| Halite | Fastest | |
| Calcite | | |
| Olivine | | |
| Pyroxenes (augite) | | |
| Calcic plagioclase | | |
| Amphiboles (hornblende) | | |
| Sodic plagioclase (albite) | | |
| Biotite | | |
| Orthoclase | | |
| Muscovite | | |
| Clay minerals | | |
| Zircon, rutile, tourmaline | | |
| Quartz | | |
| Aluminum oxides (gibbsite) | | |
| Iron oxides (hematite) | | |
| | Slowest | |

**Table 11.3**  Factors that affect the survival of resistate fragments during decomposition.

| Mineral resistance to decomposition | High | Moderate | Low |
|---|---|---|---|
| Rainfall | Low | Moderate | High |
| Temperature | Low | Moderate | High |
| Vegetation/organic activity | Sparse/low | Moderate | High |
| Erosion rate | Rapid | Intermediate | Slow |
| Relief/slope | High/steep | Intermediate/moderate | Low/gentle |
| Vegetative cover | Sparse | Moderate | Extensive |
| Surviving detrital assemblage | Many unstable components survive | Metastable and resistant components survive | Only resistant components survive |

Erosion rates depend primarily on (1) relief, (2) vegetative cover, (3) precipitation, and (4) the type of erosion agents involved. Erosion rates are generally proportional to relief. The steeper the slope, the more rapidly detrital sedimentary materials are removed from it by mass flows, running water, and even glaciers.

However, the proportional relationship between relief and erosion is affected by other factors. One of these is vegetative cover. The root systems of plants tend to hold soils in place, thus retarding rates of erosion, while aiding decomposition by the production of more acidic soil water. You may be familiar with examples of wind blowing dust from vacant lots or from freshly tilled fields but not from adjacent lots covered with grass or other vegetation. In fact, vegetation is planted to reduce erosion rates in coastal sand dunes and recently burned areas are quickly replanted to reduce erosion and mass flows.

In addition, the type of erosion agent strongly affects erosion rates. Because glaciers generally erode all sizes of material down to the bedrock, when glaciers flow over an area, vegetated or not, erosion rates are initially very rapid indeed. However, once the soil has been removed down to bedrock, rates of glacial erosion decelerate rapidly. Mass wasting processes such as landslides and debris flows erode extremely rapidly. Erosion by water increases with

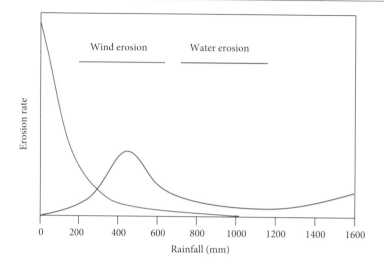

**Figure 11.10** Generalized erosion rates for wind (blue line) and water (red line) as a function of rainfall.

flow velocity, runoff, volume, and slope. Wind erosion is most effective in dry areas with minimal vegetative cover and low rainfall (Figure 11.10).

One might expect erosion rates to increase systematically with rainfall, but, as shown in Figure 11.10, this is not always the case. Where rainfall is less than 350–400 mm/yr, erosion rates generally increase with rainfall as expected. This is largely due to the lack of vegetation and relatively impermeable soils that characterize arid and semi-arid regions. As a result of dry, hardpan soils, infiltration is minimal so that most precipitation remains on the surface as overland flow, thereby increasing erosion rates of poorly vegetated soils.

Between 350–400 and 890–1015 mm/yr erosion rates actually fall. In this range, an increasingly thick vegetative cover of grasses and other plants tends to retard erosion rates. Of course one fire or land-clearing project can destroy the vegetative cover and substantially increase rates of erosion by wind, mass wasting, and surface runoff. At still higher precipitation rates, erosion rates again increase, primarily because they promote higher rates of mass wasting, which not only disperses sediments downslope but also removes much of the vegetative cover in the process.

In addition to residual minerals, new minerals develop during weathering as a result of decomposition.

### 11.3.2 New minerals

As discussed previously, the detrital fraction of residual soils generally contains a proportion of non-residual, **new minerals** that are produced by decomposition of the original rock. As discussed later in this chapter, these new minerals are commonly concentrated in the lower portions of mature soils. The most abundant group of new minerals produced by decomposition processes is clay minerals.

*Clay minerals*

Clay minerals constitute a large group of aluminum-bearing phyllosilicate minerals. Most clay crystals are of small size ($<4\,\mu m$). For this reason, the term clay has two distinct but overlapping meanings in the context of sedimentary rocks:

1   Clay is used as a compositional term for a group of phyllosilicate minerals with a specific set of chemical compositions and structures.

2   Clay is also used as a textural term for any very small particles (<0.004 mm diameter) that often are, but may not be, clay minerals.

The following discussion focuses on clay minerals defined by composition rather than texture. These phyllosilicate minerals are best understood in terms of their structures, which consist of two or more layers (sheets) connected by shared bonds, and in some cases separated by interlayer sites.

Three major types of layers (sheets) occur (Figure 11.11a):

1   **S-layers,** which are silica-rich tetrahedral layers ($Si_2O_5$) with some $Al_2O_5$ in which silicon and/or aluminum are in tetrahedral coordination with oxygen. Tetrahedral layers are also called **T-layers.**
2   **G-layers,** which are gibbsite [$Al_2(OH)_4$] octahedral layers.
3   **B-layers,** which are brucite [$(Mg,Fe)_2(OH)_4$] octahedral layers in which magnesium, iron,

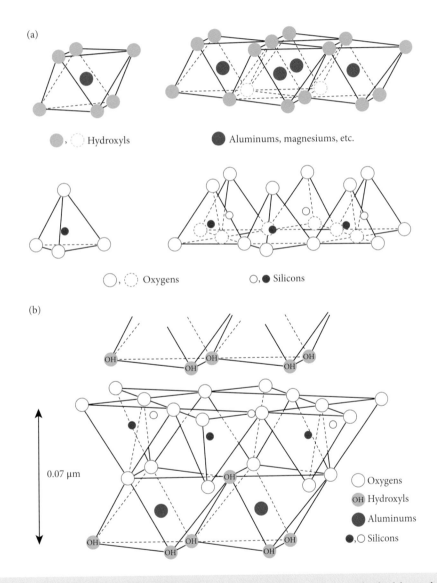

**Figure 11.11**   (a) The basic components of clay minerals: a single silica tetrahedral layer below and a single aluminum or magnesium octahedral layer above. (b) Coordination polyhedra model of a two-layer, 7 Å (0.07 μm) kandite clay mineral such as kaolinite. *Source*: Grim (1968). © McGraw-Hill.

or sometimes aluminum are in octahedral coordination with oxygen and hydroxyl ions. Octahedral G- and B-layers are also called **O-layers**.

Variations in the types and arrangements of these three types of layers permit four major groups of clay minerals to be distinguished, each with a different phyllosilicate structure. Hybrids of all four groups are common.

**Kandites** are two-layer (S-G or T-O) clays with a basic structure that consists of a single tetrahedral layer (S) bonded to a single octahedral layer (G). Pure kandites are composed of the repeated stacking of these basic structural units (S-G-S-G . . . S-G). Kandites are sometimes referred to as 7Å clays because the two layers have an aggregate thickness of approximately 7Å (0.07 μm), which is the repeat distance for the basic structural units. This repeat distance between layers is called the d-spacing and is extremely helpful in distinguishing between clay minerals using methods such as X-ray diffraction. The best known example of a kandite clay mineral is **kaolinite**, also known as China clay, an important natural resource used as the raw material in the production of fine ceramics and to produce glossy paper products. In its pure form, kaolinite consists of alternating tetrahedral sheets of $Si_2O_5$ and octahedral gibbsite sheets of $Al_2(OH)_4$ with an aggregate thickness of 7Å (Figure 11.11b). The chemical formula of pure kaolinite is written as $Al_2(Si_2O_5)(OH)_4$ or $Al_4(Si_4O_{10})(OH)_8$ and clearly reflects its two-layer structure and composition. Some kaolinites possess $H^+$ ions between the S-G layer pairs, which changes their composition slightly.

**Illites** are three-layer (S-G-S or T-O-T) clays with a basic structure that consists of a single octahedral layer (G) sandwiched between two tetrahedral layers (S) (Figure 11.12). In addition, one-fourth of the silica tetrahedra $[(SiO_4)^{-4}]$ in the tetrahedral layer are replaced by aluminum tetrahedra $[(AlO_4)^{-5}]$. This substitution creates a positive charge deficiency that necessitates the paired substitution of a similar number of positively charged cations such as potassium $(K^{+1})$ or hydrogen $(H^{+1})$. This paired cation substitution occurs in the interlayer space (//) between adjacent S-G-S units. Pure illites are composed of the repeated stacking of structural units (S-G-S//S-G-S . . . //S-G-S) with

interlayer $K^+$ ions. Illites are also referred to as 10Å clays because the three layers have an aggregate thickness or repeat distance of approximately 10Å (0.1 μm) A few water molecules can also occur in these interlayer sites. The chemical formula for illite can be written in several different ways. One common form is $(K,H)Al_2AlSi_3O_{10}(OH)_2 \cdot nH_2O$ and clearly reflects the two tetrahedral layers $(AlSi_3O_{10})$, the octahedral layer $[Al_2(OH)_4]$, and the interlayer cations and water. Clay mineral compositions are quite varied, and most illites depart from the standard composition noted above in that the silicon : aluminum (Si/Al) ratio in the tetrahedral site is a little larger than 3 : 1. This reduces the charge imbalance produced when aluminum $(Al^{+3})$ replaces silicon $(Si^{+4})$ in the tetrahedral site and thus reduces the amount of interlayer potassium $(K^{+1})$ required to electrically balance the crystal structure.

**Smectites** are three-layer (S-G-S, S-B-S or T-O-T), expandable lattice clays with a basic structure that consists of an octahedral layer (G and/or B) sandwiched between two tetrahedral layers. Interlayer sites are highly expandable and can incorporate many ions, including large amounts of water. Pure smectites are composed of the repetition of structural units such as (S-G-S//S-B-S//S-B-S . . . S-G-S) separated by expandable interlayer sites in which water and such ions as calcium $(Ca^{+2})$, sodium $(Na^{+1})$, and hydrogen $(H^{+1})$ may be absorbed. Because smectites are expandable layer clays, their repeat distance may range from 10Å (0.10 μm) to more than 21Å (0.21 μm), depending on how much interlayer absorption has occurred. Such clays tend to expand by hydration when wetted and to contract by dehydration when drying. Some sodium-rich montmorillonite can expand to ten times its normal thickness when wet (Prothero and Schwab 2013). Cation absorption is balanced by paired substitutions of cations of lower charge in the octahedral and tetrahedral sites. A common smectite clay mineral is **montmorillonite** (Figure 11.13).

Montmorillonite's composition is quite variable. Its chemical formula may be written as $(Ca,Na,H)(Al,Mg,Fe)_2(SiAl)_4O_{10}(OH)_2 \cdot nH_2O$. The first set of parentheses contains the common interlayer cations; the second set contains the octahedral G- or B-layer cations; the third set contains the tetrahedral S-layer cations; and the $nH_2O$ refers to the variable

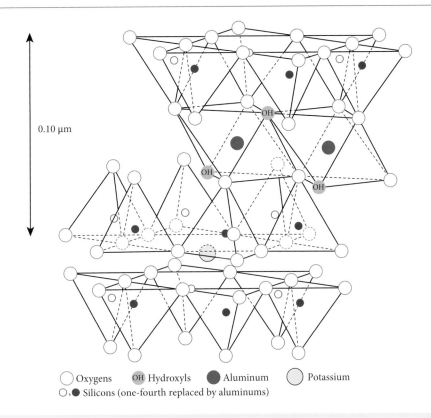

0.10 μm

⃝ Oxygens    OH Hydroxyls    ● Aluminum    ⃝ Potassium
○,● Silicons (one-fourth replaced by aluminums)

**Figure 11.12** A three-layer illite model depicting an aluminum octahedral layer (G) sandwiched between two silica tetrahedral layers (S) in the form S-G-S, with interlayer potassium ($K^{+1}$). *Source*: Grim (1968). © McGraw-Hill.

amounts of absorbed interlayer water. Soils that contain large amounts of montmorillonite are called **expansive soils**. Not only does their volume change as water is absorbed and released but also their strength and plasticity. Such soils are often involved in landslides and ground subsidence events that cause severe property damage and loss of life. Expansive soils are discussed in more detail in the section of this chapter devoted to the engineering properties of soils.

**Chlorites** are four-layer (S-B-S+B) clay minerals. The basic chlorite structure consists of an octahedral brucite layer (B) sandwiched between two tetrahedral layers (S) with an additional octahedral brucite layer (B), which may have a somewhat different composition than the other brucite layer (Figure 11.14). Pure chlorites, of which many varieties exist, have the basic stacking pattern (S-B-S+B, S-B-S+B . . . S-B-S+B). Chlorites are 14Å clays because the basic structural unit (S-B-S+B) is 14Å (0.14 μm) thick and their repeat stacking distance is 14Å. A common chlorite group mineral is **clinochlore**, whose formula can be written as $(Fe,Mg)_3(Fe_3)AlSi_3O_{10}(OH)_8$. The first set of parentheses indicates the cation content of the octahedral site sandwiched between the tetrahedral sites in which aluminum ($Al^{+3}$) substitutes for every fourth silicon ($Si^{+4}$), and the second set of parentheses shows the content of the second octahedral site. Many chlorites contain much more magnesium ($Mg^{+2}$) in the octahedral sites. Chlorite is also a common metamorphic mineral and is much more abundant in metamorphic rocks than other clay minerals.

Many clay minerals are complex hybrids of the clays discussed above. Hybrid clays, which contain stacked layer sequences characteristic of more than one type of clay mineral, are called **mixed layer clays**. For example, a clay with the structure S-B-S+B//S-G-S//S-B-S//S-B-S+B would be a mixed chlorite–smectite clay mineral, whereas

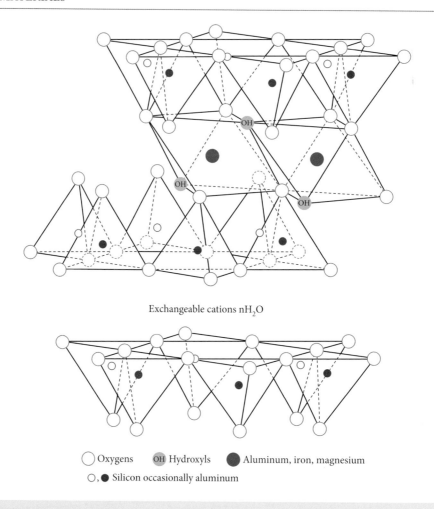

Exchangeable cations nH$_2$O

◯ Oxygens    ⬤ Hydroxyls    ⬤ Aluminum, iron, magnesium

◯, ⬤ Silicon occasionally aluminum

**Figure 11.13** A three-layer smectite clay, montmorillonite, in a partially expanded state, with absorbed water molecules and cations occupying the space between the triple layers. *Source*: Grim (1968). © McGraw-Hill.

one with the structure S-G-S-G-S-G-S-S-G-S-S-G would be a mixed kandite–illite clay mineral. The most common group of mixed layer clay minerals are illite–smectite clays.

**Degraded clay minerals** lack the interlayer constituents predicted by their general formula. In most cases, such interlayer constituents have been removed from them by pore waters during decomposition or by other types of chemical alteration. For example, illites that lack their full complement of interlayer potassium ($K^{+1}$) ions are degraded illites, and smectites that lack their full complement of interlayer calcium ($Ca^{+2}$) and/or sodium ($Na^{+1}$) ions are degraded smectites. Clay degradation occurs most commonly under acidic conditions that tend to leach

cations from the clay structures. Such leaching processes are important in the development of soils and soil horizons, as discussed later in this chapter.

It should be noted here that clay minerals form by processes other than weathering, especially by hydrothermal alteration and low temperature metamorphic processes (Chapter 15).

*Insoluble iron and manganese oxides and hydroxides*

Oxides and hydroxides of iron and manganese are significant constituents of many soils. In some soils, especially lateritic soils (oxisols) formed in warm, humid environments, oxides,

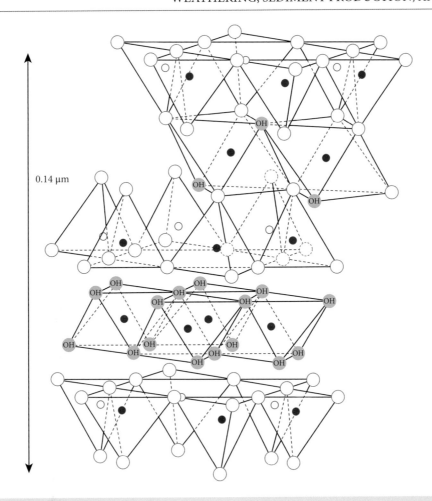

**Figure 11.14** A four-layer, 14Å (0.14 μm) structure typical of chlorites, with two tetrahedral layers that sandwich an octahedral layer in three-layer clays (S-B-S) and an octahedral brucite (B) layer below. Key as on Figure 11.13. (*Source*: Grim (1968). © McGraw-Hill.

and hydroxides are especially abundant. These minerals are largely produced by the decomposition of ferromagnesian silicates and iron-bearing sulfides by processes that include dissolution and reprecipitation, oxidation, hydrolysis, and/or hydration. The most common products are listed in Table 11.4.

Hematite is responsible for the reddish color, and limonite and goethite are responsible for the yellow-brown color of many soils and weathered rock surfaces. Manganese minerals yield dark gray to black colors in soils and on weathered surfaces. For millennia, human populations have collected these soil constituents, mixed them with water (±clay) and used them as pigments to produce artistic representations such as cave paintings (Figure 11.15a) and pictographs (Figure 11.15b).

*Highly insoluble aluminum oxides and hydroxides*

In highly acidic soils in warm, humid climates, decomposition may generate aluminum oxide and hydroxide minerals rather than aluminum silicate minerals such as clay minerals. In such cases, the silica in clay minerals is dissolved, which leaves aluminum combined with hydroxyl ions, oxygen, and some water to form a mineral suite characteristic of the major aluminum ore called **bauxite**. Most bauxite deposits consist of a variety of aluminum-bearing minerals and mineraloids that are commonly associated with the insoluble iron and manganese minerals in oxisols. Three common minerals in bauxite deposits are gibbsite [$Al(OH)_3$] and the polymorphs diaspore [$\alpha\text{-}(AlOOH)$]

and boehmite [β-(AlOOH)]. Because the aluminum in bauxite is relatively weakly bonded to the oxygen and hydroxyl ions, it is more easily and less expensively separated during refining than is the aluminum bonded to silicon in clay minerals, feldspars, and other silicates. Without the widespread formation of bauxite minerals in deeply weathered, tropical soils, the production of aluminum would be much more expensive than it is.

*Other common soil minerals*

The calcium carbonate ($CaCO_3$) minerals calcite and aragonite are abundant in some soils, especially in caliche soils called aridosols that develop in fairly arid climates, as discussed later in this chapter. The calcium sulfate minerals gypsum ($CaSO_4 \cdot 2H_2O$) and anhydrite ($CaSO_4$) also occur in some aridosols.

**Table 11.4** Common iron and manganese oxides and hydroxides produced by weathering.

| Mineral or mineraloid | Chemical composition |
| --- | --- |
| Hematite | $Fe_2O_3$ |
| Goethite | $FeOOH$ |
| Limonite[a] | $FeOOH \cdot nH_2O$ |
| Pyrolusite | $Mn(OH)_2$ |
| Manganite | $MnO_2$ |
| Romanechite | $BaMnMn_8O_{16}(OH)_4$ |

[a] Mineraloid.

Soils developed in anoxic areas with reducing environments, such as histosols, may contain sulfide minerals such as pyrite ($FeS_2$).

Under special circumstances, dozens of other minerals can be precipitated from soil pore waters, but their discussion is beyond the scope of this chapter. Precipitation of these minerals and of the more common ones discussed previously may produce anything from small clumps of partially indurated material called soil **peds** to truly solid crusts called **durisols** or **petrosols**. For this reason, soils are not completely unconsolidated materials, as will be explored in the following section.

## 11.4 SOILS

Soils are largely unconsolidated surficial deposits produced directly or indirectly by weathering processes and capable of supporting rooted plant life. Weathering of bedrock generates in situ **residual soils**. The detrital constituents of residual soils may then be eroded and dispersed by transportation to be deposited as sediments elsewhere on Earth's surface. Further weathering of such transported sediments generates **transported soils**. Most soils are relatively thin, typically up to 2 m thick according to the United States Department of Agriculture's Natural Resources Conservation Service (USDA-NRCS 1999), but compound soils produced over

(a)                                    (b)

**Figure 11.15** (a) Paintings of horses using manganese oxides and hydroxides (black) and limonite (rusty brown); from Pech-Merle Cave, France. *Source*: Photo by Anita O'Brien. © John Wiley & Sons. (b) Australian rock art using hematite (red), limonite (brown), kaolinite (white), and manganese oxides/hydroxides (black), from Nourlangie Rock, Kakadu National Park, northern Australia. *Source*: John O'Brien. © John Wiley & Sons.

several cycles of weathering may be considerably thicker. Figure 11.16 summarizes the average composition of soils, following the 2 m definition, in terms of mineral, organic, water, and air content. **Humus** is partially decayed, dead, organic matter. These percentages represent averages that do not portray the great variation in soil compositions, which depends on factors such as parent rock type, vegetative cover, seasonal moisture content, soil compaction, and soil age.

Many textural classifications of soils exist. Figure 11.17 shows one commonly used textural classification based on the percentages of sand, silt, and clay that are present. The terminology is straightforward, if elaborate. **Clay soils** contain more clay than sand or silt; **sand soils** contain much more sand than clay and silt; and **silt soils** contain much more silt than sand or clay. **Loam** is a term used for soils that contain subequal proportions of sand and silt and low to moderate amounts of clays. Various modifiers are used for clay, sand, silt, and loam soils that contain more accessory constituents than the defined limits of the four main soil types based on texture.

### 11.4.1  The importance of soils

Because we inhabit Earth's surface, soils are extremely important to many aspects of human affairs. As a result, they have been and continue to be the focus of intensive study. Soils are important as **natural resources** that support a wide range of natural vegetation, as well as agricultural and forest products. Knowing how to match crops, irrigation, and fertilizers to specific soils is an essential aspect of enhancing food production worldwide.

Soils are also important as **structure-bearing materials** that provide support for our homes, offices, factories, power plants, highways,

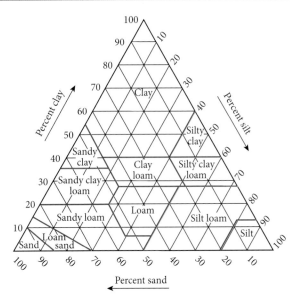

**Figure 11.17**  Textural classification of soils. *Source*: Courtesy of the US Department of Agriculture.

pipelines, reservoirs and aqueducts. When such structures fail, the consequences can be catastrophic. Understanding the load-bearing properties of soils and how their behavior might change over time as the result of human activities or natural processes is extremely important to engineers.

As discussed earlier in the chapter, water moving through unconsolidated surficial materials commonly reacts chemically with them. Where reactions involve the removal of contaminants and the purification of the water, soils act as **water filters** and as **contaminant sinks**. Groundwater flow through soils is also capable of **contaminant dispersal** over broad areas. Understanding the roles played by soils in such processes is of fundamental importance to geologists, hydrologists, and environmental scientists interested in site evaluation and remediation programs.

Soils, as the products of weathering, are also the **source of detrital sediments** and of many of the **dissolved solids** that occur in groundwater, surface water, and the oceans. The many varieties of natural mineral and spring waters (see Box 11.2) attest to the variety of chemical reactions that can take place between soils and the groundwater that passes through them.

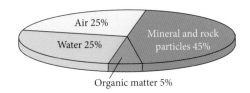

**Figure 11.16**  Proportions of the major components in average soil.

### 11.4.2 Soil layers and horizons

Many soils are clearly layered (Figures 11.18 and 11.19). When soil layers are the product of in place weathering processes they are called horizons. Each **horizon** is characterized by sets of properties produced by soil-forming processes that distinguish the horizon from the layers above and below. In many layered soils, multiple horizons occur in a vertical sequence of layers that constitutes a **soil profile**.

The classification of soil horizons is complex and has some tongue twisting jargon; please bear with us. The classification system used by USDA-NRCS (1999) divides horizons into **epipedons**, which constitute the top layer of soil profiles that have not been truncated by later erosion, and **subsurface horizons**. Eight different epipedons of diverse origins are recognized, ranging from those produced by human activities to those produced without human intervention. Eighteen different subsurface horizons are defined, each the product of a different set of soil-forming processes. Although some of these soil horizons can be tentatively identified in the field, accurate identification of many requires extensive laboratory work that involves chemical, mineralogical, and organic compound analysis.

To simplify the notion of soil horizons in a way that still carries meaning, many scientists utilize an older classification of soil horizons. One advantage of this system is that these horizons can commonly be identified in the field when a trench has been dug to expose a soil profile. This classification recognizes five major types of soil horizons, each with subdivisions, which combine to produce soil profiles. These soil horizons are generally distributed from the top down in soil profiles as (1) the O-horizon, (2) the A-horizon, (3) the E-horizon, (4) the B-horizon, and (5) the C-horizon which, in residual soils, is underlain by unaltered parent material (Figure 11.19). Each horizon may include multiple, recognizable subhorizons.

The **O-horizon**, where present, is generally a dark brown to black epipedon that occupies the upper portion of the soil in which it occurs

**Figure 11.18** Layered soil produced by the disintegration and decomposition of rock materials near Earth's surface. From top to bottom, the O-, A-, E-, B-, and C-horizons. *Source*: Courtesy of Jim Turenne, Society of Soil Scientists of Southern New England.

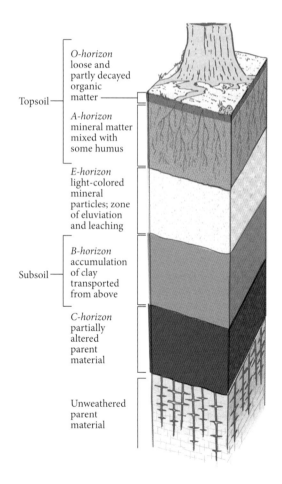

**Figure 11.19** The ideal distribution of soil horizons in a fully developed with vegetative cover. *Source*: Schoonover and Crim (2015). © John Wiley & Sons.

(Figures 11.18 and 11.19). It is characterized by being rich in organic (O) material, mostly humus. The O-horizon represents the incompletely decomposed plant debris that has accumulated near the surface over time, mostly from plants rooted in the soil.

The **A-horizon** is dominated by mineral material, with or without a significant proportion of admixed organic matter. The A-horizon is also known as the **zone of leaching** because significant amounts of material have been removed by dissolution, cation exchange, or the physical removal of fine material by the downward percolation of aqueous solutions over time. This process of downward removal of solid particles and dissolved ions from the A horizon is known as eluviation.

The lower parts of some A-horizons contain a light-colored subdivision called the **E-horizon**. The light color is caused by substantial dissolution or leaching of iron and/or aluminum from the lower, organic-poor part of the A-horizon. E-horizons are chemically resistant and often quartz-rich zones. The E-horizon is the classic expression of the zone of eluviation from which materials have been removed by leaching or eluviation.

The **B-horizon**, also known as the **zone of accumulation**, is characterized by enrichment in some of the constituents leached from the A-horizon. The process by which materials are translocated downward to be added to the lower part of a soil is known, somewhat confusingly, as *illuviation*. For this reason, the B-horizon is also known as the **zone of illuviation**. In relatively warm, humid climates, reprecipitation of amorphous or crystalline iron oxides (e.g., hematite = $Fe_2O_3$) or oxyhydroxides, e.g., limonite (HFeOOH) commonly gives the B-horizon a distinctly reddish or yellowish hue (see Figure 11.18). In humid, tropical climates, where chemical decomposition is thorough, kandite clay minerals and bauxite are also concentrated in the B-horizons to form aluminum-rich horizons.

In dryer climates, calcium carbonate ($CaCO_3$) precipitates, producing $B_k$ soil horizons. In many cases, mineral precipitation in the B-horizon binds soil particles together into hard, nodular zones, or into completely indurated sub-horizons called **duricrusts**. The most common examples of duricrusts are the calcium carbonate **calcrete** or **petrocalcic** horizons common in caliche soils. Caliche soils form in arid and semi-arid climates with seasonal deficiencies in rainfall where evaporation of soil moisture initiates calcium carbonate precipitation. Calcrete horizons occur closer to the surface in progressively dryer climates and may occur at the surface of aridosols formed in warm, arid, desert climates. Similar hard subhorizons of **silcrete** (silica) and **petrogypsic** (gypsum) layers occur less commonly in soils.

Other characteristic soil structures occur in the B-horizon interval, including:

1  **Peds**, which are partially cemented clods of soil particles of various sizes that give the soil a crumbly lump appearance.
2  **Cutans**, which are concentrations of illuviated material such as clays or iron oxides that occur as layers or that envelope less-altered cores.
3  **Glaebules**, which are prolate to equant hard lumps formed by mineral precipitation and include concretions and nodules of all sizes.

The **C-horizon**, also called the **soil mantle**, represents moderately to minimally weathered, slightly altered materials that are transitional to the underlying, unaltered parent material. Unlike the materials in the B-horizon, C-horizons are not significantly enriched in illuviated materials from above.

Where soils are developed over bedrock, the largely unweathered bedrock constitutes the so-called **R-horizon** or **regolith** horizon.

As will be seen from the discussion that follows, complete soil profiles form only in mature soils developed over $10^3–10^5$ years. Ideal conditions for the development of such soils include the presence of vegetation, sufficient precipitation, and the absence of erosion or other disturbances. The major reasons for the incomplete development of soil horizons include (1) insufficient duration of weathering processes, (2) climatic conditions that inhibit the formation of one or more horizons, and (3) soils truncated by erosional processes.

### 11.4.3 Soil classifications

Given the tremendous variety of unconsolidated materials that exist on Earth's surface and the overarching importance of surficial materials in a great variety of human enterprises, it is not surprising that the description and classification of soils is exceedingly complex. Two very different soil classifications have developed in the United States – one used by most agricultural soil scientists and a second utilized primarily by engineers. These are discussed in the sections that follow. Many other systems have evolved internationally, leading to current efforts to develop a world-wide reference system (Blum and Schad 2015).

*Agricultural classification of soils*

Soil scientists in the United States have developed a soil classification system or taxonomy that attempts to organize the thousands of individual soil types that have been recognized. This has been accomplished by organizing all soils into a hierarchy of soil categories in much the same way as the Linnaean classification system in biology organizes all life forms into a hierarchical classification system involving kingdoms, phyla, classes, orders, families, genera, and species:

Orders

   Suborders

      Great groups

         Subgroups

            Families

               Series

The purpose is to make a complex system more comprehensible and to make it easier to discern relationships between different soils. The soil taxonomy is based on the properties of the soil profile, especially the following:

1 The number and types of soil horizons present.
2 Available nutrient chemicals.
3 The distribution of organic materials.
4 Soil color.
5 Seasonal soil moisture content.
6 Overall climate.

The USDA-NRCS (1999) has organized soils into 12 major **orders**, which are fairly easy to learn (Figure 11.20). Each order is subdivided

into as many as seven **suborders** of which there are a total of 64. The suborders are subdivided into more than 300 **great groups**, which are subdivided into some 2400 **subgroups**, which are further subdivided into **families** and lastly into soil **series**. More than 19 000 different soil series have been mapped in the United States alone. Some soils can be identified on the basis of field investigations, but most require substantial laboratory analysis as well. Once a soil is placed into its appropriate soil series, a great deal of descriptive, relational, and interpretive information, gathered over many decades, is available to the practicing soil scientist. The characteristics of the 12 major soil orders are briefly described and summarized in Table 11.5. Images of examples of each major soil order are provided in Figure 11.20.

Not surprisingly, soil classifications and classification criteria vary between different countries. An international effort has evolved to facilitate communication between people using different classification systems and to provide a world-wide standard for soil data bases and maps. In 1998, a **world reference base (WRB) for soil sciences** was developed with the endorsement of the International Union of Soil Scientists (IUSS) and the Food and Agricultural Organization (FAO) of the United Nations. The WRB, revised and updated in 2006 and again in 2014, has provided a worldwide classification of soils to which soils classified by different systems can be compared. Unlike the USDA-NRSC classification scheme, climate is not considered in the WRB. Focusing on properties that can be observed and/or measured in the field, the WRB divides soils into 32 reference soil groups (RSG's), in eight categories, based on carefully-defined characteristics of the soil that include (1) soil horizons, (2) properties, and (3) materials. Table 11.6 summarizes the descriptive characteristics of each. The descriptive criteria are utilized from the top of the table downward, until a perfect match is found to identify the primary reference group. Soils in each reference group are further described using the remaining descriptive information and a set of rules that involves adding principal qualifiers in front of the soil group name (these are often other reference group names) and supplementary qualifiers behind the soil group name. This produces

Alfisol

Andisol

Aridisol

Entisol

Gelisol

Histosol

**Figure 11.20**    Examples of the major soil orders in the USDA-NRSC soil taxonomy (see Table 11.5). *Source*: Courtesy of the US Department of Agriculture.

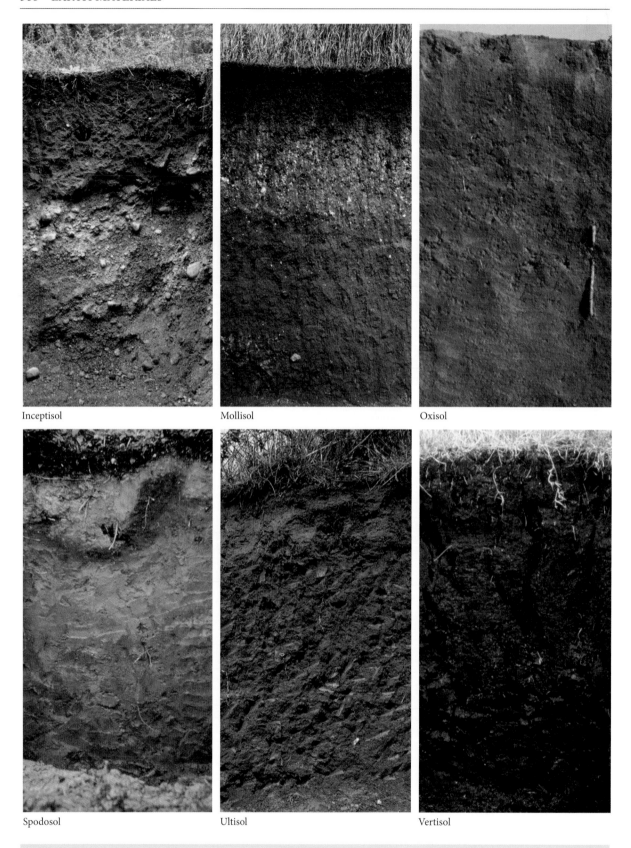

Inceptisol     Mollisol     Oxisol

Spodosol     Ultisol     Vertisol

**Figure 11.20** (Continued)

**Table 11.5**  Soil orders in the USDA-NRSC soil taxonomy; their diagnostic features and environments of formation.

| Order | Diagnostic features | Environment |
|---|---|---|
| Alfisols | Gray to brown A-horizon epipedon; B sub-horizons rich in clays with reasonably high concentrations of bases such as Ca, Na, Mg; reasonably high moisture content | Relatively humid areas with sparse forest or savannah cover; base and water content yield fertile soils |
| Andisols | Weak horizon development; rich in disordered clays and Al–humus complexes; high phosphorous retention; good moisture capacity and cation exchange capacity | Form in a wide range of non-arid climates; mostly on volcaniclastic materials; tend to be quite fertile |
| Aridosols | Sparse organic material in A-horizon epipedon; well-developed B-horizons, often rich in Ca-carbonates, even gypsum; low moisture content for long periods of time | Dominate in arid regions with sparse rainfall and vegetative cover; suitable for agriculture only if irrigated |
| Entisols | Lack significant soil horizon development; soils only because they have the capability to support rooted plants; often sand rich | Occur in any climate or setting; mostly on young surfaces; also in chemically inert parent materials or on slopes where erosion occurs |
| Gelisols | Permafrost soils and soil features; patterned ground, broken horizons and incorporation of organic matter in lower horizons produced by frost heaving and churning | In high latitude and/or high elevation areas where soils freeze for long periods |
| Histosols | Mostly very organic rich O-horizon; deeper horizons tend to be poorly developed, if at all | Mostly peat and muck from partially decomposed plant debris in swamps or bogs or water-saturated soils in areas of poor drainage |
| Inceptisols | Weak horizon development; less clay concentration in B-horizon than alfisols; carbonate and silica-rich B-horizons may occur; reasonably high moisture content | Form in a range of non-arid regions from subpolar to tropical; often with forest cover; less suitable for agriculture than alfisols |
| Mollisols | Very dark, thick, organic-rich O and A epipedon; high base content, especially calcium; clays with high cation exchange potential | Common under grasslands in semi-arid plains and steppes with seasonal moisture deficits; some under forest cover; great for grain production |
| Oxisols | Weak horizon development; extreme decomposition and base depletion; clays, mostly kaolinite, with low cation exchange capacity; bauxite under extreme conditions; quartz and iron oxides | Develop over long periods of time in tropical/subtropical settings with high rainfall and thick vegetative cover; generally infertile |
| Spodosols | Thick O-horizon; well-leached A-horizon with low Fe, Al, Ca; well-developed B-horizons with clays, reddish iron oxides or black humic material; good cation exchange | Dominate under coniferous forests; in areas with reasonable rainfall; generally suitable for agriculture |
| Ultisols | Well-leached A-horizon with some organics; clay-rich B-horizons with generally low base contents as Ca, Na, and K are largely removed which distinguishes them from alfisols | Humid climates; low base content; soils unsuitable for sustained agriculture unless fertilized with Na and K |
| Vertisols | High expansive clay content; large changes in volume associated with wetting and drying; cracks when dry and other evidence of soil movement; may have horizons | Poor soil for structures given the volume changes and tendency for strength and plasticity to change during wetting and drying |

thousands of possible combinations while providing detailed descriptions. Since a discussion of these is beyond the scope of this book, the reader is referred to the most recent rendition of the world reference base which is available on-line at www.fao.org/3/i374en.pdf. The longer range goal of these many years of effort is to promulgate a single, comprehensive, unified, worldwide soil classification system. Stay tuned!

**Table 11.6** The Reference Soil Groups (RSG) in the World Reference Base for soil sciences (2014); their properties, names, and symbols.

| Description | Reference Soil Group (RSG) | Symbol |
| --- | --- | --- |
| **1. Soils with thick organic layers** | | |
| | Histosols | HS |
| **2. Soils with Human Influence** | | |
| With long and intensive agricultural use: | Anthrosols | AT |
| Containing significant amounts of artifacts: | Technosols | TC |
| **3. Soils with limitations to root growth** | | |
| Permafrost affected: | Cryosols | CR |
| Thin or with many coarse fragments: | Leptosols | LP |
| With high content of transferable sodium (Na): | Solonetz | SN |
| Alternating wet–dry conditions, shrink–swell clays: | Veritisols | VR |
| High concentration of soluble salts: | Solonchaks | SC |
| **4. Soils distinguished by Iron–Aluminum (Fe–Al) Chemistry** | | |
| Groundwater-affected, underwater or in tidal areas: | Gleysols | GL |
| Allophanes ir Al–humus complexes; | Andosols | AN |
| Subsoil accumulations of humus and/or oxides: | Podzols | PZ |
| Accumulation and redistribution of Fe: | Plinthsolos | PT |
| Low activity clay, phosphorous (P-) fixation, many Fe-oxides, strongly structured: | Nitosols | NT |
| **5. Pronounced accumulation of organic matter in mineral topsoil** | | |
| Very dark topsoil, secondary carbonates: | Chernozems | CH |
| Dark topsoil, no secondary carbonates (unless very deep), high base status: | Phaeozems | PH |
| Dark top soil, low base status: | | |
| **6. Accumulation of moderately soluble salts or nonsaline substances** | | |
| Accumulation of, and cementation by, secondary silica: | Durisols | DU |
| Accumulation of secondary gypsum: | Gypsisols | GY |
| Accumulation of secondary carbonate: | Calcisols | CL |
| **7. Soils with clay-enriched subsoil** | | |
| Interfingering of coarser-textured, lighter-colored material into a finer-textured, stronger-colored layer: | Retisols | RT |
| Low activity clays, low base status: | Acrisolss | AC |
| Low activity clays, high base status: | Lixosols | LX |
| High activity clays, low base status: | Alisols | AL |
| High activity clays, high base status: | Luvisols | LV |
| **8. Soils with little or no profile differentiation** | | |
| Moderately developed: | Cambisols | CM |
| Sandy: | Arenasols | AR |
| Stratified fluviatile, marine, and lacustrine sediments: | Fluvisols | FL |
| No significant profile development: | Regosols | RG |

*Engineering classification of soils*

Engineers approach soil classification from a set of perspectives that differ from those of soil scientists. They are not especially focused on soil horizons and the suitability of soils for agriculture and forestry. Instead, engineers are concerned with the mechanical properties of all unconsolidated surficial deposits, regardless of origin, so that their definition of soils includes all relatively unconsolidated surface materials. This is reflected in the descriptors and classification systems of soils utilized by engineers.

Most geotechnical and engineering personnel in the United States use the **Unified Soil Classification System** (Tables 11.7 and 11.8). In this system, soils are given names and symbols according to their particle size distributions, notably the proportions of gravel, sand, silt, and expansive clays, and to the content of non-expansive clays and organic materials in the soils. Engineering definitions of gravel, sand, silt, and clay do not correspond exactly to those used by geologists, who employ the Wentworth–Udden (W–U) grade scale (Chapter 13) or to that used by the USDA-NRCS discussed above.

**Table 11.7**  Unified Soil Classification System for coarse-grained soils.

| Soil divisons | Soil characteristics | Soil group name | Soil group symbol |
| --- | --- | --- | --- |
| Clean gravel | <5% fines; continuous size variation over a range | Well-graded gravel | GW |
| Clean gravel | <5% fines; mostly one size or polymodal | Poorly graded gravel | GP |
| Dirty gravel | >12% fines; mostly silt | Silty gravel | GM |
| Dirty gravel | >12% fines; mostly clay | Clayey gravel | GC |
| Clean sand | <5% fines; continuous size variation over a range | Well-graded sand | SW |
| Clean sand | >5% fines; mostly one size or polymodal | Poorly graded sand | SP |
| Dirty sand | >12% fines; mostly silt | Silty sand | SM |

**Table 11.8**  United Soil Classification System for fine-grained soils.

| Soil divisions | Soil characteristics | Soil group name | Soil group symbol |
| --- | --- | --- | --- |
| Silt | Inorganic silts with very slight plasticity | Silt | ML |
| Silt | Inorganic silts, with mica giving soil more elasticity | Micaceous silt | MH |
| Silt | Organic silts with low plasticity | Organic silt | OL |
| Clay | Inorganic clays and silty clays of low–medium plasticity | Silty clay | CL |
| Clay | Inorganic clays of high plasticity | High plastic clay | CH |
| Clay | Organic clays with medium–high plasticity | Organic clay | OH |

In the system employed by the Unified Soil Classification System, the main groups are:

1  *Gravel (G)*, where particle diameters exceed 4.0 mm (as compared with 2.0 mm in the W–U scale).
2  *Sand (S)*, where particles range from 0.074 to 4.0 mm (as compared with 0.0625–2.0 mm in the W–U scale).
3  *Silt (M)*, where particles range from 0.004 to 0.074 mm (as compared with 0.004–0.0625 in the W–U scale).
4  *Clays (C)*, which are defined in the same way in both systems as particles smaller than 0.004 mm (4 µm). The USDA-NRCS classification system, however, defines clays as particles smaller than 2 µm.

One need not memorize these numbers, simply remember that the size classification systems are similar, but different in detail. Look them up as required or use them until they are second nature.

In the Unified Soil Classification System, **coarse-grained soils** (Table 11.7) are those that contain more than 50% total sand and gravel by weight. Gravels contain more gravel than sand, whereas sands contain more sand than gravel. Coarse-grained soils are further subdivided according to the percentage of fine-grained components (clays + silts). **Clean soils** contain less than 5% fines and **dirty soils**

contain more than 12% fines. Coarse-grained soils are subdivided further on the precise percentages of fines and on whether they are mostly silt or clay. Each soil type is represented by an appropriate symbol. In addition, transitional names can be used. For example, a well-graded gravelly soil with 5–12% fines of which the majority is silt could be called a GW–GM. Other soil characteristics are typically recorded as well by engineers. These include (1) maximum particle size, (2) color, (3) layering, (4) compactness or compressibility, (5) moisture characteristics, (6) drainage conditions, (7) structure, (8) strength, and (9) plasticity.

**Fine-grained soils** (Table 11.8) contain more than 50% total silt plus clay. Silts are defined as soils with more silt than clay, whereas clays contain more clay than silt. Fine-grained soils are further subdivided according to their mica content, organic content, and their degree of plasticity. Highly organic soils are prone to compaction, dehydration, and decomposition that result in volume loss, which makes these soils unsuitable for construction. Soil plasticity is largely determined by the soil's ability to absorb water and therefore by their smectite (expandable lattice) clay content. As will be seen in the following section, plasticity is an extremely important measure of the mechanical properties of soils and allows one to predict how they will react in different circumstances. The "L" in the group symbols stands for **loam**, a soil that contains appreciable amounts of both silt and clay in the fine fraction.

The Unified Soil Classification System contains a separate class for soils that are especially rich in organic materials. These **organic soils** are mostly peats and mucks and are roughly equivalent to the gelisols in the USDA-NRCS classification.

### 11.4.4  Soil mechanics

Soil engineers are especially concerned with the mechanical properties of soils. These properties are critical factors in the suitability of soils for use in the construction of roads, bridges, dams, and buildings, as well in many other aspects of land utilization. Many geohazards are the direct result of ignoring or failing to understand the implications of the mechanical properties of soils. Let us consider some of these, including soil strength, soil sensitivity, shrink and swell potential, and compressibility. Each is fundamental to understanding the engineering aspects of soils.

*Soil strength*

**Soil strength** is the amount of stress a soil can bear without failing by rupture or plastic flow. It is an expression of the ability of a soil to resist irreversible deformation such as inelastic changes in shape, volume, and position. Strong soils are quite resistant to stress and, along with many kinds of bedrock, generally provide excellent substrates for buildings and other structures, especially in areas prone to earthquakes. Weak soils are subject to compression, collapse or flow when stressed and therefore provide poor substrates for structures. Problems for engineers arise because soil strength can change, especially in response to changes in water content (Box 11.3), so that formerly strong soils loose strength and become weak soils that fail by rupture, flow plastically, or even flow like a liquid.

*Soil sensitivity*

The measure of a soil's tendency to change strength is expressed by **soil sensitivity**, a measure of the change in soil strength that results from changes in water content and various kinds of disturbances such as vibrations, excavations, and loading that stress soils. Soil sensitivity in response to water content is easily determined in the laboratory. It is commonly expressed by **Atterberg limits** that permit the subdivision of fine-grained soils into four classes on the basis of how they behave as their moisture content changes (Figure 11.21).

On the basis of their behavior, soils may be subdivided into four Atterberg classes: (1) brittle solids, (2) semi-solid soils, (3) plastic soils, and (4) liquid soils. The boundary between brittle solids and semi-solid soils is called the **shrinkage limit (SL)**, which is the water content below which soils do not shrink as additional moisture is lost during drying (Figure 11.21). Above the shrinkage limit, semi-solid soils shrink, and crack (think mudcracks) as they lose moisture and become progressively more stiff and brittle. Soils that remain brittle or semi-solid under all conditions of potential moisture content tend to be strong and provide excellent substrates for most construction projects so long as they are not loaded beyond their rupture strength. Generally, solid bedrock is even better.

The **plastic limit (PL)** separates semi-solid soils from plastic soils and is the water content at which soil deformation changes from rupture to plastic flow (Figure 11.21). Plastic substances change shape and/or volume in response to stress or pressure but do not

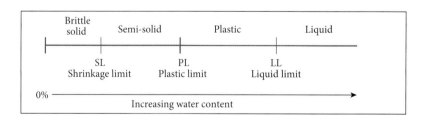

**Figure 11.21**   The major Atterberg classes of fine-grained soils and the limits that define them.

rupture visibly. Because they retain cohesive strength, they do not flow like a liquid. Plastic soils are moisture sensitive in that their strength decreases, and they deform more easily as they become progressively less cohesive with increasing moisture content. This helps to explain the many slope failure incidents that occur following heavy rainfall and the concurrent infiltration of groundwater into soils. The **plasticity** of a soil is a measure of its cohesiveness, which is sensed as a sticky, cohesive feel to the touch. It generally increases with clay content (especially expandable smectites) and water content. Soils with low clay content tend to be relatively noncohesive and therefore possess relatively low plastic limits. Soils with high clay content tend to be much more cohesive and to possess significantly higher plastic limits. Because plastic soils deform when loaded, they do not make good substrates for major construction projects.

The **liquid limit** (LL) separates plastic soils from liquid soils (Figure 11.21). It is the water content at which soils lose their shear strength and begin to flow as a liquid. When a sufficient amount of moisture has been added to a soil, it may begin to behave as a liquid; that is, it will lose cohesive strength and begin to flow under its own weight. This can have disastrous consequences for the structures placed such soils. The liquid limit tends to be relatively low for noncohesive soils such as unconsolidated sands and coarse silts and helps to explain the liquefaction of sands during an earthquake, as discussed in the succeeding paragraph. It tends to be higher for clay-rich soils, which are much more cohesive.

One important example of a sensitive soil is the tendency of water-saturated sands and coarse silts to lose their strength during an earthquake. The vibrations destroy the grain contact strength possessed when the sand grains are at rest. Vibrations cause the grains to separate. Individual grains become dispersed in the water that occupied the spaces between grains resulting in liquefaction as the soil is turned into quicksand-like material (Figure 11.22).

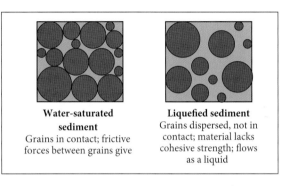

**Water-saturated sediment**
Grains in contact; frictive forces between grains give

**Liquefied sediment**
Grains dispersed, not in contact; material lacks cohesive strength; flows as a liquid

**Figure 11.22** Liquefaction produced when grains (blue) are separated; cohesive strength due to friction, present when grains are in contact is lost when grains are separated and dispersed in the liquid (white).

## Box 11.3   Liquefaction and the Van Norman Dam

As the population of Los Angeles expanded rapidly during the early years of the 20th century, the need for larger supplies of water increased. A large aqueduct system was built to bring water to the city from central California. A reservoir was required to store the water flowing from the aqueduct system. Between 1912 and 1915, the Van Norman Dam (Lower San Fernando Dam) was built to create that reservoir. The construction methods involved the building of an earthen dam, some 640 m long, composed of silty sand over a clay-rich core. A concrete parapet was placed on top, giving the dam an overall height of more than 43 m. Reservoir capacity, expanded to meet growing needs in 1930, was more than 34 billion liters of water in a 2.5 km long reservoir with a maximum depth of 40 m. The dam was built in the hinterlands of the San Fernando Valley, but it was not long before a growing population of suburbanites built their homes in the valley below.

The winter of 1970–1971 was unusually dry, even by the standards of southern California. In early February, the reservoir was at half capacity but still contained approximately 18 billion liters of water. Low reservoir levels proved fortuitous. Early on the morning of February 9, a magnitude 6.7 earthquake occurred in the San Gabriel Mountains near Sylmar, California. The earthquake destroyed two

*Continued*

## Box 11.3   *Continued*

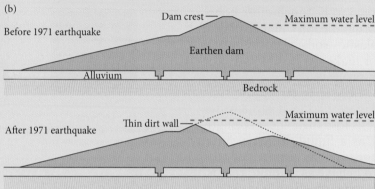

**Figure B11.2**   (a) Damage to the Van Norman (Lower San Fernando) Dam, 1971. *Source*: From the Karl Steinbrugge Collection; reproduced with the permission of the National Information Service for Earthquake Engineering, EERC, University of California, Berkeley. (b) Cross-section of the Van Norman Dam before and after failure. *Source*: Courtesy of the US Geological Survey.

hospitals, collapsed freeway overpasses, killed 65 people, and injured some 2000. But it was almost much worse. The Sylmar earthquake lasted nearly a minute. The shaking of the water-saturated sediments of which the earthen dam was constructed induced liquefaction. A major portion of the dam structure, some 550 m long, began to slide into the reservoir. Some 610 000 m³ of dam embankment were being displaced as the water-saturated reservoir side of the embankment turned to liquid. The concrete parapet was carried with it (Figure B11.2a). The dam had lost 9 m of its original height. Only a 30 cm thick portion of the downstream side of the embankment remained above water level (Figure B11.2b). Even with the lower than normal water level in the reservoir, a major disaster was about to occur.

What happened next? The earthquake ended, ground shaking stopped, and the earthen material regained its strength as the sand and silt grains regained contact and frictional resistance was restored. Of course the water in the reservoir was still exerting pressure on a flimsy 30 cm thick barrier. Fearing imminent disaster, officials evacuated 80 000 people from the valley below, while engineers drained the water from the reservoir over the next three days. Had the earthquake lasted a few more seconds, Sylmar might have a much larger place in the pantheon of historic earthquakes in California.

Los Angeles still needs water. A new dam was built close by to a much higher standard and survived the 6.7 Northridge Earthquake in 1994 with only minor damage.

(a)

(b)

**Figure 11.23** (a) Collapsed apartment buildings in Nigata, Japan, after the 1964 earthquake. (b) Neighborhood in Anchorage, Alaska, destroyed when the ground slumped, carrying structures more than 1 km, after sand lenses became liquefied during another 1964 earthquake. *Source*: The Karl Steinbrugge Collection; reproduced with the permission of the National Information Service for Earthquake Engineering, EERC, University of California, Berkeley.

Liquefied sand has no strength and therefore cannot support a load (Figure 11.23a). Liquefaction is particularly prevalent in unconsolidated sand–silt soils disturbed by earthquake ground shaking. On slopes, liquefied layers can cause entire areas to flow downhill carrying structures with them (Figure 11.23b). Water-rich, liquefied **quick clays** can also lose their strength and flow, especially when loaded. Since placing structures on the surface increases the load pressures on the substrate, careful studies of soil sensitivity must be carried out prior to the initiation of many types of construction projects.

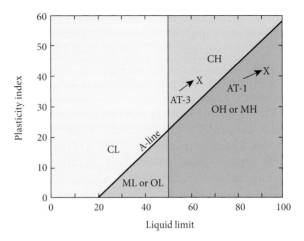

**Figure 11.24** A Casagrande plot of soil sensitivity using the plasticity index (PI) and liquid limit (LL); sand- and silt-rich soils have little or no plasticity and low liquid limits. *Source*: Courtesy of the US Geological Survey.

*Shrink–swell potential*

Other soil parameters have proven useful in soil evaluations by engineers. The **plasticity index** *(PI)* of a soil is the range of water contents over which the soil behaves as a plastic substance (Figure 11.24). It is the difference between the liquid limit and the plastic limit:

$$PI = LL - PL$$

where PI is less than 5%, small changes in water content can convert soil from a semi-solid to a liquid state. Sand- and silt-rich soils tend to possess low plastic indexes because they are not very cohesive. Clay-rich soils tend to have much higher plastic indices. Soils rich in expandable smectite clays tend to have the highest plastic indices of all (Figure 11.24). The presence of minerals that can absorb large quantities of water in the soil, such as smectite clays, can produce soils that are potentially extremely unstable. One criterion that expresses this concept is the shrink–swell potential of a soil.

**Shrink–swell potential** expresses the tendency of soils to change volume when wetted. Soils that contain large amounts of expansive clays (smectites), with their tendency to absorb water when wetted, tend to have large shrink–swell potentials. Such soils often possess large degrees of sensitivity as well and are responsible for many ground failure episodes. Ground failures include severe ground subsidence, dam failures, and major landslide events. As a general rule, such weak soils that exhibit high sensitivity, shrink–swell ratios, and plasticity indexes make poor substrates for structures unless their shrink–swell potential can be significantly reduced.

The behavior of soils can often be predicted from laboratory experiments. Figure 11.24 shows a plot of plasticity index versus liquid limit that is used extensively in laboratory testing of fine-grained soils. These plots are called **Casagrande diagrams** after the person (Casagrande 1936) who introduced them to the field. The A-line, given by PL = 0.73(LL-20), separates fine-grained soils by plasticity. The vertical line, given by LL = 50%, separates such soils by their liquid limit. Four fields result on Casagrande diagrams. Silt-rich sediments (ML or OL) tend to possess low plasticity because of their low clay contents and low liquid limits because of their minimal cohesiveness. The clay content of silty clays (CL) gives them more cohesiveness and therefore higher plasticity, while the silt content helps them retain a relatively low liquid limit. High plastic clays (CH) are rich in expansive smectite (montmorillonite) clays or somewhat less expansive illites. This results in very high plasticity and correspondingly high liquid limits. Organic clays (OH) and mica silts (MH) testify to the importance of mica in increasing the plasticity and cohesiveness of the soils in which it occurs. CH, OH, and MH soils require elevated water contents to exceed the liquid limit.

*Compressibility*

Compressible soils undergo decreases in volume when loaded. **Compressibility** expresses the tendency of soils to consolidate and lose volume. Soils with variable compressibility tend to settle unevenly causing variations in the subsidence of surfaces on which structures have been placed (Figure 11.25). If differential compaction is significant, structural damage may result.

*Porosity*

Soils hold water both by absorption in the interlayer sites of expansive clays and by storing water in the spaces between individual soil particles. The capacity of a material to hold

**Figure 11.25** Italy's famed Leaning Tower of Pisa, which was constructed on compressible, clay-rich soil. As a result of differential consolidation, the tower requires additional reinforcement (here supplied by Maureen Crowe) to maintain its equilibrium. *Source*: Courtesy of Tony Crowe, with permission.

water in its intergranular spaces is called porosity. **Porosity (Ps)** is simply the volume percentage of void spaces, called pores, in a rock. It is given by the simple formula

$$Ps = \% \text{ pore space}$$
$$= (\text{volume of pores} / \text{total rock volume}) \times 100$$

In general, porosity increases with particle size, although shape factors also play a role (Chapter 13). The primary importance of soil porosity is that it represents the total capacity of a soil to hold or store fluids such as groundwater in the void spaces between solid grains. The search for groundwater often begins in subsurface materials with substantial amounts of porosity.

*Permeability*

**Permeability (K)** expresses the hydraulic conductivity of a material which strongly influences the rate of fluid flow (Q). Permeability can be determined by using Darcy's law. In 1856, Henry Darcy, a French engineer, conducted a study of groundwater flow through a porous medium. Darcy discovered that the rate of water flow through a bed was proportional to the difference in height (h) of the water between two points and inversely proportional to the flow path length (L). In essence, the change in height divided by the distance represents the **hydraulic head** or slope of the water table. Darcy developed a "law" describing the groundwater flow where flow rate (Q; in m³/s) equals the cross-sectional area (A) of flow multiplied by the hydraulic conductivity (K) and hydraulic gradient (h/L) of the sediment or rock. Hydraulic conductivity is a measure of permeability that varies depending upon the fluid's viscosity (μ) and density (ρ), the effective permeability (K), and the acceleration of gravity (g), as given by K = ρg/μ. The hydraulic gradient (h/L) is often the slope of the water table. *Darcy's law* can be expressed as:

$$Q = A\left[K \times (h/L)\right]$$

This formula states that flow rates are proportional to hydraulic conductivity and therefore to effective permeability. Permeability increases with pore size and pore interconnectedness, both of which tend to increase with particle size and sorting. It decreases with increasing surface tension between water and the minerals with which it is in contact. As a result, sandy and gravelly soils with their larger pore spaces have higher permeability and hydraulic conductivity than clay-rich soils with their smaller pore spaces and higher surface tension, which retard fluid flow. Typical hydraulic conductivities for unconsolidated sediments are given in Table 11.9. Note the general increase in permeability with increasing particle size. Sands and gravels tend to be quite permeable and act as **aquifers**, which store and transmit water effectively. On the other hand, clay-rich layers tend to be quite impermeable and act as **aquitards**, which retard fluid flow. Rock permeability is influenced by a variety of other factors that are discussed in more detail in Chapter 13.

Permeability is also fundamentally important in aquifer studies and in determining contaminant flow behavior in the subsurface. Soil permeability is of particular importance to hydrogeologists because it determines

the rate at which water is able to flow into and through porous storage rocks in aquifers. Septic systems require permeable soils that will disperse waste materials efficiently; if such soils should become clogged with waste so that their permeability is reduced, septic system backup may occur. Permeability is also of great importance to environmental geologists interested in the rates and directions of water-borne pollutant dispersal in the subsurface. Materials with low permeability impede the dispersal of water-borne pollutants and are used as confinement barriers for hazardous waste sites.

### 11.4.5 Buried soils and paleosols

**Buried soils** are former soils that have been buried beneath the surface, usually by subsequent deposition. Buried soils occur beneath (1) glacial tills and outwash, (2) wind-blown sand, (3) river-deposited sand and mud (Figure 11.26a), (4) mass-wasting deposits produced by processes such as landslides, mudflows and debris flows, (5) marine sediments formed during transgression onto a land surface, and (6) volcanic materials produced by lava flows and pyroclastic eruptions (Figure 11.26b). Buried soils are common along regional unconformities where long-term weathering in continental environments is followed by a period of deposition. Buried soils that have been uncovered and exposed at Earth's surface by subsequent erosion are called **exhumed soils**.

**Paleosols** are ancient soils that formed under conditions not related to the present climate. They have been the subject of increasing study in recent decades as scientists have come to realize their potential as important proxies for inferring aspects of Earth's history (Retallack 2001; Driese and Nordt 2013; Tabor and Myers 2015). The study of ancient soils is called **paleopedology**.

### *Recognition of paleosols*

One major problem in recognizing paleosols is that they may be severely truncated by erosion and/or extensively altered during burial. Significant erosion may remove enough of the upper soil horizons to make soil order recognition difficult or impossible. Alteration may do the same by causing:

**Table 11.9** Generalized hydraulic conductivity of sediments.

| Sediment type | Hydraulic conductivity |
| --- | --- |
| Clays | ~$10^{-6}$ cm/s |
| Silts | ~$10^{-4}$ cm/s |
| Sands | ~$10^{-2}$–$10^{-3}$ cm/s |
| Gravels | ~$10^{-1}$–$10^{1}$ cm/s |

(a)

(b)

**Figure 11.26**   (a) Jurassic soil with plant root casts, buried by braided stream deposits, Connecticut. (b) Holocene soil with a well-defined O-horizon at the top, buried by pyroclastic deposits from the Tarawera Volcano, North Island, New Zealand. *Source*: Photo by John O'Brien. © John Wiley & Sons.

1   Decomposition of organic matter that deprives the O- and some A-horizons of a defining characteristic.
2   Oxidation of ferrous iron ($Fe^{+2}$) to ferric iron ($Fe^{+3}$), producing an oxidized layer where one did not originally exist.
3   Dehydration of goethite/limonite to hematite to produce a reddish layer where one did not originally exist.

In addition, subsequent diagenetic processes may alter other geochemical and mineralogical trends that help pedologists to recognize soil types. Despite these problems, paleopedologists have been able to recognize good examples of all the major soil orders in older rocks as well as some extinct soil types. Given Earth's long history and substantial changes in climate, biota and composition of the atmosphere, one might expect ancient soils to be very different from modern ones. Yet most ancient soils are sufficiently similar to modern ones that the USDA-NRCS soil taxonomy can be used to classify them. As Retallack (2001) states, very few of the soils found in the paleopedological record are extinct types.

Soils also yield clues to the evolving composition of Earth's atmosphere. Perhaps the most significant insight comes from the emergence of oxisols some 2.0 Ga. In rocks formed prior to this time, red ferric ($Fe^{+3}$) oxide minerals such as hematite in soils are rare to absent. Instead, the dominant soils belong to an unnamed soil order with green to gray-white soil horizons in which the iron-bearing minerals contain reduced ferrous iron ($Fe^{+2}$). What caused soils to change in this way some 2.0 Ga? Noting the widespread expansion of red-colored sediments or "red beds" at this time, most geologists believe that 2.0 Ga marks a time when the abundance of excess free oxygen increased significantly in Earth's atmosphere. This change in Earth's atmosphere from a reducing atmosphere in which methane $CH_4$ was a significant component to an oxidizing atmosphere in which $O_2$ was an important component is one of the most significant changes in Earth history, for it eventually allowed the evolution of organisms that use $O_2$ during respiration and decomposition.

Significant insights are being gained from the study of Archean (>2.5 Ga) soils concerning the emergence of life in terrestrial environments. Because living tissues selectively utilize carbon-12 ($^{12}C$) relative to carbon-13 ($^{13}C$), organic carbon possesses lower $^{13}C/^{12}C$ ratios than atmospheric carbon, as discussed in Chapter 3. The presence of organisms in a soil can be detected by such depressed $^{13}C/^{12}C$ ratios in carbonate minerals precipitated by soil waters, even when organic matter has not been preserved. Several Archean soils show such depressed $^{13}C/^{12}C$ ratios (Retallack 2001), which suggests that microbial populations inhabited terrestrial soils more than 2.5 Ga. Because the atmosphere was depleted in free oxygen ($O_2$) prior to 2.0 Ga, such microbial populations probably utilized atmospheric methane ($CH_4$) abundant in Earth's early atmosphere.

Using the ratios of carbon isotope $^{13}C$ and $^{12}C$, scientists can monitor fluctuations in the amount of the greenhouse gas $CO_2$ in the atmosphere as well. As expected, the abundance of $CO_2$ in the atmosphere generally mirrors atmospheric temperatures. When $CO_2$ levels are high, temperatures tend to be high; when they are low, temperatures tend to be low. Using these methods, scientists have been able to document a dramatic trend of global cooling that began in the late Eocene period and has continued, with many fluctuations, to the present time. This long-term cooling corresponds to a long-term decrease in atmospheric $CO_2$. What might have caused this trend? Retallack (2001) thinks he has the answer. He notes that grasses evolved following the greenhouse period that marked the early Tertiary. With the expansion of grasslands, mollisols formed for the first time in the Eocene and had spread rapidly by the Miocene to cover as much as 20% of the land surface. With their thick accumulations of slow-decaying organic material, mollisols sequester huge amounts of carbon, effectively removing $CO_2$ and another greenhouse gas, methane ($CH_4$), from the atmosphere. Mollisols also may have stimulated biological productivity in oceans and helped to increase the rate at which light is reflected from Earth's surface (albedo), accelerating the cooling trend. If Retallack is right, the rise of mollisols may well have been a major cause of the coeval long-term global cooling that eventually led to widespread glaciations during the Pleistocene.

Paleopedologists continue to push the envelope. Reflecting current concerns, they are especially interested in proxies for climate change (Costantini, 2017). They have measured very high $^{13}C/^{12}C$ ratios in Cambro-Ordovician soils that may indicate a greenhouse atmosphere with 16–18 times more $CO_2$ than exists at present (Retallack 2001) and some 10 times higher during the Cretaceous, both periods of unusually high sea level stands. They point to the soil record of vascular plants that contain lignin which begins in the Silurian and may have caused severe reductions in atmospheric $CO_2$ content. This may have eventually led to a long-term episode of global cooling that extended into the Permian. They have developed proxies for precipitation and seasonality and have analysed the pollen in soils to reconstruct vegetation patterns that are related to climate. Recent reviews of advances in paleopedology, paleoclimates, and paleoenvironments include those of Tabor and Myers (2015) and Driese and Nordt (2013). Weathering and soil development may be important in more ways than we have thought.

## CONTENT ASSESSMENT

1  Compare and contrast *weathering* and *erosion*. Then do the same for *disintegration* and *decomposition*. What role does climate play in determining the relative importance of the two sets of processes at any given location on Earth's surface? What important role does disintegration play in determining the rates of decomposition during weathering?

2  Discuss how each of the following decomposition processes changes the composition of both the original rock and the aqueous solution with which it is in contact: (a) *dissolution*, (b) *ion exchange*, (c) *hydrolysis*, (d) *oxidation*, (e) *hydration*, and (f) *chelation*.

3  Rank order the following common rock types from most to least resistant to *decomposition* and explain the reasons for your answers: (a) *quartzite* (>95% quartz), (b) *amphibolite* (40% plagioclase, 35% amphibole, 20% garnet), (c) *marble* (>95% calcite), (d) *slate* (65% clay minerals, 20% quartz, 15% fine muscovite), (e) *granite* (40% k-feldspar, 20% plagioclase, 20% quartz, 20%) biotite + hornblende, and (f) *peridotite* (60% pyroxene, 40% olivine). Which of these minerals would likely occur in mineralogically mature sediment and which might occur in mineralogically immature sediment?

4  Compare and contrast the basic crystal structures of *kandite*, *illite*, *smectite*, and *chlorite* group clay minerals in words and using both T-O and S-G-B terminology. Then compare and contrast their chemical compositions and the environmental conditions under which each is most commonly produced by weathering processes.

Under what conditions are *bauxite-rich soils* produced?

5   Distinguish between the *O, A (E), B, and C horizons* in a classic, fully developed soil profile in terms of their mineral and chemical compositions and the processes that combine to produce each. Be sure to include a discussion of *illuviation* and *eluviation*.

6   What are the differences between *porosity* and *permeability*? How, in general terms, do they affect the ability of a rock to store and transmit fluids. According to *Darcy's Law*, the rate of flow (K) of a fluid through a rock is proportional to (increases with increases in) what factors? Using the same law, to what major factor is the rate of flow is inversely proportional? This question can be a bit open-ended, so focus on Darcy's Law.

7   What is *liquefaction*? How is it related to *soil sensitivity*? What processes are important in soil liquefaction and in what ways can it damage the structures built on such soils?

## REFERENCES

Blum, H.P. and Schad, P. (2015). 90 years of soil classification of the IUSS. *IUSS Bulletin.* 6: 38–45.

Burbank, D. and Anderson, R.S. (2000). *Tectonic Geomorphology.* Oxford, UK: Blackwell Science 274 pp.

Casagrande, A. (1936). The determination of the pre-consolidation load and its practical significance. In: *Proceedings of the International Conference on Soil Mechanics and Foundation Engineering,* vol. 3, 60–64. Cambridge: Harvard University.

Costantini, E.A.C. (2017). Palesols and pedostratigraphy. *Applied Soil Ecology* 123: 1–3.

Davies, W.E. and LeGrand, H.E. (1972). Karst of the United States. In: *Karst, Important Karst Regions of the Northern Hemisphere* (eds. M. Herak and V.T. Stringfield), 467–505. Amsterdam: Elsevier Publishing.

DiPietro, J.A. (2018). *Geology of Landscape Evolution: General Principles Applied to the United States,* 2e. Amsterdam: Elsevier 611 pp.

Driese, S.G. and Nordt, L.C. (eds.) (2013). *New Frontiers in Paleopedology and Terrestrial Climatology: Paleosols and Surface Analog Systems,* Special Publication, vol. 104, 196. SEPM.

Goldich, S.S. (1938). A study in rock-weathering. *Journal of Geology* 46: 17–58.

Grim, R. (1968). *Clay Mineralogy,* 2e. New York: McGraw Hill 596 pp.

Hugget, R. (2002). *Fundamentals of Geomorphology.* London: Routeledge Publishers 336 pp.

Prothero, D.R. and Schwab, F. (2013). *Sedimentary Geology: an Introduction to Sedimentary Rocks and Stratigraphy,* 2e. New York: W.H. Freeman 557 pp.

Retallack, G.J. (2001). *Soils of the Past: An Introduction to Paleopedology,* 2e. Oxford, UK: Blackwell Science 404 pp.

Ritter, D.F., Kochel, R.C., and Miller, J.R. (2006). *Process Geomorphology.* Long Grove, IL: Waveland Press 560 pp.

Schoonover, J.E. and Crim, J.F. (2015). An introduction to soil concepts and the role of soils in watershed management. *Journal of Contemporary Water Research & Education* 154: 21–47. https://doi.org/10.1111/j.1936-704X.2015.03186.x.

Tabor, N.J. and Myers, T.S. (2015). Paleosols as indicators of paleoenvironment and paleoclimate. *Annual Review of Earth and Planetray Science* 43: 1–11.

USDA-NRCS (US Department of Agriculture, Natural Resources Conservation Service) (1999). *Soil Taxonomy: a Basic System of Soil Classification for Making and Interpreting Soil Surveys,* 2e Agriculture Handbook No. 436, 871 pp. Washington: US Government Printing Office.

Willigoose, G. (2018). *Principles of Soilscape and Landscape Evolution.* Cambridge University Press 334 pp.

# Chapter 12

## The sedimentary cycle: Erosion, transportation, deposition, sedimentary structures, and environments

### 12.1 SEDIMENTS AND SEDIMENTARY ROCKS

Sedimentary materials including soils, sediments, and sedimentary rocks, cover more than 80% of Earth's surface (Tucker 2001). They contain most of the fluid resources such as groundwater, natural gas, and petroleum on which modern societies depend. They harbor important deposits of coal, metallic ores, and the aggregate materials used in the production of products such as plaster, cement, concrete, and asphalt. They provide the substrate on which structures such as houses, commercial buildings, highways, and industrial complexes are built, and the soils in which agricultural and forest products are produced.

Because sedimentary rocks record essential information about the processes that produced them, they are also a vast repository of the history of Earth's surface over time. This repository includes the fossils of organisms that record the history of life on Earth from its emergence more than 3.5 billion years ago to the present time. It also includes rocks that preserve evidence of the specific environment and tectonic setting in which they accumulated, which provide most of the evidence on which maps depicting Earth's ancient surface (paleogeographic maps and paleotectonic maps) are based. These rocks also contain components that record major changes in Earth's climate and atmospheric composition through time. Much time, effort, and creativity has gone into interpreting this

*Earth Materials*, Second Edition. Kevin Hefferan and John O'Brien.
© 2022 John Wiley & Sons Ltd. Published 2022 by John Wiley & Sons Ltd.
Companion website: www.wiley.com/go/hefferan/earthmaterials2

repository, often with truly extraordinary results.

**Sediments** are rock materials, generally but not always unconsolidated, that form on or very close to Earth's surface. This definition excludes (1) solid volcanic materials that accumulate directly from the solidification of lava flows, (2) solid plutonic igneous and metamorphic rocks that generally form well below Earth's surface, and (3) most high temperature hydrothermal deposits. Sediments are classified genetically, according to the processes involved in their formation. Most sediment can be genetically classified into three major components: (1) detrital sediment, (2) organic sediment, and (3) chemical sediment. The latter two are sometimes grouped together as **biochemical sediment**.

**Detrital sediments**, detailed in Chapter 13, are the solid products of weathering, the breakdown of rocks at or near Earth's surface. Detrital sediments include most of the gravel, sand, and mud particles that accumulate on Earth's surface. **Organic sediments**, discussed in Chapter 14, are the solid products of organic synthesis or precipitation and include hard materials such as shells, bones, and teeth and soft materials such as cellular materials composed of organic molecules. **Chemical sediments**, also discussed in Chapter 14, are the solid products of inorganic precipitation such as mineral crystals precipitated from solution. These three types of sediment may occur in any proportions in a particular sediment or sedimentary rock.

Many other terms are used to describe major groups of sediment. **Clastic** is a general term for sedimentary particles transported as solid particles, regardless of origin. The term **epiclastic** is used for clastic sediment that is transported as clastic particles across Earth's surface. Related terms include **bioclastic**, which is used for clastic particles of organic origin such as shell fragments transported by waves or currents, **siliciclastic**, which refers to clastic particles composed of silicate minerals, and **volcaniclastic**, which is used for clastic particles produced initially by volcanic processes. Yet another related term is **terrigenous**, which refers to clastic sediment derived from a land mass.

As long ago as the 16th century, observers noticed that some solid rock bodies possessed many of the same features that are present in largely unconsolidated sediments. By the eighteenth century, geologists had begun to speculate that these solid **sedimentary rocks** had been produced by the solidification of initially unconsolidated sediments. Eventually it was recognized that for each type of unconsolidated sediment there is an equivalent sedimentary rock that is produced by its solidification. The general name for the set of processes by which unconsolidated sediment is converted into more consolidated sedimentary rock is **lithification**, literally to "turn into rock" (from the Greek work *lithos* = rock).

## 12.2 THE SEDIMENTARY CYCLE

The **sedimentary cycle** (Figure 12.1) is a simple model of the processes responsible for the production of sediments and sedimentary rocks. Even simple versions provide an excellent overview of the interrelationships between sediments, sedimentary rocks, and the processes that produce them. Simple models of the sedimentary cycle commonly illustrate five or more significant sets of processes that affect the production, dispersal, and accumulation of sediments and their conversion into solid sedimentary rocks.

The first is the production of sedimentary materials in a **source area**, largely by weathering, at or near Earth's surface. The source area has often experienced **uplift** relative to surrounding areas. **Weathering**, discussed in detail in Chapter 11, involves the physical, chemical, and/or organic breakdown of rock material into smaller mineral and rock fragments and into dissolved components. Two major types of weathering processes are (1) **disintegration**, which breaks rock materials down into smaller pieces of the original material, without changing their composition, and (2) **decomposition**, which changes the composition of the original material, producing new materials.

Weathering, by producing smaller particles, and uplift, by increasing slope angles, both enhance a second set of processes called **erosion**. Erosion is the removal of sedimentary materials from a place on Earth's surface. Erosion is an instantaneous process.

In turn, erosion initiates a period of **transportation** during which sedimentary material is moved or transported in solid or dissolved form across Earth's surface. Most transportation is by water, glaciers, mass flows or wind. Since the first three of these generally flow downhill under the influence of gravity, sedimentary processes tend to lower relatively high areas by sediment

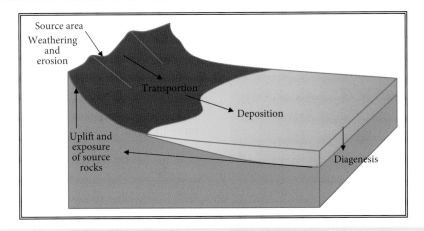

**Figure 12.1**   A simple model of the sedimentary cycle.

erosion and then disperse the sediment downhill to sites of accumulation in relatively low areas that tend to be filled in by sediment deposition.

Transportation ceases when sediments accumulate on Earth's surface by processes collectively referred to as **deposition**. It is conceptually convenient to think of erosion and deposition as opposites. Erosion involves the removal of rock material from a place on Earth's surface, whereas deposition involves the accumulation of sediment at a place on Earth's surface. Acting alone, erosion and deposition tend to decrease the relief of Earth's surface. So why, after 4.54–4.55 billion years, is Earth's surface not flat? Tectonic processes, such as folding, faulting, volcanism, and isostasy generate significant relief and prevent the development of a "flat Earth." Erosion, transportation, and deposition, discussed in detail in this chapter, normally occur multiple times during the history of a sedimentary particle, as it is dispersed from its source area to its current resting place. A grain of sand in a back-beach dune field may well have originated in a source area high in the mountains, then been eroded, transported, and deposited multiple times by glaciers, mass flows, winds, and water during its dispersal into and down a drainage system to the coastline. What might its journey be in the future?

**Diagenesis**, detailed in Chapters 13 and 14, encompasses a suite of relatively low temperature processes that affect sediments after their accumulation, typically after burial. It includes processes of lithification such as compaction and cementation. **Compaction** results from the expulsion of intergranular fluids caused by increases in confining pressure during progressively deeper burial. Compaction aids

lithification by increasing the attractive forces between grains as they move into closer contact. **Cementation** occurs when subsurface fluids precipitate minerals in the spaces between grains that bind or cement grains to one another.

A single sedimentary cycle ideally ends with burial and diagenesis. After a significant period of burial, a sedimentary rock may eventually be re-exposed at Earth's surface. This begins a new cycle of weathering, erosion, transportation, deposition, and diagenesis. These constitute a second sedimentary cycle and the products are referred to as **polycyclic** sediments and sedimentary rocks.

## 12.3 STRATIFICATION AND SEDIMENTARY ENVIRONMENTS

One of the most striking features of sediments and sedimentary rocks is their layering, which is called **stratification**. All sediments are deposited in layers called **strata**. Each stratum records a period of net sediment accumulation on Earth's surface and records information about the conditions and processes that produced it. Layer thickness tends to increase with the rate and duration of deposition and is decreased by subsequent erosion. Far more strata are eroded away than are preserved in the (therefore partial) sedimentary record. Thick strata (>1 cm thick) are called **beds**; their three-dimensional form is often that of a laterally extensive mattress, with tapered edges. Thin strata (<1 cm thick) are called **laminations** or **laminae** (Figure 12.2a); their three-dimensional form is often that of a laterally extensive, very thin yoga mat or piece of paper. Beds are commonly preserved as compound layers, defined for the purpose of describing sedimentary rocks. In

(a)

(b)

**Figure 12.2** (a) Thin laminations (above) and thicker beds (below) in the Jurassic Berlin Formation, Connecticut. (b) Cross-stratified beds (center) with initially inclined laminations, separated by a near horizontal erosion surfaces in the Jurassic Navajo Formation, Zion National Park, Utah. *Source*: Photo by John O'Brien. © John Wiley & Sons.

general, the boundaries between strata reflect changes in the type of material deposited through time or record intervals of erosion between periods of net deposition.

Most strata form when sediments come to rest on relatively flat or very gently-sloping surfaces. For this reason, most strata are horizontal or very nearly so at the time they form. This notion is summarized in the **principle of original horizontality**, which states that strata are nearly horizontal at the time they form. The major value of this principle is the recognition that very steeply inclined and/or folded strata must have been subjected to forces that have rotated them from the horizontal position after they accumulated. There is one very

significant exception to the principle of original horizontality. **Cross-strata** (Figure 12.2b) form when sediments accumulate on steeper slopes of up to 35°. They are commonly truncated by and formed over originally subhorizontal erosional surfaces and interlayered with nearly horizontal strata, as discussed later in this chapter.

Each stratum records the type of sediment that accumulated on Earth's surface and the conditions under which each type of sediment accumulated can potentially be inferred. As a result, sedimentary strata represent a vast storehouse of information concerning the changes that have occurred on Earth's surface over time. Because sedimentary layers accumulate on Earth's surface, one atop the other over time, the relative ages of sedimentary strata can be inferred from the **principle of superposition**, first stated by Niels Stensen (Nicolas Steno) in the seventeenth century. In simple terms, this principle states that in any sequence of strata that have not been overturned, strata become younger from the bottom of the layered sequence toward the top; that each layer is younger than the layer below and older than the layer above. By inferring the surface conditions that produced each layer and knowing the sequence in which those conditions occurred, we can infer a long history of how surface conditions changed at any location over time. With radiometric and other dating techniques, one can determine the absolute ages of layers and place their history into an absolute time context. With correlation techniques to relate the histories preserved in strata at many different locations, scientists have been able to develop a global history of Earth's surface, albeit one limited by the incomplete record preserved in sediments and our ignorance of how to interpret the available record completely.

Figure 12.3 illustrates several of the major surface environments in which sediments accumulate. These **depositional environments** may be broadly subdivided into terrestrial (land), transitional (paralic), and marine (oceanic) environments. Major **terrestrial** depositional environments include glacial, alluvial fan, braided stream, meandering stream and floodplain, lacustrine (lake), eolian (dune), and paludal (swamp) environments. **Transitional (paralic)** environments typically occur in the terrestrial–marine transition and include coastal–deltaic environments such

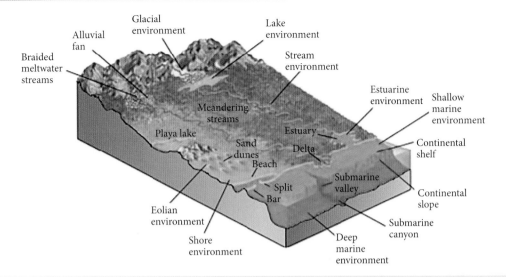

**Figure 12.3**   Major terrestrial, paralic, and marine depositional environments. *Source*: Murck and Skinner (2016). © John Wiley & Sons.

as deltas, estuaries, tidal flats, beaches, back-beach dunes, and barrier island–lagoon systems. Important **marine** depositional environments include epicontinental epeiric seas, reefs, restricted seas, continental shelves, continental slopes, submarine fans and continental rises, forearc basins, trenches, and deep ocean floors. The temporal and spatial distributions of such environments strongly depend on the existing tectonic setting and climatic conditions. Where applicable, the sediments deposited in response to processes that operate in these environments are discussed in the following sections of this chapter, and in Chapter 13 on detrital sedimentary rocks and Chapter 14 on organic and chemical sedimentary rocks. However, a detailed discussion of sedimentary environments is beyond the scope of this text. Interested readers will likely buy another book that describes sedimentary environments in moderate detail.

## 12.4 AGENTS OF EROSION, SEDIMENT DISPERSION AND DEPOSITION

As noted in the previous section, sediments are eroded, transported, and deposited by a variety of agents that flow across Earth's surface. Chief among these are water, wind, glaciers, and mass flows. These flows are either fluid flows or plastic flows, and they flow in response to the stresses that act on them.

**Fluid flows**, such as water and wind, have no shear strength; the smallest stress sets them in motion. In water on Earth's surface, the shear stress that initiates flow is typically the tangential force of gravity $(g_t)$ acting parallel to the flow surface, which causes the fluid to flow downhill, as in rivers. However, water can also be set in motion by stress provided by wind blowing across the surface, as in waves and surface currents. In addition, both wind and water flow in response to differential heating, which generates parcels of fluid with different densities, which in turn move in response to gravity, as in thermohaline and atmospheric circulation. Whatever the cause, such fluid flows move with sufficient velocity to be important agents in the transportation and dispersal of sediments across Earth's surface.

In **plastic flows**, which possess shear strength, the major stress that initiates flow is the tangential force of gravity $(g_t)$, which increases with increasing slope angle. They are set in motion only when a critical **threshold stress** is achieved. Once threshold shear stresses are achieved, glaciers flow from areas of maximum surface elevation toward areas of minimum surface elevation, whether that is downhill, as in mountain glaciers, or from areas of maximum ice thickness toward areas of minimum thickness, as in many continental glaciers. Likewise, mass flows such as debris flows and turbidity currents are initiated once critical threshold values of $g_t$ are reached, typically, but not always, on relatively steep slopes.

Flows may also be distinguished as laminar or turbulent (Figure 12.4). In **laminar flow**,

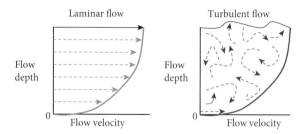

**Figure 12.4** Laminar and turbulent flow profiles; flow lines are dashed.

adjacent parcels of the flow move roughly parallel to one another in a well-organized pattern, with negligible mixing between them. In **turbulent flow**, adjacent parcels of the flow move in chaotic patterns and random mixing between parcels is common.

### 12.4.1 Water-deposited sediments

Water is the major agent by which sediments are eroded, transported, and deposited on Earth's surface. Water is a Newtonian fluid. Like all fluids, Newtonian fluids lack shear strength, but they are also characterized by a constant resistance to shear stress for a particular set of conditions, which is called **dynamic viscosity ($\mu$)** and is given by:

$$\mu = \tau_o / (dU/dY)$$

where dU/dY is the change in velocity (dU) with respect to depth (dY) in the flow and $\tau_o$ is the shear stress (e.g., $g_t$) acting on the flow's surface. Why is viscosity important? It is easier to understand the importance of dynamic viscosity if the equation is rearranged to:

$$dU/dY = \tau_o / \mu$$

It can then be seen that velocity of fluid flow is proportional to the shear stress, such as tangential force of gravity which increases with increased slope, and is inversely proportional to dynamic viscosity. The dynamic viscosity ($\mu$) is the resistance to shear stress and may be thought of as the fluid's internal resistance to flow. The viscosity of water increases with the degree of molecular linkage between water molecules, which is inversely proportional to temperature. Colder water possesses more molecular bonds between molecules (which are vibrating more slowly), thus possesses a higher viscosity or resistance to flow than warmer water. In addition to determining flow velocity at a particular shear stress, viscosity,

along with other variables, also plays an important role in determining whether flows are laminar or turbulent. These variables combine to produce several expressions for **Reynolds' number (Re)**, one of which is:

$$Re = UL\rho / \mu$$

where U is a velocity factor such a mean flow velocity, L is a length factor such as depth or particle diameter, and $\rho$ is a density factor such as fluid density.

Reynolds' number is essentially a ratio between the inertial forces and the viscous forces in a fluid medium. The Reynolds' number can be used to predict whether flow will be laminar or turbulent. Viscous forces tend to damp out turbulence and so favor laminar flow in flows of high viscosity. Inertial forces tend to favor the development of fluid turbulence, for example in flows with high velocities. For any situation, there is a critical Reynolds' number, below which flow is laminar and above which flow is turbulent. Figure 12.4 illustrates cross-sections of subcritical laminar and supercritical turbulent flow such as might occur in a stream or tidal channel. On smooth beds, a laminar flow sublayer may occur near the base of a turbulent flow; on rough beds with a larger length factor, flow will be turbulent.

Figure 12.5 illustrates the transition from subcritical laminar to supercritical turbulent flow in a natural channel, probably caused by some combination of increase in flow velocity, depth, and/or bed roughness.

**Figure 12.5** Transition from laminar flow (background) to turbulent flow (foreground) in a branch of the Delaware River, Pennsylvania. *Source*: Photo by John O'Brien. © John Wiley & Sons.

*Entrainment, transportation, and deposition*

Both laminar and turbulent flows have the ability to erode, transport, and deposit sediment. The process by which epiclastic sediment transportation is initiated by erosion is called **entrainment**. Particles are said to be entrained by the flow when their movement is initiated by the flow. As might be expected, by analogy with using a garden hose to sweep particles of different sizes from one's driveway, there is a relationship between the flow velocity of the water and the size of the particles that can be entrained by the flow. However, this relationship is not as straightforward as one might initially expect. Hjulstrom (1939) developed a diagram (Figure 12.6) that summarizes the critical velocity required for entrainment, transportation, and deposition by water of epiclastic particles of various diameters. It should be noted that the diagram applies specifically to a 20 °C, 1 m deep aqueous flow and that the details change when these conditions are changed. But, as we will see, generalizations from Hjulstrom's diagram offer powerful insights into the conditions of sediment entrainment, transportation, and deposition.

**Hjulstrom's diagram** contains two curves that divide the diagram into three fields. The upper curve is the erosion or entrainment curve that plots the **critical entrainment velocity** required to initiate particle movement against particle size. Perhaps surprisingly, the critical entrainment velocity curve (upper curved line in Figure 12.6) is lowest (~20 cm/s) for particles in the very fine sand range (~0.1 mm

diameter). The critical entrainment velocity increases progressively, as expected, for increasingly larger, more massive, particles up to >100 cm/s for pebble-size particles. This proportional relationship between critical entrainment velocity and particle size reflects the fact that progressively higher velocities and bed shear stresses are required to overcome the rest mass and initiate the movement of progressively more massive particles. However, for cohesive silt and clay particles finer than 0.0625 mm, the critical entrainment velocity increases with decreasing particle size! What is responsible for this seemingly counterintuitive relationship? The higher the amount of surface area/volume, the stronger the attractive forces, and the more cohesive the sediment will be. This **cohesiveness** is accelerated in the clay range because clays possess very large surface area: volume ratios and have large amounts of unsatisfied electric charge on their surfaces which bind the particles together. The inverse relationship between particle size and critical entrainment velocity in silts and clays reflects the increasing cohesiveness of such particles as average particle size decreases. Because clays and silts tend to stick together, they behave as larger particles with increasingly large rest masses that are increasingly difficult to entrain. The top field (in blue) on the diagram includes flow conditions under which all sizes shown on the diagram will have been eroded and entrained and so are being transported. Under such conditions, these particles will continue to be transported unless the velocity decreases substantially.

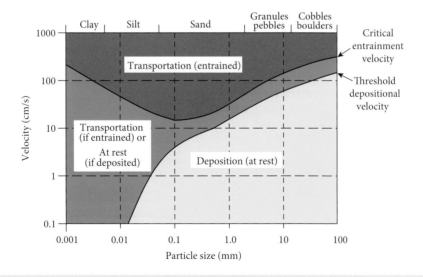

**Figure 12.6** Hjulstrom's diagram showing velocity conditions for entrainment (erosion), transportation, and deposition of sediment particles of different sizes.

The second curve on Hjulstrom's diagram represents the **threshold depositional velocity**. It illustrates the conditions under which various particle sizes cease to be transported and are deposited. The threshold velocity of deposition (lower curved line in Figure 12.6) is lower than that required for entrainment for all particle sizes. This is because once entrained, a particle has inertial momentum due to its mass and velocity. The lower velocities for deposition of a particular size reflect the need for the particle to lose momentum before it can stop moving and be deposited. The higher velocities required for entrainment of any particle size reflect the need for the particle's inertial tendencies to be overcome before it starts moving. Note that the velocity at which a particle stops moving, its threshold velocity of deposition, is proportional to particle size for all particle sizes (Figure 12.6). The bottom right field on the diagram (in yellow) includes the conditions under which sizes shown on the diagram will be at rest. Under such conditions, these particles will continue to remain at rest unless the velocity increases substantially to a value above the threshold entrainment curve.

The central field (green in Figure 12.6) lies between the critical entrainment velocity and the threshold velocity for deposition. Why is this field labeled **transportation or at rest**? Where flow velocity is increasing over a bed where all particles are at rest, the critical entrainment velocity has not yet been reached; no entrainment occurs under the particle size–velocity relationships in this field. But where flow velocity is decreasing into this field from values above the critical entrainment velocity curve, particles with size–velocity relationships in this field will continue to move because they have not yet reached the lower threshold velocity of deposition. Once entrained, coarse sand and gravel particles can be transported at velocities that are slightly lower than the velocities required for entrainment, and yet higher than those required for deposition. However, they will be deposited if the flow velocity decreases by a rather small amount to velocities below the critical depositional velocity. For example, a 10 mm pebble will be entrained at 300 cm/s and will continue to be transported as long as the velocity does not drop below 200 cm/s, a decrease in flow velocity of 33%. However, a clay particle entrained at 100 cm/s will continue to be transported unless the velocity approaches zero. This explains the much larger field for the velocity conditions under which transportation continues in fine sediments. It also explains why very calm waters are required for their deposition.

Once entrained, the **sediment load** of different particles sizes tends to be carried in different parts of the flow (Figure 12.7). The **bed load** is carried in continual or intermittent contact with the bed over which the water is flowing. It is subdivided into a traction load, which is in continual contact with the bottom, and a saltation load, which is in intermittent contact with the bottom. The **traction load** (Figure 12.7) contains the coarsest particles, those with sufficient mass that they are not lifted from the bottom by turbulence and lift forces. Particles in the traction load move down-current by (1) **rolling** along the bed, (2) **sliding** along the bed, and (3) **creep**, in which particle collisions set other particles in motion. Where the traction load is more than one particle deep, the term **traction carpet** can be used. The **saltation load** contains smaller, less massive particles that can be lifted from the bottom by turbulence, lift forces, and particle impacts, but which rapidly settle back to the bed before being lifted again. Particles in the saltation load inscribe roughly parabolic arcs as they "skip" or saltate across the bed in a down-current direction (Figure 12.7). The **suspension load** contains the smallest, least massive particles. These are particles that once lifted from the bed have such low settling velocities that they remain suspended above the bed for long periods of time. Good

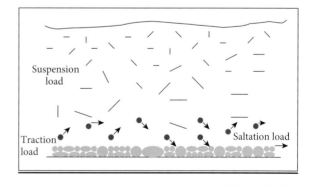

**Figure 12.7** Sediment loads in an idealized aqueous medium.

examples of suspended particles are the dust particles visible in a shaft of light shining through a window or the fine particles from a smokestack or forest fire that can remain suspended for days.

Particles are transferred between loads. As flow velocity increases, particles in the traction load may begin to saltate and particles in the saltation load may become suspended for longer periods of time. Conversely, as flow velocity decreases, particles in the suspension load may begin to saltate and particles in the saltation load may return to the traction carpet. Particles larger than coarse silt can be suspended at high velocities but generally fall back into the bed load as velocities decrease. As a result, particles larger than coarse silt are generally deposited from the traction load of aqueous flows. The very low critical depositional velocity for particles smaller than coarse silt ensures that they will remain in suspension for long periods of time. Particles finer than coarse silt are carried in and deposited from the suspension load. Only under very calm conditions, when flow velocities approach zero, are they able to settle from suspension and be deposited. Water also transports a **solution load** that consists of the dissolved solids produced by dissolution of rocks and minerals with which the water has been in contact. In the oceans, the solution load typically constitutes 3.5% of the total mass of seawater. In freshwater bodies it is considerably less, averaging about 0.015%.

Aqueous flow patterns can be very complex. However, two major types of aqueous flow are recognized in sedimentary environments: (1) unidirectional flows, and (2) oscillatory flows. If we understand how each interacts with unconsolidated sediments over which they flow, the sediments produced by each can be recognized in the sedimentary record.

### Unidirectional flow

**Unidirectional flows** are characterized by flow in one direction over a period of time. Most stream flow, sheetwash, surface and deep ocean currents, longshore currents, and flood and ebb tidal currents can be modeled as unidirectional flows. For this reason, they have been carefully studied. Engineers involved in flood control, channel diversion, dam building, irrigation, pipeline construction, and coastal erosion studies have a real need to understand the nature of unidirectional flow. Laboratory investigations and field studies have demonstrated conclusively that as unidirectional flows move over unconsolidated beds, those beds change in ways that are related to flow characteristics. One way this work has been summarized is in the **flow regime concept**, which offers insights into the origin of the primary sedimentary structures produced by unidirectional fluid flows. A simplified version of the flow regime concept for initially flat beds composed of unconsolidated medium sand is presented in Figure 12.8. Flow regimes can be varied by changing the particle

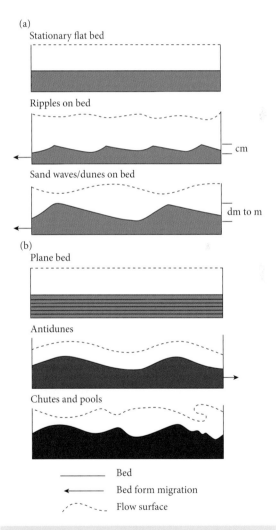

**Figure 12.8** A simplified version of the flow regime concept: (a) lower flow regime bed forms; (b) upper flow regime bed forms. Flow is from right to left; flow regime increases downward.

size of the bed, the flow velocity, the flow viscosity (e.g., temperature), the flow depth and/or the flow density. Figure 12.8 uses changes in flow regime to illustrate these concepts. Two flow regimes are recognized. These are the **lower flow regime**, where the bed surface is out of phase with the flow surface (Figure 12.8a), and the **upper flow regime**, where the bed surface is generally in phase with the flow surface (Figure 12.8b)

At very low flow regimes (e.g., very low velocity), no movement occurs on the bed, because the critical entrainment velocity for medium sand has not been achieved. Once the critical entrainment velocity is achieved, the bed is quickly deformed into small, asymmetrical **current ripples** that possess relatively gentle up-current or stoss-side slopes and steeper down-current or lee-side slopes (Figure 12.9). Straight-crested ripples typically form first, followed by various curved-crested ripples at slightly higher flow regimes. Current ripples (1) possess relatively continuous, nonbranching crests, (2) possess small ripple indices (RI = wavelength : height ratio), and (3) tend to migrate in a down flow direction over time. The asymmetry of current ripples permits them to be used to infer the direction of current flow, from the gentle stoss side toward the steep lee side, which is inclined in the flow direction. Current ripples record both lower flow regime conditions and flow directions.

As flow regime or velocity continues to increase within the lower flow regime, larger bed forms called subaqueous sand waves or dunes begin to develop on the bed. Initially these bed forms possess relatively straight crests transverse to flow, but at higher flow regimes their crests become more curved and their forms more complex. **Sand waves** are straight-crested two-dimensional bed forms where, ideally, every two-dimensional cross-section parallel to flow direction is the same. **Dunes** are curved-crested three-dimensional bed forms where cross-sections parallel to flow vary. Smaller current ripples commonly develop on the stoss side of dunes and their lee sides are inclined in the flow direction (Figure 12.10).

At still higher flow regimes that mark the transition to **upper flow regime** conditions, dunes begin to be sheared out into a flat or plane bed that is parallel to the nearly flat flow surface (Figure 12.8b). This flow regime

**Figure 12.9** Current ripples. (a) Asymmetrical current ripples on a modern beach flat, where the current flowed left to right. *Source*: Photo by John O'Brien. © John Wiley & Sons. (b) Asymmetrical current ripples in Cretaceous sandstone, where the current flowed from upper left to lower right. *Source*: Courtesy of Duncan Heron.

**Figure 12.10** Slightly wavy-crested sand waves or subaqueous dunes, Rio Hondo, southern California; the current was from right to left. *Source*: John O'Brien.

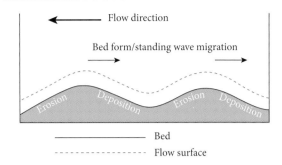

**Figure 12.11** Antidunes in phase with standing waves on a flow surface. Antidunes migrate in the opposite direction to the flow direction by lee-side erosion and stoss-side deposition.

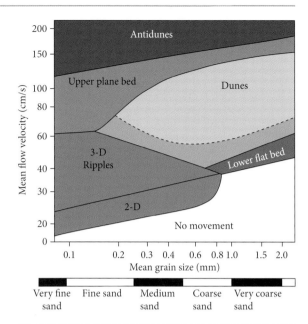

**Figure 12.12** Flow regimes with respect to mean flow velocity and grain size. *Source*: Based Harms et al. (1982). © John Wiley & Sons.

is called **plane bed** or the plane bed transition. Bed load sand moves along the plane bed at relatively high velocities. With further increases in flow regime, unconsolidated sand beds are deformed into bed forms called **antidunes** that are roughly in phase with standing waves on the flow surface (Figure 12.11). These bed forms are called antidunes because they migrate with the standing waves in an up-flow direction by lee-side erosion and stoss-side deposition. With continued flow, the standing wave collapses and antidunes are smeared out into an uneven surface characterized by shallow **chutes** and deeper **pools** (Figure 12.8b).

As noted above, the sequence of bed forms with increasing flow regime depends on the grain size of the bed over which the flow moves. Figure 12.12 illustrates the bed-form sequence for other grain sizes in the traction load. For coarse silt through fine sand, sand waves and dunes do not form. Instead the sequence of bed forms is (1) lower flow regime current ripples, (2) upper flow regime plane bed transition, and (3) antidunes. Current ripples do not form in beds composed of coarse sand or gravel. Instead the sequence of bed forms once traction movement is initiated is (1) lower flow regime flat bed, then (2) sand waves or dunes, followed by (3) upper flow regime plane bed, and (4) antidunes.

The importance of the bed forms associated with different flow regimes becomes more apparent when they are related to sediment deposition from traction. Net deposition requires that net bed aggradation (buildup) occurs as sediment is transferred from the saltation and/or suspension loads to the traction load over time. Once this important concept is understood, it is much easier to understand the relationship between bed-form migration and the sedimentary structures that form during periods of net deposition. These structures have the potential for long-term preservation and frequently record the flow regime conditions under which they formed.

Net aggradation of sediment on a flat or plane bed produces **horizontal stratification**. This occurs on upper flow regime plane beds for all sizes of sand and gravel, with minimum depositional velocities increasing with particle size as expected. It also occurs on lower flow regime flat beds for coarse sand and larger sediment (Figure 12.12). Nonetheless, as we shall see, much horizontal stratification forms by other means.

Depending on the rate of bed-form migration and bed aggradation, sediment accumulation on a lower flow regime **current rippled bed** can generate (1) **current ripples**, (2) **horizontal laminations**, (3) **small-scale cross-lamination**, or (4) **climbing ripple laminations**. One key to understanding how such a variety of structures can form from a current rippled bed is to understand how current ripples migrate. During flow, multiple current ripples (ripple

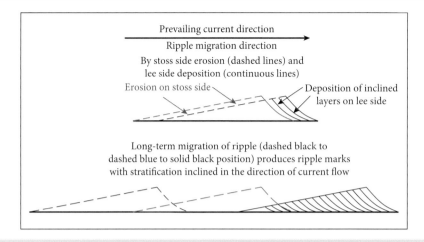

**Figure 12.13** Ripple migration by stoss-side erosion and lee-side deposition to form internal cross-laminations.

trains) migrate down-current by erosion of sediment from their up-current or stoss sides and deposition on their down-current or lee sides (Figure 12.13). Very coarse silt to medium sand is entrained on the stoss side, migrates to the ripple crest and then slides down the steep lee side to form an inclined sand layer. Repetition of this process over time causes the entire ripple form to migrate down-current with the formation of small-scale internal cross-laminations that dip in a down-current direction (Figure 12.13).

As ripple trains migrate, entire ripple forms move along the bed including crests, but also troughs. Careful examination of Figure 12.13 will persuade one that the area of maximum sediment accumulation beneath the initial crest (dashed outline to left) has become an area of no sediment accumulation under a trough (dashed outline to right). What has happened is that erosion has progressively removed sand from the stoss side, permitting the trough to migrate down-current. The trough may be visualized as an erosion surface that removes all the sand deposited by the ripple that preceded it down-current. If no sediment is added to the bed during ripple migration, no net deposition occurs because troughs erode the sand deposited by current ripples that migrate across it. Only the ripple forms of a ripple set will be preserved. Now let us add bed aggradation or "fallout from above" to the mix. If sand is being added to the bedded to the bed over time, the elevation of all parts of the ripple – including the ripple trough – will increase. As troughs migrate down-current, their elevation will increase slightly. As a result, troughs climb slightly and

do not erode all sediment deposited during the migration of the previous ripple. At very slow climb rates that result from very slow bed aggradation relative to ripple migration rates, layers a few grains thick will be preserved with the passage of each ripple trough. The passage of multiple ripple troughs will generate a sequence of nearly horizontal laminations, each bounded by trough erosion surfaces. These laminations cannot generally be distinguished from horizontal laminations generated on flat or plane beds. If, however, the aggradation and climb rates are slightly higher, but less than the slope of the stoss side, layers thick enough to preserve recognizable cross-laminations will be preserved (Figure 12.14a). The passage of multiple ripple trains will be preserved as multiple centimeter-thick sets of ripple-scale cross-laminations that dip in the direction of current flow, even though current ripple forms are not preserved. Each set is bounded by trough erosion surfaces with such gentle inclinations that the sets are nearly horizontal.

Under conditions of rapid fallout on the aggrading bed relative to ripple migration rates, troughs will climb at angles equal to or exceeding the stoss-side slope. In the former case, the entire ripple is preserved as sequences of lee-side cross-laminations separated by steep erosion surfaces produced as troughs migrated up the stoss side at the stoss-side angle (Figures 12.14b and 12.15). Such features are called **type I climbing ripple laminations**. With extreme rates of fallout from above, even the stoss side experiences net aggradation so that troughs climb at angles

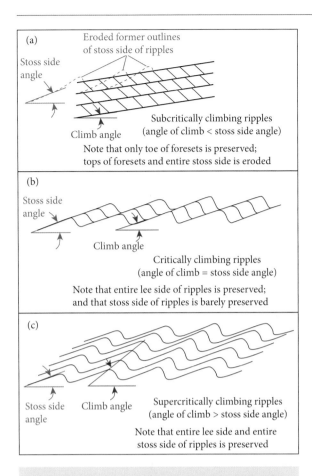

**Figure 12.14** Progressive increases in trough climb rates due to bed aggradation by fallout from above. These generate: (a) cosets of small-scale ripple laminations; (b) type I climbing ripple laminations; (c) type II climbing ripple laminations. *Source*: Courtesy of Marc Hendrix.

**Figure 12.15** Climbing ripple laminations produced by down-current (left to right) migration of ripple trains under rapid fallout and bed aggradation in the Oligocene Sacate Formation, California. Type I with no stoss-side lamination preserved can be seen to the right of the blending stub and type II with stoss-side lamination preserved is above the load-casted sand layer. *Source*: John O'Brien.

higher than the stoss-side slope angle. This produces **type II climbing ripple laminations** that preserve both stoss-side and lee-side laminations and entire ripple forms that climbed in a down-current direction (Figures 12.14c and 12.15). To summarize, as bed aggradation rates increase from zero to extreme fallout from above during the migration of ripple trains, the sequence of preserved structures evolves from (1) isolated ripple marks, through (2) horizontal laminations, (3) small-scale cross-laminations, (4) type I climbing ripple laminations, to (5) type II climbing ripple laminations. For very useful sets of visual aids, readers should refer to the many video images that relate bed-form migration and sedimentary structures (Rubin and Carter 2006; http://walrus.wr.usgs.gov/seds/).

Because lower flow regime sand waves and dunes are larger features than ripples and possess longer lee-side faces, larger scale sets of cross-strata can form as they migrate down-current. Very small rates of bed aggradation generate very small trough climb rates, which can theoretically preserve thicknesses of a few grains from the preceding sand wave or dune as horizontal stratification. Slightly larger rates of aggradation and climb generate centimeter-thick sets of small-scale cross-laminations similar to, but often coarser than, those produced by current ripple migration at lower flow regimes. Their origin from dune or sand wave migration can be documented if the sediments in such small-scale cross-laminations are coarse sand as ripples do not form in such coarse sediments. Commonly, bed aggradation rates permit widely spaced troughs to climb at angles that preserve decimeter-thick portions of longer wavelength sand waves and dunes as **medium-scale cross-stratification**. The migration of unusually large sand waves and dunes combined with rapid bed aggradation can produce meter-thick **large-scale cross-stratification**.

Laboratory experiments (see Rubin and Carter 2006; http://walrus.wr.usgs.gov/seds/), supported by video images, have shown conclusively that the three-dimensional geometry of cross-stratification generated by ripple,

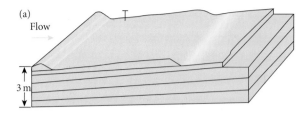

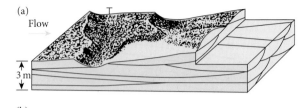

**Figure 12.16** Tabular sets of planar cross-strata. (a) Formation by aggrading, straight-crested ripples or sand waves with stoss erosion at small climb angles. *Source*: Courtesy of the US Geological Survey. (b) Single tabular set of planar cross-strata records the migration of a straight-crested bed form from left to right in the Sinian Sandstone, northwest China. *Source*: Courtesy of Marc Hendrix.

**Figure 12.17** (a) Wedge sets and trough sets of festoon cross-strata formed by aggrading curved-crested dunes with stoss erosion at moderate climb angles. (b) Trough sets of festoon cross-strata in profile at a high angle to composite dune migration. *Source*: Courtesy of the US Geological Survey. © U.S. Department of the Interior.

sand wave, and dune migration depends on the three-dimensional geometry of the migrating and evolving bed forms. Steady-state migration of straight-crested ripples and sand waves can produce **tabular sets of planar cross-strata** (Figure 12.16). Note that the tabular–planar nature of these cross-strata is most apparent in sections sub-parallel to the current direction and that the strata appear to be horizontal in sections perpendicular to flow where their apparent dip is 0°. In plan view, the traces of cross-strata are fairly straight, reflecting fairly straight sand wave crests that yield cross-strata with consistent strike.

The migration of curved-crested ripples and dunes can produce **wedge sets of planar cross-strata** when viewed in sections parallel to flow (Figure 12.17). In sections at high angles to flow, cross-strata occur in **trough sets of festoon cross-laminations**. Trough sets result from the erosion of preceding dunes by offset "scoop-shaped" troughs during bed

aggradation. Festoon cross-strata record the curvature of lee faces in cross-sections perpendicular to flow. Cross-strata traces are also curved in plan view, which reflects the curvature of dune crests. Flow direction that produces festoon cross-strata is roughly perpendicular to the place of maximum curvature in plan view (Figure 12.17). Although multiple sets of small- to medium-scale cross-strata can form by processes other than bed-form migration, most probably form by mechanisms similar to those outlined here, albeit in complicated ways for complex bed forms.

**Plane or flat bed aggradation** during a decreasing flow regime leads to the formation of **horizontal strata** (Figure 12.18b, c). Upper flow regime horizontal strata range from very fine to very coarse sand or gravel and commonly exhibit parallel orientation of elongate sand particles parallel to flow (Figure 12.18a). These are, in some cases, preserved on stratification surfaces as **parting lineations** parallel to flow direction. Parting lineation documents

**Figure 12.18** Plane bed transition. (a) View of a plane bed under shallow unidirectional flow, with a lineation on the bed. (b) Diagram showing the horizontal strata and parting lineation developed in plane bed transition. (c) Horizontal strata in sandstone produced by plane bed flow regimes. (d) Parting lineation on a sandstone bedding surface. *Source*: (a, b, and d) Courtesy of Marc Hendrix; (c) Courtesy of Duncan Heron.

formation under upper flow regime plane bed transition conditions (Figure 12.18b, d). Aggradation of coarse sand and gravel on lower flat bed surfaces produces horizontal strata that lack parting lineation. Coarse sand and gravel deposits are dominated by sometimes crude horizontal stratification.

The higher portions of the upper flow regime, under which antidunes, chutes, and pools form, are characterized more by erosion than by deposition. However, very high flow regime flows do decelerate, and under such conditions deposition is likely to occur. What kinds of stratification are produced by deposition from antidunes and chutes and pools? Field and laboratory studies show that antidunes begin as in phase stationary waves (Figure 12.19a) that migrate up-current due

to stoss-side deposition from currents that decelerate and lee-side erosion from currents that accelerate over the bed form (Figure 12.19b). Eventually antidunes shear out (Figure 12.19c) to form chutes and pools (Figure 12.19d). Figure 12.19e shows the stratification that developed in a laboratory where repeated antidune formation occurred on an aggrading bed. The resulting strata occur as bundles of low angle cross-strata separated by curved erosion surfaces. Some cross-strata are convex upward due to partial preservation of antidunes and/or chutes; others are convex downward, perhaps due to deposition in pools. These cross-strata very closely resemble the hummocky and swaley cross-strata produced by oscillatory flows, which are discussed in "Oscillatory flow."

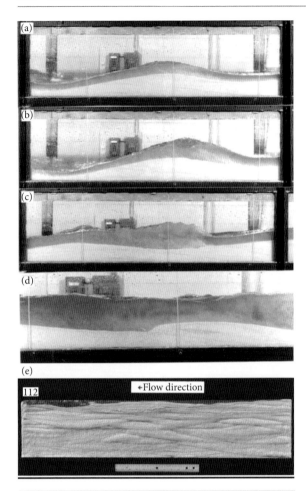

**Figure 12.19** Upper flow regime and anti-dunes. (a–d) Antidunes in laboratory flume experiments: (a) in-phase antidunes (b) migrate up the flow to the right and (c) the standing wave breaks with the development of (d) chutes and pools. (e) Antidune cross-stratification generated in a laboratory. *Source*: Photos courtesy of John Bridge.

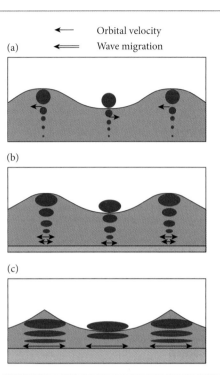

**Figure 12.20** (a) Deep water waves showing roughly circular orbitals whose diameters decrease systematically to approach zero at the wave base (L/2). (b) Transitional waves (L/2 > depth > L/20) showing progressive flattening of the orbitals as the waves interact with the bed. (c) True shallow water waves (depth < L/20) with flattened elliptical orbitals that produce oscillatory flow on the bed.

## Oscillatory flow

**Oscillatory flows** are characterized by back-and-forth flow in opposite directions over time and are generated by shallow water waves. Most wind-driven progressive waves that reach near shore environments are generated by storm winds blowing over deep water. Such waves are dispersed out of the storm area, and can then travel long distances (>10 000 km) through the body of water in which they formed. As long as they are moving through water deeper than one-half the distance between wave crests, which is called

the **wave length (L)**, they are **deep water waves.** As deep water waves move through water, water molecules inscribe roughly circular orbits in which water moves in the direction of wave motion as the crest approaches, then downward and backward as the trough approaches, then upward and forward to approach its original position as the next crest approaches, so that there is one orbit for each wave period (Figure 12.20). Orbital diameters are equal to the wave height and gradually decrease with depth until at a depth of L/2 they approach zero. Below this depth, called the **wave base,** the waves have negligible heights and essentially do not exist. The wave base (L/2) is exceedingly important in near shore environments. In water deeper than the wave base, sediments are not entrained by waves because waves do not interact with the bottom. In shallower water, waves can move some or all of the available sediment.

The position of the wave base varies with wavelength. Unusually large storm-produced waves, generated by very strong winds, begin to interact with the bottom in much deeper water than do smaller waves with shorter wavelengths. It is useful to divide shallow water coastal areas into areas of (1) shallow water, above the normal wave base, where sand is entrained and fine sediments are continually suspended by wave action; (2) deeper water, between the normal wave base and storm wave base, where sand is entrained only during storms and is deposited, along with mud, between storm events; and (3) still deeper water, below storm wave base, where sand is not entrained and muds are more or less continually deposited from suspension.

Oscillatory flows are produced by shallow water, wind-driven waves, generally in near shore environments. Waves moving through water whose depth is less than the wave base (<L/2) are called **transitional** and/or **shallow water waves**. Such waves interact with and entrain sediment from the bottom as their orbitals become progressively flattened. True shallow water waves (depth <L/20) have orbitals so severely flattened that the water essentially moves forward as the crest approaches and backward as the trough approaches. Such back-and-forth movement of water is called oscillatory flow and causes entrained sediments to move back and forth as well. Those who have been to a beach might try to visualize the back-and-forth movement of water (backwash and surge) as wave trains approach the shoreline.

Flume experiments with oscillatory flow have demonstrated that a sequence of flow regimes and bed forms develops in response to increased oscillatory flow velocity on the bed. At very low velocity, no sand is entrained, so no traction load develops. As velocity is increased, initial entrainment velocities are reached and a traction load of very coarse silt and very fine sand is produced. Subsequent entrainment velocities on noncohesive beds are proportional to particle size. Very small relief bed forms called rolling grain ripples develop early, but these have not been observed in nature. These small asperities on the bed are quickly transformed into two-dimensional vortex ripples, better known as **oscillation**

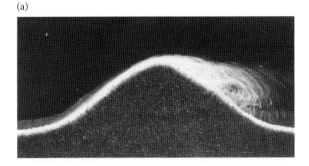

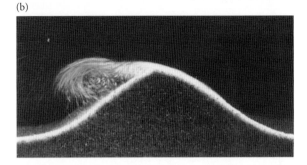

**Figure 12.21**   Oscillatory flow and sand movement (left in a, then right in b) about a symmetrical oscillation (two-dimensional vortex) ripple in a flume experiment. *Source*: Courtesy of Sergey Voropayev.

**ripples**. Oscillation ripples create forms with small ripple indices (length : height ratios) produced by the back-and-forth movement of sand across the bed and across the ripple forms (Figure 12.21). Oscillation ripples differ from current ripples in that they possess (1) fairly straight to sinuous, often branching, somewhat pointed crests, (2) symmetrical profiles (Figure 12.22), and (3) internal cross-strata that dip away from the crest in opposite directions, as sand is washed back and forth across the crest (Figure 12.23a).

Any flow process that causes one oscillatory flow component to be enhanced relative to the other so that components of both unidirectional and oscillatory flow occur produces what is called *combined flow*. Most oscillatory ripples display evidence of combined flow with one set of cross-strata more pronounced than the other. They also possess bases that are less even than those of most current ripples produced by unidirectional flow (Figure 12.23b). Two-dimensional vortex ripple heights increase with increasing oscillatory flow velocities that occur when

(a)

(b)

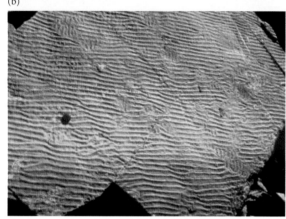

**Figure 12.22** (a) Oscillation ripple marks showing crest bifurcation, symmetry, short wavelengths and small ripple indices; modern tidal flat, Cape Cod Massachusetts. *Source:* John O'Brien. (b) Preserved in sandstone. *Source*: Photo courtesy of Peter Adderley.

(a)

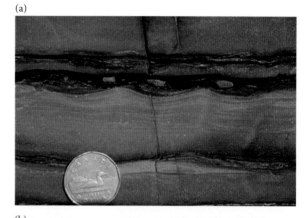

(b)

**Figure 12.23** (a) Oscillation ripples in the Carboniferous Horton Group, Nova Scotia. *Source:* Photo courtesy of John Waldron. (b) Oscillation combined flow ripples in the Jurassic Entrada Sandstone, Utah. *Source:* Courtesy of Tim Cope.

larger waves with higher orbital velocities are transformed into shallow water waves.

Rippled sands deposited in the lower parts of the lower flow regime are commonly interstratified with muds deposited from suspension under calm flow conditions. This is especially common on tidal flats where combinations of bimodal, unidirectional tidal currents and oscillatory waves produce current, oscillatory and combined flow rippled sands, and periods of slack flow allow muds to settle from suspension. Three major types of intimately interstratified sand and mud deposits, each with different sand : mud ratios, are recognized in such sequences. **Wavy stratification** is characterized by subequal amounts of relatively continuous ripple-laminated sand and mud layers with the rippled sand giving the mud layers a wavy appearance (Figure 12.24a). **Flaser stratification** is produced in less protected areas with high sand : mud ratios. Flaser are thin mud drapes deposited from suspension that partially preserve ripple forms in areas dominated by the deposition of ripple-laminated sand (Figure 12.24b). **Lenticular stratification** is produced in protected areas with low sand : mud ratios and contains isolated lenses of ripple-laminated sand deposited from traction encased in dominant muds deposited from suspension (Figure 12.24c).

What happens when oscillatory flow regimes increase, as when storm waves approach the coast? At still higher velocities, vortex ripples assume three-dimensional forms in which

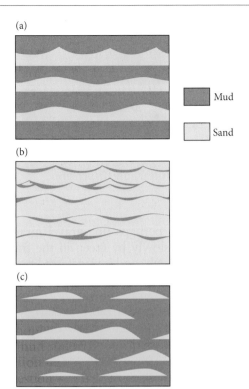

**Figure 12.24** Interlayered ripple-laminated sandstone and mudstone from tidal flat sediments, Pennsylvanian, Kentucky: (a) wavy strata; (b) flaser strata with a mud-drape flaser preserved in ripple-laminated sand; (c) lenticular strata with isolated ripples preserved in mud. *Source*: Friedman and Sanders (1978). © John Wiley & Sons.

small, elevated hummocks alternate with depressions called swales. The sequence from rolling grain ripples to two-dimensional oscillation ripples to hummocks and swales is considered to represent the lower flow regime. At higher flow regimes, the hummocks and swales are replaced by a nearly flat plane bed.

Hummocky–swaley surfaces can be seen at low tide in very shallow water but also appear to be associated with storm waves in somewhat deeper water in the geological record. Deposition on hummocks produces **hummocky cross-stratification (HCS)**, and deposition in swales generates **swaley cross-stratification (SCS)**. The two sometimes occur together in sets less than 0.5 m thick. HCS is characterized by gently inclined, convex up cross-strata, whereas SCS is characterized by gently inclined, convex down cross strata (Figure 12.25). In older rocks, these types of cross-strata generally occur in relatively fine sandstones interstratified with marine fossil-bearing mudrocks that were deposited from suspension. This suggests that they form chiefly on bottoms below the normal wave base but above the storm wave base.

### 12.4.2 Wind (eolian) and wind-deposited sediments

Winds are set in motion by a combination of differential heating, which creates air masses of different density, and the acceleration of gravity, which causes denser air masses to displace lighter ones. Because air possesses a much lower viscosity (resistance to flow) than water, it can attain much higher velocities and is generally characterized by turbulent flow. Because of its lower density, wind generally exerts smaller shear stresses on the bottom and therefore tends to entrain and transport smaller particles. In the range of normal wind speeds (<80 km/h), winds do not have the ability to entrain even small pebbles.

The erosion, transportation, and deposition of sediment by winds resemble those of unidirectional water flows in several regards (Figure 12.26). Larger grains are transported in a traction load by the process called **creep**, which is maintained by grain to grain collisions.

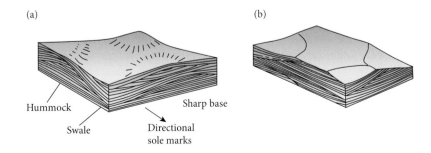

**Figure 12.25** (a) Block diagram showing hummocky cross-stratification (HCS). (b) Block diagram showing swaley cross-stratification (SCS). *Source*: From Tucker (2001). © John Wiley & Sons.

Wind direction

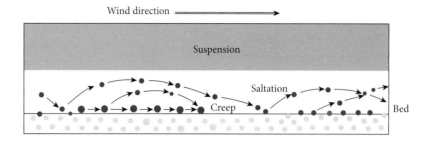

**Figure 12.26** Major modes of sediment transport by winds: creep, saltation, and suspension.

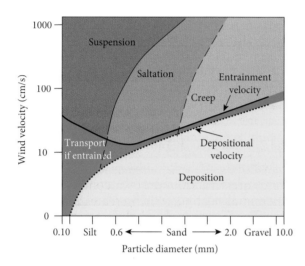

**Figure 12.27** Velocity conditions for wind erosion, transportation by suspension, saltation, and creep and deposition of various sediment sizes. *Source*: From Tucker (2001). © John Wiley & Sons.

Somewhat smaller grains are transported in intermittent suspension by saltation. During major wind storms the saltation load moves in part by **sheet flow** in which many saltating sand grains collide in mid-air, before reaching the bed, and stay suspended for longer than normal times. Large amounts of sand can be moved during such sandstorms. The finest grains are carried above the bed load in longer term suspension.

There is a critical threshold velocity where entrainment of very fine sand begins (Figure 12.27). Progressive increases in velocity entrain progressively coarser and finer (more cohesive) particles as their critical entrainment velocities are reached. As with water, grains finer than coarse silt are transported in the suspension load with fine silt and clay carried in long-term suspension. Coarse silt to medium sand is transported primarily by saltation, although it may be suspended at very high wind velocities. Fine to

coarse sand and perhaps fine gravel (granules) are transported slowly by creep in the traction load, although much sand is transferred from the creep load to the saltation load with increases in velocity. Coarser gravels are not generally entrained by wind. Even more effectively than water, wind sorts material by size during transportation. This sorting is retained during deposition so that eolian sediments are generally very well sorted. In addition, sand grains tend to undergo significant rounding during eolian transport because they are large enough to collide with considerable force, especially during saltation, and the low viscosity of air cushions such impacts less than those that occur in higher viscosity water. As a result, eolian sands tend to be well rounded and eolian lag (left behind) gravels often show evidence of significant abrasion by "sand-blasting."

As wind velocity decreases below the critical velocity for deposition (Figure 12.27), any transported fine gravel and coarse sand is deposited first, from the creep load, followed by the deposition of progressively finer sand, then coarse silt. As with water, very low wind velocities are required for fine silt and clay to settle from suspension. Think of the fine material suspended in a ray of sunlight in a calm room that slowly settles onto the top of your dresser or wardrobe over long periods of time. Where prevailing winds blow across a sediment source area over time, the suspension load is transported longer and farther than the saltation load, which is transported longer and farther than the traction load. Over time, repeated entrainment, transportation, and deposition produces a downwind sequence from (1) coarse lag gravels where the finer materials have been removed by wind erosion, a process called **deflation**; to (2) well-sorted sand deposits such as dune fields in which sand has been deposited from the traction load; and finally (3) well-sorted mud deposits formed as

**Figure 12.28** Typical loess deposit; note the paucity of stratification in comparison to that displayed by wind-blown sands. *Source*: Courtesy of the USGS. © U.S. Department of the Interior

**Figure 12.29** Wind ripples on back-beach sand dunes, Australia, with branching crests, strong asymmetry, and large ripple indices; wind from left to right. *Source*: Photo by John O'Brien.

muds settle from suspension. The extensive, poorly stratified deposits of *loess* produced by winds that entrained, transported, and deposited abundant silt from glacial deposits during the last ice age are an excellent example of the latter (Figure 12.28). Loess soils are very fertile and support significant grain production in the Great Plains of North America, Europe and Asia.

Wind blowing over noncohesive, coarse silt, and sand generates various bed forms that depend on flow velocities and the attendant shear stresses on the surface. At relatively low velocities, **wind ripples** develop in coarse silt and sand. Wind ripples are current ripples that superficially resemble aqueous unidirectional current ripples but differ from them in a number of ways. Wind ripples tend to (1) have lower heights relative to wavelengths and thus larger ripple indices (RI > 12); (2) be strongly asymmetrical, and (3) unlike continuous-crested subaqueous current ripples, show branching crests (Figure 12.29). Nothing analogous to lower flow regime dunes develops on such surfaces. Instead, at higher velocities, roughly three times the critical entrainment velocity, the ripples disappear, and the surface is transformed into a flat or plane bed.

However, much larger bed forms are commonly produced by winds in areas with sufficient sand supply. **Sand dunes** are very complex features with wavelengths of tens to hundreds of meters and heights of decimeters to tens of meters that occur in a variety of forms that are influenced by sand supply, vegetative cover, and long-term wind conditions. Figure 12.30 shows the major types of sand dune and their relationship to the prevailing wind directions.

Relatively straight-crested **transverse dunes** and curved-crested **barchan and parabolic dunes** have well-defined upwind (stoss) sides with rather gentle slopes and downwind (lee) sides with steep (20°–35°) slopes. The upwind sides experience net erosion as sand is entrained and transported in migrating wind ripples toward the dune crest. Sand that accumulates at the crest becomes unstable and slides down the steep downwind side of the dune, called the avalanche face, in the form of lobate **sand flows** whose dip decreases toward the base of the avalanche face (Figure 12.31a, b). Repeated sand flows generate multiple steeply inclined (20°–35°) cross-strata formed by lee-side deposition on and near the base of the avalanche face. As with subaqueous dunes, erosion of sand from the upwind (stoss) side and deposition on the downwind (lee) side causes wind-blown dunes to migrate downflow over time. Net aggradation of the bed, essential to long-term preservation and migration of multiple dunes, produces multiple sets of cross-strata separated by erosional surfaces that record the passage of interdune troughs. Other erosional surfaces in dune sediments record the position of the water table; wet cohesive sand resists erosion while the

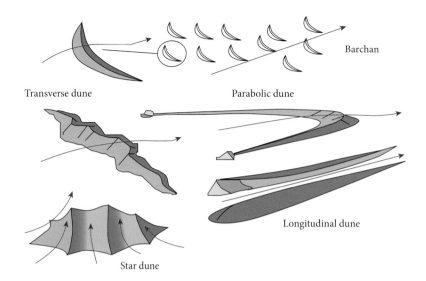

**Figure 12.30** Major types of sand dunes: transverse, barchan, parabolic, star, and longitudinal, and their relationship to wind directions. *Source*: Courtesy of Steve Dutch.

**Figure 12.31** Formation of eolian cross-strata by dune migration. (a) Dune showing stoss-side erosion and lee-side deposition by sand flows. (b) Lobate sand flows on a dune avalanche face generate large-scale, steeply inclined cross-strata. (c) Large-scale, steeply inclined, tangential cross-strata in the Jurassic Navajo Sandstone, Utah, formed by eolian dune migration with a left-to-right component. *Source*: John O'Brien.

overlying dry cohesive sand is removed. Relatively straight-crested transverse dunes migrate to produce tabular sets of planar cross-strata, whereas the migration of curved-crested dunes generates wedge sets or trough sets of festoon cross-strata. Eolian cross-strata are more tangential to the underlying trough produced erosion surfaces than are most cross-strata generated by the migration of subaqueous dunes and so tend to be more strongly concave upward in profile (Figure 12.31c).

**Longitudinal and star dunes** tend to display much more complicated bidirectional and multidirectional cross-beds. **Dras** are larger features ($2.0 \cdot 10^2$–$2.0 \cdot 10^4$ m) composed of multiple dune forms whose migrations give rise to very complex cross-stratification. Eolian sand deposits tend to be quartz rich, fine grained, and well rounded, possess excellent sorting, and contain multiple sets of large-scale cross-stratification.

### 12.4.3  Glaciers and glacial sediments

Glaciers are significant agents of sediment erosion, transportation, and deposition, especially during periods of global cooling that produce so-called ice-house conditions. At the present time, glaciers cover less than 10% of Earth's land surface and that percentage is shrinking. However, as recently as 18 000 years ago they covered nearly 30% (Paterson 1999). Glaciers form when net snow accumulation exceeds losses by ablation (melting and sublimation) over extensive periods of time. As the volume of snow increases, a combination of repeated thawing and freezing, along with increased burial pressure, slowly convert the snow into ice. As the ice volume continues to increase, the ice sheets thicken and begin to flow plastically in response to gravitational forces. Valley glaciers in mountainous areas flow down valleys. Larger continental glaciers, covering large areas ($10^5$–$10^6$ km$^2$) with thicknesses up to several kilometers, flow radially outward from their area of maximum thickness under the tangential force of gravity, in a manner roughly analogous to honey or molasses flowing away from their areas of maximum thickness. Topography still influences local flow directions. As long as the volume of glacial ice expands, the margins of the glacier tend to advance. Only when the ablation exceeds accumulation, so that glacial ice volumes contract, do glacial margins tend to retreat from areas formerly covered with ice. As these words are being written, this is happening in many parts of the world (Box 12.1).

## Box 12.1  Global warming, glaciers, and sea-level rise

One likely consequence of global warming is the global rise of sea level. This would result from the thermal expansion of sea water and the shrinkage of glacial ice volumes, especially the large ice sheets in Greenland and Antarctica. At the end of the twentieth century, the two factors appeared to be approximately equal contributors to observed sea-level rise. In the two decades since, that ratio has changed significantly. Now glacial melting appears to contribute more than twice the amount of sea-level rise caused by thermal expansion, and that change is accelerating. What has caused these changes? What does the future hold? Much attention has been given to phenomena in the Northern Hemisphere that affect global warming and sea-level rise. These include (1) accelerating melting (retreat and thinning) of the Greenland Ice Sheet (Aschwanden et al. 2019), (2) rapid decreases in the area and annual duration of sea ice cover in the Arctic Ocean (Stroeve et al. 2012), and (3) the accelerating decrease in the area and duration of snow cover in northern North America and Eurasia. What is causing these phenomena? Part of the problem is the development of a positive feedback loop between global warming and the reduction of glacial ice, sea ice cover and snow cover that it causes. Ice and snow reflect large amounts of incoming solar radiation back into the atmosphere. Areas covered by such reflective materials are said to have a high **albedo**. When such areas are converted into areas covered by soil, rocks, plants and, to a lesser extent, sea water, their albedo suddenly decreases. Instead of reflecting most of the incoming solar radiation, they absorb, and then reradiate it. This

*Continued*

**Box 12.1** *Continued*

causes surface temperatures and the temperature of the lower atmosphere to increase which in turn causes more melting and less albedo. This, in turn, leads to further melting and decreases in albedo, so that a sort of chain reaction occurs, a positive feedback loop between temperature increase and albedo decrease that adds to global warming and sea-level rise. In addition, melting of unprotected permafrost releases large amounts of methane gas which contributes to further global warming. If the Greenland Ice Sheet (GIS) were to melt completely, sea level would rise by 6–7 m (Aschwanden et al. 2019). If carbon emissions are not reduced, the authors speculate that such complete melting is likely to occur within the next millennium.

But the "elephant in the room" is not in the Northern Hemisphere. More than ten times more glacial ice is stored in the Antarctic Ice Sheet (AIS) than in the rest of the world combined. Naturally, attention has turned there, to the smaller West Antarctic Ice Sheet (WAIS) which has been retreating and thinning at an accelerating pace in this century and to the East Antarctic Ice Sheet because in holds more than 80% of the world's glacial ice (Figure B12.1).

Rignot et al. (2019) have documented net volume loss for the West Antarctic Ice Sheet since 1979 that has accelerated with each decade. Most of this loss occurred where warm salty sea water underlies sea ice shelves that are continuations of continental glaciers. The glaciers transition into sea ice

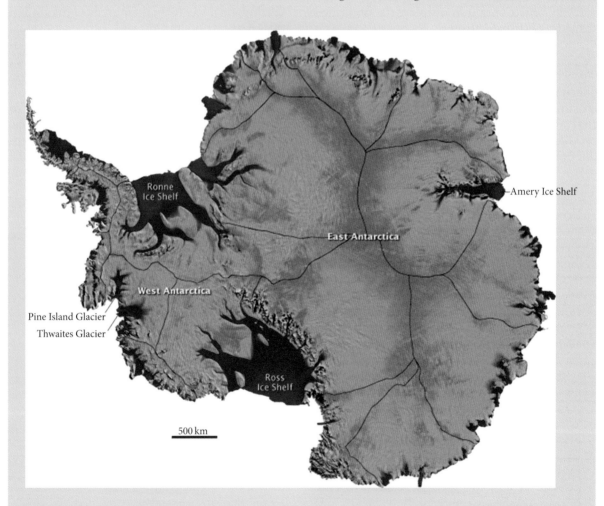

**Figure B12.1**   Map of West Antarctic Ice Sheet (WAIS) and East Antarctic Ice Sheets (EAIS). *Source*: Courtesy of NASA.

## Box 12.1  *Continued*

shelves at a **grounding line** that separates the portion of the glacier that is in contact with its base and the ice shelf portion that floats on sea water. These glacial ice sheets thicken landward, as does their weight. Because the crust is forced downward elastically in proportion glacial thickness and weight, many parts of these glaciers landward of the grounding line are well below current sea level. The thinner parts of these glaciers are underlain by more elevated land surfaces which act as a sill that separates the thickening landward portion of the glacier from the shrinking ice shelves. As the ice shelves shrink (Stroeve et al. 2012), their buttressing effect on the grounded portion of the glaciers is reduced which destabilizes the glaciers by permitting them to flow at a faster rate past the grounding line and into the sea. This sets up a feedback loop in which more sea ice melts, causing more glacial acceleration, which increases melting and so on. Furthermore, if sea-level rises to the elevation of the sill, sea water will rush in under the lower portions of the glacier and cause accelerated retreat or glacial collapse. How much of WAIS will melt in the near and intermediate future?

Garbe et al. (2020) suggest that WAIS is committed to a partial collapse sufficient to raise sea level 1.3 m/°C temperature rise relative to preindustrial temperatures. At global warming of 2 °C above preindustrial temperatures, very likely to occur before the end of this century, melting of WAIS alone would raise sea level by 2.6 m (8.7 ft). Moreover they have found that the glaciers would not regrow to their preindustrial extent unless the temperature decreased to 1 °C below preindustrial levels. This phenomenon is called **hysteresis** and occurs when the return to an original state of a system requires more than a simple reversal of the original conditions that produced the change. Of course the projected rises in sea level caused by the melting of WAIS would be augmented by GIS melting so that sea-level rise of 3 m (10 ft) or more could occur by the end of the twenty-first century. The implications are stark, given that many smaller islands and large expanses of coastal areas would be submerged as a result.

Of course the bulk of Earth's glacial ice is stored in the gigantic East Antarctic Ice Sheet (EAIS) which is fortunately extensively grounded at relatively high elevations. But even EAIS has areas that are less well grounded. The Totten and Moscow Station Glaciers have bases below sea level (Morhajerani et al. 2018). Extensive melting of these parts could produce an additional 4–5 m of sea-level rise.

Can this really happen? It seems to have happened in the not too distant past. Rapid collapse of WAIS at 125 Ka produced 6–9 m (20–30 ft) of sea-level rise in <1000 years. The extent of glacial collapse was such that several areas in the adjacent ocean stopped receiving glacial sediments at that time, because the glaciers had retreated behind sills. In addition, 400 Ka, sea water reached moraines that are now 400 km (Harwood et al. 2020) from the ocean and wave cut terraces were cut at elevations now 10–13 m (33–44 ft) above sea level (Raymo 2012). What the future holds is uncertain, but these data ought to be cause for concern.

The details of glacial erosion, transportation, and deposition depend in part on whether the glacier is a cold, dry glacier or a warm, wet glacier. **Cold, dry glaciers** occur in very cold regions where ice remains frozen most of the year. Such glaciers are dominated by processes that involve flowing ice. **Warm, wet glaciers** occur in regions that partially thaw for extended periods so that glacial ice is melted, especially along its upper, lateral, and lower margins. Warm, wet glaciers are characterized by processes that involve a mix of glacial ice flow and unidirectional aqueous flow.

How are sediments entrained by glacial ice? Because glacial ice is a plastic solid, flow rates tend to be quite slow (<1 m/day). Even at these slow speeds, glaciers tend to entrain any loose material with which they are in contact. **Glacial entrainment** occurs primarily by some combination of three processes: (1) erosion of unconsolidated sediment; (2) plucking; and (3) abrasion of bedrock. When a glacier flows over an area it entrains unconsolidated material in its path, eventually eroding the surface down to the underlying bedrock. Anyone familiar with the thin to nonexistent soils that overlay bedrock in areas subject to recent

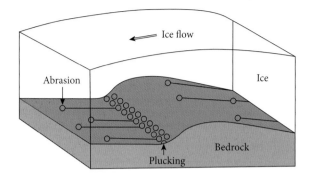

**Figure 12.32** Erosion of bedrock by glacial plucking and glacial abrasion.

**Figure 12.33** Striated bedrock surface in Cambrian dolostones, northwest New Jersey. *Source*: John O'Brien.

glacial erosion is familiar with the results of this process. As the glacier continues to flow, additional sediment is entrained from the underlying bedrock by plucking and/or abrasion. **Plucking** occurs when a glacier encounters an obstruction in an uneven bedrock surface, often where it is flowing downhill across bedrock (Figure 12.32). Pressure melting occurs; water infiltrates into joints and other cracks and then freezes and expands. Repeated freezing and expansion enlarges cracks to the place where rock fragments are dislodged from the bedrock and incorporated into the glacier. Plucking requires that glaciers be warm enough for some melting to occur. **Abrasion** occurs when rock materials carried in contact with bedrock under the pressure of the overlying ice mechanically wear away the underlying bedrock.

Medium-sized particles moving in contact with bedrock produce flow parallel scratches called **striations** (Figure 12.33a). Larger particles or clusters of particles produce larger **grooves**. Very small particles tend to **polish** the surface. Each of these processes removes particles from the underlying bedrock, which are then entrained within the glacier, including very small particles referred to as **rock flour**. Striated, polished bedrock surfaces, (Figure 12.33b) are a compelling indication of glaciation and important clues as to when, where, and to what extent glaciation has occurred in the past.

Continental glaciers entrain all sizes of sediment from huge areas, so that the entrained sedimentary particles that are eventually deposited tend to possess a great variety of compositions and sizes. These materials are carried below, within, and atop the glacier by slow, plastic (and sometimes aquatic) flow toward the terminus of the glacier. There is little tendency for sediments to be sorted by size during either transportation or deposition by glacial ice. As a result, most glacial deposits are characterized by poorly sorted debris with a wide variety of compositions. How is glacial debris deposited? Many glaciers contain a debris-rich basal layer. Some sediment is deposited beneath the glacier, often behind bedrock obstacles, by a process called **lodgment**, which occurs when the frictional resistance between the debris-rich basal layer and the bedrock is higher than that between the basal layer and the glacier that slides over it. Warm glaciers deposit sediment by local subglacial melting, which releases material from the basal layer. Even more glacial sediment is deposited at glacial margins as glaciers stagnate and/or begin to melt so that their entire load can no longer be carried. No matter which set of these depositional processes is involved, sediment deposited directly from glaciers is called **glacial till**. Glacial tills and lithified glacial tills called **tillites** tend to be very poorly sorted, boulder-bearing, polymictic (contain a wide variety of fragment compositions) and lack the well-defined internal stratification produced by agents such as wind and water that sort sediment by size during transportation (Figure 12.34a). When tills cover grooved and/or striated bedrock surfaces (Figure 12.34b), a glacial origin is, to coin a phrase, a "slam dunk."

(a)

(b)

**Figure 12.34**    (a) Glacial till, Pleistocene, Ohio; note the polymictic composition, poor sorting, and lack of stratification. *Source*: Photo courtesy of the US Geological Survey. (b) Glacial till overlying a striated bedrock surface, Switzerland. *Source*: Photo courtesy of Michael Hambrey, with the permission of www. glaciers-online.net.

**Figure 12.35**    Glacial varves from the Pleistocene, Maine: coarser, lighter colored, resistant summer layers alternate with finer, darker colored, crumbly winter layers. *Source*: Courtesy of Thomas K. Weedle, with the permission of the Maine Geological Survey.

Because ice eventually melts, especially in warm, wet glaciers, glacial sediments are intimately associated with water-laid sediments. In some cases, melt water simply modifies original glacial textures by removing many of the fines, leaving moderately–poorly sorted mixtures of gravel and/or sand. In other cases, unidirectional flow within the glacier, at the margins of the glacier or beyond the margins of the glacier deposits sediments whose characteristics are those of sediments deposited by unidirectional aqueous flow. These include sands and gravels deposited by (1) subglacial streams and preserved as **eskers**, (2) glacial margin streams preserved as **kame terraces**, and (3) braided stream channels in periglacial environments beyond the margins of the glacier and preserved as **outwash plain** deposits. Because of their enhanced porosity, these subglacial and periglacial water-laid sands and gravels act as important aquifers in many parts of the world; some older ones are significant oil reservoirs.

In addition to outwash plains, lakes commonly develop in periglacial basins near glacial margins or behind ice dams. Fine-grained deposits of clay and silt or fine sand are commonly deposited in the calm bottom waters of such lakes. Where such lakes freeze over in winter and thaw in summer, alternating dark, organic-rich winter layers and lighter, coarser silt/sand summer layers are common. Each couplet represents a year of sediment deposition. Such annual layers of sediment are called **varves** (Figure 12.35). Like tree rings, which they resemble to some extent, varves yield clues to climatic variations (e.g., summer–winter length) over time.

**Figure 12.36** Large glacial dropstone and other ice-rafted debris from the Pleistocene, Lake Spokane, Washington. *Source*: Courtesy of the US Geological Survey. © U.S. Department of the Interior.

Icebergs dislodged from the glacial margin float into adjacent lakes and marine settings carrying poorly sorted sediment. When icebergs melt, ice-rafted debris sinks to the bottom. Large clasts dropped from icebergs are called **dropstones**. Dropstones tend to disturb delicate laminations and their presence in environments so calm that muds settled from suspension is an excellent indicator of contemporaneous glaciation (Figure 12.36).

A detailed treatment of glacial sedimentary processes and features is beyond the scope of this text. For more detailed information, the reader is referred to Evans (2018), Hambrey and Alean (2016), Benn and Evans (2010), Bennet and Glasser (2009), Hambrey and others (2009), Hambrey and Alean (2004), and Mickelson and Attig (1999).

### 12.4.4 Mass (sediment gravity) flows and their deposits

In the previous sections we have discussed the entrainment, transportation, and deposition of sediment by fluid flows such as moving water and wind and plastic flows such as glaciers. Large amounts of sediment also are dispersed across Earth's surface by a fourth set of processes, variously referred to as **mass flows** and/or **sediment gravity flows**. These names reflect the fact that these flows involve relatively dense masses of sediment and/or rock material that move downslope under the tangential force of gravity, because they possess higher densities than the media that surround them and which they displace. There are many types of subaerial and subaqueous mass flows. Because of the abundance of their deposits in the geologic record, special attention will be paid to debris flows and turbidity currents.

Significant factors in most classification and nomenclature schemes developed for mass flows include the (1) the internal cohesiveness of the rock materials, (2) proportions of sediments, and ambient fluids such as water, (3) manner in which sediments are supported within the flow, and (4) maximum flow speed.

In **cohesive flows**, such as rock slides and slumps, movement occurs due to a loss of cohesion between the rock mass and its substrate. The rock mass largely maintains internal cohesion as it moves downslope. In all other mass flows, internal cohesion is not maintained as flow constituents separate and change relative positions during flow (Pierson and Costa 1987). **Granular flows** are characterized by large proportions of sedimentary particles (>80%) relative to water (<20%), though they may contain significant air. Granular flows include rapidly moving debris avalanches, slower grain and earth flows, and very slow creep. Grain-to-grain collisions and compressed air are significant in maintaining many granular flows. **Slurry flows** (Pierson and Costa 1987) are characterized by smaller amounts of sedimentary particles (~60–80%) and larger amounts of liquid (~20–40%). They are sometimes called **liquefied flows**. These include debris flows and mud flows with maximum velocities of 100 km/h, some high concentration turbidity currents and very slow flows such as those involved in solifluction. Liquids and suspended fines are important in maintaining slurry flows. **Hyperconcentrated flows** (Pierson and Costa 1987) possess lower sediment (~20–60%) to water (~40–80%) ratios and include hyperconcentrated stream flows and many, but not all, subaqueous turbidity currents. Most hyperconcentrated flows form when slurry and/or granular flows mix with and incorporate a significant amount of ambient water, which then plays a significant role in maintaining flow.

Mass flows are initiated by slope failure. They begin as masses of sediment and/or fluid with some internal cohesiveness or shear strength. In order for them to lose cohesion and be set in motion, their shear strength must be overcome. This requires that the tangential force of gravity, a shear stress, be larger than the yield stress required to overcome the internal shear strength and to cause loss of sediment cohesion. This can be accomplished by increasing the slope so that the tangential force of gravity increases or decreasing the internal shear strength of the sediment mass. Triggering mechanisms that cause slopes to oversteepen include (1) deposition, especially if rapid, (2) tectonic steepening that results from surface deformation, and (3) erosional undercutting. Mechanisms that cause loss of internal cohesion include (1) shaking by earthquake vibrations or storm waves, (2) increases in pore fluids by infiltration, (3) increases in pore fluid pressure by increased loading, and (4) liquefaction in which sediments are transformed into liquids, and which is related to the first three. Whatever the mechanism, once

the tangential force of gravity exceeds the yield strength of the material, the mass will begin to flow downhill. It will continue to do so until the tangential force of gravity decreases and/or the shear strength of the material increases, causing deposition to occur. Many mass flows that bury structures and cause loss of life are triggered by human activities including (1) oversteepening of slopes during construction projects, (2) loss of vegetative cover from construction or fire, (3) decrease in material strength from movement of Earth materials during construction, (4) increased pore fluids and/or pore fluid pressures due to water infiltration, and (5) loading from increased water use and building loads (Menzies 2002).

**Debris flows** and **mud flows** are types of slurry flows (Figure 12.37). Most geologists distinguish between the two based on a poorly defined ratio between mud and larger particles, with debris flows containing significant concentrations of larger particles, commonly

(a)  (b)

**Figure 12.37** (a) Mud flow with matrix strength sufficient to suspend boulders at the top in California. (b) Debris flow surge moving down a river valley, Mt Rainier National Park, Washington. *Source*: Courtesy of the US Geological Survey. © U.S. Department of the Interior.

boulders (Prothero and Schwab 2013). In the following discussion, we will use the term debris flow for both types of slurry flow. Lahars, detailed in Chapter 9, are debris flows with significant volumes of volcaniclastic sediment. Most debris flows form in subaerial environments, but they are well known from subaqueous environments as well. Because of their relatively low water content, debris flows have substantial yield strengths that result from frictional and/or electrical forces between grains. Most terrestrial debris flows are generated in unconsolidated material on steep slopes after periods of heavy rainfall. They are often are associated with a loss of vegetative cover that otherwise would have helped to anchor the material in place. Some form when rainwater infiltration decreases frictional forces between grains, thus lowering the shear strength of the material to where its yield strength is less than the tangential force of gravity on steep slopes. This is especially true in soils with smectite clays that expand by absorbing water (see Chapter 12). Other debris flows form when granular flows slide into rain-swollen river systems and mix with the ambient fluid to become slurry flows. Many debris flows are characterized by multiple surges (Figure 12.37b) caused by additional debris flows that enter the dispersal system (e.g., where tributary valleys enter larger ones) creating sudden increases in flow volume. Once formed, the sediment/water slurry flows as a well-mixed flow that retains surprising

cohesive strength. Sediment concentration is high and the flow so strong and "viscous" that boulders can be supported in suspension (Figure 12.37a). In many debris flows, large boulders are concentrated near the top of the flow so that during transportation, average particle size increases upward within the flow.

Debris flows begin to slow down when they reach a gentler slope. Small amounts of fluid may be expelled as grains begin to settle. These processes cause the shear strength of the debris flow to exceed the tangential force of gravity and flow ceases. This usually happens quite rapidly so that the flow is essentially frozen in place. The entire mass is deposited at once rather than grain by grain, and this **frozen bed** retains many of the characteristics the flow possessed during transportation (Figure 12.38). The debris is often reworked subsequently by normal aqueous flow processes which remove some of the finer components.

Most debris flow deposits display several characteristics useful in their identification, including (1) poor sorting of material by size, (2) crude or no internal stratification, (3) common inverse grading, (4) boulders and cobbles dispersed in and supported by a finer grained matrix, sometimes as diamictitites, and (5) many different types of gravel particles derived from different parts of a drainage basin as in polymictic conglomerates. Unlike glacial deposits, they do not occur over striated, polished bedrock surfaces. They also tend to be less polymictic than continental glacial

(a)  (b)

**Figure 12.38** (a) Debris flow deposit above an erosion surface, southern Utah. *Source*: Photo by Kevin Hefferan. © John Wiley & Sons. (b) Multiple debris flow deposits, Lone Pine, California *Source*: Photo by John O'Brien. © John Wiley & Sons.

deposits, as local drainage basin rocks may dominate the particle assemblage.

**Turbidity currents** (Natland and Kuenen 1951) form when masses of sediment begin to flow in subaqueous marine or lacustrine environments and mix with and incorporate ambient water. Turbidity current deposits are called **turbidites** and may possess any or all of the properties of slurry, hyperconcentrated and/or low concentration unidirectional flow deposits.

Turbidity currents are mostly generated by subaqueous slope failures that occur when the tangential force of gravity exceeds the shear strength of unconsolidated sediment that is usually water saturated. Triggering mechanisms include those discussed earlier that involve increasing the slope by rapid deposition, tectonic steepening and/or erosional undercutting, and/or decreasing the internal shear strength of the sediment mass by vibration, increased pore fluid pressure and/or liquefaction. Marine turbidity currents and turbidites are commonly generated on slopes. These include (1) continental slopes associated with passive margins, (2) steep slopes on the margins of pull-apart basins, (3) slopes in trench arc systems, and (4) slopes associated with forearc basins adjacent to orogenic belts. But any slope will do, and some are apparently generated on slopes of only a few degrees.

Once the sediment mass begins to move downslope, it mixes with the ambient water. The rheology of the flow strongly depends on the amount water added and/or sediment subtracted (e.g. by deposition); that is on the sediment : water ratio. If little water is added, the flow will possess the characteristics of a slurry (debris) flow. If the sediment : water ratio decreases, it will be transformed into a hyperconcentrated flow. If the ratio decreases still more, it will be transformed into a fluid flow with low sediment concentrations that resembles a unidirectional flow. Much has been learned from laboratory experiments in which liquefied sediment mixtures are released into a long tank called a flume. Figure 12.39 shows a lab-generated turbidity current.

A simplified model of turbidity currents can be used to illustrate how they evolve, recognizing that the details are both more complex than our discussion and not well understood. Repeated observations show that turbidity currents possess (1) a turbulent, multilobed region at the front of the flow, which is called the **head**, behind which is (2) a faster flowing **main body** that continually feeds sediment into the head and contains relatively high sediment concentrations, which grades back into (3) a **tail** that contains relatively low sediment concentrations (Figures 12.39 and 12.40). The turbulence developed in the head continually entrains water and/or sediment into the flow, and turbulent suspensions extend from the head back over the body of the flow (Figure 12.39).

Most erosion by turbidity currents occurs by scour concentrated in the turbulent head region. Repeated erosion by turbidity currents is responsible for the formation of most submarine canyons as well as many valley systems on continental slopes, delta slopes, trench slopes, and submarine fans. The sediment dispersion mechanism necessary to maintain flow in turbidity currents ranges from fluid turbulence in low concentration flows to cohesive matrix strength in very high concentration flows. Grain-to-grain collisions and pore pressure liquefaction may also be important. Deposition from turbidity currents is varied and complex. As one might expect, it tends to occur when flows reach gentler slopes and begin to decelerate.

Those portions of turbidity currents with high sediment concentrations are called **high concentration turbidity currents**. When they possess very elevated sediment concentrations (>60%), they behave as slurry (debris) flows. Such slurry flow turbidity currents deposit coarse-grained sandstones or conglomerates that display little internal stratification. Such deposits may contain matrix-supported gravel clasts and/or inverse grading. At lower sediment concentrations (~20–60%), high concentration turbidity currents behave as hyperconcentrated flows. Unlike slurry flows, hyperconcentrated flows lack sufficient matrix strength to suspend particles. Instead, particles are suspended primarily by fluid turbulence. The sediments in hyperconcentrated flows begin to separate by size, with larger particles moving closer to the bed and mud carried upward by fluid turbulence. Deposition occurs as individual particles accumulate on the bed, sometimes quite rapidly, as the flow decelerates. Continued deceleration generates graded beds in which grain sizes decreases upward. In hyperconcentrated flow deposits,

(a)

(b)

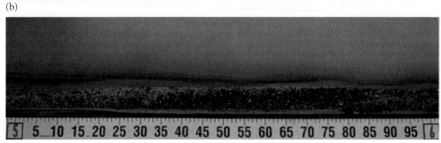

**Figure 12.39** Turbidity current in a laboratory showing the head and main body, and the turbulent suspensions that developed over both. *Source*: Sequeiros et al. (2010). Reproduced with the permission of ASCE.

gravel particles are in contact with one another, rather than matrix supported, and sandstones and conglomerates display a moderate degree of sorting. High concentration turbidity current slurry flow and hyperconcentrated flow deposits are the first sediments to be deposited from any turbidity current that contains them. As a result, such deposits are most abundant in **proximal turbidites** deposited closest to the sources of the turbidity currents near the base of prominent slopes and on the most active portions of submarine fan depositional systems.

Portions of turbidity currents with low sediment concentrations (<20%) are called **low concentration turbidity currents**. Such dilute turbidity currents behave in a manner analogous to that of unidirectional flows. The coarsest sediments, usually sand, are carried in a traction load, finer sand or coarse silt are carried in a saltation load, and finer silt and mud are transported in suspension. As such

flows decelerate, sand is deposited from the traction carpet. Plane bed transition horizontally laminated sands are commonly deposited first from the traction load of low concentrated turbidity currents. Further flow deceleration leads to the formation of current ripples in finer sand and/or coarse silt. As the flow wanes, silts and clays settle from suspension. The low concentration portions of turbidity currents may be deposited atop earlier high concentration turbidites. However, the lowest concentration turbidity currents, carrying only fine sand and suspended mud, frequently flow beyond the area where high concentration deposits form (Figure 12.40). As a result, the deposits of low concentration flows are most abundant in **distal turbidites** deposited farthest from the sources of turbidity currents. These deposits dominate turbidites some distance from the base of major slopes, on the less active and lower portions of submarine fans and on abyssal plains.

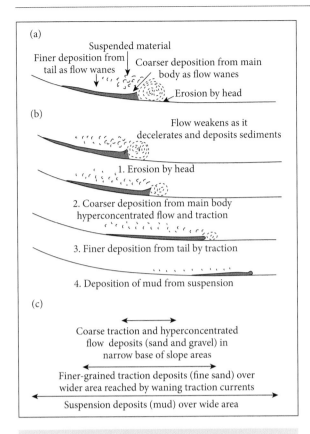

**Figure 12.40** (a) Model of a turbidity current with a head, main body, and tail. Erosion occurs primarily in the turbulent head; coarse material is deposited from the main body, finer material from the tail and very fine material from suspension. (b) Sequence of deposition during the passage of a turbidity current. (c) Distribution of sediments across an idealized basin.

A pivotal model for a classic turbidite, which contains most of the deposits discussed above, was proposed by Bouma (1962) and is now called a Bouma sequence. A complete **Bouma sequence** (Figure 12.41) consists of five major units, labeled A, B, C, D, and E, over an erosional base. Complete Bouma sequences begin with an erosional base produced by scour as the sediment-laden head and/or main body of the turbidity current passed over the bottom, often a cohesive mud bottom. Above the erosional surface is unit A, which consists of massive sandstone and/or conglomerate unit that lacks clear evidence of internal stratification. Although this unit is sometimes ascribed to deposition from the upper flow regime, it is more likely the result

of deposition from high concentration turbidity currents. If unit A is size-graded, moderately sorted, and contains any gravel clasts in contact, it is likely the product of rapid deposition from hyperconcentrated flows. Where the A unit is ungraded and/or contains matrix-supported clasts, it was likely produced from a slurry flow. Deposits from slurry flows were not included in Bouma's sequence but are common in models for high concentration turbidites (Lowe and Guy 2000). Unit B consists of sandstones with horizontal laminations that record traction load deposition by low concentration turbidity currents in the plane bed transition. Unit C, with its prominent current ripple cross-stratified finer grained sandstones, records lower flow regime deposition by low concentration turbidity currents. Unit D, which consists of weakly laminated siltstones, records small amounts of suspension deposition, perhaps by multiple waning turbidity currents. Unit E consists of very fine mudstones and is not the direct product of the turbidity currents. Instead, it is the product of normal pelagic deposition where tiny mud particles settle to the calm bottom waters of the deep sea floor. Complete Bouma sequences are rare because not all turbidity currents contain high concentration portions and because it is rare for a single turbidity current to deposit all the units at the same geographic location (see Figure 12.40).

Where erosion occurs on cohesive mud bottoms, subsequent deposition commonly fills the erosional features to produce sole marks on the sole or base of turbidite sand and/gravel beds (Figure 12.42). When turbulent eddies in the head impinge on firm mud bottoms, they produce erosional scours that are deep on the up-current side and become shallower down-current. When they are filled with sand, protuberances are produced on the sole of the turbidite bed that are large on the up-current side and become less pronounced in the down-current direction. Subsequent erosion of soft mudstones exposes these sole marks on the base of the turbidite bed; easier to see if the bed has been tilted during deformation. Such turbulent scour fill sole marks are called **flute marks** (Figure 12.42a). When stone or wood tools are dragged across cohesive mud bottoms, they produce grooves or gutters in the underlying mud that, when filled with sand, become **groove or gutter casts** on

(a)

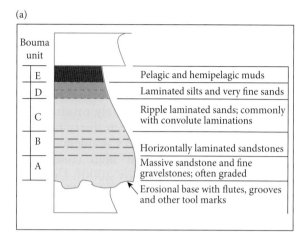

(b)

**Figure 12.41**   (a) Classic Bouma sequence showing an erosional base overlain by units A, B, C, D and E. (b) Partial Bouma sequence with a massive, graded A unit over an erosional base, horizontal laminated B unit, convolute laminated C unit and laminated D unit, from the Tertiary, Simi Hills, California. *Source*: John O'Brien.

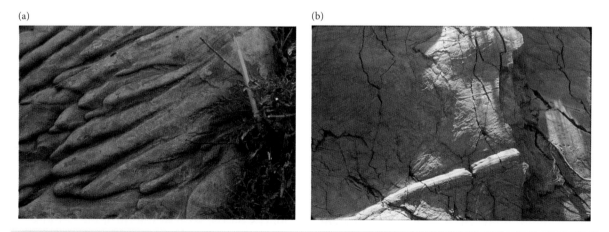

**Figure 12.42**   Sole marks on bed bases. (a) Flute marks, Austen Glen Formation, Ordovician, New York; current from lower left to upper right. (b) Groove casts, Cretaceous, Wheeler Gorge, California. *Source*: Photo by John O'Brien.

**Table 12.1**    Properties of sediments deposited by major depositional agents.

| Agent | Composition | Texture | Structures | Other |
|---|---|---|---|---|
| Water: unidirectional flow | Variable sandstone compositions | Well-sorted to moderately sorted sandstones; mudrocks; conglomerates | Asymmetrical current ripples; unidirectional cross-strata; horizontal strata common | Small ripple indices |
| Water: oscillatory flow | Variable sandstone compositions | Well-sorted to moderately sorted sandstones; mudrocks; conglomerates | Symmetrical oscillation ripples; with bidirectional cross-strata; hummocky and swaley cross-strata | Wavy, lenticular and/or flaser bedding common |
| Wind | Quartz-rich sandstones | Rounded, very well-sorted sandstones; mudrocks; negligible conglomerates | Eolian ripples; large-scale cross-stratification; tangential bases | Terrestrial flora and fauna |
| Glacier | Polymictic conglomerates with many clast compositions | Very poorly sorted, boulder-bearing, conglomerates; many matrix supported | Little or no internal stratification | Striated, polished bedrock surfaces; dropstones |
| Mass flow: mud–debris flow | Often polymictic conglomerates with many clast compositions, but less variable than glacial deposits | Poorly sorted, boulder-bearing, conglomerates; many matrix supported | Crude, if any, internal stratification; some inverse grading | |
| Mass flow: turbidity current | Sandstone, mudrocks and fine conglomerates; variable sandstone compositions | Moderate to moderately poor sorting; mud matrix common in sandstones | Bouma sequences common; massive sandstone; laminated sandstone; cross-laminated sandstone; interlayered sandstone and mudrock | Sharp, erosional base; sole marks common |

the sole of the turbidite (Figure. 12.42b). Many other sole marks are produced on the base of turbidite beds and, as might be expected, on the base of some sandstones produced by unidirectional flows. Discussion of these is beyond the scope of this text.

In this section we have focused on primary sedimentary structures that form at the time of sediment deposition and record significant information concerning the processes and conditions of deposition. Some **penecontemporaneous structures** that form shortly after sediment deposition and **secondary structures** that form well after deposition are discussed in the sections of Chapters 13 and 14 that deal with diagenesis. We have seen that each of the major agents of erosion, transportation, and deposition possesses a unique set of variable properties that impart different sets of erosional and depositional characteristics to the sediments and sedimentary rocks they produce. The major contrasts in compositions, textures, structures, and other properties of the sediments produced by each agent are summarized in Table 12.1.

Two major goals of this textbook are to help people understand how to interpret the record of Earth surface processes preserved in sedimentary rocks and how to apply the knowledge of sediments and sedimentary processes to the solution of resource needs and geohazard problems that involve such Earth materials. These will provide recurring themes in the chapters that follow.

## CONTENT ASSESSMENT

1  Sedimentary rocks occur in layers of various thicknesses called strata. What are the major factors that determine the thickness of any stratum and how do they do so?

2  What are the differences between laminar flows and turbulent flows? What factors help to determine whether a fluid flow is laminar or turbulent and how do they do so?

3   Using Hjulstrom's Diagram (Figure 12.6), please answer the following questions for a bed containing a range of particle sizes from $10^{-3}$ mm (0.001 mm) to $10^2$ mm (100 mm), remembering that the scales on the diagram are log-rhythmic).

   a   Imagine a flow whose velocity is increasing from 0 cm/s. At what velocity will it begin to entrain particles from the bed and what will their size be. Explain why this particle size is entrained rather than smaller or larger particles.

   b   If the flow velocity reaches 100 cm/s, what range of particle sizes will have been entrained? Speculate on which of these sizes will be carried in the suspension load, the saltation load and the traction load respectively.

   c   If this flow should begin to decelerate, at what velocity would it begin to deposit sedimentary particles and what would be the diameter of these particles?

   d   If the flow velocity decreased to 0.01 cm/s will any particles still be transported? If so, what will those particle sizes be?

4   What are the major criteria that permit Earth scientists to distinguish between sediments and sedimentary rocks deposited by (a) wind, (b) water, (c) glaciers, and (d) debris (slurry) flows? Please elaborate.

5   Compare and contrast *unidirectional* and *oscillatory* aqueous flows. What sedimentary structures are produced that allow the deposits of each type of flow to be recognized in sedimentary rocks? Of these, which are indicative of lower flow regimes and which are produced by upper flow regimes?

6   Using the classification of Pierson and Costa (1987), distinguish between the four major types of *sediment gravity (mass) flows*. Detail the processes that are involved in their initiation, maintenance and cessation.

7   Describe the six major elements in an ideal *Bouma sequence*. Explain how each of these features is produced over time as a result of a passing turbidity current in otherwise calm water. Then describe how and explain why the abundance of each element changes across a basin from proximal to distal turbidites.

8   Paleocurrent indicators record past directions of fluid flow. Using the information in this chapter, compile a list or table of such paleoflow indicators and how they are used to determine ancient flow patterns and, in many cases, regional slopes.

## REFERENCES

Aschwanden, A., Fahenstock, M.A., Khroulev, C. et al. (2019). Contribution of the Greenland ice sheet to sea level rise over the next millennium. *Science Advances* 5: 1096.

Benn, D. and Evans, D.A. (2010). *Glaciers and Glaciation*, 2e. London: Routledge 816 pp.

Bennet, M.B. and Glasser, N.F. (2009). *Glacial Geology: Ice Sheets and Landforms*, 2e. London: Wiley 400 pp.

Boggs, S. (2011). *Principles of Sedimentology and Stratigraphy*, 5e. Englewood Cliffs: Prentice-Hall Publishers.

Bouma, A.H. (1962). *Sedimentology of Some Flysch Deposits: a Graphic Approach to Facies Interpretation*. Amsterdam: Elsevier Publishers 168 pp.

Evans, D.J.A. (2018). *Till: A Glacial Process Sedimentology (The Cryospheric Science Series)*. Wiley-Blackwell 400 pp.

Friedman, G.M. and Sanders, J.E. (1978). *Principles of Sedimentology*. New York: Wiley 792 pp.

Garbe, J., Albrecht, T., Leuemann, A. et al. (2020). The hysteresis of the Antarctic Ice Sheet. *Nature 585*: 538–544.

Hambrey, M. and Alean, J. (2004). *Glaciers*. Cambridge, UK: Cambridge University Press 394 pp.

Hambrey, M.J. and Alean, J.C. (2016). *Colour Atlas of Glacial Phenomena*. London: CRC Press 426 pp.

Hambrey, M.T., Christopherson, P., Glasser, N.F., and Hubbard, B. (eds.) (2009). *Glacial Processes and Products*. London: Wiley-Blackwell 436 pp.

Harms, J.C., Southard, J.B., and Walker, R.G. (1982). *Structures and Sequences in Clastic Rocks*. Tulsa, OK: Society of Economic Paleontologists and Mineralogists 851 pp.

Harwood, D., Edwards, G., and Tulaczyk (2020). Biggest ice sheet on Earth more vulnerable to melting than previously thought. *National Geographic 7*.

Hjulstrom, F. (1939). Transportation of detritus by moving water. In: *Recent Marine Sediment* (ed. P.B. Trask), 5–31. Tulsa, OK: American Association of Petroleum Geologists.

Lowe, D.R. and Guy, M. (2000). Slurry-flow eposits in the Britannia Formation (Lower Cretaceous) North Sea: a new perspective on the turbidity current and debris flow problem. *Sedimentology 47*: 31–70.

Menzies, J. (2002). *Modern and Past Glacial Environments (revised student edition)*. Amsterdam: Butterworth-Heinemann Publishers 352 pp.

Mickelson, D.M. and Attig, J.W. (eds.) (1999). *Glacial Processes Past and Present*, Special Publication, vol. 337. Denver, CO: Geological Society of America 200 pp.

Mohajerani, Y., Velicogna, I., and Rignot, E. (2018). Mass loss of Totten and Moscow Station glaciers, East Antarctica, using regionally optimized GRACE mascons. *Geophysical Research Letters 45*: 7010–7018.

Natland, M.L. and Kuenen, P.H. (1951). *Sedimentary History of the Ventura Basin, California, and the Action of Turbidity Currents*, Special Publication, vol. 2, 76–104. Society for Economic Paleontologists and Mineralogists.

Nicols, G. (2009). *Sedimentology and Stratigraphy*, 2e. Oxford: Wiley-Blackwell 432 pp.

Paterson, W.S.B. (1999). *Physics of Glaciers*, 3e. Amsterdam: Butterworth-Heinemann Publishers 496 pp.

Pierson, T.C. and Costa, J.E. (1987). A rheological classification of subaerial sediment-water flows. In: *Debris Flows/Avalanches: Processes, Recognition and Mitigation*, vol. 7 (eds. J.E. Costa and G.F. Wieczorek), 1–12. Geological Society of America Reviews in Engineering.

Prothero, D.R. and Schwab, F. (2013). *Sedimentary Geology: An Introduction to Sedimentary Rocks and Stratigraphy*, 3e. New York: W.H. Freeman 557 pp.

Raymo, M. (2012). Collapse of polar ice sheets during the stage eleven interglacial. *Nature 483*: 453–456.

Reading, H.G. (1996). *Sedimentary Environments: Processes, Facies and Stratigraphy*, 3e. Oxford, UK: Blackwell Science 704 pp.

Rignot, E., Mouginot, J., Scheuchl, B. et al. (2019). Four decades of a Antarctic ice sheet mass balance from 1979–2017. *Proceedings of the National Academy of Science 116*: 1095–1103.

Rubin, D.M. and Carter, C.L. (2006). *Bedforms and Cross-bedding in Animation*. Society for Economic Paleontologists and Mineralogists Society for Sedimentary Geology Atlas Series No. 2 (DVD).

Selley, R.C. (1988). *Ancient Sedimentary Environments*, 3e. New York: Springer-Verlag 336 pp.

Sequeiros, O.E., Spinewine, B., Beauboeuf, R.T., and Sun, T. (2010). Characteristics of velocity and excess density profiles of saline underflows and turbidity currents flowing over a mobile bed. *Journal of Hydraulic Engineering 136*: 412–433.

Stroeve, J.C., Serreze, M.C., Holland, M.M. et al. (2012). The Arctic's rapidly shrinking sea ice cover. *Climatic Change 110*: 1005–1027.

Tucker, M. (2001). *Sedimentary Petrology*, 3e. Oxford, UK: Blackwell Science 272 pp.

# Chapter 13

# Detrital sediments and sedimentary rocks

Sediments and sedimentary rocks cover more than 80% of Earth's surface. A large majority of these are **detrital sediments and rocks** composed of particles that are produced by and/or survive weathering to be eroded, transported, and eventually deposited on Earth's surface. The resource value of detrital sediments cannot be overstated. Detrital sediments provide substrates on which we grow our food and timber resources and conduct our construction projects. They also provide substrates for vegetative cover that provides habitats and food sources for many animals. In addition, detrital sediments and sedimentary rocks store valuable fluid resources that include oil, gas, and water. This chapter discusses detrital sediments and sedimentary rocks with a focus on their: (1) textures, (2) compositions, (3) classification, (4) provenance, (5) tectonic implications, (6) diagenesis, and (7) uses.

## 13.1   TEXTURES OF DETRITAL SEDIMENTS

The **texture** of a sediment or sedimentary rock refers to the size, shape, and arrangement of its constituents. This includes both its solid particles and the void spaces between them. Most rocks are classified and named primarily on the basis of their texture and composition, both of which yield clues concerning their history and origin. Texture determines the ability of sedimentary materials to store and transmit fluids such as water, oil, and natural gas. The susceptibility of sediments to mass wasting processes such as landslides and debris flows is strongly influenced by their texture. Texture also plays an important role in the diagenesis of detrital sediments. Understanding the textures of sedimentary rocks provides an essential basis for their

*Earth Materials*, Second Edition. Kevin Hefferan and John O'Brien.
© 2022 John Wiley & Sons Ltd. Published 2022 by John Wiley & Sons Ltd.
Companion website: www.wiley.com/go/hefferan/earthmaterials2

classification and interpretation, for the prediction of their resource potential and for the evaluation of their potential as geohazards.

## 13.1.1   Particle size

Several different classification schemes are used to describe particle size in epiclastic rocks. The size scale used by most geologists, but not by most engineers (Chapter 11), is the **Wentworth–Udden grade scale** (Figure 13.1) created in 1918. The Wentworth–Udden grade scale defines gravel, sand, and mud particles according to their mean diameter.

1   **Gravel** particles have diameters larger than 2.0 mm, about the size of small grain of rice
2   **Sand** particles have diameters that are between 2.0 and 1/16 mm (0.0625 mm), the size range seen on sand paper or between finely granulated sugar and small grains of rice
3   **Mud** particles possess diameters smaller than 1/16 mm (0.0625 mm), like processed flour or baby powder
4   Mud particles can be further subdivided into **silt** particles with diameters between 1/16 and 1/256 mm (≈0.004 mm) and tiny **clay** particles whose diameters are less than 1/256 mm (<0.004 mm = 4 microns = 4 μm).

Each of these size classes can be subdivided further as shown in Figure 13.1. Gravel is subdivided into: (1) boulders (>256 mm), (2) cobbles (64–256 mm), (3) pebbles (4–64 mm) and (4) granules (2–4 mm).

Sand is subdivided into: (1) very coarse (1–2 mm), (2) coarse (0.5–1.0 mm), (3) medium (0.25–0.50 mm), (4) fine (0.250–0.125 mm) and (5) very fine (0.0625–0.1250 mm) sand fractions.

Silt can be subdivided into coarse, medium, fine, and very fine silt. All sizes of epiclastic grains can be described using the Wentworth–Udden grade scale.

Krumbein (1934) led a fundamental revolution in the mathematical analysis of populations of detrital particles, utilizing the fact that the Wentworth–Udden scale is a negative, base two logarithmic ($-\log_2$) scale. In order to facilitate mathematical and statistical analysis of detrital sediments, Krumbein created logarithmic phi-scale (Φ-scale) equivalents for the sizes in the Wentworth–Udden grade scale (Figure 13.1). A logarithm is just the power (x)

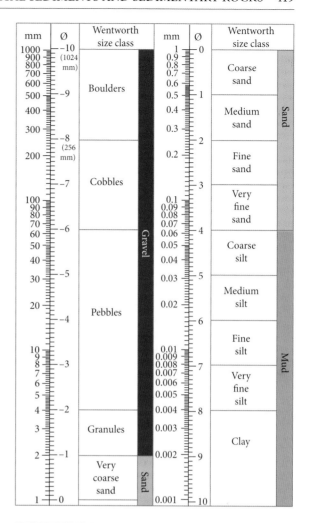

**Figure 13.1**   Wentworth–Udden grade scale with phi (ϕ) equivalents. *Source*: Lewis (1984). © Springer Nature.

to which a base (n) must be raised to produce a given number (a). Stated another way:

$$\text{if } n^x = a, \text{ then } \log_n a = x$$

A simple example may be helpful. The base 10 to the third power ($10^3$) equals 1000 and can be written in the form $n^x = a$ as $10^3 = 1000$. In this case since if $n^x = a$, then $\log_n a = x$, we can write that the base 10 logarithm of 1000 is equal to three; that is $\log_{10} 1000 = 3$. Since the Wentworth–Udden scale is a base two negative logarithmic scale, phi-values are given by the formula $\Phi = -\log_2 d$, where d is the diameter in millimeters. If this is true then the relationship $2^{-\Phi} = d$ follows from

$$\text{if } 2^{-\Phi} = d, \text{ then } \log_2 d = -\Phi$$

Although all of the boundary equivalents between mm-scales and $\Phi$-scales are given in Figure 13.1, some examples may be helpful.

1 The boundary between boulders and cobbles is 256 mm which is equal to 2 mm raised to the eighth power, that is $2^8 = 256$ mm so that $\log_2 256 = 8 = -\Phi$; $\Phi - 8$. Thus the phi-scale equivalent of 256 mm is $-8$.

2 The boundary between pebbles and granules occurs at a phi-value of $-2$ because $2^2 = 4$ mm, thus $\log_2 4 = 2 = -\Phi$; $\Phi = -2$.

3 The boundary between gravel and sand occurs at $\Phi = -1$ because $\log_2 2$ mm $= 1 = -\Phi$; and $\Phi = -1$.

4 The very coarse and coarse sand boundary occurs at $\Phi = 0$ because $\log_2 1$ mm $= 0$ because any number to the power of zero has the value of one.

5 Smaller particles have positive phi values. For example, the boundary between sand and silt occurs at 1/16 mm. Since 1/16 mm $= 2^{-4}$ mm $= -\Phi = -4$; and $\Phi = 4$. Lastly, the boundary between silt and clay occurs at a $\Phi$-value of 8 because $1/256 = 2^{-8}$, thus $\log_2 1/256$ mm $= -8 = -\Phi$; and $\Phi = 8$.

In the analysis of size distributions in sedimentary rocks, the phi-scale is almost invariably used because differences between adjacent size classes are of the same whole number magnitude, whereas in the Wentworth–Udden grade scale they are not of the same value. For example, the boundaries between very coarse, coarse, medium, and fine sand are 0, 1, and 2 phi respectively, so that each size covers an interval equal to 1 phi unit. The same boundaries are at 1 mm, and ¼ mm when expressed by the Wentworth–Udden scale, so that coarse sand covers an interval of ½ mm, medium sand an interval of ¼ mm and fine sand an interval of 1/8 mm.

### 13.1.2 Textural classification of detrital sediments

The epiclastic constituents of a sediment or sedimentary rock can be subdivided into a:

1 **Coarse fraction** or **grains** that include(s) all gravel (G) particles and all sand (S) grains and

2 **Fine fraction** or **matrix** that includes all mud (M) particles.

Detrital sediments and sedimentary rocks can then be classified and named according to the proportions of gravel (G), sand (S) and mud (M) in the detrital fraction. Many classifications exist; Folk (1974) presented a widely used classification based on triangular three-component diagrams (Figure 13.2). The first criterion used in this classification is the percentage of gravel (%G) which decreases from the top apex of the triangle (100%G) to the base (0%G). The percentage of gravel is important because it reflects the competence of the transport/depositional medium to transport large gravel particles to the site of deposition and to keep smaller particles from coming to rest. The second criterion used is

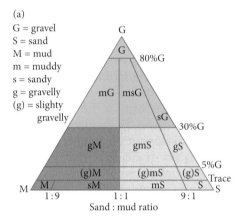

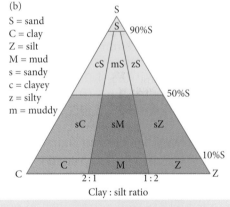

**Figure 13.2** Three-component diagrams giving the textural names of detrital sediments. (a) GSM triangle for naming sands and gravels. (b) SCZ triangle for naming muds and mud-bearing sands. Lithified equivalents are gravel-stones, sandstones, and various mudrocks including claystones, mudstones, and siltstones. *Source*: Adapted from Folk (1974). © John Wiley & Sons.

the sand/mud ratio which is important because it reflects the ability of the transport/depositional medium to keep mud in suspension and therefore to keep it from being deposited with the grain fractions. Different categories of detrital sediment defined on the proportions of gravel (G), sand (S) and mud (M) are defined by the two diagrams in Figure 13.2 and examples are discussed below.

**Gravels and gravelstones** contain more than 30% gravel in their detrital fraction. They occupy the top four sectors (G, mG, msF, sG) on the G : S : M diagram (Figure 13.2a). Detrital materials that contain more than 80% gravel fragments or clasts (G) generally possess **clast-supported textures** because the gravel clasts are in contact and therefore support one another (Figure 13.3a). Smaller sand and/or mud particles, if present, simply fill the spaces between the clast-supported frameworks. Gravels and gravelstones that contain much less than 80% gravel (G) commonly possess **sand- or mud-supported textures** because the gravel particles are generally not in contact and are supported by a sand and/or mud matrix (Figure 13.3b). If the matrix is mostly sand, they are sandy gravels or gravelstones (sG) with a sand-supported framework; if the matrix is mostly mud, they are muddy gravels or gravelstones (mG) with a mud-supported framework.

**Sands and sandstones** contain less than 30% gravel in their detrital fraction and contain more sand than mud (sand : mud ratio > 1 : 1). They occupy the six sectors in the bottom right portion of the G : S : M diagram (gS, gmS, (g)mS and (g)S, mS, and S). Epiclastic sediments that contain 5–30% gravel in the epiclastic fraction are called gravelly sands (gS) or gravelly muddy sandstones (gmS).

If a sand or sandstone contains less than 5% gravel and has a sand : mud ratio greater than 9 : 1, it is a relatively pure sand or sandstone (S). Such relatively pure sandstones are also called **arenites**. If a sand possesses a sand : mud ratio less than 9 : 1, it is a muddy sand or muddy sandstone (mS). Such mud-rich sandstones are called **wackes**. If either of these has even a trace of gravel, it is a slightly gravelly sand or sandstone [(g)S or (g)mS].

**Muds and mudrocks** contain less than 30% gravel in their detrital fraction and contain more mud than sand (sand : mud ratio < 1 : 1).

(a)

(b)

**Figure 13.3** Gravelstones. (a) A matrix-supported framework with gravel particles "floating" in and supported by a mud-rich matrix; Elatina Formation, late Precambrian, Flinders Range NP, Australia. *Source*: (a) Bahudhara, Wikimedia commons, CCBYSA, last accessed 9/21/20 https://upload.wikimedia.org/wikipedia/commons/8/86/Elatina_Fm_diamictite.JPG. https://commons.wikimedia.org/wiki/File:Elatina_Fm_diamictite.JPG. (b) A clast-supported framework with gravel particles in contact, from the Cretaceous, California. *Source*: John O'Brien. © John Wiley & Sons.

They occupy the three sectors in the bottom left portion of Folk's G : S : M diagram (Figure 13.2a). Those with more than 5% gravel are gravelly muds or gravelly mudrocks (gM). Those with less than 5% gravel are divided into sandy muds and sandy mudrocks (sM) with sand : mud ratios > 1 : 9 and relatively pure muds and mudrocks (M) with sand : mud ratios < 1 : 9. As with sands and sandstones, the prefix gravel-bearing may be used if any gravel occurs in muds or mudrocks. Muds and mudrocks may be further subdivided on the basis of clay (C), silt (Z) and sand (S) ratios using Folk's S : C : Z triangle (Figure 13.2b).

### 13.1.3 Central measures

The **central measure** of a particle population is an attempt to represent the typical particle size in the population. Detrital grain populations can be described by three central measures: (1) modal particle size, (2) median particle size and (3) mean particle size. Ideally these central measures should be calculated from diameter measurements of all of the particles in a population. In reality, there are too many grains of sand on a beach or of mud particles on a lake bottom to make this feasible. Instead, a representative or random sample of the population is collected and analyzed. For finer sediments it is more impractical to measure each particle. Instead, the sediments are sieved and/or passed through a settling column, where heavier particles tend to settle faster, and the weight percentage in each phi-size fraction is determined, after which the results are graphed in terms of weight percent vs. phi size classes (**Figures 13.4 and 13.5**).

Three types of graphs are typically used to portray the size data from which central measures may be determined:

1 A **histogram** in which the weight percent in each phi-size class is plotted in the form of a vertical bar graph (Figure 13.4)
2 A **frequency curve** in which a best fit line is plotted for similar data (Figure 13.5)
3 A **cumulative curve**, in which the cumulative weight percent of all size fractions coarser than and including the phi-size fraction under consideration are plotted against phi-size classes (Figure 13.6)

The **mode** is the most abundant particle size and is most easily determined when size data are plotted on a histogram (Figure 13.4) or a frequency curve (Figure 13.5) in which weight percentages are plotted against phi size classes. It is usually expressed in terms of the midpoint of the size class in which the largest

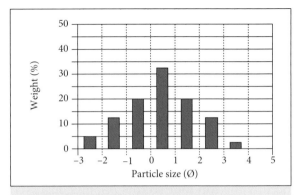

**Figure 13.4** Typical histogram of weight percent sediment versus phi size classes; the mode is between 0 ϕ and 1ϕ and the range is between +3ϕ and +4ϕ.

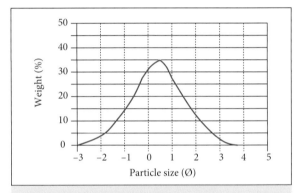

**Figure 13.5** Typical frequency curve for weight percent versus phi size classes, for data similar to those used in Figure 13.4.

weight percent occurs (e.g., 0.5Φ in Figures 13.4 and 13.5). Grain size distributions in which two size classes are more abundant than those adjacent to them, contain two modes, and are said to have **bimodal** size distributions. The more abundant mode is the **primary mode** and the less abundant is the **secondary** *mode*.

The **median size** is the particle size such that half the population is larger and half the population is smaller. Because measuring individual particles is impractical, the median is most easily estimated from a cumulative frequency curve (Figure 13.6) in which it corresponds to the 50th percentile where half the population by weight is coarser-grained and half finer-grained. This value is expressed in phi units as $\Phi_{50}$ and is obtained by locating the 50th percentile on the y-axis and finding the corresponding phi-value on the x-axis (e.g., 0.6Φ in Figure 13.6).

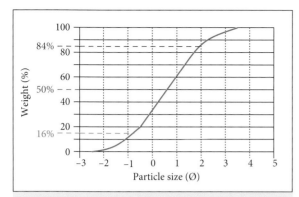

**Figure 13.6** Typical cumulative frequency curve for cumulative weight percent versus phi size classes; the 16th percentile ($\phi_{16}$) in green, 50th percentile ($\phi_{50}$) in red, and 84th percentile ($\phi_{84}$) in purple, are for data similar to those used in Figure 13.4.

The **mean** particle size ($\bar{x}$) is the average size of the particles. Ideally, mean particle size is determined by (1) measuring the mean diameter of every particle, (2) summing the particle diameters and (3) dividing the sum by the number of particles in the population. Mathematically, this is expressed by the formula:

$$\bar{x} = (\Sigma x) / n$$

where $\Sigma x$ is the sum of all diameter values ($x$), n is the number of values and $\bar{x}$ is the mean value. In reality, measuring the diameter of each particle is impractical. Using the cumulative curve (Figure 13.6), one can estimate the mean size using the *graphic geometric mean (GM)* which is calculated using the following formula:

$$GM = (\Phi_{16} + \Phi_{50} + \Phi_{84}) / 3$$

1  $\Phi_{16}$ is the size value of the 16th percentile; the size such that 16% of the population by weight is coarser,
2  $\Phi_{50}$ is the value of the 50th percentile; median such that 50% by weight is coarser
3  $\Phi_{84}$ is the value such that 84% of the population by weight is coarser.

The statistical significance of the three phi-values used is that they represent the median ($\Phi_{50}$), one standard deviation coarser than the median ($\Phi_{16}$) and one standard deviation finer

than the median ($\Phi_{84}$) for a normal probability distribution in which the mean and the median coincide. This permits the central 68% of the population to be used in estimating the mean grain size.

For a more detailed discussion of the statistics of normal probability distributions and their use in sediment studies, readers should consult Prothero and Schwab (2015), Blatt et al. (2006), Folk (1974) and Krumbein and Graybill (1966).

### 13.1.4    Sorting

**Sorting** is a measure of the degree of similarity of particle sizes in clastic sediment. As the range of particle sizes decreases, sorting increases. An everyday analogy is the sorting of eggs by size (small, medium, large, jumbo) before they are shipped to market. Sediment in which all particle grain diameters are of similar size, such as all fine sand, is well sorted. Poorly sorted deposits are those that contain a wide range of particle diameters. A good example is glacial till which commonly consists of particles that range in size from boulders to clay.

Many quantitative measures have been developed to express sorting. In statistics, one might use the standard deviation ($\sigma$) or variance ($\sigma^2$). Most studies of detrital sediments utilize the phi-scale and a simple mathematical formula to calculate a **coefficient of sorting (So)**. Where quantitative information is available concerning the percentage by weight of grains in various size classes, a cumulative curve showing these distributions can be constructed. Steeper cumulative curves represent better sorted sediments with fewer size classes, whereas more gently-sloped curves drawn to the same scale represent sediments with more varied size classes and therefore poorer sorting (Figure 13.7).

From a brief analysis of these curves, one can quickly determine the phi size value for any percentile (size at which a specific percent of the grains by weight are larger). It is easy to find values for the tenth ($\Phi_{10}$) quartile (size at which 10% of the grains by weight are larger) and for the ninetieth ($\Phi_{90}$) percentile (size at which 90% of the grains by weight are larger). The coefficient of sorting then, as given by Compton (1962), is:

$$So = |\Phi_{10} - \Phi_{90}|$$

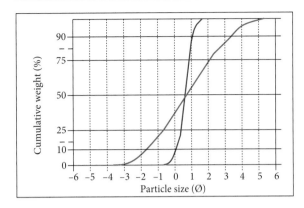

**Figure 13.7** Cumulative curves showing the median ($\phi_{50}$) and different degrees of sorting expressed by the range of sizes in the distribution. The red curve shows well-sorted sand with a narrow range of sizes (low variance). The blue curve depicts a relatively poorly sorted, gravelly, muddy sand with a similar median ($\phi_{50}$) value, but a large range of sizes (large variance).

A more complicated measure of sorting, following Folk (1974) is the **inclusive graphic standard deviation**. It involves finding values that are one standard deviation ($\Phi_{84}$ and $\Phi_{16}$) and two standard deviations ($\Phi_{95}$ and $\Phi_{5}$) above and below the mean and is given by:

$$So = \frac{\Phi_{84} - \Phi_{16}}{4} + \frac{\Phi_{95} - \Phi_{5}}{6.6}$$

General descriptive terminology for sorting based on numerical calculations following Compton (1962) and Folk (1974) are summarized in Table 13.1.

A rapid visual estimation of sorting may be made by determining the size of a particle such that only ten percent of the particles by volume are larger (roughly $\emptyset_{10}$) and the size of a particle such that only 10% by volume are smaller (roughly $\emptyset_{90}$) and using the formula in equation (13.5). This is quite satisfactory for many general purposes. Sorting can also be estimated by visual comparison with charts showing different degrees of sorting (Figure 13.8).

Transportation and depositional media of relatively low viscosity, such as wind and water, tend to sort detrital sediments according to size. Wind deposits are typically very well sorted to well sorted. Sorting in water-laid sediment is more variable, generally ranging

**Table 13.1** Descriptive terminology for sorting coefficients proposed by Compton (1962) and Folk (1974).

| Sorting description | Compton sorting coefficient (So) | Inclusive graphic standard deviation |
|---|---|---|
| Very well sorted | <1 $\phi$ | <0.35 $\phi$ |
| Well sorted | 1–3 $\phi$ | 0.35–0.50 $\phi$ |
| Moderately well sorted | NA | 0.50–0.71 $\phi$ |
| Moderately sorted | 3–5 $\phi$ | 0.71–1.00 $\phi$ |
| Poorly sorted | 5–7 $\phi$ | 1.00–2.00 $\phi$ |
| Very poorly sorted | >7 $\phi$ | >2.00 $\phi$ |

*Source*: Folk (1974) and Compton (1962). © John Wiley & Sons.

from very well sorted to moderately sorted, depending on specific flow conditions. Transport and depositional media of high viscosity such as glaciers and many mass flows such as landslides and debris flows are quite ineffective at sorting detrital sediments. Most poorly sorted and very poorly sorted sediments are deposited from glaciers or by mass flows. Sorting is extremely important in initial assessments of the likely medium of deposition for any sediment or sedimentary rock.

### 13.1.5 Particle shape

A rather complex set of quantitative terminology has been developed to describe particle shapes. Many of these are based on three essential measurements:

1 The longest dimension (a),
2 The shortest dimension perpendicular to it (c)
3 An intermediate dimension that is perpendicular to the other two (b).

In such descriptions then $a \geq b \geq c$. Figure 13.9, after Zingg (1935) illustrates four common particle shapes based on the ratios b/a and c/b and the terminology associated with each one.

1 **Equant (spheroidal)** particles are those that have similar dimensions in all directions so that $a \approx b \approx c$.
2 **Prolate (rod-shaped)** particles have an elongate cylindrical or cigar-shape such

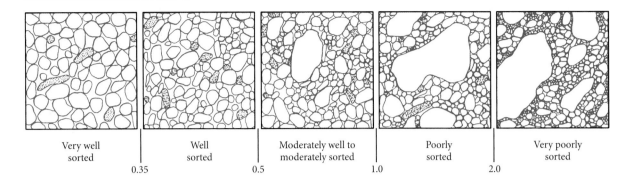

Very well
sorted
0.35

Well
sorted
0.5

Moderately well to
moderately sorted
1.0

Poorly
sorted
2.0

Very poorly
sorted

**Figure 13.8** Diagram for the determination of sorting by visual comparison, using Folk (1974) parameters. *Source*: Adapted from Folk (1974) and Compton (1962). © John Wiley & Sons.

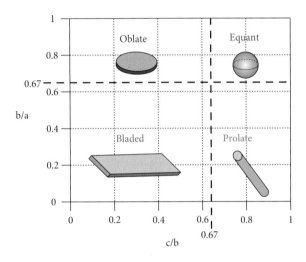

**Figure 13.9** Grain shapes defined from a, b, and c dimensions, where a ≥ b ≥ c. *Source*: Adapted from Schulz et al. (1954). © John Wiley & Sons.

that one axis that is much longer than the other two and a ≫ b ≈ c.

3  **Oblate (disk-shaped)** particles have a flattened cylindrical or disk-like shape such that one axis is much shorter than the other two with a ≈ b ≪ c.

4  **Bladed** particles have a shape that resembles a knife blade with one long axis, one short axis and one intermediate axis so that a > b ≫ c.

Prolate and bladed grains are sometimes oriented with their long axes parallel to the flow of the current, which deposited them.

In such cases they can, used with other features where possible, allow geologists to infer paleoflow directions from ancient sedimentary rocks. Care must be exercised however as in some circumstances such particles come to rest with their long axes perpendicular to current flow. Large bladed and oblate clasts sometimes show **imbrication** in which they are more or less uniformly inclined in one direction. Such imbrication commonly dips in an up current direction because large clasts come to rest on one another in a "piggy-back" manner when deposited (Figure 13.10).

*Sphericity and roundness*

Sphericity and roundness are two terms that at first glance may appear synonymous but which have different meanings. **Sphericity** is a measure of the degree to which a particle approaches the dimensions of a perfect sphere. A perfectly spherical particle would be both equant and round. An angular, equant particle will have a much higher sphericity than a rounded, but inequant particle. The particles in the lower row of Figure 13.11 have lower sphericity than the corresponding particles in the top row.

**Roundness** is measure of the degree of smoothing or curvature of grain edges. A visual comparison chart (Figure 13.12), first developed by Powers (1953), is generally used to estimate grain roundness. This chart allows one to visually compare any particle with visual standards that include relatively equant

(a)

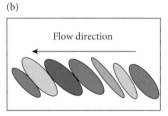

(b)

Flow direction

**Figure 13.10**    Imbricated clasts. (a) Tertiary, Montana, flow from left to right.
Figure 13.10a. Imbricated clasts in Tertiary deposits, Montana; flow from left to right. *Source*: Michael C. Rydel, Wikimedia Commons, CC, Share alike 3.0. https://commons.wikimedia.org/wiki/File:Imbricated_clasts.jpg; last viewed on January 11, 2021. (b) Imbrication diagram, flow from right to left.

grains with high sphericity and relatively inequant grains with low sphericity so that one focuses on the degree of smoothing when estimating roundness. Six roundness classes, sequenced in order of increasing roundness, were recognized by Powers (1953): (1) very angular, (2) angular, (3) subangular, (4) subrounded, (5) rounded, and (6) well rounded. Quantitative measures of sphericity and roundness are complex and rarely warrant the time and effort involved.

Most rounding of detrital grains results from abrasion caused by grain collisions during transport. Rounding therefore has the potential to yield information concerning transport history. Several factors affect the rounding of grains during transport. These include: (1) particle size, (2) hardness and durability, (3) the intensity of collisions, (4) the duration of transport and (5) the number of sedimentary cycles through which the grains have passed. Larger grains round much more quickly than smaller ones because they impact other grains with more force. As a result, pebbles and cobbles tend to round quite rapidly during transport. Sand grains round much more slowly and mud grains may not round at all. Soft particles round much more quickly than hard particles because they abrade much more rapidly. Therefore soft rocks such as

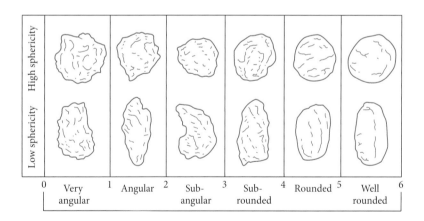

**Figure 13.11**    Diagram for the determination of rounding in grains of varying sphericity by visual estimation. *Source*: Adapted from Powers (1953), Pettijohn et al. (1987); from Tucker (2001). © John Wiley & Sons.

**Figure 13.12**    (a) Oligomictic gravel with rounded gravel and a clast-supported framework, Stromboli, Italy. (b) Polymictic conglomerate with a clast-supported framework, Kata Tjuta, northern Australia. (c) Oligomictic quartz breccia with sand-supported framework, Norlangie, Kakadu National Park, northern Australia. (d) Polymictic diamictite (a tillite) with a mud-supported framework, South Africa. *Source*: (a–c) John O'Brien. © John Wiley & Sons. *Source*: (d) Courtesy of Robert Stuhl.

mudstone, limestone, and marble round much more quickly than hard rocks such as quartzite or granitic rock fragments. Grain rounding may be inherited from an earlier sedimentary cycle.

### 13.1.6    Porosity and permeability

Porosity and permeability are critical factors in the storage and transport of fluids, such as hydrocarbons and water, in sediments and rocks. They are defined and discussed in the context of soils in Chapter 11. Here we focus on porosity and permeability in relationship to the textures of sediments and sedimentary rocks.

**Porosity (Ps)** is simply the volume percentage of void spaces, called pores, in a material and is given by the simple formula Ps = % pore space = (volume of pores/total rock volume) (100). In general, porosity increases with sorting and with grain roundness. In poorly sorted sediments, the finer constituents tend to fill the pore spaces between the larger ones.

Angular grains tend to fill poor spaces more effectively than round ones.

The primary importance of porosity is that it represents the capacity of a sediment or sedimentary rock to hold or store fluids such as groundwater, natural gas or liquid petroleum in the void spaces between solid grains. The search for groundwater or petroleum reservoirs often begins as a search for subsurface rocks with substantial amounts of porosity. Any process, such as cementation, which fills pore spaces, reduces a rock's ability to store fluids. For this reason, much attention is paid to the distribution of mineral cementation patterns in the subsurface during petroleum and hydrologic exploration surveys.

**Permeability** ($\kappa$) is related to **hydraulic conductivity** (**K**) which strongly influences the rate of fluid flow through a material. Permeability can be determined by using Darcy's Law ($Q = A(K \times h/L)$, as discussed in Chapter 12. In Darcy's equation, K is the hydraulic conductivity is given by $K = (\kappa \rho g/\mu)$, where kappa ($\kappa$) is the effective permeability. The two formulae show that flow rates (Q) through porous media increase with increasing effective permeability. Permeability increases rapidly with pore size and pore interconnectedness, both of which tend to increase with particle size and sorting. Well sorted sediments with large pore spaces (and large particles) tend to possess very high permeabilities; rocks with secondary porosity due to dissolution can possess even higher permeability; think caves and underground rivers! Permeability is extremely important in the petroleum industry because petroleum and natural gas must be able to flow into porous reservoir rocks in order to collect there and must be able to flow out of them as extraction proceeds. In addition, fluid migration is severely restricted by relatively impermeable layers that serve as petrotards that trap petroleum and natural gas in underlying permeable rocks. Permeability is also fundamentally important in groundwater aquifer studies and in determining contaminant flow behavior in the subsurface (Chapter 12). Typical hydraulic conductivities for unconsolidated sediments are given in Table 13.2. Note the general increase in permeability with increasing particle size. Sands and gravels tend to be quite

**Table 13.2** Typical hydraulic conductivity of common detrital sediments.

| Sediment type | Hydraulic conductivity |
| --- | --- |
| Clays | Low H $10^{-6}$ cm/s |
| Silts | Low to moderate H $10^{-4}$ cm/s |
| Sands | Moderate to high H $10^{-2}$–$10^{-3}$ cm/s |
| Gravels | High to very high H $10^{-1}$–$10^{1}$ cm/s |

permeable and act as **aquifers** which store and transmit water. Clay-rich layers tend to be relatively impermeable and to act as **aquitards** that retard fluid flow. Recent innovations in hydraulic fracturing ("fracking") to recover oil and gas trapped in relatively impermeable sedimentary rocks involve fracturing the rocks to increase their permeability and allow the oil or natural gas to flow more readily and to be recovered at the surface.

### 13.1.7 Textural maturity

Plumley (1948) and Folk (1951) applied the concept of textural maturity to sands and sandstones. The concept does not apply to mudrocks and gravelstones. **Textural maturity** is based on the answers to three questions regarding: (1) presence or absence of significant (>5%) mud matrix, (2) degree of sorting, and (3) degree of grain rounding.

In Table 13.3, the textural maturity increases from bottom to top as these three criteria are successively applied to the sediment or sedimentary rock in question. When applying these criteria, if the answer to any question is no, the remaining questions are not required as the textural maturity has been established.

Question #1: Is there more than 5% matrix in the sediment or sedimentary rock? Sediments and sedimentary rocks that contain more than 5% mud matrix between the sand and gravel grains are **texturally immature**. Note that Question #1 is really asking whether the mud particles were largely kept in suspension (or later winnowed out) while the sand grains accumulated or whether a significant amount of mud matrix accumulated with, or after, the coarse fraction. Since wind and many aqueous media transport mud in suspension while they deposit sand and gravel from the bed

**Table 13.3** Criteria for the textural maturity of sands and sandstones.

| (1) Is more than 5% mud matrix present? If no: | (2) Is it well sorted? If no: | (3) Are the average grains rounded? | Textural maturity |
|---|---|---|---|
| | | Yes: if yes → | Supermature |
| | Yes: if yes, go to (3) | No: if no → | Mature |
| No: if no, go to (2) | No: if no → | | Submature |
| Yes | | | Immature |

load (see Chapter 11), wind and water-laid deposits are rarely texturally immature. On the other hand, glaciers and many mass flows, such as landslides, debris flows and high-concentration sediment gravity flows (see Chapter 11) transport and deposit a wide variety of sizes together and their deposits are commonly texturally immature. If less than 5% mud matrix occurs, one proceeds to Question #2.

Question #2: Is the sediment or sedimentary rock well sorted? If the material is not well sorted, it is **texturally submature**. Question #2 assesses whether or not the medium of deposition, which was capable of keeping mud in suspension, was also capable of producing well-sorted sediment. If it is not well-sorted, it was most likely deposited by rapidly decelerating or fluctuating flows that can deposit a range of sizes in a very short time. Such flows might include river systems (e.g., braided streams and levees), many storm deposits and low concentration sediment gravity flows such as turbidity currents. If the sediment is well-sorted, then it was almost certainly deposited by wind or by aqueous media in which sediments were well washed during slow net accumulation (e.g., beaches, shallow bars, and some tidal channels). If the material is well-sorted or very well-sorted, it is at least texturally mature and one proceeds to Question #3 below.

Question #3: Are the average grains rounded? If they are rounded (≥4 on the Powers scale of roundness) the sediment is **texturally supermature**; if angular (≤3 on the Powers scale), it is **texturally mature**. This is especially significant when working with medium and finer sands as such sizes are only rounded significantly during wind transport. Care must be exercised! Wind rounded sands

can blow into any number of aqueous environments. They can even be picked up by glaciers and are frequently entrained in mass flows. In the following sections, we will describe gravelstones, sandstones, and mudstones and what we learn from their properties.

## 13.2 GRAVELSTONES

Gravelstones, such as conglomerates and breccias, constitute roughly 5% of detrital sediments and sedimentary rocks. Mudstones and sandstones are much more abundant. Still, a large majority of detrital sediments deposited directly by glaciers and debris flows and a small, but significant, proportion of water-laid sediments contain enough gravel to qualify as gravelstones.

### 13.2.1 Gravelstone classification

A **gravelstone** is a solid sedimentary rock in which at least 30% of the detrital grains by volume are gravel. Gravelstones, also called **rudites**, are classified primarily on the basis of three criteria: (1) the nature of the support framework for the gravel, (2) the roundness of the gravel fragments or clasts and (3) the composition of the clasts.

Many classifications for gravelstones exist. Table 13.4 is modified from Raymond's (2002) classification and provides an excellent basis for discriminating between the major types of gravelstone. Clast size can be used as a modifier; e.g., pebble conglomerate or boulder diamictite, as can clast composition, e.g. quartz pebble conglomerate or limestone breccia.

**Table 13.4**  Classification of gravelstones: conglomerates, breccias, and diamictites.

| Gravelstone shape | Matrix support | Gravelstone composition | Rock name |
| --- | --- | --- | --- |
| Subrounded to very rounded (conglomerate) | Gravel- or sand-supported framework | Single composition | Oligomictic conglomerate (e.g., quartz conglomerate) |
| | | Multiple compositions | Polymictic conglomerate |
| Subangular to very angular (breccia) | Gravel- or sand-supported framework | Single composition | Oligomictic breccia (e.g., limestone breccia) |
| | | Multiple compositions | Polymictic breccia |
| Any shape | Mud-supported framework (diamictite) | Single composition | Oligomictic diamictite |
| | | Multiple compositions | Polymictic diamictite |

*Source*: Adapted from Raymond (2002). © John Wiley & Sons.

## 13.2.2   Gravelstone particle shape

Gravelstones in which gravel clasts are generally rounded are called **conglomerate**. Since large clasts are rounded rapidly during transport by water, conglomerates (Figure 13.12a, b) are the most common type of gravelstone based on clast shape. Gravelstones that contain angular clasts are called **breccia**. The angular clasts of breccias (Figure 13.12c), especially clasts of soft rocks, suggest minimal transport and derivation from a nearby source area. Breccias occur in a number of environmental settings that include:

1   Fault zones where angular fragments are produced by fragmentation during slip
2   Talus slopes at the base of cliffs and ridges that result from the effects of rock weathering and mass wasting processes such as rockfall
3   **Impactites** produced when bolides (such as meteors) collide with Earth's surface
4   Karst regions where they are produced by the partial dissolution and collapse of carbonate rocks (Chapter 12)
5   The upper portions of alluvial fans where debris flows are common
6   Wherever cohesive mud clasts are eroded and transported a short distance by streams, currents or sediment gravity flows

## 13.2.3   Gravelstone framework

Conglomerates and breccias occur in clast-supported and sand or mud matrix-supported varieties. Gravelstones with a **clast-supported framework** contain gravel clasts that are generally in contact and therefore support one another to form a clast supported framework. A clast-supported conglomerate can be called an **orthoconglomerate** (Table 13.4). Clast-supported gravelstones (Figure 13.12a–c) commonly accumulate in:

1   Aqueous environments such as stream channels, beaches, and marine shoals in which sand and mud continue to be transported while gravel accumulates
2   As **sieve deposits** in alluvial fans and ephemeral braided stream systems where the fines are removed by surface water infiltrating into the ground between clasts
3   As **lag gravels** left when wind or water entrains sand and mud from the surface, leaving the gravel behind.

Many gravelstones possess a **matrix-supported framework** in which the gravel clasts are generally not in contact and are supported by the finer sand and/or mud matrix that separates them. The clasts appear to "float" in the matrix that supports them. If the matrix is mostly sand, the gravelstone possesses a **sand-supported framework**. Gravelstones with sand-supported frameworks occur wherever significant amounts of sand and gravel accumulate together. These environments include alluvial fans, river channels, braid bars and point bars in stream systems, as well as beaches, marine shoals and storm deposits in marine environments. Some high-concentration sediment gravity flows also produce sand-supported gravelstones.

If the matrix is mostly mud, the gravelstone has a **mud-supported framework** (Figure 13.12d). A gravelstone with a mud-supported matrix is called a **diamictite** or a **paraconglomerate** (Table 13.4). Because they contain significant amounts of both gravel and mud, diamictites are invariably poorly or very poorly sorted. Diamictites form by a number of different processes including:

1  Glaciers that deposit gravel and mud rich tills that are lithified to form **tillites**
2  Mudflows and many debris flows and lahars deposit mud supported gravels
3  Some gravel-bearing, sediment gravity flows, especially those that bring gravel rich sediment into areas with extensive mud accumulations where the two are mixed
4  Tectonic **mélanges** at convergent plate boundaries are diamictites produced by the mixing of gravel clasts with muds by a combination of sedimentary and tectonic processes

### 13.2.4   Gravelstone composition

Gravelstones are also classified according to clast composition. Gravelstones in which the clasts are largely of one composition are called **oligomictic** conglomerates, breccias or diamictites (Figure 13.12a, c). Fault, collapse, and rockfall breccias are commonly oligomictic, as are compositionally mature quartz-rich conglomerates.

Gravelstones that contain a variety of clast compositions are called **polymictic** conglomerates, breccias or diamictites (Figure 13.13b, d). Polymictic gravelstones can yield significant, easily accessible, information concerning the nature of the rocks in the source area. Mélange gravelstones, produced by various combinations of sedimentary and tectonic processes that mix several rock types are typically **polymictic diamictites** as are many glacial tills produced during continental glaciation where expanding glaciers erode different rock types from a large region, then deposit a large range of particle compositions and sizes together. Lahars, volcanic debris flows, commonly contain a variety of volcanic clast types, as do many other debris flows, so their deposits are also polymictic.

### 13.2.5   Gravelstone provenance

Gravelstones are by far the easiest detrital sedimentary rocks to use in hand-specimen provenance studies. Fragments of the original source rocks are often preserved intact in the pebble, cobble, and boulder populations. Many clasts can be identified macroscopically, using fresh surfaces, and almost all can be identified using microscopic methods. Large clast size allows one to identify possible source rocks and to make inferences about climate and relief, the duration and intensity of transport and the tectonic setting in which the gravelstones accumulated. Preservation of representative unaltered clasts is favored by factors that include:

1  High mechanical durability and chemical stability of clast components
2  Low precipitation and temperatures which inhibits chemical decomposition
3  High relief and low vegetative cover which promote rapid erosion and minimize the duration of chemical decomposition and clast disintegration
4  Short transportation which promotes survival of mechanically unstable clasts

Gravelstones derived from volcanic source areas are often rich in volcanic rock fragments (VRF or Lv) that record whether the source region was a volcanic arc (rich in VRF of andesitic to rhyolitic composition), or a hot spot flood plateau (rich in VRF of basaltic composition) or a bimodal suite (rich in both basaltic and rhyolitic VRF). Most non-pyroclastic volcanic rock fragments are quite hard and can survive relatively long periods of transport. Basaltic rocks, rich in ferromagnesian minerals and calcic plagioclase tend to decompose rapidly and survive weathering only in areas with relatively high relief, low rainfall, and/or low temperatures.

Gravelstones derived from plutonic igneous source areas tend to disintegrate along grain boundaries during long or intense transport. An abundance of plutonic rock fragments (Lp) suggests a relatively nearby source area. Once again, mafic fragments tend to decompose rapidly; their occurrence implies some combination of high relief and/ or rapid erosion, and/or low rainfall and/or temperature. Granitoid rock fragments (GRF), on the other hand, decompose less rapidly and are more commonly preserved in gravelstones. GRF are most commonly derived from magmatic arc and intracratonic rift settings in which granitioid rock fragments are shed from areas of reasonably high relief into nearby basins of deposition before their constituent feldspars undergo significant decomposition

Gravelstones derived from metamorphic source areas may be rich in metamorphic rock fragments (MRF or Lm). Metamorphic rock fragments can be subdivided into low-grade fragments ($Lm_1$) and higher-grade rock fragments ($Lm_2$). Hard, fine-grained MRF such as metaquartzite survive long periods of transportation, as does the vein quartz common in metamorphic terrains. On the other hand, softer MRF such as slate, phyllite, and marble do not. The occurrence of such soft MRF in gravelstone implies derivation from a nearby source. The most varied assemblages of MRF are derived from the erosion or dissection of orogenic belts. Progressive dissection of such orogenic belts generally causes the exposure of progressively higher grade MRF over time, with increases in $Lm_2$ and decreases in $Lm_1$.

Gravelstones derived from sedimentary source areas can be rich in sedimentary rock fragments (SRF or Ls). Hard, fine-grained SRF such as chert survive long periods of transportation and multiple sedimentary cycles. Softer SRF such as shale, claystone, and limestone do not commonly survive transportation and their occurrence as clasts implies derivation from a nearby source. Readily soluble SRF such as limestone and evaporates and MRF such as marble do not survive weathering unless the source area has high relief and erosion rates and/or low temperature and/or rainfall.

With the proviso that fresh material must be used, any of the techniques used by igneous petrologists to discriminate magma sources and tectonic settings (Chapter 10) and by metamorphic petrologists to determine protoliths, metamorphic facies, and tectonic settings (Chapter 18) may be applied to inferring the history of the source area from which gravelstone clasts have been derived (Box 13.1).

---

## Box 13.1 In greater depth: using conglomerate clasts to document slip on faults

For well over half a century, since Crowell (1952) used Ridge Basin conglomerates to document strike–slip on the San Gabriel Fault in southern California, geologists have utilized synkinematic deposition of conglomerates to try to unravel the timing and movement history of faults. The trick is to recognize conglomerate clasts in formations adjacent to the fault that were derived from a unique source on the other side of the fault. This source may have been adjacent to the conglomerates at the time of deposition but, if subsequent strike-slip along the fault has occurred, the source has been relocated relative to the conglomerate. The distance of relocation between the source rock and the conglomerate permits the horizontal component of slip to be documented. Knowledge of the age of the conglomerate permits average slip rates to be calculated (e.g., in km/Ma).

Geologists in New Zealand have documented the movement history of the large Alpine Fault (Figure B13.1) which has been the site of several magnitude-8 earthquakes in the past 1000 years. Like the San Andreas Fault system in southern California, the Alpine Fault is part of a major transform plate boundary system, in this case between the Australian and Pacific plates. Sutherland (1994) recognized unusual clasts derived from ultramafic rocks and greenshist–amphibolite grade schists in Pliocene (3.6 Ma) conglomerates deposited in Cascade Valley. Sutherland traced these clasts to source rocks in the Red Mountain ultramafic sequence and Haast schist that today are located 95 km to the southeast of the Miocene conglomerates. From these data, Sutherland concluded that the minimum right-lateral slip on the Alpine fault since 3.6 Ma has been 27 km/Ma (95 km/3.6 Ma) and that up to 35 km/Ma was possible.

Previously Cutten (1979) had posited that Miocene (11.5 Ma) sandstone and metamorphic clast assemblages in the Marula Basin on the northwest side of the fault had been derived from the Caples

## Box 13.1  *Continued*

and Torlesse Terranes which are now located 420 km to the southwest of the Maruia Basin (Figure B13.1). Geochemical work on both clasts and source rocks by Cutten et al. (2006) has confirmed this interpretation. The average dextral (right-lateral) slip on the Alpine Fault responsible for the present offset between the source rocks and the clasts shed from them is roughly 37 km/Ma (420 km/11.5 Ma). These slip rates were further confirmed by Lamb et al. (2016) who posited 700 km of right-lateral displacement in the past 25 Ma. All of these numbers correspond well with plate tectonic studies that imply approximately

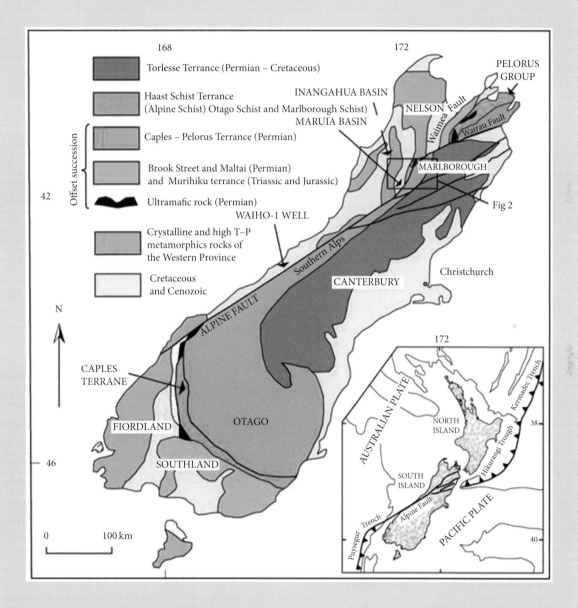

**Figure B13.1**    Geological map of South Island, New Zealand, showing the Alpine Fault, Caples Terrane and Maruia Basin. The inset map shows the plate tectonic setting. *Source*: Cutten et al. (2006). © John Wiley & Sons.

*Continued*

## Box 13.1 *Continued*

35 km/Ma of dextral slip on the Alpine Fault system transform plate boundary. Synkinematic conglomerate clasts are useful in many other tectonic interpretations. Similar procedures are utilized to determine the docking history of terranes in accretionary collisional complexes (Chapter 1). If clasts from one terrane were deposited atop another terrane, the two terranes were in close proximity (docked) at the time the conglomerates accumulated. So much critical information from so few cobbles!

## 13.3 SANDSTONES

**Sandstones** are rocks that contain less than 30% gravel in their detrital fraction and that contain more sand than mud (sand : mud > 1). They constitute roughly 20% of all sedimentary rocks.

### 13.3.1 Sandstone classification

Dozens of sandstone classifications have been proposed (e.g., Okada 1971). They use related criteria to organize sandstones into groups, use similar boundaries to separate different types, and employ related terms to name different varieties of sandstone. One widely-used sandstone classification is that proposed by Folk (1974). Folk's scheme (Figure 13.13) classifies sandstones based on the proportions of three components:

1 The percentage of quartz grains by volume (%Q) in the sand fraction,

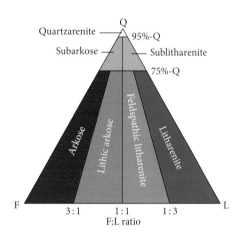

**Figure 13.13** Sandstone classification of Folk (1974). F = feldspar; L = lithic fragments; Q = quartz. *Source*: Adapted from Folk (1974). © John Wiley & Sons.

2 The percentage of feldspar grains by volume in the sand fraction (%F)
3 The percentage of polycrystalline rock fragments, called lithic fragments (%L) in the sand fraction

All other components are ignored for the purposes of basic sandstone classification. A three-component (QFL) diagram is used and is subdivided into seven sectors. Each of the seven sectors represents a different type of sandstone with a specific name that reflects its Q–F–L composition (Figure 13.14).

The first criterion used to classify sandstones is the recalculated percentage of quartz grains (%Q = [%Q/(%Q + %F + %L)] (100) in the sand fraction. Any sandstone that possesses more than 95%Q plots in the uppermost sector of the diagram and is called **quartzarenite**. Some classifications use the alternate spelling quartz arenite; others require only 90%Q. Quartzarenites are compositionally mature or supermature.

Sandstones that contain 75–95% recalculated quartz and have more feldspar than lithic fragments (F : L > 1) are called **subarkose**; those with 75%-95% quartz and less feldspar than lithic fragments (F : L < 1) are called **sublitharenite**. Subarkoses and sublitharenites are compositionally submature to mature.

All sandstones that contain less than 75%Q plot in one of the four sectors in the bottom portion of the Q–F–L triangle. These sandstones tend to be compositionally submature or immature. Feldspar-rich sandstones that contain less than 75% quartz and have an F : L ratio > 3 : 1, plot in the lower left-hand sector of the Q–F–L triangle, and are called **arkose**. A **litharenite** is a rock fragment-rich sandstone that plots in the lower right-hand sector of Q–F–L diagrams, because it contains less than 75% quartz and has an F : L ratio of <1 : 3. Many types of sandstone contain

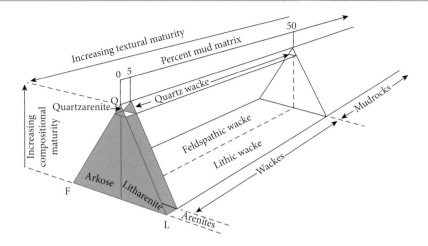

**Figure 13.14**   Four-component classification sandstones. F = feldspar; L = lithic fragments; Q = quartz. *Source*: Adapted from Blatt and Tracy (1996). © John Wiley & Sons.

substantial proportions of both feldspars and lithic fragments. These are called a **lithic arkose** when they contain less than 75% quartz and have F : L ratios between 1 : 1 and 3 : 1 and a **feldspathic litharenite** when they contain less than 75% quartz and have F : L ratios between 1 : 1 and 1 : 3.

Another valuable set of classifications for sandstones are based on a combination of sandstone composition and sandstone texture. In addition to the Q–F–L triangle, they incorporate a fourth variable, the percentage of mud matrix (Figure 13.15) which increases from zero at in the front triangle in the diagram to 50% in the back triangle of the diagram (beyond which they are mudrocks). Any sandstone with significant mud matrix is called a **wacke**; any sandstone with negligible mud matrix is called an **arenite**. Once again, boundaries and terminology vary between classifications. Blatt and Tracy (1996) put the boundary between wacke sandstones and arenites at 5% mud (Figure 13.14) so that texturally immature sandstones are wackes. They place the boundary between wackes and mudstones at 50% mud, a practice consistent with other classifications based on sand : mud percentages (Figure 13.14). Any sandstone with less than 5% mud matrix is an **arenite** of the types discussed previously (e.g., quartzarenites, arkoses, and litharenites).

An older term, **graywacke** (or greywacke) is sometimes used as a general term for matrix-rich sandstones, but has been used with so many different meanings that its value is questionable. Wackes can be subdivided into **quartz wackes** *(graywackes)*, **feldspathic wackes** *(graywackes)* or **lithic wackes** *(graywackes)* depending on the percentage of quartz, feldspar, and lithic fragments in the sand fraction percentage, as shown in Figure 13.4.

Detrital sedimentary rocks with more than 50% mud in their detrital fractions are mudrocks rather than sandstones and are classified using a different set of criteria that is discussed later in this chapter in the section on mudrocks.

### 13.3.2   Sandstone provenance

The composition of any sandstone reflects several factors that include:

1   The rock types exposed in the source area
2   Their chemical stability and decomposition history
3   The climate, relief, and rates of erosion in the source terrain
4   The proximity of the source area and the intensity and duration of transport as expressed by mechanical stability
5   The tectonic setting in which sedimentation occurred

For decades, petrologists have used various types of **discrimination diagrams** to try to distinguish igneous rocks produced in different tectonic settings, in some cases even before the advent of the formal theory of plate tectonics over fifty years ago. No one discrimination diagram makes clear distinctions in every case, but combinations of such diagrams have proven useful in distinguishing igneous rocks formed in different tectonic settings, as discussed in Chapter 10.

Dickinson (1970) and Dickinson and Suczek (1979) introduced the use of discrimination diagrams (Figure 13.15) for the determination of sandstone tectonic settings. They presented several different diagrams which, when used in combination with other data sets (such as rock texture, field relations, geochemistry, geochronology), allow one to infer the likely tectonic setting in which a particular sandstone formed. The most frequently used discrimination diagram is a Q–F–L triangular diagram, much like the ones used to classify sandstones by composition. In this case, however, Q-constituent includes all quartz grains, including monocrystalline quartz, polycrystalline quartz and chert because they are all highly resistant to weathering. The latter two are considered rock fragments in Folk's classification. The F-constituent includes all feldspars and the L-constituent includes all lithic fragments except polycrystalline quartz and chert. Effective utilization of such diagrams requires a quantitative determination of the percentages of each constituent, e.g., from detailed examination of a thin-section. In the diagrams of Dickinson and Suczek (1979) and Dickinson et al. (1983), three major provenances or tectonic settings, each with subdivisions, are recognized:

1 **Magmatic arcs** in which sediments are derived from volcanic–magmatic arcs formed over subduction zones along convergent plate boundaries
2 **Recycled orogens** in which the sediments are derived from orogenic mountain belts that developed at collisional plate boundaries
3 **Continental blocks** in which sediments are derived from stable cratonic source rocks in shields and/or platforms or in continental rift systems

A basic knowledge of plate tectonics, summarized in Chapter 1, is essential to understanding patterns of sedimentation. Detailed petrography supplemented by precise geochemical analyses, radiometric ages and cooling ages (see Box 13.2) of the constituents can allow one to test and refine hypotheses about provenance and/or

tectonic settings. This can lead to more detailed and sophisticated portrayals of the provenance and tectonic settings of sandstone deposition (e.g., Preston et al. 2002; Dickinson and Gehrels 2003; Armstrong-Altrin et al. 2004; Higgs and King 2018).

With few exceptions, sandstones from **magmatic arc source areas** plot in an area in the lower part of the diagram and are mostly litharenites, feldspathic litharenites or lithic arkoses (Figure 13.15). These rock compositions

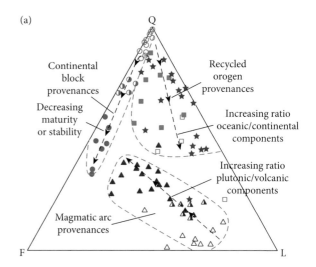

(a)

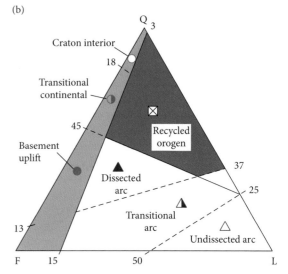

(b)

**Figure 13.15** (a) QFL (quartz, feldspar, lithic fragments) diagram after Dickinson and Suczek (1979). *Source*: Raymond (2002). © Waveland Press Inc. (b) Modified QFL diagram after Dickinson et al. (1983). *Source*: Tucker (2001). © John Wiley & Sons.

## Box 13.2   Appalachian sources for western sandstones

The Permian and Jurassic eolian quartzarenites and subarkoses of the Colorado Plateau in the western United States have long inspired geologists with their impressive, large-scale cross-stratification (Figure B13.2). Now they have another claim to fame. A large proportion of the sand grains in these wind-blown deposits were derived from the Appalachian Mountains in what is now the eastern United States.

   The story began when Dickinson and Gehrels (2003) analyzed 468 detrital zircons from five Permian eolian formations, using U/Pb dating techniques to establish their ages and delimit their provenance. Their analysis revealed six age peaks for the zircons. About half corresponded to ages of local source rocks in the Ancestral Rocky Mountains (1800–1365 Ma) and the North American Craton (3015–1800 Ma). But the other half could not have been derived from local sources. These zircons (1315–1000, 750–515 and 515–310 Ma) could only have been derived from sources located in the vicinity of the Grenville and Appalachian Orogens, now located on the eastern margin of the North American Craton. Dickinson and Gehrels concluded that a large transcontinental river system, analogous to the modern Amazon River in South America, transported detritus from the Appalachian region to the western part of the craton in Permian time. Because this area was located just north of the equator at this time, northeast trade winds did the rest, transporting and depositing the vast seas of sand that became the Permian Coconino and Jurassic Navajo Formations and their equivalents.

**Figure B13.2**   Large-scale cross-strata, Navajo sandstone, Jurassic, Utah. *Source*: John O'Brien. © John Wiley & Sons.

*Continued*

## Box 13.2 *Continued*

Rahl et al. (2003) used a more sophisticated set of analyses to support and extend these contentions. Rahl and his colleagues used U–Pb isotope methods to date zircon formation and He isotope analysis to determine zircon cooling ages which approximates their time of exposure and erosion at Earth's surface. These so-called double dating methods revealed that most of the zircons formed during the Grenville Orogeny (1200–950 Ma), but had cooling ages of 500–225 Ma that correspond to the formation and unroofing by erosion of the Appalachian Orogen. Their work confirmed the Dickinson and Gehrels hypothesis and suggested that the transcontinental river system may have existed from the Permian into the Triassic.

Campbell et al. (2005) suggest that 75% of the zircons in the Navajo sandstone were derived from Appalachian–Grenville sources. In addition Gehrels et al. (2011) have suggested that Mississippian rocks in the Grand Canyon of Arizona were partially derived from similar source rocks, thereby extending the age of the transcontinental transport system. Thomas (2011) has cautioned that the ages used are not necessarily unique and that dispersal system directions must be carefully documented in the context of a detailed knowledge of regional geology for the time periods in question. By tying together detrital sediments and source areas, detrital zircons have the potential to test ancient continental reconstructions by identifying the presence of particular source regions, the timing of their erosion and subsequent dispersal patterns. What tools we have at our disposal!

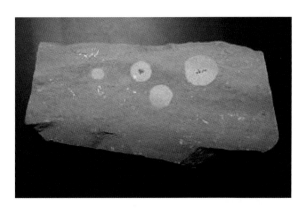

**Figure 13.16** Reduction spots around organic particles in red–purple mudrock. *Source:* Courtesy of Steve Dutch.

reflect the generally low compositional maturity of sediments shed rapidly from primary source rocks in areas of high relief. Dickinson and Suczek (1979) also noted an important trend within the magmatic arc field. Progressive erosion of a volcanic–magmatic arc produces a general decrease in lithic rock fragments (e.g., VRF) and an increase in both feldspars and quartz so that arc dissection produces the trend line from litharenites toward arkoses shown in Figure 13.16. This occurs largely because young, undissected volcanic arcs generate large numbers of fine-grained, largely aphanitic, rock fragments, whereas dissected plutonic arcs consist largely of coarser-grained granitoids whose weathering generates larger amounts of sand-size feldspar and quartz grains.

Dickinson and Suczek (1979) subdivided sandstones from **recycled orogen source areas** into:

1  **Subduction complexes** that contain sedimentary rocks, metamorphic rocks and recycled volcanic–magmatic arc sediments derived from uplifted recycled accretionary complexes
2  **Collision orogens** that contain sedimentary and metamorphic rocks, of both continental and oceanic origin, uplifted during the continental collisions that close ocean basins
3  **Foreland uplifts** that consist of diverse sedimentary, metamorphic, and plutonic igneous rock assemblages exposed in orogenic belts some distance from convergent plate boundaries

With some exceptions, sandstones from orogenic belt source areas plot in an area in the central to right upper portion of Q–F–L diagrams (Figure 13.15). These sandstones are mostly litharenites and sublitharenites along with subordinate subarkoses and a few lithic arkoses. Orogenic belt rocks are mostly recycled from sedimentary or metasedimentary source rocks that experienced at least one previous sedimentary cycle. As a result, these sediments tend to be more compositionally mature (mostly submature to mature) than those eroded from the primary source rocks in volcanic–magmatic arcs. Because recycled orogen sediments are shed relatively rapidly, they also tend to contain at least some rock fragments. One important trend in such areas is the tendency for compositions to become enriched in quartz and depleted in lithic grains and feldspars as the orogenic belt is progressively eroded down, making relief and erosion rates lower and providing more time for weathering.

**Continental block provinces** consist principally of stable cratonic areas that are subdivided into shields and platforms. **Shields** consist primarily of Precambrian plutonic and high-grade metamorphic rocks such as granitoids, gneisses, and granulites. **Platforms** are characterized by relatively thin veneer of largely mature or supermature detrital sedimentary rocks and/or carbonate sedimentary rocks that overlie shield rocks. Both shields and platforms tend to have long histories of relative tectonic stability.

Dickinson and Suczek (1979) divided continental block provenances *into*:

1 **Craton interior** in which sediments are derived from preexisting, generally mature, sedimentary rocks that overlie plutonic basement rocks in a stable platform setting with very low relief
2 **Transitional cratons** in which both platform sedimentary rocks and plutonic basement shield rocks are exposed as source rock types in settings of low to moderate relief
3 **Uplifted basement** in which the basement plutonic igneous and metamorphic rocks are exposed in an area of high relief resulting from uplift, e.g., along faults in continental rift settings

With very few exceptions, sandstones from cratonic settings plot in an area along the left side of Q–F–L diagrams (Figure 13.15).

Depending on their mineralogical maturity, these sandstones range from less mature arkoses and subarkoses to more mature quartzarenites. Dickinson and Suczek noted a significant trend within continental block sandstones that extends from feldspar-rich arkoses toward quartzarentites. They explained this trend in terms of relief; the higher the relief in the source area, the more feldspar is present in the sandstones and the lower the relief, the more quartz is present. Uplifted basement sources provide more feldspar for dispersion than do stable craton interiors.

In the past two decades, sedimentologists have utilized a variety of geochemical techniques to match detrital sediments to source rocks in their efforts to more accurately determine their most likely provenance, as discussed in Box 13.2.

## 13.4 MUDROCKS

Mudrocks dominate the sedimentary record, constituting 60–65% of all sedimentary rocks. This results primarily from the abundance of clay particles that are produced by decomposition in the source area and silt particles produced by the disintegration of larger particles in the source area and by abrasion during transportation. Despite their abundance, mudrocks generally are not well-exposed at the surface. This is because most mudrocks are fairly soft and easily eroded and because they often support abundant vegetation in nonarid climates.

**Mudrocks** contain less than 30% gravel in their detrital fraction and have more mud than sand (sand : mud < 1.0). Those with more than 5% gravel are **gravelly mudrocks**. Mudrocks with more than 10% sand (sand : mud ratio > 1 : 9) are **sandy mudrocks**. The vast majority of mudrocks contain little or no gravel and sand. This is because the traction conditions under which most gravel and sand are transported and deposited by wind and water are significantly different than the suspension conditions under which most mud is transported and deposited. As a result, the two size populations are largely separated during transportation and deposition, especially by wind and water.

**Table 13.5** Textural classification of mudrocks based on silt : clay ratios.

| Silt : clay ratio | Mudrock name |
| --- | --- |
| Predominantly silt (>2 : 1) | Siltstone |
| Abundant silt and clay (2 : 1–1 : 2) | Mudstone |
| Predominantly clay (<1 : 2) | Claystone |

## 13.4.1 Mudrock textures and structures

The two major components of mudrocks are silt (1/16– 1/256 mm = 4Φ – 8Φ) and clay (<1/256 mm ≤ 0.004 mm ≤ 4 μm ≥ 8Φ). Three common types of mudrocks are classified and named (Table 13.5) on the basis their silt : sand ratios.

**Siltstone** is a coarse mudrock that contains more than 2/3 silt in the mud fraction (silt : clay > 2 : 1). **Claystone** is fine mudrock that contains more than 2/3 clay in the mud fraction (silt : clay < 1 : 2). **Mudstone** contains substantial amounts of both clay and silt and possesses silt : clay ratios between 2 : 1 and 1 : 2. The latter term is also used as an equivalent of mudrock, at least informally.

Determination of the percentages of silt and clay in outcrop or hand-specimens is made difficult by their fine size, though coarse silt grains are still distinguishable to the unaided eye. One can use a "taste-test" as a rough guide. As a rule of thumb, clay particles feel soft, smooth, and pasty when placed between tongue and teeth, whereas most silt particles feel rather hard and gritty. This is not recommended for hazardous waste sites! Wet mud can be rolled into a string using a shearing motion with the hands; clay forms a long cohesive string while silt, being less cohesive, tends to disaggregate. The presence of mudcracks, formed by the dessication of wet mud, indicates a degree of cohesiveness in the original sediment that suggests a high clay content. For more accurate determinations, one can use laboratory methods such as settling tube analysis for unconsolidated muds and microscopic and other analytical methods for consolidated rocks.

The structure called fissility is so common in mudstones and claystones that it requires discussion. **Fissility** is the tendency of certain mudstones and claystones to split into thin layers, roughly parallel to stratification. Fissile mudstones and claystones are called **shales**.

Although the term shale is often used informally for all mudrocks, it should be reserved for those that display fissility. Fissility generally results from the sub-parallel alignment of clay and mica minerals with their characteristic sheet (phyllosilicate) structures. The rock splits parallel to the mica and clay sheets. Mudrocks such as siltstones, that contain little or no clay minerals or mudrocks in which the phyllosilicate minerals are randomly oriented do not possess fissility and are not shales.

## 13.4.2 Mudrock composition and color

The two major components of mudrocks are quartz and clay minerals. It is important to point out that the term "clay" is used in two distinctly different ways. The size term "clay" is used for any clastic particle smaller than 1/256 mm ≅ 0.004 mm ≅ 4 μm. The mineralogical term "clay" is used for a group of aluminum-bearing phyllosilicates discussed in detail in Chapter 11 which are the major products of decomposition during weathering. One relationship between the two uses of the term "clay" is that most, but not all, of the clay size fraction is composed of very small clay mineral particles.

In addition to clay minerals, the clay size fraction may contain very fine-grained quartz, feldspar, micas, and many other less abundant minerals. Silt sized minerals are predominantly angular to subangular quartz, feldspars, and mica flakes, with minor amounts of other minerals. Lithic fragments are uncommon because most polycrystalline rock fragments are too large to be included in the mud fraction. Other significant mudrock components include carbonate minerals, iron oxides and hydroxides, zeolite minerals, sulfide minerals and organic materials. The significance of organic material in mudrocks cannot be overstated. More than 95% of the cellular organic material, whose diagenetic alteration produces petroleum and natural gas, initially accumulates in detrital or carbonate muds.

Because of their tiny particle size, clay mineral analysis is difficult by standard thin-section techniques, but laboratory techniques such as x-ray diffraction can yield increasingly improved quantitative analyses. All of the major clay minerals occur in modern muds and ancient mudrocks. The general distribution of clay minerals in modern soils (Chapter 11) depends primarily on the composition of the bedrock, climate, especially rainfall and temperature, and the intensity of chemical decomposition. A brief review of the major clay minerals in mudrocks follows:

1 Chlorites are produced by the minimal decomposition of ferromagnesian minerals and commonly form in alkaline soils with impeded drainage, especially at high latitudes where precipitation and temperatures are low.

2 Smectites, such as montmorillonite, are the product of the weathering of ferromagnesian minerals plus plagioclase. Their formation is favored by impeded drainage, alkaline conditions and semi-arid climates.

3 Illites are common products of the weathering of feldspars (especially K-spars) and occur most commonly in temperate region soils with near neutral pH.

4 Mixed layer illite–smectite clays are also common in mid-latitude semi-arid to temperate soils with slightly alkaline pH.

5 Under warm, humid, acidic soil conditions, such as those common in the subtropics, cations tend to be leached from interlayer sites which gives rise to degraded illites and kandites such as kaolinite.

6 Under warm, humid conditions of unusually high acidity and low pH, intense decomposition allows silica to dissolve readily giving rise to gibbsite and other minerals of the bauxite suite.

These patterns have been traced from the source areas to adjacent areas of sediment dispersal. Chlorites are most common at high latitudes, kaolinite is most common at low latitudes, and illites, smectites, and mixed-layer clays dominate sediments elsewhere. But these patterns do not survive in the oceans or in the subsurface because clay minerals are reconstituted and transformed as they react with seawater and with pore water during diagenesis. Smectite clays, such as montmorillonite, and mixed layer smectite–illite clays dominate modern sediments and are common in younger Tertiary rocks. Older rocks are dominated by illite clays. The progressive decrease in the proportions of smectite and kandite (e.g., kaolinite) with age results from their alteration to illite and/or chlorite by diagenetic processes during burial that are detailed in the next section.

A locally important clay mineral that has not yet been mentioned is glauconite. **Glauconite** is a generally green-colored, potassium-iron rich illite that is produced in marine environments. Some glauconite is generated by slow precipitation in agitated, oxidizing, marine environments. However, glauconite also forms by the replacement of fecal pellets under marine conditions that are somewhat reducing. Although it is a clay mineral, some polycrystalline glauconite grains are of sand size. Glauconite sand grains occur as disseminated grains in sands in areas where detrital influx rates are large, but can be concentrated in areas where influx rates are smaller. In a few cases, extensive deposits occur with large concentrations of glauconite, such as the Cretaceous greensands of the New Jersey coastal plain and passive margin. Glauconite, whether it occurs in mudrocks or as grains in sandstones is a good, but not foolproof, indicator of marine sedimentation.

Mudrocks occur in a wide variety of colors that in many cases are closely related to their composition. Many mudrocks have white to pale gray to green colors that reflect the white to pale gray to green colors characteristic of clay minerals such as kaolinite (white), montmorillonite (pale gray to green), illite (pale green) and chlorite and glauconite (green). Other mudrocks possess colors that reflect the presence of nonclay coloring agents such as organic matter and iron oxides and hydroxides. Most red to purple mudrocks owe their color to the occurrence of the ferric iron ($Fe^{+3}$) oxide mineral hematite ($Fe_2O_3$). This records oxidizing conditions at the time the hematite was produced. Many red mudstones contain **reduction spots** (Figure 13.16). Reduction spots commonly form around decomposing organic matter which reduces the iron, enabling it to be removed in solution, which results in the removal of the red coloration.

Yellow to rusty-colored mudrocks contain the mineraloid limonite [$\sim FeO(OH) \cdot nH_2O$] and some brown mudrocks have colors related to the presence of the closely related mineral goethite (FeOOH). Most medium to dark gray and black mudrocks owe their color to the presence of finely-divided, carbon-rich organic material, although finely-divided pyrite ($FeS_2$) with its black streak may also contribute. The diagram in Figure 13.17 (after Potter et al. 1980; in Raymond 2002) summarizes some general relationships between color and mudrock composition.

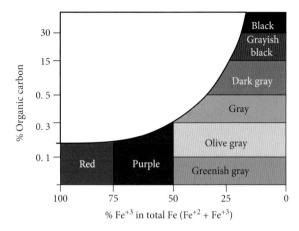

**Figure 13.17** Diagram showing the general relationships between color, organic content and oxidation state of iron in mudrocks. *Source:* Adapted from Raymond (2002) and Potter et al. (1980). © John Wiley & Sons.

### 13.4.3    Mudrock deposition

Mudrocks are the most widely distributed and abundant sedimentary rocks. Mud particles finer than coarse silt are generally transported in suspension by wind and water, whereas coarser particles are transported in the bedload (Chapter 12). This separation is maintained during deposition because small mud particles remain in suspension at very low velocities that are not capable of moving larger particles. Pure muds are deposited from suspension only in very calm environments where current velocities are exceedingly low. Mud dominated sediment depositional environments include:

1    Generally calm marine environments such as passive margin shelves below wave base, continental slopes, rises, submarine fans, abyssal plains and elsewhere on deep ocean floors; active margin trenches, forearc basins, and back-arc basins, below wave base, and in the deeper portions of marine foreland basins
2    Transitional (paralic) shorelines that are protected from significant wave and current activity such as lagoons, estuaries, fjords, bays, and tidal flats
3    Terrestrial environments, such as deltas, river floodplains, swamps, and the deeper portions of lakes, characterized by weak waves and/or currents

4    Wind-blown muds delivered to standing bodies of water where they settle out of suspension to accumulate on the bottom
5    Wind-blown loess: a silt-dominated mud produced when winds generated by the temperature differences across continental glacier margins erode fine glacial rock flour from glacial deposits and transport it away from the glacier, depositing it on land as the winds die down

### 13.4.4    Distinctive mudrock varieties

*Bentonites*

**Bentonites** are smectite-rich claystones formed by the alteration of volcanic ash deposits generated by explosive eruptions (see Chapter 9). The smectite, typically a variety of montmorillonite, forms by the alteration of glass shards and pumice fragments by diagenetic fluids. In addition to shards, clay, and silica, bentonites commonly contain the unaltered remains of euhedral crystals and angular mineral fragments such as sanadine, quartz, and plagioclase from the partially crystallized melt and various lithic fragments ripped from the walls of the vent during the explosive eruption. Because they consist of expandable clays with large shrink–swell potentials, bentonites swell when wet. When dried, they contract into small, cohesive pieces, giving dry bentonites deposits a distinctive, rubbly surface appearance that resembles "popcorn" (Figure 13.18). Their ability to swell when wet makes bentonites a potential geohazard (see Chapter 11).

Bentonite layers up to 50 m thick are known. Their thickness and the size of crystal and lithic fragments tend to decrease away from the explosive source and fragment size often decreases upward within the layer due to the more rapid settling of larger fragments. Distinctive bentonite layers can be distinguished on the basis of shard geometry, crystal composition, lithic fragment composition and bulk chemical composition. Wherever they occur, recognizable bentonites are of high stratigraphic value because each bentonite forms in a very short period of time, during short-lived explosive eruptions, yet can cover a large area. For this reason, bentonites are of great use as time horizons used to correlate geological events from one place to another. They also have many uses that are noted in Chapter 11.

**Figure 13.18**  Bentonite, with its typical, lumpy popcorn-like appearance, Mowry Formation, Cretaceous, Wyoming. *Source*: Courtesy of Wayne Sutherland, with the permission of Wyoming State Geological Survey. © Wayne Sutherland.

**Figure 13.19**  Oil shale with a dark color that results from its oil content. *Source*: Courtesy of US Department of Energy, Argonne National Laboratory.

### Carbonaceous mudrocks

**Carbonaceous mudrocks** are mudstones and claystones that contain sufficient carbon rich organic material to substantially influence their properties. A wide array of terms is used for these rocks. The terms reflect differences in properties, but also reflect the context in which the terms are used. **Black shales** are characterized by fissility and a black color which is largely the result of their elevated content (generally >2%) of incompletely decomposed, carbon-rich organic matter. The term **sapropel** (sapro = organic; pel = mud) is also used for organic rich mudstones and claystones, especially by oceanographers and paleooceanographers. **Oil shales** are characterized by organic material contents (generally 15–30%) that have the potential to yield profitable amounts of petroleum and/or natural gas when heated sufficiently. Most oil shales are black (Figure 13.19); some possess a dark brown color. Oil can be extracted from oil shales by the process of **pyrolysis** which involves crushing the shale and heating it, in the absence of air, to 500 °C. The costs of refining oil from shales, the costs of environmental protection associated with large amounts of solid waste material, and the low price of crude oil from traditional sources have inhibited large-scale production from oil shales. However, in light of the recent increases in petroleum prices and advances in hydraulic fracturing techniques, oil shales now represent a significant resource. A recent estimate from the American Association of Petroleum Geologists put the potential supply at $2.6 \times 10^{12}$ barrels, far more than the total reserves from traditional sources.

Carbonaceous mudrocks are also the **source rocks** for petroleum and natural gas. When a bituminous organic material called **kerogen** that occurs in such mudrocks is buried and heated to 100–140 °C, it is converted into petroleum and, when heated to over 160 °C, it is converted into natural gas. Because petroleum and natural gas are lighter than water, they can migrate upward from source rocks to accumulate beneath impermeable "cap rocks" in porous, permeable rocks such as limestones and sandstones. But much remains in the relatively impermeable mudrocks, from which it can now be released at reasonable cost by advanced fracking techniques.

Thick and thin accumulations of finely-laminated or fissile black shales, sapropels, and/or oil shales are widespread and locally abundant in Phanerozoic rocks. Very commonly they contain reduced iron sulfide ($Fe^{+2}S_2$) minerals in the form of the polymorphs pyrite or marcasite. These features suggest that black shales form when fine muds and organic matter accumulate together under low oxygen, reducing conditions. Several important factors interact to produce the conditions under which black shales, oil shales or sapropels accumulate. These include (1) high biological productivity of organic material in surface waters, (2) sluggish circulation which impedes both the influx of coarse detritus and oxygen replenishment of bottom waters and (3) the development of low oxygen conditions in bottom waters which impedes bacterial decomposition of organic material allowing it to accumulate in bottom sediments.

Biological productivity in near surface waters is a critical factor because high rates of biological productivity produce high rates of **biological oxygen demand (BOD)**. This occurs because the higher the rate or biological productivity, the higher the demand for oxygen by the aerobic bacteria that decompose organic material. In aqueous environments, as organisms die and the cellular material drifts toward the bottom elevated BOD begins to deplete dissolved oxygen. This is especially true in areas of sluggish circulation where the replenishment of oxygen to bottom waters is impeded. Such **anaerobic** conditions create a "dead" zone in which aerobic bacteria and organisms dependent on oxygen for respiration cannot survive. As a result, organic matter settling from surface waters is decomposed at much reduced rates by anaerobic bacteria that do not require oxygen and so tends to accumulate in bottom sediments. The sluggish circulation in these relatively stagnant waters, so important to impeding oxygen replenishment, also drastically reduces the influx of coarse detrital sediments. Instead relatively small amounts of mud accumulate with the organic matter without diluting it significantly. Because both the organic material and the mud particles are of very small size they settle out of suspension and accumulate together to produce black, organic-rich muds that become black shales or oil shales during diagenesis. Pyrite and marcasite form when reduced iron ($Fe^{+2}$) combines with sulfide produced by sulfate-reducing anaerobic bacteria to form iron sulfide ($FeS_2$).

The ideal conditions for the formation of black shales occur when large quantities of organic matter are produced in surface waters creating a high BOD, the rate of influx of detrital sediments is restricted to small amounts of mud per unit of time, and the bottom waters are depleted in oxygen because of sluggish circulation so that bacterial decomposition is inhibited. These conditions can exist in any

**Figure 13.20** Organic-rich sapropel layer (middle) between lighter colored layers in lake sediments, Jurassic Berlin Formation, Connecticut. *Source*: John O'Brien. © John Wiley & Sons.

basin in which circulation is sufficiently restricted whether it is a lake, a marginal sea, an epicontinental sea or a deep ocean basin. Examples of black shales, sapropels, and/or oil shales from all of these environments are well known. Figure 13.20 illustrates a sapropel deposited in an oxygen-depleted portion of a Jurassic lake in what is now Connecticut.

## 13.5 DIAGENESIS OF DETRITAL SEDIMENTS

As discussed in Chapter 11, all sedimentary rocks experience diagenesis. **Diagenesis** includes all sub-metamorphic, post-depositional changes that affect sediment after its accumulation. In detrital sediments, most, but not all, of these changes occur after the sediment is buried below the surface by the accumulation of additional sediments. Diagenesis occurs at relatively low temperature (up to ~$150 \pm 50\,°C$) and pressures (<1 kbar). These conditions result in the alteration of minerals and sediment textures under sub-metamorphic conditions. It is useful to distinguish stages of diagenesis that include (1) **eodiagenesis**: early, shallow diagenesis that occurs shortly after burial, (2) **mesodiagenesis**: later, deeper diagenesis, and (3) **telodiagenesis**: still later, shallow diagenesis that occurs as sedimentary rocks approach the surface due to the erosion of overlying rocks.

The major factors that control the diagenesis of detrital sediments include:

1   Temperature, which generally increases with depth
2   Confining pressure which increases with depth
3   Pore fluid chemistry, e.g. concentrations of dissolved solids, gases, pH, and Eh
4   Aqueous fluid circulation rates and patterns which depend on porosity, permeability, hydraulic head and rock structure

Diagenesis includes all of the **lithification** processes that convert unconsolidated sediments into sedimentary rocks, but involves many other changes discussed below. Because many of these processes change the porosity and permeability of sediments and sedimentary rocks, they can profoundly affect the rock's ability to hold and transmit fluids such as water, oil, and natural gas. As a result, diagenesis is of great interest to hydrologists, petroleum geologists and engineers. Because many diagenetic processes change the volume, thickness, and density of sediments, they are also of great interest to scientists who study the evolution of sedimentary basins. The discussion of diagenesis that follows is necessarily a brief overview. More comprehensive treatments, can be found in Burley and Worden (2009), MacKenzie (2005), Wolf and Chilingarian (1992, 1994), and McIlreath and Morrow (1990).

### 13.5.1   Compaction and pressure solution

Sediment **compaction** occurs as a result of increasing confining pressures as sediments are buried progressively deeper beneath the surface. The resulting increase in static pressure: (1) forces the particles closer together, (2) expels pore fluids such as air and water, (3) which results in a progressive decrease in porosity. These processes also tend to progressively reduce the volume and thickness of stratigraphic units, while increasing their density (Figure 13.21).

Rigid grains such as quartz and feldspar remain largely undeformed during compaction. In the absence of other diagenetic processes, reasonably well-sorted quartzarenite and arkosic sands can retain substantial porosity and permeability during burial (Figure 13.21).

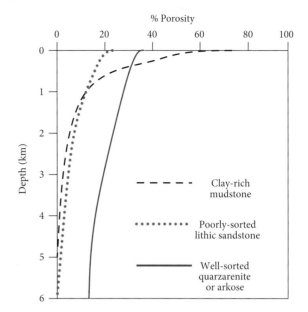

**Figure 13.21**   Idealized compaction and porosity curves for well-sorted quartzarenite or arkose, poorly sorted lithic sandstone and clay-rich mudstone.

Such sand layers are excellent fluid transmitters and generally act as aquifers or water-bearing horizons.

Wet muds commonly possess initial porosities in the range of 60–80%. During progressive burial, as pore fluids are expelled, muds undergo rapid reductions in porosity (Figure 13.21). Silt and especially plastic clay particles are squeezed together and the static charges on their surfaces cause the grains to be attracted to one another increasing their cohesion and converting muds into mudrocks. Eventually permeability may approach zero, making well compacted mudstone layers extremely impermeable. Such impermeable layers are very poor transmitters of fluids and act as aquicludes or aquitards. They form the confining layers that keep water from migrating upward from underlying **confined aquifers**. They also form **cap rocks** that keep petroleum and natural gas from migrating upward, thus effectively trapping them in underlying permeable layers.

In lithic sandstones, softer, more plastic grains such as mudrock, slate, and phyllite fragments may be substantially deformed. In thin-section, plastic grains can be seen to have been bent around more rigid grains and to have been squeezed between them to produce a diagenetic **pseudomatrix** that can be difficult to distinguish from detrital matrix constituents. The creation of pseudomatrix can drastically reduce the porosity of lithic sandstones and conglomerates, making them less efficient at storing and transmitting fluids.

As a general rule, the solubility of mineral crystals increases with increasing pressure or stress. During compaction, the highest stresses are transmitted at point contacts between rigid grains. When even a thin film of water is present, these grain contacts become the sites of dissolution, the rates of which increase with pressure. Such pressure induced dissolution is called **pressure solution** or **pressolution**. Pressolution is most common between grains in quartz sands and gravels with little or no matrix, and occurs less frequently in feldspathic sands and gravels. It occurs over a range of depths and causes significant decreases in porosity. Dissolution is controlled by the: (1) relative solubility of the two grains, (2) irregularities in the grain contacts so that not all parts of the grains are in contact and (3) presence of grain microfractures and other grain imperfections. As one or both grains dissolve, the length of their contacts tends to increase. As they evolve, such contacts may become relatively straight if both grains dissolve at similar rates, concavo-convex if one phase (concave) is more soluble than the other (convex) or sutured if microfractures or other imperfections or variable compositions lead to variable solubility between the two grains across their contact (Figure 13.22).

### 13.5.2 Dissolution and cementation

**Dissolution** occurs when crystalline solids partially or completely dissolve in pore fluids. These dissolved solid-bearing pore fluids migrate through fractures and between grains where they may eventually be precipitate fracture-filling veins and mineral cements. Dissolution is also an important process in the production of secondary porosity in sediments and sedimentary rocks, because it can produce a void space where the soluble solid previously existed. If such pore spaces are interconnected, dissolution can greatly increase permeability. Dissolution is selective; the most soluble constituents are preferentially dissolved.

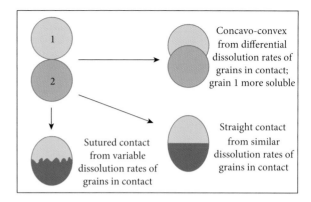

**Figure 13.22** Major varieties of long grain contacts produced by diagenetic pressure solution.

**Cementation** occurs when pore fluids precipitate intergranular mineral or mineraloid cements that bind the grains together. In order for pore fluids to precipitate mineral cements, they must first dissolve the constituents that are eventually precipitated.

Cementation generally decreases porosity by partially or wholly filling the voids between grains. It is possible of course for an earlier generation of cement to later be dissolved, increasing both porosity and permeability and making the rock an excellent host for water, natural gas and/or petroleum.

Because coarse silts, sands, and gravels commonly possess significant porosity and permeability, the precipitation of mineral cements that bind grains together is a significant process in their lithification. Given the relatively low dissolved solid content that characterizes of most subsurface waters, extraordinarily large amounts of groundwater must flow through such sediments in order to produce the significant amounts of pore-filling cements that they commonly possess.

The major cements in detrital sedimentary rocks include: (1) silica minerals, (2) carbonate minerals, (3) iron oxides and hydroxides, (4) feldspars and (5) clay minerals.

*Silica cements*

The most abundant type of **silica cement** is quartz. Quartz cement occurs chiefly in the form of **syntaxial quartz** overgrowths in

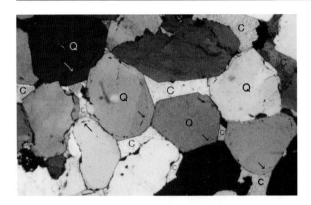

**Figure 13.23** Syntaxial quartz overgrowths (arrows) and blocky calcite cement (C) in quartzarenite (Q). *Source*: Courtesy of L. Bruce Railsback.

which the silica that precipitates from pore solutions initially nucleates on a preexisting detrital quartz grain. The quartz cement has the same crystallographic orientation as the detrital grain on which it nucleated and so the same orientation of its crystallographic axes; thus the term syntaxial. In thin section, both the syntaxial overgrowth and the detrital host grain go to extinction (Chapter 6) at the same time. This can make them difficult to distinguish. If the original quartz grain had a coating of clay or hematite or if gas bubbles (vacuoles) adhered to its surface, the boundary between the detrital quartz and the syntaxial cement will be revealed. Figure 13.23 shows rounded, detrital quartz grains surrounded by syntaxial quartz overgrowths. The overgrowths are separated from the host detrital grains by a surface with dust and vacuoles. If no dust or vacuoles are present, the boundary may be invisible and the size and shape of the original grain indistinguishable. In such cases, cement generations may develop long contacts with each other that mimic those produced by pressolution. Laboratory studies using cathodoluminescence imaging make relationships between grains and cements, including multiple cement generations, easier to distinguish. Quartz cement is particularly common in quartz-rich sandstones and gravelstones that accumulated in beach, dune, and shallow marine environments associated with intracratonic basins and passive margins.

The sources of the dissolved silica that is later precipitated during cementation are not completely understood. Quartz is rather insoluble at low temperatures, but its solubility increases with temperature and pressure. Amorphous opal is more soluble at all temperatures, even at low temperatures. Proposed sources for silica cements include the: (1) dissolution of opalline silica shells of organisms such as diatoms, radiolarian, and sponges, (2) the near-surface conversion of volcanic ash to bentonites which releases silica, (3) pressolution of silicate minerals, (4) authigenic reactions that release silica (see below), and (5) metamorphic reactions that release silica to fluids which cool as they rise through overlying sediments.

Less common varieties of silica cement include opal, chalcedony, chert, and nonsyntaxial quartz. **Opal** ($SiO_2 \cdot nH_2O$) is an amorphous mineraloid that contains various amounts of water. Opal is a common cement in volcanoclastic sediments. It is thought that the silica originates during the conversion of glass shards and pumice fragments into smectite clays (bentonites) which yields dissolved silica as a byproduct. Because opal solubility rapidly increases with temperature and depth, it is likely that opalline cements are precipitated at lower temperatures and pressures close to the surface during eodiagenesis. Opal solubility also depends on pH; opal is more soluble in mildly alkaline waters than in mildly acidic waters. Therefore, a decrease in pH can cause opal cement precipitation, whereas an increase in pH can cause dissolution. **Chert** ($SiO_2$) and **chalcedony** are cryptocrystalline to microcrystalline varieties of silica. Chert is characterized by microscopic, equant quartz crystals, whereas chalcedony crystals are bladed, commonly with a radiating to divergent habit when observed under a petrographic microscope.

*Carbonate cements*

**Carbonate cements** are the most abundant cements in sandstones and gravelstones. **Calcite** ($CaCO_3$) is by far the most abundant carbonate mineral cement. The solubility of calcite is strongly affected by the acidity-alkalinity (roughly pH) of subsurface fluids (Figure 13.24). Acidity is strongly influenced by the amount of

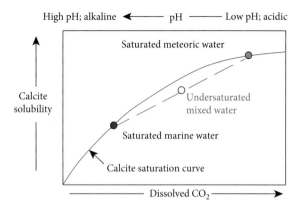

High pH; alkaline ⟵ — pH — ⟶ Low pH; acidic

**Figure 13.24** Solubility of calcite as a function of dissolved $CO_2$ content and acidity of natural pore waters; note that two saturated natural waters may mix to produce an undersaturated solution capable of dissolving calcite.

**Figure 13.25** Photomicrograph of calcite cemented sandstone with poikiloptic cement enclosing multiple quartz and feldspar grains (crossed polars). *Source*: Courtesy of Lee Phillips.

dissolved carbon dioxide. Any process such as organic respiration or bacterial decomposition of organic matter that releases carbon dioxide to pore solutions lowers their pH and increases their acidity. Any process that removes carbon dioxide from subsurface waters raises the pH, making them more alkaline.

As a result, the precipitation of calcite cements is favored by processes that increase the pH and alkalinity of subsurface solutions which causes them to become supersaturated with respect to calcium carbonate. Conversely, dissolution of calcite is favored by any process that decreases the pH and increases the acidity of subsurface waters. These processes and the most striking examples of subsurface dissolution of calcium carbonate, the formation of caves and other karst features, are discussed in more detail in Chapter 14.

Calcite cements in sandstones and gravelstones are commonly either blocky or poikiloptic. **Blocky cement** is composed of small calcite crystals that occupy pore spaces between detrital grains (Figure 13.23). **Poikiloptic cement** is composed of larger calcite crystals that nucleate and grow to fill multiple pore spaces so that they completely envelop several detrital grains (Figure 13.25) which appear as inclusions (called poikils) within a single crystal of calcite cement.

Rarer carbonate cements include **aragonite** ($CaCO_3$), the orthorhombic polymorph of

calcium carbonate, **dolomite** [$CaMg(CO_3)_2$], **ankerite** [$CaFe (CO_3)_2$], and **siderite** $Fe(CO_3)$. Because these minerals are important constituents of biochemical sedimentary rocks, their chemistry of is discussed in more detail in Chapter 14 which covers such rocks.

*Iron-rich cements*

The third most common cementing agents in sedimentary rocks are iron oxide and hydroxide minerals. Chief among these is **hematite** ($Fe_2O_3$) and a variety of water bearing minerals such as **goethite** (FeOOH) and mineraloids including **limonite** (complex, amorphous [$FeO(OH) \cdot nH_2O$]). The dissolved iron required for these minerals to form originates primarily from the alteration of ferromagnesian silicate minerals during weathering and diagenesis. The dissolved iron occurs as reduced ferrous iron ($Fe^{+2}$) which is quite soluble in waters with low oxidation–reduction potentials (Eh). Oxidation of such waters converts ferrous iron into oxidized ferric iron which is much less soluble. This leads to the precipitation of oxidized iron-bearing cements. Hematite is especially common as cement; its bright red color produces classic detrital red bed sequences. Hematite cementation is particularly common in sediments deposited in fairly arid terrestrial environment, such as alluvial fans, braided streams, meandering stream channels and flood plains and deserts. In such environments, the paucity of vegetative cover and consequent paucity of bacterial decomposition, leads to oxidizing ground waters that precipitate amorphous hydrated iron oxide [$FeO(OH) \cdot nH_2O$] which is rapidly dehydrated into hematite ($Fe_2O_3$).

*Rarer cements*

**Feldspar cements** occur in feldspar-rich detrital sedimentary rocks such as arkosic sandstones and gravelstones. Feldspar cements appear microscopically as optically-clear overgrowths that nucleated on feldspar host grains. They include orthoclase overgrowths on potassium feldspar grains and albite overgrowths on both plagioclase and potassium feldspar grains.

**Clay cements** occur in some detrital sedimentary rocks. Clay mineral stabilities are strongly controlled by temperature and pH. Kaolinite cement generally occurs as stacks of platy layers called "books" that precipitate at fairly shallow depths from low-K, acidic pore waters during eodiagenesis and telodiagenesis. Acidic pore waters are especially common in continental settings, which is where most kaolinite cements are precipitated. **Illite cement** generally forms at higher temperatures and depths from high-K, alkaline pore waters, especially in marine settings. Illite cements form mostly during late eodiagenesis and mesodiagenesis.

## 13.5.3  Additional diagenetic processes

**Mineral alteration** or **replacement** occurs when one mineral crystal is altered to another during diagenesis. Chemically stable minerals such as quartz typically display little or no alteration. Many potassium feldspar crystals are altered to clays and to the fine-grained mica called sericite which is closely related to muscovite. At higher temperatures, both calcic and potassic feldspars are commonly altered to albite. Typically such alteration begins at grain margins or along cleavage or fracture surfaces where minerals are in contact with pore solutions or aqueous films. Volcanic rock fragments commonly alter to zeolite minerals during progressive burial and diagenesis. The type of zeolite mineral yields insights into the temperature conditions and burial depth at which diagenesis occurred.

Clay mineral stability is strongly influenced by temperature. During burial, increasing temperatures lead to the transformation of lower temperature clay minerals into higher temperature mineral assemblages. In general

1  Smectites are transformed into mixed-layer clays above 100 °C
2  Kaolinite is converted into illite or chlorite above 150 °C, a metamorphic process.
3  Mixed layer clays are metamorphosed into more ordered illite above 200 °C
4  All clay minerals are metamorphosed into chlorite or micas such as muscovite above 300 °C

Many of these reactions also involve the zeolite minerals, with analcime and heulandite stable below 100 °C, laumontite stable from 100 to 200 °C, and prehnite and pumpellyite stable above 200 °C. In fact, very low-grade metamorphic rocks are often ascribed to the zeolite or prehnite–pumpellyite facies, reflecting the transitional nature between diagenesis and low-grade metamorphic processes (Chapter 15).

## 13.5.4  Diagenetic structures

Several diagenetic sedimentary structures occur in detrital sediments. The most frequently seen and/or significant of these include: (1) concretions, (2) nodules, (3) geodes, and (4) liesegang rings or bands.

**Concretions** form by the precipitation of material around a nucleation surface, such as a fossil, sand grain or shale chip. Multiple periods of precipitation cause many concretions to have a concentric structure and a roughly spherical or ellipsoidal shape (Figure 13.26a). Continued growth may cause several small concretions to coalesce into larger concretions with more complex shapes. The precipitation of mineral cement such as limonite or hematite in oxidizing conditions often causes the concretion to be harder and more resistant to weathering than the rest of the rock. As a result concretions weather out of outcrops as cannon-ball-like structures (Figure 13.26b). Siderite, pyrite or marcasite concretions form under reducing conditions. Calcite is a common component of concretions formed under a variety of oxidizing-reducing conditions.

(a)

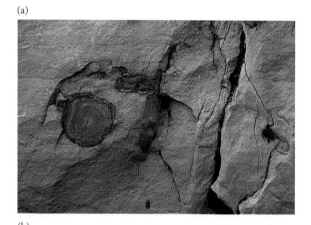

(b)

**Figure 13.26**   Images of concretions. (a) A nucleus and concentric structure in sandstone. *Source*: Courtesy of John Merck. (b) A concretion in shale, Devonian, Virginia. *Source*: Courtesy of Duncan Heron.

(a)

(b)

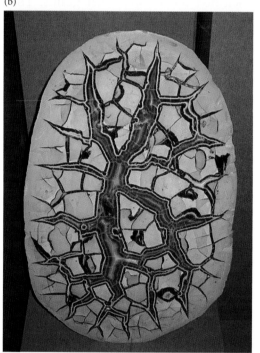

**Figure 13.27**   Nodules. (a) Three flint nodules in limestone. *Source*: John O'Brien. © John Wiley & Sons. (b) Septarian nodule; Pinch Collection, Canadian Museum of Nature, Ottawa, Ontario. *Source*: Courtesy of Keith Pomakis, cleared by the museum.

The terms nodule and concretion are sometimes used in an overlapping manner. Strictly speaking, **nodules** are similar to concretions, but lack a well-defined nucleus and generally lack concentric growth rings. Chert nodules are especially abundant in limestones and dolostones. Recall that limestones are increasingly soluble as pH decreases and acidity increases. These same conditions often favor the precipitation of opaline silica which is less soluble in acidic waters. Under such conditions, opaline silica may replace the calcite in limestones to produce nodules which later recrystallize into chert or flint (Figure 13.27a).

A spectacular group of nodules are **septarian nodules** (Figure 13.27b) which are characterized by mineral-filled cracks whose origin remains controversial.

**Geodes** are nodules that were or are partially hollow. Many geodes have roughly spherical to more irregular outer layers of chalcedony and a cavity lined with crystals (Figure 13.28) that, in some cases, have grown to fill the cavity.

Their origin remains controversial and may well involve more than one method of formation. One widely accepted explanation is that geodes originate as nodules or concretions of calcite or anhydrite. In acidic ground waters,

**Figure 13.28** Geodes showing banded chalcedony rims and quartz (plus bladed gypsum) crystal linings.

**Figure 13.29** Liesegang bands in a sandstone block; note the truncation against the joint surfaces. *Source*: Courtesy of Peter Adderley.

their exteriors are replaced by chalcedony. The interiors eventually dissolve permitting the precipitation of the crystals that line or fill their interiors.

**Liesegang bands** (Figure 13.29) are common in many detrital sedimentary rocks, especially those with carbonate or iron-rich cements. Although the bands sometimes mimic stratification, they are clearly secondary features that are often truncated against prominent joint surfaces. They form by the precipitation of various iron oxide minerals from fluids moving through bodies of rock separated by fractures. Fluids moving into such bodies of rock from the outside-in often produce ring-like patterns of bands called **Lisegang rings**.

The diagenesis of sedimentary rocks provides insights into the subsurface processes that lead to lithification. Diagenetic processes significantly alter the porosity, permeability, and density of sediments as they undergo compaction, pressolution, cementation, and chemical alteration. This in turn has a profound effect on their ability to transmit fluids such as water, natural gas and petroleum. We hope this overview of detrital sedimentary rocks and the processes that generate them will prove to be of value to you in the future.

## CONTENT ASSESSMENT

1 Detrital sediments are composed of particles of different sizes. Please distinguish between *gravel*, *sand*, *mud*, *silt*, and *clay*. When lithified, such sediments become *gravelstones*, *sandstones* or *mudrocks*. How are these three major groups of sedimentary rock distinguished from one another? What are the differences between (a) *clast supported* and *matrix supported* gravelstones, (b) *arenites* and *wackes* and (c) *claystones*, *siltstones*, *mudstones*, and *shales*?

2 Explain the difference between the *central measures* of a population of grains: (a) mode, (b) median and (c) mean. Then discuss the concept of size *sorting* and grain *roundness* and how these measures, along with grain *size*, can be used to make inferences concerning the depositional agent that deposited any detrital sediment (recognizing that sedimentary structures and field relationships would allow a more precise determination).

3 Compare and contrast *orthoconglomerates* and *paraconglomerates (diamictites)* and the environmental conditions that produce them. Then compare oligomictic and polymictic gravelstones and offer possible explanations for the similarities and contrasts between them.

4 Using Folk's classification (Figure 13.13), please give a compositional name for the sandstones whose compositions are described below. Then give each sandstone another name based on the four-component

diagram (Figure 13.14) that includes the percentage of mud matrix.

a   89% Q, 8% F, 2% L and 1% mud matrix.
b   52% Q, 11% F, 37% L and no mud matrix
c   94% Q, 4% F and 2% mud matrix
d   38% Q, 11% K-feldspar, 5% Plagioclase, 24% L and 20% mud matrix
e   73% Q, 15% F, 2% L and 10% mud matrix

5   Using Figures 13.13, 13.14 and 13.15, what are the: a. names, b. most likely provenance and c. most likely tectonic setting for each of the following sandstones whose compositions are given below.

a   quartz (97%), k-feldspar (2%), durable lithic fragments (1%) (mostly chert)
b   quartz (50%), k-feldspar (34%), plagioclase (6%), lithic fragments (3%) (mostly GRF)
c   quartz (62%), k-feldspar (12%), plagioclase (7%), lithic fragments (22%)
d   quartz (23%), k-feldpar (4%), plagioclase (15%), lithic fragments (54%) (mostly VRF)

6   Describe the major varieties of carbonaceous mudrock. Explain the environmental conditions are conducive to their formation.

7   Distinguish between the three major stages of diagenesis, eodiagenesis, mesodiagenesis, and telodiagenesis in terms of the sequence in which they occur the depths below the surface at which each occurs, and the temperature and pressure conditions under which each occurs. During which of these stages do the following most commonly occur?

a   rapid compaction of muds
b   joint development and opening
c   maximum pressolution
d   conversion of kaolinite into illite

## REFERENCES

Armstrong-Altrin, J.S., Lee, Y.I., Verma, S.P., and Ramasamy, S. (2004). Geochemistry of sandstones from the upper Miocene Kudankulam Formation, southern India: implications for provenance, weathering, and tectonic setting. *Journal of Sedimentary Research* 74: 285–297.

Blatt, H. and Tracey, R. (1996). *Petrology: Igneous, Sedimentary, and Metamorphic*, 2e. New York: W.H. Freeman Publishers 497 pp.

Blatt, H., Tracey, R., and Owens, B.R. (2006). *Petrology: Igneous, Sedimentary and Metamorphic*. New York: W.H. Freeman Publishers 530 pp.

Burley, S. and Worden, R. (eds.) (2009). *Sandstone Diagenesis: Ancient and Modern*. New York: Wiley-Blackwell 656 pp.

Campbell, I.H., Reiners, P.W., Allen, C. et al. (2005). He–Pb double dating of detrital zircons from the Ganges and Indus Rivers: implications for quantifying sediment recycling studies. *Earth and Planetary Science Letters* 237: 402–432.

Compton, R.S. (1962). *Manual of Field Geology*. New York: Wiley 378 pp.

Crowell, J.C. (1952). Probable large lateral slip faulting on San Gabriel Fault. *American Association of Petroleum Geologists Bulletin* 36: 2026–2035.

Cutten, H.N.C. (1979). Rapphannock Group: Late Cenozoic sedimentation and tectonics contemporaneous with Alpine Fault movement. *New Zealand Journal of Geology and Geophysics* 22: 535–553.

Cutten, H.N.C., Korsch, R.J., and Roser, B.P. (2006). Using geochemical fingerprinting to determine transpressive fault movement history: application to the New Zealand Alpine Fault. *Tectonics* 25: 1029–1047.

Dickinson, W.R. (1970). Interpreting detrital modes of graywacke and arkose. *Journal of Sedimentary Petrology* 40: 695–707.

Dickinson, W.R., Beard, L.S., Brakenridge, G.R. et al. (1983). Provenance of North American Phanerozoic sandstones in relation to tectonic setting. *Geological Society of America Bulletin* 94: 222–235.

Dickinson, W.R. and Gehrels, G.E. (2003). U-Pb ages of detrital zircons from Permian and Jurassic eolian sandstones of the Colorado Plateau, USA: paleogeographic implications. *Sedimentary Geology* 163: 29–66.

Dickinson, W.R. and Suczek, C.A. (1979). Plate tectonics and sandstone compositions. *American Association of Petroleum Geologists Bulletin* 63: 2164–2182.

Folk, R.L. (1951). Stages of textural maturity in sedimentary rocks. *Journal of Sedimentary Petrology* 21: 127–130.

Folk, R.L. (1974). *Petrology of Sedimentary Rocks*. Austin, TX: Hemphill Publishing 182 pp.

Folk, R.J., Andrews, P.B., and Lewis, D.W. (1970). Detrital sedimentary rock classification and nomenclature for use in New Zealand. *New Zealand Journal of Geology and Geophysics* 13: 937–968.

Gehrels, G.E., Blakey, R., Karlstrom, K.E. et al. (2011). Detrital zircon U–Pb geochronology of Paleozoic strata in the Grand Canyon, Arizona. *Lithosphere* 3: 183–200.

Higgs, K.E. and King, P.R. (2018). Sandstone provenance and sediment dispersal in a complex tectonic setting: Taranaki Basin, New Zealand. *Sedimentary Geology* 372: 112–134.

Krumbein, W.C. (1934). Size frequency distribution of sediments. *Journal of Sedimentary Petrology* 4: 65–77.

Krumbein, W.C. and Graybill, F.A. (1966). *Introduction to Statistical Methods in Geology*. New York: McGraw-Hill 475 pp.

Lamb, S., Mortimer, N., Smith, E., and Turner, G. (2016). Focusing of relative plate motion at a continental transform fault: cenozoic dextral displacement >700 km on New Zealand's Alpine Fault, reversing >225 km of Late Cretaceous sinistral motion. *Geochemistry, Geophysics, Geosystems* 17: 1197–1213.

Lewis, D.W. (1984). *Practical Sedimentology*, 229. Hutchinson Ross: Stroudsburg, PA.

MacKenzie, F.T. (ed.) (2005). *Sediments, Diagenesis and Sedimentary Rocks*. Amsterdam: Elsevier 446 pp.

McIlreath, I.A. and Morrow, D.W. (eds.) (1990). *Diagenesis*. Geological Association of Canada Geoscience Canada Reprint Series No. 4, 338 pp.

Okada, H. (1971). Classification of sandstone: analysis and proposal. *Journal of Geology* 79: 509–525.

Pettijohn, F.J., Potter, P.E., and Siever, R. (1987). *Sands and Sandstones*. New York: Springer-Verlag 586 pp.

Plumley, W.J. (1948). Black Hills terrace gravels: a study in sediment transport. *Journal of Geology* 56: 526–577.

Potter, P.E., Maynard, J.B., and Pryor, W.A. (1980). *Sedimentology of Shale*. New York: Springer-Verlag 306 pp.

Powers, M.C. (1953). A new roundness scale for sedimentary particles. *Journal of Sedimentary Petrology* 23: 117–119.

Preston, J., Hartley, A., Mange-Rajetzky, M. et al. (2002). The provenance of Triassic continental sandstones from the Beryl Field, northern North Sea: mineralogical, geochemical, and sedimentological constraints. *Journal of Sedimentary Research* 72: 18–29.

Prothero, D.R. and Schwab, F. (2015). *Sedimentary Geology: An Introduction to Sedimentary Rocks and Stratigraphy*. San Francisco: W.C. Freeman 593 pp.

Rahl, J.M., Reiner, P.W., Campbell, J.H. et al. (2003). Combined single-grain (U–Th)/He and U/Pb dating of detrital zircons from the Navajo Sandstone, Utah. *Geology* 31: 761–764.

Raymond, L.A. (2002). *The Study of Igneous, Sedimentary and Metamorphic Rocks*. Boston: McGraw-Hill 720 pp.

Schulz, E.F., Wilde, R.H., and Albert, M.L. (1954). Influence of shape on the fall velocity of sedimentary particles. In: *Sedimentation Series Report No. 5*. Omaha, NB: US Army Corps of Engineers.

Sutherland, R. (1994). Displacement since the Pliocene along the southern section of the Alpine Fault, New Zealand. *Geology* 22: 327–330.

Thomas, W.A. (2011). Detrital zircon geochronology and sedimentary provenance. *Lithosphere* 3: 304–308.

Tucker, M. (2001). *Sedimentary Petrology*, 3e. Blackwell Science: Oxford, UK 272 pp.

Wolf, K.H. and Chingarian, G.V. (eds.) (1992). *Diagenesis III*. Amsterdam: Elsevier 674 pp.

Wolf, K.H. and Chingarian, G.V. (eds.) (1994). *Diagenesis IV*. Amsterdam: Elsevier 546 pp.

Zingg, T. (1935). Beiträge zur Schotteranalyse. *Mineralogische und Petrographische Schweizerische Mitteilungen 15*: 39–140.

# Chapter 14

# Biochemical sedimentary rocks

## 14.1 INTRODUCTION

In Chapter 13, we discussed the rocks created when detrital sediments produced by weathering and erosion are dispersed and deposited by water, wind, glaciers, and gravity and lithified into detrital sedimentary rocks. In this chapter, we discuss the fate of the dissolved solids produced by decomposition, volcanism, and other surface and subsurface dissolution processes.

Organisms produce **organic (biogenic) sediments** via biochemical precipitation of skeletal materials, such as shells, bones, and teeth, and by organic synthesis of cellular materials to make organic tissues. In precipitating shells, bones and teeth, organisms extract dissolved ions from the environment and secrete skeletal materials composed of **"biomineraloids"** such as calcite, aragonite, silica, or calcium phosphate that accumulate on Earth's surface. In synthesizing organic tissues, organisms remove dissolved carbon, hydrogen, oxygen, nitrogen, sulfur, and phosphorous from solution to produce a variety of solid organic molecules. Once produced, organic sediments can accumulate in place, as **in situ** sediments, or be dispersed as **bioclastic** sediments and deposited elsewhere on Earth's surface. **Chemical sediments**, including gypsum ($CaSO_4 \bullet 2H_2O$), halite ($NaCl$), and some cherts ($SiO_2$) are formed by inorganic precipitation of minerals from solution to form solid sediments, which then accumulate on Earth's surface.

*Earth Materials*, Second Edition. Kevin Hefferan and John O'Brien.
© 2022 John Wiley & Sons Ltd. Published 2022 by John Wiley & Sons Ltd.
Companion website: www.wiley.com/go/hefferan/earthmaterials2

The boundary between organic and chemical sediments is often fuzzy. Organic activities cause changes in solution geochemistry such as oxidation–reduction potential (Eh) and acidity–alkalinity (roughly pH) that in turn trigger chemical precipitation. Are the crystals thus formed of organic or inorganic origin, or both? Because the dividing line between organic and chemical sediments is sometimes unclear, it is convenient to combine the two into a larger group called **biochemical sediments**. This is also convenient because the solid crystals produced by organic precipitation and inorganic precipitation often have the same composition. A good example is provided by the calcium carbonate ($CaCO_3$) minerals calcite and aragonite. They are the principle components of most organically precipitated shells but are also inorganically precipitated in caves, around springs, from lakes, as microcrystals from seawater and as diagenetic cements. Many limestones are composed of calcium carbonate produced by both organic and inorganic processes. This chapter discusses a variety of biochemical sedimentary rocks, with emphasis on their occurrence, composition, classification, origins, diagenesis and uses. Biochemical sedimentary rocks have been the focus of intensive study because they are extremely important as hydrocarbon reservoirs, groundwater aquifer units and are the sources of many critical industrial materials.

## 14.2 CARBONATE SEDIMENTARY ROCKS

Carbonate sedimentary rocks are by far the most abundant group of biochemical rocks. They constitute some 15% of all sedimentary rocks. They are composed chiefly of calcite and aragonite, the polymorphs of calcium carbonate ($CaCO_3$), and the calcium–magnesium carbonate mineral dolomite [$CaMg(CO_3)_2$]. Following common usage, we will employ the term minerals to describe these crystals whether or not they are of organic origin. The two most abundant carbonate sedimentary rocks are **limestone**, composed primarily of calcite and/or aragonite, and **dolostone**, composed principally of dolomite. Limestones are essential to the construction industry, because when treated, they provide the principal ingredient lime ($CaO$) used in the production of cement products.

### 14.2.1    Carbonate mineralogy

The most abundant mineral in carbonate sedimentary rocks is **calcite ($CaCO_3(r)$)**, the rhombohedral polymorph of calcium carbonate. Because limited amounts of smaller magnesium ($Mg^{+2}$) ions can substitute for calcium ($Ca^{+2}$) ions in the crystal lattice, calcite crystals can be classified according to their magnesium content. This substitution can be viewed as a limited substitution of **magnesite ($MgCO_3$)** for calcite, as explained in Chapter 2. Calcite crystals with less than 4% magnesium substitution for calcium are referred to as **low magnesium calcite (LMC)**, whereas those with more than 4% magnesium substitution are referred to as **high magnesium calcite (HMC)** (Figure 14.1).

The percentage of magnesium in calcite is significant because it plays a major role in the solubility of calcite in different pore waters. As a result, it provides important clues regarding the history of carbonate sediments and sedimentary rocks. HMC is far more abundant and low magnesium calcite less abundant in modern carbonate sediments than in ancient carbonate sedimentary rocks, in which their relative abundances are reversed. This raises questions concerning the causes of these changes in abundance with age.

The second common polymorph of calcium carbonate is the orthorhombic polymorph **aragonite ($CaCO_3(o)$)**. Together with calcite, it is an essential mineral in limestones. Because of its orthorhombic structure, small cations such as magnesium do not substitute for calcium to a significant degree. However, larger cations such as strontium ($Sr^{+2}$) substitute for calcium ($Ca^{+2}$) in much greater amounts than they do in the rhombohedral calcite structure. These large ion substitutions occur on a scale of only a few parts per thousand. Aragonite is far more abundant in modern carbonate sediments than in ancient carbonate sedimentary rocks. Why?

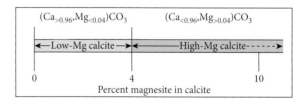

**Figure 14.1**    Compositions of low magnesium and high magnesium calcites.

In general, aragonite is most unstable during diagenesis, especially in meteoric water, whereas low magnesium calcite is most stable, especially in meteoric water. This largely explains why aragonite and HMC become progressively less common in older rocks. The composition of skeletal materials produced by organisms that secrete calcium carbonate varies between and within groups of organisms (Table 14.1).

The major mineral in dolostones and a subordinate constituent in many limestones is the double carbonate mineral **dolomite** [CaMg(CO$_3$)$_2$]. Because of their similar ionic radius, charge, and availability, a significant substitution of ferrous iron (Fe$^{+2}$) for magnesium (Mg$^{+2}$) occurs in many dolomite crystals. This can be modeled as a solid–solution substitution between **ankerite** [CaFe(CO$_3$)$_2$] and dolomite. While dolomite is a rather rare component of modern carbonate sediments, its abundance increases with age in ancient carbonate sedimentary rocks, and it is especially abundant in Precambrian rocks. Once again, this raises significant questions concerning changes in the composition of carbonate rocks with age and their probable causes.

**Siderite (FeCO$_3$)** is a rock-forming carbonate mineral common in iron-rich sediments (Section 14.4) and rarer carbonates occur in some lacustrine settings. Table 14.2 summarizes the significant minerals of carbonate sediments and sedimentary rocks.

### 14.2.2 Conditions for carbonate accumulation

Although carbonate sedimentary rocks are widely distributed in the geological record, the production of thick carbonate sediment accumulations requires a specific set of favorable environmental conditions:

1 Large rates of carbonate production.
2 Large rates of carbonate preservation.
3 Small rates of detrital sediment influx.

**Table 14.1** Common compositions of carbonate minerals for major groups of organisms that secrete or secreted calcium carbonate skeletons.

| Group | Subgroup | Skeletal composition | | |
|---|---|---|---|---|
| | | | Calcite | |
| | | Aragonite | Low-Mg | High-Mg |
| Bacteria | Blue-green | | √ | |
| Algae | Red | | | √ |
| | Green | √ | | |
| | Coccoliths | | √ | |
| Foramininera | Planktonic | | √ | |
| | Benthic | | √ | √ |
| Sponges | | | | √ |
| Annelids | Serpulids | √ | √ | √ |
| Coelenterates | Stromotoporoids | √ | √ | |
| | Rugose corals | | √ | |
| | Tabulate corals | | √ | |
| | Scleractinids | √ | | |
| | Alcyonarians | | | √ |
| Bryozoans | | √ | | |
| Brachiopods | | | √ | |
| Mollusks | Bivalves | √ | √ | |
| | Gastropods | √ | √ | |
| | Pteropods | √ | | |
| | Cephalopods | √ | | |
| | Belemonoids | | √ | |
| Arthropods | Decapods | | | √ |
| | Ostracods | | √ | √ |
| | Trilobites | | √ | |
| Echinoderms | | | √ | √ |

**Table 14.2** Common minerals in carbonate rocks.

| Mineral | Composition | Temporal distribution |
| --- | --- | --- |
| Low-Mg calcite | $(Ca_{>0.96}, Mg_{<0.04})CO_3$ | More abundant in ancient than in modern carbonate sequences |
| High-Mg calcite | $(Ca_{<0.96}, Mg_{>0.04})CO_3$ | More abundant in modern than in ancient carbonate sequences |
| Aragonite | $(CaCO_3)$; limited substitution of $Sr^{+2}$ for $Ca^{+2}$ | More abundant in modern than in ancient carbonate sequences |
| Dolomite | $[CaMg(CO_3)_2]$; limited substitution of $Fe^{+2}$ for $Mg^{+2}$ | More abundant in ancient than in modern carbonate sequences |

For reasons elaborated below, these conditions occur primarily in warm, shallow marine environments, with sufficient nutrients, in tropical and subtropical regions located in:

1 Low relief, tectonically stable areas on passive margins or in intracratonic seas where craton platforms flooded by fairly shallow water during sea level high stands.
2 Relatively shallow areas far from continents such as oceanic platforms, ocean islands, seamounts, and ocean ridges.

The majority of calcium carbonate $(CaCO_3)$ sediment is produced by organic activities such as shell secretion and precipitation by algae and bacteria. Any set of processes that increases the biomass of such organisms increases calcium carbonate production. Because sunlight is essential to photosynthesis and biological productivity, most carbonate sediment is produced in the shallow, sunlit waters of the photic zone. Because nutrients such as nitrate $(NO_3^{-1})$, phosphate $(PO_4^{-2})$, and sulfate $(SO_4^{-2})$ are essential to the synthesis of complex organic molecules, carbonate production is especially high in shallow waters with a reasonable availability of nutrients. The oversaturation and precipitation of organic or inorganic calcium carbonate are also favored by high temperature and wave agitation. These are also favored by low dissolved $CO_2$ content and elevated alkalinity (higher pH). These conditions are also common in fairly shallow water where most carbonates accumulate.

Once calcium carbonate sediments are produced, one might expect them to eventually accumulate on Earth's surface. But this is not always the case. In modern oceans, calcium carbonate dissolution occurs wherever warm, $CO_2$-poor, more alkaline, carbonate producing surface waters are underlain by colder, more acidic (lower pH), $CO_2$-rich deeper waters. In most parts of the deep ocean, as carbonate shells sink to deeper levels, they begin to undergo dissolution. This dissolution becomes significant at a depth called the *lysocline*. It becomes complete, where deeper bottom waters are sufficiently cold and acidic. Below a certain depth all $CaCO_3$ is dissolved. This depth below which calcium carbonate sediments do not accumulate is called the **carbonate compensation depth (CCD)**. In the tropics, the CCD occurs at a depth of 3000–5000 m. In subpolar regions, where cold, $CO_2$-rich water occurs close to or at the surface, the CCD does so as well. This helps to explain why carbonate sediments are abundant in warm, tropical seas where warm, alkaline waters favor $CaCO_3$ precipitation and preservation, and much less common, but not absent from, in cold, subpolar seas where more acidic waters inhibit precipitation and aid in the dissolution of $CaCO_3$.

If $CaCO_3$ is produced and preserved, it will accumulate as carbonate sediment on Earth's surface. Whether carbonates dominate sediment accumulation depends on the rate at which other types of sediment accumulate. Most importantly, it depends on the rate at which detrital mud, sand, and gravel are delivered to the site in question. If the amount of detrital sediments that accumulate in an area exceeds carbonate production and preservation, a carbonate-bearing detrital sediment (e.g., a fossil-bearing sandstone or a calcareous mudrock) will form instead of a limestone with little or no detrital sediment content. Rates of detrital sediment influx increase with proximity to areas of (1) high relief, which increases erosion rates, (2) elevated terrestrial precipitation, which generally increases erosion rates, and (3) proximity to large rivers that deliver significant volumes of detrital sediment to the coastline. This helps to explain why limestones and dolostones are rare in trench arc systems and in the portions of foreland basins adjacent to orogenic belts and are abundant along passive margins, in epicontinental seas, around small islands, and over platforms far from continents.

Much smaller volumes of carbonate sediments are formed in cool water, temperate, even subpolar, marine settings, as discussed later in this chapter. In addition, carbonate sediments are generated in terrestrial environments and include (1) stalactites, stalagmites, columns, and flowstone precipitated from groundwater in caves, (2) travertine and tufa deposits formed around springs and (3) carbonate sediments precipitated from some lake waters.

### 14.2.3 Components of carbonate rocks

Most carbonate rocks consist of various combinations of three major groups of biochemical components: (1) sand- or gravel-size clastic particles called grains or allochemical constituents (allochems), (2) mud-sized particles called mud or micrite, and (3) mineral cements. Two other types of constituents are important contributors to some carbonate rocks: (4) organically bound accumulations of carbonate called boundstones or biolithites, and (5) an array of diagenetic products that record the dissolution, replacement, and recrystallization of carbonate minerals during their low temperature alteration.

*Grains or allochems*

**Grains** or **allochems** are sand- and/or gravel-size carbonate particles. If you have ever seen a "shell" beach in Florida, Bermuda, Hawaii, or elsewhere (Figure 14.2a), you have seen a beach composed largely of carbonate grains or allochems. Allochems include the following:

1  Shells and other skeletal particles.
2  Spherical particles called ooids.
3  Clasts of carbonate sediment called limeclasts.
4  Smaller pellet-like particles called peloids.

Each grain type is an important constituent of carbonate sediments forming today, so that their occurrences and origins can be directly investigated. The properties and origins of each of these grain types are discussed in the sections that follow.

Skeletal particles: characteristics, formation, and occurrence
The most abundant sand- and/or gravel-size components of carbonate sedimentary rocks

**Figure 14.2** (a) Shell beach, Hinchinbrook Island, Queensland, Australia: an example of carbonate sediment composed largely of grains or allochems. (b) Fossil-bearing limestone, middle Devonian, New York, which records marine life of 435 million years ago. *Source*: John O'Brien. © John Wiley & Sons.

such as limestones and dolostones are **skeletal particles** that are initially precipitated by organisms. Skeletal particles (Figure 14.2b) include (1) whole and disarticulated shells, (2) shell fragments and (3) a variety of internal support structures. The vast majority of shelled macroorganisms, including mollusks, echinoderms, corals, bryozoans, brachiopods, and arthropods, secrete calcium carbonate shells composed of low magnesium calcite, HMC, and/or aragonite. Many microorganisms, including foraminifera, coccoliths, pteropods and many algae, also secrete calcium carbonate shells or internal skeletal particles. Some microorganisms are large enough to contribute to carbonate grain populations; others contribute to carbonate mud populations. Skeletal particles of bottom-dwelling (benthic) organisms may be preserved in place. These can permit inferences to be made about the environment of sediment accumulation. More commonly, shells behave as bioclastic particles, transported from the place where they formed to the site of their final accumulation. For example, the mollusk shells in Figure 14.2a were transported to their present location by a storm surge. In either case, skeletal particles record significant information about the kinds of organisms that have inhabited Earth's surface over time; when they evolved and when they became extinct.

## Ooids: characteristics, formation and occurrence

Among the most intriguing components of carbonate grain populations are ooids (Figure 14.3). Ideally, **ooids** are roughly spherical, concentrically laminated, sand-size particles that possess a nucleus. Modern ooid sands occur primarily in shallow, subtropical, and tropical environments that are agitated by waves and/or tidal currents. These warm, agitated conditions favor low dissolved $CO_2$ and high pH which favor the precipitation of calcium carbonate. Ooids form by the accretion of calcium carbonate laminae about a particle such as a shell fragment or sand grain that acts as a nucleus for precipitation. Ooids can display concentric and/or radial calcium carbonate structures (Figure 14.3b). The manner of such accretion remains controversial but seems in many cases to involve endolithic bacteria such as cyanophytes. These bacteria inhabit the surface of the particle and aid in the binding and/or precipitation of additional calcium carbonate laminae as the ooid grows. The rotation of ooids in the bed load of wave- and current-agitated environments ensures that the laminae will be of subequal thickness on all sides, thus producing approximately spherical particles. The size of the ooids is governed in part by the maximum size that can be entrained in the bed load and is generally in the sand range (<2 mm). Larger ooids are produced in exceptionally

(a)

0.50 mm

(b)

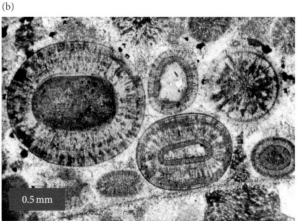

0.5 mm

**Figure 14.3**   (a) Modern ooids, Joulter Cay, Bahamas. *Source*: Photo courtesy of Jack Morelock. (b) Thin section of ooids that shows nuclei, concentric layering and radial crystals. *Source*: Mark A. Wilson, College of Wooster – Wikimedia CCO. https://commons.wikimedia.org/wiki/File:CarmelOoids.jpg; last accessed September 6, 2020.

calcium carbonate oversaturated, agitated environments. Ooids from less agitated environments tend to have fewer and less symmetrical coatings that reflect their less frequent rotation. Although ooids can be dispersed from such environments into both deeper marine environments and even terrestrial environments, ooid-rich limestones and dolostones are suggestive of warm, shallow, marine, wave- and/or tidal current-agitated environments of deposition in a tropical or subtropical setting. It should be noted in passing that small numbers of ooids also form in caves, spring deposits, and calcrete soils.

### Limeclasts: characteristics, formation, and occurrence

Another type of grain or allochem, common in limestones, is gravel-sized clasts of cohesive carbonate sediment called **limeclasts**. These carbonate clasts are produced when clasts of cohesive carbonate sediments or sedimentary rocks are produced by erosion, then transported to the site of deposition. Most are derived from nearby coeval deposits of cohesive carbonate muds, within the immediate area of deposition. Because these are derived from within the area of deposition, they are called **intraclasts** (Figure 14.4). Rarer clasts derived from the erosion of older source rocks outside the area of deposition are called **lithoclasts** or **extraclasts**.

Because penecontemporaneous cementation occurs early during carbonate diagenesis, cohesive carbonate, sediments are common in environments where carbonate sediments are being produced. Many intraclasts are generated when tidal and/or storm surges move across mud-cracked carbonate sediments, eroding the upturned edges of the surface. Others are generated by storms and tsunamis moving across partially lithified sediments in lagoonal and/or subtidal, below wave base environments. Still others are produced as mass flows moving down slopes into deeper water environments erode the cohesive bottom. In all these situations, intraclasts become a significant type of allochem in the carbonate sediments produced. Still other intraclasts are generated by the micritization (see below) of skeletal fragments or aggregates in which endolithic bacteria convert the components of gravel-size grains into micrite or lime mud. Another fascinating type of limeclast consists of cemented grains rather than cohesive muds; such limeclasts are called **aggregates**. These include **grapestones**, which consist of cemented grains such as ooids and so resemble bunches of grapes, and **botryoidal grains** that have a colloform coating of carbonate laminations (Figure 14.5). These limeclasts are produced when partially cemented grain clusters are eroded during storms and are then encrusted by cyanophytes that produce the carbonate laminae.

(a)  (b)

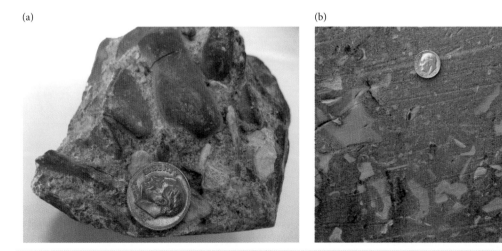

**Figure 14.4** (a) Rounded intraclasts, Jurassic, Sundance Formation, Wyoming. (b) Angular intraclasts, Cambro-Ordovician, Allentown dolostone, New Jersey. *Source*: Photo by John O'Brien. © John Wiley & Sons.

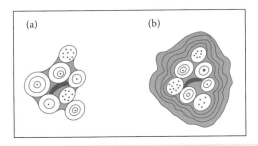

**Figure 14.5** Idealized aggregate limeclasts: (a) grapestone, (b) botryoidal aggregate.

Peloids: characteristics, formation, and occurrence

Smaller, sand-size particles composed of carbonate mud are also common as carbonate grains. Such grains are called **peloids** because they commonly resemble the ellipsoidal fecal pellets excreted by many organisms. Many peloids, particularly those with elevated levels of organic matter, are fecal pellets excreted by sediment-feeding organisms. For these the term **pellet** is appropriate. Other peloids are generated by other processes that produce sand-size particles composed of mud. These include micritization of sand-size ooids and skeletal fragments, and the production of sand-size clasts, which are otherwise similar to intraclasts, by the erosion of cohesive lime muds. That these particles are cohesive enough to be eroded and transported as clastic particles is indicated by their occurrence as grains within grain-supported frameworks, as seen in Figure 14.6.

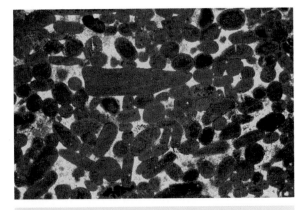

**Figure 14.6** Peloids of various shapes in grain-supported framework, with interstitial cement. *Source*: Courtesy of C. G. St. C. Kendall.

To summarize, sand- and gravel-size particles called grains or allochems are common constituents of limestones and dolostones. Major types of allochems include a variety of skeletal grains, ooids, limeclasts, and peloids – all of which yield clues concerning the production and depositional history of the carbonate rocks in which they occur. Mud-sized carbonate particles are discussed below.

*Carbonate mud or micrite: characteristics, formation, and occurrence*

The silt- and clay-size carbonate particles in limestones and dolostones are called **mud** or **micrite** (Figure 14.7a). Because the term mud is also used for detrital mud particles, we prefer to use the terms carbonate mud or micrite for clarity. Most micrite consists of rather tiny, carbonate clay-size particles, with a diameter of less than 4 μm. Coarser carbonate mud particles are sometimes referred to as **microspar** (Figure 14.7b).

The origin of carbonate mud particles or micrite has engendered a significant amount of controversy, but it is now clear that such particles are produced by a variety of organic and inorganic processes that include the following:

1  Secretion by calcareous algae and coccolithophorids.
2  Maceration of skeletal carbonate by microbes.
3  Bioerosion and micritization of preexisting grains.
4  Mechanical abrasion of preexisting carbonate grains.
5  Direct precipitation from solution.

A large percentage of modern carbonate mud is produced by calcareous green algae that secrete fine needles of aragonite in their tissues. Upon death and decomposition, the needles are released, generating carbonate mud particles. Coccolithophorids are single-celled green algae that secrete microscopic shells composed of tiny plates called coccoliths which are released during decomposition to form significant carbonate mud deposits in the pelagic realm. Other carbonate mud particles are generated by microbial **maceration** in which skeletal material composed of calcium carbonate microcrystals bound by organic material are released as bacteria decompose

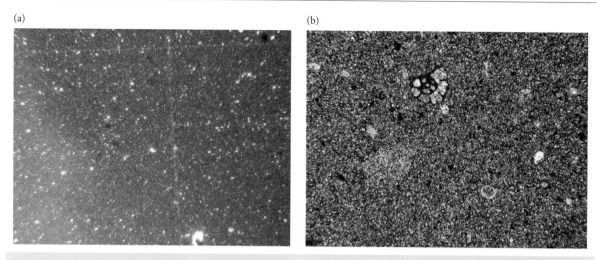

**Figure 14.7**    Carbonate mud in thin section, under plane light: (a) micrite, (b) microspar. *Source:* Photos with permission of Frederic Boulvain.

the organic material. Still others result from **bioerosion** in which the boring activity of invertebrates such as sponges and mollusks produces carbonate mud particles. The microboring activity of blue-green, cyanophyte bacteria, and algae and the filling of micropores with precipitated carbonate also converts original carbonate material into micrite. This process is called **micritization**. Still other micrite particles are generated by the mechanical abrasion of larger carbonate particles and accumulations, especially in agitated environments with vigorous wave and tidal current activity. Controversial at one time, but now widely accepted, is the hypothesis that calcium carbonate mud particles form by precipitation from seawater. In warm environments, calcium carbonate can become supersaturated with respect to alkaline seawater, especially if the salinity is slightly elevated. Under such conditions, widespread calcium carbonate precipitation, nucleated perhaps on microbes, can produce sufficient suspended carbonate mud to turn the water white. This process may have been much more common in older seas that likely were more saturated with respect to calcium carbonate (James and Jones 2016).

Clastic carbonate mud settles out of suspension under calm conditions with negligible flow velocities. Fine organic matter also settles out of suspension with micrite particles. This organic matter is an important food source for sediment-feeding organisms that extract the organic matter from the sediment and excrete fecal pellets, so that many micrites are peloidal.

*Organically bound accumulations*

Many carbonate deposits appear to consist of carbonate sediments that formed in place, as **organically bound accumulations**, rather than as separate grains and micrite. This important group of sediments includes organic reefs, stromatolites and some bioherms, all of which are discussed in the sections that follow (see also Box 14.1).

### 14.2.4  Classification of carbonate rocks

This section introduces three classification systems. One is used primarily for rapid field work, while the other two work best in conjunction with microscopic analyses of etched flat surfaces, acetate peels or thin sections.

*Field classification*

The **field classification** is based on (1) the average or modal size of the constituents and (2) whether they are composed primarily of calcium carbonate ($CaCO_3$) or dolomite [$CaMg(CO_3)_2$]. The textural terms, originally introduced by Grabau (1913), are **rudite** for gravel, **arenite** for sand and **lutite** for mud. The compositional terms, used as modifiers, are calc(i) for limestones and dol(o) for dolostones. Combining the two sets of terms produces the general descriptive terms in Table 14.3. Using this classification, it is generally easy to make identifications in the field using a hand lens, dilute HCl, which causes calcite but not dolomite to readily effervesce, and/or alizarin red-S, which stains calcite pink, but not dolomite.

Two, more sophisticated, classification systems were introduced by Folk (1959, 1962) and Dunham (1962). The latter two appeared together in an American Association of Petroleum Geologists (AAPG) memoir (Ham 1962) on the classification of carbonate rocks. Each has its advantages and disadvantages, its adherents and detractors, and each has been modified since its introduction.

### Dunham's classification system

**Dunham's classification** system emphasizes the texture of carbonate rocks, utilizing

**Table 14.3** Field classification of carbonate rocks.

| Rock name | Modal particle size | Principal composition |
|-----------|---------------------|-----------------------|
| Calcirudite or dolorudite | Gravel | Calcium carbonate or dolomite |
| Calcarenite or dolarenite | Sand | Calcium carbonate or dolomite |
| Calcilutite or dololutite | Mud | Calcium carbonate or dolomite |

easy-to-learn terminology. It recognized six major varieties of carbonate rocks (Figure 14.8). Four of these were based on rock texture, the fifth on the rock's inferred origin and the sixth on the notion that the original texture of some carbonate rocks could not be recognized due to diagenetic alteration. The first criterion used to distinguish the first four rock types is whether the rock possesses a mud-supported or a grain-supported framework (Chapter 13). Carbonate rocks with mud-supported frameworks are subdivided into **mudstone**, which contains less than 10% grains, and **wackestone**, which contains more than 10% grains. Any grains in such rocks are mud supported and therefore appear to be suspended in a mud matrix. The size of the grains is not used in the name, but the major type of grain may be used as a modifier, as in fossiliferous mudstone or peloidal wackestone. Both mudstones and wackestones accumulate in generally calm environments where significant amounts of mud settle from suspension. Carbonate rocks with grain-supported frameworks are subdivided into **packstone**, which contains mud matrix between the grains, and **grainstone**, which lacks significant mud matrix. Most grainstones contain interstitial diagenetic mineral cements that bind the grains together. As with mudstones and wackestones, the size of the grains is not used in the name, but the major type of grain may be used as a modifier, as in intraclastic packstone or ooid grainstone. Grainstones imply deposition in an

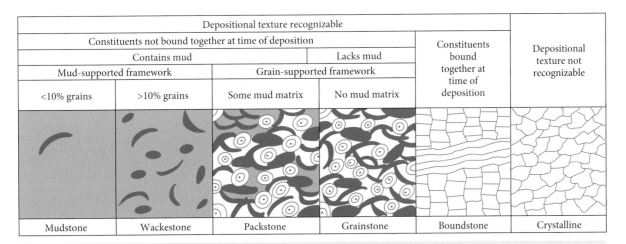

| Depositional texture recognizable | | | | | Depositional texture not recognizable |
|---|---|---|---|---|---|
| Constituents not bound together at time of deposition | | | | Constituents bound together at time of deposition | |
| Contains mud | | Lacks mud | | | |
| Mud-supported framework | | Grain-supported framework | | | |
| <10% grains | >10% grains | Some mud matrix | No mud matrix | | |
| Mudstone | Wackestone | Packstone | Grainstone | Boundstone | Crystalline |

**Figure 14.8** Dunham's classification of limestones. *Source*: Adapted from Dunham (1962). © John Wiley & Sons.

environment in which currents kept mud in suspension. Packstones suggest situations in which mud periodically settled from suspension to infiltrate grain frameworks deposited under more agitated conditions.

Two other rock types are recognized. **Boundstone** is used for *in situ* carbonate accumulations, such as reefs and stromatolites, which were organically bound at the time of accumulation. As we will see in the section that follows, the primary textures of carbonate rocks can be largely obliterated by recrystallization during diagenesis. In addition, some carbonate rocks, such as travertine, consist solely of coarsely crystalline calcium carbonate precipitated directly from solution. Dunham suggested the term **crystalline carbonate** for such carbonate rocks in which no depositional texture is recognizable.

With a sawed slab, etched by HCl, an acetate peel and/or a thin section, Dunham's classification is easy to use. For epiclastic carbonates, it involves recognition of (1) the type of framework (mud supported or grain supported), (2) the percentage of grains in rocks with mud-supported frameworks, and (3) the presence or absence of interstitial mud in grain-supported frameworks. For organically bound carbonates, recognition of their nature is sufficient.

Several significant modifications to Dunham's classification are now widely used. Embry and Klovan (1971) recommended subdividing boundstones on the basis of the process involved in binding the carbonate sediment at the time of accumulation (Figure 14.9). **Framestone** is produced by organisms that build rigid organic structures such as reefs by secreting the calcium carbonate. **Bindstone** is produced by organisms that build organic structures such as stromatolites and reefs by binding and/or encrusting preexisting carbonate material. **Bafflestone** is generated by organisms that trap carbonate sediment by acting as baffles that hinder its movement across the bottom, causing it to be trapped – an important process in reefs and bioherms. James (1984) introduced the terms **rudstone** for carbonate gravel-bearing rocks with a clast-supported framework and **floatstone** for carbonate gravel-bearing rocks with a matrix-supported framework. Other, more elaborate modifications exist (see James and Jones 2016), but are beyond the scope of this text.

Because of its emphasis on primary textures, Dunham's classification and its later modifications have been widely adopted in the petroleum industry. Rocks such as grainstones and rudstones have sufficient primary

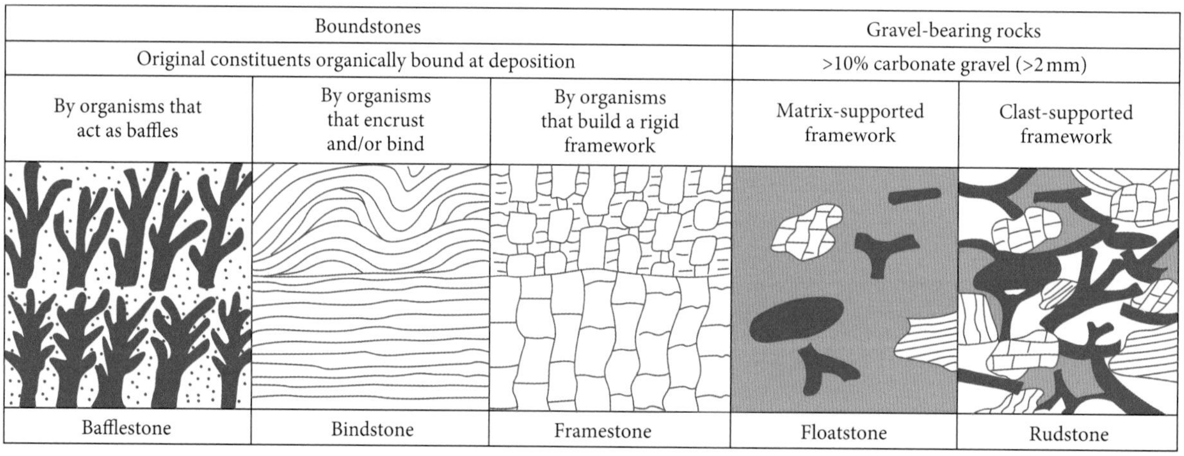

| Boundstones | | | Gravel-bearing rocks | |
|---|---|---|---|---|
| Original constituents organically bound at deposition | | | >10% carbonate gravel (>2 mm) | |
| By organisms that act as baffles | By organisms that encrust and/or bind | By organisms that build a rigid framework | Matrix-supported framework | Clast-supported framework |
| | | | | |
| Bafflestone | Bindstone | Framestone | Floatstone | Rudstone |

**Figure 14.9** Modifications of Dunham's classification. *Source*: Adapted from Embry and Klovan (1971) and James (1984). © John Wiley & Sons.

porosity and permeability to make excellent reservoir rocks, whereas mud-rich rocks such as mudstones, wackestones, and floatstones may be sufficiently impermeable and impede its flow and thus act as cap rocks.

*Folk's classification system*

**Folk's classification** revolutionized our understanding of carbonate rocks. It encouraged workers to pay careful attention to the components of carbonate rocks that lie at the heart of the classification scheme (Figure 14.10).

Because our understanding of carbonate rock components has evolved substantially since 1959, the following discussion emphasizes those aspects of Folk's classification that are still widely used.

Folk's classification (1959) recognized four major types of grains or allochems: (1) intraclasts (limeclasts), (2) oolites (ooids), (3) fossils (skeletal fragments), and (4) pellets (peloids). The term micrite, short for microcrystalline calcite, was used for carbonate mud – a valuable and widely used term that clearly distinguishes carbonate mud from

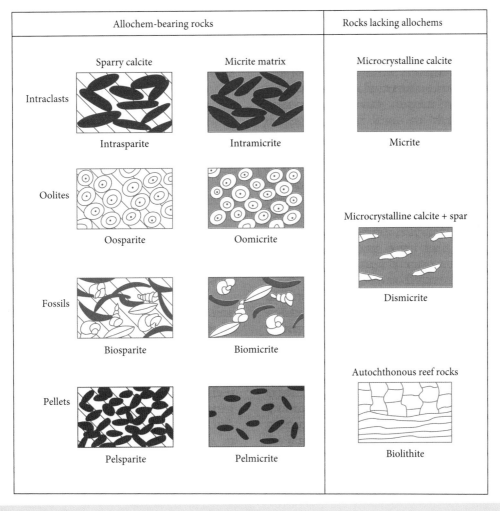

**Figure 14.10**  Folk's basic classification of carbonate rocks. *Source*: Based on Folk (1959). © John Wiley & Sons.

siliciclastic mud. Intergranular cements (see next section) were lumped together under the rubric of sparry cement or spar. Allochem-bearing carbonate rocks were classified according to the percentages of various allochems and the ratio of sparry calcite cement to micrite (Figure 14.11). Carbonate rocks with significant micrite accumulate chiefly in periodically calm environments where mud settles from suspension. Carbonate rocks with abundant sparry cement filling original pore spaces accumulate largely in continuously agitated environments where muds stay in suspension and only allochemical grains are deposited. Sparry cements are precipitated in the pore spaces between allochems during diagenesis. Rocks that lack allochems were classified separately (Figure 14.10).

In Folk's classification, allochemical rocks that contain a more than 25% intraclasts in the allochem population are given the prefix "intra." If the interstices between intraclasts are largely filled with diagenetic cement, the rock is an **intrasparite**; if filled with mud, it is an **intramicrite**. If fewer than 25% intraclasts occur, but the rock contains more than 25% ooids, the names **oosparite** and **oomicrite** are used. If neither intraclasts nor ooids exceed 25% of the allochems, then fossils or pellets (peloids in current usage) will be the dominant allochemical constituents. Where fossils dominate, the rocks are **biosparite** or

**biomicrite**. Where pellets dominate, they are **pelsparite** or **pelmicrite**. Rocks in which allochemical grains are sparse to absent are classified separately. These include **micrite** for rocks composed primarily of carbonate mud, **dismicrite** for micrites that contain small, spar-filled voids produced during diagenesis and **biolithite** for in situ carbonate accumulations (see Box 14.1) roughly equivalent to Dunham's boundstones. Folk also used the prefix "dolo" for carbonate rocks composed largely of dolomite as in dolomicrite.

Folk (1962) also developed a textural classification system (Figure 14.11) based on the proportions of micrite and sparry cement in carbonate rocks. This textural scheme permits inferences to be made concerning depositional conditions under which the carbonate particles accumulated. Carbonate rocks are subdivided into three major textural groups: (1) rocks with more than two-thirds mud matrix between the allochems (if any), called micrites, (2) rocks with more than two-thirds sparry cement between the allochems, called sparites, and (3) rocks with at least one-third of both spar and micrite between allochems, called poorly washed sparites. Micrites are subdivided according to the percentage of allochems into micrite (<1% allochems), allochemical micrite (1–10% allochems), sparse allochemical micrite (10–50% allochems), and packed allochemical micrite (>50%

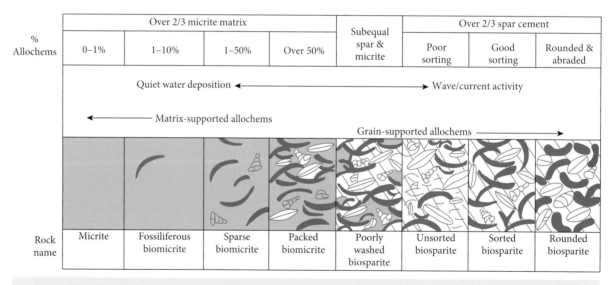

| % Allochems | Over 2/3 micrite matrix | | | | Subequal spar & micrite | Over 2/3 spar cement | | |
|---|---|---|---|---|---|---|---|---|
| | 0–1% | 1–10% | 1–50% | Over 50% | | Poor sorting | Good sorting | Rounded & abraded |
| Rock name | Micrite | Fossiliferous biomicrite | Sparse biomicrite | Packed biomicrite | Poorly washed biosparite | Unsorted biosparite | Sorted biosparite | Rounded biosparite |

Quiet water deposition ← → Wave/current activity

← Matrix-supported allochems

Grain-supported allochems →

**Figure 14.11** Folk's textural classification of carbonates. *Source*: Based on Folk (1962). © John Wiley & Sons.

allochems). The latter two are roughly equivalent to the distinction between wackestone and packstone in Dunham's classification. The name of the allochem is used as a modifier, as in Figure 14.11 in which the major allochem is assumed to be fossil grains. Sparites are subdivided according to sorting and rounding into unsorted allochemical sparites, sorted allochemical sparites, and rounded allochemical sparites. These terms echo Folk's terminology for the textural maturity of detrital sedimentary rocks (Chapter 13). The major drawback with Folk's classification is the complexity of the terminology; the strength is that it can communicate a large amount of information once it is understood.

The classification systems developed by Folk, Dunham and others have provided us with a valuable conceptual framework in which to carefully describe and interpret the components of carbonate sedimentary rocks.

### 14.2.5 Carbonate depositional environments

The conditions that favor the production of carbonate sediments include warm, fairly shallow water. As a result, the vast majority carbonate rocks originally accumulated in shallow seas in the tropics and subtropics. The extent of such seas expanded greatly during sea level highstands associated with high global tectonic activity and global warming (Chapter 1). Carbonate sedimentary rock formation also requires minimal influx of detrital sediments, so that most carbonate rocks initially accumulate in areas of low relief such as on intracratonic platforms or passive margins or in reasonably shallow oceanic environments far from land. These abundant rocks are the focus of this section. One to ten percent of calcium carbonate rocks do form in cool, temperate, or even subpolar water, mostly on rocky substrates in areas where upwelling brings many nutrients into the near surface environment. Cool water carbonates also may have been more widespread during periods of global cooling and during the Precambrian when seawater may have been more highly saturated with calcium carbonate because skeletal invertebrates had yet to evolve (James and Lukasic 2010; James and Jones 2016).

Research over the past half century, driven in part by the fact that most of the world's petroleum and natural gas reserves occur in carbonate rocks, has developed criteria that permit geologists to recognize the processes and conditions that produced particular carbonate rocks. This includes their tectonic–physiographic setting and the specific environment within that setting in which they accumulated. The sections that follow include necessarily brief discussions of the warm water carbonate rocks formed in each of these settings. Generalizations are challenging because organic evolution has completely changed the carbonate producing organisms in these environments through time. It has also played a significant role in changing the chemistry of sea water and of the atmosphere. Long-term changes in global climate also complicate the picture.

Tectonic–physiographic settings for the accumulation of such carbonate sediments include (1) carbonate ramps in a variety of settings, (2) rimmed and carbonate platforms on continental shelf passive margins, (3) epeiric sea platforms in cratonic interiors flooded during sea level highstands, and (4) isolated platforms in open oceanic environments. Many environments, subenvironments, and facies are shared between settings, but their spatial distributions depend on the tectonic–physiographic setting, local factors, and how both change with time. Within these tectonic–physiographic settings, several environments of carbonate sediment accumulation are recognized. These include nearshore, so-called **peritidal** settings that include (1) the supratidal zone, (2) the intertidal zone, (3) the subtidal zone above the normal wave base (NWB), and (4) beaches and dunes. Offshore settings include (5) the subtidal zone between the normal and storm wave bases (SWBs), (6) barrier-fringing reefs, (7) sand shoals, (8) lagoons, and (9) deeper shelf, slope, and basin environments. Their distribution in different tectonic-physiographic settings is summarized at the end of this section.

A more detailed discussion of carbonate depositional environments and the settings in which they occur is beyond the scope of this text. Readers interested in investigating this topic in more detail are referred to James and Jones (2016), Flugel and Munnecke (2010), Tucker (2001), Tucker and Wright (1990), Friedman (1981), and Bathurst (1975).

*Nearshore (peritidal) complexes*

Peritidal complexes include environments that range from

1 **Supratidal areas**, above the normal high tide line (NHTL), covered by water only during storm surge/tsunami events,
2 **Intertidal areas** between the NHTL and normal low tide line (NLTL), thus alternately submerged and subaerial at least once a day
3 **Shallow subtidal areas** that are more or less continuously submerged.

The aerial extent of these complexes is governed largely by the length of the coastline, the slope of the surface and the tidal range (difference between NHTL and NLTL). Large areal extent is favored by long coastlines, very gentle slopes and large tidal ranges and smaller areal extent by the reverse. Because many carbonate complexes develop on gently sloping platforms, continental shelves and in intracratonic settings, extensive peritidal complexes are common in the geologic record.

Mud-dominated tidal flats
Mud-dominated tidal flats form along protected, low relief coastlines where low-energy wave and current activity is the norm. This can occur because offshore islands, reefs, and/or shoals absorb most of the incoming wave energy before it reaches the shoreline or because the area is located in an embayment far from the open ocean. As a result, only fine-grained sediments are normally entrained, transported, and deposited. The extensive, flats are composed largely of mud transported from offshore areas where it is produced and then deposited near the shoreline. Roughly shore-parallel facies belts include the seaward intertidal flat, alternately submerged and exposed daily, and the landward supratidal flat which is typically subaerial and only submerged during occasional storm surges that deliver most of the sediment that accumulates on them. Intertidal flats grade seaward into the submerged subtidal zone which is frequently located in a lagoon. Supratidal flats grade landward into a variety of strictly terrestrial environments. Where tidal range is relatively large, an extensive network of **tidal channels** develops which brings water up into the flats during flood tides and drains it down off the flats during ebb tides (Figure 14.12).

The intertidal zone may be subdivided into a **lower intertidal zone** which is exposed for only a few hours a day and an **upper intertidal zone** which is exposed for longer periods of time. While details vary, the lower intertidal zone is generally dominated by muds and interlayered fine peloidal sands and, contains a normal marine fauna and therefore numerous burrow

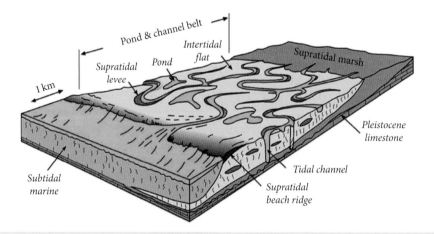

**Figure 14.12** Tidal flat depositional environments; Bahamas. *Source*: James and Jones (2016). © John Wiley & Sons.

traces. The higher intertidal zone may possesses higher salinities due to evaporation and is characterized by extensive, crinkly microbial mats and a more restricted fauna that lacks the gastropods that would otherwise feed on the mat organisms. Dessication features such as mudcracks are common in the upper part of this zone. Prior to the evolution of gastropods in the late Cambrian, microbial mats were abundant in lower intertidal and subtidal environments and produced a wide variety of layered **stromatolites**.

The supratidal zone is characterized by long-term exposure, especially at its landward edge, and periodic flooding during storm surge events. This is reflected in the characteristics of the sediments deposited here. In more humid climates, microbial mats cover large portions of the flat and generate wrinkly, laminated mud deposits, commonly with many dessication structures. In more arid climates, evaporite minerals such as gypsum and chicken-wire anhydrite are common, as discussed in the section on evaporites later in this chapter. Early diagenetic conversion of calcite and aragonite to dolomite is also common in such environments.

While the details vary, most supratidal sediments are laminated, often dolomitic, carbonates composed largely of micrite or peloidal micrite. Birdseye voids (fenestral structures) produced in algal mats are common, as are mudcracks and, in arid climates, evaporite minerals. Uncommon storm surges across supratidal flats produce significant deposition of allochems, including mud rip-up intraclasts, buoyant skeletal debris and mud that settles out of suspension onto the flat.

Muddy tidal flats are commonly cut by a shifting complex of tidal channels that can be quite deep in their outer portions, especially where tidal ranges are large. These channels may contain intertidal deposits, because flood currents flow up them during rising tides and ebb currents flow down them on falling tides. They may also contain subtidal deposits in their deeper, seaward portions. The sediments in these lower intertidal-subtidal channels are typically well washed grainstones with abundant cross-stratification and herringbone cross stratification from the back and forth migration of bedforms from alternating ebb and flow currents. Skeletal and/or intraclastic lag gravels are common over channel bases.

In the higher portions of tidal channels, where currents are weaker, fluctuating current velocity produces alternating ripple-scale cross lamination and suspension deposition leading to abundant flaser, wavy, or lenticular stratification (Chapter 12). During unusually high tides or tides associated with storm surge events, seawater rising up tidal channels may overspill the channel margins to produce finer grained, channel levee deposits and small ponds on low areas of tidal flats. Both of these areas are often quickly covered by microbial mats that generate crinkly laminated micrites. In arid climates, evaporites may form in the ponds.

In the peritidal environment, (1) shorelines and surfaces are irregular and continually shifting, (2) patterns of tidal channels are frequently changing, (3) climates evolve over time, and (4) storm surges occasionally cover large areas. As a result, long-term deposition on mud-dominated tidal flat complexes produces a complex three-dimensional mosaic of rocks generated in supratidal, intertidal, and even subtidal environments, as discussed above. Despite the complexity, peritidal shoaling upward sequences in which subtidal rocks grade upward into intertidal rocks which are overlain by supratidal rocks are common in the geologic record. They often occur as thick sequences of repeated shoaling upward cycles generated as tidal flats built outward repeatedly over subtidal deposits during multiple shoreline regressions.

Beaches and dunes

Carbonate sand and/or gravel beaches form along less protected portions of shorelines that are exposed to continuous, relatively vigorous wave action. This wave action keeps mud grains in suspension and leaves a lag deposit of sand and/or gravel grains that are lithified into grainstones and/or rudstones. These beach deposits are generally not as well sorted as detrital beach deposits. This is largely because skeletal grains are characterized by different porosities and therefore density so that particles of different sizes are hydraulically equivalent; deposited under the same hydraulic conditions.

Several **beach subenvironments** are recognized (Figure 14.13) and grade seaward from (1) backshore, (2) shoreface, (3) foreshore, and (4) foreshore subenvironments. **Backshore**

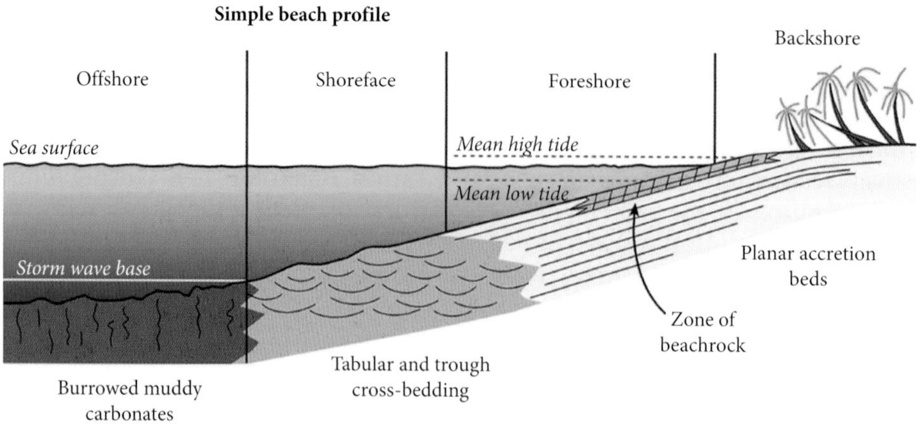

**Simple beach profile**

**Figure 14.13** Idealized sandy beach cross-section illustrating major depositional facies in beach-dune systems. *Source*: James and Jones (2016). © John Wiley & Sons.

**deposits** are above the NHTL and are quite variable. They can include dunes, sands washed over the beach during storms or supratidal flat complexes. The **shoreface** slopes seaward at changing angles of 5° to 15°, with steeper angles reflecting stronger wave regimes. Deposition on the **foreshore** or **upper shoreface** produces sets of planar laminations that dip seaward parallel to the shoreface. Changes in wave regimes and shoreface angles generate low-angle erosional surfaces that truncate sets of planar laminations that dip at different angles. These may be thought of as low-angle cross strata. These reasonably well sorted, gently dipping planar strata with low-angle truncation surfaces are characteristic of the beach shoreface. They grade seaward into **lower shoreface** grainstones that typically possess tabular and/or trough cross-laminations produced by the migration of subaqueous dunes and/or ripples. The shoreface grades, with increasing depth and carbonate mud content, into offshore deposits similar to those discussed below in the sections on lagoons and the subtidal neritic zone.

**Carbonate dunes** can form wherever winds entrain dry carbonate sand and transport it to another location, especially in relatively arid or semi-arid climates. Dunes frequently form behind beaches as winds transport sand from the beach onto the adjacent land mass or onto supratidal flats. In addition, sand initially transported onto the flats by severe storm surges or by tidal channels can be entrained, transported, and deposited by winds to form

dunes. Carbonate dune sands are not as well sorted as detrital dune sand because of the variable density of skeletal particles, as discussed earlier. As with detrital dune sands (Chapter 12), carbonate dune sands are best recognized by their large-scale tabular to trough cross-stratification formed by the migration of large straight-crested and curved-crested dunes. These can be interstratified with soil horizons, especially the calcrete soils characteristic of relatively arid climates.

*Subtidal neritic zone*

The subtidal neritic zone is the area below the low tide line, where the bottom is continually submerged under water, generally less than 200 m in depth. It is commonly subdivided into three parts, each characterized by a different set of wave regimes. These are briefly discussed below.

Subtidal above normal wave base
This zone of the seafloor is below the mean low tide line (MLTL) and above NWB, also called fair weather wave base (FWWB). The processes that characterize this zone and the characteristics of the detrital sediments that they produce are discussed in Chapter 12, in conjunction with detrital sediments. Carbonate rocks produced in this zone possess a similar set of characteristics. They are primarily grainstones, but composed of skeletal particles, less common ooids, and sparse intraclasts and coated grains. Because these sediments are

continually moved by waves, migrating bed-forms are common and produce both ripple-scale cross-laminations (in deeper and/or more protected areas) and dune-scale cross lamina-tions (in shallower and/or less protected areas).

## Subtidal between normal and storm wave base

This portion of the sea floor lies between NWB and the deeper SWB where only larger storm waves disturb the bottom at significantly greater depths. These portions of the sea floor are only periodically swept by wave action and so are characterized by alternating quiet and agitated water conditions. Over time, both quiet water muds and agitated water coarser particles are deposited on such bottoms, the muds being at least partially swept into sus-pension during storm periods. The storm pro-duced rocks are commonly called **tempestites** and are characterized by hummocky (HCS) and swaley (SCS) cross-stratification and graded bedding as discussed in Chapter 12 in connection with storm wave generated detrital sediments. Carbonate rocks ranging from mic-rites deposited under quiet conditions to rare grainstones through packstones, wackestones, and floatstones can be generated by the alter-nating quiet and storm wave swept bottom conditions that characterize this zone.

## Subtidal below storm wave base

This portion of the sea floor is swept by neither tidal currents nor storm waves, although it may on rare occasions be swept by a tsunami. Under such conditions, carbonate mud is the major material that accumulates on the bottom to form micrite, often peloidal. Since their evolu-tion in the mid-Mesozoic, the tests of calcare-ous plankton such as coccoliths and foraminifera have added significantly to such deposits. Occasional sediment gravity flows may bring coarser material into such environments pro-ducing grain-rich layers with erosive bases, graded bedding and partial Bouma sequences and contour currents may rework them to pro-duce laminated sands and muds. The processes that produce such rocks are discussed in Chapter 12, in connection with detrital rocks.

### Reefs and carbonate buildups

The term "reef" is defined in many, sometimes conflicting, ways. All workers think reefs are bioconstructions that have been built up to a significant extent by carbonate-producing organic activity that occurs *in situ*, where the reef is being formed. Much of this building involves skeletal carbonate precipitation by macroscopic invertebrates, often large colo-nial forms, and microscopic organisms. It also inorganic precipitation of calcium carbonate. As a result, the reef material is often tightly bound. Most reef buildups possess some, often significant, relief above the surrounding sea floor. Many reef structures are strong enough to be "wave resistant". Many reefs are, but increasingly workers have come to believe that some autochthonous organic buildups qualify as reefs even though they were not likely wave-resistant structures (James and Jones 2016). The latter carbonate buildups include lens-shaped **bioherms**, tabu-lar **biostromes,** and so-called **reef mounds**.

Most people think of corals as the chief reef-building organisms. They are certainly important at present, as are calcareous algae. Various groups of corals have been significant reef builders at different times in the past. However, organisms such as sponges, includ-ing stromatoporoids, mollusks, bryozoans, phylloid algae, various calcimicrobes, stroma-tolites, and extinct groups such as rudistid mollusks and archaeocyathids, have contrib-uted significantly to reef formation. Figure B14.1 highlights the groups of organ-isms that have been important reef-builders, the periods of significant reef growth and when particular reef-building organisms were prolific. These alternated with periods when large reefs were rare to nonexistent because large reef-building organisms had become extinct or had not yet evolved. Precambrian organic buildups were almost exclusively pro-duced by stromatolite-producing microbe communities which were much more wide-spread than they have been since the evolu-tion of grazing organisms such as gastropods in the late Cambrian period (see Box 14.1).

Several major types of reef exist. **Barrier reefs** form seaward of the shoreline, com-monly at the shelf edge, producing an often broad lagoon between the reef and the shore-line. The Great Barrier Reef in eastern Australia is an excellent modern example. **Fringing reefs** occur along the shoreline, with-out an intervening lagoon. Such reefs are com-mon around many Pacific Islands. **Atolls** are

## Box 14.1   Reefs through time

**Reefs** are biostromes in which organisms have built relatively rigid, wave-resistant structures over substantial periods of time. Every reef starts as a small organic accumulation, often developed on a local shoal. Baffling and binding by these organisms and by cementation eventually produces a firmer substrate on which framework-building organisms can begin to build a wave resistant structure. As organisms die and new ones build additional framework, the structure grows into the wave zone. The wave resistance results from the production of a rigid framework of framestone and bindstone during reef construction by organic activity. Large blocks of reef material are eroded from reefs, especially during storms, but many blocks are reincorporated into the reef by later encrustation to produce bindstone and by additional generation of framestone. Typical reef cores contain only 10–20% framestone. Spaces within the framework are filled with carbonate grains and mud whose eventual cementation serves to strengthen the reef structure. Auxiliary organisms that inhabit the reef add to its growth and encrusting organisms bind much of this material into bindstone. Many groups of organisms have contributed substantially to the formation of reefs and other types of biostromes since the lower Paleozoic (Figure B14.1a).

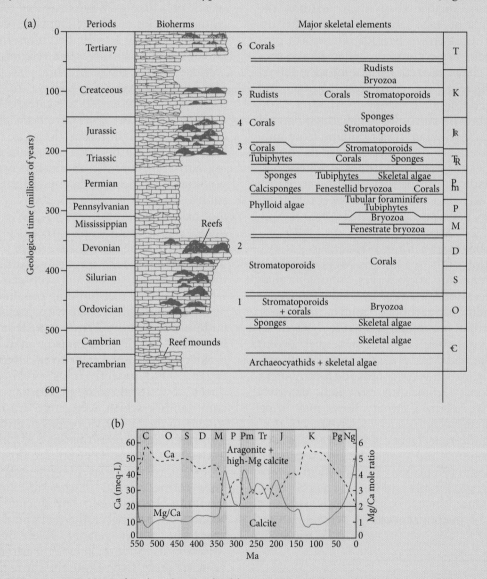

**Figure B14.1**   (a) Major reef and carbonate buildup organisms through time. Major framework builders are given in capitals, and lesser contributors in lower case letters. *Source*: After James (1983) and Raymond (2002). © John Wiley & Sons. (b) Changes in major reef and carbonate buildup organisms and their composition through time, compared with periods of calcite and aragonite seas. *Source*: After Stanley and Hardie (1999) and Prothero and Schwab (2004). © John Wiley & Sons.

## Box 14.1   *Continued*

Reef-building organisms may record significant oscillations in ocean water chemistry over time. Figure B14.1a summarizes the major periods of reef building and the organisms that contributed substantially to reef growth. Stanley and Hardie (1999) pointed out that organisms with calcite skeletons such as stromatoporoids, rugose and tabulate corals and receptaculitid algae were the dominant reef-building organisms in extensive Ordovician, Silurian and Devonian reefs. This corresponds to earlier work by Sandberg (1983), who suggested that during this time **calcite seas** favored low magnesium calcite precipitation over high magnesium calcite and aragonite (Figure B14.1b). Sandberg also suggested that from the Carboniferous through the mid-Jurassic, low calcium **aragonite seas** favored the precipitation of high magnesium calcite and aragonite. Stanley and Hardie (1999) pointed out that this period was characterized by a paucity of true reefs and that biostromal buildups were dominated by organisms with aragonite skeletons such as scleractinian corals, most sponges and some algae and by algae with high magnesium calcite skeletons. In the late Jurassic and throughout the Cretaceous, when calcite sea conditions returned, widespread construction of large reefs resumed with calcitic rudistid mollusks playing a dominant role. However, few correlations are perfect and scleractinian corals with aragonite skeletons were also important. Aragonite seas returned in the Oligocene–Miocene and coral reefs have since been built primarily by scleractinian corals and high magnesium coralline algae. Could it be that fluctuations in major reef-building organisms are controlled by and record fluctuations in seawater chemistry?

What might cause such changes in seawater chemistry and reef-building organisms? Sandberg (1983) argued that calcite seas correlated with major global warming events ("greenhouse" conditions) generated by elevated atmospheric $CO_2$ levels and aragonite seas with periods of global cooling ("icehouse" conditions) and major glaciations associated with depressed atmospheric $CO_2$. He proposed that during greenhouse periods elevated $CO_2$ increased the acidity of ocean water to the point where aragonite precipitation was inhibited so that calcite precipitation was dominant. On the other hand, during cool periods lower $CO_2$ concentration produced lower acidity that permitted aragonite to precipitate. Stanley and Hardie (1999) argued that Sandberg's mechanism was unlikely. Pointing to a correlation with plate tectonics, they suggested that the cyclic variations in ocean chemistry and skeletal composition are driven by variations in the rate of sea floor spreading. These periods of elevated sea level correspond to periods of global warming produced in part by the $CO_2$ generated by additional volcanic activity along the ridge system. On the other hand, during periods of decelerated sea floor spreading, such as those associated with the existence of Pangea (Carboniferous–Triassic) and since the collision of India with Asia (Oligocene), (1) the global rate of sea floor spreading slowed, (2) the sea level fell, (3) less magnesium was removed from seawater as hydrothermal metamorphism decreased, (4) the Mg/Ca ratio of seawater rose, and (5) aragonite seas were produced. These periods of sea level low stand correspond to periods of global cooling and glaciation produced by a combination of less $CO_2$ generated by volcanic activity along the ridge system and increased albedo. On the other hand, Kiessling (2008) argued that changes in skeletal compositions are driven by a complex set of processes, of which mass extinctions are the most important. Groups that recover well from such extinction events tend to dominate skeletal compositions in the succeeding period. Of course, all these ideas are somewhat speculative, that is how science works, but they are very interesting indeed!

ring-shaped reefs that formed as fringing reefs built up around a volcanic island. The island then slowly subsided, as upward *in situ* reef growth continued, to produce a circular barrier reef that encloses an expanding lagoon. Smaller reefs, called **patch reefs** or **pinnacle reefs**, depending on height-width ratios, also occur, most frequently in back-reef lagoons and on forereef slopes. Insight into many of the common features and processes that define

the subenvironments of reefs can be gained from a discussion of barrier reefs which follows.

Modern coral reefs that develop on the windward margin of a carbonate platform or atoll are characterized by several zones (Figure 14.14). From the platform interior toward the open ocean, these zones include the (1) backreef, (2) reef flat, (3) reef crest, (4) reef front, and (5) forereef. Ancient coral

**Zonation of a modern skeletal reef**

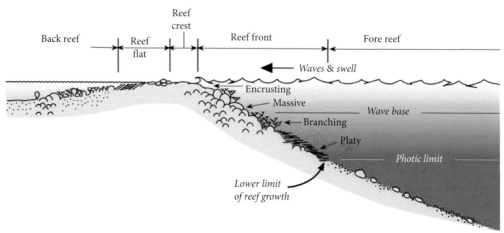

**Figure 14.14** Major environments in modern reefs. *Source*: James and Jones (2016). © John Wiley & Sons.

reefs appear to have possessed analogous zonation (Figure 14.14).

The **reef crest** lies on the seaward edge of the reef to depths of less than 10 m and receives the force of the strongest waves. This area is dominated by low relief encrusting organisms such as coralline algae and thick, stubby forms of corals. Long-term build-up of this part of the reef is preserved largely as bindstone, framestone, and cavity filling carbonate. It is sometimes referred to as the **reef core**. Behind the reef crest, on the platform side, is an extensive, slightly lower, well-cemented **reef flat**. The reef flat forms an irregular, shallow water pavement overlain by a veneer that is composed of coarse rubble eroded from the active reef by strong storm waves and dispersed toward the platform, finer sandy material deposited in slightly more protected areas and *in situ* growth of hemispherical to branching corals and other organisms that contribute to the reef flat. The sandy material may fill local depressions or form small sand shoals or even islands behind the reef crest. In the geologic record, the coarse rubble is preserved largely as rudstone, the sandy areas as, often cross-stratified, grainstone and the corals as framestone or bafflestone. The **backreef**, protected from vigorous wave action by the seaward reef flat and crest, is characterized by lower energy conditions. Many of the mud sized particles derived from erosion of the seaward portion of the reef

settle out of suspension in the backreef area where they are supplemented by carbonate mud generated by the disaggregation of calcareous algae. Sand shoals and islands can develop in backreef areas, especially behind gaps in the reef that permit higher levels of wave activity to keep mud in suspension. Corals in the backreef are largely domal and branching forms. Backreef deposits are preserved as, often peloidal, micrites, sand shoal grainstones, grain-bearing packstones, and fossil-bearing wackestones. Backreef deposits commonly grade landward over a short distance into backreef lagoonal deposits, discussed later.

The **reef front** lies below the reef crest on the seaward margin of the reef at depths roughly between 10 and 100 m. The shallower parts of the reef front are characterized by active growth of hemispherical, robust branching and columnar corals, coralline algae and many other invertebrate groups that include sponges, mollusks, and, in ancient rocks, groups such as crinoids, bryozoans, and brachiopods. The lower parts of the reef front, where currents are weaker and light penetration is limited, are characterized by delicately branching and platy corals. The **forereef** zone is dominated by coarse debris eroded from the reef and reef front and deposited on steeply sloping surfaces. This debris is composed of blocks, smaller gravel, and sand-size fragments. Because this area is well below NWB, some mud settles to the bottom as well. Forereef deposits are preserved in the geological record as steeply inclined beds of rudstone, floatstone, packstone, and wackestone that are analogous in some ways to the deposits on carbonate slopes that are discussed below.

**Fringing reefs** have similar sets of environments, but lack a backreef lagoon and commonly grade landward directly into beaches and/or dunes or peritidal to terrestrial environments.

*Sand shoals*

**Sand shoals** are composed of well-washed carbonate sand and/or fine gravel whose deposition commonly produces elevated areas on the sea floor in the form of sand bars, tidal deltas and even islands. They most commonly form (1) near platform rims, especially in intereef areas where reefs are discontinuous or absent (2) on the inner portions of carbonate ramps, or (3) as more sheet-like bodies on the shallow floors of epeiric seas. These areas are characterized by strong waves and/or tidal currents that keep mud in suspension and deposit various allochems that are preserved as grainstones in the geologic record. Packstones form where currents periodically weaken to permit mud to settle. Common constituents include ooids and skeletal debris produced both by coral reefs and elsewhere in the shallow-water "carbonate factory." This is the environment in which most ooids form, to the extent that these features are commonly referred to as "ooid shoals." Common structures include (1) ripple-scale cross strata, (2) dune-scale cross strata, (3) larger cross-strata produced by sand bar migration, and (4) subhorizontal bedding in more sheet-like sand bodies. In times when larger reefs were not being constructed, sand shoals were the major actors in forming rimmed carbonate platforms and such rimmed platforms were less common than they are today.

*Lagoons*

Lagoons typically develop seaward of shoreline peritidal complexes and landward of elevated areas on the sea floor. These elevated areas include (1) reefs, (2) sand shoals, and (3) islands, all of which provide a degree of protection from incoming waves and storms from the open ocean. They tend to develop roughly parallel to the shoreline and produce lagoonal facies belts that extend along the shore for $10-10^3$ km. Their width depends on their tectonic-environmental setting and sea level. Broad lagoons develop during sea level high stands (SLHS) when the shoreline is far from rimmed platform margins. Narrower lagoons characterize sea level low stands (SLLS) when the shoreline is near the platform margin. Narrow lagoons also develop behind sand shoals and islands near the shoreline, especially on the upper parts of ramps and on open platforms.

Under normal conditions, lagoons are characterized by relatively low-energy conditions where mud is the dominate sediment that accumulates. The mud may be pure carbonate mud, often peloidal, or contain various amounts of terrigenous mud delivered from the adjacent land mass. These muds are typically laminated, especially in the Precambrian or when the lagoon develops elevated salinity (and perhaps evaporites) or reduced oxygen levels (and undecomposed organic matter). These conditions restrict faunal diversity and suppress organic bioturbation. Bioturbation accelerated with the evolution of sediment grazing organisms in the early Paleozoic and is favoured by near-normal salinity and oxygen levels. Such pervasive burrowing may totally obscure such stratification, especially in the shoreward areas of lagoons. In the seaward portions of lagoons, storm-generated spillover sands and gravels derived from reefs and/or sand shoals or through tidal inlets may be interlayered with muds, as grainstones and packstones. Lagoons with normal salinity and oxygen levels are a highly productive part of the "carbonate factory" and a wide range of invertebrate and other fossils may be preserved in them as fossil-bearing wackestones. Small patch reefs and carbonate buildups are common features in lagoons, especially since the development of reef-building organisms, other than stromatolite-building communities, in the early Paleozoic. Lagoons that have limited connection to the open seas and are located in arid climates can be the important sites for evaporite formation, especially if their connection to the open ocean is restricted.

In the stratigraphic record, lagoonal deposits may be recognized by carbonate mud-dominated sequences that lie between shoreline peritidal and beach/dune sediments and seaward reef/sand shoal deposits. The muds generally contain a diverse assemblage of skeletal material, including infauna. However, if dissolved oxygen levels are low, black, organic rich micrites that contain an impoverished fauna will be preserved and if salinity is elevated sufficiently, a severely restricted fauna results and evaporites may form as discussed later in this chapter.

### Slope to basin carbonates

Passive margin slopes lie seaward of the shelf edge or break which as discussed above is often the location of reef and/or sand shoal platform rim. The slope grades seaward into the more level floor of the ocean basin. Similar slope to basin transitions occur seaward of isolated platforms in the open ocean. The slope and basin lie well below NWB and are therefore the site of significant mud deposition. Most of the mud is derived from the platform during storm surge relaxation, as discussed in Chapter 12, or by normal wave erosion at the shelf edge. However, significant amounts of coarser and fine debris are delivered down the slope and sometimes onto the basin floor by sediment gravity flows, also discussed in Chapter 12. These sediment gravity flows originate higher on the slope or near the shelf edge, especially through rim gaps in reefs and ooid shoals. These flows move down evolving channel systems to be deposited as the flows decelerate as the slope decreases. Rockfall generates isolated talus blocks up to 100 m in diameter that are called **olistoliths**. Coarser debris flows produce megabreccias and high concentration turbidity currents deposit massive, ungraded, poorly sorted floatstone and packstone. Finer, more dilute sediment gravity flows produce carbonate turbidites with typical ABC Bouma sequences. These products of occasional events are interstratified with the carbonate mudstones that more or less continually accumulate on the bottom. Base of slope rocks interfinger with basin floor rocks.

Some offshore basins, such as those off continental margins, are thousands of meters deep. Other, intracontinental basins, are less than 100 m deep. Basin rocks are predominately mudrocks, with a few thin, fine sand layers. If bottom conditions are anoxic, muds will be rich in undecomposed organic matter; if not they may contain an in situ benthic fauna. Since the Jurassic, a significant additional contribution to slope to basin sediments has come from the evolution of microscopic planktonic coccoliths and foraminifera in the pelagic zone. Where other sources of mud are rare, pure microfossil mud is preserved as **chalk**, as in the Cretaceous rocks of southern England and northern France. Where mud from other sources also settles to the bottom, mixtures of microscopic carbonate tests and muds form which are given the name **hemipelagite**. Basin carbonates are dominated by carbonate mudstones with sparse microturbidite rhythm rock with fine packstone to mud CDE Bouma sequences.

### Distribution of carbonate environments in different tectonic-physiographic settings

The simplest distribution of carbonate depositional environments occurs on **carbonate ramps** where, in the ideal model, the bottom slopes gently seaward over distances of $10–10^2$ km (Figure 14.15). Carbonate ramps typically develop on the margins of shallow seas. Environments occupy roughly shore-parallel bands characterized by water depths that gradually increase seaward. The ideal shore to deep ocean sequence of environments on carbonate ramps is (1) peritidal (including supratidal, intertidal, upper subtidal, and beach/dune), (2) subtidal above SWB, and (3) subtidal below SWB. The gentle slopes inhibit the development of significant mass flow deposits in the latter environments. Complications involve the local development of small patch reefs that grow upward into the subtidal zone and the local development of sand shoals in the subtidal zone above the NWB that may lead to the development of small islands (cays) behind which narrow shallow lagoons may develop.

**Continental margin platforms**, typically $10–10^2$ kilometers wide, develop over gently sloping continental shelves. The development of carbonate buildups, such as reefs and/or sand

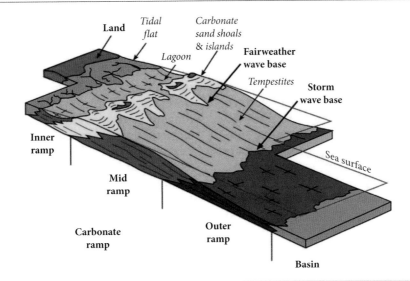

**Figure 14.15**   Major depositional environments on a carbonate ramp. *Source*: James and Jones (2016). © John Wiley & Sons.

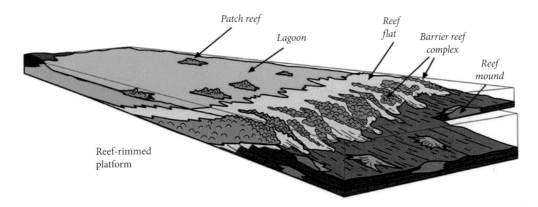

Figure 14.16   Major carbonate environments on a rimmed platform. *Source*: James and Jones (2016). © John Wiley & Sons.

shoals on the seaward edge of such platforms, early in their evolution, creates a **rimmed platform** (Figure 14.16). The rim reefs and/or shoals absorb the brunt of open ocean waves approaching from the seaward side, and a broad, quiet-water lagoon forms on the landward side. On the seaward side of the shelf edge buildups are slope to basin floor deposits produced by mass flows from the shelf edge. For rimmed carbonate platforms the ideal landward to seaward sequence of environments is (1) peritidal and its subdivisions, (2) lagoonal, (3) reef or subtidal shoal/island with local tidal channels, (4) forereef/slope, and (5) deep water mass flow/pelagic (Figure 14.16). If shelf edge reef and/or sand shoals do not develop, e.g., prior to the evolution of reef-building organisms, the unrimmed platform is called an **open platform**. Open

platforms lack the wave buffering effects of carbonate rims and therefore have higher wave energies all the way to the shoreline, thus a higher proportion of grains to mud. Facies distributions resemble those of carbonate ramps, since without rims, no large lagoon develops.

**Epeiric sea platforms and basins** develop during sea level high stands when oceans flood large portions of the craton to form shallow seas with widths on the order of $10^2$–$10^3$ km. Flooded cratons typically contain a patchwork of basins with somewhat deeper water, extensive platforms covered by shallow water and slightly elevated areas, some of which are above sea level. As a result, carbonate depositional environments show a similar patchwork pattern that depends significantly on local slope. Typically, slopes are gentle and the environments and facies that develop are those typical of carbonate ramps. However, their patterns are such that they wrap around high areas and encircle subsiding basins. As a result, environments around elevated areas are arranged in a roughly concentric pattern from terrestrial through peritidal to subtidal below wave base. Environments around subsiding basins are arranged in a bull's-eye pattern sequence: (1) peritidal and its subdivisions, (2) subtidal above NWB, (3) subtidal above SWB, and (4) subtidal below SWB.

Brief mention should be made of **isolated carbonate platforms** with widths from one to thousands of kilometers, which are very abundant in modern oceans. Such platforms develop, often on volcanic seamounts or platforms, separated from land by a considerable distance. Typically carbonate buildups and shoals develop on the platform margins and enclose a lagoon between them. Reefs (Box 14.1) typically develop on the windward side of platforms, sand shoals develop on the leeward side with lagoons and bays between them. Complications include the presence of patch reefs and islands within the lagoon. Adjacent to islands, supratidal to intertidal environments of the peritidal zone commonly occur.

## 14.2.6 Carbonate diagenesis

Carbonate rocks generally undergo extensive changes during diagenesis. One important set of controls on diagenetic reactions is pore fluid geochemistry, especially its (1) alkalinity–acidity, (2) temperature, (3) total dissolved solids (TDS) content, (4) dissolved $Mg^{+2}/Ca^{+2}$ ratios, and (5) dissolved sulfate ion $(SO_4^{-2})$ content.

The solubility of carbonate minerals is extremely sensitive to alkalinity–acidity (roughly pH in near surface environments). Small increases in dissolved $CO_2$ cause pore waters to become more acidic as their pH decreases, which leads to extensive dissolution of carbonate minerals. This is exemplified by acidification of sea water in response to increases in atmospheric $CO_2$ over the past century and a half. In contrast, small decreases in dissolved $CO_2$ cause pore waters to become more alkaline as their pH increases, which leads to extensive precipitation of carbonate minerals.

A second set of factors important in diagenesis is the mineralogy of the carbonate material, which includes aragonite, HMC, low magnesium calcite and dolomite, in a variety of crystal habits. These mineralogical and pore fluid chemistry factors interact during diagenesis to play significant roles in whether a particular species is precipitated, dissolved, or otherwise altered.

Three principle carbonate diagenetic environments (Figure 14.17) are defined by the type of water that penetrates the pore spaces: (1) **marine connate water**, mostly trapped beneath the sea floor, (2) **meteoric water**, groundwater from surface infiltration, located mostly below land surfaces, and (3) a **zone of mixing**, typically near the boundary between less dense meteoric water and a wedge of denser marine water. The meteoric zone can be subdivided into a nonsaturated meteoric **vadose zone**, above the water table, and a saturated **phreatic zone** below it. The phreatic zone, or zone of saturation, can be subdivided into a near surface zone and a deep zone. The timing of diagenesis can be (1) **synsedimentary**, coeval with deposition, (2) early (eodiagenesis), not long after burial, or (3) late, long after burial (mesodiagenesis and telodiagenesis).

The processes that occur during the diagenesis of carbonate rocks include (1) microbial micritization, (2) compaction, (3) dissolution, (4) pressolution, (5) cementation, (6) recrystallization or neomorphism, and (7) replacement by new minerals. Each of these processes is discussed in relationship to pore water

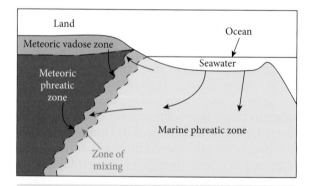

**Figure 14.17** Sketch that shows the distribution of the major zones in which carbonate diagenesis occurs; curved arrows depict groundwater flow.

chemistry, carbonate mineralogy, and diagenetic setting in the sections that follow.

*Limestone diagenesis*

**Microbial micritization** occurs largely in marine surface and near surface environments where the micro-boring activity of blue-green endolithic bacteria and algae, accompanied by precipitation of micrite into the micro-pores generated by such boring, converts original carbonate materials into micrite. Micritization of ooids, skeletal particles, and aggregate grains is common. All stages in this process have been observed, from partial micritization where the original material is recognizable to complete micritization where it is not. As discussed earlier, micritization is an important source of carbonate mud. In the context of this section, it is an important source of mud generated during eodiagenesis by endolithic microbes that inhabit carbonate sediments.

As carbonate sediments are buried, progressive increases in confining pressure cause them to undergo **compaction**. If grainstones are largely uncemented, significant breaking and crushing of rigid, but delicate, skeletal and ooid grains can occur and less brittle grains such as peloids may undergo significant plastic deformation. In any case, the porosity and permeability of carbonate sediments tend to decrease during compaction in ways similar to that which occurs during the compaction of detrital sedimentary rocks, discussed in Chapter 13.

Burial also leads to significant amounts of **pressolution** in which the portions of grains under maximum stress, especially at grain contacts, undergo selective dissolution. As in detrital sediments, pressolution leads to larger grain contacts where grains interpenetrate along concavo-convex and sutured contacts. Extensive pressolution causes interpenetration of carbonate grains producing a texture called a **fitted fabric**. Pressolution also generally decreases porosity and permeability.

Small- to large-scale **dissolution** of carbonate rock is very common. Dissolution of impure carbonate rocks leaves an insoluble residue of constituents such as clays and other siliciclastic minerals, iron oxides, and organic matter. **Stylolites** (Figure 14.18) are insoluble residue seams that cross-cut partially dissolved grains and commonly have a

**Figure 14.18** Stylolites with a "toothed" pattern (below the penny) and a thicker dissolution seam near the bottom.

toothed pattern when viewed in outcrop or microscopically. **Dissolution seams** are thicker seams of insoluble residue that often anastomose to produce a braided pattern (Figure 14.18). Many contacts between carbonate layers are marked by stylolites or dissolution seams along which substantial dissolution has occurred. It is important to remember that these are not depositional contacts and that dissolution has likely thinned the strata in question, often by a considerable amount. Especially soluble strata may disappear completely.

As discussed in Chapter 11, the existence of caves, sinkholes, and other karst features implies that carbonate rocks are extremely soluble under the appropriate conditions. Dissolution of carbonate sediments is relatively rare in warm, marine pore waters because they are approximately saturated with respect to calcium carbonate. By analogy with the CCD discussed earlier, colder, more acidic, marine pore waters are undersaturated with respect to calcium carbonate and have the ability to dissolve calcium carbonate sediments during diagenesis. However, most calcium carbonate dissolution occurs in the presence of meteoric water, which is generally much more acidic than marine pore water and generally has a lower Mg/Ca ratio as well. Although carbonates of any composition may be dissolved by meteoric groundwater, aragonite and HMC are the most susceptible. This helps to explain why HMC and, especially, aragonite become increasingly rare in older carbonate rocks exposed on land. The

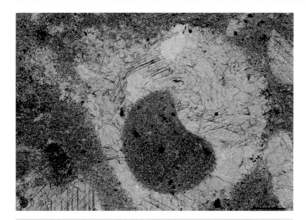

**Figure 14.19**   Moldic porosity, showing a dissolved gastropod shell, later filled with precipitated sparry cement. *Source*: Adams et al. (1984), reproduced with the permission of Wiley-Blackwell.

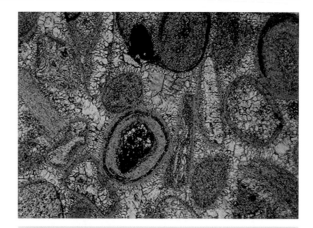

**Figure 14.20**   Marine, isopachous rim cement (brownish) on grains with pore spaces filled with second generation meteoric drusy calcite cement, Ouanamane Formation (Jurassic), Morocco. *Source*: Adams et al. (1984), reproduced with the permission of Wiley-Blackwell.

dissolution of grains composed of aragonite or HMC can produce **moldic porosity** in which the form of ooids and fossils is preserved as a cavity of similar shape. Aragonitic ooids and mollusk shells (Figure 14.19) are especially susceptible to dissolution. Later precipitation of pore-filling cements produces a cast of the original grain. Dissolution produces a variety of other void spaces that range from tiny voids to large caverns. These represent storage capacity for fluids such as water, petroleum and natural gas, as well as for aqueous fluids capable of precipitating of mineral cements.

As implied by the "sparry calcite" in Folk's classification, mineral cements are common products of carbonate diagenesis. Their composition depends largely on the pore water composition and their form depends primarily on whether cementation occurs above or below the water table, i.e., within the vadose zone or within the phreatic zone. *Marine cements* are almost exclusively aragonite and HMC. Because grains in ocean floor sediments are completely bathed in pore fluid, cements that nucleate on grains grow at similar rates to produce coatings of nearly constant thickness called **isopachous rim cements**. They occur primarily as coatings of radial, acicular to fibrous crystals and as micritic coatings of aragonite or HMC (Figure 14.20).

On the other hand, **meteoric cements** are almost exclusively composed of low magnesium calcite. In the meteoric **phreatic zone** where pore spaces are saturated with groundwater, three types of cement are common. The precipitation of **drusy calcite** cements (Figure 14.20) involves the nucleation on host grains of multiple crystals that then grow outward into pore spaces to produce a fringe of crystals with relatively straight boundaries whose size increases away from the host grains. **Syntaxial calcite** cements (Figure 14.21) involve the precipitation of low magnesium calcite that nucleates in optical continuity with a low magnesium calcite grain. These cements are especially common around single-crystal echinoderm skeletal particles. As in detrital

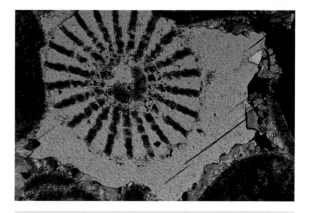

**Figure 14.21**   Syntaxial calcite in optical continuity on an echinoderm spine. *Source*: Photo courtesy of Maria Simon-Neuser, with permission of LUMIC.

rocks, **poikiloptic calcite** cement consists of a single crystal large enough to incorporate multiple grains during its growth.

**Vadose zone cements** display a very different geometry. As groundwater moves downward through the vadose zone toward the water table, its high surface tension causes it to adhere to the underside of grains and to the constricted spaces between grains. Continued precipitation on grain bottoms leads to the development of **pendant or stalactitic cements** that hang downward from the grain like a pendant on a necklace. Vadose zone water infiltrating between grains forms a meniscus, analogous to the one in a capillary tube or burette, from which precipitation of low magnesium calcite produces **meniscus cement** which may resemble an hour-glass suspended between adjacent grains.

In reality, the position of the water table varies both seasonally and over the long term so that carbonate rocks may spend multiple periods of time in both the vadose and phreatic zones. In addition, the position of the boundary or zone of mixing between marine and meteoric waters varies, especially between sea level high stands and low stands. By paying careful attention to the form and composition of calcium carbonate cements, carbonate workers can infer a complex set of processes that involve changes in pore water composition, nucleation history, and cement growth in marine environments, the transition zone and meteoric vadose and phreatic zones over long periods of time. More details on calcium carbonate diagenesis may be found in James and Jones (2016), Scholle and Ulmer-Scholle (2003), Tucker and Wright (1990), Friedman (1981), Longman (1980), and Bathurst (1975).

Carbonate sediments are also subject to recrystallization during diagenesis in which older crystals are dissolved and then precipitated as new crystals of similar composition. In carbonate rocks, the term **neomorphism** is used for processes in which new crystals form without a significant change in composition. If the crystals become larger, the neomorphism is called aggrading neomorphism; if they become smaller, it is called degrading neomorphism. The most frequently encountered neomorphism involves the progressive recrystallization of micrite into microspar and of microspar into sparry calcite by aggrading neomorphism. Such neomorphic

sparry calcite is called **pseudospar** to distinguish it from directly precipitated calcite cements. Pseudospar can be recognized where neomorphism is incomplete because patches of dusty micrite and microspar remain. In original wackestones and floatstones, pseudospar can often be inferred because grains formerly supported by mud matrix are widely separated and seem to float unsupported in sparry calcite (pseudospar).

*Dolomitization*

Dolomite is far less common in modern carbonate sediments than it is in older carbonate sequences. This fact, along with abundant evidence for calcium carbonate replacement by dolomite, supports the hypothesis that most, if not all, dolostones are of diagenetic origin – produced by a process called **dolomitization**. Dolomite is especially abundant in rocks formed during sea level high stands such as those formed during the Ordovician–Devonian and Jurassic–Cretaceous periods. It is also abundant in the Precambrian. A small number of occurrences of dolomite in modern environments are thought to be primary precipitates.

The rarity of primary dolomite is a long-standing conundrum, especially because theoretically ocean water is oversaturated with respect to dolomite. Its rarity in modern sediments suggests that something must inhibit its nucleation and block dolomite precipitation. Research suggests that dolomite nucleation and precipitation from marine waters require either elevated Mg/Ca ratios or decreases in dissolved sulfate ($SO_4^{-2}$), or both. **Primary dolomite** forms in modern intertidal to supratidal zones in the subtropics where evaporation and subsequent precipitation of aragonite ($CaCO_3$) and gypsum ($CaSO_4 \bullet 2H_2O$) produce the necessary increase in Mg/Ca ratio and decrease in dissolved sulfate ion to initiate the precipitation of dolomite. It also forms where magnesium-rich groundwater discharges into lakes, raising the Mg/Ca ratio, and where sulfate-reducing bacteria lower sulfate concentrations in saline lakes (Wright and Wacey 2004).

The origin of abundant diagenetic dolomites has been the subject of discussion and fervent controversy for more than a century (James and Jones, 2016 and Machel, 2004). The abundance of secondary dolomite requires that large volumes of water move through limestones over time. Here we will discuss major types of dolomitization processes that have had a significant number of adherents in recent decades: (1) evaporative reflux (drawdown), (2) meteoric-marine water mixing, (3) Kohout convection, (4) bacterial sulfate reduction, and (5) deep burial processes. One should remember that hybrids involving multiple processes may be significant.

It is well known that in warm, arid areas with high evaporation rates, evaporation of surface and/or ground water leads to increased salinity and the precipitation of calcium sulfate minerals such as gypsum and anhydrite (see Section 14.3). The simultaneous removal of both calcium and sulfate from such waters raises their Mg/Ca ratios and lowers their dissolved solid concentrations. At the same time, the increased salinity increases the density of such waters. Various models show that such dense brines will percolate downward through carbonate sequences by a process known as **evaporative reflux** or **evaporative drawdown**. Dense brines form beneath shallow lagoons, sabkhas and larger evaporite basins and have the requisite composition to replace calcium carbonate minerals with dolomite. The reflux mechanism gains support from the occurrence of relatively young, diagenetic dolostones, in areas where evaporative drawdown is known to occur, where groundwater shows strongly positive $\delta O^{18}$ values (Chapter 3) that reflect the preferential evaporation of $O^{16}$.

In the continually shifting **zone of mixing** (see Figure 14.13) that lies between marine and meteoric pore waters, pore waters with the appropriate composition for dolomitization can theoretically be produced. This zone could, over time, occupy a large area as it shifts during periods of prolonged sea level rise during which dolomitization is known to have been particularly common. Mixing meteoric water with seawater in proportions between 1 : 2 and 1 : 3 is sufficient to cause calcite to become soluble. The Mg/Ca ratio would remain high enough for dolomite to form, aided by the lowering of sulfate concentration due to dilution. However, this process is severely limited by the fact that the dissolution of calcite rapidly lowers the Mg/Ca ratio so that dolomite is unlikely to precipitate in large amounts. Such dolomites would have fairly

normal $\delta O^{18}$ values. This process, popular in the 1970s and l980s, is no longer considered as important in dolomitization as it once was.

Isotopic studies suggest that many dolostones have equilibrated with elevated temperatures ($50–100\,°C$), commensurate with depths of $500–2000\,m$ (Tucker 2001; Machel 2004). This suggests moderately deep circulation of dolomitizing fluids by some type of convective mechanism such as **Kohout convection** (Kohout 1965; Tucker 2001). The most common model for this process occurs along platform margins and involves the pumping of deep, cold seawater, undersaturated with respect to calcium carbonate and oversaturated with respect to dolomite, into platform margin limestones where groundwater rises, after being heated at depth. As it does so, its place is taken by seawater, perhaps aided by some combination of tidal and ocean currents. The seawater rises as it is heated, leading to convection through the limestones that is driven by heating at depth, perhaps aided locally by sinking of evaporative brines. Limestone permeability is essential to the effectiveness of convection-driven dolomitization. Because dolomite occupies a smaller volume than calcite, it is easy to envision a positive feedback mechanism in which permeability increases as dolomitization proceeds, aiding large-scale, continued dolomitization.

Many workers (James and Jones 2016) posit a **deep burial** mechanism for the formation of many dolomites, especially those that possess **saddle texture**. Saddle texture is characteristic of dolostones with curved crystals and large amounts of porosity. Some of these are likely hydrothermal, given their association with sulfide minerals such as galena and sphalerite. Others are attributed to the long-term transport of very warm, deep waters, perhaps driven by hydrostatic gradients from the hinterland toward the foreland in orogenic belts (James and Jones 2016).

Many subsurface diagenetic environments are reducing environments and some are anoxic. **Bacterial reduction of sulfate** tends to occur in many low-oxygen subsurface environments as bacteria utilize the oxygen in dissolved sulfate ions to decompose buried organic material. Such sulfate-reducing bacteria lower the dissolved sulfate contents of subsurface water, which should favor the replacement of calcite by dolomite. In conjunction with reflux or some type of convection to provide magnesium, sulfate-reducing bacteria may play a very significant role in dolomitization. The origin of secondary dolomite remains controversial. Currently, some combination of evaporative reflux, fluid mixing, deep convection, and sulfate reduction by bacteria seems to provide the most promising explanation. Stay tuned!

## 14.3 EVAPORITES

**Evaporites** are sedimentary rocks that form by chemical precipitation from highly saline waters (brines) that have become oversaturated with respect to one or more dissolved solids as the result of evaporation. Because they are nearly impermeable, evaporite rocks are also important cap rocks for petroleum reservoirs. The essential steps in the formation of evaporites are as follows:

1　The presence of surface water or shallow groundwater that contains dissolved solids.
2　Warm, dry conditions that permit net evaporation in which the progressive removal of water by evaporation exceeds the replenishment of water over time.
3　Sufficient net evaporation to progressively concentrate dissolved solids to the point where the water becomes oversaturated with respect to one or more dissolved solids.
4　That triggers the precipitation from saline brines of one or more dissolved solids in the form of evaporite minerals.

Because of their high solubility, evaporites are far more abundant in the subsurface (they underlie roughly 30% of the United States) than they are in outcrop, although the less soluble minerals are exposed at the surface in areas with arid climates. Still, the conditions under which evaporite minerals form are sufficiently limited in time and space that evaporite rocks constitute less than 1% of the sedimentary rock record. Evaporite rocks form from saline groundwater, saline lakes and highly saline, restricted seas. Evaporite rocks are climate sensitive. The vast majority formed under warm, arid climatic conditions that promote net evaporation. These conditions most commonly occur at subtropical latitudes between 10° and 30° from the equator and in rain shadows at somewhat in a broader range of latitudes. Most large evaporite deposits have formed in enclosed or restricted basins in which replenishment of water from outside sources, such as the open ocean, rainfall, groundwater inflow, and river runoff is limited. Evaporites dominate sedimentary sequences only where the influx of detrital sediment is low relative to evaporite precipitation and where the salinity severely restricts calcium carbonate production.

### 14.3.1　Marine evaporites

**Marine evaporites** form where marine water undergoes extensive net evaporation to become hypersaline brine. Continued evaporation leads to progressive concentration of dissolved solids in the remaining water. Under such conditions, hypersaline brines can become oversaturated with respect to one or more minerals, which leads to their precipitation from solution as evaporite minerals. Literally, hundreds of minerals, both primary precipitates and secondary products of diagenetic replacement and recrystallization, have been reported from marine evaporites. Most are relatively rare. The major halide and sulfate minerals in marine evaporites are listed in Table 14.4, along with their relative abundance in the context of marine evaporite rocks.

All minerals possess different degrees of solubility that depend on the chemistry of the solution and other environmental conditions. When a body of water undergoes net evaporation, increasing water salinity, and density, the least soluble minerals are precipitated first and progressively more soluble minerals are precipitated later in a specific sequence. The conditions and sequence of crystallization for average seawater are well known (Table 14.5). As seawater evaporates the following sequence occurs: (1) a small amount of calcite ($CaCO_3$) is precipitated when about 50% of the water has evaporated, especially if dissolved iron is elevated, (2) gypsum ($CaSO_4 \cdot 2H_2O$) begins to precipitate at 66% evaporation and continues to precipitate as evaporation continues, (3) halite (NaCl) precipitation is initiated at 90% evaporation, and (4) a variety of potassium and magnesium sulfates and halides (e.g., carnellite) are precipitated at >96% evaporation

484 EARTH MATERIALS

**Table 14.4** Major marine evaporite minerals and their relative abundance.

| Mineral | Chemical composition | Abundance |
|---|---|---|
| Anhydrite | $CaSO_4$ | Abundant |
| Bischofite | $MgCl_2 \cdot 6H_2O$ | Scarce |
| Carnellite | $KMgCl_3 \cdot 6H_2O$ | Common |
| Gypsum | $CaSO_4 \cdot 2H_2O$ | Abundant |
| Halite | $NaCl$ | Abundant |
| Kainite | $KMg(SO_4)Cl \cdot 3H_2O$ | Common |
| Keiserite | $MgSO_4 \cdot H_2O$ | Common |
| Langbenite | $KMg_2(SO_4)_3$ | Scarce |
| Polyhalite | $K_2Ca_2Mg(SO_4)_4 \cdot 2H_2O$ | Common |
| Sylvite | $KCl$ | Common |

**Table 14.5** Precipitation sequence of common evaporite minerals from seawater.

| Mineral | Evaporation (%) | Salinty (ppt) | Density (g/cm³) |
|---|---|---|---|
| Seawater | ~0 | ~35 | 1.04 |
| Calcite | >50 | >70 | 1.08 |
| Gypsum | >75 | >135 | 1.14 |
| Halite | >90 | >350 | 1.21 |
| Potassium and magnesium minerals | >96 | >750 | 1.27 |

from very dense brines. Saline terrestrial waters tend to have compositions that differ from marine waters. Additions of such terrestrial waters to marine waters, in various proportions, can change both the minerals that precipitate and their sequence of crystallization.

Modern marine evaporates precipitate in warm, arid, shallow, marine lagoons, and/or peritidal sabkhas on a relatively small scale. **Sabkha evaporites** form in areas of very low

relief called sabkhas that occur along arid coastal plains in the transition zone between marine and nonmarine environments. The best known sabkha occurs along the United Arab Emirates of the Persian Gulf, but many other examples are known from the subtropics. Where the influx of siliciclastic detritus is relatively small, carbonate rocks dominate sabkhas and evaporites form in the upper intertidal and supratidal zones (Figure 14.22). The essential processes begins with the evaporation of pore waters that are of either marine or mixed marine–terrestrial origin. As the water evaporates during extensive dry periods, groundwater is transformed into hypersaline brines from which gypsum ($CaSO^4 \cdot 2H_2O$) and/or anhydrite ($CaSO^4$) are precipitated.

Much of the gypsum is precipitated at the top of the water table as bladed gypsum or as gypsum rosettes; some is precipitated in the vadose zone as nodules (Figure 14.23a). Higher temperatures and salinities favor the dehydration of gypsum into anhydrite. These conditions typically exist in the middle–upper portions of the supratidal zone, some distance from normal marine influences. Here much of the gypsum is replaced by anhydrite or polyhalite, a process that involves a substantial decrease in volume. This leads to the formation of nodular or **chicken-wire anhydrite** (Figure 14.23b), so-called because the anhydrite nodules are enclosed in stringers of calcareous or siliciclastic sediment that resembles chicken-wire fencing. During burial, anhydrite generally remains the stable calcium sulfate mineral, but as rocks approach the surface in areas with sufficient rainfall, anhydrite is commonly hydrated into gypsum.

Very **large marine evaporite sequences** with thicknesses of $10^2$–$10^3$ m and aerial extents of $10^3$–$10^5$ km² occur in the geological record.

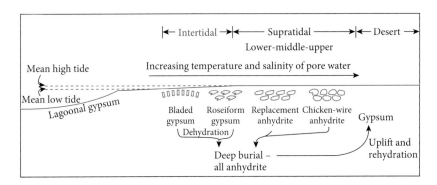

**Figure 14.22** Environments of evaporite formation in modern sabkhas and the conversions between gypsum and anhydrite. *Source*: Photo courtesy of Maria Simon-Neuser, with permission of LUMIC.

(a) 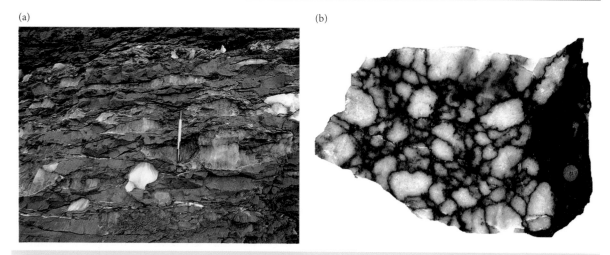 (b)

**Figure 14.23**    (a) Nodular gypsum, Triassic Mercia Group, Watchet Beach, England. *Source*: Courtesy of Nicola Scarselli. (b) Chicken-wire anhydrite, Carboniferous, Belgium. *Source*: Courtesy of Frederic Boulvain.

Some large evaporite sequences have been deposited in shallow, subsiding cratonic basins in which water depths were generally less than a few tens of meters. Examples include the Paleozoic Williston Basin centered in North Dakota and the Silurian Michigan Basin. Other large evaporites have been deposited in foreland basins of variable depth, including the Silurian Salina Group of New York and Pennsylvania in the Taconic foreland basin and the Pennsylvanian–Permian Paradox Group in a foreland basin of the Ancestral Rocky Mountains. Still other large evaporite sequences were deposited in rift basins, e.g., the Gulf Coast Basin and the proto-Atlantic Ocean Basin, of Jurassic to Cretaceous age. A few large evaporite sequences such as those in the Permian Delaware Basin in west Texas were deposited in deep basins associated with the irregular closing of ocean basins at convergent plate boundaries (Box 14.2).

There are no modern analogues for these large, ancient marine evaporite deposits. Most appear to have been formed in large barred basins (Figure 14.24a). Some of these basins were shallow (<100 m deep), including most of those on rimmed platforms and in epicontinental seas. A few were likely very deep (>1000 m) such as those associated with the irregular closing of ocean basins. The depth of water in the some evaporite basins has been the subject of considerable debate with some supporting generally deep water/

deep basin models (Figure 14.24b) and others supporting shallow water/deep basin models (Figure 14.24c).

The volume of evaporites in these basins is far too large to be explained by the simple evaporation of a body of water that filled them at one time. In many cases, the volume of evaporites implies the evaporation of volumes of seawater exceeded maximum basin volume by factors of 10–100. This can only have been accomplished by repeated flooding and dessication of such basins and/or by gradual replenishment of seawater through "leaky" barriers and/or the inflow of saline continental groundwater.

Thick sequences of anhydrite and gypsum indicate prolonged periods when salinities hovered in the appropriate range for continued precipitation of calcium sulfate. This suggests a balance between evaporation and inflow, perhaps through leaky barriers, that is consistent with the large volumes of evaporites produced. Thick sequences of halite, carnellite, and other highly soluble minerals imply severe dessication concentration of dissolved solids in dense brines. As is to be expected, these minerals are deposited from waters that covered smaller areas than those that precipitated gypsum and anhydrite in the same basin. As a result, the deposits form a bull's-eye pattern (Figure 14.25) with normal salinity carbonates and perhaps sabkha evaporites surrounding elevated salinity laminated

## Box 14.2 When the Mediterranean dried up!

In the 1960s, oceanographers using seismic reflection profilers detected a mysterious reflecting horizon known as the "M" layer in the sediments under the deeper parts (>2000–4000 m) of the Mediterranean Basin. In the summer of 1970, the Deep Sea Drilling Project sampled sedimentary rocks from the same area and a startling discovery was made (Hsu 1982). The "M" layer contains sequences up to 2000 m thick of 5.3–6.0 Ma Miocene evaporite rocks that contain anhydrite, halite, gypsum and rarer evaporite minerals. These evaporites accumulated as isolated sequences in the deeper parts of the Mediterranean Basin (Figure B14.2). In addition, these evaporites are intimately associated with stromatolite-bearing carbonate rocks that were clearly deposited in shallow water. The deep parts of the Mediterranean Basin were occupied by shallow water so saline that evaporites were precipitated. For nearly 700 000 years, the Mediterranean Sea had, at least periodically, dried up! What caused this to occur?

Prior to 6.0 Ma, one or more connections existed between the Atlantic Ocean and the Mediterranean Sea, in the area now occupied by the Straits of Gibraltar between Spain and Morocco. As long as a connection existed, any Mediterranean water removed by evaporation was replenished by the inflow of normal salinity water from the Atlantic Ocean. Just after 6.0 Ma, as Africa and Spain converged, causing uplift, the Atlantic–Mediterranean connection was apparently closed and a barrier formed to produce a restricted sea. Barrier formation was possibly aided by a 50 m fall in sea level caused by Miocene glaciation in Antarctica. Once the connection was closed, the results were inevitable. Because of the warm, arid climate, high evaporation rates removed large amounts of water from the sea surface. The low rainfall and river runoff in the region and leakage through the barrier were frequently insufficient to replenish the water lost by evaporation. Several separate saline seas formed in the deepest parts of the Mediterranean Basin while terrestrial sediments were spread across the former sea floor. Gypsum/anhydrite were precipitated from saline brines over large areas when net evaporation exceeded 70% and halite and rarer evaporites were precipitated over smaller areas when net evaporation exceeded 90%. The result is the bull's-eye pattern of evaporite minerals seen in Figure B14.2. The Atlantic–Mediterranean connection was re-established through the Straits of Gibraltar around 5.3 Ma; a giant waterfall formed and the Mediterranean basin filled rapidly.

The volume of evaporites formed in this brief interlude (5.3–6 Ma) requires the evaporation of at least 40 times the volume of the modern Mediterranean Sea. Scientists now have a recent example of the large marine evaporite basins that have formed periodically during geological history.

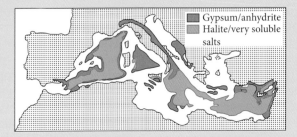

**Figure B14.2** Distribution of Miocene evaporites in the Mediterranean Basin.

calcium sulfates, which in turn surround higher salinity deposits of halite and other highly soluble evaporite minerals (Box 14.2).

Because evaporite minerals are highly soluble in most groundwater systems, evaporites are rare in Precambrian sequences. In those few instances where they do occur, they are represented by the least soluble evaporite minerals such as anhydrite and gypsum. Most evaporite rocks are Phanerozoic and halides and other highly soluble evaporites are known only from rocks of this age. As might be expected in rocks composed of highly soluble minerals, the diagenesis of evaporite rocks is exceedingly complex and can involve multiple replacement and neomorphism events. During burial,

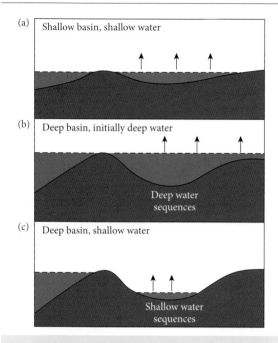

**Figure 14.24** Sketch models for large barred basin evaporite formation: (a) shallow basin, (b) deep basin with deep water. *Source*: Photo with permission of Frederic Boulvain. (c) deep basin with shallow water.

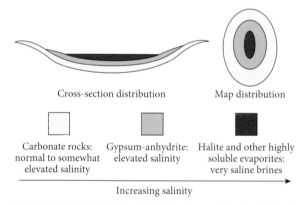

**Figure 14.25** Cross-section and sketch map view of an idealized evaporite basin. The bull's-eye map pattern shows that marine carbonates surround gypsum deposits, which in turn surround a halite bull's-eye, reflecting increased salinity and decreased aerial coverage as seas underwent increasing isolation and evaporation.

gypsum commonly loses water and dehydrates into anhydrite during eodiagenesis. The reverse process, the hydration of anhydrite to gypsum, commonly occurs during telodiagenesis as evaporite sequences approach the surface. Another very important aspect of evaporite diagenesis involves the tendency of halite and

other soluble halides to become extremely mobile when subjected to elevated temperatures and confining stresses during diagenesis. This mobility results from halite's very low density $(2.16\,g/cm^3)$ and the fact that it recrystallizes easily (Boggs 2016). Under such conditions, large masses of halide minerals may begin to flow upward by viscous plastic flow as buoyant masses called **salt diapirs** or **salt domes**. A well-known set of examples occurs in the Gulf of Mexico and is associated with the largest non-shale petroleum reserves in the United States.

### 14.3.2 Lacustrine evaporites

Lakes are filled with water that contains dissolved solids. Lakes located in interior basins and rift valleys with warm, arid climates suffer a fate similar to that of restricted seas. Small playa lakes are seasonal, filling during rainy periods, and evaporating during the dry season to leave thin crusts of evaporite minerals. Larger lakes can, over longer periods of times, produce more extensive evaporite deposits. A classic example is the Bonneville Salt Flats in northern Utah which represent the desiccated remains of Lake Bonneville. This large lake existed in this area during the Pleistocene when glacial advances made the climate cooler and wetter than it is now. In either case, **lacustrine evaporite deposits** commonly reveal a bull's-eye pattern in which extensive deposits of the least soluble minerals precipitated first when the lake occupied a larger area. Inside this, ring deposits of progressively more soluble minerals deposited as the lake shrinks to a progressively smaller area occupied by progressively more saline waters from which more soluble minerals are finally precipitated.

Because their water is derived from local interior drainage via surface water and groundwater whose dissolved solids composition reflects local source rock compositions and conditions, lacustrine evaporites tend to have more variable compositions than marine evaporites. As a result, it is more difficult to generalize about lacustrine evaporite mineralogy than about marine evaporite mineralogy. Depending on local conditions, lacustrine evaporites may be dominated by carbonates, sulfates, borates, or even nitrates. Halides, such as halite (NaCl), although present locally, tend to be less abundant than they are in marine evaporite sequences. Common carbonate evaporite minerals include calcite $(CaCO_3)$, aragonite $(CaCO_3)$, magnesite $(MgCO_3)$, and dolomite

**Table 14.6** Some important lacustrine evaporite minerals.

| Mineral | Chemical composition |
| --- | --- |
| Bloedite | $Na_2SO_4 \cdot MgSO_4 \cdot 4H_2O$ |
| Borax* | $Na_2B_4O_5(OH)_4 \cdot 8H_2O$ |
| Colemanite | $CaB_5O_4(OH)_3 \cdot H_2O$ |
| Epsomite* | $MgSO_4 \cdot 7H_2O$ |
| Gaylussite* | $CaCO_3 \cdot Na_2CO_3 \cdot 5H_2O$ |
| Glauberite* | $CaSO_4 \cdot Na_2SO_4$ |
| Kernite | $Ca_2B_4O_6(OH)_2 \cdot 3H_2O$ |
| Mirabilite | $Na_2SO_4 \cdot 10H_2O$ |
| Natron* | $NaCO_3 \cdot 10H_2O$ |
| Nahcolite | $NaHCO_3$ |
| Thenardite | $Na_2SO_4$ |
| Trona* | $NaHCO_3 Na_2CO_3 \cdot 2H_2O$ |
| Ulexite* | $Na_2B_4O_5(OH)_4 \cdot 3H_2O$ |

* Excellent indicators of lacustrine evaporites.

$[CaMg(CO_3)_2]$. Common sulfate minerals include gypsum ($CaSO_4 \bullet 2H_2O$) and anhydrite ($CaSO_4$). In addition to these minerals, the major nonmarine carbonate, sulfate, and borate evaporite minerals that form in lacustrine evaporite sequences are those listed in Table 14.6. The minerals shown with asterisks are excellent indicators of lacustrine evaporites.

Older North American examples of lacustrine evaporites include the sodium-rich trona–halite deposits in the Eocene Wilkins Peak Member of the Green River Formation in southwestern Wyoming and the Jurassic Lockatong and New Berlin Formations deposited in rift graben lakes in New Jersey and Connecticut and the Silurian age Salina Formation deposits in the Michigan Basin.

## 14.4 SILICEOUS SEDIMENTARY ROCKS

**Siliceous sedimentary rocks** occur in a wide variety of forms created in many different settings over the past 3.8 Ga. They consist primarily of silica ($SiO_2$) minerals of the chert family, such as (1) cryptocrystalline–microcrystalline **chert**, (2) cryptocrystalline, radial-fibrous **chalcedony**, and (3) amorphous **opal**. These silica minerals and mineraloids combine to form hard rocks with conchoidal fracture that are collectively referred to as the **chert family**. Details concerning the varieties of chert family minerals and mineraloids are discussed in Chapter 5.

**Modern siliceous sediments** are mostly composed of the skeletal remains of diatoms,

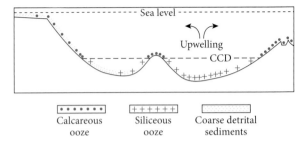

**Figure 14.26** Conditions under which modern siliceous oozes accumulate below the carbonate compensation depth (CCD), beneath the productive areas of upwelling, where detrital influx is minimal.

radiolaria, and/or silicaflagellates that secrete microscopic shells of amorphous opaline silica ($SiO_2 \bullet nH_2O$). Such deposits commonly form on modern sea floors that are below the CCD and beneath areas of upwelling where the influx of detrital sediments is minimal (Figure 14.26). Upwelling brings dissolved silica into surface waters where microorganisms extract it to secrete opaline silica shells that sink to form a significant portion of the sediment in areas where little detrital and carbonate sediment accumulates and the paucity of calcium carbonate and detrital sediment ensures relatively pure silica deposits. During diagenesis, opaline silica shells are converted to a partially crystalline form of silica called opal-CT and perhaps eventually to chert. Despite the abundance of recent siliceous sediments on the sea floor, no incontrovertible chert equivalents have been recovered by deep sea drilling.

Two contrasting types of chert deposits dominate older siliceous sedimentary rocks: bedded cherts and nodular cherts. **Bedded cherts** occur as layers in sequences up to several hundred meters thick in which chert is most commonly interstratified with mudrocks to form **ribbon rocks** (Figure 14.27) in which the remains of microscopic shells such as radiolaria are sometimes preserved in **radiolarian chert**. Many bedded cherts formed below wave base in outer shelf to deeper water marine environments, along active margins and in pull-apart basins associated with transform margins. Most bedded cherts are interlayered with mudrocks, but a few are interlayered with turbidite sandstones and even carbonates. Bedded cherts are especially

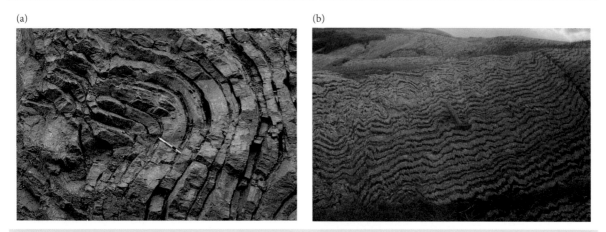

**Figure 14.27** Bedded "ribbon" chert in outcrops. (a) Radiolarian chert, Cretaceous mélange, Marin County, California. *Source*: Courtesy of Steve Newton. (b) Ribbon chert, Ordovician, Norway. *Source*: Courtesy of Roger Suthren.

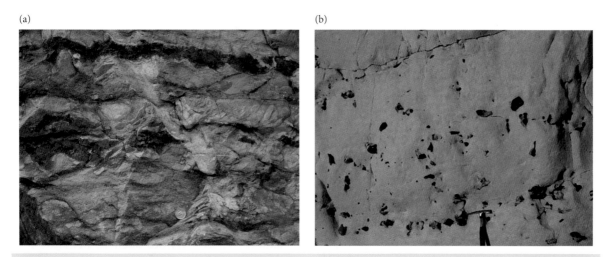

**Figure 14.28** Chert nodules. (a) Flint nodules in Kalkberg limestone, Devonian, New York. (b) Flint in chalk cliffs, Cretaceous, Seven Sisters, East Sussex, England. *Source*: John O'Brien. © John Wiley & Sons.

abundant in association with Precambrian banded iron formations (BIF) where they may have been primary precipitates and, less commonly, with phosphorites, both of which are discussed later in this chapter. They are also a significant part of the sedimentary portion of many ophiolite and alpine mélange sequences, discussed in Chapters 10 and 18.

**Nodular cherts** occur as ellipsoidal to bulbous to irregular masses less than 1 m in length that tend to be elongate, parallel to, and concentrated in certain strata (Figure 14.28). Although they occur in many different sedimentary rock types, the vast majority of

nodular cherts occur in carbonate rocks such as limestone and dolostone that formed in supratidal to subtidal marine environments. Some nodules are concentrically zoned, others are massive and still others preserve the structures of the enclosing rocks. In thin section, primary carbonate grains such as ooids, fossils, and pellets can be seen to have been replaced by silica minerals. The preservation of primary structures and carbonate grains as silica in nodular cherts strongly suggests that most nodular cherts are secondary and form by the replacement of preexisting, relatively shallow water carbonates during diagenesis.

For this reason, many nodular cherts also are referred to as **secondary** or **replacement cherts**.

Knauth (1979) pointed out that the formation of nodular cherts by limestone replacement requires pore waters supersaturated with respect to silica and undersaturated with respect to calcium carbonate. These conditions favor the dissolution of calcium carbonate and the simultaneous precipitation of silica. These conditions exist in acidic waters where the pH is decreasing. Because alkaline marine surface waters are highly undersaturated with respect to silica and nearly saturated with respect to calcium carbonate, it is unlikely that chert nodules form from marine pore waters. Nodular chert formation might occur under several sets of diagenetic conditions that involve meteoric pore waters enriched in silica from the dissolution of silicate minerals or silica skeletal particles. Wherever silica-rich, calcium carbonate-undersaturated meteoric, or mixed continental–marine waters occur, the necessary conditions of reduced pH and undersaturation with respect to calcium carbonate can be created. Most models for the formation of chert nodules involve carbonate replacement by opal or cryptocrystalline quartz during eodiagenesis in the meteoric zone or the zone of mixing.

Siliceous sediments also form in smaller amounts in alkaline lakes because the solubility of silica rapidly decreases at elevated pH (>9). It also occurs as **siliceous sinter** precipitated around hot springs (Figure 14.29) because the solubility of silica rapidly decreases as spring water cools and pH rises as carbon dioxide comes out of solution.

## 14.5 IRON-RICH SEDIMENTARY ROCKS

**Iron-rich sedimentary rocks** contain more than 15% iron by weight and occur in rocks that range in age from 3.8 Ga to the present. As the major source for iron ore, used in the manufacture of all iron and steel products, they represent a significant economic resource. Several different types of iron-rich sedimentary rocks occur. They include (1) Precambrian iron formations, (2) Phanerozoic ironstones, and (3) smaller accumulations of bog iron, iron-rich manganese nodules, and pyrite-rich black shales. Table 14.7 summarizes the significant differences between Precambrian iron formations and Phanerozoic ironstones.

### 14.5.1 Precambrian iron formations

Most, but not all, Precambrian iron formations contain iron-bearing minerals interlayered with siliceous sediments of the chert family in a way that gives them a banded appearance in outcrop (Figure 14.30). These distinctive laminated to thinly bedded iron formations are called **banded iron formations** (BIF) and contain 60% of the world's iron ores (Wenk and Bulakh 2016).

The major iron-bearing minerals (Table 14.7) include: (1) iron oxides such as hematite and magnetite, (2) iron carbonates such as siderite and ankerite, and (3) iron silicates such as the

(a)

(b)

**Figure 14.29**   (a) Siliceous sinter precipitated around hot springs, Yellowstone National Park, Wyoming. (b) Siliceous sinter terraces, Mammoth Hot Springs, Yellowstone National Park. *Source*: John O'Brien. © John Wiley & Sons.

**Table 14.7** Common minerals in Precambrian and Phanerozoic iron-rich sedimentary rocks.

| Mineral group | Mineral | Formula | Iron formations | Ironstones |
|---|---|---|---|---|
| Iron oxides and hydroxides | Hematite | $Fe_2O_3$ | Abundant | Common |
| | Magnetite | $FeFe_2O_4$ | Abundant | Rare |
| | Goethite | $FeOOH$ | | Abundant |
| Iron silicates | Stilpnomelane | $K(Fe,Mg,Fe)_8(Si,Al)_{12}(O,OH)_{27} \bullet n\ H_2O$ | Common | |
| | Greenalite | $(Fe,Mg)_6Si_4O_{10}(OH)_8$ | Common | |
| | Minnesotaite | $(Fe,Mg)_3Si_4O_{10}(OH)_2$ | Common | |
| | Riebeckite | $Na_2Fe_5Si_8O_{22}(OH)_2$ | Scarce | |
| | Chamosite | $(Fe,Mg,Fe)5AlAlSi_3O_{10}(O,OH)_8$ | | Common |
| Iron carbonates | Siderite | $FeCO_3$ | Common | Common |
| | Ankerite | $FeMg(CO_3)_2$ | | |
| Iron sulfides | Pyrite | $FeS_2$ | Common | Common |
| | Pyrrhotite | $Fe_{1-x}S$ | Scarce | |
| Common associated minerals | Chert group | Microcrystalline $SiO_2$ | Abundant | Rare |
| | Calcite | $CaCO_3$ | Rare | Common |
| | Dolomite | $CaMg(CO_3)_2$ | Scarce | Common |
| | Collophane | $Ca_5(PO_4)_3(F,OH)$ | | Common |

**Figure 14.30** Banded iron formation, Vermillion Range, Minnesota. *Sources*: Photo by Kevin Hefferan. © John Wiley & Sons. Photo by John O'Brien. © John Wiley & Sons.

iron-rich serpentine mineral greenalite, the iron-rich chlorite group mineral stilpnomelane and iron-rich talc group mineral minnesotaite. Many other, rarer minerals occur (Table 14.7).

Banded iron formations are common only in Archean and early Proterozoic rocks that range in age from 3.8 to 1.8 Ga, with a peak in abundance in the Paleoproterozoic, between 2.6 and 2.4 Ga (Bekker et al. 2010). After a hiatus of 1.0 Ga, a few examples occur in Neoproterozoic rocks formed from 0.8 to 0.5 Ga. Because there are no known modern analogues for these Precambrian rocks, their

origin remains both controversial and enigmatic. What were the conditions that permitted widespread iron formations to accumulate during the Precambrian, but not during the Phanerozoic? Why did they reappear briefly after a hiatus of a billion years?

Two major types of iron formation, each dominant during different parts of the Precambrian, are commonly recognized. **Algoma-type banded iron formations** (after deposits in the Algoma District, Ontario, Canada) dominate Archean iron-rich sedimentary rocks formed between 3.8 and 2.6 Ga. These iron formations tend to occur as fairly thin (<10–100 m), elongate lenses of limited lateral extent (<1–10 km) that occur within Archean greenstone belts formed in convergent plate settings. The iron-rich bands are composed almost exclusively of extremely fine-grained rocks called **femicrites**, by analogy with fine-grained carbonate micrites. The femicrites are interlayered with chert group minerals. Algoma-type BIF are associated with submarine ultramafic–mafic volcanic rocks, mudrocks and sparse volcanoclastic "greywacke" sandstones. These are inferred to have been formed in fairly deep water marine environments, perhaps in forearc or backarc basins or advanced intracratonic rift settings (Kappler et al. 2005). They formed well below wave base, with minimal influx of detrital sediments and probably under anoxic conditions (Bekker et al. 2010).

**Superior-type iron formations** (after deposits near Lake Superior, Minnesota) dominate Proterozoic iron-rich sedimentary rocks that formed periodically between 2.6 and 1.8 Ga and again from 0.8 to 0.5 Ga. Although their mineralogy is similar to that of Algoma-type formations, their size, textures, associations, and inferred depositional environments are quite different. Lake Superior-type iron formations tend to be much larger than Algoma-type formations. Their thickness is commonly 100–1000 m, they occur in broad belts up to $10^2$–$10^3$ km wide and have aerial extents of as much as $10^5$ km$^2$. Although they contain BIF femicrites similar to those in Algoma-type sequences, they also contain **granular iron formations** (GIF) with ooids, pisoliths, intraclasts, and pellets similar to those in shallow water carbonate sequences (Blatt et al. 2006). Stromatolites and abundant traction current features such as ripple marks and cross-strata further attest to shallow water conditions under which many GIF were formed. Lake Superior-type GIF occur in a regional context that includes quartzarenites, limestones, dolostones, and mudrocks. Taken together, these features suggest that GIF formed in fairly shallow water environments on passive margin continental shelves and platforms, with limited influx of detrital sediments. Their abundance increases in younger Superior-type sequences, and they appear to be associated with transgressive episodes associated with rising sea level and sea level highstands (Bekker et al. 2010). A substantial change in the conditions of iron formation production seems to have occurred during the transition from the Archean into the Proterozoic. What changes occurred during this transition to account for these differences? That question has been the focus of scientific investigations for the past several decades.

Perhaps because of the lack of modern analogues, many intriguing hypotheses have been put forth for the origin of both Algoma-type and Superior-type iron formations. Such hypotheses must explain the origin of the iron-rich minerals, the origin of the interlayered chert, their banded nature, and the physical and temporal contrasts between the two types of iron formation. They must also explain why their occurrence is largely restricted to rocks older than 1.8 Ga and why

they formed again in the period 0.8–0.5 Ga after a hiatus of a billion years. Over the past two decades, a broad consensus has developed concerning some aspects of their formation, but the details, especially concerning the role of microbes, remain controversial (Nealson and Myers 1990; Klein and Buekes 1992; Simonson 2003; Kappler et al. 2005; Bekker et al. 2010; Rasmussen et al. 2013; Weiqiang et al. 2015).

Several lines of evidence suggest that the Archaean–Paleoproterozoic atmosphere was depleted in free oxygen ($O_2$ gas), which would have depleted the oceans in dissolved oxygen as well. Oxidized hematite-bearing soils and red hematite-cemented sandstones do not occur until 1.8 Ga, at about the same time that reduced detrital pyrite and uraninite (unstable in the presence of free oxygen) disappear from the geological record. Although correlation is not cause, it is likely that the scarcity of free oxygen in the pre-1.8 Ga atmosphere played some role in the conditions under which iron formations were produced.

Early workers (e.g., James 1954) suggested a continental source for the iron. They argued that an early atmosphere, depleted in $O_2$ and enriched in $CO_2$, provided the ideal conditions for iron enrichment in more acidic ocean water. Under such acidic conditions, more iron would have been leached from the continents. Because ferrous iron is highly soluble in acidic water if free oxygen is not present, large quantities of dissolved iron would have entered the ocean. More recent workers (Simonson 2003, Polat and Frei 2005) suggest that surface waters at this time contained too much oxygen for ferrous iron to be transported in solution in sufficient quantities, but others, citing isotope evidence disagree (Weiqiang et al. 2015). A new consensus, supported by trace chemical and isotopic signatures, posits that the iron originated largely as dense brines from hydrothermal vents on the sea floor (Klein and Buekes 1992; Simonson 2003; Klein 2005). This certainly makes sense for the deep water Algoma-type formations whose elongate character might well be related to hydrothermal systems associated with elongate fracture systems in the sea floor, but not necessarily for the larger, shallower Superior-type formations.

The source of silica for the cherts presents a separate, but perhaps related, problem. Silica solubility [as Si(OH)$_4$] is remarkably independent of acidity for pH ranges between 2 and 10, but does increase with temperature. However, in colloidal form, the solubility of silica increases rapidly at pH levels from 4 to 5. In an acidic, CO$_2$-rich atmosphere silica would be more strongly leached from continental source rocks producing acidic oceans with far higher dissolved silica contents than modern oceans. Such oceans might on occasion become oversaturated with respect to dissolved silica, leading to its precipitation, as suggested by Rasmussen et al. (2013). However, Bekker et al. (2010) argue that the silica is diagenetic. The stage is set for banded iron formation; the remaining questions concern the mechanisms by which they originated.

Klein and Buekes (1992) proposed a widely accepted model in which Archean–early Proterozoic oceans were chemically stratified, with dense masses of anoxic, acidic, deep, ferrous iron-rich waters separated by a **chemocline** from less dense, more oxidized surface waters with little dissolved iron. Support for a stratified ocean comes from carbon isotope data that strongly suggest that Archean–early Proterozoic oceans were stratified with respect to carbon in both carbonates and organic matter (Klein 2005). A variation on this model (Cameron 1983; Simonson and Hassler 1996, cited in Simonson 2003) suggests that maximum dissolved iron concentrations occurred well above the bottom, just below the chemocline. This would certainly help to explain the origin of Superior-type iron formations in shallower water environments.

Periodic mixing of higher Eh (more oxygen) and pH (more alkaline) surface water with lower Eh (less oxygen) and pH (more acidic) deeper water – caused by changes in the position of the chemocline and/or overturning or upwelling of deep, iron-rich waters onto continental margins – could account for the cyclic precipitation of iron-bearing minerals and chert. Iron oxide minerals would develop under the highest Eh conditions, iron carbonates under the highest pH conditions and iron sulfides under the lowest Eh conditions. Hamade et al. (2003) proposed that BIF resulted from alternations between fluxes of silica derived from the continents and ferrous iron derived periodically from hydrothermal vents. On the other hand, Bekker et al. (2010) argued for a diagenetic origin for the chert layers.

Recent decades have witnessed a huge explosion in our direct and indirect (via chemical proxies) knowledge of Precambrian microbes (Konhauser et al. 2017). Nealson and Myers (1990) suggested that bacteria played a significant role in the precipitation of both silica and iron oxides, and this idea has gained many adherents. The presence of stromatolites in many GIF-type iron formations suggests that cyanophytes, photosynthetic bacteria, may have produced the oxygen necessary to cause precipitation of iron in certain shallow water Superior-type deposits. Posth et al. (2008) have also argued for a fundamental role for cyanophytes in producing the oxygen that led to the huge Superior-type deposits of the Paleoproterozoic. They have argued further that microbial-induced precipitation of iron occurred during periods of high ocean temperatures and alternated with nonmicrobial precipitation of silica at lower temperatures. Kappler et al. (2005) claimed that anoxic, photosynthetic bacteria, living in deeper water than cyanophytes, produced the oxygen that caused the precipitation of iron in Algoma-type deposits. Bekker et al. (2010) suggested that anoxygenic oxidation by protobacteria that photosynthesized without using CO$_2$, produced the oxygen for Algoma-type deposits. They also posited that oxidation by cyanophytes in the photic zone produced the oxygen for Superior-type iron formations. This correlates well with the age of the oldest, clearly documented, cyanophytes, and would explain the rapid increase in atmospheric oxygen that led to the great oxygen event (GOE) around 2.4 Ga and to the long-term disappearance of BIF around 1.8 Ga. Researchers continue to address these issues, seeking some degree of resolution.

The reappearance of iron formations in 0.8–0.5 Ga rocks corresponds roughly to the time of the Neoproterozoic "snowball Earth" (0.75–0.58 Ga). During this time, it has been suggested (Hoffman et al. 1998) that the world's oceans were frozen over for long periods of time and glaciers reached within 10° of the equator. A surface layer of ice on the oceans would have impeded the exchange of oxygen with the atmosphere, leading to the development of an anoxic ocean. The occurrence of what appear to be glacial dropstones in some BIF of this age is certainly intriguing. A snowball Earth would help to explain the reappearance of BIF at this time, but the theory remains somewhat controversial.

### 14.5.2 Phanerozoic ironstones and other iron-rich rocks

Phanerozoic iron-rich rocks are quite different from their Precambrian iron-rich counterparts in terms of mineralogy, texture, scale, associated rock types and the environment and processes under which they accumulated. The most widespread type of Phanerozoic iron-rich sedimentary rock is called **ironstone**. Most ironstone deposits are quite thin (<20–30 m thick) and are composed predominantly of goethite and hematite, with smaller amounts of the iron-bearing chlorite mineral chamosite. Siderite, magnetite, or pyrite occur in some deposits. Ironstones are commonly associated with, and even gradational into, carbonate rocks such as limestone or dolostone. In many cases, the iron minerals appear to replace carbonate grains such as skeletal particles and ooids. Quartz-rich sandstones, phosphorites, and/or cherts are less common associates. Sedimentary textures and structures, such as cross-strata, ripples, erosional scours, and abraded fossils, indicate accumulation in above wave-base, shallow water, mostly well-oxygenated, marine environments.

Ironstones range in age from Cambrian through mid-Tertiary. The vast majority of them formed during two periods of peak ironstone formation, Ordovician–Devonian and Jurassic–Cretaceous. Both of were periods of maximum global warming and sea level high stands. Periods of ironstone formation are associated with significant marine transgressions. The clear association of Phanerozoic ironstones with periods of global warming supports the idea that the iron originated from the erosion of lateritic soils formed by weathering in warm, humid, tropical climates that promoted the leaching of iron by acidic soil waters. To avoid oxidation, the iron would have to be delivered as colloidal ferrous iron attached to tiny clay particles. Once in the marine environment, the iron could then be mobilized to replace carbonate grains and/or to be precipitated as primary goethite or hematite in oxygenated near shore environments or as chamosite farther offshore.

Additional types of Phanerozoic iron formations occur, all of which have modern analogs. **Bog iron** deposits consist largely of goethite and other oxyhydroxides, siderite, and the manganese oxide minerals psilomelane and pyrolusite. They form where acidic groundwater delivers ferrous iron, dissolved mostly from pyrite, into swamps and lakes where it is oxidized, and precipitated as surface crusts. **Polyminerallic manganese nodules** (Figure 14.31) and encrustations form under oxidizing conditions on the sea floor where iron and manganese oxide minerals precipitate along with copper, cobalt, nickel, and other metals in the form of concentric nodules precipitated about a nucleus or as layered encrustations on rock outcrops. Some black shales deposited under anoxic conditions (Chapter 13) contain sufficient pyrite to be classified as iron-rich sediment. Hydrothermally precipitated chimneys around "black smoker" hot springs on the sea floor commonly contain significant amounts of iron-bearing sulfide minerals. The black color of such hot springs results from finely divided, opaque sulfide minerals, such as pyrite ($FeS_2$) and chalcopyrite ($CuFeS_2$).

Although minor rock types compared to detrital sedimentary rocks, carbonates and even evaporites, iron-rich sedimentary rocks continue to intrigue us with their variety, the mysteries surrounding their formation, and their importance as significant sources of ore for the manufacture of steel and related industrial products.

Growth rings

1 cm

**Figure 14.31** Manganese nodule. *Source*: John O'Brien. © John Wiley & Sons.

## 14.6 SEDIMENTARY PHOSPHATES

**Sedimentary phosphates,** in the form of detrital **apatite** [$Ca_5(PO_4)_3(OH,F,Cl)$], bones and teeth composed of **hydroxyapatite** [$Ca_5(PO_4)_3(OH)$], cryptocrystalline **fluorapatite** [$Ca_5(PO_4)_3(F)$], and amorphous **collophane** are minor components of many sedimentary rocks. Much rarer are Neoproterozoic to Phanerozoic phosphate-rich rocks called **phosphorites** that contain more than 50% phosphate minerals and/or 18% phosphate by weight. Because phosphate is an important nutrient element for organic synthesis and growth, these deposits have been extensively mined as sources of phosphate fertilizers. They are also attracting attention because of their sometimes high content of rare earth elements (REE's) which are essential components in many technologically sophisticate commercial products (Pufahl and Groat 2017).

The major occurrence of phosphorites is in the form of laminae and beds of cryptocrystalline fluorapatite and amorphous collophane interlayered with carbonates, siliceous sediments, and detrital mudrocks (Figure 14.32). The phosphatic strata appear nearly black in outcrop, while collophane is medium to dark brown and isotropic in thin section. The constituents of these phosphorites resemble those of carbonate rocks in that they include mud particles and grains such as ooids, fossils, peloids, and clasts. In some cases, the phosphate partially or totally replaces earlier carbonate grains, whereas in other cases its origin is less clear. Like carbonates, phosphorites tend to be concentrated in areas where the influx of detrital sediment is minimal. Additional concentration has been attributed to hydraulic processes, especially during transgressive episodes (Pufahl and Groat 2017).

Modern examples of phosphorites occur mostly as crusts near the sediment–water interface, in fairly shallow (30–500 m) tropical to subtropical waters (latitude <40°), beneath areas of upwelling. The upwelling of cold deep water brings substantial nutrients such as phosphate to near surface waters. This increases biological productivity so that phosphate is incorporated into organic matter. As these organisms die, the organic matter drifts toward the ocean floor where bacterial decomposition releases phosphate to the bottom and pore waters while utilizing sufficient oxygen to lower their oxidation–reduction potential (Eh). The details of phosphate precipitation and replacement remain uncertain, but most recent work focuses on the role of various bacteria on and within the sediment. They apparently create the geochemical conditions that allow phosphate to form by some combination of eodiagenetic replacement and/or precipitation (Brookfield et al. 2009).

Another occurrence of phosphorites is in so-called **bone beds** where various forms of apatite have been concentrated in placer deposits of bones, teeth, and other phosphatic material formed as lag deposits from which finer constituents have been removed by currents, often over long periods of time. An additional source of phosphorites, rapidly vanishing as a result of mining, is **guano deposits.** These formed where bird colonies generated thick accumulations of phosphate-rich fecal material over long periods of time.

**Figure 14.32** Black phosphate layer, Phosphoria Formation, Permian, Wyoming. *Source:* James St. John, CC BY 4.0, https://www.flickr.com/photos/jsjgeology/49192413597 last visited 09/07/2020

## 14.7 COAL AND OTHER CARBON-RICH SEDIMENTS AND MATERIALS

Sedimentary rocks typically contain very little carbonaceous organic matter. This is clearly indicated by the average amounts of carbonaceous material in sandstones (0.05%), limestones (0.3%), and even mudrocks (2.0%), cited by Tucker (2001). Carbon-rich sediments contain elevated amounts of organic carbon derived from the preservation of organic material. As pointed out in the discussions of humus-rich organic soils such as histosols in Chapter 11 and black and oil shales in Chapter 13, the ideal conditions for the preservation of organic matter occur when large quantities of organic matter are produced and then accumulate in environments that are depleted in oxygen so that bacterial decomposition is inhibited. Carbon-rich sedimentary materials that preserve significant organic matter include the coal family and solid, liquid, and gaseous hydrocarbons of the petroleum family such as asphalt, crude oil and natural gas. These are the major sources for the carbon-based fossil fuels that, for better or for worse, still provide much of the energy required by modern societies. Their use has already had significant side effects, including atmospheric pollution and global warming, the latter with its all-to-immediate potential for significant near- and longer-term global climate change and its consequences.

### 14.7.1 Coal

**Coal** is rock that consists primarily of plant materials (humus) that have been buried, compacted, heated, and biochemically altered during diagenesis. Additional constituents include small amounts of siliclastic and/or carbonate sediment, pyrite, and variable amounts of moisture. The vast majority of coal consists mostly of altered organic materials. These were derived from wood and leaves and smaller amounts of mosses, grasses, and phytoplankton that initially accumulated under anoxic conditions in swamps, bogs, and marshes. Anoxic swamps typically develop (1) in paralic shoreline environments, especially in parts of deltaic systems bypassed by distributary channels; (2) on river floodplains in the ultralow gradient parts of meandering stream systems; (3) along the shores of shallow lakes, in cratonic or continental rift valley settings; and (4) in areas of poor drainage, such as those recently covered by glaciers. In such environments, thick, dense deposits of organic **peat** may accumulate over long periods of time. Peat is converted into coal by a variety of biochemical transformations that are driven primarily by increasing temperature and, to a lesser extent, pressure during progressively deeper and longer burial. The progressive transformations that occur during coal formation are collectively referred to as **coalification**.

Because peat forms primarily from woody materials that accumulated in terrestrial and paralic environments and because woody plants did not inhabit terrestrial environments until the middle Devonian, coal deposits do not occur in rocks older than Devonian and do not become abundant until the Carboniferous. Coals are commonly associated with (1) soils, especially vertisols and histosols, (2) mudrocks such as tonsteins and bentonites, and (3) a variety of marine–terrestrial transitional (paralic) and lacustrine facies. Most coals belong to a main series that is subdivided into a number of different coal ranks. **Coal ranks** are based on the progressive changes in coal composition, texture, and appearance that occur during coalification (Figure 14.33). The sequence of coal rank varieties during progressive burial and heating of peat is (1) lignite, (2) sub-bituminous, (3) bituminous, and (4) anthracite (Figure 14.34). Other ranks, such as sub-anthracite and meta-anthracite, are sometimes recognized. Figure 14.33 shows the approximate maximum burial depths and temperatures at which each coal rank develops over sufficient periods of time.

Progressive coalification is characterized by a number of significant trends, including (1) increasing carbon content, accompanied by, (2) gradual color change from brownish to black, (3) decreasing content of moisture and other volatiles, (4) increasing hardness and

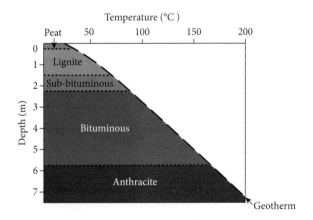

**Figure 14.33** Major ranks of coal produced during progressive coalification, with approximate burial depths and temperature ranges that produce the major ranks of coal. *Source:* Diagram by John O'Brien. © John Wiley & Sons.

Figure 14.34    Peat and the major ranks of coal derived from it during coalification (top to bottom, left to right): peat, lignite, subbituminous, bituminous and anthracite. Note the increases in compactness, hardness and luster and change in color from brown to black (carbon rich) with increasing rank. *Source*: Courtesy of Donna Pizzarella, USGS public domain. © U.S. Department of the Interior.

compactness, and (5) increasing reflectivity (Figure 14.36).

**Lignite** is a coal produced when peat is buried, compacted, and biochemically transformed by bacterial activity. Most lignite is quite soft, somewhat porous, and possesses a brownish color and rather dull luster (Figure 14.34). Carbon content in the organic fraction is relatively low (50–70%) and

volatile content (including moisture) is high (45–55%). Common volatiles include water ($H_2O$), carbon dioxide ($CO_2$), hydrogen ($H_2$), and methane ($CH_4$); less common are sulfur and nitrogen compounds. **Sub-bituminous coal** is produced by the burial and heating of lignite, which increases carbon content, drives off volatile components, and transforms some of the woody material into reflective organic compounds collectively called **vitrinite**. Sub-bituminous coal is still soft, but less porous and slightly more reflective than lignite, and generally possesses a brownish black color. Carbon content in the organic fraction is slightly higher (70–80%), and volatile contents are slightly lower (40–50%) than for lignite. Because of their relatively low carbon content and hardness, lignite, and sub-bituminous coals are considered **low rank coals** or **soft coals**. Such coals constitute almost half of the available coal reserves worldwide and are used primarily as fuel in electric power-generating stations. Their high content of volatile substances means that when burned they have an especially large potential to produce significant amounts of airborne pollutants. If these are emitted into the atmosphere, rather than being collected in the power-generating plant, they can be and are significant contributors to acid rain and global warming. For these reasons, the use of such coals to generate electricity is being phased out rapidly in many countries.

Increased coalification produces harder coals that are valued for their high carbon content. **Bituminous coal** is a harder coal produced by additional burial and heating of sub-bituminous coal. This is accompanied by additional formation of vitrinite, increased carbon content, and loss of volatiles. Bituminous coal is typically compact, hard, black, and somewhat reflective (Figure 14.34). Carbon content in the organic fraction is higher (80–90%), and volatile contents are lower (25–40%) than for sub-bituminous coals. High rank bituminous coal called **coking coal** is essential in the manufacture of steel. Relatively pure bituminous coal is first burned in an oven to drive off volatile constituents. What remains is solid coke, mostly carbon with small amounts of impurities or ash consisting largely of fused siliciclastic materials. The coke is then used as the primary fuel in the blast furnaces that smelt iron for the production of various steel products. The development of coke-fired blast furnaces for the smelting of iron in eighteenth century England was a critical event in the industrial revolution and its many consequences, both positive and negative.

With unusually deep burial and significant heating, bituminous coal is transformed into the relatively rare, vitrinite-rich coal called **anthracite**. Anthracite is very compact, hard, black and reflective (Figure 14.34). Carbon content in the organic fraction is higher (>90%), and volatile contents are lower (5–15%) than for bituminous coals. Because of its high carbon and low volatile contents, anthracite is highly valued. Its high carbon content permits anthracite to produce very large amounts of energy when burned and the low volatile content minimizes both smoke and pollution. Unfortunately, relatively clean-burning anthracite comprises less than 1% of coal deposits worldwide.

### 14.7.2 Petroleum: crude oil and natural gas

**Petroleum** is the general name for carbon-rich fluids that accumulate in the pore spaces of rock bodies, most commonly in limestone, sandstone, and dolostone. Petroleum is composed principally of different kinds of organic hydrocarbons, with various amounts of chemical impurities. As the name suggests, **hydrocarbons** are organic molecules that contain hydrogen and carbon as essential constituents. Hydrocarbons may also contain significant amounts of sulfur, nitrogen, oxygen, and phosphorous, as well as many less common constituents. Hydrocarbons are classified according to their structure, composition, and mass or density. **Natural gas** consists of lighter, more volatile hydrocarbons, whereas **crude oil** consists primarily of heavier, less volatile hydrocarbons.

Natural gas is subdivided into dry gas and wet gas components. **Dry gas** consists of very light molecules, dominated by chain-structured alkanes such as methane ($CH_4$), ethane ($C_2H_6$), and propane ($C_3H_8$) with the general formula $C_nH_{2n+2}$. **Wet gas** consists of somewhat heavier molecules, including alkanes such as butane ($C_4H_{10}$) and ring-structured cycloalkanes such as cyclobutane ($C_4H_8$) and cyclohexane ($C_6H_{12}$) with the general formula $C_nH_{2n}$. **Crude oil** is composed of heavier, generally more complex hydrocarbon molecules, many with substantial amounts of sulfur, nitrogen, oxygen, and phosphorous. Most natural petroleum deposits that contain crude oil also contain wet gas and dry gas in various proportions. All these components can be separated by distillation in an oil refinery.

The source material for petroleum is organic material that is deposited and preserved in sediments by incomplete decomposition, especially under low oxygen disoxic and/or anoxic conditions. The source materials are varied and incompletely understood. Rather than the dinosaurs suggested by certain TV advertisements, they are principally sapropels derived from marine plankton, microbial mats, and other fine organic material from marine and/or terrestrial sources. Most of this fine organic material accumulates in fine-grained sediments, including potential shales, mudstones, and micrites. During burial of these **source rocks**, progressive heating over time causes sapropels to undergo conversion to hydrocarbons by a set of process called **maturation**. Maturation is marked by three transitional stages: (1) diagenesis, (2) catagenesis, and (3) metagenesis. During shallow burial and **diagenesis**, at temperatures less than 50°C, bacteria convert some of the sapropels to methane. At the same time, increases in temperature produce thermocatalytic conversion of sapropels to kerogens, which are composed of very heavy, insoluble organic molecules. As discussed in Chapter 13, kerogens are important constituents of oil shales. The significant action begins as burial depths exceed 1.0–1.5 km and burial temperatures exceed 50°C. Under such condition, **catagenesis** begins to convert kerogens into crude oil and natural gas (Figure 14.35). Conversion to crude oil, rich in heavy hydrocarbons, continues to increase until temperatures approach 100°C, the maximum oil production temperature. In areas within the normal range of geothermal gradients, this temperature is reached at depths of 2.0–4.5 km. At higher temperatures, increasing amounts of natural gas are produced by the

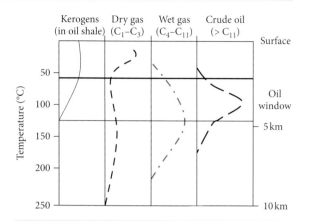

**Figure 14.35** Formation of crude oil, wet gas, dry gas and kerogens during petroleum maturation as a function of temperature and approximate burial depth in sedimentary basins. $C_n$ refers to the number of carbon atoms per organic molecule, a rough guide to the molecular density of petroleum molecules.

conversion of crude oil into lighter molecules. Conversion of crude oil to wet gas reaches a maximum near 125°C and dry gas conversion peaks near 150°C. With still higher temperatures and/or longer duration, only dry gas (methane) continues to be generated, and it is gradually lost from source rocks during **metagenesis** (Figure 14.35).

Only when organic material is cooked to temperatures of 60–120°C, the so-called **oil window**, at depths of ~2.0–4.5 km, are significant

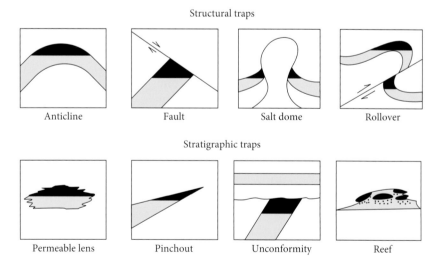

**Figure 14.36** Major types of petroleum traps: (top) structural traps produced by deformation, and (below) stratigraphic traps produced by deposition and/or diagenesis. White, impermeable rocks; stippled, permeable rocks; black, oil and/or gas. *Source*: Sketch by John O'Brien. © John Wiley & Sons.

amounts of crude oil produced. At lower and higher temperatures, only kerogens and natural gas are produced and, at still higher temperatures (>200°C), very few volatile hydrocarbons remain.

Once crude oil and natural gas are produced, their fluidity permits them to slowly migrate from the source rocks where they were produced into rocks with sufficient porosity and permeability to contain and transmit them. Their relatively low density permits fluid hydrocarbons to rise toward the surface. The rocks in which crude oil and natural gas eventually accumulate are called **reservoir rocks**. Such rocks are typically aquifers that contain significant water, in addition to petroleum. Just as relatively impermeable aquitards and aquicludes trap water in confined aquifers, relatively impermeable rocks surrounding and overlying reservoir rocks trap oil in reservoirs. Where the impermeable rocks, typically mudrocks, micrites or evaporites, overly the oil trap or reservoir and keep oil from rising toward the surface, they are called **cap rocks**. Typically, lighter crude oil rises above water and natural gas sits atop both in reservoir pore spaces. Figure 14.36 depicts several major types of petroleum traps. Petroleum traps that involve geological structures produced by deformation such as folds and faults are called **structural traps**; those produced by three-dimensional depositional patterns that trap petroleum in reservoir rocks are called **stratigraphic traps**.

More detailed discussions of biochemical rocks than can be encompassed in a book on Earth materials are to be found in Prothero and Schwab (2013), Flugel (2010), Tucker and Wright (1990), Friedman (1981), and Bathurst (1975).

## CONTENT ASSESSMENT

1   Discuss the differences between the major minerals in carbonate sedimentary rocks: (a) low-Mg calcite, (b) high-Mg calcite, (c) aragonite, and (d) dolomite. Explain how the relative abundance of these minerals related to the age of carbonate sedimentary rocks?

2   Discuss the several factors or conditions that favor the accumulation of extensive, relatively "pure," carbonate sediments on Earth's surface.

3   What are the *lysocline* and the *calcium CCD*? How and why do they influence the distribution of calcium carbonate ($CaCO_3$) sediments with depth in modern oceans? What is the likely effect of global warming on the position of each and how will that affect the distribution of calcium carbonate sediments?

4   Describe the properties of the four major types of *grains* or *allochems* preserved in carbonate sediments and sedimentary rocks. Then describe the properties of micrite, the fifth common primary component of such sediments and rocks. Lastly, discuss the processes that produce each type of the components described earlier.

5   Using Figures 14.8, 14.10 and 14.11, give an appropriate name, based on the components, for each of the following limestones, according to the classifications of Dunham (1962), Folk (1959) and Folk (1962)
    a   51% ooids, 12% skeletal fragments, 3% limeclasts, 20% spar (cement), well-sorted and rounded
    b   40% mud-supported skeletal fragments, 48% lime mud (micrite), 4% peloids and 2% spar (calcite cement)
    c   14% framework-supported skeletal fragments, 54% limeclasts, 22% mud (micrite), and 10% spar (cement)

6   Discuss the likely environmental conditions under which each of these limestones (6a, 6b and 6c) might have accumulated initially (recognizing that knowledge of their field relationships would permit a more detailed assessment).

7   Discuss the properties by which calcium carbonate cements produced in the *marine, meteoric phreatic, and meteoric vadose* zones may be distinguished from one another.

8   What is *dolomitization*? How does the abundance of dolomite change with the increasing age of carbonate rocks? Discuss the major hypotheses that have been posited to explain how dolomites form and the evidence that supports or casts doubt on each.

9   As part of an extensive coring program, your company has retrieved the following *deep* core samples from a single site and would like you to characterize the conditions under which they formed through time.

a Depth 840': 35% carnellite, 28% kainite, 22% halite, 15% polyhalite

b Depth 890'; 78% halite, 21% anhydrite, and 1% siliciclastic mud

c Depth 1020'; 7 % halite, 72% anhydrite, 3% calcite, and 18% siliclastic mud

d Depth 1275'; 50% lime mud (micrite), 46% nodular anhydrite, and 6% siliciclastic mud

10 Compare and contrast *Algoma-type BIF*, *Superior-type BIF*, and *sedimentary ironstones* in terms of mineralogy, textures/structures, size/extent, and the ages when each dominated iron-rich sedimentary rock formation.

## REFERENCES

Adams, A.E., Mackenzie, W.S., and Guilford, C. (1984). *Atlas of Sedimentary Rocks Under the Microscope*. Harlow, UK: Longman Publishers 180 pp.

Bathurst, R.G.C. (1975). *Carbonate Sediments and Their Diagenesis*, 2e. Amsterdam: Elsevier Publishers 658 pp.

Bekker, A., Slack, J.F., Planavsky, N., and Krapez, B. (2010). Iron formation the sedimentary product of a complex interplay among mantle, tectonic, oceanic and biospheric processes. *Economic Geology 105*: 467–508.

Blatt, H., Tracey, R., and Owens, B.R. (2006). *Petrology: Igneous, Sedimentary and Metamorphic*. New York: W.H. Freeman Publishers 530 pp.

Boggs, S. (2016). *Principles of Sedimentology and Stratigraphy*, 5e. New York: Pearson 688pp.

Brookfield, M.E., Hemmings, D.P., and Van Stratten, P. (2009). Paleoenvironments and origin of the sedimentary phosphates of the Napo Formation (Late Cretaceous, Oriente Basin, Ecuador). *Journal of South American Earth Sciences 28*: 180–192.

Dunham, R.J. (1962). Classification of carbonate rocks according to depositional texture. In: *Classification of Carbonate Rocks* (ed. J. Ham), 108–121. American Association of Petroleum Geologists Memoir No. 1.

Embry, A.F. and Klovan, J.E. (1971). A late Devonian reef tract on northeastern Banks Island Northwest Territories. *Bulletin Canadian Petroleum Geologists 19*: 730–781.

Flugel, E. (2010). *Microfacies of Carbonate Rocks; Analysis, Interpretation and Application*. New York: Springer Publisher.

Folk, R.L. (1959). Practical petrographic classification of limestones. *American Association of Petroleum Geologists Bulletin 43*: 1–38.

Folk, R.L. (1962). Spectral subdivisions of limestone types. In: *Classification of Carbonate Rocks* (ed. J.

Ham), 62–85. American Association of Petroleum Geologists Memoir No. 1.

Friedman, G.M. (ed.) (1981). *Diagenesis of Carbonate Rocks: Cement-Porosity Relationships*. Tulsa, OK: Society for Economic Paleontologists and Mineralogists 295 pp.

Grabau, A.W. (1913). *Principles of Stratigraphy*. New York: A.G. Seiler 1185 pp.

Ham, W.E. (ed.) (1962). *Classification of Carbonate Rocks*. American Association of Petroleum Geologists Memoir no. 1 270 pp.

Hamade, T., Konhauser, K.O., Raiswell, R. et al. (2003). Using Ge/Si ratios to decouple iron and silica fluxes in Precambrian banded iron formations. *Geology 31*: 35–38.

Hoffman, P.E., Kaufman, A.J., Halverson, G., and Shrag, D.P. (1998). A neoproterozoic snowball earth. *Science 281*: 1342–1346.

Hsu, K. (1982). *The Mediterranean Was Desert*. Princeton University Press 197pp.

James, H.L. (1954). Sedimentary facies of iron-formation. *Economic Geology 9*: 235–293.

James, N.P. (1983). Reefs. In: *Carbonate Depositional Models* (eds. P.A. Scholle, D.G. Bebout and C.H. Moore), 345–462. American Association of Petroleum Geologists Memoir 33.

James, N.P. (1984). Shallowing-upward cycles in carbonates. In: *Facies Models* (ed. R.G. Walker), 213–228. Geological Association of Canada, Geoscience Canada Reprint Series 1.

James, N.P. and Jones, B. (2016). *Origin of Carbonate Sedimentary Rocks*. Chichester, UK: Wiley 446 pp.

James, N.P. and Lukasic, J. (2010). Cool- and cold-water carbonates. In: *Facies Models*, vol. 4 (eds. N.P. James and R.W. Dalrymple), 369–398. Newfoundland: Geological Association of Canada, St. John's.

Kappler, A., Pasquero, C., Konhauser, K.O., and Newman, D.K. (2005). Deposition of banded iron formations by anoxygenic phototrophic Fe(II)-oxidizing bacteria. *Geology 33*: 865–868.

Kiessling, W. (2008). Sampling-standardized expansion and collapse of reef-building in the Phanerozoic. *Wiley VCH Fossil Record 11*: 7–18.

Klein, C. (2005). Some Precambrian banded iron-formations (BIFs) from around the world: their age, geologic setting, mineralogy, metamorphism, geochemistry, and origins. *American Mineralogist 90*: 1473–1499.

Klein, C. and Buekes, N.J. (1992). Time distribution, stratigraphy, sedimentologic settings and geochemistry of Precambrian iron formations. *Science 275*: 136–146.

Knauth, L.P. (1979). A model for the origin of chert in limestone. *Geology 7*: 274–277.

Kohout, F.A. (1965). A hypothesis concerning cyclic flow of salt water related to geothermal heating in the Floridan aquifer. *New York Academy of Science Transactions, Series 2 28*: 249–271.

Konhauser, K.O., Pllanavsky, N.J., Hardisty, D.S. et al. (2017). Iron formations: a global record of

Neoarchaean to Paleoproterozoic environmental history. *Earth Science Reviews 172*: 140–177.

Longman, M.W. (1980). Carbonate diagenetic textures from nearsurface diagenetic environments. *American Association Petroleum Geologists Bulletin 64*: 461–487.

Machel, H.G. (2004). Concepts and models of dolomitization. *Geological Society of London Special Publications 235*: 7–63.

Nealson, K.H. and Myers, C.R. (1990). Iron reduction by bacteria: a potential role in the genesis of banded iron formations. *American Journal of Science 290-A*: 35–45.

Polat, A. and Frei, R. (2005). The origin of early Archaen banded iron formations and of continental crust, Isua, southern West Greenland. *Precambrian Research 138*: 151–175.

Posth, N.R., Hegler, F., Konhauser, K.O., and Kappler, A. (2008). Alternating Si and Fe deposition caused by temperature fluctuations in Precambrian oceans. *Nature Geoscience 1*: 703–708.

Prothero, D.R. and Schwab, F. (2004). *Sedimentary Geology: an Introduction to Sedimentary Rocks and Stratigraphy*, 2e. New York: W.H. Freeman 557 pp.

Prothero, D.R. and Schwab, F. (2013). *Sedimentary Geology: An Introduction to Sedimentary Rocks and Stratigraphy*, 3e. New York: W.H. Freeman 593 pp.

Pufahl, P.K. and Groat, L.A. (2017). Sedimentary and igneous phosphate deposits: formation and exploration: an invited paper. *Economic Geology 112*: 483–516.

Rasmussen, B., Meier, D.B., Krapez, B., and Muhling, J.R. (2013). Iron silicate microgranules as precursor sediments to 2.5 billion year-old banded iron formations. *Journal of South American Earth Sciences 41*: 435–438.

Raymond, L.A. (2002). *The Study of Igneous, Sedimentary and Metamorphic Rocks*. Boston: McGraw Hill 720 pp.

Sandberg, P.A. (1983). An oscillatory trend in Phanerozoic non-skeletal carbonate mineralogy. *Nature 305*: 19–22.

Scholle, P.A. and Ulmer-Scholle, D.S. (2003). *A Color Guide to the Petrography of Carbonate Rocks*. American Association of Petroleum Geologists Memoir No. 77 474 pp.

Simonson, B. (2003). Origin and evolution of large Precambrian iron formations, in extreme depositional conditions: mega end members in geologic time. In: *Sedimentary Giants – Extreme Depositional Environments: Mega End Members in Geologic Time* (eds. M.A. Chan and A.A. Archer), 231–244. *Geological Society of America Special Paper 370*.

Simonson, B.M. and Hassler, S.W. (1996). Was the deposition of large Precambrian iron formations linked to major marine transgressions? *The Journal of Geology 104* (6): 665–676.

Stanley, M.S. and Hardie, L.A. (1999). Hypercalcification: Paleontology links plate tectonics and geochemistry to sedimentology. *GSA Today 9*: 2–7.

Tucker, M. (2001). *Sedimentary Petrology*, 3e. Oxford, UK: Blackwell Science 272 pp.

Tucker, M. and Wright, V.P. (1990). *Carbonate Sedimentology*. Oxford, UK: Blackwell Science 496 pp.

Weiqiang, L., Beard, B., and Johnson, C.M. (2015). Biologically recycled continental iron is a major component in banded iron formations. *Proceedings of the National Academy of Science of the USA 112* (27): 8193–8198.

Wenk, H.R. and Bulakh, A. (2016). *Minerals: their Constitution and Origin*. Cambridge, UK: Cambridge University Press 672 pp.

Wright, D.T., and Wacey, D. (2004) Sedimentary dolomite: a reality check. Geological Society of London Special Publications *235*, 65–74.

# Chapter 15

# Metamorphism

## 15.1 METAMORPHISM: AN INTRODUCTION

**Metamorphism** refers to a set of predominantly solid state processes by which a pre-existing "parent" rock (protolith), is transformed into a metamorphic rock. Changes in mineralogy and/or textures develop in response to: (1) changes in temperature, (2) changes in pressure and/or (3) the actions of hot fluids that serve to accelerate metamorphic reactions. Metamorphic rocks can evolve from any pre-existing rock, including an earlier formed metamorphic rock. However, by convention, the term **protolith** refers to the original igneous or sedimentary source rock, that existed prior to any later metamorphic events.

The chemical composition and mineralogy of a metamorphic rock are dependent upon: (1) the chemical composition of the protolith, (2) any components added or removed by the action of hot fluids and (3) the temperature and pressure conditions under which the rock formed. For example, silica-rich quartz arenite sandstones are most commonly metamorphosed into silica-rich metaquartzites; similarly, calcium carbonate-rich limestones are generally transformed into a calcium carbonate-rich marbles. In addition, metamorphic rocks typically contain minerals stable at the temperature and pressure conditions at which metamorphism occurred. In response to changes in the metamorphic environment, unstable minerals break down to form new stable minerals. The stable mineral suite that occurs within a metamorphic rock represents an **equilibrium assemblage** of minerals stable at the temperature and pressure conditions of metamorphism. For example, with increasing

*Earth Materials*, Second Edition. Kevin Hefferan and John O'Brien.
© 2022 John Wiley & Sons Ltd. Published 2022 by John Wiley & Sons Ltd.
Companion website: www.wiley.com/go/hefferan/earthmaterials2

temperature conditions chlorite becomes unstable and transforms to biotite which alters to garnet at even higher temperatures. Similarly, with increasing pressures augite pyroxene becomes unstable and metamorphoses into omphacite pyroxene. Only minerals stable under both the initial and final set of conditions survive in the new metamorphic rock. Quartz is a good example of a survivor mineral; although strictly speaking there are six polymorphs of quartz, each of which are stable at a particular set of temperature and pressure conditions.

Mineral transformations occur over periods varying from tens of millions of years in mountain building episodes to nearly instantaneous events in meteorite impact events. In most cases, metamorphism requires millions of years to create stable mineral assemblages that equilibrate at new temperatures and pressures from some initial state. Geologists refer to such temperature, pressure and time relationships by the abbreviation T–P–t. T–P–t conditions produce different metamorphic grades that may be characterized based on stable mineral assemblages as low to high grades or as prograde or retrograde metamorphism. **Prograde metamorphism** (also called **progressive metamorphism**) results from increasing temperature and/or pressure conditions over time that produce higher T-P metamorphic mineral assemblages and rocks. **Retrograde metamorphism** results from decreasing temperature and/or pressure so that lower temperature/pressure mineral assemblages develop which overprint earlier peak temperature/pressure mineral assemblages. Volatile components serve as catalysts in driving retrograde metamorphic reactions. Without the addition of volatiles from an external source, retrograde metamorphic conditions are difficult to attain because previously occurring prograde metamorphism already depleted the rock in volatile components.

How can we determine the conditions and paths of metamorphism? **Geothermobarometry** techniques analyze the chemistry of mineral assemblages to infer peak temperatures and pressures of metamorphism. Geothermobarometry requires a knowledge of the: (1) compositions of all equilibrium minerals that can have variable compositions; (2) thermodynamic properties of the minerals; and (3) a calibration that relates mineral chemical composition to temperature and pressure. Co-existing equilibrium mineral chemical compositions can vary based on temperature and pressure conditions. For example, equilibrium concentrations of titanium in quartz and zircon in rutile are temperature dependent, so these serve as effective thermometers to determine the temperatures at which these minerals crystallized. Yttrium concentrations in monazite and garnet are pressure dependent and serve as effective barometers, measuring the pressures at which minerals crystallized. **Pseudosections** are equilibrium phase diagrams that identify the P–T stability fields of selected equilibrium mineral assemblages likely to occur in a given bulk rock composition. For example, Inglis et al. (2017) obtained P–T estimates on garnet obtained from a metabasalt collected in the Tasriwine ophiolite of Morocco. They analyzed both the interior garnet core and outside garnet rim using a NaCaKFMASHT (Na$_2$O–CaO–K$_2$O–FeO–MgO–AlO$_2$–SiO$_2$–H$_2$O–Ti$_2$O) pseudosection. Inglis et al. (2017) determined that the garnet began to grow at 0.72 GPa and 615 °C based on the interior core analysis. The garnet rim analysis recorded pressures of 0.8 GPa and 640 °C suggesting prograde metamorphism, with increasing pressure and temperature conditions over the duration of garnet growth ~647 Ma. Increasingly, geologists are able to track PT paths of rock histories over time using sophisticated field mapping, chemical analysis and computer software. Let us consider the role of temperature, pressure and hydrothermal fluids that drive metamorphic reactions.

### 15.1.1 Temperature

Temperatures generally increase from Earth's surface (~0–35 °C), downward to the core where temperatures are ~ 6000 °C. Diagenesis (Figure 15.1) encompasses a set of post-depositional sedimentary processes such as compaction and cementation that occur at low temperatures (less than ~150 °C) and at relatively low pressures (<3 kbar or 10 km depth). The onset of low-grade metamorphism generally begins ~150 ±50 °C in silicate rocks, which corresponds to kitchen oven baking temperatures (~350 °F). The first appearance of minerals such as illite, paragonite, zeolite group, prehnite group, pumpellyite, or

(a)

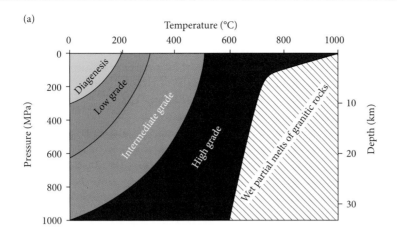

(b)

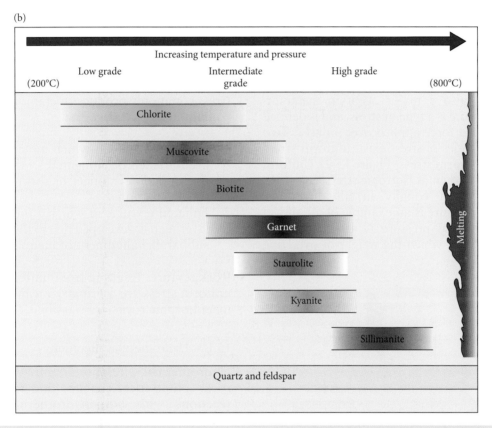

**Figure 15.1** (a) Diagram showing temperature and pressure ranges of diagenesis, different grades of metamorphic conditions, and high-temperature magmatic field. *Source*: Courtesy of Stephen Nelson. (b) Common metamorphic stability fields for aluminosilicate-rich rocks. *Source*: Levin (2006). © John Wiley & Sons.

stilpnomelane indicates the lower limit of metamorphism (Frey and Kisch 1987; Bucher and Frey 1994; Frey and Robinson 1998). Progressively higher temperatures and/or pressures result in progressively higher grades of metamorphism. At temperatures greater than 600 °C, igneous processes can be initiated due to partial melting (anatexis). If a rock contains >50% magmatic fabric it is considered to have reverted to an igneous rock. The high-temperature limit (~800–1200 °C) for metamorphism marks a transitional phase between metamorphic and igneous processes and is dependent upon factors such as the mineral composition, pressure and volatile content. Together with protolith mineral composition, temperature plays a critical role in determining the suite of minerals that occur in a metamorphic rock. Figure 15.1b illustrates progressive metamorphism for an aluminosilicate-rich protolith such as a shale. Note that chlorite and mica minerals dominate at lower to moderate grades of metamorphism. With increasing temperatures, these minerals become unstable and transform to higher-temperature stability minerals such as garnet, staurolite and sillimanite. Also, notice that quartz and feldspar persist from low grade to high grade, demonstrating their stability over a wide temperature range.

### 15.1.2 Pressure

Pressure is also a major agent in metamorphism. Pressure is a type of **stress**, and is defined as a balanced force applied over an area. Two major categories of stress exist: uniform stress and non-uniform stress (Figure 15.2). **Uniform stress**, also known **as isotropic stress**, is a stress that is applied equally in all directions. **Confining pressure** is an example of a uniform stress. Confining pressure is associated with burial depth, in which the weight of overlying

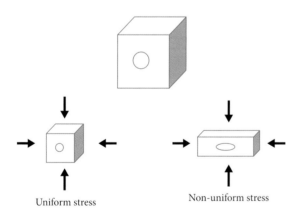

**Figure 15.2** An undeformed cube (above) subjected to uniform stress changes volume but not shape so that circles remain circles and a cube remains a cube. When the cube is subjected to non-uniform stress, the cube compresses into a rectangular shape and circles are deformed into ellipses. Non-uniform stresses produce foliations in metamorphic rocks.

rocks produces equal compressive forces directed toward a common central point. As confining stresses are equal in all directions, they produce volume changes, but not changes in shape in homogeneous materials (Figure 15.2a). As burial depth and confining pressure on rock increases, rock density increases and volume decreases as crystals are compressed closer together. Uniform stresses commonly produce metamorphic rocks with:

1 Equant grains in which individual crystals have similar dimensions in all directions.
2 Non-foliated textures that lack well-defined metamorphic layering.

**Non-uniform stress**, also known as **anisotropic** or **deviatoric** stress, occurs when stresses are not equal in all directions so that shape changes occur (Figure 15.2b). Non-uniform stress commonly produces metamorphic rocks containing:

1 Inequant grains in which mineral crystals are flattened or elongated so that at least one direction is longer or shorter than the other directions.
2 Foliated textures in which a metamorphic layering has developed due to the preferred, subparallel alignment of inequant grains.

Rock textures resulting from uniform or non-uniform stress are a primary means by which metamorphic rocks are classified.

### 15.1.3 Hydrothermal alteration

Hydrothermal alteration occurs when bulk rock composition changes due to chemical reactions with hot (hydrothermal) fluids. Hydrothermal fluids dissolve and leach primary minerals from rock, transport dissolved ions, and precipitate new minerals, in some cases producing valuable ore deposits. Fluids and vapors containing $H_2O$, $CO_2$, $CH_4$, K, Na, B, S, and Cl serve as catalysts in driving metamorphic reactions. Volatile sources include:

1 "Juvenile" fluids derived from magma, principally at ocean spreading ridges, magmatic arcs and over hot spots;
2 Seawater which infiltrates fractures and pore spaces at ocean spreading ridges and subduction zones;

3 Fluids derived by devolatilization reactions, especially in subduction zones where heating of the subducted ocean lithosphere releases: $H_2O$ from amphibole, mica and serpentine minerals and $CO_2$ from calcite in limestone and dolomite in dolostone;

4 Meteoritic fluids derived from precipitation, surface water and groundwater infiltration that are subsequently heated during burial;

5 Connate fluids, also known as formation pore fluids, stored in spaces between sediment grains during deposition and later heated during burial.

Hydrothermal alteration is indicated by the presence of:

1 Volatile-rich, secondary minerals such as calcite, epidote, muscovite, serpentine, zeolite and amphibole minerals;

2 Secondary minerals in fluid inclusions and fracture-filling veins.

While hydrothermal fluids play a significant role in most metamorphic settings, they are particularly important around ocean spreading ridges, hydrothermal vents, magma intrusions (contact metamorphism), faults and shear zones. Major types of hydrothermal alterations, mineralogy and environments are summarized in Table 15.1.

Hydrothermal alteration occurs via deuteric reactions and metasomatic reactions. **Deuteric reactions** involve reactions in which igneous rocks are "stewing in their own juices." Hot vapor-rich fluids, commonly associated with igneous intrusions, provide heat, volatiles and corrosive compounds to chemically alter minerals. Minerals produced by deuteric reactions include albite, calcite, epidote, sericite, chlorite, serpentine and talc. For example, in

**Table 15.1** Common hydrothermal alteration processes.

| Type | Major minerals | Description |
|---|---|---|
| Albitic | Albite, paragonite (Na-rich mica) | High-temperature alteration resulting in Na enrichment; common in ocean ridges. |
| Alunitic | Alunite and sulfate minerals | Occurs in hot springs and gold and copper porphyry deposits by oxidation of sulfide minerals |
| Argillic | Kaolinite, smectite, illite | Low temperature decomposition of feldspars in acidic (low pH) conditions; occurs in gold deposits hosted by sedimentary rocks. |
| Carbonatization | Carbonates such as calcite, dolomite, ankerite; accessory minerals chlorite, sericite, albite | Replacement by carbonate minerals at variable temperatures |
| Phyllic | Sericite (fine-grained white mica), quartz, pyrite | Decomposition of silicic rocks; associated with porphyry copper deposits. |
| Potassic | Biotite, K feldspar, adularia | High-temperature alteration of silicic magma resulting in K enrichment; commonly underlies phyllic zones. |
| Propylitic | Chlorite, epidote, actinolite, tremolite | Low to moderate temperature decomposition of basic and ultrabasic rocks enriched in pyroxene, amphibole, biotite, plagioclase; also occurs in gold and copper porphyry deposits. |
| Sericitic | Sericite (fine grained white mica) | Hydrothermal alteration of potassium feldspars or plagioclase feldspars |
| Serpentinization | Serpentine, talc | Low temperature alteration of basic and ultrabasic rocks |
| Silicification | Quartz, chert | Replacement by silica minerals at variable temperatures |
| Spilitization | Albite | Low temperature alteration of Ca plagioclase to Na plagioclase; common in ocean ridges. |
| Zeolite | Zeolite minerals | Low temperature replacement of glass in volcanic rocks |

propylitic reactions affecting volcanic tuff protoliths, K feldspar alters to sericite, plagioclase to epidote or sericite and micas to chlorite.

Metasomatism involves changes in rock composition due to infiltrating hydrothermal fluids that remove, add or replace elements or compounds. Metasomatism can involve: **leaching**, whereby elements are removed from minerals in the rock; **mineral precipitation**, in which elements are added to new minerals in the rock; and **mineral replacement** in which earlier formed minerals are replaced by new minerals in the rock. Metasomatism is an important process in submarine volcanic settings such as the oceanic ridge system where basaltic rocks interacts with seawater. Rocks containing calcic plagioclase (e.g., laboradorite) reacts with sodium-rich seawater; ionic exchanges convert calcic plagioclase into sodic plagioclase (e.g., albite) in a process called **albitization**. In addition, ore metals such as cobalt, copper and manganese may be leached from the rock incorporated into the hot fluids and precipitated along the ocean floor as nodular massive sulfide ore deposits. Thus, oceanic crust and ophiolites are sites of valuable metallic ore deposits due to metasomatic processes.

## 15.2 CLASSIFICATION OF COMMON METAMORPHIC ROCKS

Similar to the IUGS classification of igneous rocks, the Subcommission on the Systematics of Metamorphic Rocks (SCMR) is working to create systematic schemes for metamorphic rock terminology and definitions for international use. The SCMR classification considers properties such as minerals present, structure, protolith and genetic conditions of metamorphism related to temperature, pressure and deformation. The SCMR recognizes three broad rock classification categories: schists, gneiss, and granofels. Schists display preferred alignment of inequant grains on the scale of 1 cm or less and include the rocks schist, slate, and phyllite. Gneisses occur in which inequant mineral grains are present only in small amounts, or show a low degree of preferred alignment or, if well developed, occur in bands greater than 1 cm in thickness. Granofels include all phaneritic, equigranular

metamorphic rocks that display 0–10% foliated or linear fabric (NADMSC 2004; Schmid et al 2007). A consistent internationally accepted metamorphic naming classification continues to be a work in progress (Robertson 1999; NADMSC 2004; Schmid et al. 2007).

While a unified metamorphic rock classification remains elusive, metamorphic rocks are broadly classified into two groups based on rock fabric. Rock fabric refers to the spatial and geometrical configuration of all components that are penetratively and repeatedly developed throughout the rock (NADMSC 2004). **Non-foliated** rocks lack a metamorphic fabric characterized by preferred orientation of inequant grains. **Foliated** fabrics possess a fabric with a preferred alignment of inequant grains. Descriptions of the most common metamorphic rocks are listed in Table 15.2. Common non-foliated metamorphic rocks, most of which possess equant grains, are illustrated in Figure 15.3a. Some of the common foliated metamorphic rocks with preferred, subparallel alignment of inequant grains that produce metamorphic layering are illustrated in Figure 15.3b.

In the following section, we will consider how protolith composition influences metamorphic rock composition. In Chapter 17, we will expand upon metamorphic rock classification and address complexities within the foliated/non-foliated classification approach.

## 15.3 COMMON PROTOLITH COMPOSITIONS

We will now consider five major protolith groups and describe their general chemical and mineral composition.

### 15.3.1 Pelites

**Pelites** are rocks containing ≥40% aluminous minerals and micas. Rocks containing 20–40% aluminous minerals and micas are referred to as semi-pelites (NADMSC 2004). Pelitic protoliths include aluminum- and silica-rich rocks such as shale, mudstone, siltstone, volcanosedimentary rocks, and volcanic tuff (bentonite). Pelitic protoliths include minerals enriched in $SiO_2$, $Al_2O_3$ and $K_2O$ such as the clay minerals, micas, quartz and alkali

**Table 15.2** Protolith and major minerals of common non-foliated and foliated metamorphic rocks.

| Non-foliated metamorphic rock | Protolith | Major minerals |
|---|---|---|
| Metaquartzite | Quartz arenite | Quartz |
| Marble | Limestone, Dolostone | Calcite, dolomite and other carbonate minerals |
| Anthracite coal | Bituminous coal | None |
| Hornfels | Shale, Mudstone, Siltstone | Biotite, chlorite, epidote, zeolite, scapolite, hornblende, augite, diopside, sphene, garnet, cordierite, vesuvianite |
| Skarn | Limestone, Dolostone | Calcite, dolomite, epidote, scapolite, garnet, wollastonite, hematite, magnetite, idocrase, pyroxene, actinolite. |

| Foliated metamorphic rock | Foliated texture | Protolith | Major minerals |
|---|---|---|---|
| Slate | Very fine grains produce dull, flat, even, slaty cleavage used for pool tables, blackboards | Mudstone, shale, tuff | Muscovite, illite, quartz, feldspar, chlorite, graphite, andalusite, hematite, magnetite, zircon |
| Phyllite | Fine grains produce wavy, waxy, mildly undulating phyllitic cleavage | Shale, tuff, slate | Chlorite, muscovite, sericite, quartz, feldspar, graphite, hematite, talc |
| Schist | Medium to coarse size grains produce flaky, folded schistosity with visible minerals | Shale, tuff, slate, phyllite, granite, diorite, basalt | Muscovite, biotite, quartz, feldspar, kyanite, garnet, chlorite, actinolite, hornblende, tourmaline, glaucophane |
| Gneiss | Medium to coarse size grains produce gneissic mineral bands commonly of different colors | Granite, diorite, shale, slate, phyllite, schist | Quartz, feldspar, pyroxene, biotite |
| Mylonite | Fine to coarse grains exhibit combination of ductile flow and brittle breakage to produce wavy mylonitic cleavage | Granite, gneiss | Quartz, feldspar, garnet, pyroxene |
| Migmatite | High temperatures partially melt rock so that gneissic banding interacts with igneous magma flow producing a "mixed" texture that may appear chaotic, intensely veined or brecciated. | Gneiss | Quartz, feldspar, pyroxene, hornblende, cordierite |

feldspars. Metamorphic rocks produced from pelites inherit these protolith chemical constituents. In pelites, illite crystallinity value of <0.42D.2U, determined by X-ray diffraction analysis, defines the onset of metamorphism (Kisch 1991). Clay group minerals kaolinite, smectite/montmorillonite and illite begin to become unstable and transform to zeolite, chlorite and mica group minerals at temperature of 100–200 °C. With continued increases in temperature and pressure, progressive metamorphic mineral stability reactions produce higher-grade minerals such as the white mica group (muscovite, phengite and paragonite), chlorite group and biotite followed by garnet, staurolite, kyanite, cordierite, and sillimanite. Under non-uniform stress conditions, pelites are metamorphosed into foliated rocks such as slate, phyllite, schist and gneiss. Under uniform stress conditions, pelitic protoliths produce non-foliated rocks such as hornfels. Table 15.3 summarizes the protoliths and mineralogy of some common pelitic metamorphic rocks.

(a)

2 cm

Marble

2 cm

Metaquartzite

2 cm

Hornfels

2 cm

Skarn

2 cm

Anthracite coal

(b)

Slate

Phyllite

Schist

Gneiss

2 cm

**Figure 15.3**   Common metamorphic rocks. (a) Non-foliated rocks include marble, metaquartzite, hornfels, skarn, and anthracite coal. (b) Foliated rocks include slate, phyllite, schist and gneiss. *Source*: Photo by Kevin Hefferan. © John Wiley & Sons.

The three aluminosilicate ($Al_2SiO_5$) polymorph minerals – kyanite, andalusite and sillimanite – common in pelitic rocks are useful geothermobarometers based on their respective mineral stability fields. Andalusite is the low-pressure polymorph stable at pressures less than ~4 kbar (~11 km depth). Sillimanite is the high-temperature polymorph which becomes

**Table 15.3** Common protoliths and minerals associated with pelitic assemblages.

| | |
|---|---|
| Common pelitic protoliths | Shale, claystone, siltstone, mudstone, graywacke, tuff |
| Common pelitic (aluminosilicate) minerals | Quartz, laumontite, adularia, muscovite, chlorite, chloritoid, biotite, plagioclase, K-feldspar, pyrophyllite, corundum, phengite, paragonite, andalusite, kyanite, sillimanite, staurolite, garnet, cordierite, tourmaline. |
| Pelitic metamorphic rocks | Slate, phyllite, mica schist, pelitic gneiss, hornfels |

increasingly more common at temperatures above ~525 °C. Kyanite is a high-pressure polymorph that may occur with another high-pressure mica mineral, paragonite. The stability field for each of the three $Al_2SiO_5$ polymorph minerals is illustrated in Figure 15.4. Any two of the three polymorphs can co-exist along a line separating two fields. The only condition where all three polymorphs can co-exist defines the "triple point", where all three lines intersect. The triple point on Figure 15.4 occurs at a temperature of ~525 °C and a pressure of ~4.3 kbar.

Other common peraluminous minerals, many of which also provide useful information on temperature and/or pressure, include: tourmaline, pyrophyllite, almandine garnet (moderate temperature), corundum (high temperature), cordierite and andalusite (low pressure), staurolite and chloritoid (moderate to high pressure), and paragonite (high pressure).

### 15.3.2 Quartzofeldspathic rocks

**Quartzofeldspathic**, also known as **psammitic**, rocks contain >60% quartz and feldspar and <20% micas (NADSMC 2004). Psammitic protoliths include quartz arenite, arkose, chert and intermediate to silicic igneous rocks such as granite, granodiorite, and their volcanic equivalents. Quartzofeldspathic protoliths contain high concentrations of $SiO_2$, $Na_2O$ and $K_2O$ and relatively low concentrations of FeO and MgO. As with the metapelites, the quartzofeldspathic rocks are enriched in quartz and feldspars, such as the alkali feldspars and plagioclase feldspar minerals. Unlike the peraluminous pelitic rocks, the $Al_2O_3$ concentrations are lower and depend on the feldspar, clay and mica content in the protolith. The most common metamorphic rocks produced from quartzofeldspathic protoliths include metaquartzite (if quartz sandstone was the protolith), silicic gneisses and granulites (Table 15.4).

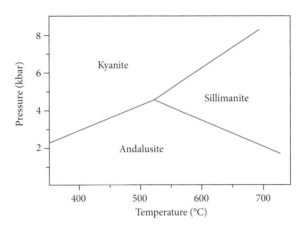

**Figure 15.4** Kyanite–andalusite–sillimanite stability fields.

Aluminous minerals, common in pelitic rocks, may occur in quartzofeldspathic rocks, especially if the protolith is a graywacke or micaceous granite. With increasing FeO concentrations, chlorite, biotite, cordierite, stilpnomelane, crossite, garnet and aegerine occur. Cordierite occurs in low-pressure or high-temperature environments. Crossite, garnet and aegirine stabilize in moderate to high-pressure environments. Under very high-pressure conditions, kyanite and the quartz polymorph coesite can form.

### 15.3.3 Calcareous rocks

Calcareous protoliths include carbonate sedimentary rocks such as limestone and dolostone (Table 15.5). Calcareous rocks are enriched in CaO, MgO, $CO_2$, and may also contain significant amounts of FeO as well as $SiO_2$ and $Al_2O_3$ where chert, siliciclastic sand and/or mud contents are significant.

Calcareous protoliths are dominated by carbonate minerals such as calcite, aragonite, and dolomite. Metamorphism of calcareous

**Table 15.4** Common protoliths and minerals associated with quartzofeldspathic assemblages.

| | |
|---|---|
| Common psammitic Protoliths | Quartz arenite sandstone, arkose sandstone quartz conglomerate, granite, granodiorite, rhyolite, dacite |
| Common psammitic minerals | Quartz, plagioclase, alkali feldspar, muscovite, chlorite, biotite, pyrophyllite, sillimanite, kyanite, andalusite, cordierite, aegerine, crossite, stilpnomelane, garnet, coesite |
| Psammitic metamorphic rocks | Metaquartzite, silicic gneiss, migmatite, granulite |

**Table 15.5** Common calcareous protoliths, minerals associated and metamorphic rocks.

| | |
|---|---|
| Common calcareous protoliths | Limestone, dolostone |
| Common calcareous minerals | Calcite, dolomite, aragonite, brucite, magnesite, quartz, graphite, pyrite, periclase, idocrase (vesuvianite), anorthite, olivine, talc, phlogopite, garnet, tremolite, wollastonite, diospide, hypersthene, coesite, lawsonite |
| Common calcareous metamorphic rocks | Marble, dolomitic marble, skarn |

**Table 15.6** Common protoliths and minerals associated with mafic and intermediate assemblages.

| | |
|---|---|
| Common mafic to intermediate protoliths | Basalt, gabbro, norite, diorite, andesite |
| Common mafic to intermediate minerals | Plagioclase, epidote, zeolite, prehnite, pumpellyite, magnesite, serpentine, talc, apatite, magnetite, ilmenite, sphene, biotite, actinolite, hornblende, garnet, crossite, riebeckite, jadeite, glaucophane, augite, enstatite, hypersthene, diopside, omphacite, olivine, zoisite, rutile, cordierite, lawsonite, quartz. |
| Common mafic to intermediate metamorphic rocks | Greenstone, greenschist, amphibolite, gneiss, granulite, blueschist, eclogite |

protoliths generates marbles rich in calcite and/or dolomite. Minor minerals include brucite and the calcium and/or the magnesium silicate minerals diopside, talc, phlogopite, periclase, grossular garnet, idocrase and olivine. Contact metamorphism of carbonate rocks produces skarn deposits containing moderate temperature minerals such as wollastonite, tremolite as well as grossular, spessartine and andradite garnet. Under high-pressure conditions, lawsonite and the $SiO_2$ polymorph coesite may form.

### 15.3.4 Mafic to intermediate rocks

Mafic to intermediate protoliths include igneous rocks such as basalt, gabbro, andesite, and diorite. These rocks are enriched in MgO, FeO and CaO and relatively depleted in $SiO_2$, $Na_2O$ and $K_2O$. As a result, amphibole, pyroxene, Ca plagioclase, mafic phyllosilicate and iron oxide minerals commonly occur as indicated in Table 15.6.

In metamorphism, minerals such as zeolite, prehnite, pumpellyite, chlorite, chrysotile serpentine and epidote occur in low-temperature and low-pressure environments such as ocean spreading ridges. Glaucophane, lawsonite, jadeite, crossite and riebeckite occur in low temperature and high-pressure rocks associated with subduction zones. Moderate temperature and pressure mineral assemblages include actinolite, hornblende, garnet. Augite, hypersthene, diopside as well as pyrope and uvarovite garnet occur in moderate to high-temperature and pressure mineral assemblages. Omphacite is a very high-pressure and high-temperature mineral associated with eclogites in the lower crust.

### 15.3.5 Ultramafic rocks

Ultramafic protoliths include pyroxenites, hornblendites, peridotites, and dunites. These rocks are highly enriched in MgO and FeO, moderately enriched in CaO, and strongly depleted in $SiO_2$, $Na_2O$ and $K_2O$. Ni and Cr

**Table 15.7**    Common protoliths and minerals associated with ultramafic assemblages.

| | |
|---|---|
| Common ultramafic protoliths | Pyroxenite, hornblendite, peridotite (harzburgite, wehrlite, lherzolite, websterite), dunite |
| Common ultramafic minerals | Olivine, zeolite, epidote, prehnite, pumpellyite, talc, antigorite, chrysotile, amosite, crocidolite, periclase, calcite, brucite, magnesite, magnetite, ilmenite, sphene, phlogopite, chromite, anthophyllite, actinolite, cummingtonite, grunerite, tremolite, hornblende, garnet, crossite, riebeckite, jadeite, glaucophane, augite, enstatite, hypersthene, diopside, lawsonite. |
| Common ultramafic metamorphic rocks | Greenstone, serpentinite, soapstone, greenschist, amphibolite, blueschist |

are especially abundant among the trace elements. Common minerals in ultramafic rocks include pyroxene, olivine and amphibole group minerals as well as biotite and iron oxide minerals (Table 15.7). Plagioclase (<10%) and the mica minerals biotite and phlogopite are common accessory minerals. Epidote, prehnite, pumpellyite, zeolite, talc and serpentine group minerals commonly occur in low temperature and pressure metamorphism at divergent plate margins and in shallow convergent margin environments. Amphibole group minerals record intermediate temperature and pressure conditions. In contrast, the pyroxene and olivine minerals indicate high-temperature conditions. The MgO-rich garnet minerals pyrope and uvarovite indicate moderate to high temperatures and pressures. Glaucophane, lawsonite, jadeite, crossite and riebeckite constitute a high-pressure and low temperature mineral assemblage associated with deeper convergent margin zones.

As with their igneous protoliths, ultramafic metamorphic assemblages are highly valued for metallic ore deposits, particularly nickel and chromium. Ultramafic metamorphic rocks also contain MgO-rich tremolite amphibole and serpentine asbestiform minerals that have both resource and environmental health hazard implications (Box 15.1).

## Box 15.1    Asbestos

Asbestos has been widely used for hundreds of years, valued for a combination of properties such as resistance to heat and friction, durability, flexibility, resistance to acids and the potential for asbestos fibers to be woven into a strong interlocking fabric. The U.S. Occupational Safety and Health Administration (OSHA) defines asbestiform elongated mineral particles as containing a type of fibrous texture with lengths greater than $5\,\mu m$, and length : width ratios $\geq 3 : 1$ (NIOSH 2011). Asbestiform is a type of fibrous texture; however, not all fibrous minerals are asbestiform. Six asbestos minerals are currently regulated by the United States Federal government. The regulated minerals include chrysotile serpentine and five amphibole minerals. However, although six minerals are regulated, approximately 400 minerals are known to exhibit asbestiform textures (Skinner et al. 1988; Strohmeier et al. 2010). Chrysotile [$Mg_3Si_2O_5(OH)_4$], also known as "white asbestos", is a widely used serpentine asbestos mineral that occurs in low grade metamorphic rocks. Chrysotile consists of soft, curly, flexible fibers and constitutes ~95% of all the asbestos used in industry. Chrysotile was widely used in construction materials for walls, flooring and ceiling tiles in homes and businesses. Manufacturing uses of chrysotile included brake linings, gaskets and insulation for appliances, ducts and pipes.

The five amphibole asbestos minerals that occur in igneous and metamorphic rocks are hard and brittle. The hard, brittle amphibole minerals are less suitable for common industrial uses and represent <5% of all asbestos used in industry. Amphibole asbestos include five different minerals. Crocidolite

*Continued*

## Box 15.1 *Continued*

$[Na_2(Fe^{2+},Mg)_3Fe_2^{3+}Si_8O_{22}(OH)_2]$, also known as "blue asbestos" is a variety of the mineral riebeckite. Crocidolite was used in pipe insulation, cements, plastics and to insulate steam engines. Amosite $[(Fe^{2+})_2 (Fe^{2+}Mg)_5Si_8O_{22}(OH)_2]$, also known as "brown asbestos" is a variety of the mineral grunerite. Amosite was used in cements, pipe insulation, ceiling tiles and for thermal insulation. Anthophyllite $[Mg_7Si_8O_{22}(OH)_2]$, actinolite $[Ca_2(Fe,Mg)_5Si_8O_{22}(OH)_2]$ and tremolite $[Ca_2Mg_5Si_8O_{22}(OH)_2]$ were used in insulation and construction materials.

Erionite $[(Na_2,K_2,Ca)_2Al_4Si_{14}O_{36}\cdot15H_2O]$ is an example of an unregulated asbestiform zeolite mineral that has caused respiratory diseases in Turkey and North Dakota (Carbone et al. 2011). Erionite occurs in volcanic ash and is widely distributed in the western USA and other locations where volcanic ash deposits are common. Erionite has been used in animal feed, pet litter, soil conditioners and for waste-water treatment.

Why are asbestiform minerals hazardous? Lightweight acicular to fibrous asbestiform minerals are easily inhaled or ingested. Individuals heavily exposed to asbestos, particularly asbestos miners and refiners, but also workers exposed to asbestos in a wide variety of industries, can develop three severe health problems. **Asbestosis** is a disease whereby lung tissue is encapsulated by asbestos particles, which harden lung tissue and decrease essential $O_2/CO_2$ exchange. This effect weakens the heart and destroys the lungs. **Mesothelioma** is a disease of the lining of lung and stomach caused by asbestos. Amazingly, mesothelioma has a 35–40 year latency between asbestos exposure and disease onset. **Lung Cancer** is the third major disease associated with asbestos and smoking.

Are all asbestos minerals equally hazardous? No. In fact, the most widely used chrysotile fibers may decompose naturally within the lungs over a period of months to years. In contrast, the less widely used amphibole asbestos minerals do not decompose over the course of decades and are serious health hazards, particularly to amphibole asbestos miners in locations such as Montana, Western Australia and South Africa (Gunter 1994; Gibbons 2000).

## 15.4 METAMORPHIC PROCESSES

Metamorphic processes producing chemical and textural changes during the transformation of protoliths into metamorphic rocks include the following:

### 15.4.1 Cataclasis

**Cataclasis** is a low temperature and pressure brittle grain fracturing process that involves grain size reduction through the mechanical grinding, rotation and crushing of rock. Cataclasis produces small, angular grains that infill between larger grains as in cement mortar. As a result, cataclasis produces a "mortar" texture in rocks such as cataclasites and fault breccias. Cataclasis occurs in high-strain rate fault zones within the upper crust as well as in meteorite impacts.

### 15.4.2 Mylonitization

**Mylonitization** is a higher-temperature and pressure ductile grain reduction process that produces oriented grains of smaller diameter.

Deformation occurs via a combination of grain fracturing, plastic bending and internal deformation and rotation of grains, which produce visible foliations in response to non-uniform stress. Although mylonitization appears to be a ductile process to the eye, microscopic grain fracturing occurs which facilitates plastic deformation. Mylonitization does not directly cause mineralogic change but commonly occurs with other processes that do so, as discussed below. Mylonitization occurs in high-temperature, high-strain shear zones of the lower crust and upper mantle.

### 15.4.3 Diffusion

**Diffusion** occurs in metamorphic rocks when individual atoms or molecules migrate in gaseous, liquid or solid phases from one location in a rock body to a new location. Diffusion is most efficient in gaseous states, where molecules are in motion and encounter minimal resistance. In the liquid state, diffusion rates are highly variable depending upon fluid viscosity: high viscosity inhibits migration whereas low viscosity enhances molecular migration. Diffusion rates are most limited in the solid state, where chemical bonding restricts the migration of atoms from one site to another. In the solid state, which characterizes most metamorphic processes, diffusion is facilitated by crystal dislocations and by the presence of small amounts of intergranular fluids that catalyze chemical reactions. This is particularly important in high-temperature metamorphic reactions where volatiles such as $H_2O$ and $CO_2$ effectively stimulate ion transfer. In addition to high-temperature environments, diffusion also plays a critical role in pressure solution reactions as described below.

### 15.4.4 Pressure solution

**Pressure solution (pressolution)** involves the dissolution of solid grains under elevated compressive stress conditions. The highest stress regimes occur at grain contacts under maximum compressive stress. Stressed grain contacts alter the crystal lattice structure so that dissolution occurs where one grain impinges upon another grain initiating a dissolution in the indented grain. Continued grain dissolution along a boundary can produce a sutured contact between the grains. Local precipitation of dissolved materials can occur in regions of lower stress producing **pressure shadows.** Pressure solution is enhanced by high concentrations of crystal defects within mineral lattice structures which facilitates dissolution (Passchier and Trouw 2005).

Pressure solution is well demonstrated in calcareous, quartzofeldspathic and pelitic rocks, where it can result in up to 50% volume loss (Wright and Platt 1982). As soluble minerals dissolve, insoluble minerals (iron oxides, micas, graphite) accumulate, commonly as a dark colored seam called a **stylolite.** Pressure solution is very important in the development of cleavage, folds, and gneissic layering where quartz and carbonate minerals dissolve in preference to clays and micas. Pressure solution involves dissolution and volume loss but also provides ions for recrystallization.

### 15.4.5 Recrystallization

**Recrystallization** occurs when existing minerals are transformed into new crystals of the same mineral without experiencing significant change in chemical composition. Recrystallization results from crystal lattice reorganization without breakage.

**Extracrystalline recrystallization** occurs by grain rotation, grain boundary migration and the dissolution and/or precipitation of grain surfaces. Extracrystalline recrystallization involves the addition or removal of mineral material by diffusion. **Intracrystalline recrystallization** occurs within individual grains due to microscopic movement of atoms related to defects, vacancies or dislocations within a crystal. Recovery describes processes whereby crystal defects are minimized by the movement of ions within a grain. Intracrystalline deformation and recovery are permanent, plastic processes that do not involve visible breakage. The net effect of these processes is to reduce the state of stress within a crystal in response to elevated temperature/pressure conditions. Both extracrystalline and intracrystalline recrystallization processes are enhanced by high volatile content, temperatures and/or pressures that can enhance crystal growth.

### 15.4.6 Neocrystallization

**Neocrystallization** refers to the nucleation and growth of new minerals as pre-existing minerals become unstable due to temperature/pressure changes. Diffusion enhances the nucleation and allows for growth of new stable minerals. Newly formed minerals distinctly larger than the minerals in the surrounding matrix are referred to as **porphyroblasts.** Continued neocrystallization is dependent upon the continuation of favorable temperature/pressure conditions and the availability of ions necessary to permit crystal growth.

### 15.4.7 Differentiation

**Differentiation** refers to the metamorphic segregation of minerals in an initially homogeneous rock into separate areas. It is due to distinctive physical or chemical processes that involve solubility, ductility, growth rate or crystallization temperature. For example, during intense folding that commonly accompanies metamorphism during non-uniform stress, more soluble minerals (e.g., calcite, quartz, feldspars) are concentrated in fold hinges while less soluble minerals (e.g., ferromagnesium minerals and clays) are preserved in fold limbs. Continued deformation and segregation can produce **transposed layering** as fold hinges and limbs become disconnected.

Through differentiation, more competent (rigid) minerals such as garnet, quartz or feldspars, can extend and grow producing elongated porphyroblast forms (Figure 15.5). Growth of these rigid grains produces pressure shadows about which more ductile minerals such as micas can accumulate and grow. As a result of their ductility contrast, segregation can occur resulting in metamorphic differentiation. Color banded gneisses and migmatites can form in this manner by developing light colored (quartz, feldspar) layers and dark colored (amphibole, pyroxene) layers. Let us now consider different types of metamorphism under which some of these mechanisms occur.

## 15.5 MAJOR TYPES OF METAMORPHISM

Metamorphism occur on a variety of scales. Local metamorphism, associated with igneous intrusions, fault/shear zones and meteorite impacts, affects areas of less than 100 km². Regional metamorphism associated with convergent and divergent plate boundaries affects areas greater than 100 km² and may encompass 1000s of km². In the following section we will briefly describe different types of local and regional metamorphism.

### 15.5.1 Impact (shock) metamorphism

**Impact or shock metamorphism** can be generated by explosive volcanic eruptions, collisions of extraterrestrial objects with Earth or nuclear explosions that generate pressures of 2–30 GPa. The high-strain rate associated with impact metamorphism produces breccia, shatter cones, shocked quartz lamellae, pseudotachylites produced by impact melting, and ultra-high-pressure (UHP) minerals such as the silica ($SiO_2$) polymorphs coesite and stishovite. A threefold IUGS classification of impactites resulting from a single impact event has been proposed consisting of shocked rock, impact melt rock, and impact breccia. Shocked rocks are non-brecciated rocks displaying unequivocal effects of impact metamorphism, but do not show evidence of whole rock melting. Impact melt rocks are crystalline, semi-glassy, or glassy rock in which >50% of the rock volume has solidified from impact melt. Impact breccias contain coarse (gravel size or larger) angular breccias with unequivocal evidence of shock metamorphism (Stöffler and Grieve 2001; Stöffler 2001a, b; NADMSC 2004).

Large numbers of meteorites bombarded our solar system around 4.3 Ga resulting in the cratered surface observed on the moon. Over time, weathering, erosion and sedimentation have largely resculpted Earth's surface

**Figure 15.5** Photomicrograph of a quartz porphyroblast that has experienced elongation and growth parallel to foliations in a mylonite. Rock sample is from Bou Azzer ophiolite, Morocco. Field of view ~2 mm. *Source*: Photo by Kevin Hefferan. © John Wiley & Sons.

masking most of Earth's meteorite crater sites. However, these impact sites are of great interest to Earth's history. The ~65 Ma Chicxulub crater near the Yucatan Peninsula is ~150 km in diameter and ranks among Earth's most famous meteorite impact sites. Breccias, shocked quartz, and ultrahigh-pressure minerals occur at the Chicxulub impact site, which has been linked with the K–T extinction event. While the Chicxulub meteorite had been proposed to be responsible for the extinction of dinosaurs, the impact is now known to predate the K–T extinction by 300 000 years (Keller et al. 2004).

Meteor Crater (Figure 15.6), located in the Arizona desert of the southwestern USA, is ~1.2 km in diameter and 180 m deep. Meteor Crater's rim has been uplifted 30–60 m due to impact debris deposited around the depression (Shoemaker 1979). Originally interpreted as a volcanic crater by respected geologist G.K. Gilbert in 1892, an impact origin was proposed by D.M. Barringer early in the twentieth century. On the basis of his work, Meteor Crater – also known as Barringer Crater – is now thought to have formed 25 000–50 000 years ago by the ~50 m diameter wide Diablo meteorite. Interestingly, when astronauts were preparing for their moon landing in the late 1960s, geologists trained NASA astronauts in Meteor Crater to simulate moon surface conditions. Debate continues regarding the distinction between magmatic and impact craters on Earth. For example, the 1.85 Ga Sudbury Complex of Ontario (Canada), long considered as a layered igneous complex renowned for metallic ore deposits, is now recognized as a meteorite impact structure that has also experienced magmatic differentiation of the impact melt sheet. In essence, the Sudbury Complex experienced pre impact metamorphism and tectonism, meteorite impact and tectonism and subsequent post-impact melting, magmatic differentiation and tectonism (Riller 2005 and references therein).

**Figure 15.6**   Meteor Crater Arizona. *Source*: Courtesy of NASA.

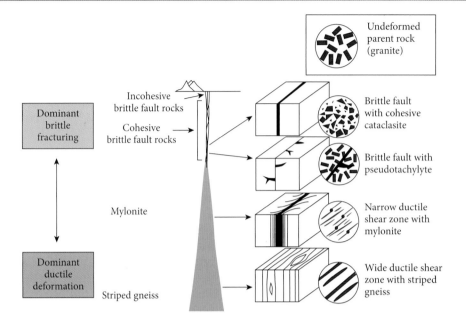

Figure 15.7 Brittle faulting produces cataclasites and high-strain rate pseudotachylites; ductile shearing produces mylonitic rocks and striped gneisses. *Source*: Cees Passchier and Rudolph Trouw; with permission of Springer Science + Business Media. © Springer Nature.

### 15.5.2 Dynamic metamorphism

**Dynamic metamorphism** is induced primarily by non-uniform stress in fault zones and shear zones (Figure 15.7). These high-strain rate environments are mostly of local extent, but may be of regional extent in large fault or shear zones. Dynamic metamorphism can occur over relatively short time intervals during which faults and shear zones experience high strain rates. Dynamic metamorphism has a high likelihood of recurring in the same fault or shear zone because of the inherent instability of these zones in response to stress.

The rock texture generated through dynamic metamorphism depends largely on temperature, pressure and strain rate conditions as well as the host rock characteristics. Relatively cold rocks in Earth's low pressure, upper crust rocks are brittle and fracture in response to stress producing cataclasites. Within the upper 5 km of the Earth's surface, the brittle crushing and grinding of rocks produces fault breccia, cataclasite, or a mélange. At depths of 5–10 km, temperatures increase so that the frictional heat and geothermal gradient combine to heat rocks within the fault zone closer to their melting point. As a result of localized rock melting, pseudotachylites develop which contain intensely sheared glasses. **Pseudotachylites** are partially melted rocks that form by quenching under high-strain rates in shear zone fractures. Ductile shear zones are zones of ductile deformation at depths greater than ~10–15 km. Ductile shear zones deform plastically producing **mylonite** rocks that contain grains reduced in size by cataclasis, plastic stretching and thinning associated with ductile deformation.

### 15.5.3 Contact metamorphism

**Contact metamorphism** develops locally where hot magma intrudes relatively cool, upper crustal (<10 km) country rock. Contact metamorphism is a variable temperature, low-pressure metamorphism. Rocks produced by contact metamorphism typically display altered mineral assemblages with non-foliated textures. Temperature variations are largely related to the size and temperature of the pluton, as well as the distance from the intrusion. Intense heat from the igneous intrusion produces a **metamorphic aureole** in the contact zone that surrounds it (Figure 15.8). Metamorphic aureoles range from centimeters to hundreds of meters in diameter. Extensional forces associated with the intrusion can hydrofracture the country rock resulting in secondary vein development and precipitation of minerals by hydrothermal fluids. Many metallic ore minerals of

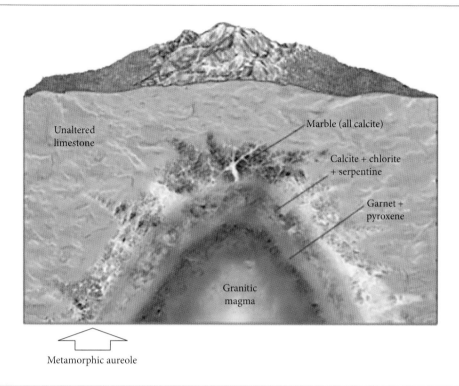

Unaltered limestone

Marble (all calcite)

Calcite + chlorite + serpentine

Garnet + pyroxene

Granitic magma

Metamorphic aureole

**Figure 15.8**    Cross-section of contact metamorphism as granite intrudes limestone creating a metamorphic aureole. *Source*: Courtesy of Murck and Skinner (2016); with permission of John Wiley & Sons.

significant economic importance precipitate in metamorphic aureoles.

Contact metamorphism produces non-foliated rock types such as hornfels, metaquartzite and skarns. Hornfels is a general term for fine grained, contact metamorphic rock rich in silicate minerals. The protolith is commonly a mudstone, but may also be a mafic or ultramafic rock. Skarns develop when igneous plutons intrude carbonate rocks. Chemical reactions with magmatic fluids result in enrichment of carbonate rocks with $SiO_2$, producing a Ca, Mg, Fe-rich calcsilicate rock. Common calcsilicate minerals produced by contact metamorphism of carbonate include wollastonite, tremolite, grossular garnet, spessartine garnet and andradite garnet. Skarn deposits are strongly affected by hydrothermal alteration. While contact metamorphism is usually considered as a local type of metamorphism, it is also prevalent in combination with other types of metamorphism at convergent and divergent boundaries as discussed below.

### 15.5.4 Ocean floor metamorphism

Hydrothermal alteration is particularly pervasive at ocean spreading ridges that experience tension, thinning and uplift. Forces within the rift valley produce extensional fracture and normal fault systems that serve as two way fluid conduits (Figure 15.9). Hot magma rises upward toward the ocean floor to cool and crystallize as mafic igneous rocks. At the same time cold, oxygenated sea water descends through the fissures and reacts with the basaltic crust. Hydrothermal fluid from seawater enriches basalt in sodium and magnesium. Hot rising fluids release helium, manganese oxide, hydrogen sulfate, methane, iron sulfide, chromium, phosphate and trace metals. Reducing conditions within the upwelling fluids result in extensive metal sulfide deposits.

Metasomatism at ocean spreading ridges involves a number of chemical reactions that include oxidation, serpentinization, and spilitization. In shallow oxidizing environments, olivine and pyroxene minerals are altered to iron oxide minerals such as magnetite and hematite. Minerals such as zeolites, epidote, chlorite, talc, brucite and magnesite are produced by similar reactions involving hydroxyl ions. Serpentinization occurs when Mg-rich olivine or pyroxene minerals are

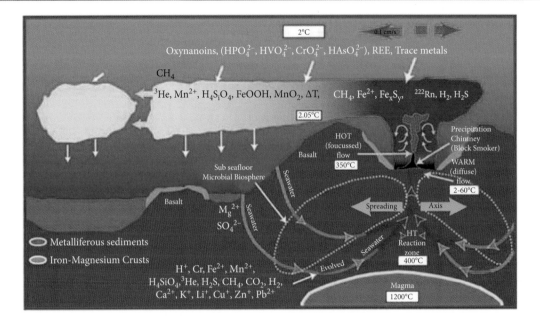

**Figure 15.9** Cross-section depicting chemical reactions at mid ocean ridges. *Source*: Courtesy of NOAA.

altered to serpentinite by seawater-derived hydrothermal fluids. Typical reactions include the following:

$$2Mg_2SiO_4 + 3H_2O$$
Forsterite

$$= Mg_3Si_2O_5(OH)_4 + \quad Mg(OH)_2 \quad .$$
Serpentine $\qquad$ Brucite (Hyndman 1985; Best 2003)

(Hyndman 1985; Best 2003)

$$6MgSiO_3 + 3H_2O$$
Enstatite $\quad$ water

$$= Mg_3Si_2O_5(OH)_4 + Mg_3Si_4O_{10}(OH)_2 .$$
Serpentine $\qquad$ Talc

(Barker 1983)

Spilitization occurs as a result of the exchange of Na from seawater for Ca in plagioclase which converts the plagioclase into albite. Spilite is the name for Na-rich basalts that form along ocean ridges and volcanic arcs. Spilites commonly occur in Precambrian greenstone belts and in ophiolites with minerals such as chlorite, epidote and actinolite that are also produced by hydrothermal alteration. Perhaps most importantly, metallic ore deposits precipitate form black smokers around ocean vents. As almost all ocean crust is generated at ocean ridges, most ocean crust is affected by low grade metamorphic reactions involving combinations of contact metamorphism and hydrothermal alteration.

### 15.5.5 Burial (static) metamorphism

**Burial metamorphism** results from increases in lithostatic stress induced by deep burial of rock and produces non-foliated textures (Coombs 1961). Burial metamorphism affects regionally subsiding basins which accumulate thick sequences of sediments and volcanic debris. Sediment rich, subsiding basins occur in a variety of environments which include: rifts, foreland basins, passive marine basins adjacent to continental margins, forearc, and back arc basins at convergent margins and thick sedimentary sequences that form in pull-apart basins associated with continental transforms.

Sediment burial initially results in diagenetic processes such as compaction, cementation and lithification. The onset of burial metamorphism, at temperatures of ~150 °C, is gradational with diagenesis. Burial metamorphism temperatures generally range from ~150 to 350 °C, producing relatively low-temperature mineral assemblages that include zeolite and prehnite group minerals, pumpellyite, chlorite and micas.

Notable modern examples of burial metamorphism include the Bay of Bengal and the Gulf of Mexico and pull-apart basins in southern California. Curray (1991) suggests that the lower part of the 22 km thick sedimentary deposits in the Bay of Bengal have experienced temperatures in excess of 350 °C. In addition to the pressure and heat associated with deep burial, warm formation pore fluids induce relatively low-temperature hydrothermal alteration. Perhaps the most significant economic aspect of burial metamorphism is in the generation of hydrocarbon deposits of oil, gas and coal.

### 15.5.6 Dynamothermal metamorphism

**Dynamothermal metamorphism** is a regional metamorphism induced by increases in both pressure and temperature. Non-uniform stress commonly produces rocks with inequant grain shapes and foliated textures (Figure 15.10).

With the exception of ocean spreading ridge hydrothermal alteration, dynamothermal metamorphism is the most aerially extensive type of metamorphism on Earth. Dynamothermal metamorphism dominates convergent margins and associated fold and thrust belts (Figure 15.11). Together, ocean floor metamorphism and dynamothermal metamorphism are by far the most volumetrically and economically important types of metamorphism on Earth, responsible for the concentration of most crustal metal deposits.

Dynamothermal metamorphism combines a complex set of processes that involve crustal shortening and thickening associated with the convergence of two lithospheric plates. As a result of crustal shortening, immense fold and thrust belts develop that characterize convergent margins throughout the world. All of the great mountain belts such as the Himalayas, Alps, Atlas, Carpathian, Cordilleran, Zagros, Andes, Appalachians, formed as a result of

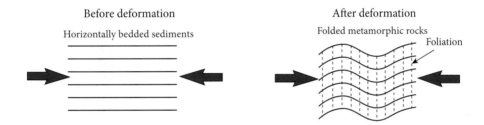

**Figure 15.10**  Compressive stress produces foliations typically at a high angle to bedding. *Source:* Courtesy of Stephen Nelson.

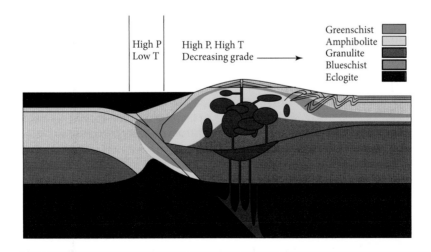

**Figure 15.11**  Dynamothermal metamorphism at convergent plate boundaries. *Source:* Courtesy of Steve Dutch.

crustal shortening, igneous activity and are the sites of intense dynamothermal metamorphism.

Metamorphic rocks are the dominant rock type in Earth's lower crust and mantle, due to the elevated temperatures and pressures deep within Earth. Metamorphic rocks are also widespread at convergent and divergent plate boundaries. This chapter serves as a brief introduction to metamorphism. In succeeding chapters we will learn more about metamorphic rock deformation, metamorphic rock classification and our understanding of the relationship between tectonics and metamorphism based upon mineral assemblages.

10  In 2004, the North American Geologic-Map Data Model Steering Committee designed a metamorphic rock classification scheme that can be viewed at: https://pubs.usgs.gov/of/2004/1451/sltt/appendixB/appendixB.pdf.

11  The British Geological Survey has a website dedicated to metamorphic rock classification that can be viewed at http://nora.nerc.ac.uk/id/eprint/3226/1/RR99002.pdf.

**12  The British Geological Survey also has a website with useful images and weblinks that can be viewed at:** http://geoscenic.bgs.ac.uk/asset-bank/action/viewHome.

## CONTENT ASSESSMENT

1   Describe the temperature conditions under which metamorphism initiates.
2   What are the three principal agents of metamorphism.
3   Name and describe on major type of hydrothermal alteration.
4   Explain the conditions that would produce a foliated versus a non-foliated metamorphic rock.
5   Identify four common foliated metamorphic rocks from low to high grade.
6   Describe conditions in which a metaquartzite might form.
7   Determine possible protolith for metamorphic rocks with the following mineral assemblages.

| Mineral assemblages | Protolith |
| --- | --- |
| Muscovite, kyanite, cordierite | |
| Chlorite, augite, epidote | |
| Calcite, aragonite, ankerite | |

8   Describe two processes by which grain changes occur in metamorphic rocks.
9   Describe metamorphic processes that commonly occur at:

| Geologic setting | Metamorphic processes |
| --- | --- |
| Ocean spreading ridges | |
| Deep sedimentary basins | |
| Convergent plate margins | |

## REFERENCES

Barker, S.D. (1983). *Igneous Rocks*. Englewood Cliffs, NJ: Prentice Hall Publishers 417 pp.

Best, M.G. (2003). *Igneous and Metamorphic Petrology*, 2e. Oxford, UK: Blackwell Publishing 752 pp.

Bucher, K. and Frey, M. (1994). *Petrogenesis of Metamorphic Rocks*, 6e. Complete Revision of Winkler's Textbook: Springer-Verlag 318 pp.

Carbone, M., Baris, Y.I., Bertino, P. et al. (2011). Erionite exposure in North Dakota and Turkish villages with mesothelomia. *Proceedings of the National Academy of Sciences* 108 (33): 13618–13623. https://doi.org/10.1073/pnas.1105887108.

Coombs, D.S. (1961). Some recent work on the lower grades of metamorphism. *Australian Journal of Science* 24: 203–215.

Curray, J.R. (1991). Possible greenschist facies metamorphism at the base of a 22-km sedimentary section, Bay of Bengal. *Geology* 19: 1097–1100.

Frey, M. and Kisch, H.J. (1987). Chapter 1. Scope of subject. In: *Low Temperature Metamorphism* (ed. M. Frey), 1–8. London: Blackie.

Frey, M. and Robinson, D. (eds.) (1998). *Low-Grade Metamorphism: Oxford*. Blackwell: Science 313 pp.

Gibbons, W. (2000). Amphibole asbestos in Africa and Australia: geology, health hazard and mining legacy. *Journal of the Geological Society of London* 157: 851–858.

Gunter, M.E. (1994). Asbestos as a metaphor for teaching risk perception. *Journal of Geological Education* 42: 17–24.

Hyndman, D.W. (1985). *Petrology of Igneous and Metamorphic Rocks*, 2e. New York: McGraw Hill 786 pp.

Inglis, J., Hefferan, K., Samson, S., and Admou, H. (2017). Determining age of Pan African metamorphism using Sm-Nd garnet whole rock geochronology and phase equilibria modeling in the Tasriwine ophiolite, Sirwa, Anti-Atlas, Morocco. *Journal of African Earth Sciences* 127: 88–98. https://doi.org/10.1016/j.jafrearsci.2016.06.021.

Keller, G., Adatte, T., Stinnesbeck, W. et al. (2004). More evidence that the Chicxulub impact predates the K/T mass extinction. *Meteoritics and Planetary Science* 39: 1127–1144.

Kisch, H.J. (1991). Illite crystallinity: recommendations on sample preparation, X-ray diffraction settings and interlaboratory standards. *Journal of Metamorphic Geology* 9: 665–670.

Levin, S. (2006). *The Earth Through Time*, 8e. New York: Wiley 547 pp.

Murck, B.W. and Skinner, B.J. (2016). *Visualizing Geology*, 4e. Hoboken NJ: Wiley Publishers 544 pp.

NADMSC (2004). North American Geologic-Map Data Model Science Language Technical Team Report on progress to develop a North American science-language standard for digital geologic-map databases; Appendix B – Classification of metamorphic and other composite-genesis rocks, including hydrothermally altered, impact-metamorphic, mylonitic, and cataclastic rocks, Version 1.0 (12/18/2004). In: *Digital Mapping Techniques '04—Workshop Proceedings: U.S. Geological Survey Open-File Report 2004-1451* (ed. D.R. Soller). 56 pp. Appendix B https://pubs.usgs.gov/of/2004/1451/sltt/appendixB/appendixB.pdf.

National Institute for Occupational Safety and Health (2011). *Asbestos Fibers and Other Elongate Mineral Particles: State of the Science and Roadmap for Research*, vol. 62. U.S. Department of Health and Human Services Centers for Disease Control and Prevention Current Intelligence Bulletin 174 pp. https://www.cdc.gov/niosh/docs/2011-159/pdfs/2011-159.pdf.

Passchier, C.W. and Trouw, R.A.J. (2005). *Microtectonics*, 2e. New York: Springer-Verlag 366 pp.

Riller, U. (2005). Structural characteristics of the Sudbury impact structure, Canada: impact-induced versus orogenic deformation – a review. *Meteoritics and Planetary Science* 40: 1723–1740.

Robertson, S. (1999). BGS Rock Classification Scheme. Classification of Metamorphic Rocks: British Geological Survey, Research Report RR 99-02, Volume 2, 24 pp.

Schmid, R., Fettes, D., Harte, B. et al. (2007). How to name a metamorphic rock. In: *Metamorphic Rocks: A Classification and Glossary of Terms. Recommendations of the International Union of Geological Sciences* (eds. D. Fettes and J. Desmons). New York: Cambridge University Press.

Shoemaker, E.M. (1979). Synopsis of the geology of meteor crater. In: *Geology of Meteor Crater, Arizona* (eds. E.M. Shoemaker and S.W. Kieffer), 1–11. Arizona State University Laboratory for Meteorite Studies No. 17.

Skinner, H.C.W., Ross, M., and Frondel, C. (1988). *Asbestos and Other Fibrous Minerals: Mineralogy, Crystal Chemistry and Health Effect*. New York, NY: Oxford University Press 94 pp.

Stöffler, D. (2001a). Towards a unified nomenclature of metamorphism: classification of impactites from single and multiple impacts. A proposal on behalf of the IUGS Subcommission on the Systematics of Metamorphic Rocks, Web version of 14.01.2001, 1 p.

Stöffler, D. (2001b). Towards a unified nomenclature of metamorphism: I. Impactites from single impacts. A proposal on behalf of the IUGS Subcommission on the Systematics of Metamorphic Rocks, Web version of 14.01.2001, 1 p.

Stöffler, D., and Grieve, R.A.F. (2001). IUGS classification and nomenclature of impact metamorphic rocks: towards a final proposal [indexed as: towards a unified nomenclature of metamorphism: IUGS classification and nomenclature of impact metamorphic rocks. A proposal on behalf of the IUGS Subcommission on the Systematics of Metamorphic Rocks, Web version of 14.01.2001], 3 p.

Strohmeier, B.R., Huntington, J.C., Bunker, K.L. et al. (2010). What is asbestos and why is it important? Challenges of defining and characterizing asbestos. *International Geology Review* 52 (7–8): 801–877.

Wright, T.O. and Platt, L.B. (1982). Pressure dissolution and cleavage in the Martinsburg Shale. *American Journal of Science* 282: 122–135.

# Chapter 16

# Metamorphism: stress, deformation, and structures

## 16.1  FORCE AND STRESS

An understanding of metamorphic rock textures requires a basic understanding of force, stress and deformation. **Force** is the push or pull applied to an object producing an acceleration. Force is defined as F = (m)(a) where force (F) is equal to mass (m) times acceleration (A). Force is represented as a vector in that it encompasses both magnitude and direction. A vector is an arrow line whose length is proportional to its magnitude and whose arrow points toward the direction in which force acts. In Figure 16.1, force (F) is applied vertically downward to an inclined plane. Force F can be resolved into two component vectors: $F_n$ is a normal component oriented perpendicular to the plane and $F_s$ is a shear component oriented parallel to the plane.

**Stresses** are balanced (equal, but oppositely directed) forces (F) applied over a cross-sectional area (A) of a plane. The Greek symbol sigma (σ) is the standard notation for stress. Stress is defined as σ = F/A. Three fundamental stress states are compression, tension, and shear (Figure 16.2). **Compressive stress** occurs where balanced forces are directed toward a plane. **Tensile stress (tension)** occurs where balanced forces are directed away from a plane. **Shear stress** occurs where balanced but opposite forces are oriented parallel to a plane. If we imagine shrinking a plane toward a vanishingly small space, we can imagine the stresses acting upon a "point."

**Deformation** is induced by stress. We discuss stress and deformation in the context of metamorphic rocks because of their importance in forming metamorphic textures. Stress and deformation are also critically important to the development of sedimentary and igneous rocks, particularly with respect to their resource potential (e.g., creating hydrocarbon reservoirs, traps,

*Earth Materials*, Second Edition. Kevin Hefferan and John O'Brien.
© 2022 John Wiley & Sons Ltd. Published 2022 by John Wiley & Sons Ltd.
Companion website: www.wiley.com/go/hefferan/earthmaterials2

and metallic ore deposits) and role in geohazards such as earthquakes and mass wasting.

The state of stress on three-dimensional rock bodies can be described in relation to three principal stress axes (Figure 16.3). The **three principal stress axes** are mutually perpendicular vectors representing normal stresses of equal (uniform) or unequal (nonuniform) magnitude that act on three principal planes at right angles to one another. Each of the three principal planes is parallel to two principal stress axes and normal (perpendicular) to the third stress axis. No shear stresses occur along the principal plane surfaces because principal planes are perpendicular to a principal stress axes. The three principal stress axes are defined by the directions of the

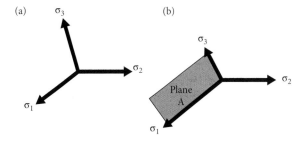

**Figure 16.3** (a) Uniform stress in which the three principal stress axes are of equal magnitude. (b) Nonuniform stress in which the three principal stress axes are not of equal magnitude as indicated by three stress axes of different lengths. Plane A is parallel to $\sigma_1$ and $\sigma_3$, and is perpendicular to $\sigma_2$. *Source*: Kevin Hefferan. © John Wiley & Sons.

maximum, intermediate and minimum compressive and/or tensional stresses (Box 16.1).

Most Earth materials are buried and experience compressive stress in all directions. As compressive stresses dominate within Earth due to the inward directed forces of the surrounding rock mass, principal stresses are usually described in relation to compressive stresses. The three principal stress axes are described as follows:

$\sigma_1$ = maximum compressive stress (or minimal tensional stress);

$\sigma_2$ = intermediate compressive stress;

$\sigma_3$ = minimum compressive stress (or maximum tensional stress).

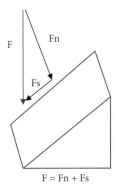

**Figure 16.1** Force acting on a two-dimensional plane can be depicted as a vector force F, which can be resolved into $F_n$ (normal force) and $F_s$ (shear force) components. *Source*: Kevin Hefferan. © John Wiley & Sons.

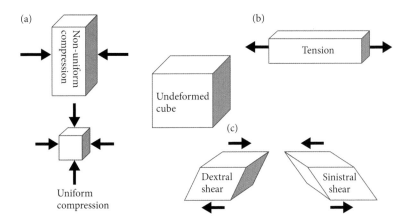

**Figure 16.2** Three-dimensional block diagrams illustrating the three main types of stress: (a) compression, (b) tension, and (c) shear stress. *Source*: Kevin Hefferan. © John Wiley & Sons.

## Box 16.1    Principal stress planes and the teeter-totter analogy

Does the concept of principal stress axes and principal planes sound confusing? Malcolm Hill suggests a playground analogy to help visualize this concept. If two friends orient a playground teeter-totter so that the board is exactly parallel to Earth's surface (Figure B16.1a), compressive gravitational stress is directed perpendicular to the board and there will be no component of shear stress on the board surface, which is to say you will not slide down the teeter-totter. Note that the vertical support on which the teeter-totter rests is a principal stress axis perpendicular to a principal plane. If your friend gradually tilts one end of the teeter-totter upwards (Figure B16.1b), the board is no longer a principal plane because it isn't perpendicular to the maximum compressive stress direction extending from your head, through your feet, to Earth's center of mass. The tilted board has a component of shear directed "downhill" along the sloping plane. If the shear stress exceeds the frictional stresses holding you in place, you may slide down the teeter-totter board.

(a)                                               (b)

**Figure B16.1**    Teeter-totter analogy of: (a) a principal plane oriented perpendicular to a principal stress axis and (b) a plane on which shear stress exists. *Source*: Kevin Hefferan. © John Wiley & Sons.

Geologists consider compressive stresses to be positive while tensional stresses are negative (Engineers, on the other hand, consider tensional stresses to be positive and compressive stresses to be negative). $\sigma_1$ may be greater or equal to $\sigma_2$, which may be greater or equal to $\sigma_3$; that is $\sigma_1 \geq \sigma_2 \geq \sigma_3$. Different states of stress can be described in relation to the three principal stress axes. These include uniform and nonuniform stresses (Figure 16.3).

### 16.1.1  Uniform (isotropic) stress

Uniform (isotropic) stress occurs when all three principal stress axes (Figure 16.3a) are

of equal magnitude ($\sigma_1 = \sigma_2 = \sigma_3$). In uniform stress, the following conditions exist:

1   The three perpendicular, principal stress axes can have any orientation as stress is equal in all directions.
2   No shear stress occurs.
3   No changes in shape occur in homogeneous materials (no shear stress results in no shear strain).
4   Volume changes may occur, but are commonly difficult to quantify.

Earth materials are overlain by rocks, sediment or water whose weights exert a uniform compressive stress referred to as confining stress or pressure (P$_c$). **Hydrostatic stress** refers to the uniform compressive force directed radially inward by a surrounding mass of water. We have a 12-ounce Styrofoam coffee cup that went to the bottom of the Mid Atlantic Ridge with the submersible Alvin. When the coffee cup returned to the ocean surface, it retained its original shape but was the size of a 3-ounce cup as a result of hydrostatic stress exerted by ocean water. **Lithostatic stress** refers to a uniform compressive force exerted radially inward due to a mass of surrounding rock. Rock stresses are commonly expressed in kilobars (kbar), megapascals (MPa), or gigapascals (GPa). Clenching teeth together in a forceful human bite produces ~1 MPa of stress. What is the relationship of rock burial to stress? Most rocks in Earth's crust have a density of ~2.6–2.9 g/cm$^3$; therefore, a relationship exists between depth of burial and lithostatic stress in which 1 kbar (0.1 GPa or 100 MPa) pressure = 3.3 km depth of burial. As a result, for each 10 km of depth, the lithostatic pressure increases by ~3 kbars. Uniform stresses tend to produce equant grains and nonfoliated textures in metamorphic rocks.

### 16.1.2 Non-uniform (anisotropic) stress

Non-uniform (anisotropic or deviatoric) stress occurs when at least one principal stress has a magnitude not equal to other principal stresses (Figure 16.3b). In non-uniform stresses, the following conditions exist:

1 Stress axes and stresses are not equal in all directions.
2 Shear stresses occur in rock bodies, on any plane that is not on a principal plane (Figure 16.4).
3 Shape changes are common, even in homogeneous materials.
4 Volume changes can occur that cause corresponding changes in density.

Non-uniform stress results in the deformation of a spherical ball into an ellipsoidal shape (Figure 16.5). Non-uniform stress also promotes the development of inequant grain growth and foliated textures. We will discuss different types of deformation in the following section.

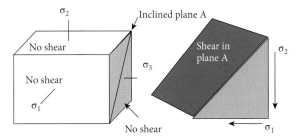

**Figure 16.4** Principle stresses directed toward a cube. Note that shear stresses do not exist on the three principal stress planes but shear stresses do exist on all other planes such as plane A, oriented 45° to a principle plane. *Source*: Kevin Hefferan. © John Wiley & Sons.

## 16.2 DEFORMATION

Stresses of sufficient magnitude produce rock deformation. **Deformation** is a physical change in the shape, volume, position and/or orientation of a rock due to an applied stress. Deformation consists of four components: distortion, dilation, translation, and rotation (Figure 16.6). Change in shape and/or volume is called **strain**.

**Distortion** is any type of deformation that produces a change in shape. **Homogeneous strain** (Figure 16.7a) occurs when strains in any particular direction are the same so that parallel lines remain parallel, perpendicular lines remain perpendicular and circles flatten to become ellipses. **Heterogeneous strain** occurs when strain intensity in any particular direction varies within a rock body. Heterogeneous strain (Figure 16.7b) causes lines that were once parallel or perpendicular to one another to be no longer parallel or perpendicular and spheres to not deform to ellipses. In the real world, heterogeneous strain predominates.

**Dilation** refers to a change in volume. Two commonplace examples of dilation in our lives are described below. If you get tested for glaucoma during an eye exam, a fluid is dropped into your eyes which makes your pupils expand. As you exit the optometrist's office, you are blinded by the brilliance of light that floods through your dilated pupils. Any woman who has given birth to a baby is also very familiar with the importance of 10 cm dilation prior to the delivery of a baby. In both of these physiological examples, dilation

**Figure 16.5** (a) Non-uniform stresses transform a sphere into (b) an ellipsoid, and (c) promote inequant grain growth and the development of foliations perpendicular to $\sigma_1$ direction. *Source*: Kevin Hefferan. © John Wiley & Sons.

refers to a volume increase (positive dilation). How does dilation affect rocks? In metamorphism, rock dilation involves a volume decrease (negative dilation) because buried rocks experience compressive lithostatic stress. Increases in confining stress during progressive rock burial reduce pore space and increase density, while reducing rock volume.

**Translation**, also known as **displacement**, refers to the movement of a body one from place to another in response to an unbalanced force. In a game such as golf, one of the measures of player skill is the length of their drive.

An unbalanced force is initiated when the golf club hits the ball. Translation is measured by the distance the golf ball travels before coming to a rest. The golf ball has not permanently changed shape or volume, but has been displaced. We can determine the displacement by measuring the distance from the tee spot, where the golf club hit the ball, to the final resting point of the golf ball. Rocks are moved from one location to another via folding, faulting, and shearing mechanisms and, on a larger scale, by plate motions. An extreme example would be the dismembered fragments

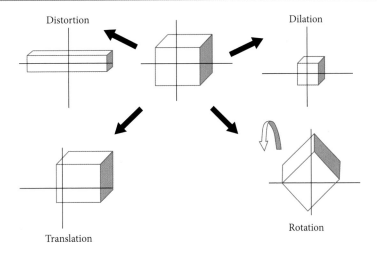

Distortion    Dilation

Translation    Rotation

**Figure 16.6**   Summary diagram of the four types of rock deformation. *Source*: Kevin Hefferan. © John Wiley & Sons.

(a)    (b)

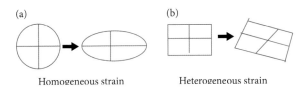

Homogeneous strain    Heterogeneous strain

**Figure 16.7**   (a) Homogeneous strain in which parallel lines remain parallel, perpendicular lines remain perpendicular and circles deform to ellipse shapes. (b) Heterogeneous strain in which parallel lines do not remain parallel and perpendicular lines do not remain perpendicular, and circles do not deform into ellipses. *Source*: Kevin Hefferan. © John Wiley & Sons.

of the Avalon terrane, which extend from Scandinavia through Ireland, Newfoundland, Massachusetts, and southward to Florida. Plate tectonics, a powerful driver, displaced these terranes thousands of kilometers.

**Rotation** occurs when an object has moved in a circular arc about an axis, sort of like tightening a screw into a wall or the way in which a wheel rotates around an axle. Rock particles also can move in a circular motion as they are displaced (Figure 16.8).

### 16.2.1   Principal strain axes

Deformation studies in metamorphic rocks focus on their strain history. Strain history refers to shape (distortion) and/or volume changes in response to one or more stress induced deformation events over time. Imagine an originally spherical mass, with three mutually perpendicular but equal length axes, deformed into an ellipsoid, with or without a new volume. The deformed ellipsoid contains three mutually perpendicular axes of unequal length: a long axis (X), a short axis (Z), and an axis of intermediate length (Y). These axes are collectively referred to as the three principal strain axes, as shown in Figure 16.9.

X = direction of maximum extension or minimum shortening

Y = intermediate direction of extension or shortening

Z = direction of minimum extension or maximum shortening.

Stress induces strain. The three principal strain axes, like principal stress axes, are lines that are perpendicular to one another and to planes of zero shear strain. In the simplest scenario, the three principal stress axes ($\sigma_3$, $\sigma_2$, $\sigma_1$) would correspond with three principal strain axes X, Y, and Z, respectively. In such cases maximum extension (X) corresponds to the $\sigma_3$ direction, the minimum compressive or maximum tensional stress. The intermediate strain direction (Y) is parallel to $\sigma_2$, the intermediate stress. The maximum shortening direction (Z) is parallel to the maximum compressive or minimum tensional stress $\sigma_1$. This is true in a given moment, that is for instantaneous strain. However, the correspondence of stress axes with strain axes is applicable in only the simplest cases, such as taking a ball of clay and squeezing it in one direction. However, in most

**Figure 16.8** Photomicrograph of a ~5 mm snowball garnet that has been rotated counterclockwise during crystal growth. *Source*: Photo courtesy of Bruce Yardley.

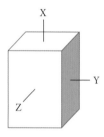

**Figure 16.9** Illustration depicting X, Y, and Z strain axes. *Source*: Kevin Hefferan. © John Wiley & Sons.

geologic studies, we cannot confidently correlate principal stress axes with principal strain axes in rocks because of complex and uncertain strain kinematic histories that involve one or more sets of shear strains.

## 16.2.2 Strain kinematics

In studying deformed rocks, geologists attempt to determine the strain path. The strain path describes the **kinematic strain** development from an initial to a final state. A kinematic study describes a series of strain increments culminating in the final strain state observed at present. **Incremental strain** refers to one or more intermediate strain steps describing separate, discrete strain conditions. Strain analysis requires knowledge of the original undeformed state of the material (rare in nature) as well as incremental stress and strain steps. In most cases, we can only study the final state of the

rock and make generalizations about the state of stress that may have produced observed deformation features. Figure 16.10 illustrates a very simple three-stage strain history.

Rarely do rocks in the field preserve evidence of their initial state, intermediate states or the stress and strain history that produced the final rock state. So one has to ask:

1  Was deformation a singular event?
2  Did the stress and strain axes remain the same throughout the deformation history?
3  Did the stress and strain axes rotate and change during the deformation history?

We usually do not have sufficient information related to stress and strain history to answer these questions. Nevertheless, later in this chapter we will be discussing rock textures that allow us to make some inferences about the state of stress and strain. Let us consider two end member examples of nonrotational and rotational strain.

**Coaxial strain** is an ideal strain in which no rotation of the incremental strain axes occurred from the initial to the final strain state. Coaxial strain produces **pure shear (irrotational strain)** in which the X, Y, and Z axes do not rotate during progressive strain. Pure shear is analogous to standing on a stationary ball so that a spherical object approaches that of an ellipse (Figure 16.11). Pure shear requires that:

1  Uniform elongation occurs in only one direction

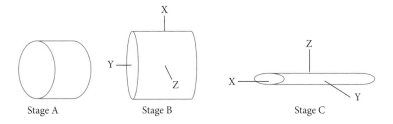

Stage A              Stage B                    Stage C

**Figure 16.10**   The equidimensional cylinder in stage A represents an undeformed state. Stress produces a deformed intermediate stage B in which the cylinder has been compressed laterally. A more recent stress event produces a deformed cylinder in stage C that has been stretched horizontally and flattened vertically. Note that significant changes in strain orientations occurred from stages A–C. The strain history for stages A and B is rarely preserved in rocks. *Source*: Kevin Hefferan. © John Wiley & Sons.

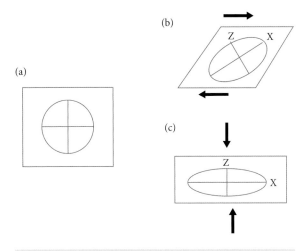

**Figure 16.11**   Diagram illustrating deformation of: (a) an initially undeformed sphere by (b) simple shear, and (c) pure shear. *Source*: Kevin Hefferan. © John Wiley & Sons.

2  Uniform contraction occurs in a perpendicular direction

3  Strain axes are parallel to principal stress axes: ($\sigma_1 = Z$, $\sigma_2 = Y$, $\sigma_3 = X$) throughout deformation

4  No change in volume occurs.

**Noncoaxial strain** is a rotational strain in which the strain axes rotate through time, so that the stationary ball described above is now experiencing shear parallel to the ground on which it rests. During incremental strain steps, the orientation of the principal strain axes change as a result of their progressive rotation (Figure 16.11). Noncoaxial strain can be modeled by **simple shear (rotational shear)** structures.

In simple shear, the following conditions exist:

1  Strain axes do not remain parallel to the principal stress axes during progressive deformation;

2  X and Z strain axes rotate during progressive deformation for a fixed single stress orientation;

3  As a result of rotation, the direction of maximum elongation is not parallel to the direction of minimum compressive stress or maximum tensional stress;

4  The direction of maximum shortening (minimum extension) is not parallel to the direction of minimum tension or maximum compressive stress.

To illustrate this, consider a card deck with circles drawn on the side of the card deck. In simple shear, as you slide the cards one on top of the other, circles on the side of the card deck deform. The total distance perpendicular to the shear plane remains constant; in other words, the thickness of the card deck remains the same. All points move parallel to a fixed direction with an amount of displacement proportional to a distance from some defined plane (e.g. parallel to the faces of cards). As slip occurs on the card deck planes, the strain axes rotate in the direction of shear.

Pure shear and simple shear are two idealized end members. In reality, rocks commonly experience general shear. **General shear** is a combination of pure shear and simple shear. Let us now consider different types of deformation that develop in response to stress.

## 16.3 TYPES OF DEFORMATION

The type or style of deformation that occurs is related to rock rheology. **Rheology** refers to the ways in which materials respond to stress. We will focus upon three types of deformation: elastic, plastic, and rupture and use them to define additional rheological terms.

1 **Elastic deformation** occurs when a body is deformed in response to a stress, but returns to its original shape and volume when stress is removed. Deformation produced by stress is totally and instantaneously reversible or recoverable.
2 **Plastic deformation** occurs when a body is irreversibly deformed in response to stress without the development of visible fractures; microfracturing may occur,
3 **Rupture deformation** creates visible fractures in response to stress. Rupture deformation results from loss of cohesion in rocks producing permanent, irreversible deformation.

### 16.3.1 Elastic deformation

All rocks possess some elasticity, even though it may be small. Many materials we use have obvious elastic properties, such as undergarments, hair accessories (e.g. ponytail holders) and rubber bands. When we apply tensional stress by pulling them apart, these items deform by stretching or changing shape so that:

1 Even a small amount of tension results in measurable stretching.
2 The amount of stretching is directly related to the amount of tension.
3 If you remove the stress, the material returns to its initial shape.

**Elastic deformation** is temporary, reversible strain in which a linear relationship exists between stress and strain (Box 16.2). Upon the application of stress, bonds temporarily stretch but return to their original position following stress removal. Rock behavior in response to stress can be depicted on a graph where stress and strain (or strain rate) are depicted on x- and y-axes. Note that in all cases a linear relationship exists between stress and strain so that the strain is proportional to the amount of stress. As this relationship is known as Hooke's Law, elastic behavior is also referred to as Hookean behavior. Note that this relationship is not time dependent: no time lag occurs so that strain begins when stress is first applied; and strain ceases immediately upon the removal of stress.

The amount of strain that occurs is inversely proportional to the **elasticity** of the material which is a measure of resistance to elastic strain (rubber bands have low, but variable elasticity). The slope of the stress-strain line is referred to as **Young's modulus of elasticity** (E), where E = stress/strain. Young's modulus of elasticity

---

## Box 16.2  Elastic deformation

Elastic deformation can be described in terms of length change, shape change (strain), and volume change (dilation). **Length change**, represented by the symbol e, refers to elongation of a linear feature. Length change is determined by dividing the change in line length (L – Lo) by the original line length (Lo), where L may be more than (lengthening) or less than (shortening) the initial line length.

$$e = \frac{(L - Lo)}{Lo} \quad Lo = \text{original length}$$

$$L = \text{final length}$$

Distortion refers to shape changes. Shear strain ($\gamma$) is a measure of angular changes to linear features that occur during distortion. The measure of a resistance to change in shape is **rigidity** (G), also known as shear modulus, For elastic materials, rigidity is the ratio of shear stress ($\sigma_s$) to shear strain such that:

$$G = \sigma_s / \gamma$$

## Box 16.2  *Continued*

**Table B16.1**  Incompressibility and rigidity values at 1 atm and 25 °C. Note that iron and copper are highly incompressible and that halite and ice are very compressible.

|  | Mineral | Incompressibility (K) | Rigidity (G) |
|---|---|---|---|
| Decreasing incompressibility | Iron | 1.7 | 0.8 |
|  | Copper | 1.33 | 0.5 |
|  | Olivine | 1.29 | 0.81 |
|  | Silicon | 0.98 | 0.7 |
|  | Calcite | 0.69 | 0.4 |
|  | Quartz | 0.3 | 0.5 |
|  | Halite | 0.14 | 0.3 |
|  | Ice | 0.07 | 0.03 |

Decreasing incompressibility

*Source*: Poirier (1985) and van der Pluijm and Marshak (2004). © John Wiley & Sons.

Dilation (D), refers to a change in volume where Vo is the original volume, V the final volume, and K represents the bulk modulus.

$$D = K\frac{(V - Vo)}{Vo}$$

The **Bulk modulus (K)** is a measure of the resistance to a change in volume. The bulk modulus, also known as incompressibility (Table B16.1), can be expressed as $K = \Delta p/\Delta v$ where K is equal to the change in pressure divided by change in volume.

The bulk modulus is related to **Poisson's ratio**, which is a measure of material "fattening" compared to its "lengthening" in response to compressive stress (Figure B16.2). Poisson's ratio = $\upsilon$ = change in object diameter divided by change in object length. Earth materials have Poisson's ratios that range from 0 to 0.5, which means that all Earth materials increase in diameter and decrease in length in response to compressive stress. Poisson's ratios for some common Earth materials are listed in Table B16.2.

Incompressibility and Poisson's ratio are very important factors in determining Earth material strength. In using Earth materials for construction,

Original dimensions

Final dimensions

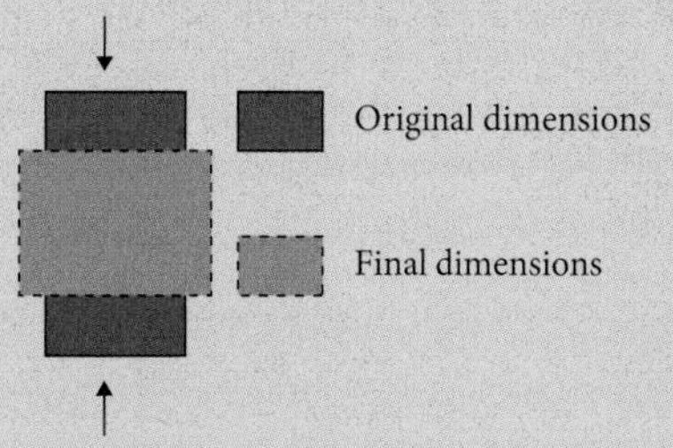

**Figure B16.2**  (a) Illustration of Poisson's ratio in which material "fattening" is compared to its "lengthening" in response to vertical compressive stress. *Source*: Poirier (1985) and Van der Pluijm and Marshak (2004). © John Wiley & Sons. (b) Poisson's ratio values for Earth materials and processed materials. *Source*: Van der Pluijm and Marshak (2004) and Hatcher (1995). © John Wiley & Sons.

engineers need information relating to the material strength and rheology so that roads, bridges, dams, tunnels, buildings, and other structures do not collapse.

*Continued*

**Box 16.2** *Continued*

**Table B16.2** Poisson's ratio values for Earth materials and processed materials.

| | Material | Poisson's ratio |
|---|---|---|
| | Fine-grained soil | 0.45 |
| | Gold | 0.44 |
| | Lead | 0.43 |
| | Crushed stone | 0.40 |
| | Concrete | 0.35 |
| | Gabbro | 0.33 |
| | Aluminum | 0.33 |
| | Limestone | 0.32 |
| | Slate | 0.30 |
| | Steel | 0.29 |
| | Gneiss, peridotite | 0.27 |
| | Shale, sandstone | 0.26 |
| | Basalt, granite | 0.25 |
| | Glass | 0.22 |
| | Metaquartzite | 0.10 |

*Decreasing incompressibility* ↓

*Source*: Van der Pluijm and Marshak (2004) and Hatcher (1995). © John Wiley & Sons.

is a constant that describes the slope of the line. The slope steepness of line E is a measure of resistance to elastic distortion (Figure 16.12). The E slope is dependent upon the stiffness or rigidity of the material. A rigid, stiff rock (high E) such as granite requires greater stress to achieve a given strain than a soft, pliable shale (low E). Young's modulus has an average value of E = $10^{-11}$ Pa for crustal rocks.

### 16.3.2 Plastic deformation

**Plastic deformation** is an irreversible strain that occurs without visible (mesoscopic) fractures. Microscopic fracturing may occur, but is not observable without the use of a petrographic microscope or other high magnification devices. Plastic behavior is preceded by reversible elastic deformation prior to the

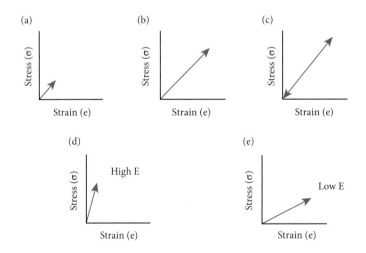

**Figure 16.12** Elastic behavior is depicted on idealized stress–strain graphs. (a) Small amount of stress produces a small amount of strain. (b) Larger amount of stress produces a larger amount of strain. (c) Stress is proportional to strain and is temporary. As stress is removed, strain diminishes. (d) Stiff rocks (e.g. granite) require high-stress values to achieve a given strain value. (e) Rocks that are not stiff (e.g. shale) deform more with a given amount of stress. *Source*: Kevin Hefferan. © John Wiley & Sons.

onset of permanent plastic strain. Ideal plastic behavior initiates after some critical stress ($\sigma_c$) value, also known as **yield stress**, is achieved and maintained over time (Figure 16.13). The critical stress is analogous to trying to inflate a balloon for the first time. An initial stress level is necessary to inflate the balloon. Once achieved, continued air pressure results in continued balloon expansion.

Plastic behavior occurs through reorientation of crystal structures within grains (intracrystalline) or along grain boundaries (intercrystalline). These processes include intracrystalline slip, twinning, kinking, dislocation, crystal defects, grain boundary sliding, dissolution, or other processes that occur in response to stress (Figure 16.14). Plastic behavior is favored by high temperatures and high confining pressures. Other important factors in plastic deformation are grain size, mineral composition and fluid pore pressure. Some of the major plastic deformation processes are described below.

**Cataclastic flow** is a mesoscopic ductile behavior facilitated by microscopic fracturing,

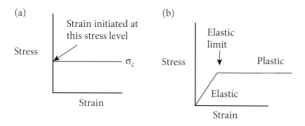

**Figure 16.13** (a) Idealized plastic deformation initiates after a critical stress or yield stress value is attained. (b) Plastic behavior preceded by elastic response at lower stress levels. *Source*: Kevin Hefferan. © John Wiley & Sons.

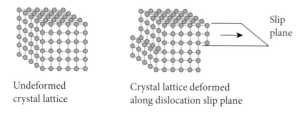

**Figure 16.14** Plastic deformation proceeds through microscopic intracrystalline mechanisms such as slip along dislocation planes or intercrystalline processes that occur along grain boundaries. *Source*: Kevin Hefferan. © John Wiley & Sons.

crushing, rolling, and frictional sliding of grains and grain fragments. An analogy is a bean bag run over by an automobile in which the bag remains intact but grains within the bag have partially pulverized and reoriented. In metamorphic rocks, a series of deformation events commonly occur so that the bean bag grains are glued back together and subsequently run over again in repeated cycles of cementation and microfracturing. Cataclastic flow is common in high strain rate faults and shear zones near the brittle–ductile transition producing rocks such as fault gouge and cataclasite.

**Crystal plasticity** is achieved by bending of the lattice through gliding along weak planes within crystalline structures. **Mechanical twinning** is common in calcite and feldspar minerals. **Kinking**, the bending of a crystal lattice into limbs and hinges, and frictional sliding along weak planes is common in micas, talc, serpentine, and other platy, phyllosilicate minerals.

**Diffusional mass transfer** is a set of plastic deformation processes that occurs in response to nonuniform stress. It involves the transfer of material from regions of maximum compressive stress to regions of minimum compressive stress which results in changes in the shape of mineral crystals. Major processes involved include pressure solution and solid-state diffusion. Although they can act in concert, each of these is favored by a particular set of temperature–pressure conditions.

**Pressure solution** is a diffusional mass transfer process that requires elevated pressures and the presence of at least a small amount of an aqueous phase. It occurs over a wide variety of temperature and confining pressures, from those in which sedimentary rocks form to moderate grade metamorphic conditions. Because the solubility of minerals phases generally increases with stress, grains begin to dissolve along grain contacts under maximum compressive stress which shortens the grains in that direction (maximum shortening is parallel to $\sigma_1$). As each mineral has different dissolution tendencies, pressure solution results in mineral differentiation whereby more soluble minerals (calcite, quartz) are removed and less-soluble minerals are concentrated (micas, clays). Dissolved minerals can be reprecipitated as beards in the areas of minimum stress producing dolphin shaped cobbles elongated in that direction as observed in Purgatory Conglomerate in Middletown, RI (Figure 16.15). Pressure solution produces volume loss and is

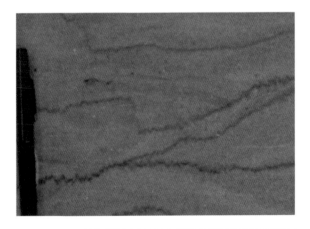

**Figure 16.15** (a) Quartz pebbles of the Purgatory Conglomerate experienced differential pressure solution resulting in dissolution on the surfaces of the cobble and precipitation on either ends, resulting in quartz beards or tails that produce elongated cobbles resembling French bread or dolphin shapes. (b) Block diagram view of quartz cobbles showing nearly circular cross sections in vertical view and distorted elongated cobbles in surface view. (c) Tectonically stretched quartz cobbles are stacked like loaves of French bread in Narragansett Basin, Rhode Island. *Source*: Kevin Hefferan. © John Wiley & Sons.

particularly important in the deformation of marble, metaquartzite, slate, limestone, dolostone, shale, and quartz sandstone (Mosher 1987). Pressure solution in carbonate rocks is commonly indicated by stylolites, which are jagged seams of insoluble mineral residue that accumulate and concentrate along a dissolution seam. Stylolites (Figure 16.16) are commonly black – due to enrichment in carbon and iron/manganese oxides – and have the appearance of wound sutures stitched together. Pressure solution in clay rich rocks helps to produce cleavage in slates, as well as embayed grains and grain overgrowths in quartz-rich rocks such as sandstone and metaquartzite.

Earth materials display Newtonian viscous creep behavior at low stresses and high temperatures. **Solid state diffusion or creep processes** include Coble creep which is a moderate to high temperature, atom and

**Figure 16.16** Stylolites in a marble slab form by concentration of insoluble residue along suture-like, black seams. *Source*: Kevin Hefferan. © John Wiley & Sons.

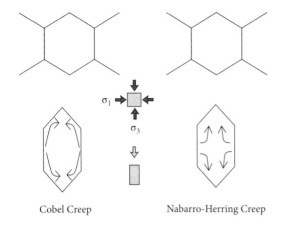

Cobel Creep                    Nabarro-Herring Creep

**Figure 16.17** Cobble creep occurs through moderate to high temperature atom and vacancy diffusion along grain boundaries; Herring–Nabarro creep is a high temperature, intracrystalline volume diffusion process. Both of these processes alter the atomic crystal lattice structure through plastic behavior. *Source*: Kevin Hefferan. © John Wiley & Sons.

vacancy diffusion along grain boundaries (Coble 1963). In contrast, Herring–Nabarro creep (Nabarro 1948; Herring 1950) is a high temperature, intracrystalline atom and vacancy diffusion that occurs within the interior of the crystal lattice (Figure 16.17). Such processes are enhanced with small grains in metamorphic and some igneous rocks. Solid state diffusion produces shape changes as atoms move along the grain surface or through the interior of the mineral. Solid state diffusion permits the migration of atoms from sites of greatest stress within or along the surfaces of grains (shortening the crystal in that direction) and relocation to regions of least stress (lengthening the mineral in that direction).

**Crystal defect migration** (dislocation creep) processes create adjustments in deformed crystal lattice structures that also lead to crystal shape changes. These defects include point defects, line defects, edge dislocations, and screw dislocations. Dislocation creep is especially important in lower temperature rock deformation reactions, whereas solid-state diffusion dominates at elevated temperatures.

All of the above plastic deformation processes and others beyond the scope of this book (e.g. superplastic flow and dynamic recrystallization) are important in mineral recrystallization and neocrystallization reactions.

### 16.3.3 Rupture deformation

Rupture refers to the visible loss of cohesion due to elevated stress. Earth's cool, low confining pressure upper crustal rocks experience reversible elastic deformation at low stress levels whereby stress is proportional to strain until the rock attains its ultimate strength. The ultimate strength is the maximum stress that a rock can attain prior to the onset of rupture. The rupture point (rupture strength) occurs when stress exceeds the rock's ultimate strength. At the rupture point, the rock loses cohesion and fractures (Figure 16.18a).

Rocks in the warm, elevated confining pressure conditions of Earth's mid to lower crust experience early reversible elastic behavior at low stress levels until they reach the elastic limit. The yield strength marks the point where the reversible, linear stress/strain relationship of elastic deformation transitions to permanent, nonlinear plastic deformation. Permanent plastic deformation continues beyond the yield point. Under some circumstances, stresses may increase until a rupture strength is reached at which point the rock loses cohesion and ruptures (Figure 16.18b). Typically, rupture deformation in the lower crust is facilitated by fluid enhanced weakening, as well as grain microfracturing, fragmentation and comminution associated with cataclasis (Petley-Ragan et al. 2019).

Elastic, plastic, and rupture deformation are fundamentally important to structural engineers

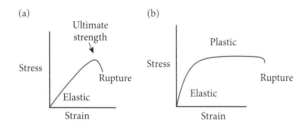

**Figure 16.18** (a) Idealized rock response to stress in which low stress levels produce elastic behavior. As stress increases, the rock reaches a maximum stress level (ultimate strength) prior to the initiation of failure. The delay between ultimate strength and visible rupture marks the development of microfracturing processes that precede visible rupture. (b) Idealized rock response in which elastic behavior is followed by permanent plastic deformation prior to rupture. *Source*: Kevin Hefferan. © John Wiley & Sons.

as well as geologists. Engineers must consider the strength and deformation properties of construction materials such as wood, metal, concrete, and stone to avoid failure of building structures. Engineers must also be aware of the geologic bedrock and soil conditions on which their structure is constructed. Material strength is also affected by earthquake faults and other geohazards. Therefore, it is essential that engineers and geologists collaborate and share their knowledge base because one size does not fit all. What do we mean by that? Whereas construction material may be somewhat standardized with respect to engineering properties, conditions that produce elastic, plastic, and rupture deformation vary greatly for different minerals, rocks, geographic locations, and stress conditions as we shall see in the section to follow.

## 16.4 DEFORMATION STYLES AND MATERIAL BEHAVIOR

Permanent, irreversible rock deformation always involves an initial period of elastic deformation. If stresses exceed the elastic limit or yield strength, rocks then deform irreversibly to produce long-term "permanent" changes in shape and/or volume. In some cases, rocks lose cohesion, and rupture immediately when the elastic limit is exceeded. This produces visible fractures. In other cases, they undergo long intervals of plastic deformation, changing shape and/or volume without developing visible fractures. This range of possibilities, which is dependent on temperature, pressure, stress states, and rock type, permits geologists to recognize two distinctive styles of deformation behavior discussed below.

### 16.4.1 Brittle behavior

**Brittle behavior** involves rupture after a period of elastic deformation, with little or no intervening period of plastic deformation. Brittle behavior in rocks is always preceded at lower stress levels by elastic behavior, and can also be preceded by a small amount of plastic deformation. The ultimate strength of a brittle material is the maximum stress level that is achieved prior to the onset of rupture. Brittle behavior is favored by low temperatures and pressures that characterizes most rocks in the

upper crust where it results in widespread fracturing and fault displacement.

### 16.4.2 Ductile behavior

Ductile behavior characterizes rocks that experience elastic, followed by large amounts of plastic deformation before rupturing. As stress increases, ductile rocks deform elastically to a point called the elastic limit. Beyond the elastic limit, plastic deformation ensues with increasing stress over a long interval prior to rupture, if any. Ductile behavior is favored by high temperatures and pressures and is typical of most rock materials in the lower crust and mantle.

### 16.4.3 The brittle–ductile transition

The **brittle–ductile transition** is the depth within Earth where the rock behavior changes from brittle to ductile (Figure 16.19). The depth brittle–ductile zone varies based on rock type, geothermal gradient, and tectonic location. For example, the brittle ductile transition is approximately 2 km depth at mid ocean ridges and may extend tens to hundreds of km in cold depths of subduction zones. However, this transition zone generally exists at depths ~10–20 km and temperatures of ~250–400 °C. As indicated in Figure 16.19, the brittle–ductile transition constitutes the depths of greatest rock strength. High confining pressure inhibits the development of fractures which increases the brittle strength of rock downward to the transition zone. Meanwhile, the relatively low temperatures of the transition zone increase ductile strength of rock compared to higher temperatures at depths greater than the transition zone. Brittle–ductile transition zones produce a mixture of fault and shear zone rocks such as cataclasites, pseudotachylites, and mylonites that we will discuss later in the textbook.

### 16.4.4 Mineral deformation behavior

The rheology of Earth materials are fundamentally important in assessing earthquake behavior, in engineering studies, and in understanding crustal and mantle processes. Factors such as depth (confining pressure), temperature, stress conditions, mineral composition, rock texture, rock competency, and strain rate affect rheology. **Strain rate** refers to the amount of strain a material experiences per

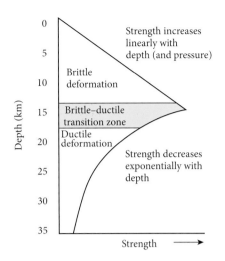
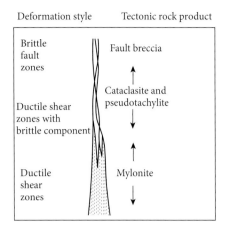

**Figure 16.19**   Brittle–ductile transition within Earth. *Source*: Courtesy of Mike Norton.

unit of time. As noted in the previous section, Earth materials tend to display:

1   Brittle behavior at shallow depths, low-temperatures, and confining pressures, and high strain rates (strain occurs more rapidly) conditions;
2   Ductile behavior at greater depths, higher temperatures and confining pressures, and low strain rates (strain occurs much more slowly) conditions.

In many rocks, minerals such as quartz, feldspars, amphiboles, garnet, and biotite are approximate geothermometers because these common minerals change strain behavior with increasing temperatures. Table 16.1 indicates approximate temperatures at which these minerals change from brittle to ductile. Actual temperatures vary depending on factors such as pressure, volatile content, and strain rate. By observing the deformation characteristics of minerals within a rock, geologists can make some educated guesses as to the metamorphic conditions.

As shown in Table 16.1, quartz and feldspars deform by brittle processes at temperatures <300 °C. At <250 °C quartz exhibits greater strength than feldspars due to its higher hardness and absence of cleavage. However, quartz progressively weakens as temperatures exceed 250 °C. At temperatures greater than 300 °C, quartz deforms via ductile processes such as dislocation creep and glide. Ductile processes produce undulose extinction, deformation

**Table 16.1**   Approximate temperatures at which some common minerals change from brittle to ductile behavior.

| Mineral | Brittle–ductile transition |
| --- | --- |
| Biotite | ~250 °C |
| Quartz | ~300 °C |
| Feldspar | ~400 °C |
| Amphibole | ~650–700 °C |
| Garnet | 600–800 °C |

*Source*: Passchier and Trouw (2005). © Springer Nature.

lamellae and recrystallization. In contrast, feldspars continue to deform by brittle processes at temperatures <400 °C, producing fractures, twinning, and deformed cleavage planes. At temperatures >400 °C, ductile processes dominate through dislocation glide and recrystallization. In assessing the deformation conditions in a granite containing quartz, feldspar, amphibole, biotite, and garnet, the following conditions are useful:

| | |
| --- | --- |
| <250 °C | All minerals exhibit brittle deformation. |
| 250–300 °C | Biotite exhibits ductile deformation; Quartz, feldspars, amphibole, and garnet remain brittle. |
| 300–400 °C | Quartz becomes ductile but feldspars, Amphibole and garnet remain brittle. |

| 400–600 °C | Quartz becomes ductile but feldspars, Amphibole and garnet exhibit brittle behavior. |
| 600–650 °C | Garnet may begin to exhibit ductile deformation. |
| >650 °C | Garnet and amphibole exhibit ductile deformation. |

In the section that follows, we describe the behavior of rocks in response to stress.

### 16.4.5 Rock deformation behavior (competency)

Different rock types deform in contrasting ways when subjected to the same conditions. One way of describing such contrast is the concept of rock competency. **Competency** is a term that describes the resistance of rocks to plastic deformation or flow. Rocks that flow easily are less competent or incompetent, rocks that flow less easily are more competent. Competency is related to strength. **Strength** refers to the amount of stress necessary to induce rock failure (rupture). In any given situation, some rocks will be more competent (stronger) and others will be less competent (weaker); a competence contrast will exist. **Incompetent** rocks are relatively weak and commonly experience ductile behavior. Common rocks that display such behavior include rock salt, shale, siltstone, slate, phyllite, and schist. These rocks contain clays, micas, evaporates, talc, chlorite, serpentine, and other relatively soft minerals with Mohr's hardness ≤3. **Competent** rocks are relatively strong and often display brittle behavior.

Metaquartzite, granite, gneiss, quartz sandstone, basalt, gabbro, and diorite are common examples. These rocks are composed of minerals with Mohr's hardness >3 such as quartz, feldspars, and ferromagnesian minerals.

Rock strength is dependent upon pressure and temperature. In general, rock competency increases with higher pressure but decreases with higher temperature. Table 16.2 lists the relative competency of some common rocks, in order of decreasing competence.

Earth materials are relatively strong under compressive stress and relatively weak under tensional stress. This explains why rocks rupture at low stresses at divergent plate boundaries producing shallow depth, relatively minor earthquakes. At convergent plate boundaries, rocks accumulate high levels of strain from compressive stress before failure. This sudden release of accumulated energy explains why convergent plate earthquakes can be of high magnitude, occur at great depths and be so destructive. Table 16.3 lists the generalized averages of rock strength under compression and tension.

In outcrop, rocks can display both brittle and ductile behavior based upon competency contrasts between rock layers. For example strong, competent rock such as a quartz sandstone or metaquartzite may rupture while interlayered shale or phyllite will flow and fold in a ductile fashion (Figure 16.20a). The rupture of competent layers, as they are stretched, produces "French bread"- or sausage-shaped structures called **boudins** (Figure 16.20b). Boudins are isolated remnants of competent rock that once formed a continuous layer surrounded by less competent rocks.

**Table 16.2** Relative competency of some common rocks at 1 atm and 25 °C.

| | Rock | Major minerals |
|---|---|---|
| Decreasing competency → | Metaquartzite | Quartz |
| | Granite | Quartz, feldspar, amphibole |
| | Gneiss | Quartz, feldspar, amphibole |
| | Quartz sandstone | Quartz |
| | Basalt, gabbro | Pyroxene, plagioclase, amphibole |
| | Dolostone | Dolomite |
| | Limestone | Calcite |
| | Schist | Mica, quartz, clay |
| | Marble | Calcite |
| | Shale | Clay |
| | Rock gypsum | Gypsum, anhydrite |
| | Rock salt | Halite |

*Source*: Davis and Reynolds (1996). © John Wiley & Sons.

**Table 16.3** Mineral and rock strength under compressive and tensile conditions at 1 atm and 25 °C.

| Mineral or rock | Compressive strength (MPa) | Tensile strength (MPa) |
| --- | --- | --- |
| Metaquartzite | 360 | — |
| Granite | 160 | 14 |
| Basalt | 100 | 10 |
| Limestone | 80 | 10 |
| Sandstone | 50 | 10 |
| Shale | 30 | 8 |
| Calcite | 27 | — |
| Halite | 14 | — |
| Clay | 10 | — |

*Source*: Handin (1966), Middleton and Wilcox (1994); and Davis and Reynolds (1996). © John Wiley & Sons.

Let us examine common brittle and ductile rock structures. Brittle deformation results in the development of fractures such as joints and faults. Ductile deformation produces crustal thinning, thickening, folds, and shear zones (Figure 16.21).

## 16.5   BRITTLE STRUCTURES

Brittle behavior commonly occurs at depths less than 10–20 km; these upper crustal low temperature and low lithostatic pressure conditions allow for rupture and the development of fractures. **Fractures** are brittle structures that develop by rupturing a previously intact rock. Fracture orientation is commonly controlled by pre-existing weak areas within the rock. Pre-existing weak surfaces include sedimentary

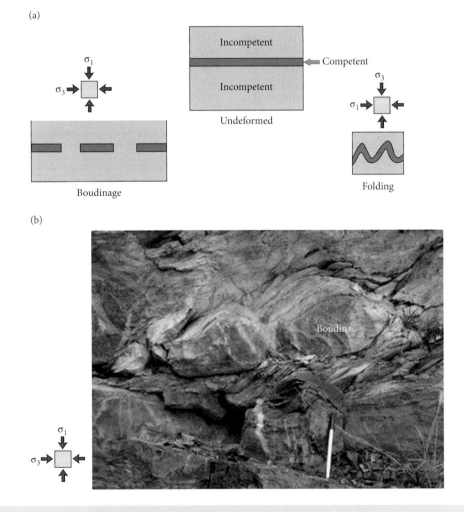

**Figure 16.20**   (a) Competence contrast results in both brittle and ductile behavior in response to the same stress conditions. *Source*: Courtesy of Robert Butler. (b) Boudinage develops within competent Baraboo metaquartzite while the incompetent phyllite flows around brittle boudin structures. *Source*: Photo by Kevin Hefferan. © John Wiley & Sons.

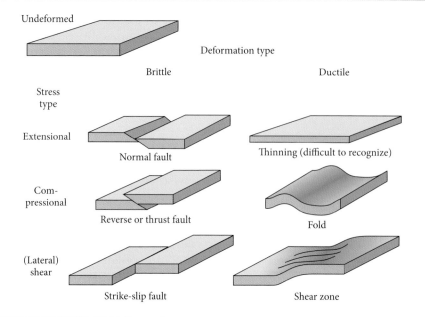

**Figure 16.21** General types of structures produced by brittle and ductile deformation. *Source*: Courtesy of Bruce Railsback.

stratification, volcanic flow contacts, weathering surfaces or metamorphic foliations. Fractures initiate at a point and migrate outward rapidly to form a discontinuity surface, which grows with time. Fractures may be random, systematic and can also occur in sets. Random fractures have many dissimilar orientations, none of which are repeated in an organized manner. Systematic fractures have similar, repeated, subparallel orientations, and commonly form in response to regional stress regimes. For example, Arches National Park in Utah contains north-striking fractures that developed due to east–west tensional stress. If regional stress patterns change over time, then two or more fracture sets may be recognized representing the product of those stress regimes.

Two major types of fracture are joints and faults. **Joints** are fractures with minimal tangential displacement; that is displacement parallel to the fracture. **Faults** are fractures that involve visible displacement parallel to the fracture. Two main types of faults exist: (1) dip–slip faults, where the slip vector (fault movement direction) is subparallel to the dip of the fault, and (2) strike–slip faults, where the slip vector is subparallel to the strike of the fault. Dip-slip faults include **normal faults**, in which the hanging (top) wall moves down relative to the underlying footwall due to the normal effects of gravity. In **reverse faults**, the

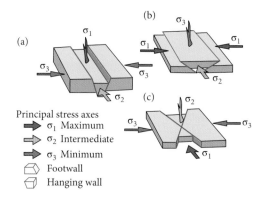

**Figure 16.22** Block diagrams of (a) two conjugate normal dip-slip faults, (b) two conjugate reverse dip-slip faults, and (c) two conjugate strike-slip faults. Principal stress axes directions are indicated. *Source*: Courtesy of Robert Butler.

hanging wall moves up relative to the footwall effectively in reverse of the gravity force. Reverse faults in which the fault angle dip is less than 45° are called **thrust faults**. Figure 16.22 illustrates the general orientation of the principal stress axes that produce normal, reverse, and strike slip faults.

**Veins** develop when fluids flow through fractures and precipitate one or more secondary minerals. Common vein minerals include quartz, calcite, zeolites, and chlorite. Ore minerals containing copper, silver, gold, and other valuable metallic elements also occur in veins. The fractures in which veins originate form as a result of tensile or shear stresses. While veins are highly associated with brittle, fractured rock, they also occur in ductile shear zones. Veins, like the fractures they fill, can occur as isolated features or in a vein arrays, in which groups of veins occur in a rock body.

As with fractures, veins can occur as random, **nonsystematic vein arrays** in which rocks randomly fracture and infill with secondary mineral material. Igneous intrusions with high volatile contents are capable of developing very high pore pressures that cause rock hydrofracturing producing nonsystematic vein arrays. (Figure 16.23).

**Systematic vein arrays** consist of veins that display consistent, subparallel orientations suggesting a common stress field origin. For example, Figure 16.24 illustrates metaquartzite beds bound by phyllite layers. In response to nearly vertical compressive stress, the metaquartzite experiences horizontal tension, resulting in brittle fracturing and the generation of extensional joints. Fluids precipitate quartz or other minerals in the joints creating tension veins parallel that are perpendicular to the minimum stress direction ($\sigma_3$) and parallel to ($\sigma_1$) to the maximum compressive stress direction.

**En echelon vein**s are parallel, slightly offset, sigmoidal veins that form due to noncoaxial rotational shear. En echelon veins develop due to tensional stress, lengthen parallel to $\sigma_1$, and rotate in response to shear stress. Fractures infill with secondary fluids and produce sigmoidal vein sets. Figure 16.25 illustrates an en echelon quartz vein array that consists of a series of offset, parallel veins that formed in response to sinistral shear within metaquartzite.

**Figure 16.23**  Nonsystematic, random veins in metabasalt, Bou Azzer, Morocco. *Source*: Photo by Kevin Hefferan. © John Wiley & Sons.

**Figure 16.24**  Parallel, regularly spaced systematic veins in competent metaquartzite boudins encased within less competent ductile phyllite, Wisconsin. *Source*: Photo by Kevin Hefferan. © John Wiley & Sons.

**Figure 16.25** En echelon quartz vein array that formed in response to sinistral shear in metaquartzite, Wisconsin. *Source*: Kevin Hefferan. © John Wiley & Sons.

Fluids that precipitate as secondary minerals in veins develop blocky or fibrous textures. **Blocky** or **sparry** minerals are equant and may display euhedral crystal faces indicating growth within an unimpeded open space (Figure 16.26).

**Fibrous veins** (Figure 16.27) are composed of numerous, subparallel acicular crystals. In many cases, crystal long axes are nearly perpendicular to vein boundaries. They fill extension fractures that formed in the X direction of maximum extension, parallel to the direction of preferred growth in the acicular minerals. In other cases, crystal long axes are not perpendicular to vein boundaries which suggests that the fracture was a shear fracture. Veins growth is often incremental in response to multiple fracture width increases. Multiple sets of fibrous veins develop by repeated cycles of "crack and seal" mechanism whereby elevated fluid pore pressures crack a vein, followed by sealing from mineralizing solutions. Fiber growth provides information regarding displacement sense as well as progressive vein growth in both brittle and ductile environments.

## 16.6 DUCTILE STRUCTURES

Ductile deformation produces many structures including folds and shear zones. Most folds form by the plastic bending of rock layers. Folded rock layers can be composed of sedimentary beds, igneous intrusions or flows, or metamorphic layers or foliations. Shear zones are essentially analogous to brittle faults in which displacement is dominated by plastic deformation processes rather than brittle rupture.

### 16.6.1 Folds

**Folds** are bends in geological surfaces that form by a variety of processes that involve the plastic bending of rock layers without producing mesoscopic ruptures. Compressive stress plays a significant role in the production of most folds, but folds can also be produced by tensional and shear stress.

Folds consist of limbs and hinges (Figure 16.28). **Limbs** are relatively straight (low curvature) layers separated by a high

**Figure 16.26** Systematic, parallel blocky spar quartz and feldspar folded granite pegmatite vein cross cutting a fine grained granite. *Source*: Photo by Kevin Hefferan. © John Wiley & Sons.

**Figure 16.27** Fibrous calcite veins indicating sinistral shear within metavolcanic rock, Bou Azzer, Morocco. *Source*: Photo by Kevin Hefferan. © John Wiley & Sons.

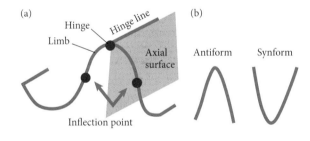

**Figure 16.28** (a) Major components of folds. (b) Terms used to describe the form of elongate folds. *Source*: Kevin Hefferan. © John Wiley & Sons.

curvature region of the hinge area. The **hinge** is a point of maximum curvature of the folded surface and separates two limbs. A **hinge line** is an imaginary line connecting a series of hinge points along the fold hinge. An **axial surface** (**axial plane or hinge plane**) is an imaginary plane that contains a series of hinge lines in multiple folded surfaces. Folds commonly occur in **fold trains**, multiple folds in a geological surface. The **inflection point** is the point at which the sense of curvature changes from

onefold to another. Elongate folds characterized by a convex upward structure (that close upward) are called **antiforms** whereas those structures with concave downward shapes (that close downward) are referred to as **synforms**. Folds that close to the side are called **recumbent folds**. When relative age relations between rock layers are known, the terms anticline and syncline are used as described below.

Fold structures in which the relative ages of rock units are known include anticlines, synclines, monoclines, basins, and domes (Figure 16.29).

1   **Synclines** are elongate folds that contain young rock in the hinge area and progressively older rock away from the hinge in either direction.
2   **Anticlines** are elongate folds that contain older rocks in the hinge and progressively younger rock further away from the hinge in either direction.
3   **Basins** are oval to circular folds in which rock layers dip toward the center and the youngest rocks are in the center.

**4  Domes** are oval to circular folds with rock layers dipping away from the center and the oldest rocks are in the center of the structure.

Folds are ductile structures that form in response to stress. In most cases, folds form in response to compressive stress. For example, synclines and anticlines commonly form where $\sigma_1$ (maximum compressive stress) is oriented perpendicular to the axial surface.

Synclines and anticlines typically occur together and are associated with low-angle reverse faults (thrust faults) in fold and thrust belts. Most of the major mountain belts on Earth are fold and thrust belts produced by orogenic (mountain building) activity at convergent

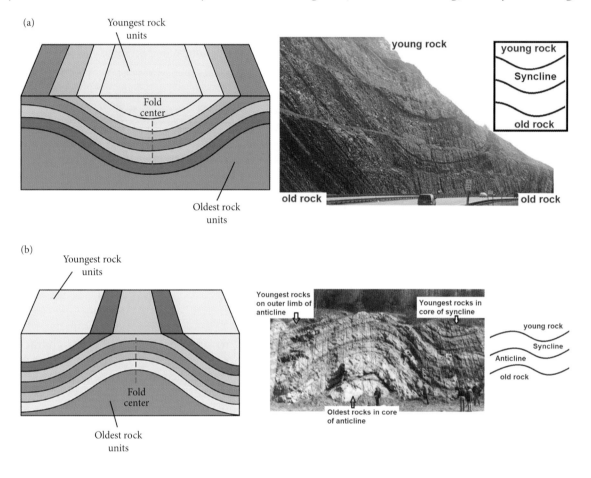

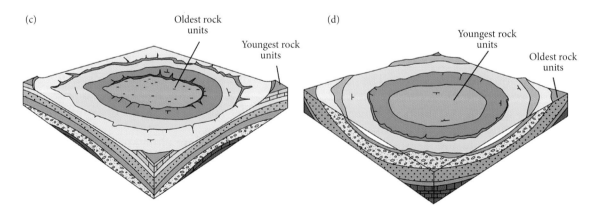

**Figure 16.29**  Idealized block diagrams illustrating (a) syncline; (b) anticline; (c) dome, and (d) basin. *Source*: Photos by Kevin Hefferan. © John Wiley & Sons.

plate boundaries. Basins and domes also commonly form in response to compressive stress, commonly landward of fold and thrust belts in continental interiors.

Folds can also form in response to tensional and shear stresses. For example, reverse drag folds occur in the hanging wall of normal fault blocks as the hanging wall rock layering rotates into the plane of the fault. Folding also commonly occurs in strike slip faults and in shear zones. However, the most magnificent folds on Earth are produced by compressional stress in the Himalayas, Alps, Rockies, and Appalachian fold and thrust belts.

Rocks may experience multiple episodes or generations of folding during one orogenic cycle or multiple orogenic pulses. The result is the formation of **refolded folds**. The result is that complex fold patterns develop that represent a culmination from a number of different fold cycles. These can produce fold interference patterns due to the successive refolding of pre-existing structures. Because younger fold structures are superimposed upon earlier fold structures, refolded folds are referred to as **superposed folds** or superimposed folds (Figure 16.30).

**Transposed folds** consist of folds in which the limbs and hinges have been separated due

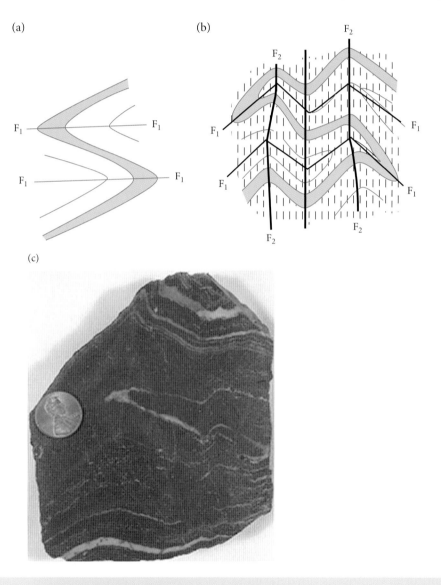

**Figure 16.30** Development of superposed folds. (a) Initial folding event with $F_1$ recumbent folds. (b) Second vertical folding event in which $F_1$ folds are refolded and axial planar $F_2$ cleavage develops. *Source*: (a, b) After Hobbs et al. 1976. © Win Means. (c) Refolded folds from the Penokean Range of Wisconsin. *Source*: Kevin Hefferan. © John Wiley & Sons.

**Figure 16.31**   Transposed folds in metamorphosed ironstone in which "rootless" fold hinges and limbs have become disassociated as a result of intense deformation. *Source*: Photo by Kevin Hefferan. © John Wiley & Sons.

to extension (Figure 16.31). Transposed folds occur with multiple fold generations, involving the replacement of an earlier tectonic fabric ($S_1$) by a more recent tectonic fabric ($S_2$) via ductile mechanisms such as recrystallization and pressure solution. Transposed folds are associated with high temperature and high-pressure metamorphism.

**Parasitic folds** are small folds occurring in the limbs and hinges of larger-scale folds. The form of the parasitic fold can provide information as to the limb or hinge location on a large fold structure. Bedding plane slip commonly produces rotation on fold limbs. The upper bed surface is displaced toward the hinge, while the lower bed surface moves away from the hinge. Rotation to the right produces clockwise bed rotation (producing a Z-shape) while bedding plane rotation to the left produces counterclockwise (or S-shape) parasitic folds. Parasitic folds in the hinge experience minimal rotation (M-shape). Figure 16.32 illustrates how parasitic fold orientation can be used in the analysis of a map-scale antiform. Geologists can determine position within a map-scale fold structure by looking down the plunge direction of the parasitic folds.

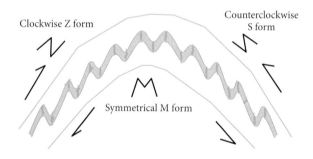

**Figure 16.32**   Parasitic folds produce Z (clockwise rotation), M (symmetrical form) or S (counterclockwise rotation) shapes indicating shear sense. *Source*: Kevin Hefferan. © John Wiley & Sons.

Faults and folds are prime targets for oil and gas exploration in sedimentary rocks, and metallic ore deposits in all rock types. Consequently, fold and fault structures are studied intensely for their potential as economically viable exploration targets.

### 16.6.2   Ductile shear zones

Ductile shear zones are essentially ductile fault zones that accommodate displacement. Shear zones develop in ductile lower crustal rocks. In the field, ductile shear zones are associated with tectonic mélanges (Figure 16.33), mylonites, and pseudotachylites (Figure 16.34). Mélange consists of larger rock blocks encased within a scaly, clay rich matrix. The matrix develops a crude layering and internal structures that can indicate the sense of motion within the shear zone. The larger rock blocks may change shape or orientation and also provide information regarding sense of shear.

**Figure 16.33** Tectonic mélange displays metabasalt encased within a muddy matrix, Bou Azzer, Morocco. *Source*: Photo by Kevin Hefferan. © John Wiley & Sons.

**Figure 16.34** Intense deformation can result in localized melting and the development of pseudotachylite rock within a mylonite, such as in the Siroua region of Morocco. *Source*: Photo by Kevin Hefferan. © John Wiley & Sons.

## 16.7 SMALL-SCALE PLANAR AND LINEAR STRUCTURES

We have summarized many of the common brittle and ductile structures that occur in deformed rocks. These structural features are part of rock fabric. **Fabric** refers to the geometric arrangement of mineral crystals and/or structures within a rock. Metamorphic rock fabric is a critical component in rock identification. Fabric can be described in a number of ways such as:

1 When did the fabric develop?
   a A **primary fabric** develops during processes that produce the original rock. In metamorphic rocks, it is inherited from the protolith.
   b A **tectonic fabric** is produced by deformation processes after the initial development of the rock. Metamorphic rocks are dominated by tectonic fabrics although some primary fabrics (e.g. sedimentary bedding, vesicles in volcanic rocks) may be preserved.
2 What is the spatial arrangement of the fabric?
   a A **continuous fabric** (penetrative fabric) is one in which the structures are continuous on a submillimeter-scale so that no undeformed parts of the rock are visible at the macroscopic scale
   b A **spaced fabric** is one in which visible spacing exists between fabric elements so that both deformed and undeformed parts of the rock are visible
   c A **random fabric** is one in which rock textures and/or structures lack a preferred orientation; they are randomly oriented.
   d A **preferred fabric** is one in which rock elements are aligned in a predictable manner.

Two main classes of preferred fabrics exist. **Foliations** are planar fabrics and **lineations** are linear fabrics (Figure 16.35).

### 16.7.1 Planar fabrics

Planar features are sheet-like structures that include primary stratification and secondary cleavage, schistosity, gneissic banding, and ductile shear zones. We have already provided brief overviews of many planar rock objects in our discussion of brittle and ductile structures. In this section, we will focus upon cleavage.

**Rock cleavage** refers to the tendency of rocks to break along sub-parallel surfaces. Cleavage is a tectonic planar fabric produced by nonuniform stresses under relatively low temperature metamorphic conditions. Processes such as

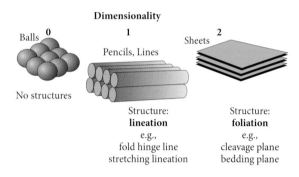

**Figure 16.35**   Random, linear, and planar fabric elements. *Source*: Courtesy of Robert Butler.

pressure solution and recrystallization produce subparallel alignment of inequant minerals so that rocks tend to split or cleave along planes. Cleavage commonly develops in response to compressive stresses, with cleavage planes developing perpendicular to $\sigma_1$, and commonly associated with dynamic or dynamothermal metamorphism. Cleavage is described on the basis of orientation with respect to other geologic structures, rock type, origin, and spacing.

Foliations are particularly well developed in folded rock bodies where it forms an axial planar cleavage. **Axial planar cleavage** consists of parallel foliations oriented nearly parallel to the axial (hinge) plane of the folds and approximately perpendicular to the maximum compressive stress. Figure 16.36 illustrates rock beds ($S_0$) that have been folded into an antiform and synform pair. Note that the axial plane cleavage ($S_1$) is steeper than the bedding angle in upright folds. In cases where axial planar cleavage is dipping less steeply than bedding, complex folding patterns have occurred such as overturned folds, recumbent

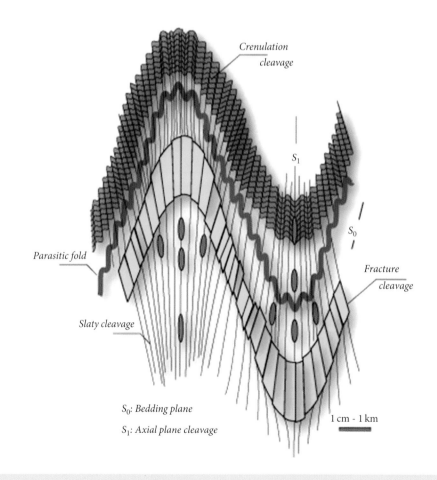

**Figure 16.36**   Note the cleavage and bedding relationships. The axial planar cleavage ($S_1$) dips at a steeper angle than bedding ($S_0$). *Source*: Courtesy of Patrice Rey.

folds or refolded folds wherein the beds have been overturned or inverted.

Other types of planar structures include slaty and phyllitic cleavage as well as schistosity, gneissic banding, and mylonitic foliations. These will be presented in our discussion of foliated textures in Chapter 17. Let us briefly consider linear fabrics.

### 16.7.2 Linear fabrics

Linear structures contain one long axis and two short axes, producing needle-like structures. Linear fabric features include intersection lineations, form lineations, stretching lineations and slip lineations (Figure 16.37).

**Intersection lineations** form by the intersection of two planar fabrics. For example, the intersection of bedding planes and cleavage planes defines a lineation. If the cleavage is axial planar, this intersection is nearly parallel to the hinge line of the associated fold. An interesting example, **Pencil cleavage**, is a continuous cleavage marked by the development of elongate pencil-like shards (Figure 16.38). Pencil cleavage commonly forms due to the intersection of two fracture sets, one parallel to cleavage and the other parallel to bedding fissilily. "Pencils" are ~5–10 cm long and generally less than 2 cm in width and height. The pencil like shape is facilitated by the internal alignment of inequant clay minerals. Pencil cleavage develops in weakly deformed shale or mudstone at an early stage in the development of slaty cleavage. **Crenulation lineations** are linear features that occur as a result of a secondary cleavage imposed upon a fine grained rock (slate or phyllite) that experienced an earlier cleavage. Crenulation lineations represent linears formed by the intersection of these two cleavages (Figure 16.39). The secondary

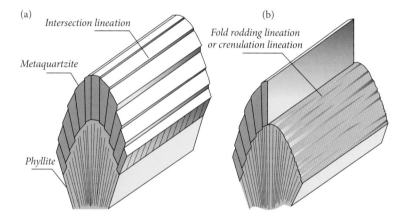

Figure 16.37   This figure illustrates: (a) Intersection lineations formed by the intersection of cleavage and bedding; (b) Form lineations such as development of rodding structures. *Source*: Courtesy of Patrice Rey.

**Figure 16.38**   Pencil cleavage from slate within the Anti-Atlas Mountains, Morocco. *Source*: Photo by Kevin Hefferan. © John Wiley & Sons.

**Figure 16.39**   Crenulation lineations in phyllite, Newport, Rhode Island. *Source*: Photo by Kevin Hefferan. © John Wiley & Sons.

cleavage commonly develops in the thinned limbs of very small-scale kink folds and the crenulation lineation is nearly parallel to the hinge lines of such folds.

**Stretching lineations** develop as elongated mineral or rock grains that define a linear fabric (Figure 16.40). Stretching lineations commonly form on metamorphic foliations, on shear surfaces or on mylonitic planes. These lineations may be due to either: (1) growth of acicular to prismatic crystals in a preferred orientation, commonly parallel to the X strain direction or (2) rotation of crystals toward the X principal strain direction in shear zones.

**Slip or fiber lineations** refer to vein mineral fibers that precipitate on rock surfaces via crack-seal processes. Slickenlines are fiber lineations produced during displacement in faults and shear zones (Figure 16.41).

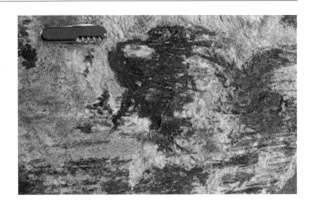

**Figure 16.40** Stretching lineations in hornblende in metamorphosed granite, Big Thompson Canyon, Colorado. *Source*: Photo by Kevin Hefferan. © John Wiley & Sons.

**Figure 16.41** Slickenlines on exposed portion of the San Andreas Fault, San Francisco CA. Notice the well-defined cylindrical rods inclined from upper right to lower left. Image view is approximately 8 m. *Source*: Photo by Kevin Hefferan. © John Wiley & Sons.

This chapter serves as an introduction to stress, strain, and deformation structures encountered in Earth materials. Metamorphic rock textures, and therefore the rocks themselves, are primarily defined by rock fabrics produced by stress. Therefore, an understanding of metamorphic rock textures requires knowledge of stress, strain, and deformation. In Chapters 17 and 18, we will discuss metamorphic textures and rocks, as well as metamorphic zones, facies, and facies series.

## CONTENT ASSESSMENT

1   Describe the type of stress generally associated with nonfoliated metamorphic rocks and foliated metamorphic rocks.

2   Name and identify characteristics of the three principal stress axes.

3   Why the three principal stress axes cannot be readily associated with X, Y, and Z strain axes.

4   Describe the three major behaviors that rocks display.

5   Why do earthquakes occur in the crust and are not generally associated with Earth's mantle.

6   What would be the likely temperature range of a deformed granite in which quartz flows producing ribbon like strands whereas feldspars fracture?

7   Identify structures observed in the Ironstone below (Figure 16.42):

**Figure 16.42** Lake Superior ironstones display plunging folds and steeply inclined faults that displace bedding.

## REFERENCES

Coble, R.L. (1963). A model for boundary diffusion controlled creep in polycrystalline materials. *Journal of Applied Physics* 34: 1679. https://doi.org/10.1063/1.1702656.

Davis, G.H. and Reynolds, S.J. (1996). *Structural Geology of Rocks and Regions*, 2e. New York: Wiley 776 pp.

Handin, J. (1966). Strength and ductility. In: *Handbook of Physical Constraints*, vol. 97 (ed. S.P. Clark Jr.), 223–289. Geological Society of America Memoir.

Hatcher, R.D. (1995). *Structural Geology: Principles, Concepts and Problems*, 2e. Englewood Cliffs, NJ: Prentice Hall 528 pp.

Herring, C. (1950). Diffusional viscosity of a polycrystalline solid. *Journal of Applied Physics* 21: 437. https://doi.org/10.1063/1.1699681.

Hobbs, B.E., Means, W.D., and Williams, P.E. (1976). *An Outline of Structural Geology*. New York: Wiley 571 pp.

Middleton, G.V. and Wilcox, P.R. (1994). *Mechanics in the Earth and Environmental Science*. Cambridge, UK: Cambridge University Press 459 pp.

Mosher, S. (1987). Pressure-solution deformation of the purgatory conglomerate, Rhode Island (USA): quantification of volume change, real strains and sedimentary shape factor. *Journal of Structural Geology* 9 (2): 221–232.

Nabarro, F.R.N. (1948). *Report of a Conference on the Strength of Solids*, 75. Physical Society of London.

Passchier, C.W. and Trouw, R.A.J. (2005). *Microtectonics*, 2e. New York: Springer-Verlag 366 pp.

Petley-Ragan, A., Ben-Zion, Y., Austrheim, H. et al. (2019). Dynamic earthquake rupture in the lower crust. *Science Advances* 5 (7): eaaw0913. https://doi.org/10.1126/sciadv.aaw0913.

van der Pluijm, B.A. and Marshak, S. (2004). *Earth Structure: An Introduction to Structural Geology and Tectonics*, 2e. New York: W.W. Norton and Company 656 pp.

Poirier, J.P. (1985). *Creep of Crystals: High-Temperature Deformation Processes in Metals, Ceramics and Minerals*, Cambridge Earth Science Series. Cambridge, UK: Cambridge University Press 285 pp.

# Chapter 17

# Texture and classification of metamorphic rocks

All rocks are classified on the basis of composition and texture. Metamorphic rock composition is largely determined by protolith chemistry. Metamorphic textures may result from the transformations induced by the temperature and pressure conditions of metamorphism or relict textures inherited from the protolith. Sedimentary protoliths contain a variety of features such as stratification, ripple marks, mudcracks, graded beds, fossils, and tool marks that can be preserved in lower grade metamorphic rocks. Relict features derived from igneous rocks include volcanic flow contacts, intrusive contacts, vesicles, phenocrysts, and other features that survive the temperatures and pressures of metamorphism.

## 17.1   TEXTURES

**Texture** refers to the shape, size, orientation, and intergranular relationships of rock constituents.

### 17.1.1 Grain shape

Metamorphic minerals form in response to high temperatures and pressures. Crystal or grain shapes are equant grains and inequant grains.

1  **Equant** grains possess nearly equal diameters in all directions and assume forms approximated by spheres or cubes. Equant grains have three mutually perpendicular axes of roughly equal length.

*Earth Materials*, Second Edition. Kevin Hefferan and John O'Brien.
© 2022 John Wiley & Sons Ltd. Published 2022 by John Wiley & Sons Ltd.
Companion website: www.wiley.com/go/hefferan/earthmaterials2

2  **Inequant** grains display at least one direction in which the grain diameter is not equal to other grain diameters. The subparallel alignment of inequant grains produces metamorphic foliations. Inequant forms include:

(a) **tabular** disc-shaped (pancake or sheet-like) grains in which one axis is significantly shorter than the other two axes.

(b) **bladed** grains in which one axis is significantly longer than the other two axes which are not equal to one another.

(c) **acicular** (needle-like) or **prolate** (cigar-shaped) grains in which one axis is significantly longer than the other two, equal length axes.

### 17.1.2  Grain size

Metamorphic grain size classification is similar to that used for igneous rocks where aphanitic grains possess diameters less than 1 mm and phaneritic grains have diameters of 1 mm or greater. Grain sizes may be fairly uniform or can vary greatly in a given rock. Rocks such as hornfels, slate, phyllite, and cataclasite generally contain uniform, fine-grained crystals. Rocks such as schist, gneiss, metaquartzite, marble, metabreccia, metaconglomerate, skarn, amphibolite, and granulite generally contain larger grain diameters and more variability in grain size.

**Porphyroclasts** are large relict grains from the protolith that have experienced deformation but have retained their original composition. Their large size relative to surrounding minerals may be inherited from the protolith, or be due to significant crushing or stretching of the surrounding minerals and/or the growth of new, smaller crystals. Common porphyroclast minerals include quartz and feldspars. **Augen** are oval shaped feldspar porphyroclasts that resemble the shape of an eye and are particularly common in gneisses, whereas **flaser** are composed of quartz. Augen and flaser grains are common in rocks that have experienced shear strain.

**Porphyroblasts** are large grains that have experienced neocrystallization and growth in response to favorable temperature and pressure conditions during metamorphism. Neocrystallization refers to the growth of new minerals stable at the temperature and pressure conditions of metamorphism. Common porphyroblast minerals include garnet, staurolite, and cordierite.

In our discussion of igneous textures the terms euhedral, subhedral, and anhedral were used to describe the degree to which crystals are bounded by crystal faces. Euhedral, subhedral, and anhedral are also appropriate terms for metamorphic crystals. Whether porphyroblasts are euhedral, subhedral, or anhedral depends upon many factors related to the conditions of metamorphism. However, as indicated in Table 17.1, metamorphic minerals possess different tendencies to develop crystal faces during solid-state growth.

### 17.1.3  Grain orientation

Metamorphic grains can be arranged in random or a preferred orientations. A rock with **random grain orientation** shows no preferred orientation of grains, resulting in a nonfoliated texture. A rock with **preferred grain orientation** occurs when inequant grains are oriented subparallel to one another producing lineations and/or foliations. Preferred orientation of acicular, bladed, or prismatic (e.g., inosilicate minerals) grains with subparallel long axes produces lineations. **Lineations** are line-like features similar to pencils all pointing in a

**Table 17.1**  The tendency of common metamorphic minerals to develop complete crystal forms.

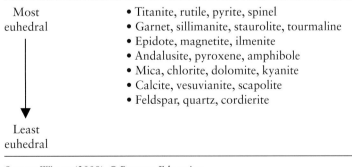

| Most euhedral ↓ Least euhedral | • Titanite, rutile, pyrite, spinel<br>• Garnet, sillimanite, staurolite, tourmaline<br>• Epidote, magnetite, ilmenite<br>• Andalusite, pyroxene, amphibole<br>• Mica, chlorite, dolomite, kyanite<br>• Calcite, vesuvianite, scapolite<br>• Feldspar, quartz, cordierite |
|---|---|

*Source*: Winter (2009). © Pearson Education.

common direction. Preferred orientation of tabular grains, especially phyllosilicate minerals, with subparallel long axes produces foliations. **Foliations** appear as metamorphic layers oriented parallel to one another, as in the pages of a book.

### 17.1.4 Intergranular relationships

Foliated textures are defined by intergranular relationships that produce a subparallel arrangement of predominantly inequant minerals (such as phyllosilicates and inosilicates) as a result of nonuniform stress conditions. **Slaty cleavage** possesses flat, dull surfaces, with parallel alignment of fine grained (<1.0 mm diameter) crystals. **Phyllitic cleavage** appears as wavy, waxy surfaces, with parallel alignment of fine grained crystals. **Schistosity** is defined by wavy to folded, subparallel alignment of medium to coarse grained (>1.0 mm diameter) crystals. **Gneissic banding** is defined by parallel alignment of medium to coarse grained crystals that define color bands.

Non-foliated textures are characterized by intergranular relationships that do not possess subparallel arrangement of minerals as a result of uniform stress conditions. **Hornfelsic textures** are characterized by predominantly fine grained (<1.0 mm diameter), equant crystals or small inequant crystals that lack preferred orientations (Figure 17.1). **Granoblastic textures** are characterized by large (>1.0 mm diameter) equant grains or large inequant crystals that lack preferred orientation (Figure 17.2). Crystals in granoblastic rocks commonly display anhedral, sutured boundaries that reflect a combination of pressure solution, recrystallization, and annealing of grains. Granoblastic textures commonly develop with minerals such as quartz, feldspar, and calcite that have low euhedral form potential (Table 17.1) and tend to crystallize in equant to subequant forms. Nonfoliated metamorphic rocks with **cataclastic textures** include fractured, angular, pulverized, and/or glassy particles. Cataclastic textures form in response to crushing, frictional sliding and rotation of grains during high strain rate deformation in upper crustal fault zones. Cataclastic rocks that exhibit cohesion in which the broken particles form a cohesive mass are called **cataclasites**. Cataclastic particles that lack cohesion are called **breccia** if they contain coarse (>2 mm diameter), angular fragments

**Figure 17.2** Photomicrograph of granoblastic texture with equant calcite crystals. *Source*: Photo courtesy of Kent Ratajeski.

**Figure 17.1** (a) Photomicrograph of hornfelsic texture with equant crystals separated by biotite-rich layers, that may represent relict sedimentary bedding. (b) Hornfels rock sample collected from the base of the Palisades Sill, New Jersey. *Source*: Photo by Kevin Hefferan. © John Wiley & Sons.

and **gouge** if they are composed of finer-sized fragments. Cataclastic textures develop during dynamic metamorphism in a wide variety of rocks within 15 km of the Earth's surface.

Metamorphic rock classification is simpler than the more complex nomenclature that has developed for igneous, and to a lesser extent, sedimentary rocks. Metamorphic rock classification is largely based on the presence or absence of foliations. However, this simple approach poses some difficulty because: (1) metamorphic rocks can retain relict layering derived from the protolith; (2) gradations exist between weakly foliated and nonfoliated textures; (3) some rocks, especially those defined primarily on the basis of composition, occur in both foliated and nonfoliated forms depending upon the stress conditions. The classification presented below focuses on the presence or absence of foliations. Other metamorphic nomenclature schemes do exist, some of which focus on the protolith. For example, a mildly metamorphosed mafic igneous rock can be called a metabasalt.

## 17.2    NON-FOLIATED METAMORPHIC ROCKS

Non-foliated textures lack metamorphic layering as defined by preferred mineral orientations. Equant (e.g., quartz, feldspars, garnet, kyanite) and inequant (e.g., mica, serpentine, amphibole) grains can occur in nonfoliated rocks; however inequant grains are not aligned subparallel to one another.

Non-foliated textures are commonly produced by dynamic metamorphism, contact metamorphism or burial metamorphism in which uniform lithostatic stress produces equant or randomly arranged minerals. However, they can also occur in dynamothermal metamorphic rocks that have experienced annealing crystallization.

### 17.2.1    Hornfels

**Hornfels** is a fine grained (<1.0 mm diameter), nonfoliated rock (Figure 17.1) produced by contact metamorphism. Hornfels is derived from fine-grained protolith rocks such as shale, siltstone, mudstone, tuff, or basalt. Hornfels display hornfelsic textures with fine grained, equant crystals. Where present, tabular to acicular crystals lack preferred alignment. Relict sedimentary fabrics such as bedding and igneous features such as phenocrysts may be preserved.

Hornfels develop in metamorphic aureoles, adjacent to igneous intrusions.

Common hornfels minerals include muscovite, biotite, andalusite, cordierite, plagioclase, K feldspar, quartz, epidote, amphibole, and pyroxene. Mineral assemblages are strongly influenced by protolith composition and the temperatures achieved during metamorphism. Recrystallization causes most hornfels to be somewhat harder and more brittle than the protoliths (e.g., mudstone) from which they commonly form. Although hornfels are composed of fine-grained crystals, spotted hornfels with larger porphyroblasts can occur.

### 17.2.2    Metaquartzite

**Metaquartzites** contain >90% quartz and are derived from quartz arenite or chert protoliths (Figure 17.3) via dynamothermal, deep burial, or contact metamorphism. Accessory minerals commonly include hematite and feldspars. In the transformation of quartz-rich protoliths to metaquartzite, recrystallization, pressure solution, and intracrystalline plastic deformation mechanisms produce interlocking quartz grains that commonly display granoblastic texture. Because the quartz grains interlock, rupture occurs through grains, rather than around grain boundaries. This gives metaquartzites a smooth, glazed appearance as opposed to the granular appearance of many quartz arenites. Because quartz is stable over a wide range of metamorphic temperatures and pressures, metaquartzite, also referred to as quartzite, is a common metamorphic rock. Metaquartzites are hard, durable rocks that produce angular surfaces when fractured. As a result of the sharp edges, metaquartzites are not ideal for road construction because of the tendency to tear rubber tires. Metaquartzites are used for rock walls, railroad ballast and drainage culverts.

### 17.2.3    Marble

**Marbles** are granoblastic metamorphic rocks rich in calcite or dolomite. Marbles are derived by recrystallization of limestone or dolostone protoliths via dynamothermal, deep burial or contact metamorphism. Marbles can retain relict sedimentary features such as bedding so some relict stratigraphic layering or fossils may be observed. Common accessory minerals include quartz, graphite, talc, diopside, forsterite,

**Figure 17.3** The Proterozoic age Baraboo metaquartzite in Wisconsin contains relict features that include ripple marks that formed during deposition of quartz sand in a coastal environment over 1 billion years ago. *Source*: Photo by Kevin Hefferan. © John Wiley & Sons.

wollastonite, serpentine, idocrase (vesuvianite), tremolite, and grossular-andradite garnet. Accessory minerals provide distinctive hues that allow marble to assume a wide variety of colors. For example, graphite produces a gray to black marble; serpentine and forsterite produce green marble. Because of their great color diversity and softness, marbles have been extensively used for sculptures, building stone and headstones for thousands of years (Figure 17.4). Artisans preferred marble for creating classic sculptures such as Venus de Milo and Michelangelo's David, sculpted from Greek and Italian marble, respectively.

### 17.2.4 Skarn

**Skarns**, also known as tactites, are granoblastic calc-silicate rocks formed by the contact metamorphism of carbonate country rocks such as limestone or dolostone (Figure 17.5). Release of silica and volatiles from the magma result in extensive metasomatism, generating calc silicate mineral assemblages and metallic ore deposits. Skarns commonly contain carbonate minerals such as calcite, dolomite, and ankerite and silica group minerals such as quartz. Common calc-silicate skarn minerals (calcium-magnesium silicates) include wollastonite, tremolite, diopside, talc, epidote, grossular garnet, phlogopite, and idocrase (vesuvianite). A number of different skarn types occur; their classification is largely related to the associated ore minerals such as gold, silver, tungsten, molybdenum, or iron.

### 17.2.5 Metabreccia

**Metabreccias** are derived from metamorphism of sedimentary or igneous breccias. Metabreccias commonly develop during burial, dynamic, or dynamothermal metamorphism. Metabreccias contain subangular to angular clasts with diameters >2 mm (Figure 17.6). Extensively recrystallized matrix grains interlock so tightly that rupture commonly occurs through grains, rather than around grain boundaries. Metabreccias can be composed of a single clast type (monomictic) or many different clast types (polymictic).

### 17.2.6 Metaconglomerates

**Metaconglomerates** are derived from conglomerate protoliths and contain subrounded to rounded relict clasts with diameters >2 mm.

**Figure 17.4** Alexander Hamilton's marble grave in Old Trinity Church graveyard. *Source*: Photo by Kevin Hefferan. © John Wiley & Sons.

Metaconglomerates can form by deep burial, dynamothermal, or contact metamorphism. Most metaconglomerates display nonfoliated textures. Extensively recrystallized matrix grains interlock so tightly that rupture occurs through grains, rather than around grain boundaries. The rock and mineral clast assemblage of metaconglomerates varies widely as a result of the wide range of protolith clast compositions, as well as the temperature and pressure conditions of metamorphism. Rock clasts may be derived from sedimentary, igneous, or metamorphic rocks. As with sedimentary conglomerates, metaconglomerates may be monomictic (one clast type) or polymictic (many clast types).

### 17.2.7 Cataclasite

**Cataclasites** are cohesive rocks with cataclastic textures produced by predominantly brittle deformation. Cataclasites form under low temperature, high strain, dynamic metamorphic conditions such as upper crustal fault zones (Figure 17.7). Cataclastic textures are classified based on the relative percentages of larger clasts and finer matrix, and their degree of cohesion (Higgins 1971; Sibson 1977). Primary cohesion distinguishes cataclasites from noncohesive fault zone rocks such as gouge and breccia. Cataclasites are commonly nonfoliated; however, phyllosilicate-rich cataclasites can exhibit foliations. These rocks are classified (Table 17.2) on the basis of percent matrix.

**Figure 17.5** Granoblastic skarn containing wollastonite, tremolite, and garnet from silicate magma intruding limestone. *Source*: Photo by Kevin Hefferan. © John Wiley & Sons.

**Figure 17.6** Quartz (monomictic) metabreccia with angular gravel size grains. *Source*: Photo by Kevin Hefferan. © John Wiley & Sons.

(a)

(b)

**Figure 17.7**  (a) Exposed fault in Trifya Basin, Bou Azzer-El Graara, Morocco. (b) Protocataclasite rocks exposed along Trifya Basin fault zone, Bou Azzer, Morocco. *Source*: Photos by Kevin Hefferan. © John Wiley & Sons.

**Table 17.2**  Cataclasite series rock terms developed by Sibson (1977).

| Percent matrix | Cataclasite series rock name |
| --- | --- |
| 10–50 | Protocataclasite |
| 50–90 | Cataclasite |
| 90–100 | Ultracataclasite |

*Source*: Modified from Sibson (1977). © John Wiley & Sons.

While cataclasites are not volumetrically significant components of the Earth's crust, they do serve as impermeable seals in fault zones, acting as cap rocks that impede the escape of resources such as oil and gas. In addition, when rocks are subjected to new stresses, cataclasites tend to rupture more easily than the surrounding host rock, so that faults are reactivated as demonstrated in Japan (Jefferies et al. 2006). As a result, cataclasites are of interest in both resource exploration and seismology.

### 17.2.8 Pseudotachylite

**Pseudotachylites** are glassy cataclastic rocks produced by high-strain rates that generate localized melting in fault zones (Figure 17.8). Near instantaneous solidification of melts produced by pressure release during fracturing produces very dark-colored, vitreous to flinty rocks. These rocks commonly occur as vein material in cataclastic rocks such as fault breccias and cataclasites.

### 17.2.9 Impactite

**Impactites** are high-strain rate cataclastic rocks created by the tremendous short term stresses associated with extraterrestrial bodies impacting Earth. Four major components of impactites are recognized: (1) **Impact breccia** and **impact shatter cones** (Figure 17.9a) produced by the fragmentation of rock upon impact; (2) Glassy spherules called **tektites** that form as rocks are locally melted due to impact. The impact of extraterrestrial rock bodies produces showers of droplets which cool very rapidly to form tectites (Figure 17.9b); (3) **Deformation lamellae** that form due to the intense stresses that deform crystal structures (Figure 17.9c). These are especially common in "shocked" quartz crystals; (d) **Ultra high pressure (UHP) mineral assemblages** such as the high-pressure polymorphs of silica including coesite or stishovite. Produced synthetically in the laboratory by Loring Coes in 1953, the high-pressure mineral coesite was first discovered in the field at Meteor Crater (Chao et al. 1960).

### 17.2.10 Anthracite coal

**Anthracites** are crystalline, high-grade coals produced by heating, compressing, and chemically altering bituminous coal. Anthracite coals form by burial metamorphism and represent a very important energy resource. Anthracite coals are vitreous, lightweight, jet black in color and commonly display conchoidal

**Figure 17.8** Pseudotachylite formed in a ductile shear zone within leucogabbro, Siroua, Morocco. *Source*: Photo by Kevin Hefferan photo. © John Wiley & Sons.

fractures (Figure 17.10). Anthracite is harder, more vitreous, and has substantially higher thermal capacity than bituminous coal due to the loss of volatiles and the concentration of carbon. Unfortunately, the mining and burning of coal has serious environmental consequences. Surface or strip mining is highly destructive to the landscape both in the removal of rock material and subsequent dumping of tailings in valleys and streams. Underground coal mining poses great dangers to the miners excavating the rock. Abandoned underground mines experience roof collapse, causing subsidence and sinkhole development on the Earth's surface. Coal burning is also a major source of greenhouse gases ($SO$, $SO_2$, $NO$, $NO_2$, $CO$, $CO_2$, $CH_4$), particulate pollution and heavy metals such as mercury and uranium. Of all the fossil fuels, coal is the most detrimental to our environment.

## 17.3 NON-FOLIATED TO FOLIATED METAMORPHIC ROCKS

The distinction between nonfoliated and foliated textures is not always well defined either in the field or in hand-specimens. In a given rock exposure, a nonfoliated rock may grade into poorly-foliated or even a well-foliated one.

Rocks such as metaconglomerates, serpentinites, soapstones, greenstones, granulites, eclogites, and amphibolites can occur with nonfoliated or foliated textures as discussed below.

### 17.3.1 Stretched pebble metaconglomerate

**Stretched pebble metaconglomerates** (Figure 17.11) form by the metamorphism of conglomerates in response to strong non-uniform stress during dynamothermal or dynamic metamorphism. During metamorphism, pebbles and cobbles are shortened and/or flattened parallel to the maximum shortening (Z) direction and relatively elongated parallel to the maximum extension (X) strain direction. Pebble alignment may define a metamorphic foliation or lineation fabric. Excellent examples of both occur in Pennsylvanian Purgatory Formation metaconglomerate in the Narraganset Basin, Rhode Island where stretched pebbles are arranged parallel to one another like loaves of French bread.

### 17.3.2 Serpentinite

**Serpentinites** (Figure 17.12) are serpentine-rich metamorphic rocks that occur in nonfoliated or foliated forms. Serpentinite forms by hydrothermal alteration of ultrabasic rocks at

(a)

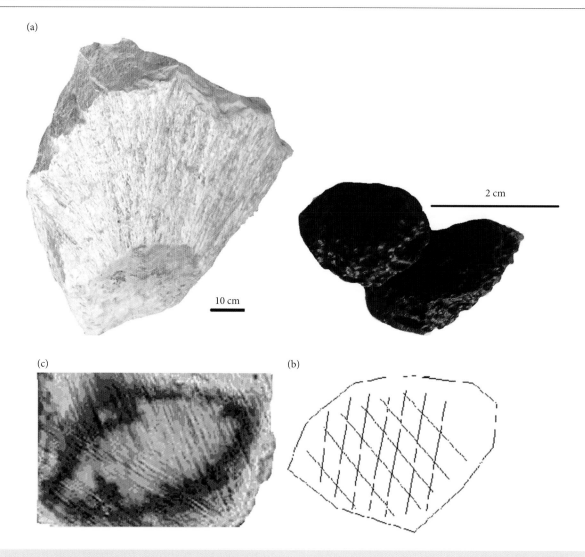

2 cm

10 cm

(c)

(b)

**Figure 17.9**  (a) Shatter impact cone produced by meteorite impact on limestone. Shatter cones display radiating fracture markings produced by shock waves and pressures >2 GPa, associated with nuclear explosions and meteorite impacts. *Source*: Kevin Hefferan. © John Wiley & Sons. (b) Glassy black tektites. *Source*: Kevin Hefferan. © John Wiley & Sons. (c) Photomicrograph of deformation lamellae in shocked quartz from the Chesapeake Bay impact site. The impact sample was collected from 250 m beneath NASA's Langley Research Center in Hampton, VA, which is located on the crater rim. Line drawing illustrates two sets of shock lamellae. *Source*: Courtesy of the US Geological Survey and Glen A. Izett. © U.S. Department of the Interior.

temperatures below ~500 °C. During metamorphism olivines and pyroxenes are hydrated to form serpentine group minerals in a process aptly named serpentinization. Serpentine group minerals include lizardite [$Mg_3Si_2O_5(OH)_4$], chrysotile [$Mg_3Si_2O_5(OH)_4$], and antigorite [$Mg,Fe]_3Si_2O_5(OH)_4$]. Lizardite and chrysotile are low-temperature serpentine minerals; antigorite is a higher temperature serpentine mineral. Common accessory minerals include calcite, talc, chlorite, brucite, and magnetite.

Serpentinization occurs at: (1) ocean spreading ridges, where hydrothermal fluids circulate through fracture systems that penetrate the ultrabasic mantle; and (2) at subduction zones where ocean water interacts with ultramafic rocks of the ocean lithosphere. Serpentinite slivers are commonly imbricated with other rocks in tectonic mélanges within subduction complexes and ophiolites. Serpentines are used for building stone and the fibrous varieties were formerly used widely as industrial and residential asbestos. Serpentinites commonly occur with soapstones, discussed below.

### 17.3.3  Soapstone

**Soapstones** are fine grained, rocks that form through the alteration of ultrabasic rocks, or magnesian-rich sedimentary rocks such as dolostone by low temperature and low pressure hydrothermal fluids. Soapstones contain talc, chlorite magnesite, serpentine, and tremolite. Talc contributes a white to green color, soapy feel and low hardness which allows soapstone to be cut easily. Soapstones are widely used in sculptures, especially Native

**Figure 17.10**   Black, glassy, lightweight anthracite coal displaying conchoidal fracture. *Source*: Photo by Kevin Hefferan. © John Wiley & Sons.

**Figure 17.12**   Green serpentinite. *Source*: Photo by Kevin Hefferan. © John Wiley & Sons.

**Figure 17.11**   Stretched pebble conglomerate in which Purgatory Conglomerate clasts have been flattened and elongated parallel to the X axis. *Source*: Photo by Kevin Hefferan. © John Wiley & Sons

American sculptors in the Arctic rim. They are also used for ornaments, kitchen sinks and countertops. Soapstone's low porosity prevents staining or seepage. Soapstones are used in the manufacture of wood stoves and fireplaces, even smoking pipes, due to their ability to withstand and evenly dissipate heat.

### 17.3.4 Greenstone

**Greenstones** are green colored rocks that form by low to moderate (200–500°C) temperature alteration of mafic, and to a lesser extent intermediate to ultramafic, igneous rocks. During metamorphism, plagioclase and primary ferromagnesian silicates such as olivine, pyroxene, and amphibole are converted into greenstone minerals such as chlorite, epidote, prehnite, pumpellyite, talc, serpentine, actinolite, and albite. Greenstones are commonly produced by hydrothermal metamorphism of basalts and gabbros in oceanic crust near divergent plate boundaries. Altered volcanic rocks such as spilites (Na-rich basalt) and keratophyres (Na-rich andesite) form greenstone belts. Greenstones are commonly incorporated into mélanges and orogenic belts along convergent plate boundaries.

Large scale Precambrian greenstone belts occur in large synclinal structures. In a typical greenstone belt sequence, ultramafic metavolcanic rocks (komatiites) and metabasalts form the basal layers and are overlain successively by intermediate and silicic metavolcanic and metavolcaniclastic sequences, which are in turn capped by graywackes and chert. Greenstone belts commonly parallel granulite belts composed of rocks of granitic to dioritic composition metamorphosed at high- temperatures and pressures. Greenstone belts are tens of kilometers wide and hundreds of kilometers long and occur in Archean (>2.5 Ga) and Proterozoic (2.5 Ga–544 Ma) cratons around the world. Phanerozoic greenstone belts are rare. The best known greenstone localities include the Barberton belt in South Africa, the Superior and Slave provinces in North America, and the Sao Francisco craton in Brazil. Greenstone belts yield valuable metallic ore deposits containing copper, gold, silver, zinc, and lead (Hoatson et al. 2006).

### Box 17.1   Greenstone belts

Greenstone belts dominate the history of continents in the Archean and Early Proterozoic Eons (>2 Ga). An estimated 260 greenstone belts occur throughout the world (de Wit and Ashwal 1997). Greenstone belts are so named because they contain green-colored minerals such as actinolite, chlorite, epidote, prehnite, pumpellyite, serpentine, and talc. These minerals formed by extensive metasomatic alteration of basic and ultrabasic rocks through that involved chemical reactions with $H_2O$ and $CO_2$. Greenstone belts are synclinal to tabular rock assemblages that contain peridotite and gabbroic intrusive rocks and ultrabasic to basic volcanic rocks called komatiites (Figure B17.1). Komatiites are overlain successively by basalt and rhyolite layers. The igneous layers are capped by metasedimentary layers that include chert, graywacke, banded iron formation and conglomerate. Archean and Proterozoic greenstones occur in association with granitic gneisses in Precambrian cratons. Two major questions remain regarding greenstone belts:

- In what tectonic setting did Precambrian greenstone belts form? The origin of Precambrian greenstone belts remains controversial (Bickle et al. 1994; de Wit 1998). Proposed environments for greenstone belt formation include ocean ridges, continental rifts (Grove et al. 1978), accreted subduction zone/island arc fragments (de Wit and Ashwal 1997), oceanic plateaus formed above deep mantle plume hot spots, or extensive lava plains unrelated to plate tectonics (Hamilton 1998).
- Why are greenstone belts uncommon in the Phanerozoic Eon? This question also continues to perplex geologists and is the subject of current research. The answer is perhaps related to lower geothermal gradients in the Phanerozoic as well as changes in plate tectonics through time. A higher Precambrian geothermal gradient may have resulted in higher-grade metamorphism at shallow depths and prevented deep subduction along steeply inclined Benioff zones. These conditions may have favored the development of shallow greenstones and prevented the development of the deep, high-pressure assemblages that characterize modern subduction zones.

**Box 17.1    *Continued***

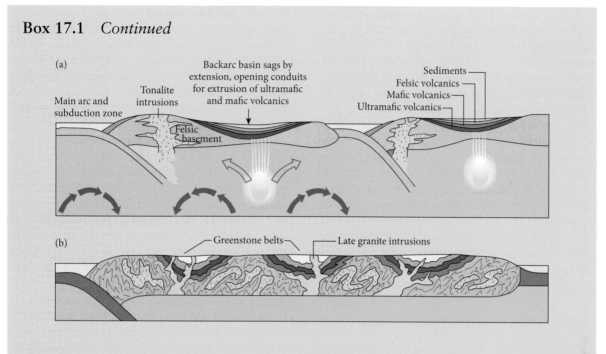

**Figure B17.1**    Cross section showing possible development of greenstone belts and associated granulite rocks. (a) Synkinematic Precambrian tectonic activity. (b) Modern erosional setting. *Source*: Levin 2006. © John Wiley & Sons.

### 17.3.5    Amphibolite

**Amphibolites** are dark colored rocks (Figure 17.13) composed largely of amphiboles, such as hornblende, and plagioclase. Garnet is also a common major mineral. Accessory minerals include quartz, K feldspar, tremolite, biotite, magnetite, chlorite, calcite, sphene, and epidote. Amphibolites form by the medium to high temperature (>550 °C) regional metamorphism of mafic igneous rocks or sedimentary rocks such as calcareous mudrocks and graywackes. Amphibolites derived from mafic igneous rocks such as basalt and gabbro are called ortho-amphibolites, whereas amphibolites produced from sedimentary protoliths are called para-amphibolites. Amphibolites are commonly, but not always, foliated.

### 17.3.6    Granulite

**Granulites** are medium to coarse grained rocks that possess granoblastic to foliated textures. Granulites form by high-temperature (>800 °C) and pressure (>1 GPa; ~33 km depth) metamorphism. High temperature granulite conditions trigger dehydration reactions resulting in the transformation of hydrous amphibole and mica minerals into anhydrous minerals such as pyroxene, K feldspar, kyanite, and garnet. The high pressure and very low water content inhibits melting, and preserves the rock's metamorphic fabric. Granulites from the type locality of Saxony, Germany are granoblastic rocks of granitic composition that contain quartz, feldspars, and minor amounts of kyanite and garnet.

2 cm

**Figure 17.13**    Dark colored amphibolite derived from mafic protolith. *Source*: Photo by Kevin Hefferan. © John Wiley & Sons.

Modern usage of the term granulite refers to rocks of the granulite facies. Such rocks may be derived from a wide variety of protoliths that include granitoids, diorite, gabbro, peridotite, and pelitic rocks such as mudstone and graywacke. Common granulite minerals include pyroxene, K feldspar, plagioclase, garnet, sillimanite, kyanite, and quartz. With increasing temperature and pressure within granulites, orthopyroxene, and plagioclase transform into clinopyroxene, garnet, and quartz (O'Brien and Rötzler 2003). Granulites form in high temperature and high pressure conditions of the lower continental crust, upper mantle and as a result of subduction of crust at convergent margins. Granulites are also common in Precambrian greenstone belts and in association with eclogites.

### 17.3.7 Eclogite

**Eclogites** are very high pressure, high temperature rocks that develop principally from basalt/gabbro protoliths. Eclogites are likely the major rock type in Earth's lower crust because they are stable at temperatures that exceed 400 °C and pressures that exceed 1.2 GPa (>40 km depth). Eclogites are commonly red and green (Figure 17.14), because they contain green omphacite (Na-Ca pyroxene) and red garnet as major minerals, composing over 75% of eclogite. Garnets in eclogite include solid solutions of pyrope, almandine, and grossular garnet. Accessory minerals consist of jadeite pyroxene, quartz, rutile, phengite mica, coesite, lawsonite, kyanite, corundum, zoisite, glaucophane, epidote, allanite, and diamond. Plagioclase is not stable and transforms to omphacite in eclogite's ultrahigh pressure conditions.

Eclogite can form by a number of processes which include: (1) High pressure recrystallization of deep, mafic crustal rocks during crustal subduction and thickening at continent-continent collisions, (2) Partial melting of the mantle followed by deep crystallization as high-pressure eclogite, or (3) High pressure metamorphism of subducted oceanic lithosphere deep within the Earth. The slab-pull effect generated by high density eclogite (3.5–4.0 g/cm³) in subducted slabs is thought to drive mantle convection and plate motion.

Let's now consider some of the more common foliated metamorphic rocks.

## 17.4 METAMORPHIC ROCKS WITH FOLIATED TEXTURES

Foliated textures are marked by the planar alignment of inequant crystals such as tabular phyllosilicates (micas, clays) and prismatic/acicular inosilicates (amphiboles, pyroxenes). Foliated textures are commonly associated with regional dynamothermal or dynamic metamorphism, in which rocks change shape in response to nonuniform stresses producing minerals aligned in a planar fabric. Foliated textures include slaty cleavage, phyllitic cleavage, schistosity, gneissic banding, migmatitic layering and mylonitic foliation. Rocks containing these textures are described below.

### 17.4.1 Slate

**Slates** are fine grained, aluminum-rich, pelitic rocks that possess flat, planar cleavage. Slaty cleavage consists of closely spaced layers along which the rock breaks or cleaves readily to produce flat surfaces with a dull luster. The layering is defined by the subparallel orientation of microscopic phyllosilicate mineral grains. Slaty cleavage develops perpendicular to the Z-direction of maximum shortening and lies in the X-Y planes of intermediate and minimum shortening that existed during metamorphism. Slaty cleavage

2 cm

**Figure 17.14** Eclogite with red garnet and green omphacite. *Source*: Photo by Kevin Hefferan. © John Wiley & Sons.

commonly forms during the metamorphism of pelitic sedimentary protoliths such as shale, silt-stone, mudstone, or altered pyroclastic volcanic rocks such as tuff. Slaty cleavage develops during metamorphism under nonuniform stress at relatively low temperatures (~150–250 °C) and low pressures. At these relatively low temperatures, relict sedimentary textures and structures may be preserved. During metamorphism, inequant phyllosilicate minerals such as clays and micas are reoriented, creating foliations that are typically at a high-angle to original bedding. With increasing metamorphic temperatures clay minerals, such as kaolinite and smectite, and zeolite minerals decrease in abundance while illite, chlorite, and mica progressively increase. Accessory minerals include quartz, graphite,

pyrite, ilmenite, plagioclase, sericite, muscovite, and iron oxide minerals such as hematite.

Slates, like the mudstones from which they commonly form, possess many colors including red, green, gray, and black which is largely inherited from the protolith. Oxidized iron imparts a brick red color. Reduced iron produces green colors. Organic matter deposited in anoxic environments and commonly recrystallized into graphite, produces dark gray and black-colored slates. Slates are actively quarried for many construction applications (Figure 17.15). The flat, planar nature of slaty foliations allows it to be split into thin sheets that make it suitable for sidewalks, roof shingles, school blackboards, pool tables, patios, and floor tiles. In rugged topography, slates

**Figure 17.15** (a) Slate with flat planar cleavage. (b) Dark gray slate with flat planar cleavage at a high angle to bedding defined by lighter sand grain layer. (c) Multi-colored slate roof tiles. *Source*: Photo by Kevin Hefferan. © John Wiley & Sons.

present a significant geohazard when steeply dipping foliations are inclined downslope which can result in unstable weathered rock faces that slide downslope by mass wasting.

As temperatures rise progressively above ~250 °C, slates are transformed into higher-grade metamorphic rocks such as phyllite, schist, and gneiss.

### 17.4.2 Phyllite

**Phyllites** display phyllitic cleavage, characterized by slightly larger crystals and more wavy surfaces than slaty cleavage (Figure 17.16). The larger, more reflective microscopic grains in phyllitic cleavage induce a silky or glossy sheen. Phyllites commonly develop by the recrystallization of slate and therefore from the same protoliths. At temperatures of ~250–300 °C, smectite and illite clays metamorphose to slightly coarser-grained minerals such as sericite, muscovite, and chlorite that align parallel to each other and define foliations. Other common metamorphic minerals include quartz, feldspar, biotite, graphite, prehnite, and pumpellyite. Major minerals are used in the rock name as a preceding adjective; for example, chlorite-sericite phyllite. Phyllitic cleavage develops in response to nonuniform stresses. As with slaty cleavage, the Z maximum shortening direction is perpendicular to the phyllitic cleavage. Many phyllites display a crenulation cleavage that

cross cuts an earlier generation of cleavage, recording multistage deformation.

Despite higher temperature conditions and levels of deformation, relict sedimentary structures can be preserved in phyllites. As temperatures increase, progressively greater amounts of recrystallization and neocrystallization occur so that grain size increases, visible porphyroblasts are produced and relict textures disappear. At temperatures above ~300 °C, phyllites are transformed gradationally into schist (Miyashiro 1973).

As with slates, phyllites present a mass wasting hazard when steeply dipping foliations are inclined downslope as these cleavage surfaces serve as slide horizons.

### 17.4.3 Schist

**Schists** are defined by a foliation called schistosity. Schistosity is a wavy foliation formed by the subparallel arrangement of macroscopic platy phyllosilicate minerals in closely spaced metamorphic layers. This foliation is commonly less regular than that of either slates or phyllites. Light-reflecting, macroscopic crystals generally impart a high sheen or sparkle to the schistosity. Even relatively rigid minerals such as quartz and feldspars are commonly stretched or flattened so as to be parallel to the foliation direction. Less competent minerals such as micas are commonly stretched and folded around more

(a)

(b)

0.25 mm

**Figure 17.16** (a) Photomicrograph depicts dark, phyllosilicate phyllite foliation. *Source*: Photo courtesy of Kent Ratajeski. (b) Field photograph of folded phyllite layers exposed on Rhode Island beach. *Source*: Photo by Kevin Hefferan. © John Wiley & Sons.

rigid porphyroclasts and porphyroblasts. Major minerals are used in the rock name as a preceding adjective; for example, sillimanite-rich schists are called sillimanite schists.

Mineral assemblages in schists depend on both metamorphic conditions and protolith composition. Schists, like slates and phyllites can develop from pelitic (shale, mudstone, graywacke) or altered tuff protoliths. Pelitic schists (Figure 17.17) are enriched in aluminosilicate minerals such as andalusite, kyanite, sillimanite, micas, and other minerals listed in Table 15.4. Schists can also be derived from many other protoliths including a wide variety of igneous rocks. For example, greenschists created from basalt protoliths commonly contain chlorite, hornblende, garnet, actinolite, and/or glaucophane. Mineral content provides key information related to rock protolith, chemistry, and the specific temperature and pressure range under which peak metamorphism occurred.

Schists are produced by dynamothermal metamorphism at convergent plate boundaries with temperatures >300 °C (Miyashiro 1973).

Under such conditions, both brittle and ductile deformation may occur. For example, quartz begins to deform plastically at ~300 °C while feldspar minerals remain rigid. Moderate to high temperatures and nonuniform stresses induce the development of visible porphyroclasts (e.g., feldspars) or porphyroblasts (e.g., garnet and/or staurolite) in a strongly foliated, schistose fabric (Figure 17.18). As a result of extensive deformation, relict textures are rarely preserved in schists.

Gemstones and industrial abrasive minerals such as garnet and corundum provide moderate resource value for some schists. As with other foliated phyllosilicate-rich metamorphic rocks, schists present a mass wasting hazard where foliations are inclined downslope.

Hydrous minerals are stable in schists as they begin to form at temperatures ~300 °C. As temperatures exceed ~400 °C, dehydration processes initiate so that hydrous minerals become unstable (e.g., talc, chlorite, amphibole). As a result, anhydrous minerals increasingly predominate in higher-grade rocks such as gneisses, migmatites, and granulites.

**Figure 17.17** Four varieties of aluminosilicate-rich schists are displayed: (a) Silliminate schist, (b) Muscovite schist, (c) Kyanite schist, (d) Garnet schist. *Source*: Photo by Kevin Hefferan. © John Wiley & Sons.

### 17.4.4 Gneiss

**Gneisses** are characterized by gneissic banding, a foliation characterized by the arrangement of minerals into distinct color bands (Figure 17.19). The color bands are commonly due to alternating light colored quartz and/or feldspar-rich layers and dark colored layers rich in biotite, amphibole, and, at increasing temperatures, pyroxenes. In addition to the major nonferromagnesian and ferromagnesian minerals listed above, accessory minerals include garnet, sillimanite, cordierite, and corundum. Rigid feldspar grains may have been sheared into oval eye shaped (augen) crystals producing augen gneiss (Figure 17.20). Quartz grains have commonly been sheared into lens shaped grains called flaser. Both are testimony to the role of shear in the development of gneissic banding in response to nonuniform stresses.

Gneisses develop from lower grade pelitic metamorphic rocks as well as from a variety of protoliths including granites, granodiorites, diorites, gabbros, and graywackes. The protolith dictates the mineral components in gneiss. Gneisses that develop from an igneous protoliths are called **orthogneiss** whereas those that develop from sedimentary protoliths are called **paragneiss**. Many gneisses with a granitic composition are referred to as granite

**Figure 17.18** Hand sample photograph of a garnet schist. *Source*: Image courtesy of Scott Horvath and the U.S. Geological Survey. https://www.usgs.gov/media/images/garnet-schist. © U.S. Department of the Interior.

**Figure 17.19** Gneissic banding defined by white quartz-feldspar layers and dark ferromagnesium-rich mineral layers. *Source*: Photo by Kevin Hefferan. © John Wiley & Sons.

**Figure 17.20** Augen gneiss with elliptical to lenticular feldspar porphyroblasts exposed in Rocky Mountain National Park, CO. *Source*: Photo by Kevin Hefferan. © John Wiley & Sons.

gneiss. Gneisses derived from gabbro can produce dark, mafic gneissic layers.

Gneissic foliations develop due to extensive layer transposition, recrystallization, and neocrystallization processes that result in segregation of minerals into separate layers. Several processes have been suggested for the development of gneissic banding:

1  Gneissic bands can originate through layer transposition. Transposition results from the pulling apart of earlier folded layers resulting in the separation of hinges and limbs. Insoluble phyllosilicate minerals (e.g., micas) and inosilicate minerals (e.g., amphibole and pyroxene) are concentrated in fold limbs; soluble quartz and feldspars are concentrated in hinge areas of tightly compressed folds resulting in alternating light and dark color bands.

2  Differentiation through partial melting (anatexis) and recrystallization. Ferromagnesian minerals have higher melting temperatures than quartzofeldspathic minerals. As a result, anatexis produces igneous layers enriched in quartz and feldspar between unmelted, refractory layers enriched in dark colored ferromagnesian minerals. Together, they produce gneissic banding.

3  "Lit par lit intrusion" which refers to thin, sill-like intrusion of magma into parallel country rock layers. Lit par lit intrusions can occur to a limited extent when granitic magma intrudes mafic county rock producing alternating light and dark color bands.

Gneisses are used as dimension stones, tiles, countertops, tombstones, and buildings. Because their hardness and crude foliation impede easy splitting, gneisses do not pose a mass-wasting geohazard. Gneisses form in dynamothermal settings at temperatures that commonly exceed ~600 °C. All relict sedimentary textures have been erased due to high temperature neocrystallization processes. However, original igneous textures and structures can be preserved in orthogneisses. At temperatures above ~700 °C, extensive melting of gneisses can lead to the disassembly of gneissic banding, generation of magmatic fluid phases and the formation of migmatites as discussed below.

### 17.4.5  Migmatite

**Migmatites** are "mixed" rocks that possess textural and structural characteristics of both igneous and metamorphic rocks. Whereas gneisses have well defined color bands, migmatites commonly display an irregular, swirling mix of colors (Figure 17.21). Migmatites typically contain zones of rock with the outward appearance of granitoid igneous rocks mixed with zones of rock that resemble typical gneiss. Light-colored segments consist of

**Figure 17.21**  Contorted migmatites exposed in Rocky Mountain National Park, CO. *Source*: Photo by Kevin Hefferan. © John Wiley & Sons.

quartz and feldspars; dark colored components are enriched in pyroxene. With additional melting, migmatites melt sufficiently so as to produce magma.

Migmatites develop under high temperature (>800 °C) conditions in the lower crust. They form via dynamothermal metamorphism at convergent plate boundaries by a number of processes that involve some combination of: (1) partial melting (anatexis); (2) magma injection; and (3) ductile deformation and plastic flow of rocks in the lower crust. Migmatites have no resource or hazard considerations.

**Figure 17.22** Folded, banded ironstones from the Precambrian Lake Superior belt. *Source:* Photo by Kevin Hefferan. © John Wiley & Sons.

### 17.4.6 Ironstones

**Ironstones** are silica and iron-rich rocks that formed primarily in the Early Proterozoic and Archean. Ironstones commonly occur as **banded iron formations (BIF)**, in which alternating hematite, magnetite, and chert layers form red and black color bands. Metamorphosed ironstones can also have gneissic banding with alternating quartz-rich layers and magnetite-rich layers which may or may not be related to the original compositional layering (Figure 17.22). Deposits in the Lake Superior region of North America and in the Hammersley Range of western Australia are among the richest iron deposits on Earth. Ironstones have been major sources of iron ore since the mid-1800s. Taconites, once considered waste rocks because of their lower iron content, are now the principal ore rocks

mined in the Lake Superior region due to depletion of richer ore deposits. Taconites are quartz-rich ironstones that contain 20–30% iron and are also commonly banded.

### 17.4.7 Mylonite

**Mylonites** are fine grained, foliated rocks produced in ductile shear zones of the lower crust and mantle. In ductile shear zones, rocks undergo intense ductile deformation that may involve both crushing and grinding (cataclasis) and plastic flow. Grains become extensively elongated as they rotate toward becoming subparallel to the shear zone. High temperatures and pressures associated with dynamic metamorphism in the lower crust and mantle result in intensely sheared and recrystallized porphyroclasts and/or neocrystallized porphyroblasts.

Mylonites can occur in a variety of rocks such as schist, gneiss, or granite. Mylonites are classified on the basis on the percentage of matrix material to porphyroclasts (Table 17.3). Higher percentages of matrix generally record greater intensity of grain size reduction and more intense shear strains during deformation.

Mylonites provide important information regarding the sense of displacement in shear zones (Figure 17.23). In the following section, we will address some shear sense indicators used in the study of mylonitic rocks. Mylonites have no significant resource or geohazard considerations.

### 17.4.8 Tectonites

A **tectonite** rock is so pervasively deformed that the original mineral composition and texture are largely obliterated. Tectonites are defined by their solid state flow fabric generated through intense ductile or brittle-ductile deformation. Foliated, planar tectonites are called **S tectonites**. Tectonites with a pronounced lineation, but no foliation, are called **L tectonites**. Tectonites with both foliation and a lineation are referred to as **L-S tectonites**.

## 17.5 SHEAR SENSE INDICATORS

Shear sense indicators are rock structures or textural elements that provide information concerning the relative sense of displacement of rock components within fault zones or shear zones of all scales (Simpson 1986). Shear sense indicators include a number of microscopic and megascopic structures discussed below.

### 17.5.1 Grain tail complexes

**Grain tail complexes** are asymmetric porphyroclasts or poryphyroblasts with mineral tails that "point" in the direction of shear. Tail complexes form commonly about minerals such as feldspars and quartz. Mineral tails may form by a combination of:

1 Plastic flattening of preexisting mineral grains;
2 Pressure solution of material from the grain surface;
3 Dynamic recrystallization of material at the rim of the grain;

**Table 17.3** Mylonite series as defined by Sibson (1977).

| Percent matrix | Mylonite series rock name |
| --- | --- |
| 10–50 | Protomylonite |
| 50–90 | Mylonite |
| 90–100 | Ultramylonite |

Note that this terminology follows that used for the cataclasite series in Table 17.2.
*Source*: Modified from Sibson (1977). © John Wiley & Sons.

4 Neocrystallization in pressure shadows around the grain.

Grain tail complexes are particularly well developed in schists, gneisses, and mylonitic rocks. Mylonitic rocks experience intense ductile shearing which favors plastic deformation processes required to create grain tail complexes. Sigma and delta grain tail complexes are distinguished by the relationship between grain tails and an imaginary reference plane of shear that passes through the grain center.

$\sigma$ (**Sigma**) **grain tail complexes** (Figure 17.24) consist of wedge-shaped tails that do not cross the reference plane of shear. Through pressure solution and rotation, mineral material is removed from the outer part of the grain and precipitated in pressure shadows parallel to the minimum stress direction. Incremental tail growth occurs in the direction of minimum compressive stress ($\sigma_3$), located at 45° to the reference plane (and shear plane). Previously formed parts of the tail rotate progressively toward the shear plane while new portions of the tail continue to grow at 45° to the reference plane. During rotational shear, $\sigma$-grain tail complexes develop by slow grain rotation. As tail growth continues, the classic curved sigma tail shape is produced.

$\delta$ (**Delta**) **grain tail complexes** (Figure 17.25) are produced by more rapid grain rotation relative to tail growth rate. Rapid grain rotation results in a significant bending of the earlier

(a)

(b)

**Figure 17.23** (a) Deformed rock displaying protomylonite and ultramylonite layers. (b) Photomicrograph (~4 mm across) of a mylonite with an asymmetric feldspar porphyroclast. *Source*: Photos by Kevin Hefferan. © John Wiley & Sons.

formed parts of the tail so that it crosses the reference plane (Passchier and Trouw 2005).

### 17.5.2 Fracture patterns

Fracture patterns can develop preferred orientations that indicate sense of shear. **Synthetic fractures** are preferentially oriented in the direction of shear. The fractures are inclined at low-angles (<45°) to foliations. Synthetic fractures show displacement consistent with overall sense of shear (Figure 17.26). **Antithetic fractures** are preferentially oriented in a direction opposite to the sense of shear. Antithetic fractures are inclined at high-angles (>45°) to the foliation plane and display an opposite

sense of movement to overall sense of shear (Figure 17.27).

### 17.5.3 S-C foliations

**S-C foliations** develop in mylonitic, schistose, and gneissic rocks subjected to ductile shear. The letter "S" represents schistosity (foliation) and the letter "C" is for **cisaillement**, a French term for shear direction, which lies in the C plane (Figure 17.28).

Figure 17.29 illustrates S-C structures developed within granite mylonite subjected to dextral shear. This granite mylonite is used as floor tile in the Denver International Airport, Colorado. Figure 17.30 illustrates a stunning

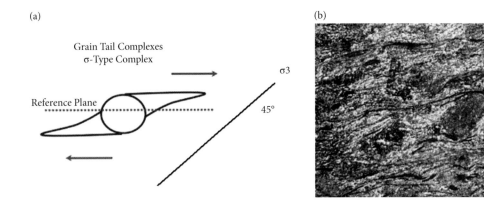

**Figure 17.24**   (a) Cross-section indicating σ grain tail complex in dextral shear zone, oriented 45° to σ3. (b) Photomicrograph if a sigma tail grain complex within mylonitic rock subjected to sinistral shear, Bou Azzer, Morocco. *Source*: Photo by Kevin Hefferan. © John Wiley & Sons.

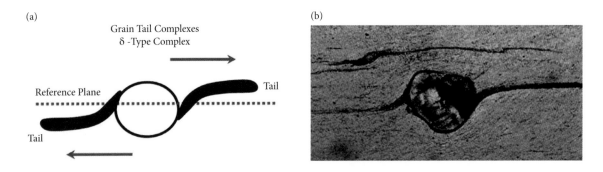

**Figure 17.25**   (a) Cross section illustrating dextral shear as indicated by a delta grain tail complex which crosses the imaginary reference plane. (b) Delta grain tail complex in an ultramylonite that has experienced dextral shear. *Source*: Photo courtesy of Cees Passchier and Rudolph Trouw; with permission of Springer Science+Business Media. © Springer Nature.

**Figure 17.26**   A set of synthetic fractures, all of which are inclined to the right and parallel to the sense of movement, indicate dextral shear in metamorphosed cherts, Bou Azzer, Morocco. *Source*: Photo by Kevin Hefferan. © John Wiley & Sons.

**Figure 17.27**   Outcrop-scale antithetic en echelon, antithetic fractures in metavolcanic rock recording dextral shear in Bou Azzer, Morocco. *Source*: Photo by Kevin Hefferan. © John Wiley & Sons.

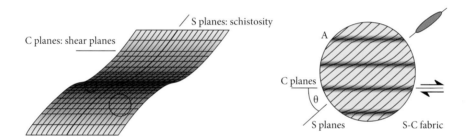

**Figure 17.28**   S-C structures in a dextral shear zone. *Source*: Courtesy of Patrice Rey.

**Figure 17.29**   S-C structures in Proterozoic mylonitic granite from Colorado. This rock is used as floor tiles in the Denver Airport. *Source*: Photo by Kevin Hefferan. © John Wiley & Sons.

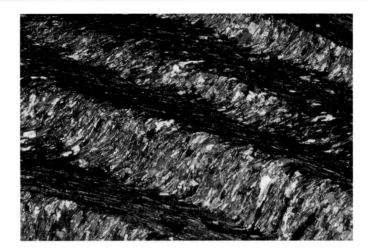

**Figure 17.30**   S-C structures in phyllosilicate-rich mylonite. Note the metamorphic segregation that has produced gneissic layers, wherein ferromagnesian minerals are concentrated in the C planes which represent fold limbs. The light-colored mineral bands of the S plane occur in the S fold hinges. *Source*: Photo courtesy of Cees Passchier and Rudolph Trouw; with permission of Springer Science+Business Media. © Springer Nature.

photomicrograph of S-C structures in which phyllosilicates form the schistosity and dissolution seams occur within the cisaillment zone.

All of these different shear sense indicators can be useful in determining sense of shear in deformed rocks. Only under relatively rare circumstances can displacement be accurately measured. In the following chapter we will focus upon mineral and rock assemblages that develop in response to changes in temperature and pressure.

## CONTENT ASSESSMENT

1   Describe conditions in which a hornfels develops.
2   How does a granoblastic texture differ from a cataclastic texture?
3   What is the economic value of skarns?
4   Why are cataclasites important with respect to geohazards?
5   Explain why greenstones were more common in the Archean and Proterozoic than in the Phanerozoic.
6   Name and describe four common foliated metamorphic rocks.
7   What are the conditions in which mylonites form and how are they classified?

8   Describe two shear sense indicators used in the field analysis of metamorphic rocks.
9   For further interest reading, Dave Waters has an excellent series of weblink resources on thermobarometry at https://www.earth.ox.ac.uk/~davewa/pt/index.html.

## REFERENCES

Bickle, M.J., Nisbet, E.G., and Martin, A. (1994). Archean greenstone belts are not oceanic crust. *Journal of Geology* 102: 121–138.

Chao, E.C.T., Shoemaker, E.M., and Madsen, B.M. (1960). First natural occurrence of coesite from Meteor Crater, Arizona. *Science* 132: 220–222.

de Wit, M.J. (1998). On Archean granites, greenstones, cratons and tectonics: does the evidence demand a verdict? *Precambrian Research* 91: 181–226.

de Wit, M.J. and Ashwal, L.D. (1997). *Greenstone Belts*. Oxford: Clarendon Publishers 836 pp.

Grove, D.I., Archibald, N.J., Bettenay, L.F., and Binns, R.A. (1978). Greenstone belts as ancient marginal basins or ensialic rift zones. *Nature* 273: 460–461.

Hamilton, W. (1998). Archean magmatism and deformation were not products of plate tectonics. *Precambrian Research* 91: 143–179.

Higgins, M.W. (1971). Cataclastic Rocks. United States Geological Survey Professional Paper No. 687.

Hoatson, D.M., Jaireth, S., and Jazues, A.L. (2006). Nickel sulfide deposits in Australia: characteristics, resources and potential. *Ore Geology Reviews* 29: 177–241.

Jefferies, S.P., Holdsworth, R.E., Shimamoto, T. et al. (2006). Origin and mechanical significance of foliated cataclastic rocks in the cores of crustal-scale faults: examples from the Median Tectonic Line, Japan. *Journal of Geophysical Research* 111: B12303. 17 p. doi:https://doi.org/10.1029/2005JB004205.

Levin, S. (2006). *The Earth Through Time*, 8e. New York: Wiley 547 pp.

Miyashiro, A. (1973). *Metamorphism and Metamorphic Belts*. New York: Wiley 479 pp.

O'Brien, P.J. and Rötzler, J. (2003). High-pressure granulites: formation recovery of peak conditions and implications for tectonics. *Journal of Metamorphic Geology* 21: 3–20.

Passchier, C.W. and Trouw, R.A.J. (2005). *Microtectonics*, 2e. New York: Springer-Verlag 366 pp.

Sibson, R.H. (1977). Fault rocks and fault mechanisms. *Geological Society of London Journal* 133: 191–213.

Simpson, C. (1986). Determination of movement sense in mylonites. *Journal of Geological Education* 34: 246–261.

Winter, J.D. (2009). *Principles of Igneous and Metamorphic Petrology*, 2e. Upper Saddle River, NJ: Prentice Hall 720 pp.

# Chapter 18

# Metamorphic zones, facies, and facies series

Metamorphic mineral assemblages effectively serve as geothermobarometers, accurately recording elevated temperature and pressure conditions of metamorphism. Ideally, metamorphic reactions proceed slowly over time to produce minerals in equilibrium with the temperature and pressure conditions of metamorphism. In addition to protolith composition, key factors in determining metamorphic mineral assemblages are temperatue, pressure, and volatiles as described below.

1   Temperatures (T) of metamorphism vary from as little as 100 °C to over 1200 °C. This range can be broadly subdivided into low temperature (~100–400 °C), moderate temperature (400–600 °C), and high temperature (>600 °C) conditions (Fyfe et al. 1958).

2   Pressures (P) of metamorphism vary from near atmospheric conditions in some contact metamorphic reactions to extremely high pressures in lower crust, mantle, and subduction zone environments. This pressure range can be broadly subdivided into low pressure (0–2 kbars ≈ 0–6 km depth), moderate pressure (2–6 kbars), or high pressure (>6 kbars ≈ 18 km depth). Well into the second half of the twentieth century, metamorphic petrologists considered the highest metamorphic pressures to be ~15 kbars (Fyfe et al. 1958). Remember that 3.3 kbars (330 MPa) corresponds to ~10 km depth within the Earth, so the maximum depth of formation of metamorphic rocks exposed at the surface was considered to be 45–50 km. With a greater understanding of convergent margins and the discovery of ultrahigh-pressure rocks, geologists now

*Earth Materials*, Second Edition. Kevin Hefferan and John O'Brien.
© 2022 John Wiley & Sons Ltd. Published 2022 by John Wiley & Sons Ltd.
Companion website: www.wiley.com/go/hefferan/earthmaterials2

recognize that metamorphic pressures may exceed 30 kbars (3 GPa) which is equivalent to depths >100 km. In fact, some ultrahigh-pressure xenoliths form at depths ~400 km (Pearson et al. 2003).

3 Volatiles serve as catalysts in hydrothermal and metasomatic reactions, thereby altering mineral composition and in some cases, creating ore deposits. Without volatiles, metamorphic reaction kinetics slow or cease. This explains why anhydrous high T and P assemblages return to Earth's surface relatively unchanged. The absence of volatile catalysts inhibits retrograde metamorphic reactions.

Despite the wide variety of possible mineral components, most metamorphic rocks contain one to six essential minerals. The mineral assemblages provide critical information regarding the protolith and the conditions of metamorphism. In the following discussion, we will consider key minerals and mineral assemblages that provide constraints on the conditions of metamorphism. We will also introduce some chemical reactions by which unstable minerals are converted into minerals stable under new temperature/pressure conditions. It is on the basis of our knowledge of these reactions that we can infer the increasing temperature and/or pressure conditions of progressive metamorphism or, less commonly, the decreasing temperature and/or pressure of retrograde metamorphism. A historical perspective of metamorphic petrology involves the concept of metamorphic zones and facies, a discussion of which follows.

## 18.1 METAMORPHIC ZONES

Since the late nineteenth century, geologists have recognized that some metamorphic minerals crystallize in, and effectively record, specific ranges of temperature and pressure. Minerals such as quartz and calcite, stable over a wide range of temperature and pressure, are of little value in this regard. However, other minerals crystallize within a more limited temperature and/or pressure range. These **index minerals** provide critical information because they approximate indicate the T–P conditions of metamorphism. Using index minerals, George Barrow (1893, 1912) and C.E.

Tilley (1925) pioneered research in regional metamorphism by field mapping the appearance, and disappearance, of index minerals in metamorphosed pelitic (mudstone) rocks exposed in the Scottish Highlands. The index minerals used by Barrow and Tilley were, in order of increasing temperature: chlorite, biotite, almandine garnet, staurolite, kyanite, and sillimanite. George Barrow defined isograds and metamorphic zones mapping the field locations of these six index minerals (Figure 18.1). **Isograds** are lines drawn on geologic maps that mark the first appearance of a particular index mineral. For example, the chlorite isograd marks the first appearance of chlorite and the biotite isograd marks the first appearance of biotite.

A **metamorphic zone** denotes a region marked by the appearance or disappearance of index minerals and is bound by two isograd lines.

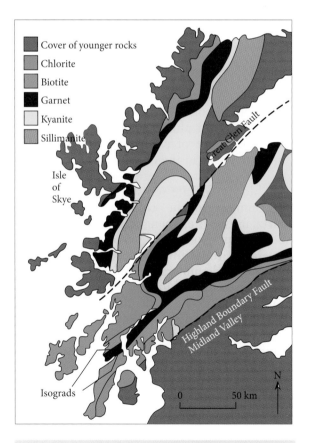

**Figure 18.1** Barrovian zones of metamorphism based on index minerals of Scotland's Highlands. Note that the staurolite zone is not shown in this figure. *Source*: After Tilley (1925) and Murck et al. (2010). © John Wiley & Sons.

For example, the chlorite zone (Figure 18.1) is bound on one side by the chlorite isograd that marks where chlorite first appears and on the other side by the biotite isograd which marks the higher temperature conditions where biotite first appears. The biotite isograd marks the transition to the higher grade biotite zone (in which chlorite may still occur). Barrow, Tilley, and others recognized that isograds record important changes in temperature during metamorphism.

The six metamorphic zones (Figure 18.1) based on index mineral isograds are called **Barrovian zones** in honor of Barrow's insightful work. In essence, Barrow discovered a key to interpreting progressive metamorphism whereby progressively higher grades of metamorphic minerals and rocks are produced with increasing temperature in an orogenic belt. The Barrovian zones developed by Barrow (1912), Tilley (1925), and Harker (1932) are discussed below, from lowest grade to highest grade of metamorphism.

### 18.1.1 Chlorite zone

The **chlorite zone** is bound by the chlorite and biotite isograds. The chlorite isograd records metamorphic grades sufficient for the first appearance of chlorite group minerals. Common rocks in the chlorite zone are chlorite-bearing slate, chlorite–sericite phyllite, and chlorite–sericite schist. Accessory minerals common to the chlorite zone include quartz, muscovite, phengite, K feldspar, albite (sodium plagioclase), pyrophyllite, stilpnomelane, tournamline, zircon, monazite, and iron oxide minerals. Chlorite group minerals such as clinochlore and chamosite commonly form by the hydrothermal alteration of ferromagnesium minerals. Chlorite becomes unstable and begins to be replaced by biotite at the high temperature limit of the chlorite zone. Chlorite to biotite transition occurs over a range of temperatures and pressures by a number of different chemical reactions which include the following (Best 2003):

$$Chlorite + K\ feldspar => biotite + water$$

$$Ankerite + muscovite => biotite + calcite$$

$$Phengite + chlorite => biotite + muscovite \\ + quartz + water$$

### 18.1.2 Biotite zone

The **biotite zone** occurs between the biotite isograd, marking the first appearance of biotite, and the almandine garnet isograd. Common rocks in the biotite zone include sericite–biotite phyllite and biotite schist. Other minerals common to pelitic rocks in the biotite zone include quartz, sodium plagioclase, chlorite, and muscovite. Biotite forms by chemical reactions that transform minerals such as chlorite, muscovite, quartz, magnetite, and rutile. Note that lower-temperature index minerals are not confined to the Barrovian zone that bears their name. Index minerals such as chlorite, biotite, and almandine garnet (which we will discuss next) persist in higher temperature zones. Why is this so? The lower-grade minerals persist because the transformation from chlorite to biotite, or biotite to iron rich almandine garnet is not attained at a single temperature/pressure, nor by a simple transformation of one index mineral into another; instead, multiple chemical reactions can occur incrementally over a temperature range producing several coexisting minerals. Almandine garnet producing chemical reaction include the following:

$$Chlorite + muscovite => almandine\ garnet \\ + Mg - chlorite + biotite + water\ (Best\ 2003)$$

### 18.1.3 Almandine (garnet) zone

The **almandine garnet zone** occurs between the almandine garnet isograd, marking the first appearance of almandine garnet, and the staurolite isograd. Common rocks in the almandine garnet zone include chlorite-garnet schist or garnet-mica schist. Garnet commonly occurs as porphyroblasts. Accessory minerals common in pelitic rocks of the almandine zone include chlorite, biotite, muscovite, magnetite, quartz, and sodium plagioclase minerals such as albite or oligoclase. Almandine forms through the chemical transformation of minerals such as:

$$Chlorite + magnetite + quartz => \\ almandine\ garnet + water$$

As temperatures increase and approach the staurolite isograd, almandine garnet begins to alter to staurolite.

## 18.1.4 Staurolite zone

The **staurolite zone** lies between the staurolite isograd, marking the first appearance of the higher temperature mineral staurolite, and the kyanite isograd. Staurolite-mica schist and staurolite-garnet-mica schist are common rock types in this zone. Accessory minerals common to the staurolite zone include quartz, almandine, potassium feldspar, biotite, and muscovite. Potassium feldspar forms through the alteration of muscovite. Staurolite forms through the chemical transformation of minerals such as almandine garnet, chlorite, and muscovite as indicated by the reaction below:

$$Fe_3Al_2(SiO_4)_3 + (Mg,Fe,Al)_6(Si,Al)_4O_{10}(OH)_2$$
Almandine        Chlorite

$$+ KAl_2(AlSi_3O_{10})(OH,F)_2$$
Muscovite

$$=> Fe_2Al_9O_6(SiO_4)_4(O,OH)_2$$
Staurolite

$$+ K(Mg,Fe)_3(AlSi_3O_{10})(OH,F)_2 + SiO_2$$
Biotite        Quartz

$$+ H_2O \quad (Philpotts\,1990)$$
Water

Unlike the lower temperature index minerals (chlorite, biotite, and almandine garnet which exist in multiple zones), staurolite exists only within the staurolite zone. Similarly, the other higher grade index minerals such as kyanite and sillimanite, also only occur as stable minerals in their namesake zone.

## 18.1.5 Kyanite zone

The **kyanite zone** exists between the kyanite and sillimanite isograds that mark the first appearances of kyanite and sillimanite, respectively. Kyanite is the high-pressure member of the $Al_2SiO_5$ (otherwise written as $AlAlOSiO_4$) polymorph series andalusite, kyanite, and sillimanite. Common rock types include kyanite schist and kyanite-mica schist. Other minerals commonly occurring in pelitic rocks of the kyanite zone include biotite, muscovite, almandine garnet, cordierite, and quartz. Kyanite forms by transformation of aluminosilicate minerals such as andalusite (a low-pressure polymorph) or through a dehydration

reaction involving staurolite or pyrophyllite as indicated below:

$$6Fe_2Al_9O_6(SiO_4)(O,OH)_2 + 11SiO_2$$
Staurolite        Quartz

$$=> 23Al_2SiO_5 + 4Fe_3Al_2(SiO_4)_3 + 3H_2O$$
Kyanite     Almandine     Water

$$2Al_2Si_4O_{10}(OH)_2 => Al_2SiO_5 + 3SiO_2$$
Pyrophyllite     Kyanite     Quartz

$$+ H_2O \quad (Haas\ and\ Holdaway\,1973)$$
Water

## Sillimanite zone

The **sillimanite zone** occurs inside the sillimanite isograd and marks the highest temperature zone defined by Barrow and Tilley (Figure 18.1). Common rock types include sillimanite schist, sillimanite gneiss, and cordierite gneiss. Sillimanite is the high temperature $Al_2SiO_5$ polymorph mineral. Other minerals that commonly occur in the sillimanite zone include biotite, muscovite, cordierite, quartz, oligoclase, and orthoclase. Sillimanite, and K feldspar can also develop by dehydration of muscovite in the presence of quartz as in the reaction:

$$KAl_2(AlSi_3O_{10})(OH,F)_2 + SiO_2$$
Muscovite        Quartz

$$=> K(AlSi_3O_8) + Al_2SiO_5 + H_2O$$
K feldspar    Sillimanite    Water

(Philpotts and Ague 2009)

The early mapping of Barrow (1893, 1912), Tilley (1925), and others focused on pelitic (mudstone) rocks of Scotland. Subsequently, Barrovian zones have been recognized in many orogenic belts that contain both pelitic and nonpelitic assemblages (Harker 1932; Wiseman 1934; Kennedy 1949). Geologists now recognize approximate equivalents of Barrovian zones in nonpelitic rocks. Table 18.1 presents a list of both pelitic and nonpelitic metamorphic protoliths and common index minerals associated with Barrovian zones.

Barrovian zones and metamorphic isograds remain in use today. However, metamorphic zones are less useful in nonpelitic rocks, or in rocks that form in subducting plates or contact metamorphic environments. How can we infer the conditions that produced metamorphic rocks of all compositions and environments?

**Table 18.1** Barrovian zones in pelitic, calcareous, and mafic protoliths.

| Zone | Pelitic protolith | Calcareous protolith | Mafic protolith |
|---|---|---|---|
| Chlorite | Chlorite, muscovite, quartz, pyrophyllite, albite, graphite | Calcite, albite, biotite, zoisite, quartz | Chlorite, albite, epidote, sphene, calcite, actinolite |
| Biotite | Biotite, muscovite, chlorite, quartz | | |
| Almandine | Almandine, biotite, magnetite, muscovite, quartz | Garnet, andesine, zoisite, biotite | |
| Staurolite | Staurolite, biotite, muscovite, almandine, quartz | Bytownite, anorthite, diopside, garnet | Hornblende, plagioclase, epidote, almandine, diopside |
| Kyanite | Kyanite, biotite, muscovite, almandine, quartz | | |
| Sillimanite | Sillimanite, biotite, cordierite, muscovite, almandine, quartz, oligoclase, orthoclase | Bytownite, anorthite, diopside, garnet | |

*Source*: After Harker (1932) and Turner (1958). © John Wiley & Sons.

Eskola (1915, 1920, 1939) introduced the concept of metamorphic facies, a more comprehensive approach to assessing the conditions recorded by metamorphic rocks. Mapping in Finland, Eskola noticed that certain metamorphic rock and mineral assemblages of various compositions recurred together in time and space, which he called "facies." Metamorphic facies are defined by a group or assemblage of critical minerals, rather than a single index mineral as used for Barrovian zones. While Barrovian zones were defined for pelitic rocks, metamorphic facies were named for common mafic rocks and minerals.

## 18.2 METAMORPHIC FACIES

**Metamorphic facies** are distinctive mineral assemblages in metamorphic rocks that form in response to a particular range of temperature and/or pressure conditions. Eskola (1920) initially identified five metamorphic facies: sanidinite, hornfels, greenschist, amphibolite, and eclogite. Additional facies proposed since Eskola's early work included the blueschist and granulite facies (Eskola 1939; Turner 1958), the zeolite facies, (Turner 1958) and prehnite–pumpellyite facies (Coombs 1961). The original hornfels facies proposed by Eskola (1920) was expanded by Eskola (1939) and Fyfe et al. (1958) into four facies that include in order of increasing temperature: (1) albite–

epidote hornfels, (2) hornblende hornfels, (3) pyroxene hornfels, and (4) sanidinite hornfels. In addition to the four hornfels facies, the other seven widely recognized metamorphic facies include the: (1) zeolite, (2) prehnite–pumpellyite, (3) greenschist, (4) amphibolite, (5) granulite, (6) blueschist, and (7) eclogite facies (Figure 18.2).

We will begin our discussion with metamorphic facies that contain low pressure mineral assemblages associated with contact metamorphism and shallow dynamothermal metamorphism. Then we will consider higher pressure and temperature mineral assemblages associated with burial and deeper dynamothermal metamorphism.

### 18.2.1 Hornfels facies

**Hornfels facies** include nonfoliated, fine grained hornfels rocks, and coarser grained rocks with granoblastic textures. Hornfels facies form by heat induced metamorphism in aureoles surrounding igneous intrusions. The contact metamorphic aureoles are localized around the intrusion, commonly having widths of 100 m or less. Ocean spreading centers represent regionally extensive zones of hydrothermal metamorphism which produces hornfels facies rocks on a large scale. Sea floor spreading of hornfels facies rocks effectively encompass entire ocean basins as these altered rocks move away from ocean spreading ridges over time.

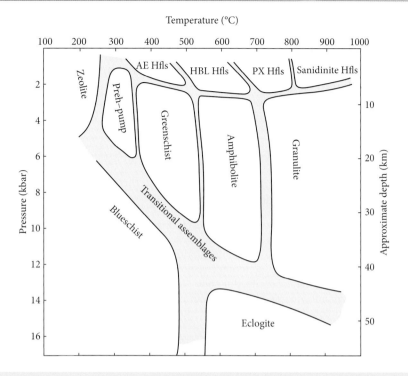

**Figure 18.2** Common metamorphic facies depicted on a temperature and depth/pressure graph. Hfls, hornfels; Preh–pump, prehnite–pumpellyite. *Source*: Yardley 1989.

The four types of hornfels facies (Figure 18.2) are defined based on their nonfoliated textures and critical minerals. Note that these critical minerals are not unique to the hornfels facies. These minerals also occur in greenschist, amphibolite, and granulite facies rocks of equivalent temperature but higher pressure conditions of dynamothermal metamorphism which tend to produce foliated metamorphic rocks. The combination of nonfoliated textures and the distinctive mineral assemblages define the (1) albite–epidote hornfels facies, (2) hornblende hornfels facies, (3) pyroxene hornfels facies, and (4) sanidinite hornfels facies.

The **albite–epidote hornfels facies** is the low temperature hornfels facies, with temperatures generally <450 °C and pressures <2 kbar (depth < 6 km). Albite–epidote hornfels minerals develop in the outer fringes of many metamorphic aureoles, farthest away from the magmatic heat source. The characteristic minerals of this assemblage are albite and epidote which most commonly occur in metabasaltic tuffs and lavas thermally metamorphosed at ocean ridges, hot spots, and in volcanic–magmatic arcs. The minerals and rocks associated with the albite–epidote hornfels facies depend upon the protolith composition as indicated in Table 18.2. The albite–epidote hornfels facies is roughly the low pressure equivalent of the greenschist facies discussed later in this chapter.

**Hornblende hornfels facies** rocks compose the bulk of many metamorphic aureoles, forming at temperatures generally between 450 and 600 °C and at pressures <2.5 kbar (<8 km). The minerals and rocks associated with the hornblende hornfels facies depend upon the protolith composition as indicated in Table 18.3. Chlorite, albite, epidote, and actinolite which are common in albite–epidote hornfels facies, are notable by their absence in the hornblende hornfels assemblage. At temperatures above 450°, dolomite is replaced by diopside via the following chemical reaction:

$$\underset{\text{Dolomite}}{CaMg(CO_3)_2} + \underset{\text{Quartz}}{2SiO_2} => \underset{\text{Diopside}}{CaMgSi_2O_6}$$
$$+ \underset{\text{Carbon dioxide}}{2CO_2} \quad (Turner\,1981).$$

The hornblende hornfels facies may be loosely considered as the low pressure equivalent of the amphibolite facies, discussed later in this chapter.

**Pyroxene hornfels facies** are less common than the lower temperature hornfels facies previously discussed because they require very high temperatures not commonly attained by granitoid magmas. Where they occur, commonly adjacent to higher temperature mafic intrusions, pyroxene hornfels facies rocks develop at temperatures of 600–800 °C and pressures <2.5 kbar (<8 km). The minerals and rocks associated with the pyroxene hornfels facies depend upon the protolith composition (Table 18.4). Note that with the exception of biotite, hydrous minerals do not occur. At the higher temperatures

**Table 18.2** Mineral assemblages and rocks in albite–epidote hornfels facies related to protolith composition.

| Protolith | Common albite–epidote facies minerals | Common rocks |
|---|---|---|
| Pelitic | Chloritoid, albite, epidote, muscovite, chlorite, biotite, andalusite, quartz | Hornfels |
| Mafic | Epidote, albite, chlorite, actinolite, biotite, talc, sphene | Hornfels |
| Ultramafic | Antigorite (serpentine), actinolite, tremolite, talc, chlorite, biotite, albite, magnesite, brucite, dolomite | Hornfels |
| Quartz-feldspathic | Albite, epidote, quartz, microcline, muscovite, biotite | Hornfels, metaquartzite |
| Calcareous | Calcite, dolomite, epidote, tremolite, idocrase (vesuvianite), magnesite, brucite | Marble, skarn |

*Source*: Turner and Verhoogen (1951) and Turner (1958). © John Wiley & Sons.

**Table 18.3** Mineral assemblages and rocks in hornblende hornfels related to protolith composition.

| Protolith | Common hornblende hornfels facies minerals | Common rocks |
|---|---|---|
| Pelitic | Andalusite, cordierite, anthophyllite, plagioclase, muscovite, biotite, quartz, microcline | Hornfels |
| Mafic | Hornblende, plagioclase, biotite, diopside | Hornfels |
| Ultramafic | Talc, forsterite, cummingtonite, grunerite, hornblende, lizardite, anthophyllite, plagioclase, diopside, tremolite, clinochlore, brucite, spinel, bronzite | Hornfels |
| Quartz-feldspathic | Hornblende, microcline, quartz, plagioclase, muscovite, biotite | Hornfels Metaquartzite |
| Calcareous | Calcite, diopside, tremolite, calcium plagioclase, wollastonite, grossular garnet, idocrase (vesuvianite), brucite | Marble, skarn |

*Source*: Turner and Verhoogen (1951) and Turner (1958). © John Wiley & Sons.

**Table 18.4** Mineral assemblages and rocks in pyroxene hornfels related to protolith composition.

| Protolith | Common pyroxene hornfels facies minerals | Common rocks |
|---|---|---|
| Pelitic | Cordierite, andalusite or sillimanite, plagioclase, orthoclase, quartz, biotite, corundum, spinel, hypersthene | Hornfels |
| Mafic | Hypersthene, diopside, cordierite, plagioclase, biotite, augite | Hornfels |
| Ultramafic | Forsterite, hypersthene, diopside, augite, grossular garnet, enstatite | Hornfels |
| Quartz-feldspathic | Augite, hypersthene, diopside, orthoclase, andalusite or sillimanite, quartz, plagioclase, biotite | Hornfels Metaquartzite |
| Calcareous | Diopside, wollastonite, calcium plagioclase (anorthite, bytownite), grossular garnet, forsterite, periclase | Marble, skarn |

*Source*: Turner and Verhoogen (1951) and Turner (1958). © John Wiley & Sons.

at which pyroxene hornfels form, dehydration reactions produce a largely anhydrous suite of minerals. The pyroxene hornfels facies is roughly the low pressure equivalent of the granulite facies, discussed later in this chapter.

**Sanidinite hornfels facies** rocks are very rare, forming in very high temperature (>800 °C) and low pressure (<2.5 kbar ≈ <8 km) conditions in association with mafic and ultramafic intrusions. Sanidinite hornfels develop where the country rock is in contact with the intrusion or in country rock inclusions (xenoliths) within the intrusion. The minerals and rocks associated with the sanidinite hornfels facies depend upon the protolith composition as indicated in Table 18.5. Note the absence of hydrous minerals, the result of dehydration reactions such as the following:

$$KMg_3AlSi_3O_{10}(OH)_2 + 3SiO_2$$
$$\underset{\text{Phlogopite}}{} \qquad \underset{\text{Quartz}}{}$$

$$=> K(AlSi_3O_8) + MgSiO_3 + H_2O$$
$$\underset{\text{Sanidine}}{} \quad \underset{\text{Enstatite}}{} \quad \underset{\text{Water}}{}$$

(Wones and Dodge 1968)

The contact metamorphic facies described above record temperature driven mineral changes as a result of igneous intrusions into country rock, or hydrothermal metamorphism related to shallow magmatic activity as occurs along oceanic ridge systems. Contact metamorphism can occur anywhere within Earth's crust, but predominates in the upper 5-10 km where pressures are relatively low.

The next two facies we will discuss, zeolite and prehnite–pumpellyite, contain low to moderate temperature and pressure mineral assemblages (Figure 18.2) and are commonly produced by burial metamorphism in thick sedimentary basins. Zeolite and prehnite–pumpellyite minerals can also form at ocean ridges, hot spots, and in association with volcanic arc complexes.

### 18.2.2 Zeolite facies

Turner (1958) defined the **zeolite facies** as a low grade metamorphic facies produced by temperatures between ~150 and 300 °C and pressures less than 5 kbar (500 MPa ~ 15 km depth). The zeolite facies is named after the zeolite mineral group (Table 18.6). Zeolites are a hydrous sodium and calcium aluminum tectosilicate mineral group formed by diagenetic or low temperature metamorphic reactions.

Critical zeolite facies minerals, which commonly co-exist with quartz, include analcime, laumontite, heulandite, and wairakite. Accessory minerals in the zeolite facies may include: albite, kaolinite, vermiculite, adularia, pumpellyite, sphene, epidote, prehnite, montmorillonite, smectite, muscovite, chlorite, and calcite (Turner 1958; Coombs 1961). The minerals and rocks associated with the zeolite facies depend upon the protolith composition as indicated in Table 18.7.

Zeolite facies minerals originate from the hydrothermal alteration of volcanic protoliths such as basalt and andesite, the devitrification of basaltic glass and tuff, and the reaction of pelites and graywackes with saline waters. Zeolite minerals occur in isolated vesicles and veins or as pervasively disseminated minerals within the metamorphic rock. The low temperature and low pressure conditions of zeolite facies metamorphism commonly preserve relict structures inherited from the protolith.

Turner (1958) created the zeolite facies based upon D.S. Coomb's (1954) geologic mapping of metamorphosed volcanic graywackes in Southland, New Zealand. Southland's metagraywakes

**Table 18.5**  Mineral assemblages and rocks in sanidinite hornfels related to composition.

| Protolith | Common sanidinite hornfels facies minerals | Common rocks |
|---|---|---|
| Pelitic | Cordierite, sillimanite, corundum, sanidine, calcium plagioclase (anorthite, bytownite), tridymite. | Hornfels |
| Mafic | Hypersthene, pigeonite, augite, enstatite, forsterite, magnetite, ilmenite, hematite, diopside, plagioclase, corundum, spinel | Hornfels |
| Ultramafic | Forsterite, diopside, pigeonite, enstatite, hypersthene, cordierite, augite, spinel, ilmenite, magnetite | Hornfels |
| Quartz-feldspathic | Sanidine, tridymite, calcium plagioclase (anorthite, bytownite), diopside, hypersthene, pigeonite | Hornfels Metaquartzite |
| Calcareous | Wollastonite, diopside, calcium plagioclase (anorthite, bytownite) | Marble, skarn |

*Source*: Turner and Verhoogen (1951) and Turner (1958). © John Wiley & Sons.

**Table 18.6** Zeolite minerals and chemical formulas.

| Zeolite mineral | Chemical formula | Occurrence |
|---|---|---|
| Natrolite | $Na_2Al_2Si_3O_{10} \cdot 2H_2O$ | Diagenesis and zeolite facies |
| Stilbite | $NaCa_2Al_5Si_{13}O_{36} \cdot 14H_2O$ | Diagenesis and zeolite facies |
| Chabazite | $Ca_2Al_2Si_4O_{12} \cdot 14H_2O$ | Diagenesis and zeolite facies |
| Heulandite | $CaAl_2Si_7O_{18} \cdot 6H_2O$ | Diagenesis and zeolite facies |
| Thompsonite | $NaCa_2(Al_5Si_5)O_{20} \cdot 6H_2O$ | Zeolite facies |
| Analcime | $Na(AlSi_2)O_6 \cdot H_2O$ | Zeolite facies |
| Laumontite | $Ca(Al_2Si_4)O_{12} \cdot 4H_2O$ | Zeolite facies |
| Wairakite | $Ca(Al_2Si_4)O_{12} \cdot 2H_2O$ | Zeolite facies |

**Table 18.7** Common minerals and protoliths of the zeolite facies.

| Protolith | Common zeolite facies minerals | Common zeolite facies rocks |
|---|---|---|
| Pelitic | Kaolinite, zeolite, quartz, montmorillonite, vermiculite, phengite, epidote, muscovite | Metapelite, argillite, slate |
| Quartzo feldspathic | Quartz, zeolite, albite, sphene, epidote, quartz, muscovite | Metaquartzite or metagraywacke |
| Mafic | Zeolite, albite, quartz, phengite, sphene, epidote, chlorite, prehnite, pumpellyite | Metabasite or greenstone |
| Ultramafic | Lizardite serpentine, talc, olivine, chlorite, prehnite, pumpellyite | Serpentinite or greenstone |
| Calcareous | Zeolite, calcite, quartz, epidote, dolomite, lawsonite, talc, muscovite | Marble |

contain andesitic volcanic lapilli and glass particles affected by low temperature diagenesis and zeolite facies metamorphism. Metagraywacke rocks subjected to low temperature diagenetic alteration are enriched in stilbite and heulandite. As temperatures and pressures increase within the zeolite facies, stilbite and heulandite are replaced by laumontite and wairakite:

Stilbite => heulandite

=> laumontite => wairakite

(Miyashiro and Shido 1970).

$$CaAl_2Si_7O_{18} \cdot 6H_2O$$
$$\text{Heulandite}$$

$$=> Ca(Al_2Si_4)O_{12} \cdot 4H_2O + 3SiO_2 + H_2O$$
$$\text{Laumontite} \qquad \text{Quartz} \quad \text{Water}$$

(Coombs et al. 1959)

$$Ca(Al_2Si_4)O_{12} \cdot 4H_2O$$
$$\text{Laumontite}$$

$$=> Ca(Al_2Si_4)O_{12} \cdot 2H_2O + 2H_2O$$
$$\text{Wairakite} \qquad \text{Water}$$

(Bird and Helgeson 1981)

Zeolite facies metamorphism develops by hydrothermal alteration at divergent margins, hot spots, and convergent margins or during burial metamorphism at depths less than 5 km. Laumontite and heulandite are particularly common in the zeolite facies. At temperatures approaching 250 °C and depths of 3–5 km, zeolite facies minerals begin to alter to prehnite and pumpellyite facies minerals which occur in the upper zeolite facies but are the dominant minerals in the prehnite–pumpellyite facies, discussed below.

### 18.2.3 Prehnite–pumpellyite facies

Coombs (1961) defined the **prehnite–pumpellyite facies** based on metasedimentary basin deposits in New Zealand. Prehnite–pumpellyite facies minerals are produced by hydrothermal alteration and burial metamorphism at temperatures and pressures that exceed zeolite facies conditions. Zeolite facies and prehnite–pumpellyite facies rocks formed by hydrothermal alteration are widespread at oceanic ridges and therefore affect the oceanic crust generated at spreading ridges. Although any protolith can be metamorphosed to prehnite–pumpellyite facies, the most common protoliths include basalt, graywackes, and mudstones (pelites).

The prehnite–pumpellyite facies generally forms under low temperature (250–350 °C) and fairly low pressure (<6 kbar or ~20 km depth) conditions. In addition to prehnite and

**Table 18.8**  Common minerals and protoliths of the prehnite–pumpellyite facies.

| Protolith | Common prehnite–pumpellyite facies minerals | Common prehnite–pumpellyite facies rocks |
| --- | --- | --- |
| Pelitic | Chlorite, quartz, illite, phengite, smectite, sphene, titanite, epidote, stilpnomelane | Slate or phyllite |
| Quartzofeldspathic | Quartz, epidote, feldspar | Metaquartzite or metagraywacke |
| Mafic | Prehnite, pumpellyite, albite, quartz, phengite, sphene, titanite, epidote, chlorite, actinolite, zeolite | Metabasite or greenstone |
| Ultramafic | Lizardite serpentine, talc, forsterite, tremolite, chlorite, zeolite | Serpentinite, soapstone, or greenstone |
| Calcareous | Prehnite, calcite, quartz, epidote, dolomite, lawsonite, talc, muscovite, zeolite | Marble |

pumpellyite, other common minerals include: quartz, albite, chlorite, muscovite, illite, phengite, smectite, sphene, titanite, epidote, lawsonite, and stilpnomelane (Table 18.8). Although some zeolite minerals occur in the prehnite–pumpellyite facies, laumontite, and heulandite are restricted to the zeolite facies. The minerals and rocks associated with the prehnite–pumpellyite facies depend upon the protolith composition as indicated in Table 18.8.

Due to the relatively low temperature and low pressure conditions, prehnite–pumpellyite facies rocks commonly retain relict textures and structures. As with zeolite facies rocks, prehnite–pumpellyite facies rocks are commonly referred to by referencing the protolith as in metabasites, metagraywackes, and metapelites. Higher temperature alteration of prehnite and pumpellyite results in the neocrystallization of actinolite and epidote, two minerals that mark the transition to the higher grade albite–epidote hornfels facies discussed previously and the greenschist facies discussed below. The higher temperature assemblage containing pumpellyite and actinolite has been called the transitional pumpellyite–actinolite facies (Hashimoto 1966; Turner 1981).

The facies discussed in the following sections largely form in response to higher temperatures and pressures than those that produce the zeolite and prehnite–pumpellyite facies (Figure 18.2). They are typically produced by dynamothermal metamorphism under conditions of nonuniform stress, particularly at convergent plate boundaries.

### 18.2.4 Greenschist facies

**Greenschist facies** assemblages form under medium temperature (350–550 °C) and pressure (3–10 kbar = 0.3 – 1.0 GPa ≈ 10–30 km depth)

conditions associated with dynamothermal metamorphism at convergent plate boundaries. At these higher temperature and pressure conditions, pervasive recrystallization and/or neocrystallization may result in the obliteration of relict textures. Greenschist facies metamorphism can affect rocks of any composition, but the most common protoliths include mafic and ultramafic igneous rocks, tuff, sandstones, mudrocks, and limestone. Key minerals in the greenschist facies include epidote, chlorite, and actinolite (green amphibole). These minerals impart the green color to both foliated (schistose) greenschist rocks and nonfoliated greenstone rocks in this facies. Minerals and rocks common to the greenschist facies are listed in Table 18.9.

Increasing temperatures associated with greenschist facies metamorphism results in the liberation of volatile components such as $CO_2$ and $H_2O$, as shown in the chemical reactions below, which serve as flux catalysts driving igneous and metamorphic processes at convergent plate boundaries.

$$\underset{\text{Kaolinite}}{Al_4\left(Si_4O_{10}\right)\left(OH\right)_8} + \underset{\text{Quartz}}{4SiO_2}$$

$$\Rightarrow \underset{\text{Pyrophyllite}}{2Al_2Si_4O_{10}\left(OH\right)_2} + \underset{\text{Water}}{2H_2O}$$

(Hower et al. 1976)

$$\underset{\text{Dolomite}}{3CaMg\left(CO_3\right)_2} + \underset{\text{Quartz}}{4SiO_2} + \underset{\text{Water}}{H_2O}$$

$$\Rightarrow \underset{\text{Tremolite}}{Ca_2Mg_5Si_8O_{22}\left(OH\right)_2} + \underset{\text{Calcite}}{3CaCO_3} + \underset{\text{Carbon dioxide}}{7CO_2}$$

(Metz and Winkler 1963)

$$\underset{\text{Dolomite}}{5CaMg\left(CO_3\right)_2} + \underset{\text{Quartz}}{8SiO_2} + \underset{\text{Water}}{H_2O}$$

$$\Rightarrow \underset{\text{Talc}}{Mg_3\left(Si_4O_{10}\right)\left(OH\right)_2} + \underset{\text{Calcite}}{3CaCO_3} + \underset{\text{Carbon dioxide}}{3CO_2}$$

(Metz and Winkler 1963)

**Table 18.9** Common minerals and protoliths of the greenschist facies.

| Protolith | Common greenschist facies minerals | Common greenschist facies rocks |
| --- | --- | --- |
| Pelitic | Pyrophyllite, andalusite, kyanite, quartz, albite, epidote, chlorite, chloritoid, biotite, phengite, muscovite, garnet (almandine, spessartine), chloritoid, sphene, tourmaline | Slate, phyllite or schist |
| Quartz feldspathic | Quartz, albite, muscovite, epidote, chlorite, | Metaquartzite |
| Mafic | Actinolite, albite, epidote, chlorite, quartz, biotite, magnetite, sphene, stilpnomelane, calcite | Greenschist or greenstone |
| Ultramafic | Antigorite serpentine, talc, epidote, chlorite, actinolite, tremolite, magnetite, almandine or spessartine garnet, sphene | Serpentinite, soapstone or greenstone |
| Calcareous | Calcite, talc, tremolite, dolomite, stilpnomelane, scapolite, magnesite, quartz, pumpellyite | Marble |

*Source*: Turner and Verhoogen (1951) and Turner (1958). © John Wiley & Sons.

Greenschist facies assemblages are abundant in orogenic fold and thrust belts, where they record regional, moderate temperature/pressure metamorphic conditions at convergent plate boundaries. Classic greenschist facies rocks are extensively exposed in orogenic belts such as the Appalachians, Cordilleran, Alps, and the Otago fold and thrust belt of Southern New Zealand.

Earlier in this chapter, we discussed index minerals in pelitic rocks used to map Barrovian zones. Where metapelites occur, the greenschist facies can be subdivided into three Barrovian zones:

1  The chlorite zone corresponds to lower greenschist facies conditions with minerals such as chlorite, dolomite, stilpnomelane, calcite.
2  The biotite zone corresponds to upper greenschist facies conditions and contains biotite, tremolite.
3  The lower part of the almandine garnet zone corresponds to the uppermost greenschist to epidote–amphibolite facies.

Figure 18.3 illustrates how Barrovian zones relate to the metamorphic facies common in orogenic belts.

If pressures become more uniform (less deviatoric), greenschist facies metamorphism can grade into the albite–epidote hornfels facies. If nonuniform pressures continue and progressively higher temperature conditions develop within orogenic belts, the greenschist facies grades into higher grade amphibolite facies rocks, as discussed below.

### 18.2.5  Amphibolite facies

**Amphibolite facies** assemblages form at high temperatures (~550–750 °C) and moderate to high pressures (4–12 kbar = 0.4–1.2 GPa ≈ 12–40 km depth) in regional orogenic belts at convergent margins. Amphibolite facies metamorphism can affect rocks of any composition, although the most common protoliths include mafic and ultramafic igneous rocks, tuff, sandstone, mudstone, and limestone. Minerals and rocks common to the amphibolite facies are listed in Table 18.10. The transition from greenschist to amphibolite facies is marked by an increase in hornblende, garnet, and anthophyllite and a decrease in actinolite, chlorite, biotite, and talc in mafic and ultramafic rocks. In addition, plagioclase minerals become less sodic and more calcic. The amphibolite facies also marks the appearance of staurolite in pelitic rocks, where staurolite may occur with kyanite (Figure 18.3). Increasing temperature in amphibolite facies rocks results in the transformation from kyanite to sillimanite in pelitic rocks where sillimanite commonly occurs with muscovite. In calcareous rocks, dolomite breaks down and releases carbon dioxide via the following chemical reactions:

$$\underset{\text{Dolomite}}{CaMg(CO_3)_2} + \underset{\text{Quartz}}{2SiO_2} => \underset{\text{Diopside}}{CaMgSi_2O_6}$$

$$+ \underset{\text{Carbon dioxide}}{2CO_2} \quad (\text{Turner 1981}).$$

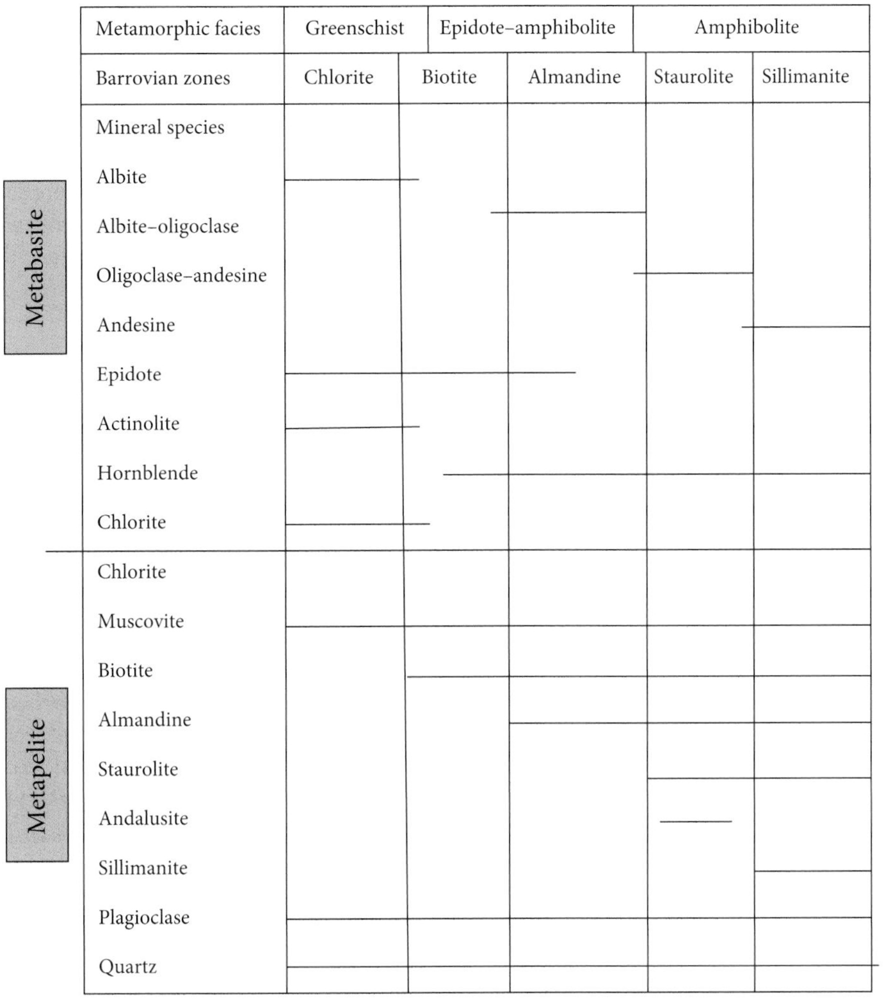

| Metamorphic facies | Greenschist | Epidote–amphibolite | | Amphibolite | |
|---|---|---|---|---|---|
| Barrovian zones | Chlorite | Biotite | Almandine | Staurolite | Sillimanite |
| **Metabasite** Mineral species | | | | | |
| Albite | | | | | |
| Albite–oligoclase | | | | | |
| Oligoclase–andesine | | | | | |
| Andesine | | | | | |
| Epidote | | | | | |
| Actinolite | | | | | |
| Hornblende | | | | | |
| Chlorite | | | | | |
| **Metapelite** Chlorite | | | | | |
| Muscovite | | | | | |
| Biotite | | | | | |
| Almandine | | | | | |
| Staurolite | | | | | |
| Andalusite | | | | | |
| Sillimanite | | | | | |
| Plagioclase | | | | | |
| Quartz | | | | | |

**Figure 18.3**  Common metapelite and metabasite minerals in greenschist, epidote–amphibolite, and amphibolite facies. *Source*: Yardley 1989.

**Table 18.10**  Common minerals and protoliths of the amphibolite facies (Turner and Verhoogen 1951; Turner 1958).

| Protolith | Common amphibolite facies minerals | Common amphibolite facies rocks |
|---|---|---|
| Pelitic | Kyanite, sillimanite, staurolite, muscovite, almandine garnet, quartz, plagioclase, biotite, epidote | Schist, gneiss |
| Quartzofeldspathic | Quartz, microcline, hornblende, biotite, muscovite, almandine garnet | Metaquartzite, gneiss |
| Mafic | Hornblende, calcium plagioclase (oligoclase–anorthite), epidote, almandine garnet, diopside, cummingtonite, grunerite, augite, quartz, ilmenite, magnetite | Amphibolite, schist |
| Ultramafic | Talc, anthophyllite, olivine, diopside, cummingtonite, serpentine, spinel, calcium plagioclase (oligoclase–anorthite), enstatite, phlogopite, tremolite | Schist |
| Calcareous | Calcite, dolomite, tremolite, anorthite plagioclase, almandine to grossular garnet, diopside, augite, phlogopite | Marble |

Increasing intensity of metamorphism results in devolatization reactions resulting in the release of $H_2O + CO_2$ as shown by the two examples below:

$$7Mg_3Si_4O_{10}(OH)_8 \atop \text{Talc}$$

$$\Rightarrow \underset{\text{Anthophyllite}}{3Mg_7Si_8O_{22}(OH)_2} + \underset{\text{Quartz}}{4SiO_2} + \underset{\text{Water}}{4H_2O}$$

$(\text{Chernosky } 1976)$

$$\underset{\text{Tremolite}}{Ca_2Mg_5Si_8O_{22}(OH)_2} + \underset{\text{Calcite}}{3CaCO_3} + \underset{\text{Quartz}}{2SiO_2}$$

$$\Rightarrow \underset{\text{Diopside}}{5CaMgSi_2O_6} + \underset{\text{Water}}{H_2O} + \underset{\text{Carbon dioxide}}{CO_2}$$

$(\text{Turner } 1981)$.

The amphibolite facies encompasses a number of different Barrovian zones (Figure 18.3) that, with increasing temperature, include the upper part of the almandine zone, all of the staurolite and kyanite zones and the lower part of the sillimanite zone. The low temperature part of the amphibolite facies that corresponds with the almandine zone is also known as the epidote-amphibolite facies ("transitional" facies) because both epidote and hornblende can coexist under these conditions. With decreasing pressure, the lowest pressure field of the amphibolite facies metamorphism grades into the hornblende hornfels field. As temperatures increase, the amphibolite facies grades into the higher temperature granulite facies discussed below.

## 18.2.6 Granulite facies

The **granulite facies** consists of high temperature (~700–900 °C), and moderate to high pressure (3–15 kbar = 0.3–1.5 GPa ≈ 10–50 km depth) mineral assemblages. Granulite facies metamorphism can affect rocks of any composition, although the most common protoliths include granitic to ultramafic igneous rocks, schists, gneisses, pelites, sandstones, and limestones. The minerals and rocks associated with the granulite facies largely depend upon the protolith composition as indicated in Table 18.11.

Granulite facies rocks are commonly, but not always, dominated by rocks with granoblastic textures arranged in a non-foliated to weakly-foliated fabric. Foliated granulite rocks are less common because many of the inequant, hydrous phyllosilicate minerals stable at lower temperatures (e.g. muscovite) have been largely transformed to anhydrous minerals (e.g. K feldspar) with equant forms. Granulite facies minerals are predominantly anhydrous (Table 18.11), due to dehydration reactions at high temperatures. Hydrous minerals hornblende and biotite, but not muscovite, can occur in lower part of granulite facies, sometimes referred to as granulite I. The upper part of the granulite facies, sometimes referred to as granulite II, is characterized entirely by anhydrous minerals.

Amphibole minerals (tremolite, anthophyllite, hornblende) dehydrate to pyroxene (enstatite, diopside, and hypersthene). Phyllosilicate minerals such as muscovite dehydrate to anhydrous orthoclase in response to high

**Table 18.11** Common minerals and rocks in granulite facies based on protolith composition (Turner and Verhoogen 1951; Turner 1958).

| Protolith | Common granulite facies minerals | Common granulite facies rocks |
|---|---|---|
| Pelitic | Sillimanite, kyanite, orthoclase, quartz, almandine garnet, calcium plagioclase (anorthite, bytownite, andesine), cordierite, sapphire, magnetite, ilmenite, rutile | Gneiss, granulite |
| Mafic | Hypersthene, calcium plagioclase (anorthite, bytownite, andesine), diopside, garnet, cummingtonite, grunerite, magnetite, ilmenite | Gneiss, granulite |
| Ultramafic | Enstatite, hypersthene, diopside, olivine, calcium plagioclase, corundum, spinel | Gneiss, granulite |
| Quartz-feldspathic | Orthoclase, quartz, hypersthene, almandine garnet, plagioclase | Gneiss, charnockite, and granulite |
| Calcareous | Calcite, diopside, almandine garnet, forsterite, scapolite, corundum | Marble |

temperatures. Such reactions are completed either in the transition from amphibolite to granulite facies or in the higher temperature/pressure transition within the granulite facies (granulite I to granulite II). Examples of these dehydration reactions include:

$$2Ca_2Mg_5Si_8O_{22}(OH)_2$$
Tremolite
$$=> 4CaMgSi_2O_6 + 3MgSiO_3$$
Diopside    Enstatite
$$+ 2SiO_2 + 2H_2O$$
Quartz    Water
$$(Boyd 1954)$$
$$Mg_7Si_8O_{22}(OH)_2$$
Anthophyllite
$$=> 7MgSiO_3 + SiO_2 + H_2O$$
Enstatite    Quartz    Water
$$(Greenwood 1963; Chernosky 1979)$$
$$Ca_2Mg_4Al_2Si_7O_{22}(OH)_2$$
Hornblende
$$=> CaMgSi_2O_6 + 3MgSiO_3$$
Diopside    Hypersthene
$$+ CaAl_2Si_2O_8 + H_2O$$
Anorthite    Water
$$KAl_2(AlSi_3O_{10})(OH,F)_2 + SiO_2$$
Muscovite    Quartz
$$=> K(AlSi_3O_8) + Al_2SiO_5 + H_2O$$
Orthoclase    Sillimanite    Water

Quartz and orthoclase are common granulite facies minerals in pelitic and quartz-feldspathic rocks. Calcareous rocks are marked by the appearance of wollastonite and the absence of hydrous minerals such as phlogopite. Meta-mafic and meta-ultramafic granulites commonly contain both orthopyroxene (hypersthene) and clinopyroxene (diopside). This two pyroxene association that characterizes many granulites does not occur in amphibolites or other lower temperature facies.

Granulite facies rocks are dominated by gneisses, granulites, charnockites, and migmatites. **Granulites** are medium to coarse grained, granoblastic rocks with feldspars, quartz, and anhydrous ferromagnesium minerals such as the garnets and ortho- and clino-pyroxenes. **Charnockites** are hypersthene bearing granitic gneisses. Migmatites are mixed rocks, meaning they display textural relations of both metamorphic and plutonic igneous rocks. The

igneous component is due to varying degrees of partial melting driven by the release of volatile flux components that enhance rock melting and the generation of magma.

Granulite facies metamorphism occurs in the highest temperature dynamothermal metamorphism region at (1) convergent plate boundaries, (2) at the base of thick continental crust, and (3) in the uppermost part of the mantle. Some mafic granulites may represent the refractory residual rock material following partial melting at the base of continental lithosphere. Granulites occur in rocks of all ages, but are especially common in Precambrian shields and associated anorthosite complexes where long-term erosion has exposed rock formed deep below the surface.

The granulite facies corresponds with the upper parts of the Barrovian sillimanite zone and, at still higher temperature, the cordierite-garnet zone (Figure 18.3). With decreasing pressure, the low-pressure field of granulite facies metamorphism grades into the pyroxene and sanidinite hornfels facies. As progressively higher temperature conditions develop within orogenic belts, migmatites of the granulite facies are produced. So far, we have focused largely on facies changes affected primarily by temperature or temperature-pressure conditions. The facies described below are defined largely on changes in pressure.

### 18.2.7 Blueschist facies

The **blueschist facies** consists of moderate to high pressure ($4$–$20$ kbar $= 0.4$–$2.0$ GPa $\approx 12$–$70$ km depth), low temperature ($150$–$500°C$) mineral assemblages. Although blueschist facies metamorphism affects rocks of any composition, common protoliths include mafic to ultramafic igneous rock and sedimentary graywackes and mudstones. The blue amphibole mineral glaucophane provides the distinctive color for which this facies is named. In addition to glaucophane, other common minerals include magnesio-riebeckite, lawsonite, jadeite pyroxene, aegirine, crossite, and kyanite. Minerals and rocks occurring in the blueschist facies depend upon the protolith composition as indicated in Table 18.12.

Blueschists form in subduction zones where oceanic lithosphere is forced downward to great depths at geologically rapid rates. As cold oceanic lithosphere is dragged downward

**Table 18.12** Common minerals and rocks in blueschist facies based on protolith composition (Turner and Verhoogen 1951; Turner 1958).

| Protolith | Common blueschist facies minerals | Common blueschist facies rocks |
|---|---|---|
| Pelitic | Kyanite, chlorite, epidote, titanite, almandine garnet, quartz, phengite, coesite | Schist |
| Quartzo feldspathic | Kyanite, quartz, coesite, jadeite, magnesio-riebeckite, stilpnomelane | Kyanite metaquartzite |
| Mafic | Glaucophane, jadeite, crossite, lawsonite, albite plagioclase, chlorite, epidote, titanite, magnesio-riebeckite, magnesio-arfvedsonite, aegerine, augite, pumpellyite, stilpnomelane | Schist |
| Ultramafic | Lizardite, glaucophane, magnesio-riebeckite, magnesio-arfvedsonite, jadeite, aegerine, lawsonite, pyrope garnet, acmite | Schist |
| Calcareous | Calcite, epidote, chlorite, lawsonite | Marble |

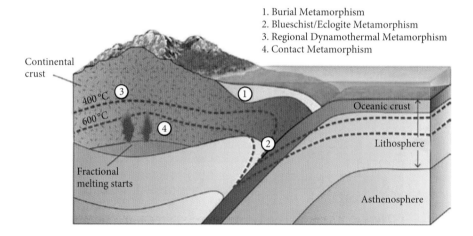

**Figure 18.4** Generalized cross-section indicating depressed temperatures in the subduction zone. *Source*: Furukama (1993) and Murck et al. (2010) © John Wiley & Sons.

into Earth's interior it absorbs heat from the surrounding asthenosphere very slowly, reaching great depths while remaining relatively cool. The subduction of cold lithosphere lowers the temperature in the subduction zone environment to temperatures not found anywhere else on Earth at comparable depths. This has the effect of depressing the isotherms (lines of equal temperature) in the vicinity of the subducted slab (Figure 18.4). Subduction of oceanic lithosphere serves as a proverbial icicle in the Earth's interior. The result is the creation of a unique set of high pressure, low temperature conditions that produces blueschist facies metamorphism. In fact, blueschist facies rocks are a critical indicator for subduction zones in convergent margins.

Blueschist facies rocks present several interesting scientific dilemmas. Two critical questions and possible solutions are listed below:

1 How do blueschist mineral assemblages produced at great depth return to the Earth's surface? Suggested mechanisms include:
   a Extreme extension perpendicular to the arc (Lister et al. 1984; Platt 1986; Shermer 1990) which allows deep seated rocks to rise toward the surface;
   b Extension associated with oblique subduction (Avé Lallemant and Guth 1990);
   c Uplift associated with rising, buoyant serpentinites (Pilchin 2003);
   d A reverse loop within the base of an accretionary wedge that regurgitates deeply buried (underplated) subducted

rock fragments upwards to the Earth's surface (Platt 1986);

e    Uplift along thrust fault systems combined with hanging wall erosion (Unruh et al. 1995).

2    Why are blueschists relatively common in the Phanerozoic but relatively rare in the Precambrian? Explanations include:

a    Their precarious positions at the leading edges of subduction zones makes them vulnerable to further subduction and crustal recycling (Möller et al. 1995);

b    Uplift and removal by erosion during continental collisions;

c    Retrograde metamorphism to greenschist facies assemblages during later tectonic events with more normal P/T ratio conditions (Zwart 1967; Ernst 1988)

d    Elevated Precambrian geotherms which lowered P/T ratios by raising temperatures during metamorphism (De Roever 1956; Burke et al. 1977);

e    More limited upper mantle convection and thinner lithosphere in the Precambrian (Liou et al. 1990) which caused shallower subduction.

Research regarding the scarcity of Precambrian blueschists and the uplift of blueschists continues. At temperatures between 350 and 450 °C and pressures of ~8 kbar, a blueschist and greenschist facies transition occurs (Maruyama et al. 1986). With increasing temperatures and pressures exceeding 12 kbar, the blueschist facies transforms to the eclogite facies.

## 18.2.8  Eclogite facies

The **eclogite facies** develops at high temperatures (400–900 °C) and very high pressures (12–25 kbar = 1.2–2.5 GPa ≈ 40–80 km). Ultrahigh-pressure (UHP) rocks, also considered part of the eclogite facies, form at even higher pressures and greater depths, as discussed in the next section. The rock after which the eclogite facies is named, eclogite is chemically similar to a silica undersaturated, anhydrous basalt and generally develop from mafic protoliths. Eclogite rocks are fine to coarse grained, dense, dark green rocks, commonly with reddish brown garnet porphyroblasts. The two key minerals are garnet and omphacite, a sodium rich, jadeitic clinopyroxene.

Garnet group minerals stable at high-pressure eclogite conditions include majorite, pyrope garnet, and lesser garnet grossular; at lower pressures almandine garnet can also occur. While the rock eclogite most commonly forms from mafic protoliths, the eclogite facies includes rocks of other compositions. For example, kyanite rich quartz eclogite rocks developed from pelitic protoliths in the Sanbagawa belt of Japan (Miyamoto et al. 2007). Other minerals that can occur within the eclogite facies include enstatite, jadeite, rutile, zoisite, coesite, phengite, lawsonite, corundum, and diamond. The eclogite facies constitutes a high pressure facies in which plagioclase is unstable as shown by the reaction:

$$\underset{\text{Albite}}{Na\left(AlSi_3O_8\right)} => \underset{\text{Jadeite}}{Na\left(Al, Fe\right)Si_2O_6} + \underset{\text{Quartz}}{SiO_2}.$$

Minerals and rocks occurring in the eclogite facies depend upon the protolith composition as indicated in Table 18.13.

Eclogite facies rocks occur in three major environments:

1    In the lower continental crust and mantle (>40 km depth), later exposed on the Earth's surface in deeply eroded fold and thrust belts;

2    At convergent margin ophiolite complexes and subduction zone mélanges; and

3    As xenoliths in diamond bearing kimberlite pipes.

Baldwin et al. (2004) studied the youngest documented eclogites in the world, produced along the Papua New Guinea oblique subduction zone. Oblique subduction involves components of both thrusting and strike-slip faulting. The 4.33 ±0.36 Ma eclogites were metamorphosed at depths of 75 km. Geothermobarometric analysis indicates that the minimum rate of uplift was ~1.7 cm/yr, comparable to spreading rates at some ocean ridges. The 1.7 cm/yr estimate assumes vertical uplift which is unlikely in the oblique shear zones of Papua New Guinea. More likely, transport rates associated with the strike-slip component of motion ranged between 3.5 and 5.1 cm/yr – geologically blazing speeds! Now we will consider ultrahigh-pressure minerals, which geologists consider as part of the eclogite facies.

**Ultrahigh-pressure (UHP) minerals** occur within the eclogite facies (Figure 18.5) at pressures > 25 kbars (=>2.5 GPa = >80 km depth)

and temperatures >600 °C. UHP conditions are indicated by critical minerals such as:

1 Coesite, a high pressure polymorph of silica;
2 Diamond, the high pressure polymorph of carbon; and
3 Majorite, a high pressure mineral ($Mg_2Al_3Si_2O_{12}$–$MgSiO_3$).

Coesite was first observed in a laboratory, where it was synthetically created by Loring Coes, Jr. in 1953. Naturally occurring coesite was first noted at Meteor (Barringer) Crater, Arizona (USA) where it formed by high pressure impact metamorphism (Chao et al. 1960). Christian Chopin (1984) discovered the first dynamothermally produced coesite in Alpine

**Table 18.13** Common minerals and rocks in eclogite facies based on protolith composition (Turner and Verhoogen 1951; Turner 1958).

| Protolith | Common eclogite facies minerals | Common eclogite facies rocks |
| --- | --- | --- |
| Pelitic | Kyanite, almandine garnet, grossular garnet, pyrope garnet, ompacite, phengite, rutile, quartz, coesite, diamond | Kyanite quartz eclogite |
| Quartzo feldspathic | Kyanite, quartz, coesite, phengite, jadeite, diamond | Kyanite metaquartzite; metaquartzite |
| Mafic | Omphacite, pyrope garnet, glaucophane, rutile, titanite, zoisite, jadeite, crossite, lawsonite, diamond | Eclogite |
| Ultramafic | Omphacite, pyrope garnet, glaucophane, rutile, titanite, zoisite, jadeite, aegerine, lawsonite, diamond | Eclogite |
| Calcareous | Calcite, lawsonite, pyrope-almandine garnet | Marble |

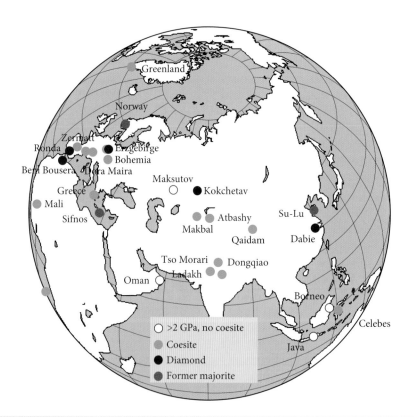

**Figure 18.5** Eurasian–African view of coesite, diamond, and majorite ultra high pressure mineral occurrences. *Source*: Courtesy of Bradley Hacker.

rocks. Since that time, researchers have found UHP assemblages in many orogenic belts (Figure 18.5). In addition to coesite, diamond, and majorite, which form at pressures >25 kbar (2.5 GPa) and depths of ≥80 km, other minerals that commonly occur in the UHP facies include potassium rich clinopyroxene, magnesium rich garnet (pyrope), lawsonite, aluminum rich rutile, glaucophane, phengite, diopside, kyanite, jadeite, and ellenbergite. UHP minerals such as coesite and majorite can occur in a number of environments that include:

1 Meteorite impact sites such as at Meteor Crater, Arizona;
2 Kimberlite pipes in which UHP minerals such as coesite and diamond occur in brecciated ultramafic host rock.
3 Deepest levels of subduction zones where quartz transforms to coesite;
4 Convergent margins involving two continental lithospheric plates that collide and partially subduct. Although early plate tectonic proponents suggested that continental crust is too thick and buoyant to subduct, UHP rocks indicate that continental lithosphere can indeed be subducted to depths of 150 km. UHP rocks are distributed along continent–continent collision zones such as in the Alpine-Himalayan belt, the Ural Mountains, the Western Gneiss Region of Norway and in Pan African belt rocks (Figure 18.5).

It is a matter of current debate as to whether UHP minerals define a separate metamorphic facies or represent the high-pressure end of the eclogite facies. Given that increasingly higher pressure suites are being recognized, it is likely that a separate UHP facies will be formally recognized in the future. UHP research is one of the dynamic aspects of current research, particularly in relation to the depth to which continental crust can subduct and reemerge on Earth's surface as UHP assemblages.

A wide variety of temperature and pressure conditions exist during metamorphism producing distinct metamorphic zones and facies. Individual facies can occur alone, but are more commonly associated in time and space with other facies. These multiple, associated facies form through temperature and pressure conditions that vary laterally and vertically in an area during metamorphism. In recognition of the existence of patterns of multiple metamorphic facies within a given region, Miyashiro (1961, 1973, 1994) proposed the concept of metamorphic facies series, which we will now address.

## 18.3 METAMORPHIC FACIES SERIES

A **metamorphic facies series** is a sequence of facies aerially distributed across a metamorphic terrane that results from spatial and temporal differences in temperature and pressure conditions. In order to describe such a sequence of changing metamorphic conditions, geologists refer to Pressure–Temperature–time (P–T–t) relations in which the history of pressure and temperature changes over a time range are inferred from the rock record. Each facies series is characterized by the development of a particular sequence of individual facies, with each facies stable at a specific range of temperature and pressure conditions. Why is the concept of a facies series so important? Facies series provide key information concerning the progressive T-P-t conditions as well as the tectonic setting in which metamorphism occurred.

Metamorphic facies series were defined (Miyashiro 1994) on the basis of pressure and temperature gradients, both of which are related to the conditions of metamorphism and tectonic setting (Figure 18.6). Five metamorphic facies series, assigned to three major groups, are recognized. The low P/T series group (shaded zone A in Figure 18.6) consists of the contact facies series (Figure 18.7) associated with hot magma induced contact metamorphism. The moderate P/T series group (shaded zone B in Figure 18.6) contains the Buchan and Barrovian facies series (Figure 18.7) associated with dynamothermal metamorphism in orogenic belts at convergent plate margins. The high P/T series group (shaded zone C in Figure 18.6) includes the Sanbagawa and Franciscan facies series (Figure 18.7) associated with dynamothermal metamorphism produced by convergent plate boundary subduction.

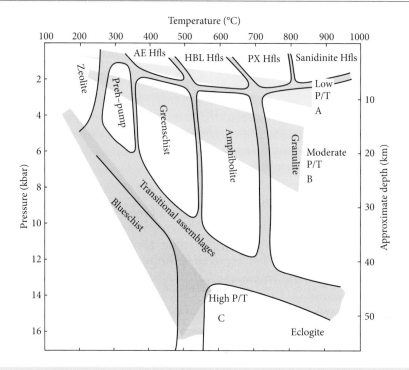

**Figure 18.6** Five metamorphic facies series grouped into three (A–C) trends based on temperature and pressure relations. AE, albite epidote; HBL, hornblende; Hfls, hornfels; PX, pyroxene. *Source*: Courtesy of Bradley Hacker.

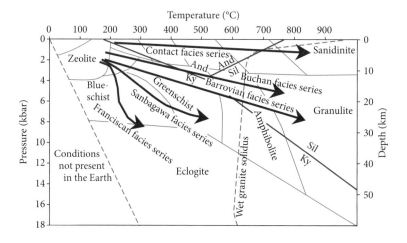

**Figure 18.7** Five metamorphic facies series defined by Myashiro (1961) and others. And, andalusite; Ky, kyanite; Sil, sillimanite. *Source*: Courtesy of Stephen Nelson.

### 18.3.1 Contact facies series

The **contact facies series** forms at relatively low pressure (<2.5 kbars = <250 MPa ≈ 8 km depth) and moderate to high temperature conditions. The geothermal gradient for the contact facies series is very high, >80°/km.

This facies series develops by contact metamorphism in aureoles adjacent to hot igneous intrusions. While contact metamorphism commonly occurs at shallow depths, it may also develop in moderate pressure conditions of the lower crust. Low pressure contact metamorphism produces hornfelsic and/or granoblastic rather than foliated textures. However, the rocks may display (1) older, relict foliated fabrics due to deformation that preceded contact metamorphism or (2) younger, overprinted foliated fabrics from deformation that occur after contact metamorphism.

The contact facies series consists of several individual hornfels facies, each of which is defined by a low pressure assemblage of minerals stable at specific temperature ranges (Section 18.2.1). For example, pelitic rocks commonly contain the low temperature polymorph andalusite in low temperature contact facies and the high temperature polymorph sillimanite in high temperature contact facies. With increasing temperature, the contact facies series progresses through the sequence: (1) zeolite facies, (2) albite–epidote hornfels facies, (3) hornblende hornfels facies, the less common (4) pyroxene hornfels facies and, at very high temperatures the still rarer (5) sanidinite hornfels facies. On a pressure–temperature diagram, a line marking the trajectory of progressive contact metamorphism (Figure 18.7) is nearly horizontal which reflects the progressive increase in temperature at near constant pressure experienced as the protolith underwent contact metamorphism. Higher temperature facies occur in close proximity to the intrusion with progressively lower temperature facies toward the outer margins of the contact metamorphic aureole.

The contact facies series can occur anywhere hot plutons intrude rock: convergent margins, ocean spreading ridges, hot spots, and large intraplate intrusions. In contrast, the Buchan, Barrovian, Sanbagawa, and Franciscan facies series are associated with orogenic belts at convergent plate boundaries.

## 18.3.2 Buchan facies series

The **Buchan facies series** records high geothermal gradients that range from 40 to 80 °C/km. This facies series is named after the Buchan area of northeastern Scotland (Figure 18.8). The Buchan facies series is also known as the Abukuma facies series, after the Abukuma region in Japan (Miyashiro 1961, 1994). The individual facies within the Buchan facies series are defined by low to moderate P/T mineral assemblages. As with the contact facies series, pelitic rocks may contain the low temperature polymorph andalusite or the high temperature polymorph sillimanite; the high

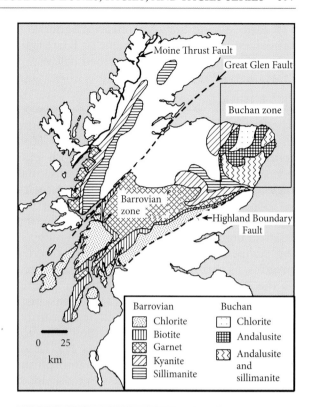

**Figure 18.8**  Generalized geologic map of northern Scotland showing Barrovian and Buchan zones. *Source*: Adapted from Miyashiro (1961) and Best (2003). © John Wiley & Sons.

pressure polymorph mineral kyanite is typically absent. However, because of its higher P/T ratio, the trajectory of progressive metamorphism followed by the Buchan series on a P/T diagram is somewhat steeper than that of the contact series, reflecting the increase in both pressure and temperature that occurs during progressive, dynamothermal metamorphism.

The Buchan facies series progresses through: (1) zeolite, (2) prehnite–pumpellyite, (3) greenschist, (4) amphibolite to (5) the high temperature, moderate pressure granulite facies. Buchan facies series metamorphism reflects relatively higher rates of temperature increase that accompanied increases in pressure during regional metamorphism, often due to associated regional magmatism. Because of the nonuniform stress states produced in orogenic belts, rocks of the Buchan facies series are commonly, but not always, foliated. Nonfoliated Buchan facies series rocks can occur in regions that experience crustal thinning and heating which lowers the

P/T ratio during metamorphism. Higher P/T ratios result in the development of Barrovian facies series assemblages, described below.

### 18.3.3 Barrovian facies series

The **Barrovian facies series** develop in response to geothermal gradients of ~20 to 40 °C/km, reflecting the progressive increase in both temperature and pressure during dynamothermal regional metamorphism. The Barrovian facies series is named for George Barrow, the geologist who first mapped isograd zones in the Scottish highlands (Section 18.1). With increasing temperature, the Barrovian facies series progresses through the same facies sequence as the Buchan facies series, from: (1) zeolite, (2) prehnite–pumpellyite, (3) greenschist, (4) amphibolite to (5) the high temperature, moderate to high-pressure granulite facies (Figure 18.7). Because of its higher P/T ratio, the trajectory of progressive metamorphism followed by Barrovian series metamorphism on a P/T diagram is somewhat steeper than that of the Buchan series (Figure 18.7), reflecting the lower geothermal gradient as pressures increase with greater depths. As a result, the Barrovian equilibrium mineral assemblage commonly differs from the Buchan facies series. For example, pelitic rocks may contain not only the low temperature polymorph andalusite and the high temperature polymorph sillimanite, but also the high-pressure polymorph kyanite. Barrovian facies series rocks commonly occur in orogenic belts and display foliated textures as a result of nonuniform stresses at convergent plate boundaries. However, nonfoliated rocks such as metaquartzites and marbles may also occur.

### 18.3.4 Sanbagawa facies series

The **Sanbagawa facies series** are produced by geothermal gradients in the range of 10 to 20 °C/km. Miyashiro (1994) named this facies after the Sanbagawa belt of Japan (Figure 18.9). The Sanbagawa facies series progression includes: (1) zeolite, (2) prehnite–pumpellyite, (3) blueschist facies followed in some cases by (4) greenschist and/or (5) amphibolite facies. Sanbagawa facies series metamorphism reflects the rapid increase of pressure relative to temperature during progressive dynamothermal regional metamorphism at convergent plate boundaries. Rocks within this facies series are commonly foliated and highly deformed due to nonuniform compressive stress. The P/T trajectory of the Sanbagawa series metamorphism is steeper than that of the Barrovian facies series (Figure 18.7) reflecting higher P/T ratios associated with subduction zones. The Sanbagawa and the Franciscan facies series, described below, document conditions in which the cool slabs of downgoing lithosphere (low temperature) are subducted to great depths (high

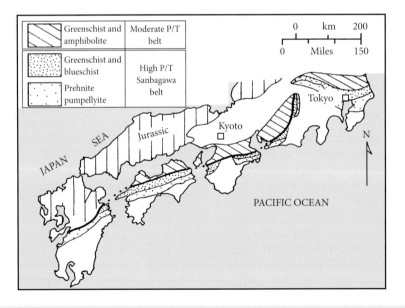

**Figure 18.9** Metamorphic facies map of Japan's Sanbagawa belt. *Source*: Adapted from Miyashiro (1994) and Best (2003). © John Wiley & Sons.

pressure). The slightly higher temperatures in the Sanbagawa facies series may be due to (1) slower subduction that allows rocks to be heated over longer periods of time or by the adjoining magmatic arc complex, or (2) higher geothermal gradients during subduction.

### 18.3.5 Franciscan facies series

The **Franciscan facies series** document very low geothermal gradients of <10°C/km. Franciscan facies rocks are characterized by extremely high P/T ratios found only in subduction zones. Because of its very high P/T ratio, the trajectory of progressive metamorphism during Franciscan series metamorphism on a P/T diagram is very steep (Figure 18.7). Miyashiro (1994) named this facies series after the Franciscan Complex (Figure 18.10) of California (USA). The Franciscan facies series progresses from: (1) zeolite, (2) prehnite–pumpellyite, (3) blueschist and locally into the (4) eclogite facies (Figure 18.7). Franciscan series metamorphism reflects rapid subduction

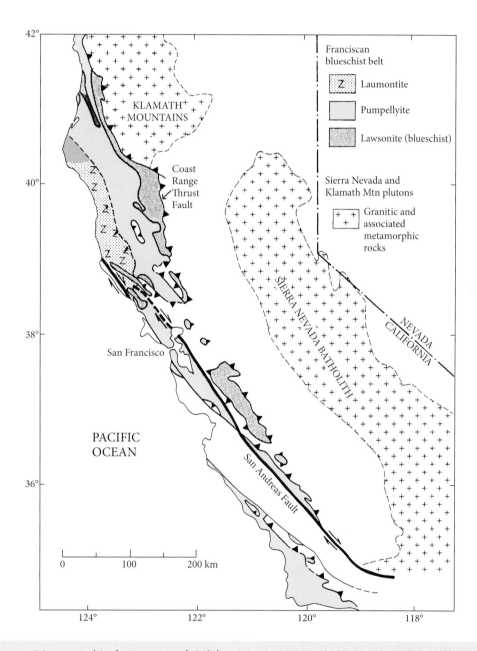

**Figure 18.10** Metamorphic facies map of California's Franciscan belt. *Source*: Adopted from Murck and Skinner (2016). © John Wiley & Sons.

of old, cold crustal rocks shoved into Earth's deep subduction canal, thereby lowering the geothermal gradient. The high pressure minerals jadeite, glaucophane, and lawsonite are particularly important indicators of the high pressure, low temperature conditions. Kyanite, the high pressure polymorph of aluminum silicate, and phengite are common in Franciscan pelitic rocks. Franciscan facies series rocks are commonly fragmented, highly deformed and foliated.

In the following section, we will briefly introduce the study of equilibrium suites of minerals through the use of ternary diagrams, a powerful tool in the interpretation of metamorphic mineral assemblages.

## 18.4  PHASE RULE, CHEMICAL REACTIONS, AND THREE-COMPONENT PHASE DIAGRAMS

In this section, we pose two major questions:

1  What are the factors that determine the equilibrium assemblage of metamorphic minerals that develop during closed system metamorphism?
2  Can mineral assemblages be graphically represented for ease of understanding?

When rocks experience high temperatures and/or high pressures over long periods of time, in the presence of appropriate catalysts that permit chemical reactions to proceed, pre-existing mineral assemblages (the reactants) become unstable and are converted into new mineral assemblages that are stable in the new T/P environment. Equilibrium conditions occur when chemical reactions that produce stable minerals have gone to completion in a system. If temperature and/or pressure conditions or any other variable in the system should change, additional reactions may occur over time to reestablish chemical equilibrium with respect to the new conditions. The concept of equilibrium mineral assemblages is the basis for the Barrovian zones, metamorphic facies, and metamorphic facies series described earlier in this chapter.

How can we determine whether a mineral assemblage records equilibrium or disequilibrium conditions? Microscopic analysis of thin-sections can provide information. In thin section, disequilibrium conditions can be indicated by: reaction rims on minerals, coexisting minerals that cannot exist in equilibrium, or incomplete replacement of minerals. Equilibrium conditions are indicated by the absence of the features stated above as well as planar grain contacts between mineral crystals. If equilibrium conditions appear to have been achieved based on thin section or hand sample analysis of rocks, metamorphic petrologists can make some inferences regarding the conditions that produced the equilibrium mineral assemblages using the phase rule.

### 18.4.1  The phase rule and metamorphic minerals

The phase rule Gibbs (1928) is a rule used by petrologists to predict equilibrium assemblages of minerals in many systems, including both igneous and metamorphic rocks. Recall the phase rule which states that:

$$P = C + 2 - F$$

where $P$ represents the number of phases present in a system; $C$ designates the minimum number of chemical components; and $F$ refers to the number of degrees of freedom or variance. A **phase diagram** is a visual means by which mineral stability fields can be displayed. Let us examine an example of how the phase rule can be applied to phase diagrams in metamorphic petrology (Figure 18.11). Consider the one chemical component ($C = 1$) system $Al_2SiO_5$ (also written as $AlAlOSiO_4$) which can exist as three different mineral phases ($P$) or polymorphs: (1) andalusite (low temperature polymorph), (2) kyanite (high pressure polymorph), and (3) sillimanite (high temperature polymorph). Temperature–pressure stability fields for the three polymorph minerals are separated by lines. At any point within the kyanite stability field (e.g. point A), $Al_2SiO_5$ exists as kyanite as a single phase ($P = 1$). Using the phase rule $P = C + 2 - F$, since $1 = 1 + 2 - 2$, $F = 2$. This means that both temperature and pressure can be varied independently of one another while the system remains within the kyanite stability field. This stability field is called a **divariant field** because the two variables can change independently of one another without changing the phase composition of the system. As suggested by

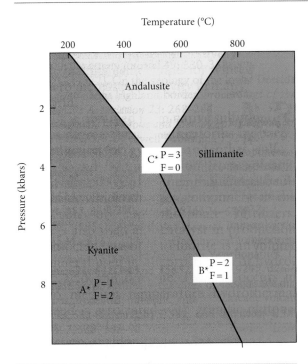

**Figure 18.11**  Phase rule as applied to the Al$_2$SiO$_5$ polymorph mineral group.

Figure 18.11, similar large divariant fields occur at higher temperatures for sillimanite and at lower temperatures for andalusite. The divariant fields for each polymorph phase are separated from one another by phase boundary lines.

However, on the boundary line between the kyanite and sillimanite stability fields, such as at Point B, kyanite and sillimanite co-exist (P = 2). From the phase rule (P = C + 2 − F or 2 = 1 + 2 − F), it is clear that along this line, only one independent variable exists (F = 1) in this one component system. An independent change in either the temperature or pressure requires a change in the other dependent variable in order for the system to remain on the kyanite–sillimanite boundary line. All phase boundary lines in this system are **univariant lines** because only one variable can change independently. Any change in one variable requires a compensating change in the second variable in order to maintain the two phases in equilibrium. Univariant lines are fundamentally important in a one-component system because they indicate:

1  Boundaries between divariant fields in which only one mineral species is stable;

2  Conditions under which two phases can coexist;

3  Conditions across which changes in equilibrium mineral assemblages occur.

Point C, the only point on the diagram where three univariant phase boundaries intersect, defines the unique temperature–pressure conditions under which all three Al$_2$SiO$_5$ mineral phases coexist (P = 3). From the phase rule (P = C + 2 − F), since 3 = 1 + 2 − 0, the co-existence of three phases at a single point implies that F = 0. Only at this fixed temperature and pressure, called an **invariant point**, can all three phases coexist. In the Al$_2$SiO$_5$ system, kyanite, andalusite, and sillimanite all coexist in equilibrium only at a temperature of 510°C and pressure of 3.8 kbar. Note that univariant lines intersect at invariant points. Invariant points are critically important because they indicate:

1  Unique points where three univariant lines intersect and three mineral phases coexist in equilibrium.

2  Specific temperature and pressure conditions at which three stable phases coexist in a one-component system.

Most metamorphic mineral assemblages have compositions that require more than one component in order to adequately describe phase compositions. These are discussed in the section that follows.

### 18.4.2  Equilibrium mineral assemblage grids

The phase rule can be applied in metamorphic systems that involve more than one component. By displaying a number of phase stability relationships on a temperature–pressure diagram, we can graphically demonstrate the temperature and pressure conditions at which specific mineral transformations occur, and the minerals likely to occur within specific temperature and pressure ranges. By overlaying such information onto the temperature–pressure fields for metamorphic facies, we can demonstrate which facies and facies series involve specific reactions and mineral stability fields (Figure 18.12). Such equilibrium mineral reaction and assemblage diagrams are referred to as **petrogenetic or paragenetic grids**. Petrogenetic refers to the conditions under which the rock originated, whereas the term paragenesis refers to the formation sequence of an equilibrium set of minerals that formed at different times, e.g. along a metamorphic T–P trajectory. Both are important in understanding the formation and evolution of metamorphic rocks.

Figure 18.12 illustrates a paragenetic grid for minerals derived from a pelitic protolith. Note the many divariant fields, univariant lines, and invariant points for a number of different aluminosilicate mineral transformations. Phase transformations across univariant lines (dashed) that separate mineral stability fields are labeled with the mineral reactions that occur. This information is overlain on the metamorphic facies stability fields (solid lines). Progressive metamorphic paths or metamorphic trajectories for three different geothermal gradients that approximately range from Franciscan through Barrovian facies series (10 to 30°/km) are also shown. This grid provides an example of the various types of data that can be succinctly summarized in a single diagram. By following any metamorphic

trajectory, one can see metamorphic transformations that can occur during progressive metamorphism, compare it with the different reactions that characterize other trajectories, and gain insights into the variations in mineral assemblages between metamorphic facies.

### 18.4.3 Ternary diagrams

Equilibrium mineral assemblages can be graphically displayed on **three-component (ternary) diagrams**. The major chemical components in metamorphic reactions include: $SiO_2$, $Al_2O_3$, $FeO$, $Fe_2O_3$, $MgO$, $CaO$, $K_2O$, $Na_2O$, $H_2O$, and $CO_2$. Ternary diagrams display three sets of oxide compounds at the X, Y, and Z apices of the triangular diagram respectively. The most common X, Y, and Z oxide compounds selected include $Na_2O$, $K_2O$, $CaO$, $Al_2O_3$, $FeO$, $MgO$. Chemically similar components, such as ferrous iron and magnesian oxides, are grouped as a single component ($FeO + MgO$) for simplification. As suggested by the phase rule ($3 + 2 = 5$), the

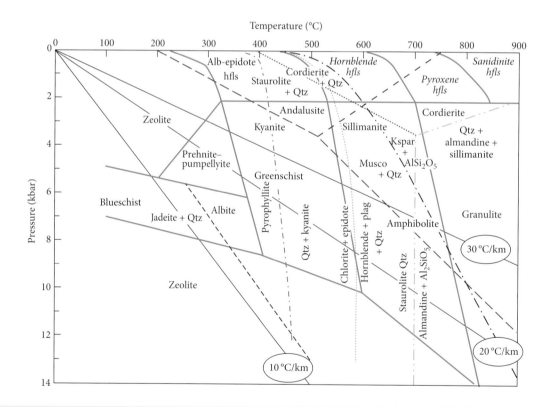

**Figure 18.12** Equilibrium assemblage grid for minerals derived from politic protoliths. Hfls, hornfels; Qtz, quartz; Musco, muscovite. *Source*: Courtesy of Stephen Nelson.

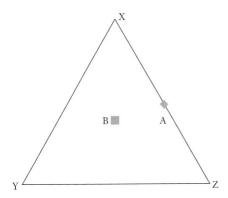

**Figure 18.13** Generalized ternary diagram with X, Y, and Z representing end member oxide compounds. Point A has a normalized composition of 50% X and 50% Z. Point B has a normalized concentration of 33% X, 34% Y, and 33% Z.

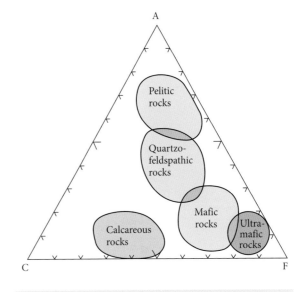

**Figure 18.14** ACF ternary diagram for protolith composition where: A = (Al$_2$O$_3$ + Fe$_2$O$_3$) – (Na$_2$O + K$_2$O), C = (CaO – 3.33 P$_2$O$_5$), and F = (FeO + MgO + MnO). *Source:* Courtesy of Stephen Nelson.

three major chemical components combine to produce up to five major coexisting minerals in a given metamorphic rock. Ternary diagrams depict: (1) stable mineral assemblages that exist under specific temperature and pressure conditions; and (2) how mineral assemblages react to changes in temperature and pressure. As such, ternary diagrams are a powerful tool in understanding the evolution of metamorphic mineral assemblages, rocks, and series. Pentti Eskola (1915) used ternary diagrams (Figure 18.13) to display the composition of common metamorphic mineral assemblages in terms of their concentration of common oxide components. Mineral concentrations are normalized so that the concentrations of the three oxide components in a given mineral total 100% (e.g., X + Y + Z = 100%). As shown in Figure 18.13 a mineral that plots at point A has a composition expressed as 50% X, 50% Z, and 0% Y, whereas a mineral represented by point B has a composition expressed as 33.3% X, 33.3% Y, and 33.3% Z. Several types of ternary diagrams are discussed below.

The **ACF ternary diagram** proposed by Eskola (1915) is based on the molecular amounts of three components ACF where: A = (Al$_2$O$_3$ + Fe$_2$O$_3$) – (Na$_2$O + K$_2$O), C = (CaO – 3.33 P$_2$O$_5$) and F = (FeO + MgO + MnO). The ACF diagram depicts average compositional differences between the five major compositional groups of metamorphic rocks:

(1) the aluminum-rich pelitic rocks, (2) the calcium-magnesium rich, aluminum-poor calcareous rocks, (3) the magnesium/iron-rich ultramafic rocks, (4) the iron/magnesian/calcium-rich mafic rocks, and (5) quartz-feldspathic rocks based on average amounts of all three components in each. The ACF diagram is particularly useful for displaying common equilibrium mineral assemblages that occur in rocks derived from the quartzo-feldspathic, mafic, calcareous, and pelitic protoliths (Figure 18.14).

Figure 18.15 illustrates common metamorphic minerals whose compositions can be plotted on an ACF diagram. Only a small subset of these minerals will occur under any given temperature and pressure conditions, depending composition and mineral stability ranges. Minerals with fairly constant compositions are shown by points or black squares. Minerals with extensive substitution solid solution can have variable compositions (e.g., chlorite, hornblende, anthophyllite) that occupy larger areas within ternary diagrams.

The **A'KF ternary diagram** (Eskola 1915) discriminates equilibrium mineral assemblages derived from pelitic and quartzo-feldspathic protoliths, with excess Al$_2$O$_3$ and

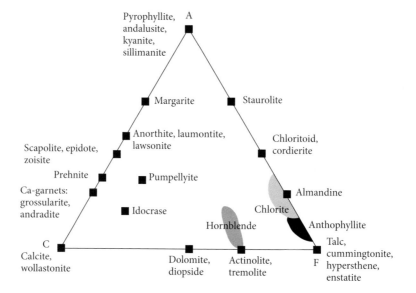

**Figure 18.15** ACF ternary diagram illustrating approximate chemical composition of common metamorphic minerals where: A = $(Al_2O_3 + Fe_2O_3) - (Na_2O + K_2O)$, C = $(CaO - 3.33\ P_2O_5)$ and F = $(FeO + MgO + MnO)$. Boxes indicate fairly constant mineral composition. Shaded zones indicate variable mineral composition. *Source*: Courtesy of Stephen Nelson.

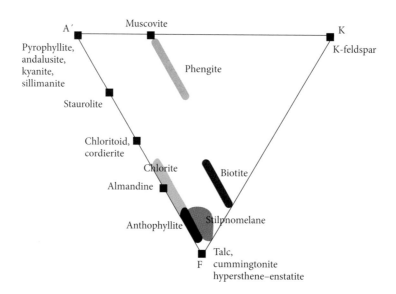

**Figure 18.16** A'KF ternary diagram indicating approximate chemical composition of common pelitic and quartzofeldspathic metamorphic minerals where endpoints: A' = $(Al_2O_3 - (Na_2O + K_2O + CaO))$, K = $K_2O$ and F = $(FeO + MgO + MnO)$. Boxes indicate fairly constant mineral composition. Shaded zones indicate variable mineral composition. *Source*: Courtesy of Steve Dutch.

$SiO_2$ (Figure 18.16). A'KF ternary diagram includes endpoints:

A' = $(Al_2O_3 - (Na_2O + K_2O + CaO))$, K = $K_2O$ and F = $(FeO + MgO + MnO)$. The A'KF ternary diagram is useful for equilibrium mineral assemblages from a continental area.

While ACF and A'KF diagrams combine MgO, FeO, and MnO as F, Thompson (1957) utilized an **AFM ternary diagram** for metamorphic rocks where A = $Al_2O_3$, F = FeO, and M = MgO. The AFM diagram is particularly useful in discriminating equilibrium mineral compositions in ferromagnesian-rich mafic and ultramafic rocks, as well as some pelitic rocks. Figure 18.17 illustrates compositions of aluminosilicate minerals associated with pelitic rocks such as biotite, cordierite, andalusite, sillimanite kyanite, chloritoid, stilpnomelane, garnet, and staurolite on the Thompson AFM diagram.

**CMS ternary diagrams** are used for equilibrium mineral assemblages of calcareous rocks (Figure 18.18), in which the three components are CaO, $SiO_2$, and MgO, respectively. The CMS diagram has powerful applications in the depiction of mineral assemblages in calcsilicate rocks (containing minerals such as calcite, dolomite, wollastonite, diopside) and ultramafic (containing minerals such as tremolite, anthophyllite, forsterite) rocks. In using these diagrams, it is assumed that sufficient $CO_2$ is present to make the carbonate minerals such as calcite and dolomite and sufficient $H_2O$ is available to produce hydrous minerals such as talc, tremolite, antigorite, and brucite.

Ternary diagrams such as these depict the equilibrium assemblage of minerals that may occur. However the constraints imposed by the phase rule ($F = C + 2 - P$) limit ($P = C + 2 - F$) the number of minerals (5 or less) that coexist in equilibrium at a given set of temperature and pressure conditions ($F = 0$) in a three-component system.

**Tie lines** are drawn on ternary diagrams between the endpoint composition of equilibrium minerals that coexist under specific temperature and pressure conditions. Compositions located on tie lines contain the two minerals joined by the tie line while points on either end of a tie line represent only one stable mineral. Mineral stability fields that contain three equilibrium minerals are defined by three tie lines. Figure 18.19 depicts a simplified model of a CSM ternary diagram with CaO (C), MgO (M), and $SiO_2$ (represented by Q for quartz) components. Figure 18.19a illustrates the stable minerals in this system at ~400 °C, in which tie lines connect mineral pairs. Some tie lines occur along the margins of the triangle (e.g., Q–C), others cross the interior (e.g., D–Tc). Tie lines divide the ~400 °C ternary diagram in Figure 18.19a into three separate fields, each of which is completely bound by tie lines between stable mineral phases. Mineral assemblages within any stability field contain the minerals located on the tie lines that bound the stability field. For example, compositions in field "x" contain quartz, calcite, and talc. Field "y" contains calcite, dolomite, and talc. Field "z" contains talc, dolomite, and magnesite. This permits prediction of minerals that can coexist at 400 °C for any composition expressed by the three end-member components.

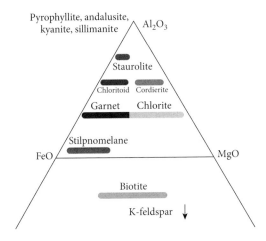

**Figure 18.17** AFM ternary diagram illustrating approximate chemical composition of common silicate minerals. *Source*: Courtesy of Steven Dutch.

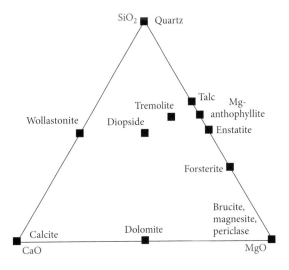

**Figure 18.18** Ternary CMS diagram illustrating approximate chemical composition of common metamorphic minerals that occur in rocks enriched in CaO, $SiO_2$ or MgO. *Source*: Courtesy of Steve Dutch.

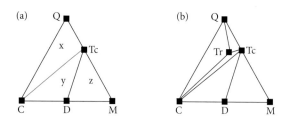

**Figure 18.19** A simplified CMS ternary diagram illustrating common minerals assemblages at (a) 400 °C, and (b) 500 °C where C, calcite; D, dolomite; M, magnesite; Q, quartz; T, talc; Tr, tremolite. *Source*: Courtesy of Steve Dutch.

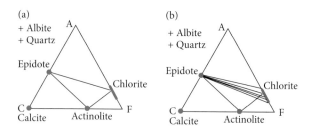

**Figure 18.20** (a) ACF ternary diagram in which the different chemical compositions of the chlorite group minerals are represented by a blue bar line. (b) ACF ternary diagram in which a series of sub-parallel divergent (radiating) tie lines connect various compositions of chlorite group minerals. *Source*: Courtesy of Stephen Nelson.

Figure 18.19b illustrates the same ternary diagram at >500 °C. Note the occurrence of tremolite which documents higher temperature conditions. The Tr–Tc and Tr–Q tie lines cause the ternary diagram to be divided into five distinct fields. Compositions that lie within the x-field on this diagram are composed of quartz-calcite-talc at 400 °C, but at 500 °C are composed of (1) quartz–calcite–tremolite in the field is bound by the Q–Tr–C tie lines, (2) quartz–talc–tremolite in the field is bound by the Q–Tr–Tc tie lines, and (3) tremolite–talc–calcite in the field is bound by the Tr–Tc–C tie lines. Tremolite forms at ~500 °C by the chemical reaction of quartz, talc, and calcite, until one of these phases disappears and a new chemical equilibrium is achieved. The equilibrium mineral assemblages whose compositions plot in stability fields y and z do not change over this temperature range. Figure 18.19 illustrates a simple example of how mineral assemblages change as a result of chemical reactions during progressive metamorphism.

As noted earlier, many minerals have compositions that vary considerably as a result of solid solution. These include plagioclase, garnet, chlorite, and amphibole minerals such as hornblende and actinolite–tremolite. In many of the ternary diagrams presented in this chapter bold, black lines are used to indicate the compositional range of minerals such as chlorite with variable composition due to solid solution (Figure 18.20a). Position on the tie line corresponds to the chemical composition of the mineral. In scientific journals and other research documents, you may see a series of

sub-parallel, diverging lines directed towards the solid solution mineral (Figure 18.20b). These represent tie lines between other minerals and various compositions of the mineral in question. In Figure 18.20b, multiple tie lines are shown between epidote and different compositions of chlorite.

Ternary diagrams are used as a predictive model for mineral assemblages in rocks. In the real world, where the number of components is generally more than three, mineral assemblages are more complex than the simple diagrams presented above. Nevertheless, three component models serve as a very useful starting point for determining the likely assemblage of major metamorphic minerals produced from different protoliths. For these reasons, ternary diagrams continue to be widely used in the study of metamorphic rocks.

Figure 18.21 presents a series of ACF, A'KF, and AFM diagrams for the greenschist, amphibolite, granulite, and eclogite facies for metabasic/calcsilicate (left column) and pelitic (right column) protoliths. Note that several of the diagrams indicate other minerals that occur such as quartz, muscovite, and K feldspar, which require additional components. One can take any composition on one of these sets of diagrams and trace the changes in equilibrium mineral assemblages that occur as temperature and/or pressure conditions change during prograde or retrograde metamorphism. Of course, different diagrams will be used for Franciscan facies series and

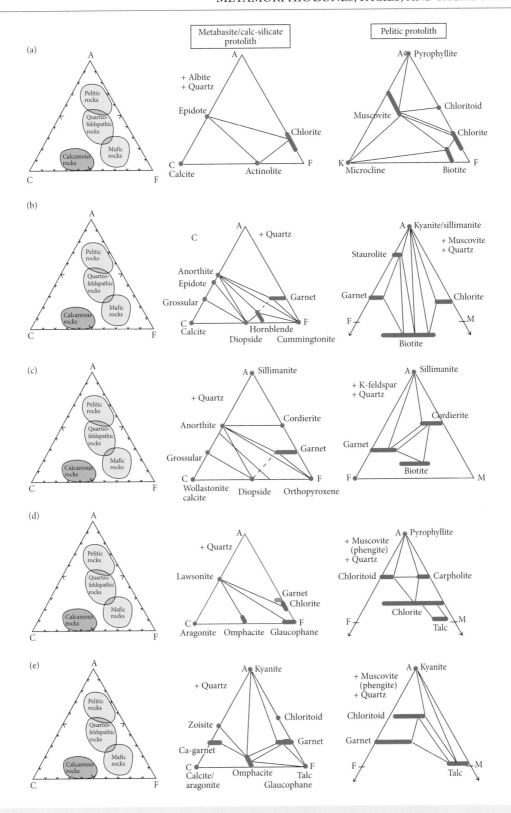

**Figure 18.21**   Ternary diagrams for rocks derived from basites (mafic) and calcsilicates (ACF, A'KF) and pelitic (AFM) rocks. (a) Greenschist facies; (b) amphibolite facies; (c) granulite facies; (d) blueschist facies, and (e) eclogite facies. Note the ACF inset diagram to the left of each diagram indicating a generalized composition of protolith rock type on an ACF diagram. *Source*: Courtesy of Stephen Nelson.

Barrovian facies series depending on the sequence in which metamorphic assemblages and facies develop along their metamorphic trajectory. The above discussion is an abbreviated introduction to ternary diagrams.

While the pioneering work on Barrovian zones, metamorphic facies, metamorphic facies series, mineral assemblage grids and ternary diagrams occurred primarily in the early twentieth century, the second half of the twentieth century provided a context by which we were able to incorporate these mineral and rock assemblages into a global tectonic framework. For more detailed information, refer to a Mineralogical Society of America text by Frank Spear (1995) on metamorphic phase equilibria and Temperature-Pressure-time paths. In the following section, we will relate metamorphic rocks, facies, and facies series to the context of plate tectonics.

## 18.5 METAMORPHIC ROCKS AND PLATE TECTONICS

Different types of metamorphism – including meteorite impacts, dynamic metamorphism, contact metamorphism and burial metamorphism – can occur anywhere on Earth under the appropriate conditions. However, most large-scale metamorphism is inextricably linked to (Figure 18.22):

1 Divergent plate boundaries associated with continental rifts, oceanic rifts, and seafloor spreading, where hydrothermal processes dominate.
2 Convergent plate boundaries associated with subduction, magmatic arcs, and with continental collisions, where dynamothermal metamorphic processes dominate.

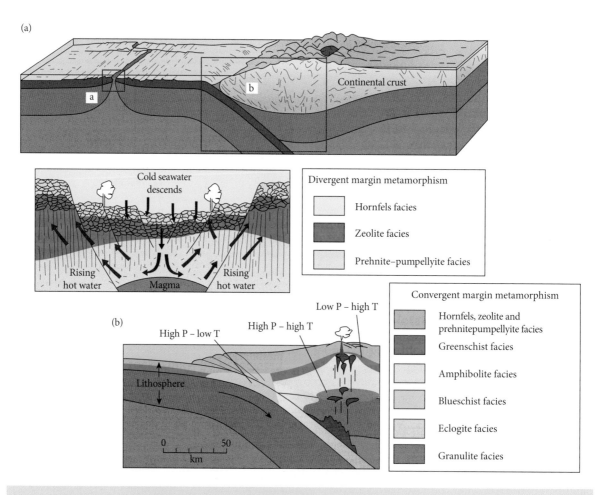

**Figure 18.22** Cross-section view of divergent and convergent plate boundaries and corresponding metamorphic facies. P, pressure and T, temperature.

In the section that follows we relate metamorphic processes, mineral assemblages, rocks, facies, and facies series to the tectonic processes that occur at divergent and convergent plate boundaries.

## 18.5.1 Metamorphism at divergent plate boundaries

Divergent plate boundaries are the site of widespread metamorphism at continental rifts and ocean ridge systems. The rise of warm asthenosphere produces uplift and generates tensional stress. Tensional stresses and extensional deformation result in the thinning and fracturing of lithosphere. Buoyant, hot basaltic magma produced by decompression melting of the asthenosphere rises upward through extensional fissures where it reacts with volatiles and surface waters. Continued extension and magmatism generates rift valleys that evolve to ocean spreading ridges.

### Continental rift basin metamorphism

**Continental rift basin metamorphism** can occur by a number of processes:

1  Contact metamorphism from the intrusion of shallow dikes and sills, and flood basalts;
2  Hydrothermal alteration associated with hot magmatic and wall rock volatile fluids;
3  Dynamic metamorphism due to brittle extensional faulting in the upper crust and ductile shearing in the lower crust;
4  Burial metamorphism producing zeolite and prehnite–pumpellyite facies assemblages due to the deposition of thick, nonmarine detrital sediment sequences in rift basins.

### Ocean ridge metamorphism

**Ocean ridge metamorphism** occurs as continued rifting evolves into sea floor spreading. In sea floor spreading ridge environments, cold, saline seawater flows down through fractures and interacts with hot mafic and ultramafic components of the newly formed ocean lithosphere (Figure 18.22). As a result, extensive hydrothermal metamorphism of basalt, gabbro, and peridotite occurs at ocean ridges resulting in albite–epidote hornfels, zeolite, prehnite–pumpellyite facies mineral assemblages. At deeper levels within oceanic crust and upper mantle, gabbro, and peridotite are altered to higher temperature hornblende hornfels facies assemblages. As a result, mostly hydrous minerals such as zeolite, prehnite, pumpellyite, chlorite, calcite epidote, actinolite, albite, talc, serpentinite, brucite, magnesite, biotite, and hornblende are formed. Because hydrothermal metamorphism of the sea floor occurs in an environment with minimal compressive stress and low confining pressures, foliations, and high-pressure minerals such as garnet are not formed. Metasomatism via the reaction of cold seawater infiltrating into fractured ocean lithosphere, results in the addition of Na, Cl, Br, $CO_3$, $SO_4$, $H_2O$, and $O_2$ to oceanic crust, generating spilitized (Na rich) basalts. At the same time, elements such as Mn, Ni, Zn, Cu, Co, Au, Ag, and Fe are leached from mafic and ultramafic rocks and precipitate onto the seafloor via black smokers producing metallic ore deposits. Eventually, hydrothermally altered oceanic crust, which covers ~70% of Earth's surface, moves away from the spreading ridge and encounters a convergent plate boundary, discussed below.

## 18.5.2 Convergent plate boundaries

As noted earlier, the concepts of metamorphic zones, facies, and facies series predated our understanding of plate tectonics. Miyashiro (1961) noted the existence of paired metamorphic belts, consisting of high P/T terranes adjacent to low P/T terranes – in Japan and other regions within the circum-Pacific region. The question arose as to why paired metamorphic belts exist in the circum-Pacific, Caribbean, Scotia, Eastern Indian Ocean regions, and not elsewhere? Miyashiro's observation was particularly timely as 1960s research revolutionized our knowledge of Earth's dynamic nature. Within a few years, the paired metamorphic belts that encircle the Pacific Ocean were recognized as subduction zone (outer metamorphic belt) and magmatic arc (inner metamorphic belt) assemblages that developed in response to lithospheric plate convergence and subduction.

**Paired metamorphic belts** align with the two major components of convergent margins (Figure 18.22):

1  The outer metamorphic belt occurs on the ocean or trench side and consists of Sanbagawa or Franciscan facies series

rocks, created by lithospheric subduction with low thermal gradients, high pressures and high to very high P/T trajectory assemblages. Rapid, steep subduction favors the development of very high P/T ratios of the Franciscan facies series, whereas slower, shallower subduction favors the development of the moderate to high P/T ratios of the Sanbagawa facies series. The outer metamorphic belt contains hornfels, zeolite, and prehnite–pumpellyite facies metamorphism, in part from early ocean ridge alteration. However, these rocks are commonly overprinted by more recent greenschist facies and high-pressure assemblages such as blueschist and eclogite facies rocks due burial in the subduction zone. Notably absent are the high temperature assemblages of the amphibolite and granulite facies. The outer metamorphic belt is exposed in subduction zone complexes, accretionary wedges, and mélanges.

2   The inner metamorphic belt occurs on the continent or arc side and consists of Buchan or Barrovian facies series rocks, characterized by moderate to high temperature gradients and moderate P/T to low P/T mineral assemblages. The inner metamorphic belt contains hornfels, zeolite, prehnite–pumpellyite, greenschist, amphibolite, and granulite facies metamorphic facies produced by arc magmatism and dynamothermal metamorphism. Notably absent are high P/T assemblages of the blueschist and eclogite facies. The inner metamorphic belt occurs along the magmatic arc complex.

Convergent plate boundaries generate immense compressive stresses, producing widespread, regional belts of dynamothermally metamorphosed rock. Three different convergent margins settings occur based on the type of lithosphere at plate leading edges: (1) ocean–ocean convergence, (2) ocean-continent convergence, and (3) continent–continent collision.

### Ocean–ocean convergence

**Ocean–ocean convergence** is marked by trenches, subduction zones, forearc regions, and volcanic island arcs. **Trenches** (Figure 18.23) are deep curvilinear troughs in ocean basins that mark the surface expression of inclined subduction zones. **Subduction zones** occur along the Pacific rim, the eastern Indian Ocean, the Caribbean, the Mediterranean Sea, and around the Scotia plate in the southern Atlantic Ocean. The subduction of relatively cold lithosphere to depths as great as 700 km produces high P/T Sanbagawa and Franciscan facies that occur only in association with ocean plate subduction. While these high P/T blueschist rocks are generated in the subduction zone, they can later be incorporated into forearc accretionary wedges and mélanges discussed below.

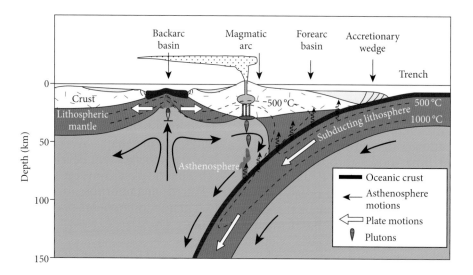

**Figure 18.23**   Cross-section illustrating trench, accretionary prism wedge, forearc basin, and underlying basement, magmatic arc, and backarc basin.

**Forearc regions,** located on the overriding plate between the trench and the volcanic arc, may include the accretionary wedge (prism), forearc basin, and forearc basement. The forearc **accretionary wedge** (Figure 18.23) contains a diverse assemblage of deformed rocks that are tectonically dismembered through faulting, folding, shearing, and mass wasting. The ocean lithosphere that undergoes subduction contains ocean plateaus, seamounts, guyots, ocean ridges, and sediment layers of varying thickness. Much of the ocean lithosphere is subducted to great depths with the downgoing plate. However, in a set of processes collectively referred to as **obduction,** some ocean lithosphere is offscraped, underplated, folded, and sheared into tectonically deformed rock material that accumulates as an imbricated accretionary mélange package on the overlying plate. As a result of subduction zone tectonism and offscraping, a diverse suite of rock assemblages are incorporated into the accretionary wedge and include the following:

1 Basalt, gabbro, and peridotite from the downgoing ocean lithosphere, ocean plateaus, ocean islands, and seamounts.
2 Limestone and chert from pelagic sediment that accumulated on the ocean floor;
3 Muds, mudrocks, and coarser volcanodetrital material derived from the island arc; Intermediate to silicic volcanic and plutonic rocks derived from the adjacent volcanic arc complex or from far traveled arcs that have migrated towards the subduction zone.
4 Metamorphic rocks that are highly variable in composition, origin, and facies. The metamorphic suite includes zeolite, prehnite–pumpellyite, greenschist, amphibolite, granulite.
5 High P/T assemblages such as blueschist, and in rare instances eclogite and UHP, facies rocks derived from the underlying subduction zone.

**Tectonic mélanges** develop within the forearc accretionary prism. Mélange, from the French word for chaotic mixture, commonly contains fragments of rocks that include: basalt, spilite, gabbro, peridotite, chert, limestone, sandstone, serpentinite, phyllite, zeolite, prehnite–pumpellyite, greenschist, amphibolite, blueschist, and eclogite facies rock fragments encased in a scaly mudrock matrix. Some of these rock materials are offscraped at the inner trench wall where the downgoing lithospheric slab is stripped of overlying rock material. Other rock material is transferred from the subducted slab to the base of the overriding slab accretionary wedge by **underplating,** then rises upward as regurgitated rock components metamorphosed at greater depths by enigmatic processes involving faulting and uplift. The accretionary mélange matrix is enriched in pore fluids that are expelled upon further compression to produce a scaley matrix (Figure 18.24). Heated pore fluids enhance hydrothermal alteration and metasomatic reactions. Widespread compressional shortening produces fold and thrust belts in the accretionary wedge whereby rock slices are stacked vertically, sometimes producing an elevated **forearc high,** separated from the magmatic arc by a lower forearc basin.

Highly faulted, partially dismembered **ophiolite complexes** commonly occur in accretionary wedges. Because of their location above subduction zones (Figure 18.25), these are referred to as supra-subduction zone (SSZ) ophiolites. SSZ ophiolites form by the offscraping and offslicing of oceanic lithosphere resulting in the emplacement of oceanic lithosphere fragments onto the overriding hanging wall plate through obduction.

Blueschist facies mineral assemblages, the diagnostic facies of subduction zones, may be exposed in subduction mélanges within the accretionary wedge. The subduction mélange is characterized by a diverse suite of metamorphic facies including high P/T blueschist and eclogite assemblages of the Franciscan and/or Sanbagawa metamorphic facies series (Figure 18.26). In contrast, the forearc basin is largely affected by burial metamorphism and the arc complex is characterized by low to moderate P/T assemblages as discussed below.

**Forearc basins** (Figure 18.23) develop between the high P/T rocks of the accretionary wedge/subduction zone complex and the low P/T rocks of the magmatic arc. Forearc basins are commonly affected by burial metamorphism due to deposition of massive volcaniclastic sediment derived from the adjacent volcanic arc. Total basin fill thicknesses can exceed 10 km, resulting in zeolite to prehnite–pumpellyite facies assemblages. Coombs'

**Figure 18.24**   Franciscan mélange with a mixture of blueschist and greenschist facies rocks from clay to boulder size.

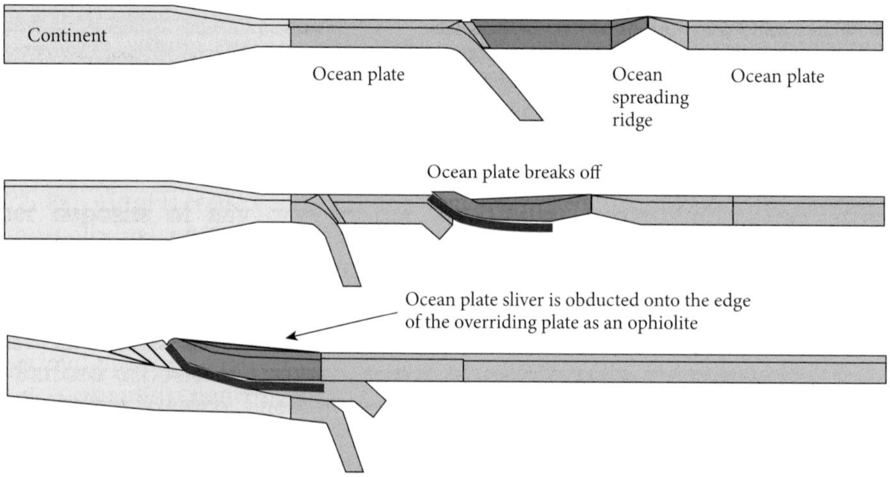

Continent–ocean convergent margin with incipient slab flaking

Continent

Ocean plate

Ocean spreading ridge

Ocean plate

Ocean plate breaks off

Ocean plate sliver is obducted onto the edge of the overriding plate as an ophiolite

**Figure 18.25**   Simplified development of an ophiolite by obduction of ocean lithosphere above a subduction zone. *Source*: Courtesy of Bradley Hacker.

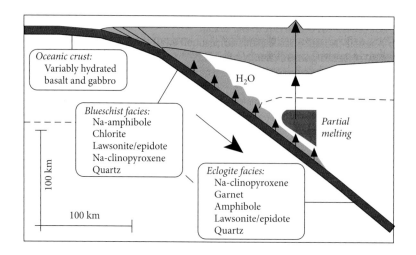

*Oceanic crust:*
    Variably hydrated
    basalt and gabbro

$H_2O$

*Blueschist facies:*
    Na-amphibole
    Chlorite
    Lawsonite/epidote
    Na-clinopyroxene
    Quartz

*Partial melting*

*Eclogite facies:*
    Na-clinopyroxene
    Garnet
    Amphibole
    Lawsonite/epidote
    Quartz

100 km

100 km

**Figure 18.26**   Cross section of high-pressure mineral assemblage within the subducting slab.
*Source*: Courtesy of Simon Peacock.

research in forearc basins of New Zealand served as the pioneering work in understanding the role of burial metamorphism at sub-greenschist facies conditions (Coombs et al. 1959; Coombs 1961). Forearc basins, such as the Great Valley basin (Figure 18.27) in California (USA) provide for excellent agriculture due to the rapid deposition of nutrient rich sediment derived from nearby volcanic arc complexes.

The forearc basin is underlain by **forearc basement**, consisting of overriding plate rocks that existed prior to subduction, younger volcanic–magmatic arc material that formed after subduction began, and partially subducted lower plate rocks that underplate basement rocks. The forearc basement may experience thickening by lower plate underplating or subduction erosion as the arc system migrates towards the trench. In **subduction erosion**, forearc basement rock is eroded and cannibalized into deeper levels of the subduction zone. The loss of the overriding plate can occur by: (1) frontal erosion at the upper, outer toe margin of the plate or (2) by basal erosion where the base of the overlying plate is sliced and transported to great depths in the subduction channel. Whereas subduction erosion can result in the loss of upper plate rocks to great depths, **suprasubduction zone ophiolites** (SSZ) develop by tectonic slicing and subsequent uplift of the forearc basement onto the upper plate where they are exposed and studied to reveal their enigmatic origin.

The **magmatic island arc complex** (Figure 18.23) consists of predominantly intermediate to silicic plutons overlain by composite volcanoes. High geothermal gradients (≥40 °C/km) associated with magmatism generate Buchan, Barrovian, and contact facies series assemblages. Progressive Buchan to Barrovian facies series metamorphism

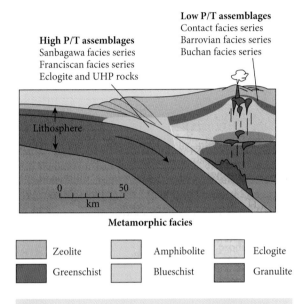

**Figure 18.28** Cross-section of paired (dual) metamorphic belt at convergent margin with high P/T assemblages in the subduction zone and Low P/T assemblages associated with the magmatic arc complex.

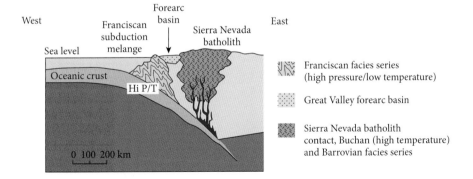

**Figure 18.27** Cross-section showing Mesozoic subduction of ocean crust and development of Franciscan subduction mélange, Great Valley forearc basin, and Sierra Nevada batholith. Note the paired metamorphic belts associated with the Franciscan mélange and the Sierra Nevada batholith complex. *Source*: Levin (2006). © John Wiley & Sons.

produces regionally developed zeolite, green-schist, amphibolite, and granulite facies metamorphism. High geothermal gradients also create localized contact metamorphic aureoles around arc plutons (Figure 18.28). Modern magmatic arc complexes with extensive Buchan and Barrovian facies series rocks are well developed around the Pacific rim, the eastern Indian Ocean and in the Caribbean Sea and Scotia Arc. Ancient examples occur in the Caledonian, Appalachian, Ural, and the Alpine-Himalayan fold and thrust belts. Magmatic underplating and the tectonic underplating of subducted rock slices results in long-term thickening of arc complex litho-sphere. In extreme instances, increasing thick-ness can lead to the development of granulite, eclogite, and UHP assemblages.

In some convergent margin settings, back-arc basins (Figure 18.23) and pull apart basins (Figure 18.29) develop in response to local stresses within the arc complex. Backarc extension in an ocean–ocean convergent mar-gin, such as in the western Pacific Ocean, gen-erates actively spreading marine **backarc basins** that can evolve into marginal seas. Backarc basins can also form in ocean-conti-nent convergent margins, as for example in the Tyrrhenian Sea and in the Mediterranean Sea. Metamorphism in backarc basins is simi-lar to the processes occur in ocean ridges because both systems involve rifting and metasomatism involving cold seawater react-ing with hot magma in fractured mafic rock. Zeolite and prehnite–pumpellyite facies as well as hornfels facies metamorphism occur at relatively shallow depths. Greenschist facies temperatures can be attained at greater depths.

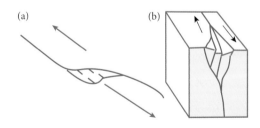

**Figure 18.29**   (a) Map view of pull-apart basin developing by tension along a bend in a sinistral strike-slip fault. (b) Block diagram of a pull-apart basin with normal faults. *Source*: Kevin Hefferan. © John Wiley & Sons.

**Pull-apart basins** form in response to tran-stensional stress in strike-slip fault systems (Figure 18.29). Local extension produces a down-dropped block which infills with sedi-ment thicknesses capable of producing burial metamorphism. Pull-apart basins mostly occur along transform as well as within intra plate settings. As with backarc basins and forearc basins, pull-apart basins generally experience burial metamorphism conditions in the zeolite and prehnite–pumpellyite facies. Pull-apart basins occur notably along California's San Andreas fault zone.

*Ocean-continent convergence: fold and thrust belts and foreland basins*

**Fold and thrust belts** and foreland basins form where the overriding plate is composed of continental lithosphere. They are among the most distal tectonic features created in the overlying plates of many convergent margins. Formerly gently dipping to horizontal conti-nental slope and continental shelf sedi-mentary rocks (such as shales, limestones, dolostones, and sandstones deposited in pas-sive marine settings) are subjected to intense subhorizontal compressive stress. The hori-zontal shortening is accommodated by folds and thrust faults that produce the telescop-ing or "piggybacking" of multiple thrust slices. The net effect is lithospheric thicken-ing by vertically stacking thrust sheets. Metamorphism is driven by the higher tem-peratures and pressures associated with rocks being buried progressively deeper as the orogenic belt is thickened by folding and telescoping of thrust slices. For example, the Sevier fold and thrust belt and Rocky Mountain foreland basin (Figure 18.30) developed in the Mesozoic and Early Cenozoic in response to subduction activity. In the fold and thrust belt, Precambrian, Paleozoic, and Mesozoic sedimentary rocks were folded and displaced eastward by thrust faults. Syn- to post-orogenic magmatism intruded the fold and thrust belt. Lithospheric loading associated with the thickened pile of folded and faulted terranes produced a sub-siding foreland basin eastward of the thrust sheets into which sediments were deposited. The combination of moderate to high tem-peratures, intense compression, thickened

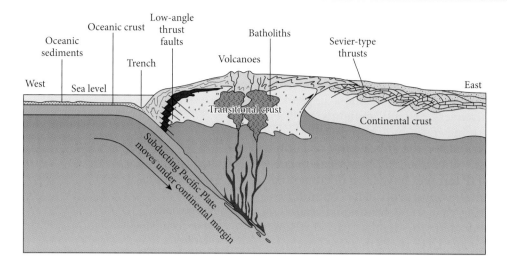

**Figure 18.30**    Mesozoic (~100 Ma) convergent margin activity in the western United States. *Source:* Levin (2006). © John Wiley & Sons.

lithosphere, and late silicic intrusions produces regional Buchan to Barrovian facies series assemblages in fold and thrust belts.

Orogenic thickening of the lithosphere causes an isostatically depressed lithosphere on the continentward side of the fold and thrust belt to form a **foreland basin**. In the foreland basin, an initially deep basin fills with marine deposits producing alternating shale, sandstone, chert, and carbonate layers producing what are referred to as **flysch** deposits. With continued thrusting and infilling, fine-grained marine rocks are succeeded by sandstones and conglomerates in what are referred to as **molasse** deposits. Earlier formed deposits are uplifted and eroded as the fold and thrust belts migrates continentward. Essentially, early deposits are cannibalized and reincorporated into younger basins that develop in the direction of thrusting. Total layer thicknesses in foreland basins can range from a few kilometers to tens of kilometers resulting in burial metamorphism in the zeolite or prehnite–pumpellyite facies. Fold and thrust belts and foreland basins provide reservoirs for oil and gas deposits as well as extensive lignite, bituminous and anthracite coal deposits. For these reasons, fold and thrust belt and foreland basin environments are of great interest with respect to fossil fuel energy production.

*Continent–continent collision*

**Continent–continent collision** marks the end result of continued subduction of the elimination of ocean lithosphere, the accretion of far traveled arc complexes and microcontinents, the closing of an ocean basin and the cessation of subduction as two continental lithospheric plates collide. A suture zone develops, marking the former site of the subduction zone, within an accretionary mélange containing intensely deformed rocks that include ocean plateau, ophiolite, forearc, blueschist, and UHP assemblages. Continued horizontal shortening generates vertical uplift and intense lithospheric thickening accommodated by folding and thrust faulting. The elevated temperatures and pressures create greenschist, amphibolite, and granulite facies with thermal gradients ranging ~20–40 °C/km, in the Barrovian facies series. This situation is well displayed in the Himalayan belt which contains the highest mountain peaks on Earth. The regionally elevated terrain is due to a double lithosphere thickness because of the partial subduction, continental underplating, and telescoping of the Indian lithosphere with the Eurasian plate. The Himalayan Mountains contain an uplifted Tibetan plateau, suture zone, and fold and thrust belt that formed by continent–continent collision. The Himalayan

Mountains constitute the primary modern UHP study area.

Traditionally, plate tectonic theory did not account for continental crust subduction because continental crust (density ~2.8 g/cm³) is more buoyant than Earth's mantle (~3.3 g/cm³). Until the discovery of UHP minerals such as coesite in Alpine rocks by Chopin (1984), no evidence existed for the subduction of continental crust. Since the 1990s, global occurrences of UHP terranes have been documented at continental convergent plate boundaries demonstrating that some continental crust is indeed subducted to depths of ~150 km. Amazingly, some UHP rocks migrate upwards and are exhumed on Earth's surface providing unique portals to deep subduction cycling (Liou et al. 2004; Hacker et al. 2013; Froitzheim et al. 2016). The limited subduction of continental crust provides yet another mind bending component of our understanding of plate tectonic processes.

Metamorphism, igneous activity, and sedimentation are intricately related so that numerous themes recur when discussing Earth Materials. Hopefully, this chapter has elucidated key connections from the early work of Barrow to the modern study of UHP assemblages and continued twists in our understanding of plate tectonis. Many questions remain unresolved such as the origin of greenstone belts and the processes by which blueschists and UHP rocks return to Earth's surface. Geology is a journey without end. We hope you have enjoyed the ride thus far.

## CONTENT ASSESSMENT

1  Why was the work of George Barrow and C.E. Tilley groundbreaking?
2  Describe limitations associated with the metamorphic zone concept
3  Identify the Barrovian zone based upon the following mineral assemblages

| Mineral assemblage | Barrovian zone |
|---|---|
| Chlorite, muscovite, pyrophyllite | |
| Sillimanite, biotite, cordierite, quartz | |
| Staurolite, biotite, almandine | |

4  How has Pentti Eskola's metamorphic facies concept improved upon the early work of Barrow and Tilley?
5  Identify the metamorphic facies based on the following mineral assemblages

| Mineral assemblage | Metamorphic facies | Protolith |
|---|---|---|
| Wollastonite, grossular garnet, tremolite, calcite | | |
| Albite, prehnite, pumpellyite, epidote, chlorite | | |
| Actinolite, chlorite, epidote, sphene, biotite | | |
| Glaucophane, jadeite, lawsonite | | |
| Kyanite, coesite, diamond, jadeite | | |
| Enstatite, hypersthene, olivine, diopside | | |

6  Identify likely tectonic environments associated with the following facies series:

| Metamorphic facies series | Likely tectonic environment |
|---|---|
| Franciscan | |
| Buchan | |
| Barrovian | |

7  Describe the origin of the paired metamorphic belts identified by Miyashiro (1961) and why that discovery was so important with relation to plate tectonics.
8  What does the discovery of Ultrahigh-pressure rocks infer regarding the fate of continental crust in subduction zones?

## REFERENCES

Avé Lallemant, H.G. and Guth, L.R. (1990). Role of extensional tectonics in exhumation of eclogites and blueschists in an oblique subduction setting, northeastern Venezuala. *Geology* 18: 950–953.

Baldwin, S.L., Monteleone, B.D., Webb, L.E. et al. (2004). Pliocene eclogite exhumation at plate tectonic rates in eastern Papua New Guinea. *Nature* 431: 263–267.

Barrow, G. (1893). On an intrusion of muscovite–biotite gneiss in the southeast Highlands of Scotland, and its accompanying metamorphism. *Geological Society of London Quarterly Journal 49*: 330–358.

Barrow, G. (1912). On the geology of the lower Deeside and the southern highland border. *Proceedings of the Geologist's Association 23*: 268–284.

Best, M.G. (2003). *Igneous and Metamorphic Petrology*, 2e. Oxford, UK: Blackwell Publishing 752 pp.

Bird, D.K. and Helgeson, H.C. (1981). Chemical interaction of aqueous solutions with epidote–feldspar mineral assemblages in geologic systems: II. Equilibrium constraints in metamorphic/geothermal processes. *American Journal of Science 281*: 576–614.

Boyd, F.R. (1954). *Amphiboles. Annual Report of the Director*, vol. 53, 108–111. Geophysical Laboratory, Carnegie Institution of Washington Year Book.

Burke, K., Dewey, J.F., and Kidd, W.S.F. (1977). World distribution of sutures – the sites of former oceans. *Tectonophysics 40*: 69–99.

Chao, E.C.T., Shoemaker, E.M., and Madsen, B.M. (1960). First natural occurrence of coesite from Meteor Crater, Arizona. *Science 132*: 220–222.

Chernosky, J.V. (1976). The stability of anthophyllite – a reevaluation based on new experimental data. *American Mineralogist 61*: 1145–1155.

Chernosky, J.V. (1979). The stability of anthophyllite in the presence of quartz. *American Mineralogist 64*: 294–303.

Chopin, C. (1984). Coesite and pure pyrope in high-grade blueschists of the western Alps: a first record and some consequences. *Contributions to Mineralogy and Petrology 86*: 107–118.

Coombs, D.S. (1954). The nature and alteration of some Triassic sediments from Southland, New Zealand. *Royal Society of New Zealand Transactions 82 (I)*: 65–109.

Coombs, D.S. (1961). Some recent work on the lower grades of metamorphism. *Australian Journal of Science 24*: 203–215.

Coombs, D.S., Ellis, A.J., Fyfe, W.S., and Taylor, A.H. (1959). The zeolite facies with comments on the interpretation of hydrosynthesis. *Geochemica et Cosmochimica Acta 17*: 53–107.

De Roever, W. (1956). Some differences between post-paleozoic and older regional metamorphism. *Geologie en Mijnbouw 18*: 123–127.

Ernst, W.G. (1975). *Metamorphism and Plate Tectonic Regimes: Benchmark Papers in Geology*. New York: Halsted Press 440 pp.

Ernst, W.G. (1988). Tectonic history of subduction zones inferred from retrograde blueschist P–T paths. *Geology 16*: 1081–1084.

Eskola, P. (1915). On the relations between the chemical and mineralogical composition in the metamorphic rocks of the Orijarvi region. *Bulletin of the Commission of Geology Finlande 44*: 109–143.

Eskola, P. (1920). The mineral facies of rocks. *Norsk Geologisk Tidskrift 6*: 143–194.

Eskola, P. (1939). Die Metamorphen Gesteine. In: *Die Entstehung der Gesteine* (eds. T.F.W. Barth, C.W. Correns and P. Eskola), 263–407. Berlin: Julius Spring Publishers.

Froitzheim, N., Miladivona, I., Janak, M. et al. (2016). Devonian subduction and syncollisional exhumation of continental crust in Lofoten, Norway. *Geology 44*: 223–226.

Furukama, Y. (1993). Magmatic processes under arcs and formation of the volcanic front. *Journal of Geophysical Research 98*: 8309–8319.

Fyfe, W.S., Turner, F.J., and Verhoogen, J. (1958). Metamorphic reactions and metamorphic facies. *Geological Society of America Memoir 73* 259 pp.

Gibbs, J.W. (1928). *The Collected Works of J. Willard Gibbs*, vol. I. New Haven, CT: Thermodynamics: Yale University Press 438 pp.

Greenwood, H.J. (1963). The synthesis and stability of anthophyllite. *Journal of Petrology l4*: 317–351.

Haas, H. and Holdaway, M.J. (1973). Equilibria in the system $Al_2O_3–SiO_2–H_2O$ involving the stability limits of pyrophyllite. *American Journal of Science 273*: 449–464.

Hacker, B.R., Gerya, T.V., and Gilotti, J.A. (2013). Formation and exhumation of continental UHP terranes. *Elements (Quebec) 9*: 289–293.

Harker, A. (1932). *Metamorphism: A Study of the Transformation of Rock Masses*. London: Methuen Publishers 376 pp.

Hashimoto, M. (1966). On the prehnite–pumpellyite metagraywacke facies. *Geological Society of Japan Journal 72*: 253–265.

Hower, J.E., Eslinger, M.E., Hower, M.E., and Perry, E.A. (1976). Mechanism of burial metamorphism of argillaeous sediments: I. Mineralogical and chemical evidence. *Geological Society of America Bulletin 87*: 725–737.

Kennedy, W.Q. (1949). Zones of regional metamorphism in the Moine schists of the western Highlands of Scotland. *Geological Magazine 86*: 43–56.

Levin, S. (2006). *The Earth Through Time*, 8e. New York: Wiley 547 pp.

Liou, J.G., Maruyama, S., Wang, X., and Graham, S. (1990). Precambrian blueschist terranes of the world. *Tectonophysics 181*: 97–111.

Liou, J.G., Tsujimori, T., Zhang, R.Y. et al. (2004). Global UHP metamorphism and continental subduction/collision: the Himalayan model. *International Geology Review 46*: 1–27.

Lister, G.S., Banga, B., and Feenstra, A. (1984). Metamorphic core complexes of Cordilleran type in the Cyclades, Aegean Sea, Greece. *Geology 12*: 221–225.

Maruyama, S., Cho, M., and Liou, J.G. (1986). Experimental investigations of blue schist–green schist transition equilibria: pressure dependence of $Al_2O_3$ contents in sodic amphiboles – a new geobarometer. *Geological Society of America Memoir 164*: 1–16.

Metz, P. and Winkler, H.G.F. (1963). Experimentelle gesteinsmetamorphose – VII. Die bildung von talc aus kieseligem dolomite. *Geochemica et Cosmochimica Acta 27*: 431–457.

Miyamoto, A., Enami, M., Tsuboi, M., and Yokoyama, K. (2007). Peak conditions of kyanite-bearing quartz eclogites in the Sanbagawa metamorphic belt, central Shikoku, Japan. *Journal of Mineralogical and Petrological Sciences 102*: 352–367.

Miyashiro, A. (1961). Evolution of metamorphic belts. *Journal of Petrology 2*: 277–311.

Miyashiro, A. (1973). *Metamorphism and Metamorphic Belts*. New York: Wiley 479 pp.

Miyashiro, A. (1994). *Metamorphic Petrology*. New York: Oxford University Press 416 pp.

Miyashiro, A. and Shido, F. (1970). Progressive metamorphism in zeolite assemblages. *Lithos 6*: 13–20.

Möller, A., Appel, P., Mezger, K., and Schenk, V. (1995). Evidence for a 2 Ga subduction zone eclogites in the Usagaran belt of Tanzania. *Geology 23*: 1067–1070.

Murck, B.W. and Skinner, B.J. (2016). *Visualizing Geology*, 4e. Hoboken NJ: Wiley Publishers, 544p.

Pearson, D.G., Canil, D., and Shirey, S.B. (2003). Mantle samples included in volcanic rocks: xenoliths and diamonds. In: *Treatise on Geochemistry*, vol. 2 (eds. R.W. Carlson, H.D. Holland and K.K. Turekian), 171–275. New York: Elsevier.

Philpotts, A.R. (1990). *Principles of Igneous and Metamorphic Petrology*. New York: Prentice Hall 498 pp.

Philpotts, A.R. and Ague, J.J. (2009). *Principles of Igneous and Metamorphic Petrology*, 2e. New York: Cambridge University Press 667 pp.

Pilchin, A. (2003). The role of serpentinization in exhumation of high- to ultra-high-pressure metamorphic rocks. *Earth and Planetary Science Letters 237*: 815–828.

Platt, J.P. (1986). Dynamics of orogenic wedges and the uplift of high pressure metamorphic rocks. *Geological Society of America Bulletin 97*: 1037–1053.

Shermer, E.R. (1990). Mechanisms of blueschist facies metamorphism and preservation in an A-type subduction zone, Mount Olympos region, Greece. *Geology 18*: 1130–1133.

Spear, F. (1995). *Metamorphic Phase Equilibria And Presure–Temperature–Time–Paths*. Mineralogical Society of America, 799 pp. ISBN: 0-939950-34-0.

Thompson, J.B. Jr. (1957). The graphical analysis of mineral assemblages in pelitic schists. *American Mineralogist 42*: 842–858.

Tilley, C.E. (1925). A preliminary survey of metamorphic zones in the southern Highlands of Scotland. *Quarterly Journal of the Geological Society of London 81*: 100–112.

Turner, F.J. (1958). Mineral assemblages of individual metamorphic facies. In: *Metamorphic Reactions and Metamorphic Facies*, vol. 73 (eds. W.S. Fyfe, F.J. Turner and J. Verhoogen), 199–239. Geological Society of America Memoir.

Turner, F.J. (1981). *Metamorphic Petrology: Mineralogical, Field and Tectonic Aspects*. New York: McGraw-Hill 524 pp.

Turner, F.J. and Verhoogen, J. (1951). *Igneous and Metamorphic Petrology*. New York: McGraw-Hill 602 pp.

Unruh, J.R., Loewen, B.A., and Moores, E.M. (1995). Progressive arcward contraction of a mesozoic – tertiary fore-arc basin, southwestern Sacramento Valley, California. *Geological Society of America Bulletin 107*: 38–53.

Wiseman, J.D.H. (1934). The Central and South-west Highland epidiorites: a study in progressive metamorphism. *Quaterly Journal of the Geological Society of London 90*: 354–417.

Wones, D.R. and Dodge, F.C.W. (1968). *On the Stability of Phlogopite*, Special Paper, vol. 101. Geological Society of America 242 pp.

Yardley, B.W.D. (1989). *An Introduction to Metamorphic Petrology*. London, UK: Longman Publishers 248 pp.

Zwart, H.J. (1967). The duality of orogenic belts. *Geologie en Mijnbouw 46*: 283–309.

# Chapter 19

# Mineral resources and hazards

Minerals have many economic applications and have been broadly grouped into gemstones, industrial minerals, and metallic ores. More recently, we have identified critical or strategic minerals essential Earth materials necessary to modern society. Recent developments in telecommunication, healthcare, information technology, energy production, and climate change create breakthrough uses and Herculean demands for Earth materials.

**Gem minerals** have been used throughout human history for currency, ornamentation, jewelry, and prestige. Gem mineral value relates to factors such as hardness, brilliance, beauty, rarity and social status. **Industrial minerals** are widely used due for physical and/or chemical properties allowing them to be used in fertilizers, aggregates, abrasives, fillers, fluxes, refractory and construction materials, and many manufactured items increases. Industrial mineral use grows in the twenty-first century as our human population and resource demands continue to grow. An **ore** is a naturally occurring mineral or rock deposit sufficiently enriched in metal that it may be economically mined, processed, and sold for a profit. Metallic ores have been so important in human history since the Stone Age (2.5 million years ago) that civilizations are largely classified based upon the metals developed during that time such as the Bronze Age (beginning ~5300 years ago) and the Iron Age (beginning ~3200 years ago). In what now may be termed the silicon age, metals continue to play critical roles due to the electromagnetic properties associated with metallic bonds. **Critical minerals** are those essential minerals that directly impact economic viability and national security, have no viable substitutes, and face potential supply disruption based upon scarcity or limited geographic access. The US considers 35 elements and/or mineral groups, most of which are metals, to be critical minerals. Critical minerals include

*Earth Materials*, Second Edition. Kevin Hefferan and John O'Brien.
© 2022 John Wiley & Sons Ltd. Published 2022 by John Wiley & Sons Ltd.
Companion website: www.wiley.com/go/hefferan/earthmaterials2

aluminum (bauxite), antimony, arsenic, barite, beryllium, bismuth, cesium, chromium, cobalt, fluorspar, gallium, germanium, graphite, hafnium, helium, indium, lithium, magnesium, manganese, niobium, platinum group metals, potash, the rare-earth-elements group, rhenium, rubidium, scandium, strontium, tantalum, tellurium, tin, titanium, tungsten, uranium, vanadium, and zirconium (National Resource Council 2008; Schulz et al. 2017; USGS 2020).

## 19.1 ORE MINERALS

Of ~4500 known minerals, less than 100 are important metallic ore minerals. Earth's metallic minerals are widely distributed in low concentrations. In relatively rare occurrences, enriched mineral deposits constitute an ore body that can be mined, processed, and sold for a profit. Enriched ore bodies commonly occur with **gangue minerals** that are not currently economically valuable. Common gangue minerals include quartz, calcite, dolomite, barite, gypsum, feldspar, garnet, chlorite, clay, fluorite, apatite, pyrite, marcasite, pyrrhotite, and arsenopyrite. An ore is deemed valuable due to a combination of physical and chemical properties, availability, uses, cost of processing, impact on human health, market, and demand. Several examples of these factors are briefly described. Aluminum was first isolated as a metal in 1824 by Hans Christian Orsted; its limited production through much of the nineteenth century resulted in aluminum's value exceeding that of silver and gold. However, when Carl Joseph Bayer discovered a method to extract aluminum from bauxite in 1889, its value diminished thereafter as production increased. Conversely, lead was once widely used in gasoline, paint, batteries, glassware, and many other items. However, since the 1970s, lead use has declined due to its adverse neurological impact on human health. Prior to the 1980s, rare earth metal value was low due to limited market demand. However, with the advent of telecommunications (cell phones and computers), renewable energy expansion (wind turbines, solar photovoltaics, fuel cells), and electric and hybrid vehicles, strategic rare earth element (REE) metals are now highly valued. Until the mid-twentieth century, only ~15 metallic elements had significant practical use. Today, nearly all the naturally occurring elements in the periodic table are utilized. For example, the production of a modern high capacity computer chip incorporates approximately 60 of the 92 naturally occurring elements (Schulz et al. 2017). In short, Earth's scientists and mineral resources are more important than ever before.

Ore minerals contain metallic elements such as platinum, gold, silver, nickel, cobalt, iron, lead, palladium, tungsten, aluminum, chromium, zinc, molybdenum, and copper. As cations, metals commonly bond with anions which are removed by processing. As a result, most metallic ore minerals are oxides, sulfides, or native elements from which the metallic components are easily separated during refining processes releasing oxygen or sulfur-bearing waste material. Sulfur-bearing waste material is a particular hazard because in reacting with water, sulfuric acid is produced which contaminates groundwater and surface water. Relatively few ores are silicate minerals due in part to the difficulty in separating metallic elements from the tightly bonded silica and the large amounts of slag waste material generated. For example, aluminum is the third most common element in Earth's crust and is an essential element in many feldspar minerals, the most common mineral group in Earth's crust. However, it is not economically feasible to extract aluminum ore from feldspars at current or foreseeable future market prices.

Ores can be classified on mode of development. **Syngenetic** ore deposits represent primary mineralization where ores form at the same time (synchronous) as the rock. **Epigenetic** ore deposits form by secondary mineralization wherein ores concentrate after the rock has formed. Epigenetic mineralization can post-date the host rock by hundreds of millions of years. Hot silicate fluids, water, and other volatile fluids play critical roles in dissolving, transporting, and precipitating metals in concentrated deposits. Hydrothermal fluids leach and concentrate metallic ores in two fundamental ways. **Hypogene enrichment** is a primary or syngenetic process that occurs as deep, upwelling magmatic fluids concentrate ore synchronous with rock development. These primary ore deposits are commonly later altered deep within Earth to create metamorphic ore deposits or reworked by near surface processes to produce sedimentary ore deposits. **Supergene enrichment** occurs as surface waters percolate downward, leaching near surface metals and concentrating them at deeper levels within Earth's crust. Supergene enrichment is an epigenetic process that commonly develops below oxidizing zones near the water table.

Ores may possess tabular, cylindrical, or irregular forms depending upon the stress and tectonic conditions, rock characteristics, and mode of emplacement. Ores can accumulate in narrow, concentrated zones or be broadly disseminated in low concentrations over large areas. Tabular orebodies include veins or layered, stratiform bodies. Tabular ore bodies commonly form along planar fracture systems, igneous layers, metamorphic foliations, or sedimentary beds. Cylindrical orebodies, appropriately referred to as pipes or chimneys, commonly form in response to ore enriched magma or hydrothermal solutions that rise buoyantly toward Earth's surface. Lensoid ore deposits are referred to as podiform, meaning that they have a foot-like shape. Irregular or broadly disseminated ore deposits typically develop within or in close proximity to large igneous intrusions in which ore-bearing fluids infiltrate, and in many cases locally metamorphose, surrounding rock. Intense forces associated with igneous intrusions result in elevated fluid pressures. These fluid pressures are directed outward into the rock body resulting in extensive hydrofracturing of the surrounding host rock. We will now address some of the major ore forming environments, many of which are associated with tectonic plate boundaries.

### 19.1.1 Igneous ore forming environments

Igneous processes create massive ore deposits at convergent and divergent plate boundaries as well as within intraplate settings. Magmatic ore forming processes develop from eruptions of lava on Earth's surface, as in volcanogenic massive sulfides, as well as from the intrusion of magma within Earth.

**Volcanogenic massive sulfide (VMS)** deposits are copper–zinc–lead sulfide deposits concentrated on the ocean floor, primarily at either divergent or convergent plate boundaries (Figure 19.1a). Other VMS ores include silver, gold, cobalt, nickel, iron, tin, selenium, manganese, cadmium, bismuth, germanium, gallium, indium, and tellurium. While plate margin magmatism at ocean spreading ridges or volcanic arcs is the driving force behind the development of VMS deposits, extensive hydrothermal alteration and metamorphism play critical roles in altering the igneous rocks and concentrating ores through both hypogene and supergene enrichment. How does this occur? Tensional forces at ocean ridges, back arc basins, and intra-arc basins produce extensional fractures through which hot magmatic fluids rises upward and interact with descending cold seawater. As first documented by Francheteau et al. (1979), upwelling hydrothermal fluids release plumes of black, metal laden "smoke" from ocean ridge vents. The metal laden ~360°C black smokers create chimney-like structures enriched in chalcopyrite, sphalerite, pyrite, and anhydrite (Figure 19.1b). Metasomatic mixing of seawater and hydrothermal fluids also produce "white smokers," erupting ~300°C plumes that precipitate quartz, calcite, anhydrite, pyrite, and barite on the sea floor. Other common sulfide and oxide ore minerals at VMS include galena, chalcocite, bornite, cobaltite, magnetite, hematite, pyrolusite, and enargite (Herzig and Hannington 1995; Humphris et al. 1995; Galley et al. 2007).

Black smokers produce mounds and nodules enriched in manganese, zinc, iron, cobalt, copper, and nickel. Beneath the overlying mound, hydrothermal fluid flow and steeply inclined chemical and thermal gradients produce cylindrical pipe zones with a higher temperature, chalcopyrite rich inner core and lower temperature, sphalerite-galena rich outer zones (Galley et al. 2007). While ancient VMS deposits on land (Figure 19.2) have been mined for thousands of years, it was not until the discovery of actively forming, VMS deposits at 21°N Latitude of the East Pacific Rise that the genesis of these metal deposits was understood (Herzig and Hannington 1995). VMS deposits accumulate by the growth and subsequent collapse of black smoker chimneys that results in layered (stratiform), lens shaped (podiform) metal deposits and/or collapse breccias overlying magma feeder, stockwork sulfide-silicate dike structures. Modern VMS deposits occur globally and are associated with hydrothermal vents at divergent and convergent boundaries (Figure 19.3).

(a)

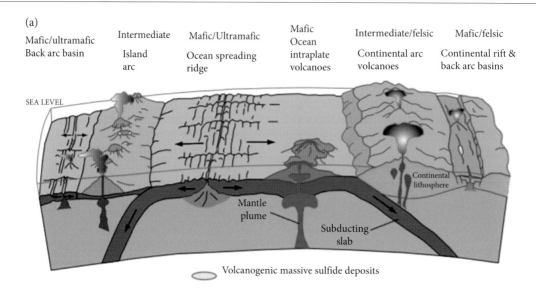

Mafic/ultramafic
Back arc basin

Intermediate
Island
arc

Mafic/Ultramafic
Ocean spreading
ridge

Mafic
Ocean
intraplate
volcanoes

Intermediate/felsic
Continental arc
volcanoes

Mafic/felsic
Continental rift &
back arc basins

(b)

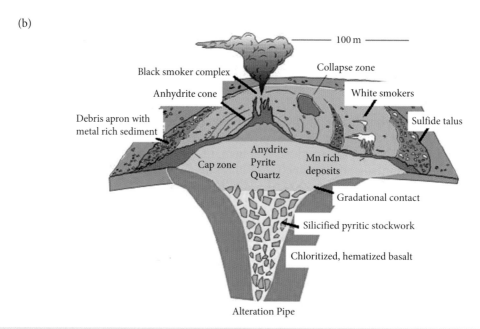

**Figure 19.1** (a) Block diagram illustrating tectonic locations where VMS deposits originate and their associated rock suites. *Source*: Morgan and Schulz (2012). Public Domain. © The United States Department of the Interior. (b) Schematic VMS cross section illustrating lower temperature cap rock enriched in anhydrite, pyrite, and quartz overlying higher temperature sulfide and oxide ore assemblages within a highly altered pipe vent system. *Source:* Galley et al. (2007). © GAC.

Various VMS deposit classification schemes have been developed based on genetic origin (Cox and Singer 1986; Prokin and Buslaev 1999) and/or dominant rock assemblages (Franklin et al. 2005; Mosier et al. 2009; Morgan and Schulz 2012), as indicated in Table 19.1. The Cox and Singer (1986) classification focused on famous VMS localities such as the Island of Cyprus and the Besshi and Kuroko regions of Japan. More recent VMS classifications focus on rock type and tectonic setting distinguishing three (Morgan and Schulz 2012) to five groupings (Mosier et al. 2009).

Whereas Morgan and Schulz (2012) define five major groupings based on lithology and tectonic setting, Mosier et al. (2009) recognize three groups based primarily on rock type.

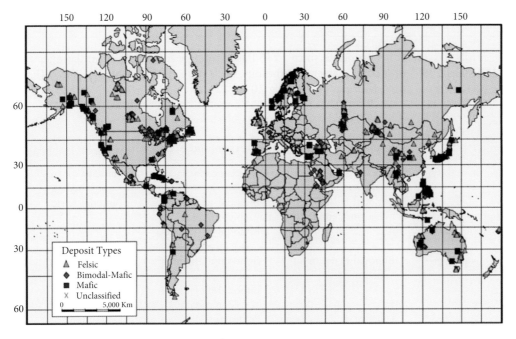

Mosier, Berger and Singer, 2009 USGS Open File Report 2009-1034
VMS deposits ofthe world- database and grade and tonnage models

**Figure 19.2** Global distribution of known land-based VMS deposits (Mosier et al. 2009) based on type of igneous rock suite. *Source*: Mosier et al. (2009). Public Domain. © The United States Department of the Interior.

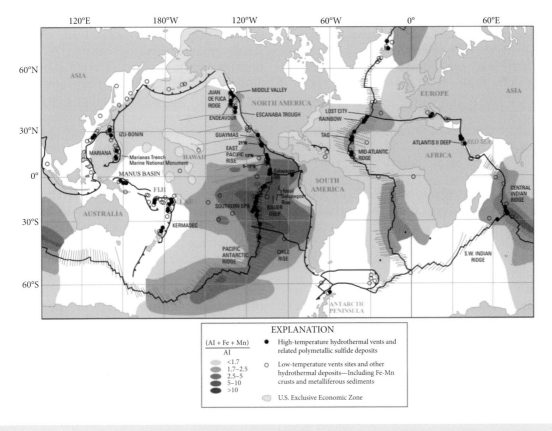

**Figure 19.3** Global distribution of modern hydrothermal vents and related polymetallic sulfide deposits identified on the ocean floor. *Source*: Boström et al. (1969), Hannington et al. (1999), and Shanks and Thurston. (2012). © John Wiley & Sons.

**Table 19.1**   VMS classifications proposed since 1986 based on type locality (Cox and Singer 1986; Prokin and Buslaev 1999) or a combination of lithology, rock suite, and tectonic boundary (Franklin et al. 2005; Mosier et al. 2009; Morgan and Schulz 2012).

| Cox and Singer (1986) | Prokin and Buslaev (1999) | Franklin et al. (2005) | Morgan and Schulz (2012) | | Mosier et al. (2009) |
|---|---|---|---|---|---|
| Kuroko | Baimak-type | Siliciclastic-felsic in a mature epicontinental arc | Siliciclastic-felsic in a mature epicontinental arc | Felsic to intermediate | Felsic |
| | | Bimodal-felsic in an epicontinental arc | Bimodal-felsic in an epicontinental arc | Bimodal-felsic | |
| | | | Bimodal-felsic in an oceanic arc | | |
| | Urals-type | Bimodal-mafic in an oceanic arc | Bimodal-mafic in an oceanic arc | Bimodal-mafic | Bimodal-mafic |
| | | | Bimodal-mafic in an epicontinental arc | | |
| Besshi | Besshi-type | Pelite-mafic in a mature oceanic backarc | Pelite-mafic in a mature oceanic backarc | Pelite-mafic | Mafic |
| Cyprus | Cyprus-type | Mafic in a primitive oceanic backarc | Mafic in a primitive oceanic backarc | Mafic | |
| | | | Mafic in a midoceanic ridge | | |

*Source*: Mosier et al. (2009). Public Domain. © The United States Department of the Interior.

1   Felsic group consists dominantly of rhyo-dacite to rhyolite volcanic rocks. Dacite to andesite may be abundant; basalt is usually less than 10%. Siliciclastic rocks (sandstone or quartzite) may be associated with the felsic volcanic rocks. In a felsic-dominated bimodal sequence, basalt units constitute 10–50%; and intermediate rocks (dacite and andesite) are rare to absent.

2   Bimodal-mafic group consists predominantly of mafic volcanic rocks (basalt), with rhyolite to dacite constituting 10–40%.

3   Mafic group consists dominantly of mafic volcanic rocks (andesitic basalt to basalt). Mafic volcanic rocks may be associated with gabbro, diabase, and ultramafic rocks of an ophiolite sequence. Pelitic rocks may also be abundant and dominate some sequences. Felsic volcanic rocks are rare (less than 5%) or absent.

Mosier et al. (2009) relate the three major rock types to the tonnage, type, and grade of ore such that (i) felsic and bimodal-mafic groups are significantly larger in ore tonnage than the mafic group; (ii) the felsic group has significantly higher zinc, lead, and silver grades than the bimodal-mafic group; and the bimodal-mafic group has significantly higher zinc grades than mafic group; (iii) the mafic group has significantly higher copper grades than both the felsic and bimodal-mafic groups.

Within intracontinental **rift basins**, ore deposits form by the initial precipitation of metals from magma, with later reworking of metallic elements and their incorporation into overlying sedimentary rocks. Basalt and rhyolite lavas erupt in continental rift basins producing vesicular textures associated with high gas content. In response to hydrothermal fluids and oxidation processes, metals such as iron, copper, zinc, nickel, and platinum group elements, precipitate in gas vesicles forming ore deposits. In some locations, rift basin ores are leached from their igneous host rocks and concentrated in surrounding sedimentary rocks. The Bethlehem, Pennsylvania iron deposits and the Keweenaw, Michigan iron–copper–nickel deposits represent two important rift deposits involving sedimentary deposits bearing ore minerals derived from underlying rift basalts.

The Keweenaw basin in the Lake Superior region of Michigan is the most famous example of a rift basin deposit (Figure 19.4). Keweenaw ore, principally copper, was mined from the 1840s to the 1950s. The Keweenaw belt represents a 1.1 billion year old continental rift that formed synchronous with the Grenville Orogeny in Eastern North America. Continental rifting produced a downdropped basin which infilled with 2.5–5 km thick basaltic lava flows and underlying gabbro, troctolite, and granite intrusions. Copper, nickel, iron, titanium, plati-num group elements (platinum and palladium) and silver precipitated in the igneous rocks. Weathering resulted in dissolution of metals and subsequent secondary enrichment in over-lying conglomerate, sandstone, and shale lay-ers resulted in the world's largest concentration of native copper (Bornhorst and Mathur 2017). The Keweenaw rift is underlain by the Duluth complex (Figure 19.5) which uplifted, frac-tured, and concentrated valuable metallic ores

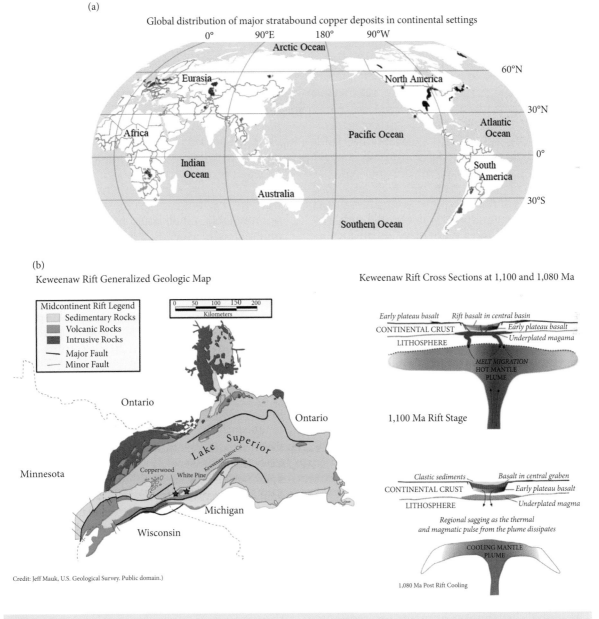

**Figure 19.4**    (a) Generalized geologic map of the Keweenaw rift, which is part of the Midcontinent rift around Lake Superior. Red stars correspond to native copper deposits. (b) Conceptual cross sections indicate magmatic heating and rifting 1.1 billion years ago followed by cooling and subsidence 1.080 billion years ago. *Source*: Images courtesy of Jeff Mauk and the U.S. Geological Survey.

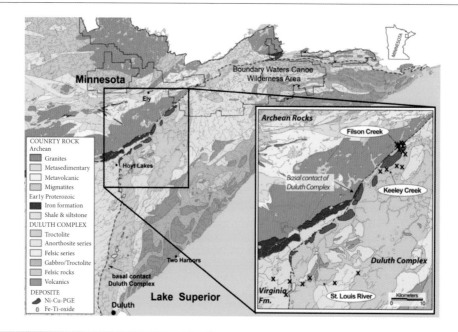

**Figure 19.5** Geologic map of the Duluth Complex, part of the Keweenaw rift, illustrating platinum group elements (PGE), copper, nickel, titanium, and iron deposits associated with rifting and magmatism. *Source*: Piatak et al. (2016). Public domain. © The United States Department of the Interior.

in the overlying crust. The Duluth gabbro-peridotite complex is recognized as an important platinum group element ore deposit based on exploratory drill core results (Koerber and Thakurta 2019).

At depth within Earth's interior, **immiscible metal sulfide** liquids separate from metal-rich, silicate magma. Evidence for immiscible liquid separation includes fluid inclusions of immiscible fluids in a host material such as occurs with oil and water in salad dressing. Immiscible liquid separation in silicate magma results in the concentration of metallic sulfide deposits containing copper, iron, nickel, chromium, vanadium, palladium, and platinum. Common ore minerals include chalcocite, bornite, chalcopyrite, chromite, pentlandite, nickelline, magnetite, and hematite. These valuable ores occur in layered igneous intrusions, which consist of stratiform gabbroic and pyroxenite rocks that crystallized in layered intrusions. Exceptional examples of ore-bearing layered igneous intrusions include the Stillwater mine in Montana, the Greenland's Skaergaard intrusion, and South Africa's Bushveld Complex. The global distribution of major layered igneous intrusions is illustrated in Figure 19.6.

Platinum group element (PGE)-enriched sulfide mineralization occurs due to exsolution of immiscible sulfide liquid from mafic-ultramafic magma as it ascends toward Earth's surface. Magma composition changes that promote sulfide exsolution include the assimilation of crustal country rock, magma mixing, fractional crystallization, and venting of the magma chamber associated with eruptive phases (Ripley 1999; Barnes et al. 2006; Holwell and McDonald 2010; Naldrett 2010; Zientek 2012).

**Porphyry deposits,** the source of 74% of the world's copper, develop primarily from granite or granodiorite intrusions at convergent plate boundaries (Hammarstrom et al. 2019). Forces associated with magma injection, coupled with hydrothermal fluid pressures, result in the widespread metamorphic alteration (Lowell and Guilbert 1970) and diffuse infiltration of ore-bearing fluids into a complex network of fractures and pore spaces of the pluton and the surrounding rock at temperatures >500 °C (Figure 19.7). Cooling and crystallization results in massive, low concentration (<2%) deposits of copper, molybdenum, gold, zinc, mercury, silver, lead, lithium, and tin disseminated throughout a zone of alteration within the host rock. Common porphyry ore minerals include chalcocite, chalcopyrite, bornite, molybdenite, sphalerite, cinnabar, enargite,

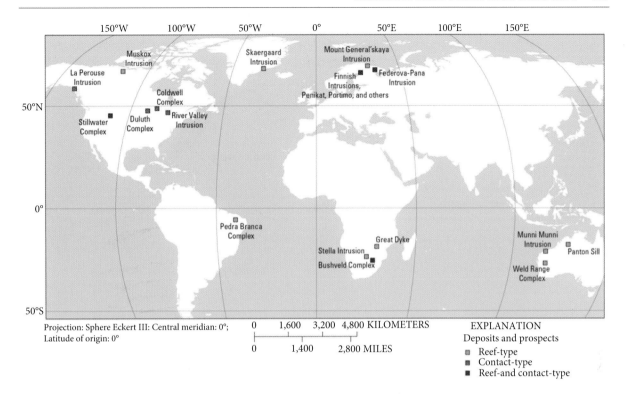

**Figure 19.6**   World map displaying major PGE ore deposits. *Source*: Zientek (2012). Public Domain. © The United States Department of the Interior.

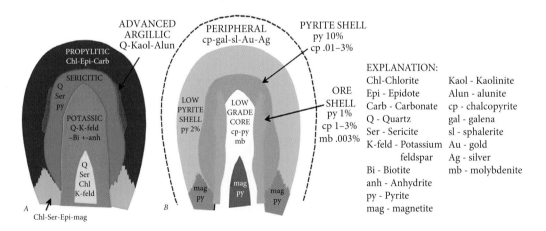

**Figure 19.7**   (a) Cross section of porphyry ore deposits illustrating metamorphic alteration classification (Table 15.1). (b) Cross section of porphyry deposit indicating major ore mineral occurrences. *Source*: John et al. (2010). Public Domain. © The United States Department of the Interior.

spodumene, cassiterite, and galena. Large tonnage (tens to thousands of million metric tons of ore), low grade porphyry deposits are commonly mined in large open pits.

Porphyry copper deposits occur around the Pacific rim, such as in western North America and the Andes Mountains of western South America (Figure 19.8). Chile contains the

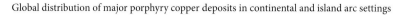

Global distribution of major porphyry copper deposits in continental and island arc settings

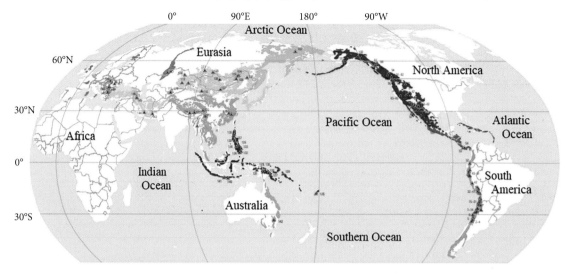

**Figure 19.8** This global map illustrates 150 of the major porphyry deposits. Note that these deposits are concentrated along modern and ancient convergent plate boundaries. *Source*: Hammarstrom et al. (2019) and Singer (1995). © John Wiley & Sons.

largest copper deposits of any type in the world. Other sites include the controversial and undeveloped Pebble deposit in Alaska, the Bingham deposit in Utah, the Continental/ Butte deposit in Montana, the Morenci-Metcalf and Safford deposits in Arizona, and the Cananea deposit in Mexico. The Southeast Asia Archipelagos region also hosts immense porphyry deposits in the Philippines, Indonesia, and Papua New Guinea. Other copper porphyry deposits include the Grasberg Group deposit in Indonesia, the Oyu Tolgoi deposit in Mongolia, and the Reko Diq Group in Pakistan (Hammarstrom et al. 2019).

**Vein deposits** are commonly associated with magma intrusions due to extensional hydrofractures that develop as hydrothermal fluids escape upward from the magma. Metals such as gold are dissolved in volatilized gases that subsequently cool and crystallize producing highly concentrated metal vein deposits (Figure 19.9). Vein deposits occur at plate boundaries as well as intraplate settings. Because of the high concentration of metals, vein deposits are referred to as lode deposits and can contain gold, silver, copper, or metal sulfides that occur with gangue minerals such as quartz or calcite. California's 1849 gold rush and Canada's Klondike gold rush are

familiar examples of vein-filling fracture deposits produced by granitic intrusions at convergent margins.

Pegmatites are coarse grained igneous textures that develop due to slow cooling of a low viscosity magma containing a high volatile content (Figure 19.10). Volatiles such as OH, fluorine, boron, and $H_2O$ promote ion diffusion and the development of large crystals. **Pegmatite deposits** are closely associated with vein deposits described above. Granitic pegmatites (containing quartz, feldspars, amphiboles, and micas as major minerals) occur within continental plates and at convergent plate boundaries. Minor and accessory minerals in pegmatites include beryl, apatite, lepidolite, spodumene, cassiterite, wolframite, wulfenite, molybdenite, scheelite, tourmaline, topaz, uraninite, lithiophillite, columbite, tantalite, gold, and silver. Pegmatites are important producers of tin, molybdenum, tungsten, gold, and silver and the primary source for beryllium, lithium, tantalum, niobium, and REE ores. Lithium–cesium–tantalum (LCT) pegmatites are of special interest because they host most of the world's supply chain of rare and strategic elements, accounting for about one-third of world lithium production, most of the tantalum, and all of the cesium. LCT

(a)

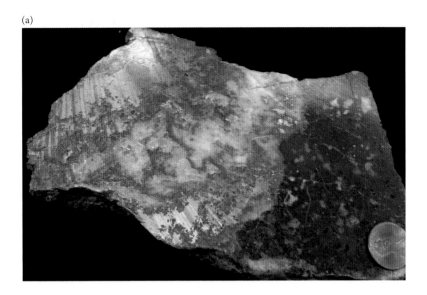

**Figure 19.9** (a) Copper and quartz vein within metamorphosed basalt.

(a)

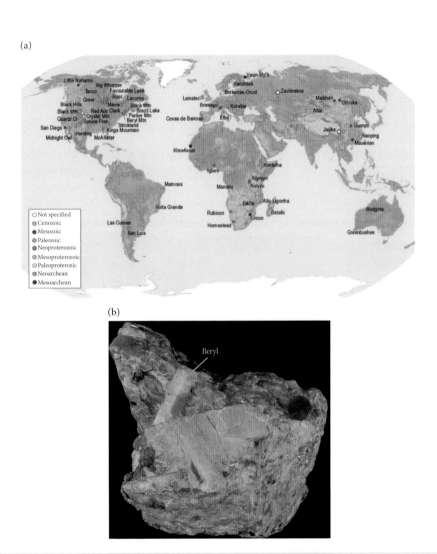

**Figure 19.10** (a) Known LCT pegmatite deposits on Earth, color coded with respect to age. *Source:* Bradley and McCauley (2013). Public Domain. © The United States Department of the Interior.); (b) granite pegmatite from the Black Hills of South Dakota containing large crystals of beryl, quartz, and feldspar. *Source:* Photo by Kevin Hefferan. © John Wiley & Sons.

pegmatites formed throughout Earth's history but are most concentrated in Archean-Proterozoic age cratons. LCT pegmatites are associated with greenschist to amphibolite metamorphism ($\approx$350–550 °C; $\approx$3 kb) acquired during late-stage collisional tectonic activity, rather than subduction. In the USA, King's Mountain, North Carolina, and the Black Hills, South Dakota, were formerly important producers of lithium (London 2008; Bradley and McCauley 2013). Let us further consider metamorphic processes that occur in regions surrounding igneous intrusions.

### 19.1.2 Metamorphic ore forming environments

Metamorphic processes concentrate metallic elements into ore deposits through solid state changes and hydrothermal fluid reactions. Hydrothermal fluids catalyze metamorphic reactions and concentrate ore metals. When we think of hydrothermal processes, Yellowstone or Iceland may come to mind where hot magma reacts with subsurface fluids to create geysers and hot springs. In fact, hydrothermal deposits originate by a number of different means that may involve meteoric (surface) waters, seawater, groundwater, formation pore fluids, or deep magmatic fluids.

**Greenstone belts** are hydrothermally altered assemblages that contain thick sequences of metamorphosed volcanic and sedimentary layers. They are called greenstones because of green-colored metamorphic minerals such as chlorite, epidote, and serpentine. Greenstone belts are particularly abundant in Archean cratonic belts where they contain among the world's greatest concentrations of copper, chromium, nickel, gold, cobalt, silver, and iron ore deposits. Common ore minerals include chalcocite, chalcopyrite, chromite, nickelline, pentlandite, cobaltite, erythrite, enargite, hematite, and magnetite (Figure 19.11). Greenstone belts consist of downwarped basinal deposits in which peridotite rocks are overlain successively by basaltic layers, silicic igneous rocks, and marine sediments. Many of these greenstone ore deposits are derived from altered komatiite, basalt, and peridotite rocks that form at very high temperatures (>1400 °C) and occur the base of the greenstone assemblage. Although they occur on all continents, greenstone belts are particularly extensive in the Precambrian cratons of Africa, Canada, and Australia.

**Skarns** are contact metamorphosed rocks enriched in calc-silicate minerals. Skarns form through the high-temperature alteration of country rocks, usually carbonate rocks, by the

**Figure 19.11** Cobalt mine in greenstone belt in Bou Azzer Morocco. *Source*: Photo by Kevin Hefferan. © John Wiley & Sons.

intrusion of silicate magmas. Hot magma intruding carbonate rock produces ion exchange via hydrothermal solutions. Minerals such as calcite and dolomite release $CO_2$ and obtain $SiO_2$ from the magma; as a result, a distinctive suite of calc-silicate minerals form that include Ca-pyroxenoid (e.g. wollastonite), Ca-amphibole (e.g. tremolite), Ca pyroxene (e.g. diopside), and Ca garnet (grossular, andradite, or uvarovite). Other associated minerals include quartz, calcite, phlogopite, brucite, talc, serpentine, and periclase. In addition, dissolved volatiles such as $H_2O$, $CO_2$, $SO_2$, $H_2S$, and HCl serve as catalysts in promoting the dissolution of metallic ions and transporting them in solution. Ion exchange commonly results in the concentration of copper, lead, zinc, iron, molybdenum, tin, tungsten, cobalt, manganese, silver, and gold ore deposits. Common ore minerals include chalcopyrite, chalcocite, sphalerite, galena, hematite, magnetite, siderite, molybdenite, cassiterite, wulfenite, wolframite, cobaltite, rhodochrosite, rhodonite, and enargite. Not all skarns form in carbonate rock. **Exoskarn** refers to skarns that develop in any sedimentary country rock, whereas **endoskarns** occur in igneous country rock. Modern skarns form in geothermal systems, hot springs, hydrothermal vents on the sea floor and at convergent and divergent plate boundaries. In these environments, skarns are commonly associated with porphyry, pegmatite, vein, and VMS deposits (Figure 19.12).

### 19.1.3 Sedimentary ore forming environments

Sedimentary ore forming environments develop through hydrothermal, depositional, and weathering processes. Let us begin by considering the role of hot fluids in leaching, transporting, and concentrating ore minerals in sedimentary deposits.

Sedimentary **iron ore deposits** consist of quartzite, chert, or shale alternating with magnetite, hematite, and lesser amounts of iron-rich carbonate, sulfide, or oxide minerals. Most banded iron formations (BIF) formed 1.8 to 3.7 Ga before free oxygen became abundant in Earth's atmosphere. They were deposited largely in marine basins (Figure 19.13). Three major types of iron ore deposits are Superior, Algoma, and oolitic ironstones.

**Superior-type banded iron deposits** are the largest and most important iron deposits and formed as a result of chemical precipitation in shallow marine environments 1.8–2.5 billion years (Ga) ago. These deposits are well developed in the Lake Superior region of North America, where they have represented over 80% of US production since 1900, and are referred to as Superior-type deposits. Similar massive deposits also occur in Australia's Hamersley Basin, Brazil's Minas Gerais deposits, and Russia's Kursk region. **Superior-type banded iron formations (BIF)** consist of alternating iron-rich and silica-rich layers. The iron-rich layers contain both ferrous and ferric iron. Ferrous iron minerals include magnetite and siderite, whereas ferric minerals include hematite and goethite. These shallow marine deposits formed in the Early Proterozoic when iron-rich, deep seawater mixed with shallow oxygenated shelves (Figure 19.13). As a result, BIF layers several hundred meters thick and encompassing >100 km$^2$ in area created the greatest iron deposits on Earth.

**Algoma-type deposits** consist of iron ore deposits in metasedimentary greenstone basins, most of which are Archean (>2.5 Ga) in age. Algoma-type ironstones contain hematite and magnetite interbedded with volcanic rocks, graywackes, turbidites, and pelagic sedimentary rocks. Algoma ironstones formed in deep abyssal basins heated by submarine volcanic activity (Figure 19.14). Volcanic hot springs release heated waters containing dissolved iron which precipitate in downwarped sedimentary basins. Algoma iron deposits form concentrated iron-rich layers ~30–100 m thick and extending a few square kilometers in area. Algoma-type deposits are associated with volcanogenic massive sulfide deposits and likely form at ocean ridges or volcanic arc environments. These deposits have been extensively mined in South Africa and in Canada's Algoma region for which these deposits are named (Kesler 1994).

**Oolitic ironstones** were deposited in the Early Paleozoic Era from the Cambrian to Devonian Periods (~540–350 Ma). Oolitic ironstones consist of rounded, iron rich sand sized grains that were deposited with carbonates in shallow marine environments. Minerals in oolitic ironstones include hematite, magnetite, siderite,

(a)

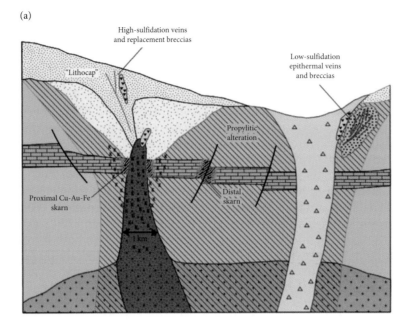

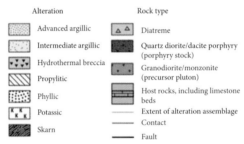

(b)

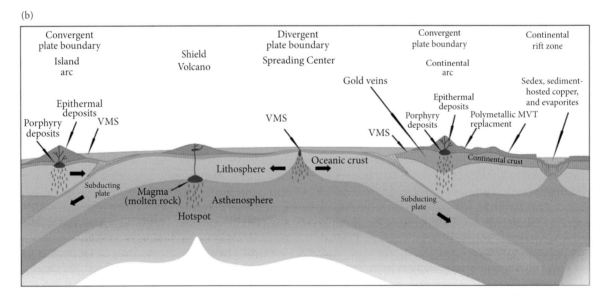

**Figure 19.12**   (a) Cross section illustrating skarns developing in association with porphyry deposits that commonly occur together at convergent plate boundaries. Metamorphic grades are discussed in Table 15.1. *Source*: Hammarstrom et al. (2019). Public Domain. © The United States Department of the Interior. (b) Cross section illustrating common ore forming processes at lithospheric plate boundaries. *Source*: Modified from Taylor et al. (2009). © John Wiley & Sons.

(a)

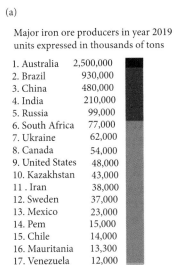

Major iron ore producers in year 2019
units expressed in thousands of tons

1. Australia    2,500,000
2. Brazil        930,000
3. China         480,000
4. India         210,000
5. Russia         99,000
6. South Africa   77,000
7. Ukraine        62,000
8. Canada         54,000
9. United States  48,000
10. Kazakhstan    43,000
11. Iran          38,000
12. Sweden        37,000
13. Mexico        23,000
14. Pem           15,000
15. Chile         14,000
16. Mauritania    13,300
17. Venezuela     12,000
Countries    1,000–10,000

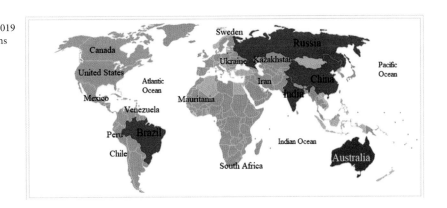

(b)

(c)

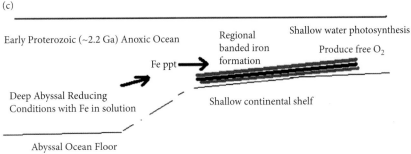

Early Proterozoic (~2.2 Ga) Anoxic Ocean

Regional banded iron formation

Shallow water photosynthesis

Produce free $O_2$

Fe ppt

Deep Abyssal Reducing Conditions with Fe in solution

Shallow continental shelf

Abyssal Ocean Floor

**Figure 19.13** (a) Global production of Iron ore in 2019. Image authored by PMJ and courtesy of wikimedia. Iron ore data from US Geological Survey. *Source*: Taylor et al. (2015) and Bekker et al. (2010). © John Wiley & Sons. (b) Banded iron formation rock. *Source*: Photo Kevin Hefferan. © John Wiley & Sons. (c) Anoxic ocean model for BIF development.

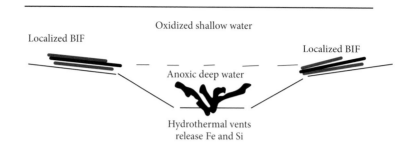

**Figure 19.14** Algoma-type deposits that form due to hydrothermal fluids interacting in deep sedimentary basins near the oxic-anoxic boundary. *Source*: Kevin Hefferan. © John Wiley & Sons.

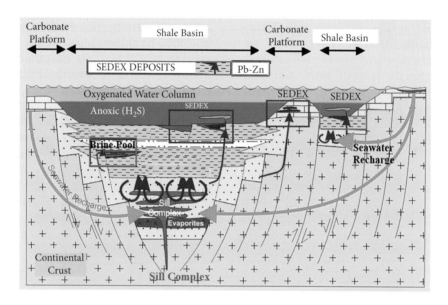

**Figure 19.15** Cross section of Sedex deposits formed by an external, magmatic heat source "exhaling" into a sedimentary basin. *Source*: Goodfellow and Lydon (2007). © GAC.

goethite, kaolinite, and chamosite. Whereas the Superior and Algoma-type banded iron formations have iron concentrations of 50% or more, oolitic ironstones generally contain less than 35% iron (Taylor et al. 2015).

**Sedimentary exhalative deposits (sedex)** are similar to Algoma-type deposits in that hydrothermal fluids (100–200 °C) leach and concentrate metallic ore. Whereas Algoma deposits are enriched in iron, sedex deposits contain lead–zinc–iron sulfides precipitated by submarine hot springs in extensional basins. Hydrothermal fluids containing dissolved metals rise upward and are "exhaled" into marine sedimentary basins releasing metal rich brine solutions into the overlying ocean water

(Figure 19.15). The brines precipitate as hot spring sedimentary deposits creating massive lead, zinc, and iron sulfides in sedimentary rocks such as shale, chert, and carbonate rocks. Sedex deposits typically consist of stacked tabular lenses of sulfide ore a few tens of meters thick that can extend hundreds of meters. Common minerals include galena, sphalerite, pyrite, and pyrrhotite. Sedex deposits are mined for zinc, lead, and silver in regions such as Australia, British Columbia, northern Europe, and the Yukon (Kesler 1994). The global distribution of sedex deposits is illustrated in Figure 19.16. Sedex deposits provide over 50% of the world's zinc and lead reserves a significant amount of silver. Over 129 Sedex

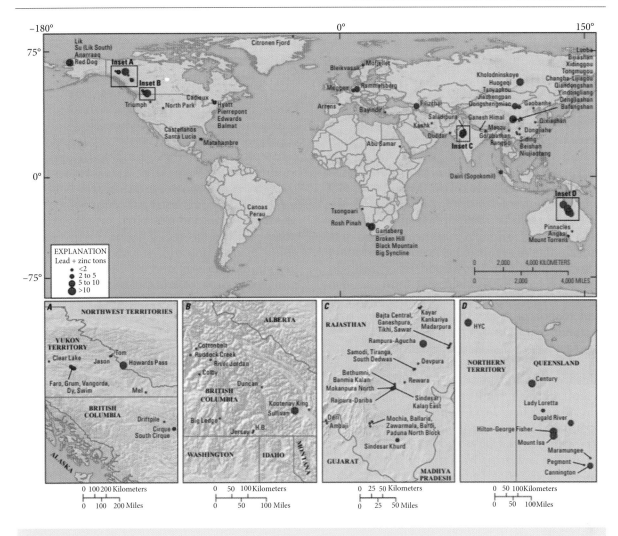

**Figure 19.16**   Global map of major sedex deposits in continental settings. Note that Sedex deposits occur in continents around the world but are concentrated in Canada (a, b), Europe (c), and Australia (d). *Source*: Emsbo et al. (2010); Courtesy of the U.S. Geological Survey.

deposits are recognized around the world (Tikkanen 1986; Goodfellow and Lydon 2007).

Unlike the Algoma and sedex deposits that have an external heat source, **Mississippi valley type (MVT) deposits** form by warm (75–200 °C) saline solutions that flow in the pore spaces within permeable carbonate or sandstone rocks in deep sedimentary basins. MVT deposits precipitate lead and zinc in thick limestone, dolostone, or sandstone deposits (Figure 19.17). Sulfide ore bearing minerals form by replacement of carbonate minerals affected by dolomitization and dissolution. These stratified deposits most commonly occur in large, distal Phanerozoic foreland basins and passive margin environments. Basal brines with 10–30 wt. % salt flow through permeable sedimentary rock producing large-scale epigenetic deposits (Leach et al. 2005). MVT deposits produce ~24% of the world's zinc and lead (Leach et al. 2010). In addition to lead and zinc from minerals, such as galena and sphalerite, the saline, brine formation waters precipitate pyrite, marcasite, dolomite, calcite, halite, sylvite, gypsum, calcium chloride, barite and minor amounts of gold, silver, copper, mercury, and molybdenum. Mississippi valley type deposits are named for deposits occurring from Missouri to Wisconsin in the central USA. MVT deposits also occur in Poland, Spain, Ireland, the Alps, and the Northwest Territories of Canada (Kesler 1994). The global distribution of MVT deposits is illustrated in Figure 19.18.

Mississippi Valley-Type Lead-Zinc Deposits

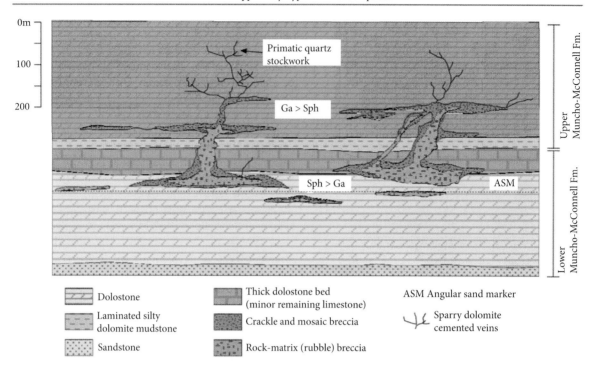

**Figure 19.17** Cross section of Canada's Robb Lake Zn-Pb sulfide bodies in brecciated MVT deposits. Ga, galena; Sph, sphalerite. *Source*: Paradis et al. (2007). © GAC.

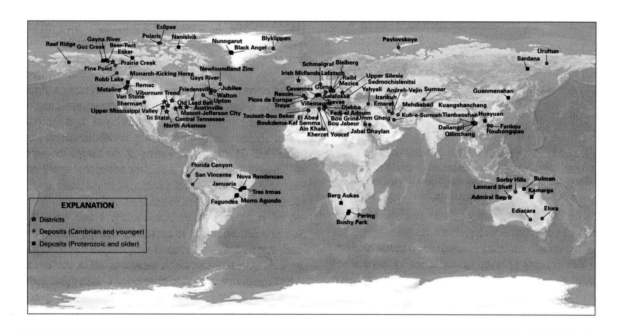

**Figure 19.18** Global map of the major MVT deposits. Note that these deposits only occur in continental settings. *Source*: Taylor et al. (2009); Courtesy of the U.S. Geological Survey.

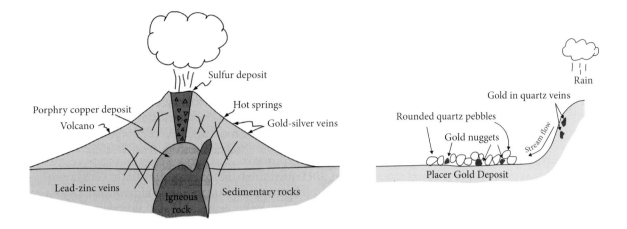

**Figure 19.19** Precious metals derived from magmatic veins are exposed, eroded, and transported by streams to form placer ore deposits in stream beds and beaches. *Source*: Taylor et al. (2009). Public Domain. © The United States Department of the Interior.

As most metallic elements have a high specific gravity, they tend to be concentrated as **placer deposits** in higher energy stream or beach environments (Figure 19.19). Placer deposits contain heavy minerals that may have originated in igneous intrusions or metamorphic belts. Upon uplift and weathering, the metals are eroded and transported by streams, waves, or currents to a high-energy depositional site. Strong water action transports lower density minerals down current and tends to deposit moderate to high specific gravity minerals as lag deposits on the Earth's surface.

Placer mining usually is conducted by dredging or panning methods by which sediments are sorted by, density using water. Common high specific gravity minerals in placer deposits include gold (20), platinum (14–19), silver (10), uraninite (7.5–10), cassiterite (6.8–7.1), columbite (5.2–7.3) ilmenite and zircon (4.7), chromite (4.6), rutile (4.2), and diamond (3.5). Uranium mining largely occurs in placer deposits (e.g. Athabasca Basin) within sandstones deposited in stream and beach environments. Canada and Australia are the two largest uranium exporters in the world, and many of these deposits are associated either with placer deposits or with major unconformity deposits.

**Unconformity deposits** occur along erosional surfaces representing time gaps between depositional cycles. In Precambrian (> 1 Ga), cratons consisting of gneisses and granites,

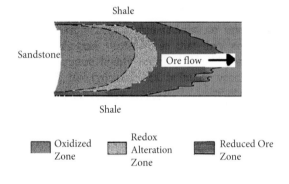

**Figure 19.20** Roll-front, tongue-shaped uranium deposits form in stream channels. *Source*: Modified from Reynolds and Goldhaber (1978). © John Wiley & Sons.

100—300° C hydrothermal fluids leach metals from underlying rocks and deposit them as ores in tabular to podiform deposits along unconformity surfaces. Major unconformity-derived deposits of uranium were discovered in Canada's Rabbit Lake deposit in 1968 and in Australia's East Alligator River field in 1970. In addition to uranium, other ores that occur in these unconformity deposits include nickel, copper, arsenic, silver, molybdenum, and selenium (Evans 1993).

Sedimentary deposits of uranium also occur in tongue-shaped **roll-front deposits** (Figure 19.20). Roll-front deposits form in fluvial sandstones as dissolved uranium is transported in stream channels. The uranium is derived from granitic rocks and silicic tuffs. In the presence of oxygen, uranium is soluble and moves downstream as a dissolved phase. Under reducing conditions, uranium precipitates within the pores of sandstone. Variations in oxygen concentrations result in cycles of dissolution and precipitation producing irregular, tongue-shaped deposits. In addition to uranium, other roll-front ores include vanadium, copper, silver, and selenium (Evans 1993).

**Weathering processes**, involving the chemical breakdown of preexisting rocks, can produce residual ores by concentrating aluminum, nickel, manganese, and iron. Through the action of surface waters and shallow groundwater, leaching processes result in the decomposition and oxidation of igneous, metamorphic, and sedimentary rocks on Earth's surface. Generally, decomposition rates correlate with precipitation and temperature so that tropical environments – regions of high temperature and high precipitation rates – are sites of intense chemical weathering capable of producing laterite soils (Figure 19.21). **Laterite soils** are intensely leached soils in which soluble components have been removed from the decomposed rocks and insoluble residues have been concentrated. Laterites are commonly enriched in clay minerals – produced by hydrolysis of feldspars – and can contain substantial ore metal ore concentrations. The specific ore type is determined by the chemical composition of the residual rock from which the laterite is derived. Laterite soils enriched in aluminum hydroxide minerals collectively referred to as bauxite. Bauxite, the primary source of aluminum and gallium, is largely produced by the intense leaching of granitic rocks containing aluminum-rich feldspar minerals, and in some cases gold ore. Laterites derived from the breakdown of basic-ultrabasic igneous rocks such as gabbro, pyroxenite or peridotite also produce iron, nickel, cobalt, chromium, titanium or manganese ores.

Bauxites also form in paleokarst depressions as aluminum-rich clay deposits within carbonate beds (Bárdossy 1982, 1994; Pajović 2009). Karst bauxites are black to gray due to organic carbon and pyrite

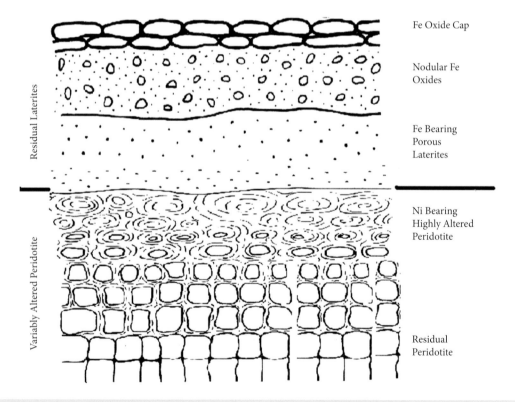

**Figure 19.21** Intense weathering of peridotite igneous rock concentrates iron and nickel deposits in overlying sediment. *Source*: Modified from Chetelat (1947). © John Wiley & Sons.

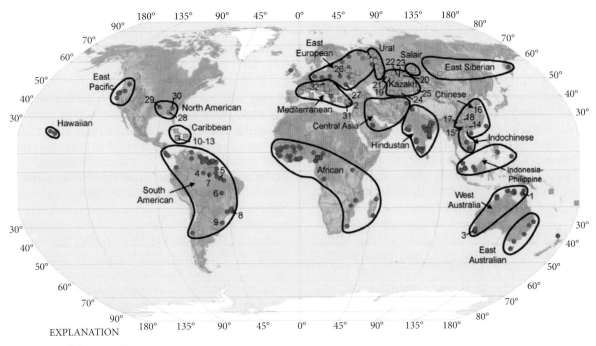

**Figure 19.22**   Global map of bauxite deposits from karstic and laterite sources. *Source*: Schulte and Foley (2014). Public Domain. The United States Department of the Interior.

(Retallack 2010). Approximately 88% of global bauxite deposits are from laterites and 12% are derived from karst environments (Bárdossy and Aleva 1990). Major bauxite resources occur in Africa, Australia, South America, and the Caribbean (Figure 19.22). Let us now consider some of the important metals and their uses in our society. Let us now consider important metals and their use in human history.

### 19.1.4  Metals and alloys

Over the last 8,000 years, people have learned to value metals for cooking utensils, tools, buildings, gems, currency, and weapons. Initially, people used pure base metals such as copper, zinc, or tin. Over time, it was discovered that tool strength and durability increase when two metals are alloyed or forged together using heat. As a result, the Bronze Age developed over 5000 years ago when copper and tin were alloyed together. Subsequently brass was created by combining copper and zinc. **Metal alloy** use continues to serve three primary

functions: (i) increasing hardness, (ii) reducing corrosion; and (iii) increasing high-temperature strength. Metals continue to be critically important in many aspects of our existence, from bodily function and health, to our homes, industry, transportation, and agriculture. Over the past several decades, metal use has expanded to play critical roles in medicine, energy, telecommunication, and information technology. Metals are nonrenewable resources that impart a significant financial and environmental cost in mining. Conservation, recycling, and finding cost-effective substitutes for these materials are increasingly recognized as a requirement. Metals are commonly grouped into nonferrous and ferrous metals. Nonferrous metals include (i) precious metals, (ii) light metals, and (iii) base metals. Ferrous metals are alloyed with other elements such as manganese, cobalt, nickel, chromium, silicon, molybdenum, and tungsten to make steel. Let us consider nonferrous metals.

The **precious metals** include the platinum group elements, gold and silver, which are in great demand. They are primarily used (i) as catalysts in chemical reactions, (ii) as conductors of electricity, (iii) for the production of nitric acid, (iv) in the fabrication of laboratory equipment, (v) as fillings and caps in dental restoration, (vi) as currency, (vii) in electronic equipment such as cellphones, and (viii) for jewelry. **Platinum group elements (PGE)** are hard, durable metals that include platinum, palladium, rhodium, ruthenium, iridium, and osmium. PGE minerals occur as native elements and as compounds with other metals (copper, iron, mercury, nickel, silver, bismuth, lead, and tin), metalloids (antimony, arsenic, and tellurium), and nonmetals (selenium and sulfur). Platinum and palladium are the "most abundant" of the rare PGE, occurring in concentrations of ~5 ppb in Earth's crust. PGE have similar chemical characteristics, serving primarily as catalysts in chemical reactions. PGE are essential in catalytic converters to reduce carbon monoxide, hydrocarbon, and nitrous oxide emissions in automobile exhaust. PGE serve as catalysts to refine crude oil and create high-octane gasoline. Platinum and platinum–rhodium alloys are needed to manufacture nitric oxide used in explosives, fertilizers, and nitric acid. PGE are in the production of fiberglass and high-purity single crystals used in the production of light-emitting diodes (LEDs). PGE are essential in computer hard disks, integrated circuits, and ceramic capacitors. PGE are used in medical implants (pacemakers) and cancer-fighting drugs. Platinum alloys are prized in jewelry for their white color, strength, and tarnish resistance. PGE serve as investment commodities and in currency exchanges. South Africa and Russia control over 90% of PGE primary production. The secondary supply through recycling catalytic converters, jewelry, and electronic equipment provides an essential component of the world's PGE availability and must continue to expand (Zientek et al. 2017).

PGE are obtained primarily from three-layered intrusions: South Africa's layered mafic to ultramafic Bushveld Complex; Zimbabwe's layered mafic to ultramafic Great Dyke intrusion; and sill-like intrusions associated with Russia's Noril'sk-Talnakh flood basalt region. Montana's Ultramafic Stillwater complex and Minnesota's Duluth Gabbro Complex and Ontario's Sudbury Complex are North American sources where PGE occur with chromite and nickel deposits. PGE also occur in modern ocean ridge black smokers, ancient VMS deposits, and in placer deposits (Kesler 1994; Zientek et al. 2017).

**Gold** occurs in a number of different forms which include (i) native element state, (ii) bonding with silver to form electrum, or (iii) bonding with tellurium to form telluride minerals such as calaverite. Gold is valued in jewelry because of its ductility, softness, and resistance to tarnish. Gold is also utilized in dentistry, building ornamentation, as currency. Because of its very high conductivity, gold is also widely sought in the electronics industry. **Silver** occurs in its native state, with gold in electrum and with base metals in sulfide minerals such as argentite, tennantite, and tetrahedrite. Unlike gold, silver readily tarnishes dull gray brown due to oxidation reactions. Silver is widely used in electronics because of its high conductivity; and in jewelry, coins and dentistry due to its ductility and softness. Mercury and silver are the main components of dental amalgam and their use inhibits recurring tooth decay. Silver is also used for automobile catalytic converters, photovoltaic solar cells, water purification, antimicrobial bandages, pharmaceuticals, plastics, batteries, electroplating, inks, and mirrors (USGS 2020). Gold and silver originate via magmatic processes and are concentrated by hydrothermal reactions in a number of igneous environments. The largest gold deposits on Earth – the Witwatersrand deposits of South Africa – occur in an Archean age (>2.5 Ga) placer deposits that formed by weathering granite source rocks and depositing high-density gold and uranium in a clastic sedimentary basin. The Witwatersrand deposit was discovered in 1886 and mining continues to this day. Notable gold rushes in the nineteenth century, which included the California and the Klondike gold rushes, also involved placer deposits in which high specific gravity gold flakes and nuggets weathered from vein deposits, settled to the base of streams and were later extracted by panning techniques.

The gold and silver found in placer deposits typically originate via granitic intrusions, Precambrian greenstone belts, deep turbidite sedimentary basins, and as finely disseminated grains within porphyry copper deposits and VMS deposits. Porphyry deposits such as Utah's Bingham copper mine and the Grasberg copper mine in Indonesia produce significant tonnage of both gold and silver (Kesler 1994). The leading producers of gold include China, Russia, United States, Australia, South Africa, Peru, Indonesia, Ghana, Mexico, and Canada. The leading producers of silver include Peru, Mexico, China, Australia, Chile, Poland, United States, and Canada (USGS 2020).

**Light metals** consist of low-density elements such as magnesium ($1.7\,g/cm^3$), beryllium ($1.85\,g/cm^3$), aluminum ($2.7\,g/cm^3$), and titanium ($4.5\,g/cm^3$). Their lightweight and relatively high strength are particularly prized in the aerospace and transportation industries (aircraft fuselages and wings, rockets, trains, automobiles, and trucks) as well as for common uses such as aluminum foil and baseball bats. Aluminum baseball bats are alloyed with scandium and cored with other lightweight materials such as graphite. Compared to wooden bats, aluminum bats have a greater elastic response upon impact with a baseball, resulting in the ball traveling higher velocities which can endanger infielders.

Aluminum and magnesium are the third and seventh most abundant elements in Earth's crust, respectively. Despite the fact that aluminum constitutes ~8% of Earth's crust by weight, it is a difficult metal to obtain and process. Aluminum is mined primarily from laterite soils that have experienced extreme leaching in tropical environments. Laterite soils derived from the weathering of granitic or clay-rich rocks commonly contain bauxite group minerals diaspore, gibbsite, and boehmite. Although bauxite is the only commercial source of aluminum, this light metal could be processed from clay or feldspar group minerals at a significantly higher cost. Whether derived from bauxite, clay, or feldspars, aluminum production requires an enormous amount of energy. Bauxite is strip mined with the largest deposits occurring in Australia, China, Brazil, India, Papua New Guinea, and Jamaica (USGS 2020).

Earth's crust contains about 2% magnesium by weight. **Magnesium** occurs in minerals such as olivine, pyroxene, amphibole, dolomite, magnesite, and brucite. Economic deposits are derived largely from seawater and evaporate brines and deposits. Magnesium is used in the automotive industry and as an alloy with aluminum to increase aluminum's hardness and resistance to corrosion. Magnesium is also used as a refractory material for furnaces and molds in the steel industry. Other uses include wastewater treatment, animal feed supplements, fertilizer for agricultural, and industrial applications in the form of caustic-calcined magnesia, magnesium chloride, magnesium hydroxide, and magnesium sulfates. Leading producers of magnesium include China, Brazil, Turkey, Russia, and Slovakia (USGS 2020).

**Titanium** represents ~1% by weight of the Earth's crust and occurs in ore minerals such as ilmenite, rutile, and anatase. Titanium ore occurs in layered gabbroic intrusions and in coastal placer deposits derived from the weathering of gabbroic intrusions. Titanium is a lightweight, low corrosion metal with a very high melting temperature ($1678\,°C$). Titanium is widely used in high speed turbines and the aerospace industry, representing up to 30% of modern aircraft weight (Kesler 1994). Ocean research submersibles such as the Woods Hole Oceanographic Institute submersible Alvin also consist largely of titanium. Lightweight and high strength allow titanium to be widely used in armor, marine hardware, space stations, and recreational sports equipment such as bicycles and golf clubs. Titanium dioxide whiteness, opacity, and chemical inertness combine for widespread use as pigment for paints, paper, plastic, sunscreen, toothpaste, and wallboard. Because of its high strength, low weight, low toxicity, and hypoallergic properties, titanium is also widely used for joint replacements and prostheses in the medical industry. For people with allergic reactions to gold or silver metal jewelry such as earrings, titanium is also the metal of choice. Significant producers of titanium include Australia, Canada, Norway, and South Africa; additional amounts are produced in Brazil, India, Madagascar, Mozambique, Sierra Leone, and Sri Lanka (Woodruff et al. 2017; USGS 2020).

**Beryllium** is an expensive, relatively rare element that occurs in the minerals beryl, spodumene, and bertrandite ($Be_4Si_2O_7(OH)_2$). Ore deposits of these minerals occur in silicic pegmatite intrusions, associated hydrothermal veins, and rhyolite tuffs, for example Utah's Spor Mountain deposit (USA). Beryllium is alloyed with copper to increase hardness and is widely used in computer, telecommunications, aerospace, military, and automotive electronics industries due to its high conductivity, lightweight, stability at high temperature, stiffness, and resistance to corrosion. Beryllium is critical to aircraft and satellite structural components that have a high stiffness-to-weight ratio. Beryllium–nickel alloys produce durable, deformation resistant, high-temperature springs. Beryllium is also used in the medical industry, for X-ray windows and treatments, as well as in nuclear fusion reactor experiments. Beryllium oxide is used in missile guidance systems, radar applications, and cellphone transmitters. Beryllium is also essential for magnetic resonance imaging (MRI) machines, medical lasers, and portable defibrillators. Beryllium is a known carcinogen and inhalation of beryllium dust is particularly lethal, which is the reason why beryllium is no longer used in fluorescent light fixtures. The leading producers of beryllium are the United States, Brazil, China, Madagascar and Mozambique, and Portugal (Foley et al. 2017; USGS 2020).

**Base metals** are nonferrous metals that oxidize easily. Base metals include copper, zinc, lead, tin, lithium, uranium, mercury, arsenic, cadmium, antimony, germanium, rhenium, tantalum, zirconium, hafnium, inidium, selenium, bismuth, tellurium, and thallium (Table 19.2).

Copper occurs in the native state as well as in sulfide, oxide, hydroxide, and carbonate minerals. Copper ore minerals include chalcopyrite, chalcocite, bornite, cuprite, enargite, tetrahedrite, malachite, and azurite. Copper ores form in many different environments. These include porphyry copper deposits, massive volcanogenic sulfide deposits, rift basin deposits, hydrothermal vein deposits, and oxidized secondary supergene enrichment environments. Copper is perhaps the earliest metal used by humans and continues to be among the most widely used, particularly in electronics, due to its high conductivity. Demand and value of copper is rising dramatically in response to the industrialization of China and growth in India. Major producers of copper include Chile, Peru, China, Congo, United States, Indonesia, Australia, Russia, Mexico, Kazakhstan, and Zambia (USGS 2020).

Zinc occurs primarily in the sulfide mineral sphalerite, and to a lesser degree in the zinc silicate mineral willemite. Zinc oxide minerals, zincite and franklinite, were exclusively mined at the Sterling Mine in Franklin, New Jersey (U.S.), which ceased operations in the 1980s. Sphalerite occurs with the copper minerals listed above in VMS deposits, skarns and other hydrothermal environments. Sphalerite occurs with galena in MVT and sedex deposits. Zinc is widely used as a metal alloy to prevent oxidation and corrosion in galvanized sheet metal, nails, bolts, and other construction equipment. Zinc is also used in the production of brass, bronze, and in ammunition. Like copper, zinc is greatly in demand due to economic expansion in China and India. The leading producers of zinc include Australia, China, South Africa, Peru, United States, Canada, Mexico, and Kazakhstan (Kesler 1994; USGS 2020).

Lead occurs primarily in the sulfide mineral galena. Minor lead minerals include anglesite, cerussite, and crocoite. As noted previously, lead occurs with zinc in MVT and sedex environments. Lead has been used since ancient times in lead crystal glassware, lead glass, and as a sweetener for wine. Physical and mental debilitation due to ingestion of lead has been cited as one of the possible causes for the fall of the Roman Empire. Until recently, lead was commonly used in paints, gasoline and in fishing equipment (among other uses). Since the 1970s, restrictions have been placed on lead use due to severe health risks. Lead is now used primarily in lead-acid batteries, as starting-lighting-ignition (SLI) batteries for automobiles and aircraft and as industrial-type batteries for standby power for computer and telecommunications networks. Lead is also utilized in ammunition, radioactive shields, glass, pigments, and ceramics. Lead remains a very significant health risk in our environment as will be discussed later in this chapter. The principal producers of lead include Australia, China, Ireland, Mexico, Peru, Portugal, Russia, and the United States (Kesler 1994; USGS 2020).

Tin is derived primarily from the mineral cassiterite and is associated with silicic igneous intrusions in sedimentary rock. Tin is closely associated with tungsten and molybdenum ore deposits in granite plutons and associated hydrothermal vein networks in continental crust overlying subduction zones. Tin also occurs in VMS, MVT, and in placer deposits. In the past, tin was widely used for canned food and beverages. However, its market share has dramatically been replaced by glass, aluminum, and plastic. Tin continues to be used for tinplate, solder, and as an alloy for brass. The major producers of tin include China, Indonesia, Peru, Bolivia, Brazil, Russia, Malaysia, and Australia (USGS 2020).

Uranium is obtained from oxide and phosphate minerals such as uraninite and carnotite in unconformity and placer deposits derived by weathering granite source rocks. Uranium is the primary fuel for nuclear reactors and is a major energy source. According to the International Atomic Energy Agency, nuclear energy provides ~10% of the world's electricity from ~440 nuclear reactors in over 30 countries. While the production of nuclear energy does not emit contaminants or greenhouse gases associated with fossil fuels, the mining and disposal of uranium material pose major environmental problems. Uranium mining is associated with high lung cancer incidence due to inhalation of radioactive radon gas, produced as a radioactive decay product of uranium. Radium, another uranium daughter product, is a major contaminant of groundwater. As radium is chemically similar to calcium, it is readily absorbed into bones causing cancer. Uranium waste from nuclear reactors poses two long-standing problems: (i) uranium ore can be enriched to weapons grade material which can be used in nuclear weapons and (ii) permanent

**Table 19.2**   Some additional base metal occurrences and uses are presented.

| Element and minerals | Occurrence | Uses and/or hazards |
| --- | --- | --- |
| Antimony: in stibnite, tetrahedrite and jamesonite | Hydrothermal vein deposits, MVT deposits and in Kuroko-type VMS deposits | Antimony (Sb) imparts strength, hardness, and corrosion resistance. Used as fire retardant in safety equipment, household materials (mattresses). Sb used in batteries, ceramics, and glass. China produces >60% of world's supply. Bolivia, Canada, Mexico, Russia, South Africa, and Tajikistan mine antimony with Pb, Zn, Ag, Sn, Wo. No USA production. Sb causes liver, respiratory, cardiovascular, skin diseases. |
| Arsenic: in realgar, orpiment, enargite, arsenopyrite, and tennantite | Hydrothermal veins with Cu, Ni, Ag, Au, and in Cu porphyry deposits | Arsenic (As) toxicity used in copper chromate arsenic (CCA) wood preservatives. InGaAs semiconductors used in biomedical, communications, computer, electronics, and photovoltaic instruments. Producers include China, Morocco, Belgium, Japan, Russia, Namibia, Iran. No USA production since 1985. Toxicity impacts respiratory and cardiovascular system as well as causes bladder, lung, and skin cancer. |
| Bismuth: in bismuthinite | By-product of Wo, Mo, and Pb mining in porphyry deposits | In 2019, a stable bismuth-based perovskite oxide semiconductor was discovered that could be used in thin-film solar technology. Bi is used for stomach remedies, foundry equipment, pigments. As a nontoxic replacement for lead, Bi is increasingly being used in plumbing, fishing weights, ammunition, lubricating grease, soldering alloy. Because of its low melting temperature, Bi is used as an impermeable low-temperature coating on fire sprinklers. Bi is mined in China, Bolivia, Vietnam, Korea, Mexico, Kazakhstan, and Japan. |
| Cadmium: in greenockite but is primarily derived from sphalerite | By-product of Zn mining in VMS and MVT type deposits | Cadmium (Cd) is alloyed with other metals such as CdTe for thin-film solar-cell panels and NiCd for rechargeable batteries in alarm systems, cordless power tools, medical equipment, electric cars, and semiconductors. Used in steel and PVC pipes for corrosion resistance and durability. Previously used as a yellow, orange, red, and maroon pigment. Cd interferes with Ca, Cu, and Fe metabolism resulting in softening of the bones and vitamin D deficiency causing health problems. Producers include China, Australia, Canada, Peru, and USA |
| Gallium | By-product in processing of zinc and bauxite ores | Gallium is used as a dopant in semiconductors. Gallium is a critical element widely used in electronics such as in microwave circuits, LEDs, and lasers. Gallium alloys are used in thermometers as a nontoxic alternative to mercury. |

*(Continued)*

**Table 19.2** (*Continued*)

| Element and minerals | Occurrence | Uses and/or hazards |
|---|---|---|
| Germanium: rarely forms its own mineral but occurs with Zn and Cu | By-product from MVT deposits and in Cu ore deposits | Germanium (Gm) is used for fiber-optic cables where it has replaced Cu in wireless communication, solar panels, semiconductors, microscope lenses, and infrared devices for night-vision applications in luxury cars, military security, and surveillance equipment. Gm is also used as a catalyst in the production of polyethylene terephthalate (PET) plastic containers and has potential for killing harmful bacteria. Producers include the USA, China, and Russia. |
| Hafnium: in ilmenite and rutile | Placer deposits | Used in the construction of nuclear rods because Hf does not transmit neutrons. Major producers of zirconium and hafnium include Australia and South Africa |
| Indium: in sphalerite, cassiterite, and wolframite | VMS, MVT, hydrothermal veins with Sn, Wo, and Zn | Indium–tin oxide thin-film coatings used for electrical conductivity in solar cells and flat-panel displays such as touchscreens, liquid crystal displays (LCDs). Indium (In) used in alloys and solders, electrical components, windshield glass, semiconductors, breathalyzers, dental crowns. Producers include China, Canada, Korea, and Taiwan. |
| Lithium: in spodumene, lepidolite, and lithiophilite | Granite pegmatite and alkali brines of playa basins | Lithium (Li) batteries show dramatic growth in electric cars, cellphones, electric tools, computers, energy grid storage, rechargeable batteries, glass, ceramics, grease. Li replaced Pb and Cd due to adverse health issues. Producers include Bolivia, Argentina, Chile, China, USA, Canada, Mexico, Congo, Germany, and Zimbabwe |
| Mercury: in cinnabar | Low temperature (<200 °C) hydrothermal vein deposits. | Only metal liquid at room temperature for which it is commonly referred to as quicksilver. Previously used for automotive switches, cosmetics, pigments, gold processing, and hat making. Hg is highly soluble and readily enters the bloodstream where it attacks the nervous system causing psychotic behavior ("mad as a hatter"), irritability, tremors and can lead to death. Hg continues to be used in thermometers, batteries, fluorescent lights, dental amalgam fillings and is used as a catalyst in paper production. Hg use in these and other capacities will continue to decline because of its adverse health effects. Hg is obtained from mines in China, Peru, Kyrgyzstan or obtained as a secondary ore in copper, zinc, lead, and gold mines throughout the world |
| Rhenium substitutes for molybdenum: in molybdenite | Cu porphyry deposits | Used with Pt as a catalyst in oil refining and the generation of high-octane, lead-free gasoline. Rhenium improves the high-temperature (1000 °C) strength properties of some nickel-base superalloys for aerospace, turbine engines, crucibles, electrical contacts, electron tubes and targets, electromagnets, heating elements, ionization gauges, mass spectrographs, semiconductors, thermocouples, and vacuum tubes. Producers of rhenium include Chile, USA, Kazakhstan, Peru, Canada, Russia, Poland, and Armenia |
| Selenium: in selenite gypsum but more commonly occurs with sulfide minerals such as FeS and CuS minerals | By-product of Cu ore deposits | Selenium (Se) is photovoltaic, converting light into electricity and also photoconductive in that its electrical conductivity increases with increasing light. Se is deemed an energy critical element in photovoltaic cells and solar panels. Se is used for energy-efficient plate glass windows that limit heat transfer. Se also used in thin-film photovoltaic cells that convert solar energy into electricity. Se is also used in the electrolytic process to increase current efficiency and manganese metal deposition rate. Se is a salt used as a vitamin supplement to animal feed and as a fertilizer for agricultural crops. Se has both positive and negative health aspects: deficiencies increase the incidence of stroke while excess selenium is related to deformities. Major producers of selenium include Japan, Canada, and Belgium. |

**Table 19.2** (*Continued*)

| Element and minerals | Occurrence | Uses and/or hazards |
|---|---|---|
| Tantalum is in microlite, pyrochlore and tantalite | Pegmatites, hydrothermal veins and placer deposits | Tantalum (Ta) are used in electrical capacitors, automotive electronics, personal computers, and cellphones. Major producers of tantalum include Australia, Brazil, Canada, Congo (Kinshasa), and Rwanda. |
| Tellurium in calaverite | Mafic and ultramafic layered intrusions, VMS, porphyry, skarns, veins. | Tellurium (Te) is used in cadmium telluride (CdTe) for thin-film solar cells. Bismuth telluride (BiTe) is used in thermoelectric equipment for both cooling and energy generation. Te also increases heat resistance of rubber. Te serves as a metal alloy to increase ductility and strength. Alloyed with copper, Te improves machinability without reducing conductivity. Te is alloyed with selenium in copying machine photoreceptors and as a coloring agent in ceramics and glass. Te is a semi-conductor in integrated circuits, laser diodes, and medical instruments. Te is among the rarest elements in Earth's crust. Producers include Canada, China, Germany |
| Thallium occurs in association with sphalerite | VMS and MVT deposits | Thallium (Tl) is used in photoelectric cells, infrared optical materials, and low melting glasses. Tl is also used an activator in gamma radiation detection equipment, infrared detectors, and in high-temperature superconductors used for wireless communication. Tl was previously used in rat and ant poison until its toxicity to humans became apparent. Tl interferes with the metabolism of K and can cause death. Major producers include China, Kazakhstan, and Russia. Since 2005, major Tl deposits have been identified in Brazil and Macedonia. |
| Zirconium is in rutile and ilmenite | Placer deposits | Used as a fuel in nuclear reactors due to the ease with which it transmits neutrons; also used in ceramics and as an abrasive, metal alloy, and refractory material due to its very high melting temperature (2550 °C). Major producers include Australia, South Africa, China, Senegal. |

*Source*: Kesler (1994), Schulz et al. (2017), and USGS (2020). © John Wiley & Sons.

nuclear waste repositories for safe disposal do not yet exist. Major producers of uranium include Canada, United States, Australia, Russia, Namibia, and France.

The origins, uses and hazards of other base metals are listed in Table 19.2.

**Rare Earth metals** include scandium, yttrium, and the 15 lanthanide elements (Table 19.3). Rare Earth metals occur in minerals such as monazite, which occurs in granite pegmatites, hydrothermal veins, and in placer deposits. Rare Earth metals are widely used as telecommunication and hybrid/electric car batteries, as catalysts in oil refining, in chemical synthesis, catalytic converters in automobiles, glass additives, glass polishing, fiber optic lasers, phosphors for fluorescent lighting, color televisions, electronic thermometers, X ray screens, pigments, superconductors, as dopants and more. A **phosphor** exhibits the phenomenon of phosphorescence and is utilized in equipment such as fluorescent lights and cathode ray tubes. REEs yttrium, europium, and terbium are the red-green-blue phosphors used in light bulbs, computer panels, and televisions. A **dopant** is an impurity that alters the optical and electrical properties of semiconductors. Major sources of rare Earth metals include China, India and Malaysia. China controls ~90% of REE and has placed limits on these strategic mineral exports that have impacted economies throughout the world. The restriction of REE supply can have a major impact on telecommunications and hybrid automobile development (Schulz et al. 2017; USGS 2020).

**Ferrous metals** and **Ferrous alloys** are also very important. Strictly speaking, the term

**Table 19.3** Common rare Earth metal (REE) uses.

| REE | Use |
| --- | --- |
| Cerium | Phosphors for cathode ray tube, flat panel display screens, incandescent, fluorescent, and light-emitting diode lighting. Also used in steelmaking to remove impurities and as catalysts in automotive catalytic converter |
| Dysprosium | Dysprosium is used in lasers, commercial lighting, and in computer data storage. Dysprosium magnets used in hybrid and electric car motors, wind turbine generators. |
| Erbium | Erbium is used in amplifiers and lasers and serves as neutron absorbers in nuclear reactor control rods. Erbium is used for its pink color in glasses. |
| Europium | Europium is used in euro currency as it glows red under UV light; forgeries can be detected by the absence of a red glow. Also used in low-energy light bulbs and in lasers. Europium is in super-conducting alloys. Europium absorbs neutrons, making it valuable in nuclear reactor control rods. |
| Gadolinium | Phosphors used in X-ray imaging and MRI. |
| Holmiun | Dopant in laser crystals; also used in magnets and as a yellow or red coloring for glass and cubic zirconia. |
| Lanthanum | Major use in hybrid and electric car batteries; serve as catalyst in petroleum refining; used in steel making to remove impurities and to produce superalloys; major component of digital camera lenses and cell phone cameras. |
| Lutetium | Phosphor; also used as a catalyst for cracking hydrocarbons in the oil industry. Lu is used in cancer therapy and as a detector in positron emission tomography (PET). PET scan creates a three-dimensional cellular image of a body. |
| Neodynium | Neodymium–iron–boron magnets are the strongest magnets known, also used as glass additive, dopant in laser crystals |
| Praseodymium | Yellow pigment in ceramics; steel alloy removes impurities; and produces superalloys. |
| Promethium | Promethium is used in luminous paint, watches, pacemakers, and as a light source for signals. Promethium has replaced carcinogenic radium for luminous purposes, although it too is radioactive. |
| Samarium | Samarium is used in optical lasers, infrared absorbing glass and as a neutron absorber in nuclear reactors. It is also used in the chemical industry for decomposing plastics, dechlorination of polychlorinated biphenyls (PCBs) pollutants. |
| Scandium | Scandium is used as a lightweight strong metal for aerospace industry components and recreational items such as bicycle frames, fishing rods, golf clubs, and baseball bats. It is also used in film industry mercury vapor lamps. |
| Terbium | Terbium is used in fuel cells, electronic devices, naval sonar systems, and sensors. Terbium also serves as a green phosphor in fluorescent lighting. |
| Thulium | Thulium used in military, medical, and meteorological laser technology; X-ray technology in medicine and dentistry; and as a high-temperature superconductor. |
| Yttrium | As an alloy, yttrium increases the strength of aluminum and magnesium. Yttrium is used in LEDs, microwave filters, radar, and lasers. |
| Ytterbium | Ytterbium is used as doping agent and amplifier in fiber optic cable technology. Ytterbium may serve as a radiation source for portable X-ray machines. |

*Source*: Kesler (1994), Schulz et al. (2017), and USGS (2020). © John Wiley & Sons.

"ferrous" refers to iron in the +2 oxidation state, having lost 2 electrons ($Fe^{+2}$). Iron is derived from (i) oxide minerals such as magnetite and hematite, (ii) hydroxide minerals such as goethite and limonite, and (iii) carbonate minerals such as siderite. **Iron** is largely mined from Superior-type and Algoma-type deposits throughout the world. Iron also occurs in magmatic deposits and skarns.

Demand for iron is steady due in part to continued economic growth in China. Major producers of iron include Australia, Brazil, Russia, China, Ukraine, Canada, India, United States, and Iran. Iron is used principally in the production of steel through combining iron with various metal alloys. **Steel** is an alloy consisting mostly of iron with 0.02 to ~2.0% carbon content by weight. Carbon is the most

**Table 19.4**  Common steel alloys (Kesler 1994; USGS 2020).

| Alloy element and minerals | Origin | Attributes and major producers |
|---|---|---|
| Aluminum-diaspore, boehmite, gibbsite | Bauxite in laterite deposits derived from weathering silicic igneous rocks | Lightweight and high strength metal. Major producers include Australia, China, Brazil, India, and Papua New Guinea. |
| Beryllium-Beryl | granite pegmatites, hydrothermal veins around silicic igneous rocks | Lightweight metal stable with high-temperature strength. Major producers include the USA, China, and Brazil. |
| Chromium-Chromite | Layered gabbroic intrusions, ophiolites, VMS, laterites, placer deposits | Corrosion resistance; important alloy in "stainless steel." Major producers include South Africa, Kazakhstan, and India |
| Carbon-Graphite | Organic or inorganic sources | Alloyed for hardness. Producers are widespread globally. |
| Cobalt-Cobaltite | Layered gabbroic intrusions, hydrothermal veins; evaporite brines in desert basins. | Corrosion and abrasion resistance; used in industrial and gas turbine engines. Major producers include Congo (Kinshasa), Zambia, Australia, Russia, and Canada. |
| Magnesium-Magnesite, olivine | Marine carbonate rocks and layered igneous intrusions. | Hardens Al and corrosion resistance. Magnesium producers are widespread globally. |
| Manganese-Pyrolusite, psilomelane, rhodonite, and rhodochrosite | Supergene weathering of Mn rich rocks in laterites; Mn nodules in VMS; marine limestones and shales | Corrosion resistance in Al; alloyed with Cu for strength; replaced lead in fuel. Major producers include South Africa, Gabon, Australia, Brazil, China. |
| Molybdenum-Molybdenite | Porphyry-type deposits and granite pegmatites | Hardness; corrosion resistance, high-temperature strength. Producers include the United States, Chile, China, and Peru |
| Nickel-nickelline, pentlandite, millerite | Layered gabbroic intrusions ophiolites, komatiites, and laterites derived from ultrabasic rocks | High-temperature strength; corrosion resistance; used in jet engines and turbines. Major producers include Russia, Canada, Australia, Indonesia, New Caledonia, Columbia, and Brazil. |
| Silicon-Quartz | Clastic sedimentary rocks, pegmatites, and hydrothermal veins | Deoxidant in steel; increases strength and corrosion resistance. Silicon producers are widespread globally. |
| Titanium-ilmenite and rutile | Placer sand deposits (beach dunes) or from layered gabbroic igneous intrusions | Lightweight, strong; strategic mineral with limited reserves; 6 times stronger than steel; aircraft commonly contain 1/3 Ti. Producers include Australia, Canada, Norway, and South Africa. |
| Tungsten-scheelite and wolframite | Pegmatites and hydrothermal veins derived from granitic intrusions. Also skarns, hot springs | High temperature strength; high speed drills, armor plating, light filaments. Major producers: China, Russia, Kazakhstan, Austria, Portugal, and Canada |
| Vanadium-Vanadinite, uraninite | Pegmatites, veins, layered igneous intrusions, phosphate sedimentary rock. | High strength, ductility, toughness; used in high rises, oil platforms. Major producers: South Africa, China, Russia |
| Zinc-Sphalerite, zincite | VMS or MVT | Corrosion resistance. Major producers include China, Peru, Australia, Ireland |

*Source*: Adapted from Kesler (1994) and USGS (2020). © John Wiley & Sons.

cost effective alloy, serving as a hardening agent and preventing dislocations in the iron crystal lattice structure. In addition to carbon other steel alloys include aluminum, chromium, cobalt, magnesium, manganese, molybdenum, nickel, silicon, titanium, tungsten, and vanadium. Important aspects of these alloys are briefly described in Table 19.4.

## 19.2 INDUSTRIAL MINERALS AND ROCKS

Industrial minerals and rocks constitute economically important Earth materials in addition to metals, fuels, or gems. Industrial minerals and rocks serve a number of vital roles in the global economy and are widely used: (i) as fertilizers and chemicals, (ii) in construction, and (iii) in manufacturing.

Fertilizers are widely used to increase crop production. Fertilizers contain primary nutrients such as nitrogen, phosphorous, and potassium, and secondary nutrients such as calcium, magnesium, and sulfur. In addition to serving as nutrients, industrial minerals containing arsenic (realgar and orpiment) and mercury (cinnabar) have been widely used as pesticides. While the use of arsenic and mercury compounds has diminished due to the detrimental health effects on humans and other animals, arsenic continues to be used in pressure treated lumber used for outdoor recreation purposes.

The chemical industry relies on minerals to produce large quantities of sulfur, fluorine, and chlorine for diverse purposes such as pharmaceuticals, petrochemical refining, cosmetics, purifiers, and various manufacturing processes. The construction of roadways, buildings, dams, and other structures requires aggregate minerals that include quartz sand, calcite, gypsum, dimension stone, bricks, cement, clay, sand, and gravel. Manufacturing includes a wide array of uses such as abrasives, absorbents, fillers, and filters. In the following section, we will address the three major categories of industrial minerals and demonstrate their significance to the global economy.

### 19.2.1 Fertilizers and chemicals

Fertilizers and industrial chemicals are derived from a number of sources that include evaporite minerals, marine sediments, and organic deposits derived from the remains of animals and bacterial alteration. **Fertilizers** improve nutrient level within soils and serve as soil conditioners to adjust pH. Fertilizers include nitrates, phosphates, potash, limestone, dolostone, sulfur, and salts. Nitrogen is a key element widely used in agriculture. Although the nitrates can be extracted from the evaporite

mineral niter, most nitrogen is obtained from chemical processes and not from the mining of nitrates. In addition, bat excrement, known as guano, has been used as a minor nitrogen source. Excess nitrates in waterways cause eutrophication and excess algal growth, diminishing water quality. Excess nitrates in drinking water cause "blue baby" syndrome. Blue baby syndrome results when infants drink nitrate rich water. Their body convert nitrates into nitrites, which bind to hemoglobin in the blood, forming methemoglobin. Methemoglobin reduces oxygen causing baby skin to appear blue and potentially results in impaired brain development.

Phosphate is derived from the calcium phosphate mineral apatite and represents a critical fertilizer element. Phosphate is a component of DNA and many other critical organic compounds, forming the skeletal material of many vertebrates. Approximately 95% of world phosphate production is used as fertilizer and ~5% is used in products such as detergents, fire retardants, and toothpaste. Phosphate occurs in three major settings: (i) continental shelves rich in organic material such as bones and teeth; (ii) igneous intrusions (syenite, carbonatite) at continental rifts; and (iii) guano deposits from bat and bird excrement in caves and tropical islands. Excess phosphates in waterways cause eutrophication and excess algal growth, diminishing water quality.

Halide elements such as the evaporite minerals halite, sylvite, polyhalite are used for potash fertilizer (sylvite) as well as for food additives (salt), water purification (chlorine), and in the production of sodium (caustic soda) and chlorine gas. In the days before refrigeration, salting meat was essential to prevent food from going rancid. Salt continues to be used as a very common additive in food and still serves as a preservative. Sodium and lime are used to adjust the pH of soils. Lime is produced from limestone which forms in the marine environment, beach systems, continental hot springs, and fresh water lakes. Lime is produced by heating limestone to 700–1000 °C, effectively driving off $CO_2$.

$$\underset{\text{Calcite}}{CaCO_3} + heat \rightarrow \underset{\text{lime}}{CaO} + \underset{\text{carbon dioxide}}{CO_2}$$

Lime neutralizes soil acidity, purifies water, and controls pH of natural lake systems. For example, the liming of lakes increases the pH (less acidic, more basic) and reduces acidification. Industrial processes involving lime include making cement and for whitewashing houses to increase reflectivity in many parts of the world.

Sulfur is obtained from native sulfur and, to a lesser degree, pyrite. Sulfur forms by bacterial reduction of gypsum and other sulfate minerals and by sublimation from volcanic vents. Sulfur is used for the production of sulfuric acid, which is a key component in the generation of superphosphate fertilizers. Sulfur is also used as a catalyst in production of high octane gasoline. Fluorite is an important mineral used in the production of hydroflouric acid and fluorine. Fluorite occurs in low to medium temperature hydrothermal veins and is associated with granitic plutons. Fluorite is used in the enrichment of uranium, production of chlorofluorocarbons (CFC) for refrigerants and propellants, as a flux in steel making, in dentistry for cavity protection, and in glass and ceramic manufacturing.

Chemicals are also widely used in pharmaceuticals, health products, and cosmetics. Barium is used to enhance x-ray images of organs and blood vessels; compounds containing calcium are used to prevent bone loss; clays (kaopectate), magnesium, and calcium are widely used for digestion. Other dietary supplements contain zinc, iron, selenium, cobalt, and sodium. Shampoos and soaps contain nitrogen compounds, zinc, potassium, chlorine, and titanium. Sunscreens are composed largely of zinc oxide and titanium oxide. Cosmetics are largely composed of clays, talc, micas with color additives containing oxides of iron, titanium, zinc, chrome, aluminum, and copper. Micas provide the sheen in lipsticks. Healthcare products rely heavily on naturally occurring elements occurring in minerals.

### 19.2.2 Construction material

Common materials used in construction include dimension stone, aggregate, concrete plaster, and other clay building materials (Figure 19.22). **Dimension stone** consists of rock slabs of limestone, marble, sandstone, granite, gneiss, and slate. Dimension stone has been used since at least the construction of the pyramids over 4000 years ago and until recently, dimension stone remained the primary material used in sidewalks, roads, bridges, dams, building foundations, walls and roofs, as well as interior uses such as countertops and fireplaces. Dimension stone was acquired by quarrying operations throughout the world. Michelangelo used marble quarried locally in Carrara, Italy. Vermont slate was the rock of choice for sidewalks, blackboards, and roof tiles in the eastern United States; slate from Wales provided the same service in Wales and England. Homes in Ireland and England were largely constructed of limestone or sandstone. The Palisades sill basalt studied by Norman Bowen was the "trap" rock commonly used for road construction in the New York metropolitan area. The rock of choice in any given location was determined by the underlying geology at that site. Dimension stone continues to be used, but it has largely been displaced by less costly aggregate and cement materials.

**Aggregate** consists of sand, gravel, and crushed stone. Aggregate material is most commonly derived from unconsolidated sedimentary layers deposited in beaches, deltas, deserts, stream systems, or by glaciers. In the absence of available sand and gravel, aggregate is produced by mechanically crushing massive rock. The equipment and energy required to crush rock results in greater expense. Due to their low hardness and abundance, limestone and dolostone are preferred for crushed aggregate. Rocks such as granite, basalt, sandstone, and metaquartzite are also readily available in many locations but their greater hardness results in higher costs. Aggregate quality is lowered by the presence of pyrite, which oxidizes to produce rust, shale which weathers easily and chert which tends to produce "pop outs" in freeze–thaw conditions, fracturing the aggregate.

**Concrete**, developed by the Romans 2000 years ago, is the dominant construction material in use. Concrete consists of Portland cement mixed with an aggregate of sand or gravel. Portland cement is made by heating limestone and clay in a kiln to produce clinker. Clinker is then ground to a powder and mixed with a little gypsum. Portland cement is the most common hydraulic cement, meaning that it hardens with water. In many construction projects, cement has replaced dimension stone because of its ease of transport, workability, and the ability to create forms to fit site-specific construction projects. Cement quality is affected by the presence of chert (promotes pop out cracks in concrete), magnesium content (MgO must be <5%), and pyrite concentrations (produces $SO_2$ gas). The production of cement, through heating calcite, is responsible for ~8% of the global $CO_2$ emissions.

Gypsum is an evaporite mineral that commonly forms in sabkhas or evaporite basins of marine systems. In addition to use in cement, plaster of Paris (gypsum cement) is produced by heating and dehydrating gypsum at temperatures ~150°C. Gypsum is also widely used in wallboard, also known as sheetrock. Other minerals such as clay, quartz, hematite, and calcite are major components of common construction materials such as bricks, adobe, stucco, roof tiles, drains, and sewer pipes. Bricks, adobe, and stucco are widely used where dimension stone and lumber are not readily available. Again, in the absence of ostentatious wealth, construction utilizes the resources at hand, largely due to the cost associated with transporting rock material great distances. In these times of rapidly expanding populations and urbanization, construction represents the single greatest growth area for nonfuel mineral resources.

### 19.2.3 Manufacturing minerals

**Manufacturing materials** are used in a wide array of mineral applications, such as abrasives, absorbents, ceramics, drilling fluids, fillers, filters, fluxes, glass, insulators, lubricants, pigments, and refractory materials. All of these applications are used in manufacturing and in our daily lives throughout the world. Many minerals are adept at multitasking. For example, clay minerals serve as absorbents, fillers, in drilling mud, and in construction.

**Abrasives** grind, polish, abrade, scour, and clean surfaces. Abrasive minerals display high hardness and rigidity, retain their grain shape and size, and do not contain softer impurities. Abrasives are used for saws, drills, sandpaper, and grinding wheels. Common abrasive minerals include diamond, emery (corundum), garnet, and quartz. Large, high-quality mineral samples of diamond and corundum are considered gemstones, but the varieties used as abrasives have imperfect, flawed, and/or small crystals so that they are considered to be of industrial grade rather than gem quality. Abrasives are used on industrial scales for cutting and polishing manufactured goods. We also use these for cleaning kitchen utensils, finishing wood products, in emery boards for fingernails, and in toothpaste for brushing teeth.

**Absorbents** remediate spills and remove waste products from liquids. Absorbents are used for industrial scale removal of wastes and contaminants as well as in household items such as baby's diapers, feminine hygiene pads, antiperspirants, and cat litter. Silica serves as an absorbent in packaging electronics and footwear. Minerals commonly used as absorbants include the hydrous aluminosilicate zeolite minerals and phyllosilicate minerals such as expansive smectite clays and vermiculite. These hydrous aluminum silicate minerals have crystal structures that allow them to incorporate molecules such as water between silica tetrahedral sheets. Vermiculite is used to increase soil aeration and retain moisture and chemicals in agricultural soils.

**Ceramics** utilize clays and other minerals such as feldspars, hematite, bauxite, bentonite, pyrophyllite, borax, wollastonite, barite, lepidolite, and spodumene. Clay forms the basis of pottery and other ceramic materials. Feldspars, borax, and wollastonite are used to produce ceramic glazes on pottery that increases hardness, provides a vitreous luster, and preserves color. Hematite serves as a pigment. Barite hardens ceramics and serves to preserve pigments. Bauxite and pyrophyllite provide high-temperature strength. Lithium minerals lepidolite and spodumene are added to ceramics to prevent volume change with temperature variations. Each of these minerals serves to improve the quality of clay-based ceramic materials (Chang 2002).

Minerals used as **detergents** include evaporate minerals such as borax, halite, and trona. Together with baking soda ($HNO_3$) or ammonia, evaporite minerals are very effective in cleansing soiled items. Depending on the application, detergents can be used in combination with abrasive materials for effective scrubbing action. Phosphate minerals were used in the past but have largely been removed from detergents. Phosphates promote eutrophication of water bodies by deteriorating water quality due to excessive nutrient input.

**Drilling fluids** are used in well drilling for oil gas and groundwater, and in geophysical studies such as earthquake studies. Inert minerals with high specific gravities such as barite are combined with water to produce drilling muds, which allows rock chips to float to the surface for analysis of borehole geology. Drilling muds also control well pressure, maintain borehole stability, lubricate the drilling apparatus, and protect the drill target from contamination. Barite, which has a specific gravity of 4.5–5, is combined with bentonite (smectite) clay to produce an expansive mud capable of supporting rock chips and cuttings. After having constructed a borehole well using drilling fluids, swelling bentonite pellets pack the top of the well to prevent surface fluids from flowing into the borehole contaminating the well. Upon exposure to water, swelling clays can expand up to 20 times their original volume. In this capacity, the bentonite pellets serve as fillers discussed below.

Fillers include barite and phyllosilicate minerals such as kaolinite, smectite, mica, talc, and pyrophyllite. Fillers are inert, inexpensive materials that extend volume of material at low cost. The ability of fillers to expand volume without reducing material quality can substantially reduce manufacturing costs. The most effective fillers may in fact improve material quality. For example, in paper production, fillers such as lime and kaolin improve print and appearance of paper. Fillers are used in the production of food, rubber, paint, wallpaper, cosmetics, plastics, roofing, textiles, soaps, ceramics, and paper. For example, vermiculite is used as a lightweight filler in cement, concrete, and plaster. Barite is used as a filler in chocolate, and kaolinite is used as a filler in ice creams and shakes. In a sense, water can be considered as a filler when it is used to dilute juices or other drinks to reduce their sugar content or simply to extend the volume.

Filters purify, clarify, and clean liquids by removing solids or other contaminants. Common mineral filters include zeolites, graphite, clays, charcoal, quartz, and diatomite. Diatomite is a siliceous rock composed of microscopic marine organisms called diatoms. Filters serve a wide variety of industries that include water softeners in homes, beer filters in breweries, and hydrocarbon filters in oil, and gas refining. Filters are also essential in the production of medicines, fruit juices, and the remediation of contaminated ground water and surface water. Zeolites are multipurpose industrial minerals in that they serve as absorbents as well as filters. Zeolites filter lead, cadmium, and other potentially hazardous elements from potable water. Zeolites also serve as filters in oil refining and petrochemical industry.

A **flux** breaks chemical bonds thereby lowering melting temperature of materials. A common analogy is the fluxing or melting of ice by adding salt. In manufacturing, a lower melting temperature drastically reduces energy demands. Common industrial fluxes include fluorite, limestone, and borate minerals. Limestone is widely used as flux in steel industry so that metals melt at a lower temperature requiring less heat input. Fluorite and boron act as fluxes in manufacture of glass, glass fibers, and insulation. Cryolite deposits from Greenland have been used as a flux in aluminum refining.

**Insulators** reduce the transmission of heat or sound. Common insulating materials include asbestos, micas, silica, and vermiculite. Insulators are commonly used in flooring, walls and ceiling tiles, brake linings, roof shingles, around pipes, and in the production of fire-resistant clothing. Lead is also an insulator in preventing the transmission of radioactive energy as anyone who has had an X-ray can attest. Chrysotile asbestos was among the most popular insulating materials in homes and buildings due to its ability to retain heat, ease of use, and its fireproof nature. However, since the 1980s, chrysotile asbestos has largely been replaced due to health concerns about mesothelomia, asbestosis and lung cancer – despite the fact that illnesses are generally caused by amphibole asbestos minerals that are less commonly used. Asbestos has been replaced by other materials, such as fiberglass and vermiculite. Although commonly used as rolled insulation, fiberglass is a health concern because the silica glass fibers lacerate lung tissue when inhaled. Vermiculite is blown into walls, crawl spaces, and attics as an unconsolidated fluffy material. Although vermiculite is inert, it commonly contains amphibole (tremolite) asbestos which is a known carcinogen as well as a known contributor to mesothelomia and asbestosis.

**Lubricants** are low hardness minerals used for a variety of purposes ranging from petroleum jelly use to automotive engines in transportation and heavy machinery in agricultural and industrial applications. Lubricants may be wet or dry. In all cases, lubricants reduce friction and abrasion,

**Table 19.5** Common mineral pigments used for providing color to materials.

| Pigment color | Pigment minerals or elements |
|---|---|
| Yellow | Limonite, goethite, spinel, orpiment cadmium, uranium |
| Blue | Azurite, lazurite, zircon, spinel, cobalt |
| Green | Malachite, glauconite, epidote, celadonite, atacamite [$Cu_2Cl(OH)_3$], barite, chromium |
| Red | Hematite, cinnabar, iron, mercury, strontium |
| Orange | Realgar |
| Turquoise | Azurite, chrysocolla, turquoise |
| Copper | Malachite, dioptase, gaspeite [$(Ni,Mg)CO_3$], copper |
| Blue Gray | Vivianite [$Fe_3(PO_4)_28(H_2O)$], aegerine |
| Purple | Hematite, pyrolusite, cuprite, purpurite ($Mn_3PO_4$) |
| Lapis Lazuli | Lazurite [$(Na,Ca)_8(AlSiO_4)_6(SO_4, S,Cl)_2$] |
| Pink | Montmorillonite, lepidolite, rhodonite, rhodochrosite, titanium, manganese |
| Gray-Black | Graphite, magnetite, pyrolusite |

minimize heat generation, and extend the usability of materials. Molybdenite is widely used in manufacturing oils and greases to reduce friction and abrasion of machinery. Graphite is widely used in grease, brake linings, and pencils. The hardness of a pencil (#1 being soft and #4 being hard) is related to the ratio of graphite to clay. Soft pencil lead contains more graphite and hard pencil lead contains more clay. Other uses for these soft, lightweight, and strong minerals include sports equipment such as bicycles, tennis rackets and golf clubs, batteries and military purposes such as in Stealth aircraft.

**Pigments** provide color variations in paints, glass and other materials and have been used for centuries. Minerals used as pigments commonly include metallic elements that provide vibrant hues. Table 19.5 lists common mineral pigments used in art and construction materials. Some of the metallic elements, such as arsenic, cadmium, uranium, mercury and lead, are harmful and potentially lethal.

**Refractory** minerals are heat-resistant materials that have very high melting temperatures and are chemically resistant to breakdown. Because refractory minerals do not melt readily

they can be used for high-temperature industrial applications such as in furnaces or crucibles. Refractory minerals are also used in jet engines, turbines, and high-speed drills that generate extremely high temperatures. Refractory minerals include magnesite, wolframite, Mg-chromite, olivine, zircon, bauxite, kyanite, sillimanite, andalusite. Many of these minerals are key alloys in the production of high strength steel discussed earlier.

Quartz is the primary mineral used in the production of **glass**. The purest, highest quality quartz sand occurs in beach or desert dune deposits. Fresh water deposits are preferable as salts corrode metal. High-quality quartz must lack color (no Fe) and lack refractory elements (Fe, Al, Sn, Zr, Cr). High-quality quartz sand, such as from the St. Peters Sandstone, is used: in glass making, as foundry sand for making metal molds, and as an abrasive. While quartz is the dominant glass component, other minerals and elements include feldspar (reduces breakage), trona soda ash (glaze and pigment), fluorite (pigment), barite (scratch resistance), selenium (reduces tint and infrared rays), spodumene and zircon (reduces thermal expansion), celestite (television glass), borax (optical lenses), and lead in artisian glass works.

## 19.3 GEMS

Gems form an important part of our lives, culture and economy. Although over 4500 minerals are known to exist, fewer than 100 minerals are considered gems. Exactly what are gems? **Gems** are durable, beautiful, somewhat rare solid substances that, with proper cut and polish, may be used as jewels or for ornamentation. Gems include naturally occurring, hard, crystalline minerals that are the most valuable precious stones. Gems also include mineraloids and manufactured gems produced in a laboratory. **Mineraloid** gems are naturally occurring solids that lack one of the critical aspects of minerals such as crystal form or inorganic origin. The gems amber, ivory, and pearls are mineraloids. Amber is both noncrystalline and organic, having formed from plant material. Ivory is noncrystalline and organic, forming the tusks of large animals brutally slaughtered for these gem materials. Pearls are organic substances produced by oysters in the marine environment. **Manufactured gems** consist of treated gems, synthetic gems, and imitation gems as well as doublets and triplets. **Treated gems** constitute natural stones "improved" by artificial means such as artificial coloring or staining, heat treatment, irradiation, or special mountings. **Synthetic gems** are solid substances produced in a laboratory. One well-known synthetic gem is the isometric zirconium oxide crystal called cubic zirconia. Imitation gems resemble gemstones but are composed of inferior materials such as glass or plastics and are commonly referred to as "costume jewelry." **Doublets and triplets** consist, respectively, of two or three layers fused together to resemble a single coherent crystal. The doublet or triplet may consist of natural gem stone material or a less expensive imitation substance.

Why are gems deemed so valuable? First and foremost, gems are treasured for their beauty. However, gem value varies somewhat with fashion. Minerals valued today as precious stones may have at one time been considered a nuisance! Platinum, the most precious of metals, was in earlier times cursed for its difficult-to-isolate occurrence with the more highly valued gold and silver. We should all be so cursed.

Gem minerals may be classified based upon relative value. "Precious" gems are those perceived to have the greatest value; these include the nonmetals diamond, emerald, ruby, sapphire, and alexandrite. Precious metals such as platinum, gold, and silver may also be considered precious gemstones. These metallic minerals are commonly used for gem mountings. However, as these metals bear hardnesses ranging from 2.5 to 4.5, they must be alloyed with other metals to increase their hardness and wearability. Pure **platinum** is soft and flexible, with a hardness of 4–4.5. Platinum hardness increases substantially when alloyed with 5–10% iridium. Pure 24 k **gold** is relatively soft (H = 2.5–3); alloying 18 parts gold

**Table 19.6**  Some of the more common gems are described in this table.

| Mineral | Chemical formula | Gem variety |
|---|---|---|
| Aragonite | $CaCO_3$ | Pearls |
| Beryl | $Be_3Al_2(Si_6O_{18})$ | Emerald (Cr-green), Aquamarine (blue), Helliodore (yellow) |
| Chrysoberyl | $BeAl_2O_4$ | Alexandrite (green by day; red by night);Cats Eye (yellow-green-brown) |
| Chrysocolla | $Cu_4H_4O_{10}(OH)_8$ | Green |
| Corundum | $Al_2O_3$ | Ruby (red) Sapphire (blue) |
| Diamond | C | Diamond |
| Garnet | $A_3B_2(SiO_4)_3$<br>A = Ca, Mn, Mg, Fe<br>B = Al, Fe, Cr | Demantroid (Andradite Garnet-green-yellow)<br>Hessonite (Grossular Garnet-green-yellow)<br>Uvarovite-green (chromium) Almandine, Pyrope, Spessartine (Red) |
| Jadeite | $NaAlSi_2O_6$ | Jade |
| Lazurite | $(Na, Ca)_8(AlSiO_4)_6(SO_4, S, Cl)_2$ | Lapis lazuli (lazurite with calcite, pyroxene) |
| Malachite | $Cu_2CO_3(OH)_2$ | Green |
| Microcline | $KalSi_3O_8$ | Amazonite |
| Olivine | $(Fe, Mg)_2SiO_4$ | Peridot |
| Plagioclase | $(Na, Ca)AlSi_3O_8$ | Labradorite (iridescent)<br>Moonstone (opalesque) Sunstone (golden shine) |
| Quartz | $SiO_2$ | Agate (banded chalcedony)<br>Amethyst (purple)<br>Aventurine (mint green)<br>Jasper (red chert)<br>Chrysoprase (green chalcedony)<br>Citrine (yellow-orange),<br>Moss Agate (mottled gray)<br>Onyx (layered chalcedony)<br>Rock Crystal Quartz<br>Rose Quartz<br>Smoky Quartz<br>Tigers eye |

*(Continued)*

**Table 19.6** (*Continued*)

| Mineral | Chemical formula | Gem variety |
|---|---|---|
| Rhodochrosite | $MnCO_3$ | Pink |
| Sodalite | $Na_4(AlSiO_4)_3Cl$ | Violet blue color |
| Spinel | $MgAl_2O_4$ | White, red, all colors |
| Topaz | $Al_2SiO_4(F,OH)_2$ | Topaz (yellow, clear, pink, red, green, blue) |
| Tourmaline | $(Na,Ca)(Li,Mg,Al)(Al,Fe,Mn)_6$ $(BO_3)_3(Si_6O_{18})(OH)_4$ | Various colors |
| Tremolite-Actinolite | $Ca_2Mg_5Si_8O_{22}(OH)_2$ | Nephrite Jade (amphibole) |
| Turquoise | $CuAl_6(PO_4)_4(OH)_8*4H_2O$ | Turquoise |
| Zoisite | $Ca_2Al_3(Si_3O_{12})(OH)$ | Tanzanite-violet blue |
| Zircon | $ZrSiO_4$ | Zircon (diamond like) |

with 6 parts metal alloy yields 18 k gold which is significantly harder than 24 k gold. Alloyed gold results in a color change. Yellow gold, alloyed with silver and copper, becomes progressively lighter in color with increasing silver to copper ratio and becomes darker colored with increasing copper to silver content. White gold is produced by alloying 18 k gold with nickel, copper, and zinc. Green gold is generated by alloying 18 k gold with silver, nickel, and copper. Rolled gold or gold plated refers to a thin veneer of gold overlying a base metal (Kraus and Slawson 1947). Silver is another soft, flexible precious metal with a hardness of 2.5. Silver must be alloyed with 7.5–10% copper to produce a harder "sterling silver" or "coin silver" (Kraus and Slawson 1947). Table 19.6 lists some of the more common nonmetal gemstones.

Gem beauty and value results from the specimen's color, play of color, brilliance, purity, transparency, luster, wearability, cut and rarity; many of these properties are interrelated such that when one aspect is diminished, a cascading effect occurs reducing the gem's beauty and value. **Color** refers to the degree of hue, tone, and intensity of light within a substance. **Hue** is a function of the frequency of light and is described as red, orange, yellow, blue, green, indigo, and violet. **Tone** varies from very light to very dark. **Intensity** refers to the saturation or purity of a color. It is generally true that for colored gems, the greater the hue, tone and intensity, the greater the value of the stone. Notwithstanding this general rule, some colorless gems are more prized for exactly that attribute – the absence of color. This is particularly true with diamonds.

**Table 19.7** Metals commonly provide vibrant color to gemstones.

| Element | Gem |
|---|---|
| Chromium | Ruby, emerald, pyrope, grossular, uvarovite, tourmaline |
| Copper | Dioptase, malachite, azurite |
| Iron | Sapphire, aquamarine, olivine, almandine garnet, amethyst quartz |
| Manganese | Morganite, pink tourmaline, spessartite garnet |
| Vanadium | Green beryl, blue zoisite, garnet |
| Titanium | Blue sapphire |

Gems consist of both metallic and nonmetallic minerals. Metallic minerals include gold, platinum, silver, titanium, copper, brass (copper and zinc), and bronze (copper and tin). Even though most gems consist of nonmetals, metals play a vital role in imparting color to nearly all gems. Metallic elements behave as brilliant dyes in transforming colorless minerals into minerals exhibiting hue, tone, and saturation. For example, the metals listed in Table 19.7 are responsible for the following gemstone colors.

**Refractive Index** (R.I.) refers to the ratio between the velocity of light through air and the velocity of light through a mineral substance. Minerals with a high refractive index (diamond R.I. is ~2.4) tend to be more brilliant than minerals with lower refractive indices (quartz R.I. is ~1.5). **Brilliance** refers to the tendency of a mineral to sparkle or radiate transmitted and reflected light outward in all directions, producing a brilliant display of color and light. Diamonds, with their adamantine luster, are most noted for brilliance.

Brilliance is dependent upon naturally occurring physical properties such as dispersion. **Dispersion** refers to the ability to split light into its component colors based upon the varying wavelengths and velocities of light. The rainbow effect produced by white light penetrating a prism is an example of dispersion.

Gems may be isotropic or anisotropic. Isotropic minerals, which crystallize in the isometric system, have the same refractive index in different directions and tend to have a constant color. Anisotropic minerals crystallize in the nonisometric crystal systems and are characterized by having different refractive indices in different directions. Anisotropic gems tend to display **pleochroism**, characterized by changes in color in different directions and or **photochroism**, whereby colors change under different lighting conditions.

**Play of color** refers to the ability of some anisotropic gems to display multiple changing colors when subjected to the transmission (light passing through a mineral) and reflection (light bouncing off a surface) of light. Gems such as pearls and opals display a **"mother of pearl"** opalescence; a play of color characterized by an intermingled rainbow-like assortment of green, yellow, orange, and blue. Acicular-fibrous gems such as tiger's eye (yellow-orange quartz) and cat's eye (chrysoberyl) display **chatoyancy**, which is a translucent play of yellow, orange, brown, and gray colors. Another precious variety of chrysoberyl, alexandrite, displays **photochroism**; that is, alexandrite appears green in natural light (sunlight) but turns ruby red in artificial (incandescent) light. Why? Incandescent light is enriched in red wavelengths of light and relatively depleted in blue-green wavelengths.

**Purity** is a measure of the lack of impurities (other minerals or substances) and flaws in gemstones. Flaws include fractures, inclusions, notched corners, cleavage or earthy, granular, rough, or splintery textures. Impure, flawed gems are characterized by reduced transparency and brilliance. Fractures, bubbles, and inclusions of other substances reduce the transmission of light through the gemstone, thus reducing their transparency. Impurities and flaws also reduce luster by reducing the quality and intensity of reflected light. In contrast, pure, unflawed, transparent minerals with high refractive indices display an adamantine or highly vitreous luster. Even

the most valuable gems have flaws. The most valuable gems, however, are marked by extraordinary color, clarity, transparency, luster, cut, brilliance, hardness, and rarity

**Wearability** refers to the ability of a gem to be worn and exposed to "wear and tear." A mineral with high wearability retains its luster, brilliance, polish and is not chemically or physically diminished through prolonged use. Wearability is dependent upon the **durability** of a gem. Durable gems are hard and tough. **Hardness** is defined as the ability to resist abrasion or scratching. **Toughness**, the resistance to breakage, is distinctly different than hardness. Very hard minerals may be deficient in toughness due to properties such as cleavage. For example, diamonds have a hardness of 10, but also contain 4 sets of cleavage planes that reduce its toughness. Toughness may also be reduced by susceptibility to chemical weathering, such as the tendency of some minerals to break down in acidic solutions and others to oxidize when exposed to the atmosphere.

No story of gem properties and value would be complete without a description of the modifying role played by humans. Cut, brilliance, and rarity are at least partially due to the impacts of humans. **Cut** refers to the manner in which a jeweler has sliced the mineral specimen to enhance its natural beauty and increase transparency, luster, and brilliance.

**Rarity** is a measure of the lack of availability of gem quality specimens. Certainly, most gemstones are rare in nature. The most precious gems – diamond, emerald, ruby, and sapphire as well as precious metals such as platinum – are deemed exceedingly rare. However, in some cases, rarity is due to highly effective marketing.

## 19.4 MINERALS AND HEALTH

Just as minerals play pivotal roles in the health of the global economy, minerals also serve vital roles in human health. How are minerals incorporated into our bodies? Rocks containing minerals break down on the Earth's surface in the presence of water to produce soils. Soils contain organic material, weathered rock material, and dissolved ions in pore waters. With continued weathering, dissolved ions are distributed throughout the soil profile and are absorbed by plants. Animals such as cows, sheep, fowl, and humans eat plant materials – or other animals – and incorporate these dissolved nutrient ions in their bodies. Diet and location determine nutrient intake. Climate and the underlying geology play a critical role in determining the concentration of soil nutrients. Plants grown in limestone soils will have more calcium but less potassium than those grown in granite. Leafy vegetables provide elements such as potassium, iron, calcium, magnesium, phosphorous, and copper. Cereal grains provide manganese, iron, zinc, and chromium to our diet.

Nutritionists classify "minerals" – we would consider them elements – with regard to the amount needed for physiological function. "Macrominerals" are required in relatively large amounts while only trace amounts (parts per million) of "microminerals" are required to satisfy physiological needs within the body. Table 19.8 lists some key "macrominerals" and "microminerals." Elements provide nutrients and serve as catalysts in chemical reactions within our bodies. For example, vitamin B12 is found in cobalt; zinc is a vital component of enzymes, DNA, RNA,

protein synthesis, carbohydrate metabolism, and cell growth; copper allows for our bodies to process iron; iron allows our blood to carry oxygen; magnesium promotes growth and stabilizes human behavior as well as provides a vital role in chlorophyll production in plants (Kesler 1994). Without these essential ingredients, life as we know it would cease to exist.

On the other hand, excessive exposure to some elements can cause bodily harm. Exposure to amphibole asbestos causes potentially lethal diseases such as lung cancer, mesothelioma, and asbestosis due to the ability of asbestos fibers to encapsulate and contaminate lung tissue. Radioactive elements such as uranium, radium, radon, polonium, and thorium are known carcinogens. Overexposure to copper can be indicated by green hair, and intense overexposure can lead to death. Excess aluminum in drinking water has been correlated to Alzheimer's disease. Metals such as arsenic, mercury, cadmium, and chromium are known carcinogens that also attack the nervous system. Prolonged inhalation of beryllium dust causes a chronic lung disease

**Table 19.8**   Key elements, referred to by nutritionists as macro- and micro-minerals, required for human health.

| "Macrominerals" | Physiological function |
| --- | --- |
| Calcium | Bones, teeth, cartilage, neural transmission, muscle function |
| Chlorine | Water and electrolyte balance, digestive acid |
| Magnesium | Regulates chemical reactions, nerve transmission, blood vessels, chlorophyll production in plants. Magnesium sulfate helps prevent cerebral pals in infants. |
| Phosphorous | Bones, teeth, cartilage, cell function, blood supply |
| Potassium | Growth, body fluids, muscle contractions, neural transmissions |
| Sodium | Regulates acid–base balance, neural transmission, blood pressure |
| Sulfur | Proteins, thiamine, hair, skin |
| **"Microminerals"** | **Physiological function** |
| Chromium | Glucose metabolism |
| Cobalt | Vitamin B12, red blood cells |
| Copper | Red blood cells, nervous system, metabolism, prevents anemia |
| Fluorine | Tooth decay, strengthens bones |
| Iron | Hemoglobin, immune system |
| Iodine | Thyroid hormones, reproduction |
| Manganese | Tendon and bones, central nervous system, enzyme reactions |
| Molybdenum | Growth and enzyme reactions |
| Selenium | Antioxidant, detoxifies pollutants, prevents cardiovascular disease, cancer |
| Zinc | Enzymes, red blood cells, sense of taste and smell, immune system, metabolizes alcohol, protects liver |

*Source*: Adapted from Dunn (1983) and Wenk and Bulakh (2004). © John Wiley & Sons.

called berylliosis. Prolonged exposure to silicon dust causes the lung disease silicosis. Lead causes both physiological and neurological damage such as nausea, anemia, decreased nerve function, encephalitis, loss of coordination, coma, brain damage, and death. Lead causes anemia by substituting for iron in blood and causes bone loss by substituting for calcium and blocking zinc enzymes. While lead has been removed from paint and gasoline since the 1970s in the United States, many products still contain lead. In 2007, toys manufactured in China were noted for high lead concentration and many were belatedly pulled from shelves. Perhaps most importantly, the burning of fossil fuels, such as coal, oil, and gas releases gases (carbon oxides, nitric oxides, and sulfur oxides), metals (Hg, U, Pb), and particulate materials that contaminate our air, water, and soils. Table 19.9 summarizes the detrimental health aspects of elements based on whether they cause cancer (carcinogenic), birth defects (teratogenic), or are radioactive.

In addition to the ill effects caused by the use and overexposure of potentially dangerous minerals, the mining of minerals poses significant risks. Mining can be conducted in deep open pits, shallow strip mines, or underground mining. Underground mining poses a significant hazard due to underground explosions and mine collapse as well as subsidence on the Earth's surface. Strip mines and mountain top mining remove vast quantities of near surface rock resulting in loss of topography

**Table 19.9** Naturally occurring toxic elements segregated based on whether they cause cancer, birth defects, or are radioactive.

| Toxic effect | Element |
| --- | --- |
| Carcinogenic | Arsenic, beryllium, cadmium, chromium, cobalt, copper, iron, lead, nickel, tin, zirconium |
| Teratogenic | Aluminum, arsenic, beryllium, cadmium, copper, indium, lead, lithium, manganese, mercury, molybdenum, selenium, tellurium, thallium, zinc |
| Radioactive | Radon, radium, thorium, uranium |

*Source*: Smith and Huyck (1999) and Wenk and Bulakh (2004). © John Wiley & Sons.

and arable land use. Waste material is commonly dumped in low lying valleys and river systems. Open pit mining involves the excavation of holes several kilometers wide and over 1 km deep. Each of these types of mining operations results in immense waste piles. In metal mining and processing, rocks are exposed to cyanide leaching to segregate metallic ores. Pulverized rock contains sulfur that reacts with ground water and surface water to produce sulfuric acid, resulting in the contamination of water resources. High-temperature smelters vaporize metals such as lead, nickel, mercury, and copper during the processing, resulting in release to our atmosphere and precipitation into our soils and waters.

Prior to the 1970s mining operations in most countries were largely unregulated with respect to the disposal of hazardous waste materials and impact to land and water. Beginning in the 1970s, environmental regulations were instituted that required mining operations to minimize contamination to the environment. However, these regulations are not universally applied and are challenged by mining companies. As a result, mining has increased in those countries where environmental regulations are relatively lax, resulting in contamination in developing countries. Indeed, the majority of our mineral resources are imported, rather than domestically mined and produced. The United States imports 100% of 20 mineral commodities and over 50% of another 50 mineral commodities, some of which are listed in Figure 19.23 from countries depicted in Figure 19.24.

Global concerns exist on the availability of minerals to meet increased demand for a growing population and the expansion of renewable energy and telecommunication fields. In 2010, China, which produces over 80% of REE metals, drastically reduced the export of these critical metals to protect its domestic needs. China's REE export limits rattled world markets and highlighted the tenuous global economic reliance on mineral commodities controlled by a limited number of countries. In addition to REE, China is also the leading producer of antimony, bismuth, fluorspar, germanium, graphite, and indium (Price 2013). Similarly, the Democratic

2016 U.S. NET IMPORT RELIANCE

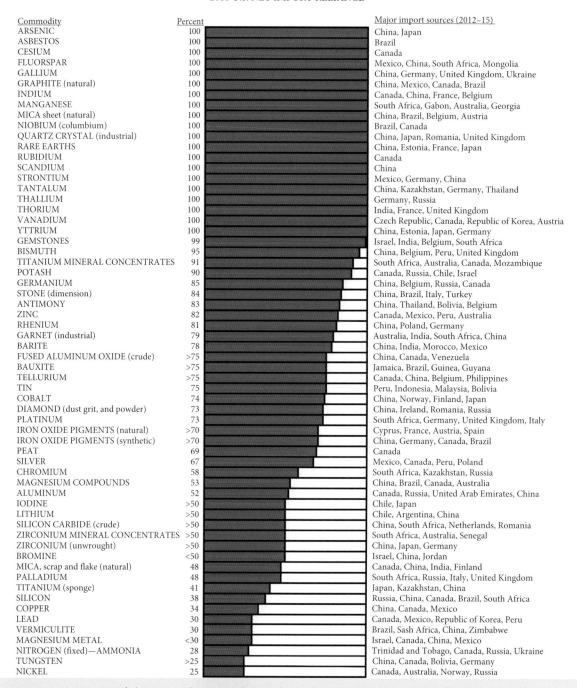

| Commodity | Percent | Major import sources (2012–15) |
|---|---|---|
| ARSENIC | 100 | China, Japan |
| ASBESTOS | 100 | Brazil |
| CESIUM | 100 | Canada |
| FLUORSPAR | 100 | Mexico, China, South Africa, Mongolia |
| GALLIUM | 100 | China, Germany, United Kingdom, Ukraine |
| GRAPHITE (natural) | 100 | China, Mexico, Canada, Brazil |
| INDIUM | 100 | Canada, China, France, Belgium |
| MANGANESE | 100 | South Africa, Gabon, Australia, Georgia |
| MICA sheet (natural) | 100 | China, Brazil, Belgium, Austria |
| NIOBIUM (columbium) | 100 | Brazil, Canada |
| QUARTZ CRYSTAL (industrial) | 100 | China, Japan, Romania, United Kingdom |
| RARE EARTHS | 100 | China, Estonia, France, Japan |
| RUBIDIUM | 100 | Canada |
| SCANDIUM | 100 | China |
| STRONTIUM | 100 | Mexico, Germany, China |
| TANTALUM | 100 | China, Kazakhstan, Germany, Thailand |
| THALLIUM | 100 | Germany, Russia |
| THORIUM | 100 | India, France, United Kingdom |
| VANADIUM | 100 | Czech Republic, Canada, Republic of Korea, Austria |
| YTTRIUM | 100 | China, Estonia, Japan, Germany |
| GEMSTONES | 99 | Israel, India, Belgium, South Africa |
| BISMUTH | 95 | China, Belgium, Peru, United Kingdom |
| TITANIUM MINERAL CONCENTRATES | 91 | South Africa, Australia, Canada, Mozambique |
| POTASH | 90 | Canada, Russia, Chile, Israel |
| GERMANIUM | 85 | China, Belgium, Russia, Canada |
| STONE (dimension) | 84 | China, Brazil, Italy, Turkey |
| ANTIMONY | 83 | China, Thailand, Bolivia, Belgium |
| ZINC | 82 | Canada, Mexico, Peru, Australia |
| RHENIUM | 81 | China, Poland, Germany |
| GARNET (industrial) | 79 | Australia, India, South Africa, China |
| BARITE | 78 | China, India, Morocco, Mexico |
| FUSED ALUMINUM OXIDE (crude) | >75 | China, Canada, Venezuela |
| BAUXITE | >75 | Jamaica, Brazil, Guinea, Guyana |
| TELLURIUM | >75 | Canada, China, Belgium, Philippines |
| TIN | 75 | Peru, Indonesia, Malaysia, Bolivia |
| COBALT | 74 | China, Norway, Finland, Japan |
| DIAMOND (dust grit, and powder) | 73 | China, Ireland, Romania, Russia |
| PLATINUM | 73 | South Africa, Germany, United Kingdom, Italy |
| IRON OXIDE PIGMENTS (natural) | >70 | Cyprus, France, Austria, Spain |
| IRON OXIDE PIGMENTS (synthetic) | >70 | China, Germany, Canada, Brazil |
| PEAT | 69 | Canada |
| SILVER | 67 | Mexico, Canada, Peru, Poland |
| CHROMIUM | 58 | South Africa, Kazakhstan, Russia |
| MAGNESIUM COMPOUNDS | 53 | China, Brazil, Canada, Australia |
| ALUMINUM | 52 | Canada, Russia, United Arab Emirates, China |
| IODINE | >50 | Chile, Japan |
| LITHIUM | >50 | Chile, Argentina, China |
| SILICON CARBIDE (crude) | >50 | China, South Africa, Netherlands, Romania |
| ZIRCONIUM MINERAL CONCENTRATES | >50 | South Africa, Australia, Senegal |
| ZIRCONIUM (unwrought) | >50 | China, Japan, Germany |
| BROMINE | <50 | Israel, China, Jordan |
| MICA, scrap and flake (natural) | 48 | Canada, China, India, Finland |
| PALLADIUM | 48 | South Africa, Russia, Italy, United Kingdom |
| TITANIUM (sponge) | 41 | Japan, Kazakhstan, China |
| SILICON | 38 | Russia, China, Canada, Brazil, South Africa |
| COPPER | 34 | China, Canada, Mexico |
| LEAD | 30 | Canada, Mexico, Republic of Korea, Peru |
| VERMICULITE | 30 | Brazil, Sash Africa, China, Zimbabwe |
| MAGNESIUM METAL | <30 | Israel, Canada, China, Mexico |
| NITROGEN (fixed)—AMMONIA | 28 | Trinidad and Tobago, Canada, Russia, Ukraine |
| TUNGSTEN | >25 | China, Canada, Bolivia, Germany |
| NICKEL | 25 | Canada, Australia, Norway, Russia |

**Figure 19.23** Some of the mineral commodities of which the USA imports >50%. *Source*: USGS (2020). Public Domain. © The United States Department of the Interior.

Republic of Congo is the sole major producer of cobalt; Brazil controls the niobium market; South Africa and Russia dominate the platinum market. Many of these mineral commodities are essential for telecommunications, renewable energy growth and national security of countries throughout the globe. Clearly, international cooperation, education and expanded research are essential to wisely use existing resources and to seek alternative

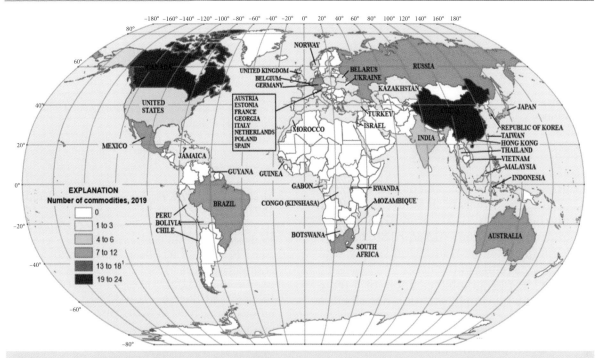

**Figure 19.24**   Major import sources of nonfuel mineral commodities for which the US net import reliance exceeded 50% in 2019. *Source*: USGS (2020). Public Domain. © The United States Department of the Interior.

sources for those critical minerals in high demand and short supply.

Minerals have been critically important to human endeavors from ancient times to the present. From the moment you get up in the morning and flush a toilet you begin a daily routine that relies upon the availability of minerals: clay and feldspars in porcelain toilets, stainless steel in plumbing, copper in electrical wiring, water stored in sedimentary rock and quartz sand aquifers, cement sidewalks, gravel roads, metallic bicycles and automobiles, cement and stone buildings and homes – the list goes on and on. Conservation and recycling of existing mineral resources, and the search for substitute sources, are vital. We hope this textbook has provided a framework on the global importance of minerals and rocks, as well as their daily role in our lives. Cherish our blue planet, our loved ones, and the minerals, rocks, water, and atmospheric resources that allow life to continue on Earth. No viable substitute exists.

## CONTENT ASSESSMENT

1   Name and describe the three major uses of minerals.
2   Explain a major ore forming process at ocean spreading ridges
3   Describe a major ore forming process at convergent plate boundaries.
4   Describe three different uses of industrial minerals. Refer to the figures in the first 20 pages from the USGS Mineral Commodities Yearbook 2020 at https://pubs.usgs.gov/periodicals/mcs2020/mcs2020.pdf
5   From which two countries does the United States import the most nonfuel minerals?
6   Which US states produce the most metallic ore minerals?
7   Which state are you from and what nonfuel mineral deposits are mined from that locality?
8   Identify specific elements, minerals, and geographic mineral sources required for your cell phone using the USGS document at https://pubs.usgs.gov/gip/0167/gip167.pdf

| Element | Applicable properties | Mineral | Geographic source |
|---|---|---|---|
| LED | | | |
| Display | | | |
| Electronics | | | |
| Battery | | | |
| Speaker | | | |

Refer to the U.S. Geological Survey Critical Mineral Resources – an Introduction (2017) located at: https://pubs.usgs.gov/pp/1802/a/pp1802a.pdf

9  Comment on the US mineral supply situation currently in the news.

10  Provide a geographic example of a tenuous mineral supply situation described in the text.

11  Define critical or strategic elements/minerals and explain why they are so important.

12  Describe three changes that have occurred since the 1970s with respect to our study of ores.

## REFERENCES

Bárdossy, G. (1982). *Karst Bauxites – Bauxite Deposits on Carbonate Rocks*, 441 p, Developments in Economic Geology 14. Amsterdam: Elsevier.

Bárdossy, G. (1994). Carboniferous to Jurassic bauxite deposits as paleoclimatic and paleogeographic indicators: Canadian Society of Petroleum Geologists. *Memoir 17*: 283–293.

Bárdossy, G. and Aleva, G.J.J. (1990). *Lateritic Bauxites*, 624 p, Developments in Economic Geology 27. Amsterdam: Elsevier.

Barnes, S.J., Cox, R.A., and Zientek, M.L. (2006). Platinum-group element, Gold, Silver and Base Metal distribution in compositionally zoned sulfide droplets from the Medvezky Creek Mine, Noril'sk, Russia. *Contributions to Mineralogy and Petrology 152* (2): 187–200. https://doi.org/10.1007/s00410-006-0100-9.

Bekker, A., Slack, J.F., Planavsky, N. et al. (2010). Iron formation – the sedimentary product of a complex interplay among mantle, tectonic, oceanic, and biospheric processes. *Economic Geology 105*: 467–508.

Bornhorst, T.J. and Mathur, R. (2017). Copper isotope constraints on the genesis of the Keweenaw Peninsula native copper district, Michigan, USA. *Minerals 7* (10): 185. https://doi.org/10.3390/min7100185.

Boström, K., Peterson, M.N.A., Joensuu, O., and Fisher, D.E. (1969). Aluminum-poor ferromanganoan sediments on active ocean ridges. *Journal of Geophysical Research 74*: 3261–3270.

Bradley, D. and McCauley, A. (2013). A preliminary deposit model for lithium-cesium-tantalum (LCT) pegmatites (ver. 1.1, December 2016). U.S. Geological Survey Open-File Report 2013–1008, 7 p. https://doi.org/10.3133/ofr20131008.

Chang, L.L.Y. (2002). *Industrial Mineralogy: Materials, Processes and Uses*, 472 pp. Englewood Cliffs, NJ: Prentice Hall Publishers.

Chetelat, E. (1947). La genese et l'evolution des gisements de nickel de la Nouvelle-Caledonie. *Bulletin de la Société Géologique de France 5*: 105–160.

Cox, D.P. and Singer, D.A. (1986). *Mineral Deposit Models*, 379 pp. US Geological Survey Bulletin No. 1693.

Dunn, M.D. (1983). *Fundamentals of Nutrition*, 581 pp. Boston, MA: CBI Publishers.

Emsbo, P., Seal, R.R., Breit, G.N., et al. (2010). Sedimentary exhalative (sedex) zinc-lead-silver deposit model. U.S. Geological Survey Scientific Investigations Report 2010–5070–N, 57 p., doi: https://doi.org/10.3133/sir20105070N. https://pubs.usgs.gov/sir/2010/5070/n/sir20105070n.pdf

Evans, A.M. (1993). *Ore Geology and Industrial Minerals*, 3e, 389 pp. Oxford: Blackwell Publishing.

Foley, N.K., Jaskula, B.W., Piatak, N.M., and Schulte, R.F. (2017). Beryllium. In: *Critical Mineral Resources of the United States – Economic and Environmental Geology and Prospects for Future Supply*, 797 p (eds. K.J. Schulz, J.H. DeYoung Jr., R.R. Seal II and D.C. Bradley). U.S. Geological Survey Professional Paper 1802 https://doi.org/10.3133/pp1802.

Francheteau, J., Needham, H.D., Choukroune, P. et al. (1979). Massive deep-sea sulfide ore deposits discovered by submersible on the East Pacific Rise: Project RITA, 21°N. *Nature 277*: 523–528.

Franklin, J.M., Gibson, H.L., Galley, A.G., and Jonasson, I.R. (2005). Volcanogenic massive sulfide deposits. In: *Economic Geology 100th Anniversary Volume* (eds. J.W. Hedenquist, J.F.H. Thompson, R.J. Goldfarb and J.P. Richards), 523–560. Littleton, CO: Society of Economic Geologists.

Galley, A.G., Hannington, M.D., and Jonasson, I.R. (2007). Volcanogenic massive sulphide deposits. In: *Mineral Deposits of Canada: A Synthesis of Major Deposi Types, District Metallogeny, the Evolution of Geological Provinces, and Exploration Methods* (ed. W.D. Goodfellow), 141–161. Geological Association of Canada, Mineral Deposits Division, Special Publication No. 5.

Goodfellow, W.D. and Lydon, J. (2007). Sedimentaryexhalative (SEDEX) deposits. In: *Mineral Deposits of Canada: A Synthesis of Major Deposit Types, District Metallogeny, the Evolution of Geological Provinces, and Exploration Methods* (ed. W.D. Goodfellow), 163–183. Geological Association of Canada, Mineral Deposits Division, Special Publication No. 5.

Hammarstrom, J.M., Zientek, M.L., Parks, H.L., et al. (2019). Assessment of undiscovered copper resources of the world, 2015 (ver. 1.1, May 24, 2019). U.S. Geological Survey Scientific Investigations Report 2018–5160, 619 p. (including 3 chap., 3 app., glossary, and atlas of 236 page-size pls.), doi: https://doi.org/10.3133/sir20185160.

Hannington, M.D., Poulsen, K.H., Thompson, J.F.H., and Sillitoe, R.H. (1999). Volcanogenic gold in the massive sulfide environment. *Reviews in Economic Geology 8*: 325–356.

Herzig, P.M. and Hannington, M.D. (1995). Polymetallic massive sulfides at the modern seafloor: a review. *Ore Geology Reviews 10*: 95–115.

Holwell, B.D.A. and McDonald, I. (2010). A review of behavior of platinum group elements within natural magmatic sulfide ore systems. *Platinum Metals Review 54* (1): 26–36.

Humphris, S.E., Herzig, P.M., Miller, D.J. et al. (1995). The internal structure of an active sea-floor massive sulfide deposit. *Nature 377*: 713–716.

John, D.A., Ayuso, R.A., Barton, M.D., et al. (2010). Porphyry copper deposit model, chapter B of Mineral deposit models for resource assessment. U.S. Geological Survey Scientific Investigations Report 2010–5070–B, 169 p. https://pubs.usgs.gov/sir/2010/5070/b

Kesler, S.E. (1994). *Mineral Resources, Economics and the Environment*. New York: Macmillan Publishers 391 pp.

Koerber, A.J. and Thakurta, J. (2019). PGE-enrichment in magnetite-bearing olivine gabbro: New observations from the Midcontinent Rift-related Echo Lake Intrusion in Northern Michigan, USA. *Minerals 9*: 19 p. doi: https://doi.org/10.3390/min9010021.

Kraus, E.H. and Slawson, C.B. (1947). *Gems and Gem Materials*, 332 pp. New York: McGraw Hill Publishers.

Leach, D.L., Sangster, D.F., Kelley, K.D. et al. (2005). Sediment-hosted lead-zinc deposits: a global perspective: Society of Economic Geologists. *Economic Geology One Hundredth Anniversary Volume 1905–2005*: 561–607.

Leach, D.L., Taylor, R.D., Fey, D.L., et al. (2010). A deposit model for Mississippi Valley-Type lead-zinc ores, chap. A of Mineral deposit models for resource assessment. U.S. Geological Survey Scientific Investigations Report 2010–5070–A, 52 p. https://pubs.usgs.gov/sir/2010/5070/a/pdf/SIR10-5070A.pdf

London, D. (2008). *Pegmatites*, 347 p. The Canadian Mineralogist Special Publication 10.

Lowell, J.D. and Guilbert, J.M. (1970). Lateral and vertical alteration-mineralization zoning in porphyry ore deposits. *Economic Geology 65*: 373–408.

Morgan, L.A., and Schulz, K.J. (2012). Physical volcanology of volcanogenic massive sulfide deposits in volcanogenic massive sulfide occurrence model. U.S. Geological Survey Scientific Investigations Report 2010–5070 –C, chapter 5, 36 p.

Mosier, D.L., Berger, V.I., and Singer, D.A. (2009). Volcanogenic massive sulfide deposits of the world; database and grade and tonnage models: U.S. Geological Survey Open-File Report 2009–1034. http://pubs.usgs.gov/of/2009/1034. https://pubs.usgs.gov/of/2009/1034/of2009-1034_text.pdf

Naldrett, A.J. (2010). From the mantle to the bank – the life of a Ni-Cu-(PGE) sulfide deposit. *South African Journal of Geology 113* (1): 1–32. https://doi.org/10.2113/gssajg.113.1-1.

National Research Council (2008). *Minerals, Critical Minerals, and the U.S. Economy*, 245 p. Washington, D.C.: The National Academies Press.

Pajović, M. (2009). Genesis and genetic types of karst bauxites. *Iranian Journal of Earth Sciences 1*: 44–56.

Paradis, S., Hannigan, P., and Dewing, K. (2007). Mississippi Valley-type lead-zinc deposits. In: *Mineral Deposits of Canada: A Synthesis of Major Deposit Types, District Metallogeny, the Evolution of Geological Provinces, and Exploration Methods* (ed. W.D. Goodfellow), 185–203. Geological Association of Canada, Mineral Deposits Division, Special Publication No. 5.

Piatak, N.M., Seal, R.R. II, Jones, P.M., and Woodruff, L.G. (2016). Copper toxicity and dissolved organic matter: Resiliency of mineralized watersheds in northern Minnesota and Michigan: Institute on Lake Superior Geology Proceedings, 62, Part 1 – Program and Abstracts (4–8 May 2016), Duluth, MN, pp. 117–118. https://www.lakesuperiorgeology.org/Volumes.html#2010

Price, J.G. (2013). The challenges of mineral resources for society. In: *The Impact of the Geological Sciences on Society* (ed. M.E. Bickford), 1–19. Geological Society of America Special Paper 501 https://doi.org/10.1130/2013.2501(01).

Prokin, V.A. and Buslaev, F.P. (1999). Massive copper–zinc sulphide deposits in the Urals. *Ore Geology Reviews 14*: 1–69.

Retallack, G.J. (2010). Laterization and bauxitization events. *Economic Geology 105*: 655–667.

Reynolds, R.L. and Goldhaber, M.B. (1978). Origin of a South Texas roll-type uranium deposit: I. Alteration of iron titanium oxide minerals. *Economic Geology 73*: 1677–1689.

Ripley, E.M. (1999). Sulfur and oxygen isotopic evidence of country rock contamination inthe Voisey's Bay Ni-Cu-Co deposit, Labrador, Canada. *Lithos 47*: 53–68.

Schulte, R.F. and Foley, N.K. (2014). Compilation of gallium resource data for bauxite deposits: U.S. Geological Survey Open-File Report 2013–1272, 14 p., 3 separate tables, doi: https://doi.org/10.3133/ofr20131272.

Schulz, K.J., DeYoung, J.H. Jr., Seal, R.R. II, and Bradley, D.C. (eds.) (2017). *Critical Mineral Resources of the United States – Economic and Environmental Geology and Prospects for Future Supply*, 797 p. U.S. Geological Survey Professional Paper 1802 https://doi.org/10.3133/pp1802.

Shanks, W.C. Pat, III and Thurston, R. (eds.) (2012). Volcanogenic massive sulfide occurrence model. U.S.

Geological Survey Scientific Investigations Report 2010–5070–C, 345 p. https://pubs.usgs.gov/sir/2010/5070/c/SIR10-5070-C.pdf

Singer, D.A. (1995). World class base and precious metal deposits – a quantitative analysis. *Economic Geology* *90*: 88–104.

Smith, K.S. and Huyck, H.L.O. (1999). An overview of the abundance, relative mobility, bioavailability and human toxicity of metals. In: *The Environmental Geochemistry of Mineral Deposits. Society of Economic Geologists*, vol. *A* (eds. G.S. Plumlee and M.J. Logsdon), 29–70. Littleton, CO.: Society of Economic Geologists.

Taylor, R.D., Leach, D.L., Bradley, D.C., and Pisarevsky, S.A. (2009). Compilation of mineral resource data for Mississippi Valley-type and clastic-dominated sediment-hosted lead-zinc deposits: U.S. Geological Survey Open-File Report 2009–1297, 42 p.

Taylor, C.D., Finn, C.A., Anderson, E.D., et al. (2015). Algoma-, Superior-, and oolitic-type iron deposits of the Islamic Republic of Mauritania (phase V, deliverable 83), chap. O of Taylor, C.D., ed., Second projet de renforcement institutionnel du secteur minier de la République Islamique de Mauritanie (PRISM-II): U.S. Geological Survey Open-File Report 2013–1280-O, 94 p., doi: https://doi.org/10.3133/ofr20131280

Tikkanen, G.D. (1986). World resources and supply of lead and zinc. In: *Economics of Internationally Traded Minerals* (ed. W.R. Bush), 242–250. Society of Mining Engineers, Inc.

U.S. Geological Survey (2020). *Mineral Commodity Summaries 2020*, 200 p. U.S. Geological Survey: https://doi.org/10.3133/mcs2020.

Wenk, H.-R. and Bulakh, A. (2004). *Minerals: Their Constitution and Origin*, 646 pp. Cambridge: Cambridge University Press.

Woodruff, L.G., Bedinger, G.M., and Piatak, N.M. (2017). Titanium. In: *Critical Mineral Resources of the United States – Economic and Environmental Geology and Prospects for Future Supply*, 797 p (eds. K.J. Schulz, J.H. DeYoung Jr., R.R. Seal II and D.C. Bradley). U.S. Geological Survey Professional Paper 1802 https://doi.org/10.3133/pp1802.

Zientek, M.L. (2012). Magmatic ore deposits in layered intrusions – descriptive model for reef-type PGE and contact-type Cu-Ni-PGE deposits: U.S. Geological Survey Open-File Report 2012–1010, 48 p. https://pubs.usgs.gov/of/2012/1010/contents/OF12-1010.pdf

Zientek, M.L., Loferski, P.J., Parks, H.L. et al. (2017). Platinum-group elements. In: *Critical Mineral Resources of the United States – Economic and Environmental Geology and Prospects for Future Supply*, 797 p (eds. K.J. Schulz, J.H. DeYoung Jr., R.R. Seal II and D.C. Bradley). U.S. Geological Survey Professional Paper 1802 https://doi.org/10.3133/pp1802.

# Index

Page numbers in **bold** refer to tables, page numbers in *italic* refer to figures.

*Earth Materials*, Second Edition. Kevin Hefferan and John O'Brien.
© 2022 John Wiley & Sons Ltd. Published 2022 by John Wiley & Sons Ltd.
Companion website: www.wiley.com/go/hefferan/earthmaterials2